Soil and Groundwat
Remediation Technolo

Soil and Groundwater Remediation Technologies

A Practical Guide

Edited by

Yong Sik Ok

Jörg Rinklebe

Deyi Hou

Daniel C.W. Tsang

Filip M.G. Tack

CRC Press
Taylor & Francis Group
Boca Raton London New York

CRC Press is an imprint of the
Taylor & Francis Group, an **informa** business

First edition published 2020
by CRC Press
6000 Broken Sound Parkway NW, Suite 300, Boca Raton, FL 33487-2742

and by CRC Press
4 Park Square, Milton Park, Abingdon, Oxon OX14 4RN

First issued in paperback 2023

© 2020 Taylor & Francis Group, LLC
CRC Press is an imprint of Taylor & Francis Group, an Informa business

No claim to original U.S. Government works

ISBN 13: 978-1-03-257082-2 (pbk)
ISBN 13: 978-0-367-33740-7 (hbk)

DOI: 10.1201/9780429322563

Visit the Taylor & Francis Web site at
http://www.taylorandfrancis.com

and the CRC Press Web site at
http://www.crcpress.com

Typeset in Times
by Lumina Datamatics Limited

Contents

Editors

Yong Sik Ok, PhD, is a Full Professor and Global Research Director of Korea University in Seoul, Korea. His academic background covers waste management, the bioavailability of emerging contaminants, and bioenergy and value-added products (such as biochar). Professor Ok also has experience in fundamental soil science and the remediation of various contaminants in soils and sediments. Together with graduate students and colleagues, Professor Ok has published over 700 research papers, 60 of which have been ranked as Web of Science ESI top papers (56 nominated as "Highly Cited Papers" [HCPs] and 4 nominated as "Hot Papers"). He has been a Web of Science Highly Cited Researcher (HCR) since 2018. In 2019, he became the first Korean to be selected as an HCR in the field of Environment and Ecology. He maintains a worldwide professional network through his service as an Editor (former Co-Editor in Chief) of the *Journal of Hazardous Materials*, Co-Editor for *Critical Reviews in Environmental Science and Technology*, Associate Editor for *Environmental Pollution* and *Bioresource Technology*, and as a member of the editorial boards of *Renewable and Sustainable Energy Reviews, Chemical Engineering Journal, Chemosphere*, and *Journal of Analytical and Applied Pyrolysis*, along with several other top journals. Professor Ok has served in a number of positions worldwide including as Honorary Professor at the University of Queensland (Australia), Visiting Professor at Tsinghua University (China), Adjunct Professor at the University of Wuppertal (Germany), and Guest Professor at Ghent University (Belgium). He currently serves as Director of the Sustainable Waste Management Program for the Association of Pacific Rim Universities (APRU). He has served as chairman of numerous major conferences such as Engineering Sustainable Development, organized by the APRU and the Institute for Sustainability of the American Institute of Chemical Engineers (AIChE).

Jörg Rinklebe, PhD, is Professor for Soil and Groundwater Management at the University of Wuppertal, Germany. His academic background covers environmental science, bioavailability of emerging contaminants, and remediation of contaminated soils using value-added products such as biochar. He works at various scales, ecosystems, and spheres including pedosphere, hydrosphere lithosphere, biosphere, and atmosphere and has a certain experience in fundamental soil science. His main research is on soils, sediments, waters, plants, and their pollutions (trace elements and nutrients) and linked biogeochemical issues with a special focus in redox chemistry. He also has a certain expertise in soil microbiology. Professor Rinklebe is internationally recognized particularly for his research in the areas of biogeochemistry of trace elements in wetland soils. He has published a number of scientific papers in leading international and national journals. According to SCOPUS (January 22, 2020), Professor Rinklebe has published over 214 research papers, among which 23 were nominated as "Highly Cited Papers" and 1 as "Hot Paper." He is a highly cited researcher. His *h*-index is 41. He has published two books entitled *Trace Elements in Waterlogged Soils and Sediments* (2016) and *Nickel in Soils and Plants* (2018), as well as numerous book chapters. He is Editor of the international journal *Critical Reviews in Environmental Science and Technology* (*CREST*), Chief Editor for special issues of *Environmental Pollution* and the *Journal of Hazardous Materials,* and Guest Editor of the international journals *Environment International, Chemical Engineering Journal, Science of the Total Environment, Chemosphere, Journal of Environmental Management, Applied Geochemistry*, and *Environmental Geochemistry and Health*. He is a member of several editorial boards (*Ecotoxicology; Geoderma, Water, Air, & Soil Pollution; Archive of Agronomy;* and *Soil Science*) and reviewer for many leading international journals. He has organized many special symposia at various international conferences such as "Biogeochemistry of Trace Elements" (ICOBTE) and the "International Conference on Heavy Metals in the Environment" (ICHMET). He has also been an invited speaker (plenary and keynote) at many international conferences. In 2016, he received an appointment as Honorable Ambassador for Gangwon Province, South Korea, which was renewed in June 2018. In addition, he received an appointment as Adjunct Professor at the University of Southern Queensland, Australia; Visiting Professor at the Sejong University, Seoul, South Korea; and Guest Professor at the China Jiliang University, Hangzhou, Zhejiang, China. Recently, Professor Rinklebe was elected as Vice President of the International Society of Trace Element Biogeochemistry (ISTEB).

Deyi Hou, PhD, is an Associate Professor at the School of Environment of Tsinghua University in Beijing, China. Professor Deyi's research interests include sustainability of soil use and management, regional to global sources and distribution of heavy metal pollution, contaminant transport and fate in porous media, and green and innovative remediation materials and technologies. He received his BE from Tsinghua University, MS from Stanford University, and PhD from the University of Cambridge. He has published over 100 papers in international journals including *Nature, Science, Nature Climate Change,* and *Nature Sustainability*. He is currently serving as Editor-in-Chief for *Soil Use and Management* and Associate Editor for *Science of the Total Environment*. He also serves as an editorial board member for *Critical Reviews in Environmental Science and Technology, Environmental Pollution*, and *Remediation Journal*. He has also served as co-chair, session chair, and scientific committee member for numerous international conferences.

Daniel C.W. Tsang, PhD, is currently an Associate Professor in the Department of Civil and Environmental Engineering at the Hong Kong Polytechnic University and Honorary Associate Professor at the University of Queensland. He was an IMETE Visiting Scholar at Ghent University in Belgium, Visiting Scholar at Stanford University in the United States, Senior Lecturer and Lecturer at the University of Canterbury in New Zealand, and Postdoctoral Fellow at Imperial College London in the United Kingdom and the Hong Kong University of Science and Technology. Dr. Tsang's research group strives to develop low-impact solutions to ensure sustainable development and foster new ways in which we utilize biomass waste, contaminated land, and urban water. He has published over 350 SCI journal papers with an *h*-index of 45 (SCOPUS), and currently serves as Associate Editor of the *Journal of Hazardous Materials, Science of the Total Environment, Critical Reviews in Environmental Science and Technology*, as well as editorial board member of *Environmental Pollution, Bioresource Technology, Chemosphere,* and *Advanced Sustainable Systems*. He has received the Excellence in Review Award from *Environmental Science and Technology; Resources, Conservation & Recycling*; and *Chemosphere*. He is the chair and organizer of multiple international conferences including the Fifth Asia Pacific Biochar Conference (APBC 2020).

Filip M.G. Tack, PhD, is Professor in Biogeochemistry of Trace Elements at the Department of Green Chemistry and Technology of Ghent University. He is Head of the Laboratory of Analytical Chemistry and Applied Ecochemistry of Ghent University. Recent research topics include the study of the occurrence, chemical speciation, and behavior of trace metals in riparian zones and dredged sediment disposal sites, treatment of wastewater using plant-based systems, and management/remediation of moderate metal contamination using phytoremediation and phytostabilization. He is author or co-author of more than 200 publications in international journals (*h*-index 65), 20 chapters in books, and more than 240 contributions to international congresses. Professor Tack teaches in topics related to aquatic and soil chemistry, environmental chemistry, and soil remediation. He was one of the initiators and currently serves as Coordinator at Ghent University of the Erasmus Mundus Master program International Master in Environmental Science and Engineering, which is organized jointly by the IHE Delft Institute for Water Education and the University of Chemistry and Technology Prague. He also has a teaching assignment at Ghent University Global Campus in Songdo, Republic of Korea and is Guest Professor at China Jiliang University in Hangzhou, China. He is currently Associate Editor of the *Science of the Total Environment* and *Journal of Plant Nutrition and Soil Science*, and Coordinating Editor for *Environmental Geochemistry and Health*. He has served as Guest Editor for special issues in *Science of the Total Environment, Chemosphere, Applied Geochemistry*, and *Journal of the Cleaner Production*. He is a member of the International Society of Trace Element Biogeochemistry, the Belgian Society of Soil Science, and IE-NET, the Flemish Society of Engineers.

Contributors

Yaser A. Almaroai
Department of Biology
Umm Al-Qura University
Makkah, Saudi Arabia

Andon Vasilev Andonov
Department of Plant Physiology and Biochemistry
Agricultural University Plovdiv
Plovdiv, Bulgaria

Vasileios Antoniadis
Department of Agriculture Crop Production and Rural
    Environment
University of Thessaly
Volos, Greece

Rabia Amen
Institute of Soil and Environmental Sciences
University of Agriculture Faisalabad
Faisalabad, Pakistan

Paul Bardos
r3 Environmental Technology Ltd.
Department Soil Science
University of Reading
Reading, United Kingdom

and

School of Environment and Technology
University of Brighton
Brighton, United Kingdom

Hamna Bashir
Institute of Soil and Environmental Sciences
University of Agriculture Faisalabad
Faisalabad, Pakistan

Jingzi Beiyuan
School of Environmental and Chemical Engineering
Foshan University
Foshan, P.R. China

Irshad Bibi
Institute of Soil and Environmental Sciences
University of Agriculture Faisalabad
Faisalabad, Pakistan

Nanthi S. Bolan
Global Centre for Environmental Remediation (GCER)
Advanced Technology Centre
The University of Newcastle
Callaghan, New South Wales, Australia

Julian Bosch
Intrapore GmbH
Essen, Germany

Elio Brunetti
Intrapore GmbH
Essen, Germany

Jochen Bundschuh
School of Engineering and Surveying
University of Southern Queensland
Queensland, Australia

and

UNESCO—Chair on Groundwater Arsenic within the
    2030 Agenda for Sustainable Development
University of Southern Queensland
Toowoomba, Queensland, Australia

Liang Chen
Department of Civil and Environmental Engineering
The Hong Kong Polytechnic University
Hong Kong, P.R. China

David O'Connor
School of Environment
Tsinghua University
Beijing, P.R. China

Xiaomin Dou
College of Environmental Science and Engineering
Beijing Forestry University
Beijing, P.R. China

Yinqing Fang
Research Center for Eco-Environmental Sciences
Chinese Academy of Sciences
Beijing, P.R. China

Alina Gawel
Intrapore GmbH
Essen, Germany

Nan Geng
College of Water Conservancy and Environmental
    Engineering
Zhejiang University of Water Resources and Electric
    Power
Hangzhou, P.R. China

and

Korea Biochar Research Center
O-Jeong Eco-Resilience Institute (OJERI)
Division of Environmental Science and Ecological
    Engineering
Korea University
Seoul, Republic of Korea

Viraj Gunarathne
Ecosphere Resilience Research Center
University of Sri Jayewardenepura
Nugegoda, Sri Lanka

Udaya Gunarathne
Ecosphere Resilience Research Center
University of Sri Jayewardenepura
Nugegoda, Sri Lanka

Ziyu Han
Research Center for Eco-Environmental Sciences
Chinese Academy of Sciences
Beijing, P.R. China

Kiran Hina
Department of Environmental Science
University of Gujrat
Gujrat, Pakistan

Deyi Hou
School of Environment
Tsinghua University
Beijing, P.R. China

Jian Hu
Research Center for Eco-Environmental Sciences
Chinese Academy of Sciences
Beijing, P.R. China

Muhammad Mahroz Hussain
Institute of Soil and Environmental Sciences
University of Agriculture Faisalabad
Faisalabad, Pakistan

Longjie Ji
Research Center for Eco-Environmental Sciences
Chinese Academy of Sciences
Beijing, P.R. China

Wentao Jiao
Research Center for Eco-Environmental Sciences
Chinese Academy of Sciences
Beijing, P.R. China

Yuanliang Jin
School of Environment
Tsinghua University
Beijing, P.R. China

Jurate Kumpiene
Waste Science & Technology
Luleå University of Technology
Luleå, Sweden

Sang Soo Lee
Department of Environmental Engineering
Yonsei University
Wonju, Republic of Korea

Efi Levizou
Department of Agriculture Crop Production and Rural
    Environment
University of Thessaly
Volos, Greece

Debao Lu
College of Water Conservancy and Environmental
    Engineering
Zhejiang University of Water Resources and Electric Power
Hangzhou, P.R. China

and

Korea Biochar Research Center
O-Jeong Eco-Resilience Institute (OJERI)
Division of Environmental Science and Ecological
    Engineering
Korea University
Seoul, Republic of Korea

Li Lyu
Department of Environmental Science and Engineering
Zhejiang Gongshang University
Hangzhou, P.R. China

Reza Mahinroosta
CSU Engineering
Charles Sturt University
Wagga Wagga, New South Wales, Australia

Nabeel Khan Niazi
Institute of Soil and Environmental Sciences
University of Agriculture Faisalabad
Faisalabad, Pakistan

and

School of Civil Engineering and Surveying
University of Southern Queensland
Toowoomba, Queensland, Australia

Yong Sik Ok
Korea Biochar Research Center
O-Jeong Eco-Resilience Institute (OJERI)
Division of Environmental Science and Ecological
    Engineering
Korea University
Seoul, Republic of Korea

Kumuduni Niroshika Palansooriya
Korea Biochar Research Center
O-Jeong Eco-Resilience Institute (OJERI)
Division of Environmental Science and Ecological
    Engineering
Korea University
Seoul, Republic of Korea

Jin Hee Park
Department of Environmental and Biological Chemistry
Chungbuk National University
Chungbuk, Republic of Korea

Jason Prior
Institute for Sustainable Futures
University of Technology Sydney
Sydney, New South Wales, Australia

Zhirui Qin
Research Center for Eco-Environmental Sciences
Chinese Academy of Sciences
Beijing, P.R. China

Jörg Rinklebe
University of Wuppertal
School of Architecture and Civil Engineering
Institute of Foundation Engineering, Water- and Waste
    Management
Laboratory of Soil- and Groundwater Management
Wuppertal, Germany

Ajit K. Sarmah
Civil and Environmental Engineering Department
The University of Auckland
Auckland, New Zealand

Christian Schemel
Intrapore GmbH
Essen, Germany

Volker Selter
Geobau GmbH Company–Beratende Geologen und
    Ingenieure
Bochum, Germany

J.A.I. Senadheera
Ecosphere Resilience Research Center
University of Sri Jayewardenepura
Nugcgoda, Sri Lanka

S.T.M.L.D. Senevirathna
CSU Engineering
Charles Sturt University
Wagga Wagga, New South Wales, Australia

Sabry M. Shaheen
School of Architecture and Civil Engineering
Institute of Foundation Engineering, Water and Waste
    Management
Laboratory of Soil and Groundwater Management
University of Wuppertal
Wuppertal, Germany

and

Department of Arid Land Agriculture
King Abdulaziz University
Jeddah, Saudi Arabia

Muhammad Shahid
Department of Environmental Sciences
COMSATS University Islamabad
Vehari, Pakistan

Muhammad Bilal Shakoor
Department of Environmental Sciences and Engineering
Government College University Faisalabad
Faisalabad, Pakistan

Zhengtao Shen
Department of Earth and Atmospheric Sciences
University of Alberta
Edmonton, Alberta, Canada

Yinan Song
School of Environment
Tsinghua University
Beijing, P.R. China

Yuqing Sun
Department of Civil and Environmental Engineering
The Hong Kong Polytechnic University
Hong Kong, P.R. China

Filip M.G. Tack
Department of Green Chemistry and Applied Ecochemistry
Ghent University
Ghent, Belgium

Sofie Thijs
Centre for Environmental Sciences
Hasselt University
Diepenbeek, Belgium

Daniel C.W. Tsang
Department of Civil and Environmental Engineering
The Hong Kong Polytechnic University
Hong Kong, P.R. China

Jaco Vangronsveld
Centre for Environmental Sciences
Hasselt University
Diepenbeek, Belgium

and

Department of Plant Physiology
Maria Curie-Skłodowska University
Lublin, Poland

Meththika Vithanage
Ecosphere Resilience Research Center
University of Sri Jayewardenepura
Nugegoda, Sri Lanka

Di Wang
Department of Civil and Environmental Engineering
The Hong Kong Polytechnic University
Hong Kong, P.R. China

Hailong Wang
School of Environmental and Chemical Engineering
Foshan University
Foshan, P.R. China

Lei Wang
Department of Civil and Environmental Engineering
The Hong Kong Polytechnic University
Hong Kong, P.R. China

and

Department of Materials Science and Engineering
The University of Sheffield
Sheffield, United Kingdom

Linling Wang
School of Environmental Science and Engineering
Huazhong University of Science and Technology
Wuhan, P.R. China

Shiyu Wang
Research Center for Eco-Environmental Sciences
Chinese Academy of Sciences
Beijing, P.R. China

Malgorzata Wojcik
Department of Plant Physiology
Maria Curie-Skłodowska University
Lublin, Poland

James Tsz Fung Wong
Department of Civil and Environmental Engineering
The Hong Kong University of Science and Technology
Hong Kong, P.R. China

Yinfeng Xia
College of Water Conservancy and Environmental Engineering
Zhejiang University of Water Resources and Electric Power
Hangzhou, P.R. China

and

Korea Biochar Research Center & Division of
    Environmental Science and Ecological Engineering
Korea University
Seoul, Republic of Korea

Dong Zhang
Department of Environmental Science and Engineering
Hangzhou Dianzi University
Hangzhou, P.R. China

Chaosheng Zhang
International Network for Environment and Health (INEH)
School of Geography and Archaeology & Ryan Institute
National University of Ireland
Galway, Ireland

Yiyun Zhang
Department of Engineering
University of Cambridge
Cambridge, United Kingdom

1 Overview of Soil and Groundwater Remediation

Filip M.G. Tack and Paul Bardos

CONTENTS

1.1 INTRODUCTION

Our environment remains the victim of industrial growth. The number of recognized waste sites continues to grow daily (Panagos et al., 2013; Brombal et al., 2015; Carré et al., 2017). Past and continuing abuses require high remediation efforts in terms of technical, financial, and social resources. There is now international recognition of the problems of environmental damage, especially to soil. Land contamination is one of the eight threats mentioned in the EU Thematic Soil Strategy (European Commission, 2002), and is listed as one of the ten key threats that hamper the achievement of sustainable soil management (FAO, 2017). Managing soil pollution has been explicitly recognized as one of the needs to achieve sustainable development (UNEP, 2018). Accordingly, management and remediation of contaminated land is a rapid growth area for applications and technological innovations.

Cleaning up contaminated soils in an effective way is very difficult, even more so as it needs to be economically feasible at the same time. Increasingly, broader considerations with respect to overall environmental impact, long-term concerns, and sustainability further affect and change the way soil remediation is developing. Soil remediation and management of contaminated soil has evolved and matured over the past half-century toward promoting sustainable approaches that (Ellis and Hadley, 2009)

- Minimize consumption of energy or other natural resources
- Maximally avoid releases to the environment
- Harness or mimic natural processes

- Result in the reuse or recycling of land and materials
- Favor approaches that lead to effective destruction of contaminants

In many countries, the management of contaminated land has matured, and it is developing in many others. China, for example, recently has put a comprehensive policy and legislative framework in place (Li et al., 2015). In other countries, the process is yet to start (Sam et al., 2016; Arias Espana et al., 2018). This chapter aims to provide an overview of the issue of soil contamination and soil remediation approaches.

1.2 SOIL DEGRADATION AND SOIL CONTAMINATION

Natural soils in the environment are the result of thousands to millions of years of soil forming processes (Targulian and Krasilnikov, 2007). Time is one of the five major soil formation factors, next to climate, organisms, relief, and parent material (Jenny, 1994). As such, the natural soil is a unique resource that, once disturbed, cannot be reverted to a pristine state in the short term. In this sense it is non-renewable.

Soils provide numerous services to humanity. Soils are a substrate for biomass production, providing food, fodder, renewable energy, and raw materials. These functions are the basis of human and animal life (Blum, 2005). Soil has important filtering, buffering, and transformation capacities, providing a stable environment for supporting life functions, and largely contributing to the quality of air, food, and water, and thus to human health (Carré et al., 2017). It is a biological habitat for a large variety of organisms and thus hosts a large gene reserve. Soil is the physical support for human activities, and provides a source for raw materials such as clay, gravel, sand, and minerals. Humanity must move to a truly sustainable use of the soil as a limited resource.

Soil degradation refers to a decrease of the soil's actual or potential capability to produce quantitative and qualitative goods and services as a result of degradative processes. These degradation processes may be of physical, chemical, or biological nature (Lal and Stewart, 2012). Compaction and hardsetting, laterization, and erosion/desertification by wind and/or water deteriorate the physical properties of the soil. A decline of organic matter and reduction of soil biological life reflect biological soil degradation. Chemical soil degradation is related to depletion of nutrients/fertility and establishment of imbalances in chemical composition. Contamination with toxic compounds is but one of the several threats to soil quality, and its remediation is the scope of this book.

Chemical contaminants are emitted into the environment by human activities related to mining, urbanization, industry, and agriculture (Horta et al., 2015). In Europe, waste disposal and treatment has been estimated to account for 37% of the contaminated sites, followed by industrial and commercial activities (33.3%), storage (10.5%), and transport spills on land (7.9%) (Panagos et al., 2013). *Local contamination* occurs when inadequate handling of wastes or accidents during human activities introduces excessive contaminants in the soil. *Diffuse contamination* is defined where the relationship between the contaminant source and the level and spatial extent of soil contamination is indistinct (Jones et al., 2012). Dispersed sources and transformation and dilution in other environmental compartments may have occurred before the contaminants ended up in the soil over wider areas and typically in more moderate concentrations.

The number of contaminated and potentially contaminated sites has been estimated at about 10 to 20 million worldwide (Carré et al., 2017) (Table 1.1). About 60% of contamination in soils in Europe is by metals (34.8%) and petroleum hydrocarbons (23.8%). Contamination with CHC (8.3%), BTEX (10.2%), and PAH (10.9%) constitutes another 30% of soil contamination cases (Panagos et al., 2013) (Figure 1.1). Although many countries actioned on their urgent sites, overall progress of remediation is slow. In the European Union, 5% of all identified sites were cleaned up in 2013. In Australia, the progress is at a rate of 5% per year. However, each year, about 5% new sites are identified. The United States had 1322 superfund sites in 2014. These are the most severely contaminated sites that are being monitored at the federal level. While 375 sites were cleaned up, 53 new sites were proposed for being designated superfund sites (Horta et al., 2015; Carré et al., 2017).

Soil remediation may entail substantial costs. In 2011, total costs for remediation of a site in Europe ranged from less than 5000 € to more than 50,000,000 €. More than

TABLE 1.1
Estimated Number of Contaminated and Potentially Contaminated[a] Sites Worldwide

	Number of Sites	Population	Total
European Union	0.005 per capita	1×10^9	5×10^6
Emerging economies	0.0025 per capita	2×10^9	5×10^6
Developing economies	0.001 per capita	4×10^9	4×10^6

Source: Carré, F. et al., Soil Contamination and Human Health: A Major Challenge for Global Soil Security, In *Global Soil Security*, Springer, pp. 275–295, 2017.

[a] "Contaminated site" refers to a well-defined area where the presence of soil contamination has been confirmed and this presents a potential risk to humans, water, ecosystems. or other receptors. "Potentially contaminated site" refers to sites where unacceptable soil contamination is suspected but not verified, and where detailed investigations need to be carried out to verify whether there is an unacceptable risk of adverse impacts on receptors.

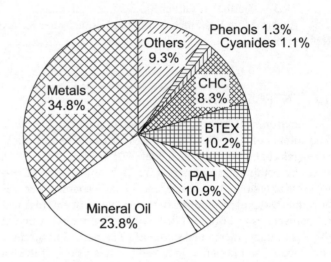

FIGURE 1.1 Distribution of contaminants in contaminated sites in Europe. (Based on Panagos, P. et al., *J. Environ. Public Health*, 1–11, 2013; original and current figure licensed under Creative Commons Attribution 3.0 Unported, https://creativecommons.org/licenses/by/3.0/legalcode.)

40% of the remediation cases costed between 50,000 and 500,000 € for site investigation and remediation (Figure 1.2) (Van Liedekerke et al., 2014). These costs may be offset in the longer term by the revalorization of the cleaned land and added value of the remedial actions. Remediation can create opportunities for recycling materials and potentially energy recovery. Revalorization of contaminated derelict land may create a wide range of beneficial opportunities including renewable energy, renewable feedstocks, water resource improvement, flood capacity and drainage management, carbon management, and societal benefits from amenity land and landscaping (Bardos et al., 2000; Dickinson et al., 2009; Bardos et al., 2011; Cundy et al., 2016).

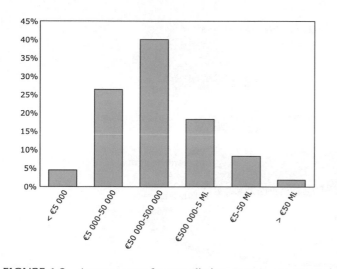

FIGURE 1.2 Average costs for remediation measures as reported in 2011. Costs include site investigations and remediation measures. (Reproduced from Van Liedekerke et al. (2014), https://www.eea.europa.eu/dataand-maps/daviz/average-cost-categories-for-site#tab-chart_2 (last retrieved 2019-06-24), licensed under Attribution 2.5 Denmark, https://creativecommons.org/licenses/by/2.5/dk/legalcode.)

1.3 THE PROCESS OF SOIL AND GROUNDWATER REMEDIATION

Every contaminated site has its own characteristics in terms of soil and contaminant properties, soil stratification, hydrology, current land use, land use and pollution history, ownership, social and economic needs, etc. (Table 1.2). Accordingly, every new case of soil remediation requires a specific approach where the remediation strategy will be designed accounting for site-specific needs and requirements (Khan et al., 2004; Scullion, 2006; Vik et al., 2001).

Over the past 40 years, awareness of issues surrounding land contamination has grown and many efforts have since

TABLE 1.2

Factors Affecting the Design of a Soil Remediation Project

Site-specific factors
    History of the site
    Location of the site
    Current use
Features of the contamination
    Type of contaminants
    Combination of contaminants present
    Effective hazard of the contaminants as a function of soil and
      contaminant chemistry
Socioeconomic factors
    Economic value of the site
    Connection to neighborhood
    Acceptability by the public

been done to remediate contaminated sites. While in the beginning (1980s, 1990s) focus was on an effective removal of the pollutants through digging, pumping, burying, and burning, a paradigm shift occurred where focus moved to risk management involving more sustainable approaches based on recycling, reusing, transforming, and biodegrading (Smith, 2019; Ellis and Hadley, 2009). In *sustainable risk-based contaminated land management* (SRBLM), decisions on contaminated land should be made based on risks to environmental and public health (Vegter and Kasamas, 2002; NICOLE and COMMON FORUM, 2013; Smith, 2019). The risk management should also meet sustainable development principles (Rizzo et al., 2016) and consider all of the three elements of sustainability, i.e., environmental, economic, and social (Bardos et al., 2016). Large efforts across many countries led to the publication of the sustainable remediation ISO standard, ISO 18504:2017 (ISO, Geneva, Switzerland).

For a risk to be present, there must be a source, a receptor, and a pathway (Figure 1.3) (Bardos et al., 2012). The source is a hazardous compound or property, whereas the receptor is a target that could be adversely affected by the contamination. It might be a human, an ecosystem, water resources, a building, or an ecological "good or service" provided by the wider environment. The pathway links the source to the receptor. *Remediation* involves the actions involved to manage source and/or pathway in order to eliminate the risks for the receptor (Nathanail and Bardos, 2004). Risk management interventions can take place at the level of the source, the pathway, the receptor, or a combination of these. Source management involves removal or immobilization of the source of the pollution. In pathway management, the pollution is prevented from migrating along pathways, e.g., groundwater flow. Receptor management involves actions to prevent receptor access to a pathway. A common approach is "institutional control," for example, closing access to a site, temporarily prohibiting use of water from an impacted well, etc.

A conceptual site model (CSM) is a key component in risk based management of contaminated sites. It is a representation of the site which forms the basis for the mathematical modeling of contaminant fate and transport (Thomsen et al., 2016), and describes the significant site-specific features and processes, including sources, pathways, receptors, and their linkages. Comprehensive risk assessment relies on the use of a CSM to perform modeling-based environmental impact assessment (Rügner et al., 2006). The CSM, which is evolving as more information is collected, is used in all stages of the remediation process, from site investigation over risk assessment, risk management, and verification to aftercare.

Once risk management goals have been set and indicate remediation should take place, four phases can be distinguished in the soil and groundwater remediation process:

- Option appraisal
- Implementation

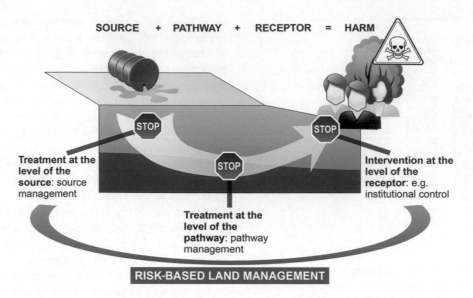

FIGURE 1.3   Risk management along a contaminant (S-P-R) linkage. (Figure licenced by the authors under a Creative Commons Attribution 4.0 International License.)

- Verification and monitoring
- Aftercare and stewardship

Option appraisal is the phase during which potential strategies for controlling the risks of contaminants are identified and evaluated. Mainly the technical suitability and feasibility of available remediation options to meet the risk management objectives agreed for a site are considered. These that appear most optimal for the site concerned will be planned in detail and carried out during the **implementation** phase. **Verification** involves all activities aimed to assess the effectiveness of a particular operation on-site, in order to confirm whether risk management objectives have been met or identify where failures in compliance have occurred. The verification typically needs to meet the requirements set out by the regulatory bodies (Nathanail et al., 2013). **Monitoring** of soil, water, and air occurs during the remediation operations to control and optimize the process, and continues after the remediation work as part of the verification that the remediation objectives have been met. It eventually serves as an early warning of adverse trends. **Aftercare and stewardship** are used to describe the site management once remediation, including monitoring, is finished. In case of containment-based remediation or pump and treat, also maintenance is involved. This phase is of particular importance in gentle remediation options, where treatments may last for decades. Sound institutional controls are needed to ensure adequate aftercare, verification, and reporting, and to ensure the availability of information in the long term.

1.4   OVERVIEW OF APPROACHES FOR SOIL AND GROUNDWATER REMEDIATION

A range of approaches is possible to effectuate risk control and remediation of a contaminated site. Intervention can take place on excavated soil or extracted water (*ex situ*) or in soil and water in the subsurface (*in situ*) (Figure 1.4). Because no soil is excavated, *in situ* treatment is less disrupting and generally less expensive than *ex situ* methods. Treatment times, however, may extend from several months to years. The major drawback of *in situ* treatment is the limited control on the uniformity of the treatment. Heterogeneities in the subsoil may cause treatment not to proceed equally fast or effective in all zones. Slow release or diffusion limitation in specific areas may cause lenses of pollution to remain after a long treatment time. This is why practitioners have to account for the risk of *rebound effects* (O'Connor et al., 2018). After a certain treatment period, concentrations in the monitoring wells may have decreased to acceptable levels, suggesting that treatment is complete. However, after a while, a slow release of contaminants from remaining lenses of pollution may cause the contaminant levels to increase again.

While *ex situ* treatment is often more complicated and expensive, a fundamental advantage is that the contamination

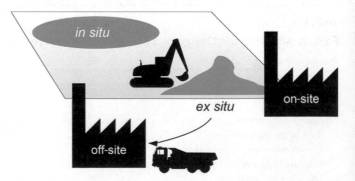

FIGURE 1.4   Soil remediation approaches classified according to whether soil is excavated for treatment (*ex situ*) or is treated in place (*in situ*). Soil that is being excavated can be treated nearby and then typically is returned to the original location (on-site) or can be transported for treatment in specialized soil treatment centers (off-site). (Figure licensed by the authors under a Creative Commons Attribution 4.0 International License.)

effectively is taken away by excavation of the contaminated soil. The practitioner has full control on the contaminated excavated material, which can be subject to a fully controlled sequence of treatments to screen, homogenize, and separate or neutralize contaminated fractions or contaminants.

Soil may be treated in mobile facilities "on-site," in which case the treated material is returned to the site. This approach would be suited if the treatment is relatively simple and fast. Alternatively, this also suits mega projects, where the scale of the project permits to establish more sophisticated temporary installations, with the benefit of supplying cleaned soil materials for regenerating the affected site on location. An example is the cleanup of the Stratford site in London to establish the Olympic Park for the 2012 Olympic Games (Hellings et al., 2011; Hou et al., 2015). "Off-site" treatment implies that the excavated soil is shipped to specialized soil treatment centers. As such, all contaminated soil is quickly taken away from the contaminated site.

Soil remediation approaches may also be classified according to their operational principle. Five categories can be discerned; biological, physical, chemical, thermal treatment, or stabilization (Martin and Bardos, 1996) (Table 1.3). The distinction is not always sharp, and approaches may combine different principles.

1.4.1 BIOLOGICAL

In bioremediation and composting, organic contaminants are broken down into products that are harmless to the environment. Biological approaches rely on the naturally occurring microorganisms in the soil, or can involve the introduction of microorganisms with an enhanced activity (Crawford and Crawford, 1996; Luka et al., 2018; Elekwachi et al., 2014).

Bioremediation is implemented in a variety of ways (Table 1.4). *Ex situ* techniques include different approaches all aimed at the stimulation of biological organic contaminant breakdown. Optimal conditions for microbial activity are established by controlling aeration and moisture, and

eventually by adding nutrients and/or microbial inocula. In *landfarming*, soil is spread over a surface and managed using conventional agricultural plowing, tiling, etc. *Windrow turning* borrows from techniques used in composting, where soil is placed in windrows where it can be periodically turned to enhance aeration. The *biopile* is a more technical implementation aimed at being more space efficient than the two previous *ex situ* methods. Soil is piled in layers on large heaps with tubings in between for aeration through air extraction or air injection, and for injection of water, and optionally nutrients and/or inocula. An industrial implementation involves the use of *bioreactors*, where technology is borrowed from industrial fermentation processes to achieve fast and specific conversion of contaminants to harmless compounds.

Many biological treatment techniques are applied *in situ*. *Natural attenuation* essentially involves leaving the system alone, relying on the natural tendency of pollutants to be broken down or to be increasingly sorbed/sequestered with time (Rügner et al., 2006). A prerequisite of the approach is that the existing pollution poses no risk for the surrounding environment. If the natural buffering capacity of the immediate surroundings does not sufficiently protect, temporary isolation measures including hydraulic control or the establishment of physical barriers may be required.

Bioventing, biosparging, and *bioslurping* are known as *pump-and-treat* methods. A system of tubings and wells is installed on the site to enhance circulation of air and oxygen (bioventing and biosparging) or groundwater (bioslurping) using pumps. Bioventing involves enhancing the air flow in the unsaturated zone, whereas biosparging refers to introducing air or other gasses under the groundwater table (Mueller et al., 1996).

Phytoremediation has been defined as "the use of green plants to remove pollutants from the environment or to render them harmless" (Cunningham and Berti, 1993). Phytoremediation of metal contaminated soils can be based on transferring the metals to the above ground part of the plant, and subsequently removing the metals from the site through

---

TABLE 1.3

Categories of Soil Remediation According to Main Operating Principles

Biological	*Contaminant degraded/immobilized/destroyed by the activity of living organisms*
	in situ: natural attenuation; bioventing; bioslurping; biosparging; phytoremediation/management; reactive barrier…
	ex situ: landfarming; biopile; windrow; bioreactors
Physical	*Contaminants separated based on physical, chemical, and thermal differences*
	in situ: soil flushing; air venting; electrokinetic remediation
	ex situ: size separation; hydrodynamic separation; gravity separation; froth flotation; magnetic separation; solvent extraction; supercritical extraction; vapor extraction
Chemical	*Contaminants converted, immobilized, or mobilized following chemical reactions*
	in situ: chemical soil washing; active treatment wall
	ex situ: chemical soil washing; dehalogenation; chemical reduction/oxidation
Stabilization	*Contaminants rendered immobile by physical containment or binding agents that change physicochemical conditions*
	in situ: hydraulic control, impermeable screens; cement and puzzolan-based systems; lime-based systems; vitrification
	ex situ: vitrification
Thermal treatment	*Contaminants destroyed or removed using elevated temperatures*
	in situ: hot air injection; steam stripping; electrical resistance heating; conductive heating; smoldering
	ex situ: thermal desorption; incineration; vitrification

TABLE 1.4
Biological Remediation Techniques

Ex situ
    Landfarming
    Windrow turning
    Biopile/composting
    Bioreactor/slurry reactor
In situ
    Natural attenuation
    Bioventing
    Biosparging
    Bioslurping
    Phytoremediation/phytomanagement

TABLE 1.5
Techniques for the Physical Separation of Contaminants or Contaminated Soil Fractions

Soil washing	
Physicochemical soil washing	Separation according to solubility
Hydrocyclones, screens	Separation according to grain size
Jigs, spirals, shaking tables	Separation according to differences in density
Froth flotation	Separation based on differences in surface properties
Magnetic separation	Separation based on differences in magnetic susceptibility
Electrokinetic remediation	Separation based on charge; mobilization through electroosmosis
Soil venting	Separation based on volatility

the harvest of the biomass. Another approach involves using plants that have a low tendency to accumulate metals to aim for a decrease in metal mobility. This decrease is effectuated by sequestration of the metals in the top soil by soil organic matter, accumulated and maintained by the established vegetation (root development, litter fall) and the associated biological activity in the root zone (Tack and Meers, 2010). With respect to organic contaminants, plants may directly take up organic contaminants or contribute to sorption and an accelerated breakdown in the root zone by excreting degradative enzymes and/or stimulating microbial activity in the root zone (Alkorta and Garbisu, 2001). In place of "phytoremediation," *phytomanagement* emerged as a broader term that includes all biological, chemical, and physical technologies in connection to vegetated sites. It involves the manipulation of soil-plant systems to affect the fluxes of trace metals in the environment, with the goal of remediating contaminated soils, recovering valuable metals, or increasing micronutrient concentrations in crops (Robinson et al., 2009), and establishes synergies with other needs such as biomass energy, improved biodiversity, watershed management, soil protection, carbon sequestration, and improved soil health (Dickinson et al., 2009).

1.4.2 PHYSICAL

Physical methods of soil remediation rely on differences in physical properties of the contaminants and different soil fractions and particles to effectuate a separation (Table 1.5). Contaminated particles may be separated based on differences in size, density, surface properties, magnetic properties, and/or charge. Properties of the contaminants include volatility, density, charge, and solubility.

Soil venting, a group of *in situ* physical remediation techniques, includes soil vapor extraction and steam stripping (which also can be classified as a thermal soil remediation technique). Volatilization of the contaminant from the soil is enhanced based on the relatively high vapor pressure of many semi-volatile and volatile organic contaminants. The collected off-gas from these processes is treated above ground in a conventional waste gas treatment unit. Non-volatile organic contaminants are not eligible for cleanup using soil venting.

Ex situ physical soil remediation involves the use of soil washing systems that exploit differences in size, density, surface properties, and magnetic properties between contaminants, contaminated soil particles, and uncontaminated soil particles to separate clean from contaminated fractions. A typical soil washing scheme is presented in Figure 1.5. Much of the technology is borrowed from the mineral processing industry. At present, commercially available techniques are limited to soils relatively low in organic matter or clay content. Fine soil fractions are often the most contaminated fraction and are usually separated for confined disposal. Physical separation usually is cost effective only for soils with a sand content higher than 50%–70%, although some approaches, including wet screening, hydrocyclones, or froth flotation, can be used for the remediation of fine-grained materials such as sediments (Dermont et al., 2008).

Extraction technologies transfer contaminants from the soil into a mobile medium where they are concentrated for further treatment or disposal. Approaches including soil flushing (Atteia et al., 2013), non-aqueous extraction (Silva et al., 2005; Chu and Kwan, 2003), and supercritical extraction (Saldaña et al., 2005) rely on differences in solubility. The extracting liquid before extraction, the leachant, can be aqueous solution (acids, alkalis, chelating agents, surfactants), organic solvents (e.g., triethylamine, TEA), or supercritical fluid (SCF, e.g., carbon dioxide or propane). These techniques essentially do not destroy the contaminant. The concentrate must therefore be further treated or safely disposed.

Electroremediation involves the application of direct currents across electrodes inserted in the soil to generate an electrical field for mobilization and extraction of contaminants. The driving mechanisms are transport under electrical fields, coupled with electrolysis and other reactions. In addition, the electrical field will cause soil-water movement through the mechanism of electroosmosis. Thus, electroremediation provides a tool to move contaminants through treatment zones in soils where the low hydraulic conductivity impairs the use of wells and pumps to circulate soil-water (Ho et al., 1999; Roulier et al., 2000).

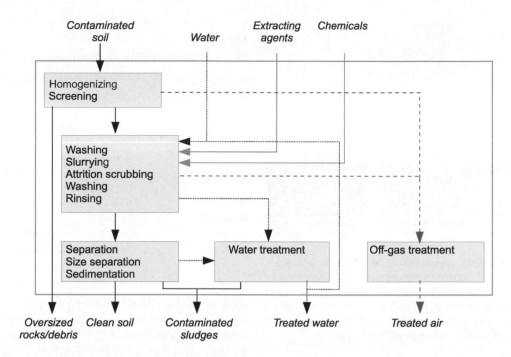

FIGURE 1.5 A typical soil washing scheme. (Figure licenced by the authors under a Creative Commons Attribution 4.0 International License.)

1.4.3 CHEMICAL

Chemical soil remediation methods rely on initiating chemical reactions that cause the contaminant to be destroyed, change solubility, or decrease toxicity (Martin and Bardos, 1996). In extraction or leaching, chemical treatment is aimed at moving the contaminant into the liquid phase, to concentrate contaminants for further treatment or safe disposal. In solidification and stabilization techniques, treatments are aimed at reducing the environmental availability of soil contaminants. Examples of detoxification are the reduction of highly toxic hexavalent chromium to insoluble and much less toxic trivalent chromiumoxides (Dhal et al., 2013), or the oxidation of As(III) to less soluble, less toxic As(V) (Miretzky and Cirelli, 2010). Organic contaminants can be subject to hydrolysis to break down the compounds in smaller, more polar, and more soluble compounds. Polymerization involves the combination of molecules to form a larger, more complex molecule. A polymer consists of a combination of a single compound, while a co-polymer is a combination of different molecules. Contaminants that are amenable for potential polymerization include styrene and vinyl chloride. Dehalogenation involves the removal of halogen atoms (chlorine, bromine) from halogenated hydrocarbons (Stroo et al., 2003).

In situ chemical soil washing requires extensive control on the collection of leachates to prevent dispersion of contaminants and reagents, which is difficult and expensive to implement in the field. Therefore, chemical soil remediation most commonly will be performed *ex situ* in soil washing installations. Large-scale application of chemical extraction processes has several issues. The use of chemical reagents significantly increases the costs; physicochemical properties of the soil may strongly be altered, not allowing revegetation without intensively adding amendments; residues of toxic chemical reagents may still be present in the treated soil and thus hamper both reuse and disposal unless subject to further expensive treatment; washing waters require intensive treatment (Dermont et al., 2008).

1.4.4 STABILIZATION

Stabilization technologies rely on eliminating hazardous effects of the contaminants by preventing their dispersion into the surrounding environment. Contaminants are not removed, but through containment, immobilization, or solidification (Figure 1.6), the source-pathway-receptor chain, i.e., the potential pathways for exposure to the target organism/environment/human is broken (Kumpiene 2020). Metal/metalloid contamination is a main target of immobilization/solidification approaches because, unlike many organic contaminants, chemical elements cannot be broken down.

The contaminated environment may be isolated from the surroundings by establishing physical screens or exerting hydraulic control on the groundwater table to prevent contaminants from moving out of the contaminated area (*containment*). *Immobilization* of the contaminant involves changing the physico/chemical environment of the soil through the use of amendments to favor a reduction in the solubility, and as a consequence also the mobility and bioavailability, of the contaminant. Soil amendments stabilize metals in soils by enhancing processes including cation exchange, adsorption, surface complexation, and precipitation. Organic amendments that can be used include sewage sludge, organic refuse, manure compost, and bagasse, whereas inorganic amendments include lime, phosphate salts, and fly ashes among others (Guo et al., 2006).

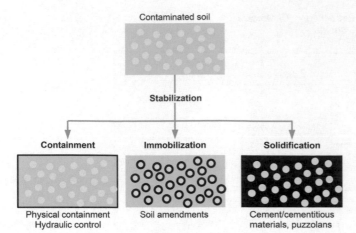

FIGURE 1.6 Conceptual approaches in stabilization techniques. The contaminant may be contained using physical screens/barriers or by effectuating hydrological control, or is rendered chemically less mobile through the addition of amendments. In solidification, the contaminant is incapsulated in a solid matrix. (Figure licensed by the authors under a Creative Commons Attribution 4.0 International License.)

Solidification involves a conversion of the contaminated environment into a solid durable matrix where contaminants are encapsulated (Hunce et al., 2012) and water flows through the matrix, which may solubilize and leach out contaminants, are strongly reduced or eliminated. Binders such as Portland cement and other cementing or puzzolanic materials (fly ash, rice husk ash, furnace slag, etc.) are mixed into the contaminated soil in order to obtain a solid material with low leachability of the contaminants (Voglar and Leštan, 2011; Zhou et al., 2006).

1.4.5 THERMAL TREATMENT

Thermal treatment technologies involve using heat to render contaminants more mobile for subsequent removal, to volatilize and separate the contaminant, or to break it down through oxidation or pyrolysis (Zhao et al., 2019; Vidonish et al., 2016). Techniques include thermal desorption, radio frequency/microwave heating, hot air injection, steam injection, smoldering, incineration, pyrolysis, and vitrification (Table 1.6).

Thermal desorption is used for the treatment of most volatile or semi-volatile contaminants, including polycyclic aromatic hydrocarbons (PAHs), polychlorinated biphenyls (PCBs), dichlorodiphenyltrichloroethane (DDT), total petroleum hydrocarbon (TPH), and Hg (Zhao et al., 2019). In *ex situ* treatment installations, the "thermal desorber" is the unit operation that heats the soil to a sufficient temperature to volatilize organic contaminants. The contaminant is removed through a gaseous exhaust stream for further treatment. The desorption units typically operate between 150°C to 350°C for petroleum contaminations, although some systems can operate up to 650°C (Troxler et al., 1993).

Alternatively, incineration involves the destruction of the contaminants by high temperature (600°C–1600°C) combustion of the impacted soils. Incineration typically is applied *ex situ* (Vidonish et al., 2016). Although these techniques

TABLE 1.6
Thermal Remediation Techniques

Technique	Main Mechanism
Thermal desorption	Desorption
Radio frequency/ microwave heating	Desorption, enhanced contaminant mobility, enhanced biodegradation
Hot air injection	Increase mobility by decreasing viscosity and increasing vapor pressure
Steam injection	Increase mobility by decreasing viscosity and increasing vapor pressure
Smoldering	Oxidation, pyrolysis, desorption
Incineration	Oxidation
Pyrolysis	Pyrolysis (thermal cracking), desorption
Vitrification	Entrapment in solid amorphous matrix, desorption, pyrolysis, oxidation

allow to quickly and reliably meet cleanup standards, they are high energy consuming and damage soil properties (Vidonish et al., 2016). Pyrolysis (400°C–500°C), where contaminants are exposed at high temperature in a low oxygen environment, may provide an equally effective approach for breakdown of contaminants while preserving nutrients and soil properties, but still is in the experimental phase (Vidonish et al., 2016).

In *in-situ* thermal treatment, the treatment area is heated by means of electrical currents (electrical resistance heating), periferal heat sources (conductive heating), steam injection, radio-frequency, or *in situ* soil mixing with steam and hot air injection. The increased temperature causes a reduction in viscosity and an increase in vapor pressure, facilitating the removal by either liquid pumping or gas phase extraction (Triplett Kingston et al., 2010). *In situ* smoldering involves establishing a self-sustained smoldering wave to destroy hydrocarbons without excavation. Via air injection and heating, combustion is started. After ignition, the injection of heat can be stopped, while air injection continues throughout the remediation process (Vidonish et al., 2016).

1.5 TOWARD MORE SUSTAINABLE SOIL AND GROUNDWATER REMEDIATION AND MANAGEMENT

Soil remediation is often associated with high costs, and the environmental footprint of the remedial actions can be significant compared to the reduction of environmental risks (Rosén et al., 2015). More awareness has arisen about the contradictory effects of remediation over recent years. There is increasing concern not only about the extent by which land can be cleaned, but also about how the cleaning is performed. There is concern on extensive use of energy and resources, disruptive effects of soil remediation on-site, noise pollution, traffic and contamination of air and water during soil remediation operations, CO_2 footprint, requiring space for disposal elsewhere when cleaning a location, etc. Several networks, including two European networks, NICOLE and COMMON FORUM, and many national initiatives (Sustainable Remediation Forum

(SURF) in the United States, SuRF-UK, SuRF-NL, SURF-China S) and many others in the International Sustainable Remediation Alliance (ISRA) have been instrumental in the rapid achievement of ISO 18504:2017. Essentially, concepts of sustainable remediation are based on achieving a net benefit overall across a range of environmental, economic, and social concerns that are judged to be representative of sustainability (Bardos, 2014). Any remedial treatment should achieve a balance between protecting the environment now and not limiting use of the environment in the future, acceptance by the general population, and not being too expensive (Hodson, 2010). Sustainable remediation goes beyond risk control and must consider the overall benefits and impacts of remediation. The end result of sustainable remediation should be either the removal of contaminants or their immobilization, in a way that the soil can be used once again for the benefit of society (Hodson, 2010). This movement has paved the way to (1) have more attention for sustainability aspects in conventional soil cleaning, (2) increase the acceptance of "gentle" remediation (phytomanagement, stabilization) options as feasible alternative approaches to conventional soil cleaning, and (3) to adapt legislation moving to a pragmatical risk-based approach in the management of contaminated land. Sustainable remediation should not only help accelerating an efficient cleanup of contaminated land, but also eliminate the potential of future land contamination (Hou and Al-Tabbaa, 2014).

REFERENCES

Alkorta, I and C Garbisu. 2001. "Phytoremediation of Organic Contaminants in Soils." *Bioresource Technology*, Reviews Issue, 79 (3): 273–276. https://doi.org/10.1016/S0960-8524(01)00016-5.

Arias Espana, VA, AR Rodriguez Pinilla, RP Bardos, and R Naidu. 2018. "Contaminated Land in Colombia: A Critical Review of Current Status and Future Approach for the Management of Contaminated Sites." *Science of the Total Environment* 618: 199–209. https://doi.org/10.1016/j.scitotenv.2017.10.245.

Atteia, O, E Del Campo Estrada, and H Bertin. 2013. "Soil Flushing: A Review of the Origin of Efficiency Variability." *Reviews in Environmental Science and Bio/Technology* 12 (4): 379–389. https://doi.org/10.1007/s11157-013-9316-0.

Bardos, RP. 2014. "Progress in Sustainable Remediation." *Remediation Journal* 25 (1): 23–32. https://doi.org/10.1002/rem.21412.

Bardos, RP, B Bone, R Boyle, D Ellis, F Evans, ND Harries, and JWN Smith. 2011. "Applying Sustainable Development Principles to Contaminated Land Management Using the SuRF-UK Framework." *Remediation Journal* 21 (2): 77–100. https://doi.org/10.1002/rem.20283.

Bardos, RP, BD Bone, R Boyle, F Evans, ND Harries, T Howard, and JWN Smith. 2016. "The Rationale for Simple Approaches for Sustainability Assessment and Management in Contaminated Land Practice." *Science of the Total Environment* 563–564: 755–768. https://doi.org/10.1016/j.scitotenv.2015.12.001.

Bardos, RP, M Knight, and S Humphrey. 2012. "Sustainable Remediation." *Environmental Scientist* 21 (3): 46–49.

Bardos, RP, P Morgan, and RPJ Swannell. 2000. "Application of In Situ Remediation Technologies – 1. Contextual Framework" Land Contamination & Reclamation 8 (4): 22.

Blum, WEH. 2005. "Functions of Soil for Society and the Environment." *Reviews in Environmental Science and Bio/Technology* 4 (3): 75–79. https://doi.org/10.1007/s11157-005-2236-x.

Brombal, D, H Wang, L Pizzol, A Critto, E Giubilato, and G Guo. 2015. "Soil Environmental Management Systems for Contaminated Sites in China and the EU. Common Challenges and Perspectives for Lesson Drawing." *Land Use Policy* 48: 286–298. https://doi.org/10.1016/j.landusepol.2015.05.015.

Carré, F, J Caudeville, R Bonnard, V Bert, P Boucard, and M Ramel. 2017. "Soil Contamination and Human Health: A Major Challenge for Global Soil Security." In *Global Soil Security*, pp. 275–295. Springer. Cham, Switzerland.

Chu, W and CY Kwan. 2003. "Remediation of Contaminated Soil by a Solvent/Surfactant System." *Chemosphere* 53 (1): 9–15. https://doi.org/10.1016/S0045-6535(03)00389-8.

Crawford, RL and DL Crawford. 1996. *Bioremediation: Principles and Applications*. Vol. 6. Biotechnology Research Series 6. Cambridge, UK: Cambridge University Press.

Cundy, AB, RP Bardos, M Puschenreiter, MJ Mench, V Bert, W Friesl-Hanl, I Müller et al. 2016. "Brownfields to Green Fields: Realising Wider Benefits from Practical Contaminant Phytomanagement Strategies." *Journal of Environmental Management* 184: 67–77. https://doi.org/10.1016/j.jenvman.2016.03.028.

Cunningham, S and W Berti. 1993. "Remediation of Contaminated Soils with Green Plants: An Overview." *In Vitro Cellular & Developmental Biology–Plant* 29 (4): 207–212.

Dermont, G, M Bergeron, G Mercier, and M Richer-Laflèche. 2008. "Soil Washing for Metal Removal: A Review of Physical/Chemical Technologies and Field Applications." *Journal of Hazardous Materials* 152 (1): 1–31. https://doi.org/10.1016/j.jhazmat.2007.10.043.

Dhal, B, HN Thatoi, NN Das, and BD Pandey. 2013. "Chemical and Microbial Remediation of Hexavalent Chromium from Contaminated Soil and Mining/Metallurgical Solid Waste: A Review." *Journal of Hazardous Materials* 250–251: 272–291. https://doi.org/10.1016/j.jhazmat.2013.01.048.

Dickinson, NM, AJM Baker, A Doronila, S Laidlaw, and RD Reeves. 2009. "Phytoremediation of Inorganics: Realism and Synergies." *International Journal of Phytoremediation* 11 (2): 97–114.

Elekwachi, CO, J Andresen, and TC Hodgman. 2014. "Global Use of Bioremediation Technologies for Decontamination of Ecosystems." *Journal of Bioremediation & Biodegradation* 5 (4). https://doi.org/10.4172/2155-6199.1000225.

Ellis, DE and PW Hadley. 2009. "Sustainable Remediation White Paper-Integrating Sustainable Principles, Practices, and Metrics into Remediation Projects." *Remediation Journal* 19 (3): 5–114. https://doi.org/10.1002/rem.20210.

European Commission. 2002. "Towards a Thematic Strategy for Soil Protection. Communication from the Commission to the Council, the European Parliament, the Economic and Social Committee and the Committee of the Regions. Com 179 Final."

FAO. 2017. *Voluntary Guidelines for Sustainable Soil Management*. Rome, Italy: Food and Agriculture Organization of the United Nations. http://www.fao.org/3/a-bl813e.pdf.

Guo, G, Q Zhou, and LQ Ma. 2006. "Availability and Assessment of Fixing Additives for the In Situ Remediation of Heavy Metal Contaminated Soils: A Review." *Environmental Monitoring and Assessment* 116 (1–3): 513–528.

Hellings, J, M Lass, J Apted, and I Mead. 2011. "Delivering London 2012: Geotechnical Enabling Works." *Proceedings of the Institution of Civil Engineers: Civil Engineering; London* 164 (6): 5–10.

Ho, SV, C Athmer, P Wayne Sheridan, B Mason Hughes, R Orth, D McKenzie, PH Brodsky et al. 1999. "The Lasagna Technology for In Situ Soil Remediation. 1. Small Field Test." *Environmental Science & Technology* 33 (7): 1086–1091. https://doi.org/10.1021/es980332s.

Hodson, ME. 2010. "The Need for Sustainable Soil Remediation." *Elements* 6 (6): 363–368. https://doi.org/10.2113/gselements.6.6.363.

Horta, A, B Malone, U Stockmann, B Minasny, TFA Bishop, AB McBratney, R Pallasser, and L Pozza. 2015. "Potential of Integrated Field Spectroscopy and Spatial Analysis for Enhanced Assessment of Soil Contamination: A Prospective Review." *Geoderma* 241–242: 180–209. https://doi.org/10.1016/j.geoderma.2014.11.024.

Hou, D and A Al-Tabbaa. 2014. "Sustainability: A New Imperative in Contaminated Land Remediation." *Environmental Science & Policy* 39: 25–34. https://doi.org/10.1016/j.envsci.2014.02.003.

Hou, D, A Al-Tabbaa, and J Hellings. 2015. "Sustainable Site Clean-up from Megaprojects: Lessons from London 2012." *Proceedings of the Institution of Civil Engineers – Engineering Sustainability* 168 (2): 61–70. https://doi.org/10.1680/ensu.14.00025.

Hunce, SY, D Akgul, G Demir, and B Mertoglu. 2012. "Solidification/Stabilization of Landfill Leachate Concentrate Using Different Aggregate Materials." *Waste Management* 32 (7): 1394–1400. https://doi.org/10.1016/j.wasman.2012.03.010.

Jenny, H. 1994. *Factors of Soil Formation: A System of Quantitative Pedology.* New York: Dover.

Jones, A, P Panagos, S Barcelo, F Bouraoui, C Bosco, O Dewitte, C Gardi, M Erhard, J Hervás, and R Hiederer. 2012. *The State of Soil in Europe.* Luxembourg: Publications Office of the European Union. https://core.ac.uk/download/pdf/38625630.pdf.

Khan, FI, T Husain, and R Hejazi. 2004. "An Overview and Analysis of Site Remediation Technologies." *Journal of Environmental Management* 71: 95–122.

Kumpiene, J. 2020. "Basic Principles of Risk Assessment of Contaminated Sites." Chapter 3 In *Soil and Groundwater Remediation Technologies: A Practical Guide*, edited by YS Ok, et al. 1: 23–29; Boca Raton, FL: CRC Press.

Lal, R and BA Stewart. 2012. "Soil Degradation: A Global Threat." In *Soil Degradation*, edited by R Lal and BA Stewart, xiii–xvii. Advances in Soil Science: Soil Degradation, Vol. 11. New York: Springer-Verlag.

Li, XN, WT Jiao, RB Xiao, WP Chen, and AC Chang. 2015. "Soil Pollution and Site Remediation Policies in China: A Review." *Environmental Reviews* 23 (3): 263–274. https://doi.org/10.1139/er-2014-0073.

Luka, Y, BK Highina, and A Zubairu. 2018. "Bioremediation: A Solution to Environmental Pollution-A Review." *American Journal of Engineering Research* 7 (2): 101–109.

Martin, I and RP Bardos. 1996. *A Review of Full Scale Treatment Technologies for the Remediation of Contaminated Soil.* Richmond, UK: EPP Publications.

Miretzky, P and A Fernandez Cirelli. 2010. "Remediation of Arsenic-Contaminated Soils by Iron Amendments: A Review." *Critical Reviews in Environmental Science and Technology* 40 (2): 93–115. https://doi.org/10.1080/10643380802202059.

Mueller, JG, CE Cerniglia, and PH Pritchard. 1996. "Bioremediation of Environments Contaminated by Polycyclic Aromatic Hydrocarbons." In *Bioremediation: Principles and Applications*, edited by Ronald L Crawford and Don L Crawford, 6: 125–194. Biotechnology Research Series 6. Cambridge, UK: Cambridge University Press.

Nathanail, CP and RP Bardos. 2004. *Reclamation of Contaminated Land.* Chichester, UK: John Wiley & Sons.

Nathanail, CP, RP Bardos, A Gillett, R Ogden, D Scott, and J Nathanail. 2013. "International Processes for Identification and Remediation of Contaminated Land." Report No. 1023-0. Nottingham, UK: Land Quality Management Ltd. randd.defra.gov.uk/Document.aspx?Document=11863_1023DefraInternational22combined(2).pdf.

NICOLE and COMMON FORUM. 2013. "Risk-Informed and Sustainable Remediation. Joint Position Statement by NICOLE and COMMON FORUM, June 9, 2013." http://www.nicole.org/uploadedfiles/2013%20NICOLE-Common-Forum-Joint-Position-Sustainable-Remediation.pdf.

O'Connor, D, D Hou, Y Sik Ok, Y Song, AK Sarmah, X Li, and FMG Tack. 2018. "Sustainable in Situ Remediation of Recalcitrant Organic Pollutants in Groundwater with Controlled Release Materials: A Review." *Journal of Controlled Release* 283: 200–213. https://doi.org/10.1016/j.jconrel.2018.06.007.

Panagos, P, M Van Liedekerke, Y Yigini, and L Montanarella. 2013. "Contaminated Sites in Europe: Review of the Current Situation Based on Data Collected through a European Network." *Journal of Environmental and Public Health* 2013: 1–11. https://doi.org/10.1155/2013/158764.

Rizzo, E, RP Bardos, L Pizzol, A Critto, E Giubilato, A Marcomini, C Albano et al. 2016. "Comparison of International Approaches to Sustainable Remediation." *Journal of Environmental Management* 184: 4–17. https://doi.org/10.1016/j.jenvman.2016.07.062.

Robinson, BH, G Bañuelos, HM Conesa, MWH Evangelou, and R Schulin. 2009. "The Phytomanagement of Trace Elements in Soil." *Critical Reviews in Plant Science* 28: 240–266.

Rosén, L, P-E Back, T Söderqvist, J Norrman, P Brinkhoff, T Norberg, Y Volchko, M Norin, M Bergknut, and G Döberl. 2015. "SCORE: A Novel Multi-Criteria Decision Analysis Approach to Assessing the Sustainability of Contaminated Land Remediation." *Science of the Total Environment* 511: 621–638. https://doi.org/10.1016/j.scitotenv.2014.12.058.

Roulier, M, M Kemper, S Al-Abed, L Murdoch, P Cluxton, J-L Chen, and W Davis-Hoover. 2000. "Feasibility of Electrokinetic Soil Remediation in Horizontal Lasagna(TM) Cells." *Journal of Hazardous Materials* 77 (1–3): 161–176.

Rügner, H, M Finkel, A Kaschl, and M Bittens. 2006. "Application of Monitored Natural Attenuation in Contaminated Land Management—A Review and Recommended Approach for Europe." *Environmental Science & Policy* 9 (6): 568–576. https://doi.org/10.1016/j.envsci.2006.06.001.

Saldaña, MDA, V Nagpal, and SE Guigard. 2005. "Remediation of Contaminated Soils Using Supercritical Fluid Extraction: A Review (1994–2004)." *Environmental Technology* 26 (9): 1013–1032. https://doi.org/10.1080/09593332608618490.

Sam, K, F Coulon, and G Prpich. 2016. "Working towards an Integrated Land Contamination Management Framework for Nigeria." *Science of the Total Environment* 571: 916–925. https://doi.org/10.1016/j.scitotenv.2016.07.075.

Scullion, J. 2006. "Remediating Polluted Soils." *Naturwissenschaften* 93 (2): 51–65. https://doi.org/10.1007/s00114-005-0079-5.

Silva, A, C Delerue-Matos, and A Fiúza. 2005. "Use of Solvent Extraction to Remediate Soils Contaminated with Hydrocarbons." *Journal of Hazardous Materials* 124 (1): 224–229. https://doi.org/10.1016/j.jhazmat.2005.05.022.

Smith, JWN. 2019. "Debunking Myths about Sustainable Remediation." *Remediation* 29: 7–15. https://doi.org/10.1002/rem.21587.

Stroo, HF, M Unger, CH Ward, MC Kavanaugh, C Vogel, A Leeson, JA Marqusee, and BP Smith. 2003. "Remediating Chlorinated Solvent Source Zones." *Environmental Science & Technology* 37 (11): 225–230.

Tack, FMG and E Meers. 2010. "Assisted Phytoextraction: Helping Plants to Help Us." *Elements* 6: 383–388.

Targulian, VO and PV Krasilnikov. 2007. "Soil System and Pedogenic Processes: Self-Organization, Time Scales, and Environmental Significance." *Catena* 71 (3): 373–381. https://doi.org/10.1016/j.catena.2007.03.007.

Thomsen, NI, PJ Binning, US McKnight, N Tuxen, PL Bjerg, and M Troldborg. 2016. "A Bayesian Belief Network Approach for Assessing Uncertainty in Conceptual Site Models at Contaminated Sites." *Journal of Contaminant Hydrology* 188: 12–28. https://doi.org/10.1016/j.jconhyd.2016.02.003.

Triplett Kingston, JL, PR Dahlen, and PC Johnson. 2010. "State-of-the-Practice Review of In Situ Thermal Technologies." *Ground Water Monitoring & Remediation* 30 (4): 64–72. https://doi.org/10.1111/j.1745-6592.2010.01305.x.

Troxler, WL, JJ Cudahy, RP Zink, JJ Yezzi, and SI Rosenthal. 1993. "Treatment of Nonhazardous Petroleum-Contaminated Soils by Thermal Desorption Technologies." *Air & Waste* 43 (11): 1512–1525. https://doi.org/10.1080/1073161X.1993.10467224.

UNEP. 2018. "United Nations Environment Programme – UNEP (2018) United Nations Environment Assembly of the United Nations Environment Programme Third Session Nairobi, December 4–6, 2017 3/6. Managing Soil Pollution to Achieve Sustainable Development." UNEP /EA.3/Res.6. Nairobi, Kenya: United Nations. https://papersmart.unon.org/resolution/uploads/k1800204.english.pdf.

Van Liedekerke, M, G Prokop, S Rabl-Berger, M Kibblewhite, and G Louwagie. 2014. "Progress in the Management of Contaminated Sites in Europe." Reference Report by the Joint Research Centre of the European Commission. Copenhagen, Denmark: European Environment Agency. https://www.eea.europa.eu/data-and-maps/data/external/eionet-nrc-soil-data-collection-1.

Vegter, J and H Kasamas, Eds. 2002. *Sustainable Management of Contaminated Land: An Overview.* Umweltbundesamt GmbH (Federal Environment Agency Ltd). http://rgdoi.net/10.13140/RG.2.2.20348.03204.

Vidonish, JE, K Zygourakis, CA Masiello, G Sabadell, and PJJ Alvarez. 2016. "Thermal Treatment of Hydrocarbon-Impacted Soils: A Review of Technology Innovation for Sustainable Remediation." *Engineering* 2 (4): 426–437. https://doi.org/10.1016/J.ENG.2016.04.005.

Vik, EA, P Bardos, J Brogan, D Edwards, F Gondi, T Henrysson, BK Jensen et al. 2001. "Towards a Framework for Selecting Remediation Technologies for Contaminated Sites." *Land Contamination & Reclamation* 9 (1): 120–128.

Voglar, GE and D Leštan. 2011. "Efficiency Modeling of Solidification/Stabilization of Multi-Metal Contaminated Industrial Soil Using Cement and Additives." *Journal of Hazardous Materials* 192 (2): 753–762. https://doi.org/10.1016/j.jhazmat.2011.05.089.

Zhao, C, Y Dong, Y Feng, Y Li, and Y Dong. 2019. "Thermal Desorption for Remediation of Contaminated Soil: A Review." *Chemosphere* 221: 841–855. https://doi.org/10.1016/j.chemosphere.2019.01.079.

Zhou, Q, NB Milestone, and M Hayes. 2006. "An Alternative to Portland Cement for Waste Encapsulation—The Calcium Sulfoaluminate Cement System." *Journal of Hazardous Materials* 136 (1): 120–129. https://doi.org/10.1016/j.jhazmat.2005.11.038.

2 Sustainable Remediation and Socio-Environmental Management at Contaminated Sites

Deyi Hou, David O'Connor, and Yuanliang Jin

CONTENTS

2.1 INTRODUCTION

The remediation industry has evolved through three stages (Figure 2.1). Initially, under pressure from regulators and the public, remediation practitioners attempted to remove all contaminants from cleanup sites. By the 1990s, it was recognized that there are bio-geophysical constraints that hinder the capacity to remediate all sites to pristine conditions. Moreover, many nations found the cost of removing all contaminants to be considerably higher than the perceived benefits to society. Therefore, a compromise was embraced, favoring risk-based remediation that makes land suitable for an intended use, but no further (O'Connor and Hou, 2018). In the most recent decade, researchers, remediation practitioners, and policy makers have also come to recognize the environmental, social, and economic impacts of remediation. A new imperative has arisen, demanding that the industry conduct remediation in a way that sustainability is maximized.

This led to the advancement of green and sustainable remediation (GSR), which is defined as a holistic approach, where the environmental, social, and economic benefits of remediation are maximized for all stakeholders, inside and outside of the site boundary, in both current and future generations. In this regard, sustainable remediation should meet the following five criteria:

1. All viable remediation alternatives are evaluated by an evidence-based sustainability assessment of environmental, social, and economic impacts;
2. The sustainability benefits of the chosen remedial alternative exceed the local and wider detrimental impacts on a life cycle basis;
3. Relevant and up-to-date best management practice is applied to minimize secondary emissions, waste, energy and resource use, and ecological impacts;
4. The social impacts to workers and local communities are considered and addressed by stakeholder engagement; and
5. The remediation minimizes life cycle project costs and maximizes gains in the wider economy.

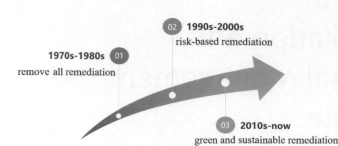

FIGURE 2.1 The three evolutionary stages of remediation.

The environmental aspect of sustainable remediation stresses the importance of minimizing the life cycle secondary emissions associated with remediation operations, for example, the greenhouse gases (GHGs) that are released during the manufacture of treatment reagents (Diamond et al., 1999; Volkwein et al., 1999; Blanc et al., 2004). There are also tertiary impacts to consider, which are associated with post-remediation site usage. The magnitude of tertiary impacts can far exceed secondary impacts, but they are more difficult to estimate, and only a few studies have considered this dimension (Hou et al., 2018a). The social aspect takes into account the social impacts of remediation projects on affected people's lives, including remediation workers, the local community, and vulnerable groups. This includes, for example, the assessment of the impacts on remediation worker's health and safety, neighborhood impacts, stakeholder satisfaction, social inclusion, etc. The economic aspect includes life cycle project costs and benefits, and the effects that remediation can have on local economies, e.g., workers' wages, local employment.

As such, the emergence of GSR has created an imperative for (i) robust sustainability assessment and (ii) socio-environmental management to the benefit of the three sustainability pillars. These aspects form the focus of this chapter. The adoption of GSR has been aided by the introduction of International Organization for Standardization (ISO) 18504:2017 on sustainable remediation (Nathanail et al., 2017; ISO, ISO 18504:2017) and an ASTM standard on GSR (ASTM, 2013), to which readers are directed for further guidance.

2.2 REMEDIATION SUSTAINABILITY

2.2.1 SUSTAINABILITY ASSESSMENT

Remediation sustainability impacts can be evaluated by various sustainability assessment (SA) tools. Life cycle assessment (LCA), as standardized by the ISO 14040 series, is considered the most comprehensive approach to SA. It is beneficial to perform quantitative SAs so that the impacts of different remediation alternatives can be objectively compared. On the other hand, qualitative assessments, such as multi-criteria assessment (MCA), offer reduced cost and complexity (Bardos et al., 2016). Moreover, when large uncertainties exist that cannot be expressed by LCA, SAs will need to be

conducted qualitatively (Suèr et al., 2004). Some assessors follow a tiered approach to SA, which begins with simplistic qualitative methods, and only proceeds to full quantitative LCA if a robust sustainability decision cannot be made. It should be noted that although LCA tools have recently become more holistic by incorporating social and economic metrics, the environmental aspect remains their predominant focus. In practice, environmental LCA is sometimes combined with socioeconomic MCA for easier assessment. Various technical guidelines for remediation sustainability assessment and LCA have been published by organizations including SuRF-UK and SuRF-US (Surf-UK, 2010; Favara et al., 2011). Moreover, proprietary SA tools based on MCA are available from various companies and professional organizations, e.g., GoldSET by Golder Associates, VHGFM by the Swedish Geotechnical Association (Brinkhoff, 2011; Golder, 2012), and Sitewise by Battelle (NAVFAC, 2013).

The increasing use of SA has brought new insights about the sustainability impacts of the remediation industry, particularly the large amount of greenhouse gas (GHG) emissions. This was exemplified by a single remediation project in New Jersey, USA, which was calculated to have the potential to emit 2.7 million tons of CO_2 if dig and haul (D&H) was employed as the remediation alternative (Garon, 2008). This would have been equivalent of ~2% of the total annual emissions for the entire state. Based on a (small) number of LCA studies, it has been revealed that the remediation of each kg of contaminants can result in up to 5000 kg of CO_2 emission (average 15 kg CO_2). Moreover, the cleanup of each kg of contaminants in groundwater may result in 130,000 kg of CO_2 emission (average 1300 kg CO_2) (Hou and Al-Tabbaa, 2014). Remediation GHG emissions are mostly associated with traditional technologies such as D&H and pump and treat (P&T). The D&H method is also known to generate large quantities of hazardous waste, while the P&T method disposes of large quantities of valuable water resources. Other remediation technologies are also associated with adverse impacts, such as acidification, eutrophication, ozone depletion, and ecological damage.

A recent study assessed the life cycle environmental impacts of a remediation project on a wider city-level scale. Hou et al. (2018a) examined the primary impacts associated with the physical state of San Francisco's brownfield and greenfield sites, the secondary impacts associated with remediating the brownfield sites, and the tertiary impacts associated with post-remediation usage (Hou et al., 2018a). It was found that redeveloping brownfield sites to avoid greenfield land use could lead to a potential net GHG reduction of 0.74 Mt CO_2 per year, the equivalent of 14% of San Francisco's GHG emissions (Hou et al., 2017).

Based on the philosophy of GSR maximizing the net environmental benefit of remediation, quantitative SA can also provide a basis to set optimal cleanup targets. For this, researchers have combined LCA with human health risk assessment (Hou et al., 2017). A case study was conducted at a lead contaminated site where it was found that an optimum cleanup target of 800 mg kg^{-1} could increase the net environmental benefit by ~3% and reduce the economic costs by more than a third, compared to the regulatory value of 255 mg kg^{-1} applicable at the time (Hou et al., 2017).

2.2.2 Sustainable Remediation Drivers

The pressure for sustainable remediation derives from the following three sources (Hou and Al-Tabbaa 2014): (1) secondary environmental impacts from remediation operations (e.g., life cycle GHG emission, air pollution, energy consumption, waste production), (2) stakeholder demand for economically sustainable brownfield restoration and green practice, and (3) institutional pressures, e.g., social norms and public policy, that promote sustainability practice.

There are promoting forces and barriers to the adoption of GSR. Such forces can be broadly grouped as stakeholder or institutional pressures. For example, regulators may act as a coercive stakeholder force if they communicate the GSR message to contaminated site managers and workers, as well as providing technical guidance. The regulators may also relieve withholding pressures (i.e., eliminating impeding forces) for adopting GSR. Institutional pressures are less tangible than stakeholder pressures, which may arise from the following: (1) society as a whole increasingly viewing sustainability as socially desirable and watches for such sustainability labels and certifications; (2) a growing number of principles and guidance promulgated by government agencies as well as professional organizations; and (3) expanded reporting practices associated with corporate social responsibility and carbon accounting.

Companies are often under corporate social responsibility pressures. When it comes to remediation, companies will often place a high priority on avoiding a negative public perception of a lack of action, or inappropriate action. There are also incentives to adopting sustainability initiatives such as GSR. For instance, during the preparations for the London 2012 Olympic Games, an independent assurance body known as the Commission for a Sustainable London 2012 (CSL) was established. The CSL influenced decision making in favor of a more sustainable remediation alternative, which saved approximately £68 million compared to the D&H alternative.

Among developing countries, China is notable as having severe soil and groundwater pollution issues and a fast growing remediation market (Hou et al., 2018b; O'Connor et al., 2018a; Peng et al., 2019). A 2014 national soil survey indicated that 16.1% of China's soil contains contaminants exceeding soil quality standards (since replaced) (MEP, 2014). In response to social and environmental pressure, the Chinese government unveiled their action plan to curb soil pollution and clean up contaminated land in May 2016. The plan, entitled the "Soil Pollution Prevention and Control Action Plan," creates a demanding schedule for national and local governments. It is expected to lead to the cleanup of 700,000 ha of contaminated land by 2020 and ensure that 95% of contaminated land is used in a safe manner by 2030. It was estimated that this will generate RMB 450bn (~US$65bn) revenue for the environmental industry by 2020 and will stimulate RMB 2.7 trillion (~US$392bn) of economic growth (People's Daily, 2016). However, based in literature and questionnaire survey findings, it is apparent that the adoption of GSR among industrial practitioners is very limited.

With improved design and cautious implementation, GSR could help optimize the use of China's resources, improve environmental quality, enhance public health, curb social inequality, act as a springboard for remediation technology innovation, and help establish a mature remediation market. In 2018, the Chinese Ministry of Science and Technology launched a major research-funding theme in soil pollution prevention and control. The 25 billion RMB (~US$3.7 billion) pot includes funds to develop green remediation materials, sustainable remediation technologies, and GSR technical standards. It is expected to stimulate a large number of studies on GSR. Recently, the first GSR technical standard, drafted by a team led by Tsinghua University, Beijing, has been made available for public comment by the China Association of Environmental Protection Industry (CAEPI). Overall, the development of GSR in China is lagging behind the West, but holds much promise going forward.

2.2.3 Adaption to Future Changes

Sustainability addresses both current and future generations (see Section 2.1); therefore, it is important to take a future perspective in GSR considerations. The long-term sustainability of a coupled human-nature system requires both change and persistence, according to complexity theories (Holling, 2001). Persistence regards the effectiveness of remediation; change regards the adaptability to new concepts by remediation practitioners and flexibility of remediation systems to incorporate new technologies.

In the context of sustainable remediation, resilience to future change is a critical consideration, including the following aspects: (1) capability to meet evolving human health and environmental standards; (2) adaptability to a variety of future land use choices; and (3) resistance to changing geophysical conditions (e.g., global climate change). As a shift in regulatory standards can reshape the balance between pollution prevention and contamination remediation (Hou, 2011), a similar trend exists for choosing various remediation strategies. As shown in Figure 2.2, as regulatory standards become more stringent, an initially costlier solution (BCC'B') may become more economical under the new social condition (ACC''A' versus ACC'A'').

In addition, geophysical changes, like climate change or pollution migration, may add further life costs to a less resilient remediation strategy. The need for resilient remediation can be important for certain industrial activities. For instance, in shale gas exploitation, surface spills and deep fracture fluid injection can lead to contamination of aquifers (Hou et al., 2012; Jiang et al., 2001) over tens to hundreds of years. Therefore, it is imperative to develop resilient strategies for these types of contamination.

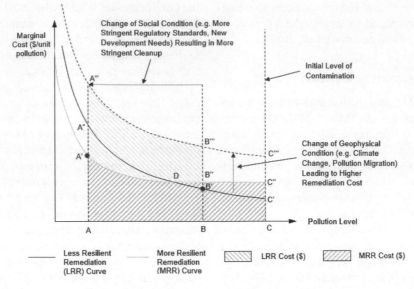

FIGURE 2.2 A more resilient remediation (MRR) strategy may initially render higher cost (i.e., BCC''B'') than a less resilient remediation (LRR) strategy (i.e., BCC'B'); however, MRR cost over the long term (i.e., ACC''A') can be much lower than LRR cost (i.e., ACC'B'B'''A''').

2.3 SOCIAL-ENVIRONMENTAL MANAGEMENT

2.3.1 PRINCIPLES

The need to protect site workers and members of the public from the social-environmental impacts of remediation first received attention in the 1980s due to high-profile cases. This is exemplified by the Corby case in the United Kingdom, one of the largest remediation projects in Europe at the time, entailing 680 acres of soil (>750 thousand m³) contaminated by dioxins, heavy metals, and polycyclic aromatic hydrocarbons. The soils were dug up and hauled through the town of Corby to a disposal site several miles away (Jiang et al., 2001). However, as an exemplar of unsustainable practice, it was reported that large quantities of contaminated soils were dropped onto public roads from uncovered trucks. Particulate dust contaminants were also reportedly emitted to the local atmosphere. As such, a group action alleging negligence, breach of statutory duty, and public nuisance was brought forward (Benning et al., 2000), thus emphasizing the cost of deficient social-environmental management.

Owing to cases like Corby, aspects of social-environmental management (SEM) are encompassed in long-standing best-practice procedures used by the established remediation industry, like sheeting trucks. Other SEM aspects, particularly regarding social management, have only recently been introduced, as the remediation industry's cleanup paradigm shifted toward GSR (Liu et al., 2018; O'Connor et al., 2018b, 2018c). SEM is now commonplace among developed remediation markets (Hou and Al-Tabbaa, 2014), with relatively systematic procedures in place, such as the US EPA's "Risk Assessment Guidance for Superfund" and "Superfund Community Involvement Handbook." It should be noted that SEM incorporates economic aspects. Based on standard

practice, socio-environmental management should abide by the following principles:

- People-oriented: The social impacts to workers and local communities should be considered
- Safety-first: The safety of workers and local communities should be the first priority in site remediation
- Green and sustainable: In order to realize GSR, various environmental and social impacts of remediation should be managed
- Life cycle: Realizing a "cradle to grave" life cycle management from site investigation to closure

2.3.2 ENVIRONMENTAL MANAGEMENT

The environmental management aspect of SEM aims to prevent the secondary polluting impacts caused during the various remediation phases, from site investigation to site closure. SEM actions may relate to the mitigation and regulation of impacts on local water resources and water quality (water); management of waste produced in site remediation (waste); dust and odor (dust); natural and man-made disasters in the process of site remediation, such as debris landslide and hazardous waste leakage (disasters); impacts on air quality (air); energy conservation and carbon reduction (energy); noise (noise); traffic (traffic); and biodiversity (biodiversity) (Hou and Al-Tabbaa, 2014). O'Connor et al. (2019) suggested these could be categorized as (1) on-site/local impacts and (2) widespread impacts.

2.3.2.1 Local Impacts

Local environmental impacts are experienced on-site, or near to the site, because of remediation activity. For example, dust and odor impacts are associated with remedial processes that involve earth moving due to suspension or volatilization of organic contaminants. Suitable ways to mitigate the dust and odor impacts of remediation works may include water-spraying, plastic sheeting over bare soil, wind shielding, and stopping work in high winds or in extremely dry conditions. Management of dust and odor is a vital part of managing local impacts at remediation sites. Although rarely experienced, the especially hazardous nature of natural and manmade disasters as a result of remediation activity demands remediation engineers to use suitable factors of safety in the engineering designs, including for remediation excavations, soil stockpiles, or contaminated water container bunding. Such requirements are often legally required and form part of general construction design and management regulations. Remediation impacts on local air quality may include dust and odor, as discussed above, as well as vehicle and machinery emissions. The impacts on air quality are experienced acutely locally, as compared to widespread atmospheric impacts. SEM steps for vapor phase contaminants include removal or destruction approaches, such as catalytic oxidation or granulated activated carbon. Vehicle/machinery exhaust emissions may be managed by using exhaust treatment technology (i.e., catalytic converters), maintenance of vehicle engines, and using clean fuel (e.g., ultra-low sulfur diesel), and monitoring atmosphere emissions regularly, as well as the step to control dust and odor discussed above.

Traffic impacts may have consequences for local air quality, noise, and dust. Traffic levels can significantly increase around remediation sites due to site workers commuting to and from the site, equipment and material deliveries, and truck movements to haul excavated soils, etc. Local residents soon become aware of increased traffic in residential areas. Remediation site operators may consider that this aspect is beyond their control, as it occurs off-site. However, some traffic management steps can be incorporated in SEM to help alleviate this impact. These include encouraging or incentivizing car sharing or public transport use, scheduling deliveries and truck movements for times of low traffic congestion, or remediating/reusing soils on-site rather than moving it off-site. It should be noted that fossil fuel burning has both local and wider ranging impacts. It is mainly associated with vehicle emissions or remediation equipment. Vehicle emissions may be reduced by reducing traffic volume, as discussed above. The reduction of fossil fuel use for remediation equipment can be achieved by using alternative renewable energy sources, employing efficient technology, or by using existing technologies in ways that are more efficient.

Noise can be caused at remediation sites by vehicles or remediation equipment. Noise mitigation measures include banning the use of vehicle horns when entering or leaving sites, scheduling noisy operations to suit local residents, selecting low noise equipment, and installing noise isolation devices and sound insulation. Monitoring noise levels at boundary points can help ensure that noise impacts are identified and managed. Biodiversity can be impacted by remediation in terms of regional land utility patterns and surface ecological processes, eco-system types and integrity, habitat destruction,

pest invasion, and damage to sensitive areas including sites of special scientific interest. Biodiversity management aims to identify species and habitats that need protection and use minimally invasive remediation techniques. For example, remediation at the London 2012 Olympic Games site required the relocation of 4000 smooth newts, 100 toads, and 330 common lizards (Rowan and Roberts, 2006).

2.3.2.2 Widespread Impacts

Widespread impacts relate to water, waste, and energy. Impacts on water quality may derive from inadequately treated contaminated groundwater, leaks, wastewater from remediation equipment/machinery use (e.g., drilling fluid), contaminated soil or waste stockpiles leaching, equipment and tool decontamination water, and sewage. The focus of remediation involves dealing with contaminated soils and groundwater, and, therefore, the remediation industry may be considered adept in managing water quality. On-site water management measures may include recycling water for beneficial use. For water quality protection, the amount and concentration of contaminants in waters produced during remediation can be reduced by treatment technology. Solid waste production at remediation sites mainly arises from treatment processes, including waste oil, chemicals, contaminated soils and sludges, discarded materials, and other wastes. Ways to deal with wastes may include minimizing the excavation of hazardous soils by using *in situ* technology or maximizing the efficiency of *ex situ* treatments, reusing treated soils on-site as backfill/made ground, and reusing or recycling other materials, as far as is practical.

As discussed in Section 2.3.2.1, energy conservation and carbon reduction is associated with both the local and widespread impacts of remediation. Widespread impacts include CO_2 emissions, which, as a GHG, have global implications. This type of impact tends to be poorly managed at remediation sites, being a low priority among the remediation industry (O'Connor et al., 2019). Traditionally, the industry has much greater experience and competence in handling polluted water and hazardous wastes than reducing GHG emissions, which have only recently become a concern with the introduction of GSR. Energy conservation and carbon reduction targets may achieve reduced fossil fuel consumption. Maintaining and updating vehicles/equipment can also reduce this impact. Life cycle analysis can be used to help to identify key activities associated with GHG emissions (US EPA, 2004).

2.4 SOCIAL MANAGEMENT

The social aspect of SEM aims to improve the impacts of remediation projects on people such as workers, the local community, and vulnerable groups. Social management actions may relate to protection of site workers' health, safety, and employment rights (safety), environmental monitoring to safeguard site workers' and the local community's health (monitoring), identification of affected groups and reduction of local community risk (groups), addressing complaints (complaints), encouraging public participation and stakeholder involvement (participation), information disclosure to the public (disclosure), increasing property values (property), gender and ethnic equality (gender), cultural heritage (heritage), involuntary migration (migration), enhancing local employment (employment), and bringing prosperity to disadvantaged communities (increase tax revenue, education, security, etc.) (prosperity). O'Connor et al. (2019) suggested these could be categorized as (1) community inclusion; (2) economic gain; and (3) health, safety, and welfare.

2.4.1 Community Inclusion

Community inclusion as a part of SEM involves addressing complaints, encouraging public participation and stakeholder involvement, information disclosure to the public, gender and ethnic equality, and cultural heritage. Among these, addressing complaints is often critical to companies due to the reputational risk of mishandling complaints, or because they have established compliant mechanisms that are required by law/ organizational policy. For instance, the World Bank asserted an appeal mechanism for the process of preparation and implementation of an environmental impact assessment at a remediation project in Shenyang, China. The public complaint procedure incorporated three phases: Phase 1: The affected person dissatisfied with an indirect impact can submit a written complaint. Phase 2: If the affected person is still dissatisfied with the decision of Phase 1, he/she may appeal. Phase 3: If the affected person is still dissatisfied with the decision of the appeal, he/she can appeal to the court after receiving the decision. Disclosure and stakeholder engagement are important for managing complaints, both as a means to providing information and a way to provide a route for complaints to be made. Project disclosure transfers important information to stakeholders during remediation, such as project progress, problems, and response to questions and complaints. Information disclosure systems/procedures can be used to promptly discern any negative aspects of remediation and uphold the interests of relevant stakeholders. An information disclosure system/procedure may include records of disclosure implementation, management measures for public complaints/appeals, response system of complain/appeal events, and records of complaints/appeal events.

Community inclusion has gained a prominent role in the GSR movement. In this context, remediation decisions should regard the views of stakeholders with a clear process by which they can participate (Roberts et al., 1996). Planning for public participation involves identifying the people affected by remediation, with a focus on the local area. Local communities will want to avoid negative impacts on them and to see benefit from remediation. To garner information specifically about how the local communities normally become involved in SEM, the following question was asked to remediation practitioners by O'Connor et al. (2019): "How has your team disclosed information about remediation details to the public?" With responses given as a choice of four options, or "other." The responses are illustrated in Figure 2.3. The survey participators were also asked: "How has your team engaged in

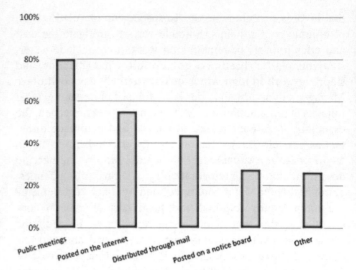

FIGURE 2.3 How teams disclose information about remediation details to the public.

public participation/dialogue?" With responses given as a choice of five options, or "other." The responses received are illustrated in Figure 2.4. As Figures 2.3 and 2.4 both show, public meetings are a key way to include local communities. Public consultation planning may include questionnaire surveys mailed to houses or house-to-house surveys to check whether sufficient SEM measures are being put in place to acquire the opinions and suggestions from the local community and to answer any relevant questions. Posting on the Internet is another important way of disclosing information to local communities, as well as traditional ways like posting information on notice boards. Other disclosure methods may include disclosure via local authorities or NGOs.

Community involvement is a major deficiency for SEM in developing countries. For example, Wang et al. (2014) pointed

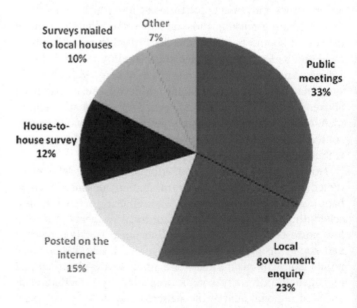

FIGURE 2.4 How teams engage in public participation/dialogue.

out that public participation in the Chinese remediation field is in need of significant improvement as the concerns of local communities are usually neglected during remediation decision making, which leads to negative social impacts (2016). The Changzhou Foreign Languages School controversy exemplified this issue, when the school was subject to relocation without the parents or students being informed about the new site being a former chemical manufacturing facility.

The gender and ethnic equality aspects of community inclusion aim to determine whether certain groups, vulnerable people, ethnic minorities, or indigenous peoples are unfairly affected by remediation. For example, pregnant women, disabled people, or low-income people may foreseeably be disproportionately impacted by remediation due to greater sensitivity to contaminated dust and odor, poor air quality, and increased traffic, etc. Management of cultural heritage takes into account the local community's culture by considering whether there are cultural relics or culturally important natural landscapes that need to be protected from remediation activity. Community engagement can help determine the value of cultural heritage to a community, which may otherwise be difficult to ascertain. Historic cultural resources may be identified by local governmental departments, who can help determine the degree of protection required. Cultural resource impact evaluation can help identify how site remediation processes may affect heritage, but this will usually require specialist knowledge. For example, remediation of the London 2012 Olympic Games site uncovered, among other artifacts, a nineteenth century boat, an eighteenth century roadway, iron age skeletons, and a bronze age hut (Rowan and Roberts, 2006), which needed to be protected.

2.4.2 Economic Gain

Economic gain in SEM covers property values, involuntary migration (migration), enhancing local employment (employment), and bringing prosperity to disadvantaged communities (increased tax revenue, education, security, etc.). Remediation makes land fit for use that could not otherwise be put to economic use, e.g., removal of legal restrictions. Therefore, increased land value is intrinsically associated with remediation but should not be overlooked. Involuntary migration may sometimes be necessary so that sites can be remediated. Compulsory purchase/use of contaminated land with the removal of occupants is not used in most remediation projects, but when it is used, it has potential to affect people's livelihoods significantly. A relocation settlement plan should be implemented if willingness exists, and sufficient compensation is paid for the land and property. Migrants' needs and concerns should be fully considered. It is desirable for resettlement to be finalized before remediation construction begins and monitored throughout the remediation process. For instance, migrants can be consulted whether they want to return to the original site as soon as possible. Among the economic gain factors, enhancing local employment and bringing prosperity to disadvantaged communities are the least directly

associated with remediation, the most difficult to measure or quantify, and are not within the traditional scope of remediation projects. With the advancement of GSR, managing employment and investment opportunities from contaminated site remediation, promoting local sustainable development, enhancing local employment, and bringing prosperity may gain much greater prominence going forward.

2.4.3 Health, Safety, and Welfare

Health, safety, and welfare covers protection of site workers' health, safety, and employment rights; environmental monitoring to safeguard site workers' and the local community's health; identification of affected groups and reduction of local community risk; and addressing complaints. Among the social components of SEM, health, safety, and welfare is generally perceived to be well managed by the industry (O'Connor et al., 2019). This may be due to long standing regulations and legal requirements and the high economic, legal, and moral cost of injuries and deaths, or, perhaps, because of the nature of remediation projects in managing health risks.

2.5 A PATH FORWARD FOR SUSTAINABLE REMEDIATION AND SOCIO-ENVIRONMENTAL MANAGEMENT

Sustainable remediation is not only a technological issue, but also a behavioral issue. The decisions made by practitioners in every step of a remediation project affect its sustainability. Decisions made in different geographic areas at different times are influenced by social, economic, cultural, and regulatory differences. There is higher awareness and adoption of sustainable remediation in North America and Europe than Asia and Latin America (Hou et al., 2016). Practices pertaining to risk control, e.g., "reduce local community risk" and "reduce site workers' risk," are of paramount importance among remediation practitioners in all regions. Socioeconomic values such as "enhancing local employment" and "bringing prosperity to a disadvantaged community" are not well adopted by remediation professionals in any particular area (Hou et al., 2016; ITRC, 2011).

The emergence of GSR represents a critical intervention point when the remediation field can be re-shaped with new norms and standards established for practitioners for years to come. GSR is becoming a new imperative for the remediation field, with implications for consultants, contractors, liability owners, regulators, and technology vendors. The industry may be viewed as a complex self-organizing adaptive system. In the ongoing sustainability movement, it is evident that there is an attempt to make a transition toward a more sustainable future. To accelerate this transition, it is necessary to develop SA tools and better understand sustainable behavior in order to catalyze interactions among researchers and actors at various levels in the industry.

GSR presents a great opportunity. The new norm should not only assist in accelerating the sustainable remediation of existing contaminated sites, but should also assist systematical planning for efficient future land remediation. GSR is not simply a technological solution or options appraisal method. It represents a new way of thinking. Practitioners should incorporate sustainability considerations across all phases of remediation. Moreover, the long-term success of GSR depends on adaptive management and institutional learning. The key is to sustain the skills and learning necessary for professionals to understand how to achieve greater sustainability, which requires an expanded scientific knowledge base. In a sense, concerted action by academia, governments, and the remediation industry is needed to fulfill the promise of the ongoing GSR movement.

The GSR movement demands that informed and integrated management steps are taken, with social-environmental management becoming a key part of sustainable risk-based remediation. Social-environmental management could also become increasingly important in developing countries such as China, where soil pollution can be extremely severe and the remediation industry is in its infancy. A global survey of remediation participators was undertaken by O'Connor et al. (2019) who found that impacts that the industry are historically most familiar with generally scored higher than those introduced by the GSR movement recently and may require different skills to manage. For example, managing risks to site workers is well established among the remediation industry, whereas there is considerably less experience in bringing prosperity to disadvantaged communities. It may be beneficial to focus efforts in mature remediation markets to bolster the management of widespread environmental impacts and to maximize gains in the wider economy. In developing countries like China, much can be learned from the strides that have made in the GSR movement in the recent decade. At present, community involvement is a major lack in the SEM of many remediation projects conducted in developing countries. In China, community inclusion by disclosure and public engagement is often very limited and requires much improvement.

REFERENCES

ASTM, Standard guide for integrating sustainable objectives into cleanup (E2876-13). 2013, West Conshohocken, PA: ASTM International.

Bardos, R.P., B.D. Bone, R. Boyle, F. Evans, N.D. Harries, T. Howard, and J.W. Smith, The rationale for simple approaches for sustainability assessment and management in contaminated land practice. *Science of the Total Environment*, 2016. **563–564**: 755–768.

Benning, L.G., R.T. Wilkin, and H.L. Barnes, Reaction pathways in the Fe–S system below 100°C. *Chemical Geology*, 2000. **167**(1–2): 25–51.

Blanc, A., H. Métivier-Pignon, R. Gourdon, and P. Rousseaux, Life cycle assessment as a tool for controlling the development of technical activities: Application to the remediation of a site contaminated by sulfur. *Advances in Environmental Research*, 2004. **8**: 613–627.

Brinkhoff, P., Multi-Criteria Analysis for Assessing Sustainability of Remedial Actions. 2011. Göteborg, Sweden: Chalmers University of Technology.

Diamond, M.L., C.A. Page, M. Campbell, S. McKenna, and R. Lall, Life-cycle framework for assessment of site remediation options: Method and generic survey. *Environmental Toxicology and Chemistry*, 1999. **18**: 788–800.

Favara, P.J., T.M. Krieger, B. Boughton, A.S. Fisher, and M. Bhargava, Guidance for performing footprint analyses and life-cycle assessments for the remediation industry. *Remediation Journal*, 2011. **21**: 39–79.

Garon, K.P., Sustainability Analysis for Improving Remedial Action Decisions. In *2008 State Superfund Managers Symposium*, 2008. Association of State and Territorial Solid Waste Management Offices: Scottsdale, AZ.

Golder, GoldSET: Fast and Reliable Sustainability Decision Support Tool. 2012.

Holling, C.S., Understanding the complexity of economic, ecological, and social systems. *Ecosystems*, 2001. **4**: 390–405.

Hou, D. and A. Al-Tabbaa, Sustainability: A new imperative in contaminated land remediation. *Environmental Science and Policy*, 2014. **39**: 25–34.

Hou, D., G. Li, and P. Nathanail, An emerging market for groundwater remediation in China: Policies, statistics, and future outlook. *Frontiers of Environmental Science & Engineering*, 2018a. **12**(1): 16.

Hou, D., J. Luo, and A. Al-Tabbaa, Shale gas can be a double-edged sword for climate change. *Nature Climate Change*, 2012. **2**(6): 385–387.

Hou, D., P. Guthrie, and M. Rigby, Assessing the trend in sustainable remediation: A questionnaire survey of remediation professionals in various countries. *Journal of Environmental Management*, 2016. **184**: 18–26.

Hou, D., S. Qi, B. Zhao, M. Rigby, and D. O'Connor, Incorporating life cycle assessment with health risk assessment to select the "greenest" cleanup level for Pb contaminated soil. *Journal of Cleaner Production*, 2017. **162**(9): 1157–1168.

Hou, D., Vision 2020: More needed in materials reuse and recycling to avoid land contamination. *Environmental Science & Technology*, 2011. **45**: 6227–6228.

Hou, D., Y. Song, J. Zhang, M. Hou, D. O'Connor, and M. Harclerode, Climate change mitigation potential of contaminated land redevelopment: A city-level assessment method. *Journal of Cleaner Production*, 2018b. **171**: 1396–1406.

ISO, ISO 18504:2017 Soil quality – Sustainable remediation. 2017, International Organization for Standardization (ISO). p. 23.

ITRC, Green and Sustainable Remediation: State of the Science and Practice. 2011. Washington, DC: Interstate Technology & Regulatory Council.

Jiang, W.T., C.S. Horng, A.P. Roberts, and D.R. Peacor, Contradictory magnetic polarities in sediments and variable timing of neoformation of authigenic greigite. *Earth and Planetary Science Letters*, 2001. **193**(1–2): 1–12.

Little, M.E., N.M. Burgess, H.G. Broders, and L.M. Campbell, Mercury in little brown bat (*Myotis lucifugus*) maternity colonies and its correlation with freshwater acidity in Nova Scotia, Canada. *Environmental Science & Technology*, 2015. **49**(4): 2059–2065.

Liu, K., D. Huisingh, J. Zhu, Y. Ma, D. O'Connor, and D. Hou, Farmers' perceptions and adaptation behaviours concerning land degradation: A theoretical framework and a case-study in the Qinghai–Tibetan Plateau of China. *Land Degradation & Development*, 2018. **29**(8): 2460–2471.

MEP, National Soil Contamination Survey Report. 2014. Beijing, China: Ministry of Environmental Protection.

Nathanail, C.P., L.M. Bakker, P. Bardos, Y. Furukawa, A. Nardella, G. Smith, J.W. Smith, and G. Goetsche, Towards an international standard: The ISO/DIS 18504 standard on sustainable remediation. *Remediation Journal*, 2017. **28**(1): 9–15.

NAVFAC, SiteWise Version 3 User Guide. 2013.

O'Connor, D. and D. Hou, Targeting cleanups towards a more sustainable future. *Environmental Science: Processes & Impacts*, 2018. **20**: 266–269.

O'Connor, D., D. Hou, J. Ye, Y. Zhang, Y.S. Ok, Y. Song, F. Coulon, T. Peng, and L. Tian, Lead-based paint remains a major public health concern: A critical review of global production, trade, use, exposure, health risk, and implications. *Environment International*, 2018a. **121**(Pt 1): 85–101.

O'Connor, D., D. Hou, Y.S. Ok, Y. Song, A.K. Sarmah, X. Li, and F.M.G. Tack, Sustainable in situ remediation of recalcitrant organic pollutants in groundwater with controlled release materials: A review. *Journal of Controlled Release*, 2018b. **283**: 200–213.

O'Connor, D., D. Müller-Grabherr, and D. Hou, Strengthening social-environmental management at contaminated sites to bolster Green and Sustainable Remediation via a survey. *Chemosphere*, 2019: 295–303.

O'Connor, D., T. Peng, J. Zhang, D.C.W. Tsang, D.S. Alessi, Z. Shen, N.S. Bolan, and D. Hou, Biochar application for the remediation of heavy metal polluted land: A review of in situ field trials. *Science of the Total Environment*, 2018c. **619**: 815–826.

Peng, T., D. O'Connor, B. Zhao, Y. Jin, Y. Zhang, L. Tian, N. Zheng, X. Li, and D. Hou, Spatial distribution of lead contamination in soil and equipment dust at children's playgrounds in Beijing, China. *Environmental Pollution*, 2019. **245**: 363–370.

People's Daily, "Soil Ten" Propose to Curb Deteriorating Soil Pollution by 2020 (2016) [in Chinese]. *People's Daily Press*, Beijing China 2016.

Prior, J., The norms, rules and motivational values driving sustainable remediation of contaminated environments: A study of implementation. *Science of the Total Environment*, 2016. **544**: 824–836.

Roberts, P., R.L. Reynolds, K.L. Verosub, and D.P. Adam, Environmental magnetic implications of greigite (Fe_3S_4) formation in a 3 my lake sediment record from Butte Valley, northern California Andrew. *Geophysical Research Letters*, 1996. **23**(20): 2859–2862.

Rowan, C.J. and A.P. Roberts, Magnetite dissolution, diachronous greigite formation, and secondary magnetizations from pyrite oxidation: Unravelling complex magnetizations in neogene marine sediments from New Zealand. *Earth and Planetary Science Letters*, 2006. **241**(1–2): 119–137.

Suèr, P., S. Nilsson-Påledal, and J. Norrman, LCA for site remediation: A literature review. *Soil & Sediment Contamination*, 2004. **13**(4): 415–425.

Surf-UK, A Framework for Assessing the Sustainability of Soil and Groundwater Remediation. 2010. London, UK: Contaminated Land: Applications in Real Environments.

US EPA, Guidance for Monitoring at Hazardous Waste Sites: Framework for Monitoring Plan Development and Implementation. OSWER Directive 9355.4-28. 2004.

US EPA, Draft Investigation of Ground Water Contamination near Pavillion, Wyoming. in US EPA. 2011. Washington, DC: United States Environmental Protection Agency.

Volkwein, S., H.W. Hurtig, and W. Klöpffer, Life cycle assessment of contaminated sites remediation. *The International Journal of Life Cycle Assessment*, 1999. **4**: 263–274.

Wang, Z., C. Fang, and M. Megharaj, Characterization of iron–polyphenol nanoparticles synthesized by three plant extracts and their fenton oxidation of azo dye. *ACS Sustainable Chemistry & Engineering*, 2014. **2**(4): 1022–1025.

3 Basic Principles of Risk Assessment of Contaminated Sites

Jurate Kumpiene

CONTENTS

3.1 INTRODUCTION

Mining, agriculture, urbanization, industrialization, and other human activities have left millions of sites worldwide affected by contamination. The full extent of this problem around the world is yet unknown. FAO encourages the world countries to place prevention of soil pollution among the top priorities (FAO, 2018). Human health and the quality of the environment are threatened by the amount of harmful substances in soil. Risks that contaminated sites pose to our health and the environment have to be assessed, and the ones that are found to be unacceptable need to be managed.

Risk is defined as the combined effect of the probability of a harmful event to occur and the magnitude of the consequence (Whyte and Burton, 1980; Andersson and Lindvall, 1995). A distinction should be made between a risk and a hazard. Hazard in this context means an event that can have adverse consequences and is related to the intrinsic properties of a substance (e.g., toxicity).

Risks related with contaminated sites occurs when contaminants from a source (contaminated soil) can spread through a pathway (usually water, but also air and dust), reach a receptor (plants, animals, and ultimately humans), and cause negative effects in any of the exposed receptors. For the risks to occur, the source-pathway-receptor chain should be unbroken, i.e., the pollutant linkage should exist (Figure 3.1).

The process of risk assessment comprises a step-wise evaluation of a site in order to determine whether it is contaminated or not, if risks exist, and if remediation to reduce those risks is required (NRC, 1996). Thus, *risk assessment* is

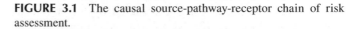

FIGURE 3.1 The causal source-pathway-receptor chain of risk assessment.

a process of estimating the potential impact of a hazard (e.g., contaminant) on environment or a specified population under specific site conditions. It is based on objective scientific findings. When an action is taken, e.g., to confine or eliminate identified risks through remediation, the *risk management* is applied. This process comprises risk perceptions and policy driven decisions. That is, social, economic, and political factors are taken into consideration (Davies, 1996), which can lead to situations where even high risks are perceived as acceptable. For example, if risks to human health or the environment are defined (e.g., emissions of a factory causing soil contamination and health deterioration of inhabitants through increased rates of human cancer cases in the surrounding villages), the willingness of the society to support (or pay for) the elimination of those risks (closure of the factory and improvement of the environmental quality for everyone) might be very low if this is the only source of people's incomes (personal economic losses).

A number of initiatives have been implemented in order to review, summarize, and standardize risk assessment and analysis efforts (e.g., by the European networks NICOLE, CARACAS, the NATO/CCMS Pilot Study program, ISO

Technical Committee 190, etc.). Numerous publications by scientists and responsible authorities are available describing risk assessment procedures (e.g., Ferguson et al., 1998; Swedish EPA, 2009; US EPA, 2019). The aim of this chapter is to provide a generalized overview of the concepts of risk assessment of contaminated sites.

3.2   RISK ANALYSIS

All activities aimed at understanding, analyzing, and evaluating risks are jointly called the *risk analysis*. Generically, a risk analysis framework is composed of three stages (Covello and Merkhofer, 1993; Figure 3.2):

1. *Hazard identification* – Identification of risk compounds, the conditions and events under which they potentially cause adverse effects to humans or the environment
2. *Risk assessment* – Description and quantification of the risks
3. *Risk evaluation* – Comparison and judgment of the significance of the risk

These activities provide input to *risk management* – identification, selection, and implementation of proper actions to control the identified risks.

The National Research Council (NRC, 1996) only distinguishes two phases: risk assessment and risk management. The hazard identification becomes a part of the risk assessment, while risk evaluation is considered by Hartlén et al. (1999) to be closer to the risk management rather than to risk analysis as defined by Covello and Merkhofer (1993) in Figure 3.2. Risk management is often complemented with risk communication, which aims at increasing the public awareness and knowledge on contamination-related risks (Davies, 1996).

Risk analysis framework can be more specific and have a focus on either human health risk assessment or on ecological risk assessment. Background and detailed guidelines for each of these frameworks can be found, for example, as described by the US National Research Council (NRC, 1983, 1996); US EPA (2019), etc.

FIGURE 3.2   Risk analysis framework as an input for risk management. (Modified from Covello, V.T. and Merkhofer, M.W., *Risk Assessment Methods: Approaches for Assessing Health and Environmental Risks*, Plenum Press, New York, 1993.)

3.3   RISK ASSESSMENT

A contaminated site can pose a risk to humans and the environment through exposure to toxic substances occurring within the contaminated area or in its vicinity. The aim of the risk assessment is to find out how much damage can be expected as a result of a specific event of a contaminant release, transport and uptake by living organisms today and in the future, and how much the risk needs to be reduced to avoid unacceptable effects. It comprises a set of analytical techniques able to provide a scientific and legally defendable basis for the support of risk management decisions. The majority of European countries use a risk assessment-based approach in decision making on remediation of contaminated sites (Perez, 2012).

Classical risk assessment comprises several steps or stages, namely hazard identification, hazard characterization, exposure assessment, and risk characterization. The structure of the ecological risk assessment and the human health risk assessment is similar as they both contain the core components of the risk assessment. The difference is that the human health risk assessment is directed toward protecting the health of individual human beings, while the ecological risk assessment is intended for protection of populations and communities of organisms (these can be individual species of organisms in their habitat) and ecological integrity (impact on changes of certain types of species in their habitat over time). Although both types of assessments use exposure models, the ecological risk assessment only considers exposure to terrestrial animals through ingestion pathways, while the human health risk assessment considers exposure to humans via ingestion, inhalation, and dermal contact. In both assessments, concentrations of contaminants are compared with environmental quality guidelines (threshold values). In risk analysis, both assessments are concluded with risk management and communication.

The Technical Guidance Document on Risk Assessment of the European Commission (CEC, 2003) describes the risk assessment process for both human health and the environment as follows:

1. Assessment of effects, comprising: (a) hazard identification, which is the identification of the adverse effects that a substance has an inherent capacity to cause; and (b) dose (concentration) – response (effects) assessment, which includes estimation of the relationship between dose (or level of exposure) and the incidence and severity of an effect.
2. Exposure assessment, which includes estimation of the concentrations/doses to which human populations or environmental compartments (aquatic environment, terrestrial environment, and air) are or may be exposed.
3. Risk characterization, which includes estimation of the incidence and severity of the adverse effects likely to occur in a human population or environmental

compartment due to actual or predicted exposure to a substance, and may include "risk estimation," i.e., the quantification of that likelihood.

The National Research Council of the Academy of Sciences model of the human health risk assessment contains the same steps as the CEC model (NRC, 1996).

A common issue that is discussed in the context of risk assessment of contaminated sites is the bioavailability of a contaminant in soil. A site might contain contaminant concentrations exceeding background and regulatory values, i.e., considered as contaminated, but the risks to the ecosystem functions and human health might be negligible. One of the main reasons is that the toxicity criteria of a substance derived based on the toxicological and epidemiological studies, usually with animals, might be different from the toxicity of the substance that is observed in a contaminated soil. For example, spilled contaminants age in soil, leading to changes in their speciation from highly soluble ionic metal forms to stable minerals (e.g., Oorts et al., 2007). These changes in soil over time may decrease the actual risks for a contaminant to cause adverse effects in living organisms. Hence, chemical speciation, mineralization, sorption, and uptake can significantly modify the actual availability of a contaminant for living organisms. Although relative bioavailability may be taken into account when calculating site-specific guideline or threshold values, decisions based on the generic soil target values assuming 100% bioavailability of substances in soil are common in generic models.

3.3.1 Human Health Risk Assessment

People are exposed to chemical substances through several pathways and in various environments (via air, food, water, medicines, at home and at work, etc.). Contaminated sites compose only a part of the total contaminant exposure to humans. The Swedish environmental protection agency (Swedish EPA, 2009) considers that the exposure from a contaminated area to a single individual should not exceed 50% of the tolerable daily intake of a chemical (TDI or corresponding toxicological reference value). This applies to substances, including non-genotoxic carcinogens or non-mutagens, that are believed to have a threshold below which no adverse health effects occur. For specific contaminants, for example metals without any known physiological function in humans, such as lead, cadmium, and mercury, the level of exposure to contaminants from contaminated sites should be below 20%, while for persistent organic pollutants, such as dioxins and polychlorinated biphenyls (PCBs), it can only be less than 10% (Swedish EPA, 2009). The proportion of the tolerable daily intake that is acceptable from the polluted area varies in different countries.

For substances without a threshold effect (genotoxic carcinogens or mutagens), the risk for human health decreases with decreasing exposure, but no exposure is safe. For genotoxic carcinogens in Sweden, as in many other European countries, it is assumed that one additional cancer case per 100,000 exposed individuals over a lifetime can be accepted (Swedish EPA, 2009). The acceptable level of an additional cancer case varies in various countries from per 10,000 to per 1,000,000 individuals (Ferguson et al., 1998).

The toxicology data that is used in risk assessment of contaminated sites are mainly derived from animal tests. The tested chemicals are usually administered orally in the feed or by gavage in a soluble form (Ferguson et al., 1998). Hence, the chemical form and exposure conditions are quite different from those occurring to humans at a contaminated site; therefore the defined permissible intake values or TDI may be considered as very conservative.

3.3.2 Ecological Risk Assessment

The ecological risk assessment aims at evaluating the likelihood of harmful effects to occur in biological organisms at higher levels, i.e., ecosystems or their components (populations, communities) through exposure to a specific concentration of a chemical. The European Union requires conducting the ecological risk assessment for industrial and agricultural chemicals, as the US Toxic Substances Control Act (Zeeman and Gilford, 1993) does for industrial chemicals and the Federal Insecticide, Fungicide, and Rodenticide Act for pesticides (Touart, 1995). But testing the effects of chemicals on all species is impossible. Therefore a bottom-up approach is used, meaning that the effects of chemicals are tested on the survival, growth, and reproduction of a few selected test species on an individual level and the results are extrapolated to the higher, ecologically relevant levels (Forbes and Forbes, 1993; Calow, 1998; Forbes et al., 2008). The concentration that is unlikely to cause adverse effects in exposed ecological systems is determined, which is then compared with an estimated exposure concentration (Zeeman and Gilford, 1993; Van Leeuwen and Hermens, 1995).

The extrapolation of the single-species toxicity test results to the ecosystem level can be done applying an assessment factor method (e.g., Chapman et al., 1998; Duke and Taggart, 2000). In this approach, the sensitivity of the most sensitive species is assessed and then assessment factors are applied to account for uncertainties in extrapolation from laboratory toxicity test data for a limited number of species to the natural environment (OECD, 1992; CEC, 2003; Verdonck et al., 2005). In the EU Technical Guidance Document on Risk Assessment of existing and new substances (CEC, 2003), data on endpoints from at least three trophic levels are considered. Endpoints from acute tests are generally expressed as concentrations affecting 50% of the test population (for sublethal effects – median effective concentration EC50; for lethal effect – LC50 lethal concentration). Endpoints from chronic tests are presented as the highest-observed concentration below which no detectable effect compared to an unexposed control is observed and is called no observed effect concentration (NOEC). Recommended extrapolation factors are then used to derive a predicted no-effect concentration (PNEC) for ecosystems by dividing the lowest short-term L(E)C50 or long-term

TABLE 3.1

Assessment Factors Presented in the EU Technical Guidance Documents for Existing and New Substance Legislation to Derive a PNEC$_{aquatic}$ in Ecological Risk Assessment

Available Data	Assessment Factor
At least one short-term EC50 or LC50 from each of three trophic levels of the base-set (fish, *Daphnia*, and algae)	1000
One long-term NOEC (either fish or *Daphnia*)	100
Two long-term NOECs from species representing two trophic levels (fish and/or *Daphnia* and/or algae)	50
Long-term NOECs from at least three species (normally fish, *Daphnia*, and algae) representing three trophic levels	10
Species sensitivity distribution (SSD) method	5-1 (to be fully justified case by case)
Field data or model ecosystems	Reviewed case by case

Source: CEC (Commission of the European Communities), Technical guidance document in support of commission directive 93/67/EEC on risk assessment for new notified substances and commission regulation (EC) No 1488/94 on risk assessment for exiting substances, Part 2, Luxembourg, 2003.

NOEC value by an assessment factor (Table 3.1, CEC, 2003). PNEC is considered as the concentration of a chemical below which adverse effects will most likely not occur.

The most reliable assessment of the environmental risk is considered to be feasible only for the aquatic environment, and most of the available effect data that exists is for aquatic organisms (CEC, 2003).

Another way to extrapolate toxicity data from single-species toxicity tests to multiple-species ecosystems is by composing a Species Sensitivity Distribution (SSD) curve (Posthuma et al., 2002). In this approach, a large toxicity data set from long-term tests for different taxonomic groups is log transformed and fitted using a distribution function, defined percentiles of which are used as criteria (Figure 3.3). NOEC and LOEC (lowest-observed effect concentration) are defined as well. It is presumed that all organisms are equally important, but they have different sensitivity for contaminants. Such statistical extrapolation approach is applied in several countries (e.g., United States, Australia, Europe) when deriving regulatory environmental quality criteria and estimating ecological risks (Aldenberg and

Slob, 1993; Hose and Van den Brink, 2004; Solomon et al., 1996; Sijm et al., 2002; Stephan, 1985; Stephan, 2002).

3.3.3 SOIL QUALITY TRIAD APPROACH

Recently, an international standard has been released describing the procedure of a site-specific ecological risk assessment of soil contamination following a TRIAD approach (ISO 19204:2017). The term TRIAD represents three lines of evidence: chemistry, toxicology, and ecology, and means that a tiered (step-wise) approach is used combining information from these disciplines for a site-specific ecological risk assessment of contaminated soil. The TRIAD starts when a decision that an ecological risk assessment needs to be performed is already taken and is completed with a decision on how to proceed (Figure 3.4). Each consecutive tier of TRIAD has

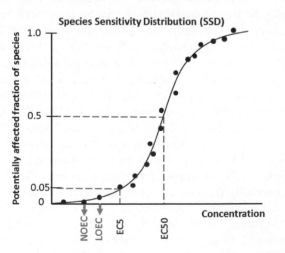

FIGURE 3.3 Example of a Species Sensitivity Distribution (SSD) curve and derived concentrations for quality criteria used in ecological risk assessment. NOEC – no observed effect concentration; LOEC – lowest-observed effect concentration; EC5 and EC50 – effective concentration affecting 5% and 50% of tested species in an ecosystem.

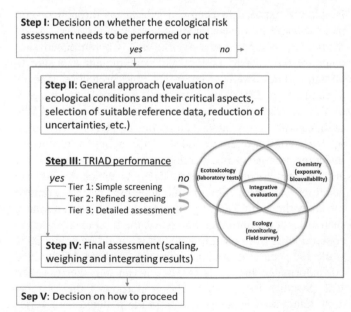

FIGURE 3.4 Decision tree of the five steps of integrating soil quality TRIAD into a site-specific ecological risk assessment. (Modified from ISO 19204:2017 (E), Soil quality – Procedure for site-specific ecological risk assessment of soil contamination [soil quality TRIAD approach], International Standardization Organization, Geneva, Switzerland.)

increasing complexity and is adjusted to the site-specific conditions. Data obtained from these lines of evidence are scaled and weighted, and the results are integrated into one risk number of the soil quality. The risk number indicates whether the ecological risk at a given site exists or not. The approach gives a possibility to stop further investigations after each tier and either decide to remediate the site or redefine the land use (i.e., redefine the level of the acceptable risks).

It is believed that this approach allows for an efficient, ecologically robust, practical and cost-effective ecological risk assessment. The actual performance of the three lines of evidence is provided in additional technical standards (e.g., ISO 15799, ISO 17616).

3.4 GUIDELINE VALUES

3.4.1 GENERIC GUIDELINE VALUES

If there is any suspicion that a site can be contaminated, sampling and analysis of soil is needed to confirm or reject the suspicion. If the background values of substances are known, they can be compared with the measured concentrations in soil. According to the Swedish EPA (2009), a contaminated site is a relatively well-defined area that contains one or more contaminants. Contaminants are defined as substances derived from human activity and present in concentrations exceeding background levels. Thus, if the measured concentrations are higher than the background concentrations, the area is considered contaminated. Nevertheless, the exceedance of the background values does not directly imply that there is a risk to human health and the environment. Also, it is generally agreed that a certain degree of exposure to contaminants is acceptable.

For the assessors implementing risk assessments, it is convenient to have a reference value which can be used as a threshold indicating non-negligible risks to people and the environment during a long-term exposure. Many countries have put efforts in developing such national or regional generic values for soil (called threshold, guideline, trigger, intervention, or regulatory values). The values are often derived based on the epidemiological studies and (eco)toxicological reference values focusing mainly on human health effects, but also on biota. In some countries, e.g., Flanders (Belgium) (VLAREBO, 2007) and China (State EPA, 1995), soil factors such as clay and organic matter contents, soil pH_{KCl}, and/or cation exchange capacity (CEC) values are also taken into account when deriving soil threshold values.

Some countries, e.g., Sweden, differentiate their soil guideline values to those that are applied for land with sensitive use (agricultural, residential, recreational areas) and for land with less sensitive use (commercial and industrial areas). While others, e.g., China (GB 15618-1995) and Latvia (Cabinet of Ministers Regulation No 804) define three levels of threshold values indicating different levels of soil contamination. The Netherlands has defined so-called intervention values for soil indicating when the functional properties of soil for humans and biota are or can be seriously impaired (RIVM, 2001).

The generic guideline values are one of the tools in risk assessment of contaminated sites that can be used to determine the need and extent of risk management efforts. Although exceedance even of the generic guideline values may not necessarily imply negative effects on humans and the environment, the values are often used in simplified risk assessment as the sole basis for a decision on-site remediation. The generic guideline values are very conservative, and to meet these criteria in all cases of site contamination may not be possible. There are cases when background concentrations of metals in certain areas are higher than the generic guideline values. Using these generic guidelines to make decisions on soil remediation in such sites would be nonsense. It is therefore necessary to have tools allowing for adjustment of guideline values for specific scenarios, that is, to be able to derive site-specific guideline values.

3.4.2 SITE-SPECIFIC GUIDELINE VALUES

Site-specific guideline values are developed taking into account the specific conditions of the assessed area. A site-specific risk assessment is required to estimate risks that deviate from those assumed in the initial risk assessment stage using the generic model. Although site-specific values usually allow for a higher contaminant concentrations to be accepted at the site, contrary scenarios are also possible where, for example, several different media (soil, groundwater, surface water, sediment) are contaminated and contaminant spreading constitutes a significant risk.

Calculations of site-specific guideline values are performed applying various models. The Swedish EPA has developed an Excel-based tool that is available for any assessor for this purpose. The model allows for adjustments of contaminant sources, release and spreading mechanisms, exposure pathways, as well as objects or receptors to be protected (people, environment, or natural resources). For example, exposure pathways of contaminants through intake of vegetables can be excluded in areas that are and will be used for industrial purposes. In this stage, adjustments of the contaminant bioavailability can also be made (Swedish EPA, 2009).

Calculations of site-specific guideline values require extensive knowledge and understanding of the basic assumptions made in the generic model. If the underlying assumptions are not clearly described, the calculated site-specific guideline values can be dismissed by a responsible authority. There can always be a case that deviates so much from the generic model's basic assumptions that the model can become inappropriate to calculate the site-specific guideline values. It is the user's responsibility to make an assessment whether the calculation tool is suitable for the given situation or not (Swedish EPA, 2009).

3.5 UNCERTAINTIES

Uncertainties exist in all parts of the risk assessment process, from problem description to the risk characterization. The uncertainties occur as a result of (i) lack of knowledge

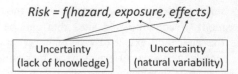

FIGURE 3.5   Uncertainties in risk assessment of contaminated sites.

or understanding and (ii) variability due to natural variation of inherent properties of ecosystem components (Figure 3.5). Extrapolating chemical effects from laboratory (eco)toxicological test results to natural ecosystems also contain uncertainties.

Risks related to the ingestion of contaminated soil, skin absorption, and inhalation of dust are usually easier to assess, and uncertainties are lower compared to the risks related to exposure as a result of the contaminant spreading from soil to other media (spreading by vapor to indoor air, spreading to groundwater wells or surface water, or through uptake by plants) (Swedish EPA, 2009). If such exposure pathways dominate, a more thorough risk assessment may need to be performed.

When derivation of guideline values are used in risk assessment using the Swedish Excel-based calculation model, a sensitivity analysis may be applied to identify the parts of the system where uncertainty is the greatest and where susceptibility to errors in the model will have the greatest impact on the final result. Uncertainties can also be estimated for future scenarios to account for long-term changes and episodic events that could increase the risks, e.g., climate change, changes in land surface topography, etc. (Swedish EPA, 2009).

3.6   RISK MANAGEMENT

Once risks are assessed using objective scientific data, options for managing the risks (policy driven process) need to be defined. Risks can be perceived by general public and stakeholders very differently as they can have different perspectives on the risks. The objective components of risk assessment need to be integrated with the subjectivity of the risk perception. Situations of land contamination are usually very complex (high complexity of the site contamination, multiple site owners, shared responsibilities, etc.). It is therefore important to communicate risks using various tools and approaches to be able to reach the stakeholders and engage them in the decision making process, get their support and acceptance for the taken decisions (Ellen et al., 2007). Although risk assessment of contaminated sites is complex and contains uncertainties, the stakeholders need to receive clear and easily understood information (SNIFFER, 2010). Guidance and recommendations on how to develop and apply effective communication strategies have been developed by various organizations and networks (e.g., NICOLE, 2004; SNIFFER, 2010).

If the defined risks are unacceptable, they should be appropriately managed. Management options are often proposed using various decision support tools, e.g., life cycle analysis (LCA), cost benefit analysis (CBA), multi-criteria analysis (MCA), and so on (Rosén et al., 2015; Söderqvist et al., 2015). The purpose of these tools is to support the decision making process in selecting the most suitable management option for a given site. However, application of most of these tools is time consuming and costly, and requires large amounts of data and competence. These tools also contain a certain degree of arbitrariness, while different tools can provide different outcomes. Similarly, application of soil remediation techniques, especially *in situ*, requires skills, competence, and knowledge of different disciplines. Time pressure and willingness to get rid of the problem once and for all often leads to situations where excavation and landfilling is selected despite questionable environmental benefits and high costs.

All parts of risk analysis and management require competent specialists to implement the objective risk assessment and to clearly communicate the message to stakeholders. This is often associated with time-consuming and costly procedures. But if done properly, it can promote the most sustainable options for management of contaminated sites and actually save resources in the long term.

REFERENCES

Aldenberg, T. and W. Slob. 1993. Confidence limits for hazardous concentrations based on logistically distributed NOEC toxicity data. *Ecotoxicology and Environmental Safety* 25:48–63.

Andersson, I. and T. Lindvall (Eds.). 1995. Risk assessment – Health – Environment (in Swedish) Swedish Environmental Protection Agency Report 4409. Swedish Environmental Protection Agency's Publishers, Stockholm.

Cabinet of Ministers Regulation No 804 "Regulation of the Quality Normatives for Soil and Subsoil" (issued on October 25, 2005), Latvia. http://likumi.lv//ta/id/120072.

Calow, P. 1998. Ecological risk assessment: Risk for what? How do we decide? *Ecotoxicology and Environmental Safety* 40:15–18.

CEC (Commission of the European Communities). 2003. Technical guidance document in support of commission directive 93/67/EEC on risk assessment for new notified substances and commission regulation (EC) No 1488/94 on risk assessment for exiting substances. Part 2. Luxembourg.

Chapman, P.M., A. Fairbrother, D. Brown. 1998. A critical evaluation of safety (uncertainty) factors for ecological risk assessment. *Environmental Toxicology and Chemistry* 17:99–108.

Covello, V.T., M.W. Merkhofer. 1993. *Risk Assessment Methods: Approaches for Assessing Health and Environmental Risks.* Plenum Press, New York.

Davies, J.C. (Ed.). 1996. *Comparing Environmental Risks: Tools for Setting Government Priorities.* Resources for the future, Washington, DC.

Duke, L.D. and M. Taggart. 2000. Uncertainty factors in screening ecological risk assessments. *Environmental Toxicology and Chemistry* 19:1668–1680.

Ellen, G.J., L. Gerrits, A.F.L. Slob. 2007. Risk perception and risk communication. *Sustainable Management of Sediment Resources* 3:233–247.

FAO (Food and Agriculture Organisation of the United Nations). 2018. The solution to soil pollution. Global symposium on soil pollution, May 2–4, 2018, Rome, Italy. Outcome document.

Ferguson, C., D. Darmendrail, K. Freier, B.K. Jensen, J. Jensen, H. Kasamas, A. Urzelai, J. Vegter (Eds.). 1998. *Risk Assessment for Contaminated Sites in Europe.* Volume 1. Scientific Basis. LQM Press, Nottingham, UK.

Forbes, T.L. and V.E. Forbes. 1993. A critique of the use of distribution-based extrapolation models in ecotoxicology. *Functional Ecology* 7:249–254.

Forbes, V.E., P. Calow, R.M. Sibly. 2008. The extrapolation problem and how population modeling can help. *Environmental Toxicology and Chemistry* 27:1987–1994.

Hartlén, J., A-M. Fällman, P-E. Back, C. Jones. 1999. Principles for risk assessment of secondary materials in civil engineering work: Survey. AFR-Report 250. AFN, Naturvårsverket. Stockholm.

Hose, G.C. and P.J. Van den Brink. 2004. Confirming the species-sensitivity distribution concept for endosulfan using laboratory, mesocosm, and field data. *Archives of Environmental Contamination and Toxicology* 47:511–520.

ISO 15799. Soil quality – Guidance on the ecological characterization of soils and soil materials. International Standardization Organization, Geneva, Switzerland.

ISO 17616. Soil quality – Guidance on the choice and evaluation of bioassays for ecotoxicological characterization of soils and soil material. International Standardization Organization, Geneva, Switzerland.

ISO 19204:2017 (E). Soil quality – Procedure for site specific ecological risk assessment of soil contamination (soil quality TRIAD approach). International Standardization Organization, Geneva, Switzerland.

NICOLE. 2004. Risk communication on contaminated land. NICOLE, Network for Industrially Contaminated Land in Europe, Report. DOW Benelux BV. http://www.nicole.org/uploadedfiles/2004-communication-contaminated-land.pdf

NRC (National Research Council). 1983. *Risk Assessment in the Federal Government: Managing the Process.* National Academy Press, Washington, DC.

NRC (National Research Council). 1996. *Understanding Risk: Informing Decisions in a Democratic Society.* National Academy Press, Washington, DC.

OECD. 1992. Report of the OECD Workshop on the extrapolation of laboratory aquatic toxicity data on the real environment. Organisation for Economic Cooperation and Development (OECD), OECD Environment Monographs No. 59, Paris.

Oorts, K., U. Ghesquiere, E. Smolders. 2007. Leaching and aging decrease nickel toxicity to soil microbial processes in soils freshly spiked with nickel chloride. *Environmental Toxicology and Chemistry* 26:1130–1138.

Perez, J. 2012. The soil remediation industry in Europe: The recent past and future perspectives. The Environment Directorate-General of the European Commission hosted conference 'Soil remediation and soil sealing', May 10–11, 2012, Brussels, Belgium.

Posthuma, L., G.W. II Suter, T.P. Traas (Eds.). 2002. *Species Sensitivity Distributions in Ecotoxicology.* Lewis Publishers, CRC Press, Boca Raton, FL.

RIVM. 2001. Technical evaluation of the intervention values for soil/sediment and Groundwater. Human and ecotoxicological risk assessment and derivation of risk limits for soil, aquatic sediment and groundwater. National Institute of Public Health and the Environment (RIVM) report 711701 023.

Rosén, L., P.-E. Back, T. Söderqvist, J. Norrman, P. Brinkhoff, T. Norberg, Y. Volchko, M. Norin, M. Bergknut, G. Döberl. 2015. SCORE: A novel multi-criteria decision analysis approach to assessing the sustainability of contaminated land remediation. *Science of the Total Environment* 511:621–638.

Sijm, D.T.H.M., A.P. van Wezel, T. Crommentuijn. 2002. Environmental risk limits in the Netherlands. In: Posthuma L, Suter GW, Traas TP (Eds.), *Species-Sensitivity Distributions in Ecotoxicology.* Lewis, Boca Raton, FL, pp. 221–254.

SNIFFER. 2010. Communicating understanding of contaminated land risks. Project UKLQ13, Final report. Scotland & Northern Ireland Forum for Environmental Research (SNIFFER), Edinburgh, UK. https://www.folkestone-hythe.gov.uk/moderngov/documents/s23615/rcabt20170419%20App%202%20contaminated%20land.pdf

Solomon, K.R., D.B. Baker, R.P. Richards, D.R. Dixon, S.J. Klaine, T.W. La Point, R.J. Kendall et al. 1996. Ecological risk assessment of atrazine in North American surface waters. *Environmental Toxicology and Chemistry* 15:31–74.

Stephan, C.E. 1985. Are the "Guidelines for deriving numerical national water quality criteria for the protection of aquatic life and its uses" based on sound judgments? In: Cardwell, R.D., Purdy, R., Bahner, R.C. (Eds.), *Aquatic Toxicology and Hazard Assessment: Seventh Symposium,* ASTM STP 854. American Society for Testing and Materials, Philadelphia, PA, pp. 515–526.

Stephan, C.E. 2002. Use of species sensitivity distributions in the derivation of water quality criteria for aquatic life by the US Environmental Protection Agency. In: Posthuma, L., Suter, G.W., Traas, T.P., (Eds.), *Species-Sensitivity Distributions in Ecotoxicology.* Lewis, Boca Raton, FL, pp. 211–220.

State EPA. 1995. Environmental quality standard for soils. GB15618–1995. Beijing, China: SEPA; 1995. State Environmental Protection Administration of China.

Swedish EPA. 2009. Risk assessment of contaminated sites. Guidance for simplified and in deep risk assessment (in Swedish). Swedish Environmental Protection Agency, Report 5977.

Söderqvist, T., T. Norberg, L. Rosén, P.-E. Back, J. Norrman. 2015. Cost-benefit analysis as a part of sustainability assessment of remediation alternatives for contaminated land. *Journal of Environmental Management* 157:267–278.

Touart, L.W. 1995. The Federal Insecticide, Fungicide, and Rodenticide Act. In: Rand, G.M. (Ed.), *Fundamentals of Aquatic Toxicology,* 2nd ed. Taylor & Francis Group, Philadelphia, PA, pp. 657–668.

US EPA. 2019. Risk assessment guidelines. https://www.epa.gov/risk/risk-assessment-guidelines#tab-1.

Van Leeuwen, C.J. and J.L.M. Hermens. 1995. *Risk Assessment of Chemicals: An Introduction.* Dordrecht, The Netherlands, Kluwer Academic Publishers.

Verdonck, F.A.M., P.A. Van Sprang, P.A. Vanrolleghem. 2005. Uncertainty and precaution in European environmental risk assessment of chemicals. *Water Science and Technology* 52:227–234.

VLAREBO. 2007. The Order of the Government of Flanders of 14 December 2007 establishing the Flemish regulation on soil remediation and protection ("VLAREBO").

Whyte, A.V.T. and I. Burton (Eds.). 1980. Environmental Risk Assessment. SCOPE, book 15. *The Scientific Committee on Problems of the Environment (SCOPE) of the International Council of Scientific Unions (ICSU),* John Wiley & Sons, 157 p.

Zeeman, M.G. and J. Gilford. 1993. Ecological hazard evaluation and risk assessment under EPA's Toxic Substances Control Act: An introduction. In: Landis, W.G., Hughes, J.S., Lewis, M.A. (Eds.), *Environmental Toxicology and Risk Assessment.* ASTM (American Society for Testing and Materials), Philadelphia, PA, pp. 7–21.

4 International Trace Element Regulation Limits in Soils

Vasileios Antoniadis, Sabry M. Shaheen, Efi Levizou, and Jörg Rinklebe

CONTENTS

4.1 INTRODUCTION

Trace elements (TEs) are those elements found in small concentrations in soil; they may become toxic to organisms, including microbes, plants, animals, and humans. Trace elements range in natural pristine soils mostly below the level of 100 mg kg^{-1} (with the exception of Fe and Mn, which are more abundant). Some of the TEs are of metallic nature and are commonly also called "heavy metals" due to their density of higher than 5 g cm^{-3}. These typically include Cd, Co, Cr, Cu, Fe, Mn, Mo, Ni, Pb, Sn, V, and Zn (Page, 1974; Hodson, 2004; Pourret and Bollinger, 2018; Wu et al., 2018). However, some metalloids (e.g., As, Sb) and non-metals (e.g., Se), can also be harmful. Of the TEs, some are known to have certain important physiological roles in plants; these include Mo, Cu, Zn, Ni, Mn, Co, Se, and Cr (Brady and Weil, 2002). Nevertheless, these elements and those without any essential biological role (e.g., Zr, Sb, Pb, Hg, and Cd) can negatively influence biochemo-morphological traits in microbes, plants, animals, and humans beyond their critical exposure level in environmental media.

Trace elements may occur in soil both from natural and anthropogenic processes (Liu et al., 2018; Bolan et al., 2014). As for the former, the levels of geogenically derived soil TEs are linked to the corresponding concentrations in parent materials from which soils have developed. As for the latter, it has the highest possibility of resulting in enhanced TE loads. Anthropogenic inputs may include industrial, mining, and agricultural activities, as well as disposal of wastes of communal sources. Such inputs often modify the natural level, biochemical balance, and geochemical cycling of TEs in the environment.

Soil contamination with TEs over the past decades has drawn a wide interest due to the potential health risks associated with it. Governments and health associations have long realized that TEs can cause serious health damages if critical exposure levels are exceeded, either through the food-to-human or through the soil-to-human pathways (Bolan et al., 2014; Beckers and Rinklebe, 2017; Gupta et al., 2019). This is why various regulatory organizations have attempted to set up safe limits of TEs in soil (and the same applies to TE levels in water and food). However, there is a problem associated with the regulation limits: they are greatly variable across countries. Although it is understandable that legislations cannot be identical at national and even at regional scale within countries, there is a need for minimizing discrepancies. One reason of this necessity is the fact that, due to globalization, products being traded among countries may fall in and out of certain limits depending on a given country. In other words, in case of a product traded across several countries, it may comply with TE permissible limits in some countries but breach the legislations of some other countries. Thus, in order to better understand the critical exposure concentrations of TEs, it is highly important to compare these values among different countries and organizations. This book chapter serves this necessity, as it collects, compares, and critiques regulation limit values of major countries and organizations around the globe.

4.2 TRACE ELEMENT BEHAVIOR IN SOIL

4.2.1 FACTORS AFFECTING TRACE ELEMENT AVAILABILITY AND MOBILITY

Soil-to-plant TE transfer is linked to their soil bioavailability (Shahid et al., 2016). Inorganic elements are generally found in soil in the following pools: (a) as free hydrated metal ions, soluble in soil solution; in that pool often various other soluble metal species are also included; (b) bound by organic colloids or complexed as stable insoluble organic phases; and (c) sorbed onto, and/or complexed with, inorganic phases (this includes, among others, electrostatic sorption onto clay colloids, interlayer fixation, complexation with carbonates, and retention by oxides) (Pourrut et al., 2011). It is agreed that the former has the highest potential for plant uptake (Natasha et al., 2018). Thus TE solubility and bioavailability govern their soil-to-plant-to-human transfer (Liu et al., 2017a). Apart from that, it should be noted that TE availability is dependent on plant species and root exudates (Puschenreiter et al., 2003; Antoniadis et al., 2017a). Such soil processes are governed by soil factors such as soil pH, particle size distribution, organic matter, clay content and mineralogy, redox conditions, and competing cations (Shahid et al., 2016; Antoniadis et al., 2017a).

Also, it is generally agreed that TE soil-to-plant mobility decreases with ageing (Antoniadis et al., 2017b). Indeed, anthropogenic inputs tend to initially cause higher mobility of TEs, with bioavailability decreasing with time (Antoniadis et al., 2017a). Geogenically originating TEs are sparingly soluble, and therefore of very low mobility in soil. This is caused by the fact that added TEs are found mostly in residual soil pools by being strongly retained in inert, non-exchangeable forms (Shaheen et al., 2017a). Fractionation studies have repeatedly exhibited this fact (Rinklebe and Shaheen, 2017; Shaheen et al., 2019). For example, Cu and Pb mainly occur in soil in complexed forms (Pourrut et al., 2011), with Cu also having a high affinity for the formation of soluble organo-metallic complexes. Thus, soluble soil Pb and Cu fractions generally constitute a minor percentage the total soil concentrations (Natasha et al., 2018). On the other hand, Cd, Ni, and Zn soluble soil fractions are typically of higher percentage than that of Cu and Pb, and this results in higher soil-to-plant dynamics (Shahid et al., 2016).

Although there is ample evidence to this effect, there are certain works exhibiting some worse-case-scenarios concerning re-mobilized TEs, in case certain soil properties change over time. For example, McBride (1995) reported that decreasing soil pH (known to enhance the solubility of cationic species) over time could increase TE mobility and thus plant uptake. Another possible adverse effect could be caused if soil organic matter (SOM) would decrease; such decrease could likely occur in case global warming is realized, an effect that would induce enhanced SOM mineralization rates. The same could occur if land use shifts from intact ecosystems (e.g., forest) to intense-tillage agriculture. In that case, previously bound TEs onto organic colloids could be solubilized due to decreased soil retention capacity.

On the other hand, plants deal with high TE concentrations mainly by compartmentalizing absorbed TEs to less-vital organs so that important physiological processes may not be significantly harmed (Antoniadis et al., 2017a). This means that plant toxicity induced by enhanced TE concentrations may not become evident until highly toxic levels are already well established (Chaney, 1980).

4.2.2 CHEMICAL BEHAVIOR AND USES OF THE STUDIED TRACE ELEMENTS

Among the studied TEs in this work, As and Pb are known to be particularly toxic; they are recognized as having both carcinogenic and non-carcinogenic health effects to humans (Sun and Chen, 2018). Arsenic (As) has two dominant valencies, As(III) and As(V), with As(III) being more mobile and toxic than As(V) (Niazi et al., 2018). Arsenic causes numerous human toxicity symptoms (Pan et al., 2018). Lead (Pb^{2+}) is not very bioavailable because it is usually well-retained by soil colloids. However, it also causes serious health effects, including nervous, skeletal, and circulatory damages of various degrees (Li et al., 2014).

Among the studied elements, Cd and Hg are both metallic and also known for their highly toxic effects. Cadmium occurs in soil solution as Cd^{2+} and is known to be highly mobile: very often it exists in its exchangeable and soluble fractions (El-Naggar et al., 2018), and it is reported as being poorly sorbed (Shaheen, 2009; Elbana et al., 2018). As for Hg (inorganic form, Hg^{2+}), it is easily methylated, and this increases its mobility (Frohne et al., 2012; Beckers and Rinklebe, 2017). Also for Hg, the gaseous emissions in the soil-plant-atmosphere pathway are very important in its chemical behavior (e.g., Böhme et al., 2005; During et al., 2009; Rinklebe et al., 2010; Beckers and Rinklebe, 2017).

Concerning Ni and Zn, they do have a highly important and essential role in plants and other organisms. Molybdenum and Tl are both metals that are used in a variety of industrial processes; Mo is used extensively in steel factories for improving steel quality and hardness, while Tl is used in electronic devices (Liu et al., 2017b); it was also used widely in past decades as rodenticide (Evans and Barabash, 2010). The two metals are indeed found in many works to be interrelated (e.g., Antoniadis et al., 2019a). As for Se, it is a nonmetal, and found in the wastes of coal-powered industries; it is also often used as fungicide (He et al., 2010). Chromium and Ni are among the most widely used metals, with many industrial applications: Cr is used in electronics and pigments, and Ni in batteries and catalysts. As for V and Co, they are often recognized as being of lithogenic origin in soils, especially as part of the clay lattice structure.

4.3 REGULATION LIMITS

4.3.1 THE NECESSITY OF SETTING UP LIMITS

In order to combat the adverse effects caused by soil contamination with TEs, countries and organizations have set up regulations that impose certain maximum limits of certain target TEs. Such limits result from the definition of the deficient, essential, optimal, and critical exposure limits of TEs for different organisms under various exposure situations. It is important to note that various age groups are under different levels of risk, with children being the most vulnerable (Jiang et al., 2017a; Tang et al., 2017; Kusin et al., 2018; Yang et al., 2018). Thus TE contamination has received substantial attention worldwide due to their toxic effects (Antoniadis et al., 2017a, 2019a), resulting in the development of regulation limits due to the discrepancies that are discussed immediately below.

4.3.2 DISCREPANCIES IN TRACE ELEMENT REGULATION LIMITS

4.3.2.1 Land Uses and Soil Properties

Regulations are often based on different land uses. These are mostly residential and industrial; however, commercial (in most countries treated as identical to industrial) and agricultural are also cited. On the other hand, some countries have developed regulations based on different soil properties. Such properties usually include pH (China; Wang and Shan, 2013), texture (Latvia), and both pH and texture (Russia; Chernova and Beketskaya, 2011); some countries have a range of values as maximum allowable limits instead of a single value (Finland and the EU Directive). However, some other countries regulate TEs based on more advanced properties, such as water table depths and/or potable/non-potable groundwater: these include Poland (Carlon, 2007) and the Province of Ontario, Canada (Ontario, 2011).

Residential areas and gardens ("allotments" in the United Kingdom) are given the lowest values, while industrial/commercial the highest. This concerns the Belgium regions of Flanders, Brussels, and Wallonia, as well as Austria, Czech Republic, France, Germany Italy, and Poland, while for countries outside Europe, Australia, Canada, Korea, South Africa, and the United States (Table 4.1).

4.3.2.2 Regulated Trace Elements

There is an evident disagreement as to the minimum number of TEs that are regulated. Thus various countries regulate different combinations of TEs, some even leaving important toxic elements such as As, Hg, and Cr unregulated. One can note that As (Figure 4.1), Cd (Figure 4.2), Pb (Figure 4.3), Hg (Figure 4.4), and Zn, Cu, and Ni (Table 4.1) are regulated in most inspected countries (but not all these TEs in all studied countries), while Co, Cr, Mo, Se, V, and Tl (Table 4.1) are rather rarely cited.

4.3.2.3 Names of Regulation Categories

In regulation limits, different terms exhibit different levels of potential risk, although the order from the lower to the higher risk is never identical. In Europe, eleven countries have identified limits as action value (AV), critical value (CV), guideline value (GV), intervention value (IV), precautionary value (PV), screening value (SV), and trigger value (TV). Among these, it is not certain which term represents what level of exposure and risk. Thus for some countries, TV is the safest value (Finland, the Netherlands, Latvia, the three Belgium regions – Wallonia, Flanders, and Brussels – Lithuania, and Germany), while for the Czech Republic the lowest is PV. The highest risk is also represented in different terms: AV (Czech Republic), CV (Latvia), GV (Sweden and Finland), and IV (Lithuania, Belgium, Netherlands, and Austria). Some European countries – Italy, Poland, and the United Kingdom – use SV differently: they have this single level of risk only (Figures 4.1 through 4.4, Table 4.1).

On the other hand, some countries use completely different terms, identical to none other: Denmark ("Soil Quality Criteria," "Ecotoxicology SQC," "Indicative Value of Serious Contamination," and TV as well; DEPE, 2002), Slovakia ("Maximum Allowable Concentrations" and "Value for Decontamination Measures"; Carlon, 2007), France ("Soil Statement Definition Value" and "Impact Statement Value"; Carlon, 2007), and Russia ("Maximum Permissible Concentration" and "Provisional Permissible Concentration"; Chernova and Beketskaya, 2011).

Concerning countries outside Europe, India has low and high limits, as is also the case with Finland and the EU Directive. The United States has one single name for limits, SV, with two levels assigned to residential and industrial uses (US EPA, 2016). Korea identifies its limits as "Soil Contamination Warning Limits" and "Soil Contamination Counterplan Limits," and within these there are three different limits for various land uses (categorized as Zones 1 to 3; MoE, 2018); Australia identifies its limits as "Ecological Investigation Levels" and "Health Investigation Levels" (HIL), and within HIL there are further differentiations for land uses (residential with and without gardens, parks-schools, and industrial-commercial; DoEC, 2010). South Africa does not have any specific name for the limits, but distinguishes limits among various land uses or water protection regimes (water protection, informal and standard residence, commercial-industrial, and a value named "Protection of Ecological Health"; GGS, 2012). As for Canada, apart from one set of national regulation values, there also exist different regulations in various Provinces. The two studied here are as follows: Alberta differentiate TEs in land uses (natural, agricultural, residential/park, industrial, and commercial; Alberta, 2016), while Ontario has many different limits distinguishing for categories such as uniform/stratified soils, deep/shallow water table, and potable/non-potable groundwater. Also within each category, there are two land uses, residential and industrial (Ontario, 2011). As for the national Canada limits ("Soil Quality Guidelines," SQG), TEs are put in three categories according to "agricultural," "industrial," and "residential" soil uses (CCME, 2018).

TABLE 4.1

Limits for Co, Cr, Cu, Mo, Ni, Se, V, Tl, and Zn (mg kg^{-1} dry soil) for Various Land Uses (Residential, Industrial, Agricultural, and Non-Specified Use) in Various Countries

	Co	Cu	Ni	Mo	Se	V	Zn	References
							Residential	
Europe								
Italy (SV)	2	120	120	NE	3	90	150	Carlon (2007)
UK (SV)	NE	NE	130	NE	35	NE	350	Carlon (2007)
Wallonia (IV)	NE	110	150	NE	NE	NE	230	CO (2008)
Germany (TV)	NE	NE	140	NE	NE	NE	NE	BBodSchV (1999)
Czech R. (AV)	NE	600	250	100	NE	450	2500	Carlon (2007)
Flanders (IV)	NE	400	470	NE	NE	NE	1000	Chernova and Beketskaya (2011)
Brussels	NE	400	470	NE	NE	NE	1000	Carlon (2007)
Non-Europe								
USA (SV)	0.3	3100	NE	390	390	NE	23000	Ontario (2011)
Ontario-Ca (SS)	8	140	100	6.9	2.4	86	340	Ontario (2011)
Alberta-Ca (Res/Park)	0.4	63	45	4	1	1	200	Alberta (2016)
Canada (SQG)	0.4	63	45	10	1	130	250	CCME (2018)
S. Africa (StandRes)	13	2300	620	1500	NE	300	19000	MoE (2018)
Korea (SCCL, Z. 1)	15	450	300	NE	NE	NE	900	MoE (2018)
Australia (HIL, Res/No-gard)	400	4000	2400	NE	NE	NE	28000	DoEC (2010)

	Co	Cu	Cr	Ni	Mo	Se	V	Tl	Zn	References
									Industrial	
Europe										
Italy (SV)	15	600	800	500	NE	15	250	350	1500	Carlon (2007)
Poland (SQS, Deep/LHC)	NE	200	800	500	200	NE	300	NE	3000	Carlon (2007)
France (ISV)	NE	950	7000	900	1000	NE	NE	NE	NE	Carlon (2007)
Czech R. (AV)	NE	1500	800[a]	500	240	NE	550	NE	5000	Carlon (2007)
Germany (TV)	NE	NE	1000	900	NE	NE	NE	NE	NE	BBodSchV (1999)
Flanders (IV)	NE	800	800	700	NE	NE	NE	NE	700	Chernova and Beketskaya (2011)
Brussels	NE	800	800	700	NE	NE	NE	NE	3000	Carlon (2007)
Wallonia (TV)	NE	500	165	210	NE	NE	NE	NE	1300	CO (2008)
UK (SV)	NE	NE	5	1800	NE	8	NE	NE	NE	Carlon (2007)
Non-Europe										
USA (SV)	63	47,000	1,800,000	NE	5800	5800	NE	NE	350000	US EPA (2016)
Ontario (SS)	40	5,600	11,000	510	1200	1200	NE	160	15000	Ontario (2011)
Alberta	1.4	91	87	89	40	2.9	NE	1	360	Alberta (2016)
Canada (SQG)	1.4	91	87	89	40	2.9	130	NE	410	CCME (2018)
Australia (HIL)	500	5,000	600	3000	5,100	NE	7,200	NE	35000	DoEC (2010)
Mexico	NE	NE	NE	NE	NE	NE	NE	NE	NE	Perez-Vazquez et al. (2016)
S. Africa (Comm/Ind)	40	19,000	790,000	10000	12,000	NE	2,600	NE	150000	GGS (2012)
Korea (SCCL, Z.3)	120	6,000	NE	1500	NE	NE	NE	NE	5000	MoE (2018)

	Co	Cu	Cr	Ni	Mo	Se	V	Zn	Tl	References
									Agricultural	
Europe										
Austria (SV)	NE	100	100	60	NE	NE	NE	300	1	CF (2009)
Wallonia (IV)	NE	50	85	65	NE	NE	NE	155	NE	CO (2008)
France (ISV, Gard/playgr)	NE	190	130	140	200	NE	560	9000	10	Carlon (2007)
UK (SGV, Allotment)	NE	NE	NE	230	NE	120	NE	NE	NE	CLAIRE (2018)
Flanders (IV)	NE	200	130	100	NE	NE	NE	600	NE	Chernova and Beketskaya (2011)
Poland (SQS, Deep/LHC)	NE	100	380	210	210	NE	300	720	NE	Carlon (2007)
Czech R. (AV, Garden)	NE	190	380	210	100	NE	NE	720	NE	Carlon (2007)
Germany (AV, Agriculture/garden)	NE	1300[a]	NE	1900	NE	NE	NE	NE	15a	BBodSchV (1999)
Non-Europe										
Korea (SCCL, Z. 2)	15	1500	NE	600	NE	NE	NE	1800	NE	MoE (2018)
Alberta-Ca	0.4	63	64	45	4	1	NE	200	1	Alberta (2016)
Canada (SQG)	0.4	63	64	45	5	1	130	250	NE	CCME (2018)
Australia (HIL, Res/Gard)	100	1000	120,000	600	390	NE	550	700	NE	DoEC (2010)
Taiwan	NE	1000	NE	600	NE	NE	NE	7000	NE	Lai et al. (2010)
Mexico	NE	NE	NE	NE	NE	NE	NE	NE	NE	Perez-Vazquez et al. (2016)
S. Africa (PEH)	NE	16000	NE	1400	NE	NE	NE	240	NE	GGS (2012)

(Continued)

TABLE 4.1 (*Continued*)

Limits for Co, Cr, Cu, Mo, Ni, Se, V, Tl, and Zn (mg kg⁻¹ dry soil) for Various Land Uses (Residential, Industrial, Agricultural, and Non-Specified Use) in Various Countries

	Co	Cu	Cr	Ni	Mo	Se	V	Zn	Tl	References
					Non-Specified Use					
Europe										
Russia (PPC, clay/pH > 5.5)	NE	250	NE	80	NE	NE	NE	200	NE	Chernova and Beketskaya (2011)
Denmark (TV)	NE	1000	1000	30	NE	0.7	42	1000	1	DEPE (2002)
Spain	NE	NE	NE	405	NE	NE	NE	80	NE	Nadal et al. (2016)
Sweden (GV, Non-sens)	35	200	150	120	100	NE	200	500	NE	INSURE (2017)
Latvia (CV, clay)	NE	150	350	200	NE	NE	NE	700	NE	INSURE (2017)
Slovakia (VDM)	NE	500	800	500	200	20	500	3000	NE	Carlon (2007)
Austria (IV)	150	600	250	140	25	NE	NE	NE	10	CF (2009)
Netherlands (IV)	240	190	380	210	200	100	250	720	15	MvV (2000)
EU Dir (High)	NE	140	NE	75	NE	NE	NE	300	NE	CEC (1986)
Lithuania (IV)	120	200	600	300	15	4.5	450	1200	NE	CF (2009)
Finland (GV, High)	250	200	300	150	NE	NE	250	400	NE	MotE (2007)
Non-Europe										
China (Grade III, pH > 6.5)	NE	500	400	200	NE	NE	NE	400	NE	Wang and Shan (2013)
India (High)	NE	600	NE	150	NE	NE	NE	270	NE	Saha et al. (2015)
Taiwan	NE	2000	NE	200	NE	NE	NE	400	NE	Lai et al. (2010)
Armenia	NE	220	NE	80	NE	NE	NE	132	NE	Tepanosyan et al. (2017)

[a] Concerns grassland only.

Abbreviations: **SV** = Screening Value; **IV** = Intervention Value; **TV** = Trigger Value; **AV** = Action Value; **SS** = Stratified Site Condition Standards in a Non-potable Groundwater Conditions; **Res/Park** = Residential/Parks; **Res/No-gard** = Residence with no garden; **SQG** = Soil Quality Guidelines; **StandRes** = Value for "Standard Residential Area"; **SCCL** = Soil Contamination Counterplan Limit; **Z.1** = Zone 1 (includes residential area); **HIL** = Health Investigation Levels; **Gard/playgr** = Garden/playground; **SQS** = Soil Quality Standards; **Deep/LHC** = Agricultural use/deep soil (depth > 15 m with low hydraulic conductivity (<10⁻⁷ m s⁻¹); **ISV** = Impact Statement Value; **Comm/Ind** = Commercial/Industrial; **Z.3** = Zone 3 (includes industrial area); **SGV** = Soil Guideline Value; **Z.2** = Zone 2 (includes agricultural land use); **Res/Gard** = Residence with garden; **PEH** = "Protection Ecological Health"; **PPC** = Provisional Permissible Concentration; **VDM** = Value for Decontamination Measures; **High** = High value in a range of values; **Grade III** = Applying for High Adsorption Capacity Soils Only; **Non-sens** = Non-sensitive soils; **NA** = Non-applicable; **NE** = Non-existent.

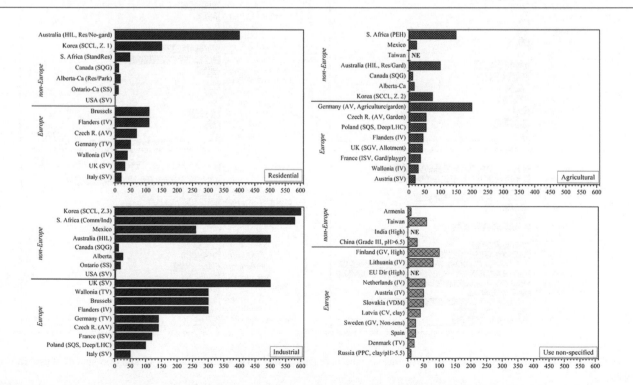

FIGURE 4.1 Limits for As (mg kg⁻¹ dry soil) for various land uses (residential, industrial, agricultural, and non-specified use) in various countries.

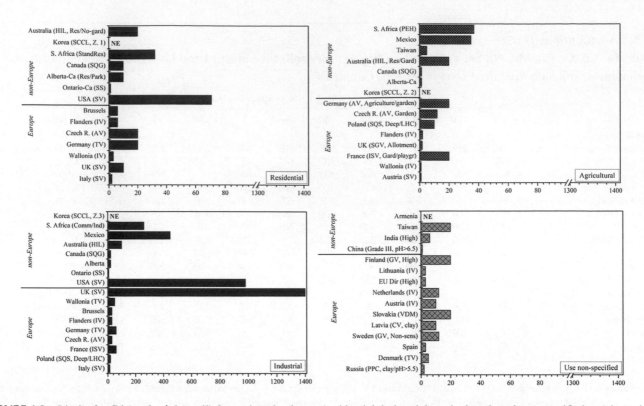

FIGURE 4.2 Limits for Cd (mg kg^{-1} dry soil) for various land uses (residential, industrial, agricultural, and non-specified use) in various countries.

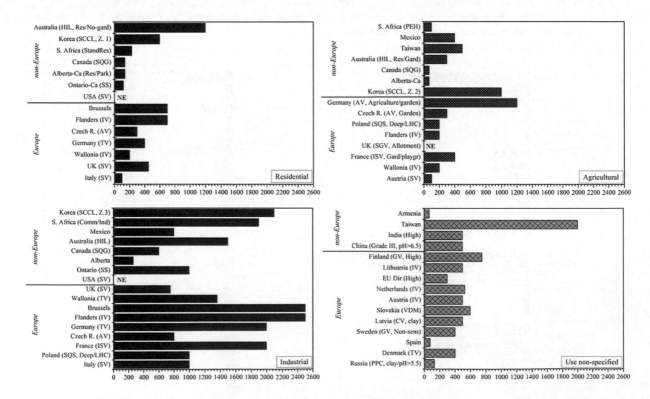

FIGURE 4.3 Limits for Pb (mg kg^{-1} dry soil) for various land uses (residential, industrial, agricultural, and non-specified use) in various countries.

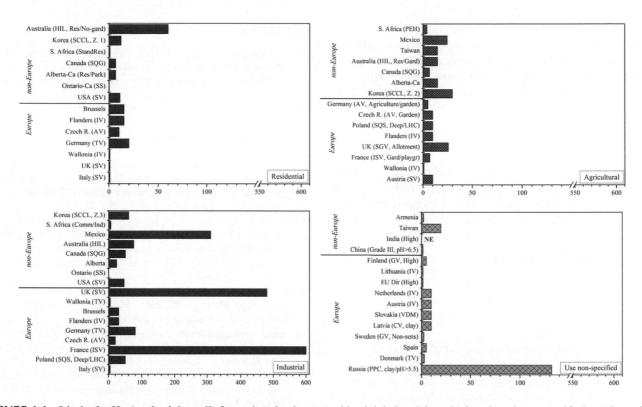

FIGURE 4.4 Limits for Hg (mg kg⁻¹ dry soil) for various land uses (residential, industrial, agricultural, and non-specified use) in various countries.

4.3.2.4 Maximum Values in Regulation Limits

Maximum values in the regulations are very different among countries. For As, residential limits range from 20 (SV, Italy; Carlon, 2007) to 110 (IV, Flanders, Chernova and Beketskaya, 2011); for non-European countries: from 0.68 (SV, USA; US EPA, 2016) to 400 (HIL, Australia; DoEC, 2010) (Figure 4.1). Similar is the case with Cd: residential limits range from 2 (Italy; Carlon, 2007) to 20 in Germany (TV; BBodSchV, 1999) and Czech Republic (AV; Carlon, 2007). In non-European countries limits range from 1.2 (SS, Ontario-Canada; Ontario, 2010) to 71 (USA; US EPA, 2016) (Figure 4.2). For Pb, the diversity is also high: the lowest among the scrutinized countries for residential use is 100 (Italy) and the highest 700 (Flanders). In the non-European countries the variability is similar (lowest value is 120 for Ontario-Canada, highest 1200 for Australia) (Figure 4.3; all above values in mg kg⁻¹). It is noteworthy that As is not even regulated in the EU Directive; because of that, As is absent from those European countries that have complied with the EU Directive (such as Greece); outside Europe, similar is the case with India. Likewise, Pb is not regulated in the United Kingdom and the United States. Great is also the variability for Hg: in residential use it ranges from 1 to 15 in Europe and 0.27 to 60 outside Europe, in industrial use from 5 to 600 in Europe, and 0.27 to 75 outside Europe, and it also ranges from 1.5 to 132 in Europe and 1.5 to 20 outside Europe when land use is not specified (Figure 4.4). In India, Hg is not even

regulated. The diversity concerning limits of Zn, Cu, and Ni (Table 4.1) is also great. All this exhibits the great differences in environmental policy across countries.

4.3.2.5 Countries with No Soil Regulation

Although there is a problem of countries with similar climate and soil conditions having greatly diverse limits, some other countries face an even more serious problem: they lack regulations, an often-occurring situation in African and Asian countries (Antoniadis et al., 2019b). In such countries, scientific works that need to utilize limits for comparison reasons in monitoring works present guidelines of other countries. These comparisons and recommendations are very complicated due to expected differences in ambient conditions.

4.4 CONCLUSIONS

This review exhibited the diversity in regulation limits: Limits of various countries do not include the same TEs, while some countries and organizations fail to regulate highly toxic elements, such as As. Also land uses determine the TE limits, but not all countries use them and no uniform land use categories are drafted. Moreover, TE values presented in regulation limits are vastly different among countries. This leads to the realization that some TEs are overly protective: this results in high costs of legislation enforcement. At the same time, some regulations are not sufficiently protective concerning other TEs.

In that case, health risks may occur. We conclude that there is a great necessity of dramatically reducing the existing variability of TE limits. There should be a serious attempt at (a) taking into consideration plants as part of the soil-to-plant inter-relation of the TEs and (b) considering multi-element contamination cases.

ACKNOWLEDGMENTS

Dr. Antoniadis would like to thank the German Academic Exchange Service (Deutscher Akademischer Austauschdienst; DAAD) for awarding the program "Research Stays for University Academics and Scientists 2018," during which this project was conducted with Prof. Rinklebe, Wuppertal University, Germany.

LIST OF ABBREVIATIONS

AV	action value
CCOL	contamination cutoff level
CUL	cleanup levels
CV	critical value
EIL	ecological investigation levels
ESQC	ecotoxicology soil quality criteria
FD	full-depth generic site condition standard in non-potable groundwater conditions
Grade I	BG concentration
Grade II	upper permissible levels for safe food, healthy plant growth, and no expected effect on water
Grade III	applying for high adsorption capacity soils only.
GV	guideline value
HIL	health investigation levels
ISV	impact statement value
IV	intervention value
IVSC	indicative value of serious contamination
Lim Agr	limited agricultural use
MAC	maximum allowable concentration
MPC	maximum permissible concentration
PPC	provisional permissible concentration
PV	precautionary value
SCCL	soil contamination counterplan limits
SCLW	soil contamination warning limits
SGV	soil guideline value
SQC	soil quality criteria
SQG	soil quality guidelines
SS	stratified site condition standard in non-potable groundwater conditions
SSDV	soil statement definition value
SV	screening value
TV	target value
VDM	value calling for decontamination measures
Zone 1	fields, paddies, orchards, pastures, building lots (residential), school sites, ditches, parks, and children's playgrounds, and is subject to the strongest limits
Zone 2	forests, salterns, building lots (non-residential), warehouses, rivers, historic ruins, physical sites, recreational sites, and miscellaneous sites
Zone 3	factory sites, gas station sites, roads, parking lots, railway sites, and national defense and military facilities

REFERENCES

Alberta Government, 2016. Alberta Tier 1: Soil and Groundwater Remediation Guidelines. Available at https://www.alberta.ca/part-one-soil-and-groundwater-remediation.aspx?utm_source=redirector (Last accessed on June 3, 2019).

Antoniadis, V., Golia, E.E., Liu, Y., Wang, S., Shaheen, S.M., Rinklebe, J. 2019a. Soil and maize contamination by trace elements and associated health risk assessment in the industrial area of Volos, Greece. *Environ. Int.* 124, 79–88.

Antoniadis, V., Golia, E.E., Polyzois, T., Petropoulos, S., 2017b. Hexavalent chromium availability and phytoremediation potential of *Cichorium spinosum* as affected by manure, zeolite and soil ageing. *Chemosphere* 171, 729–734.

Antoniadis, V., Levizou, E., Shaheen, S.M., Ok, Y.S., Sebastian, A., Baum, C., Prasad, M.N.V., Wenzel, W.W., Rinklebe, J., 2017a. Trace elements in the soil-plant interface: Phytoavailability, translocation, and phytoremediation–A review. *Earth Sci. Rev.* 171, 621–645.

Antoniadis, V., Shaheen, S.M., Levizou, E., Shahid, M., Niazi, N.B., Vithanage, M., Ok, Y.S., Bolan, N., Rinklebe, J. 2019b. A critical prospective analysis of the potential toxicity of trace element regulation limits in soils worldwide: Are they protective concerning health risk assessment? – A review. *Environ. Int.* 127, 819–847.

BBodSchV, 1999. Bundes-Bodenschutz- und Altlastenverordnung (BBodSchV) vom 12. Juli 1999. Bundesgesetzblatt I 1999, 1554 (English Version) [Federal Soil Protection and Contaminated Sites Ordinance, dated July 12, 1999].

Beckers, F., Rinklebe, J., 2017. Cycling of mercury in the environment: Sources, fate, and human health implications: A review. *Crit. Rev. Environ. Sci. Technol.* 47, 693–794.

Böhme, F., Rinklebe, J., Staerk, H.-J., Wennrich, R., Mothes, S., Neue, H.-U., 2005. A simple field method to determine mercury volatilisation from soils. *Environ. Sci. Pollut. Res.* 12, 133–135.

Bolan, N., Kunhikrishnan, A., Thangarajan, R., Kumpiene, J., Park, J., Makino, T., Kirkham, M.B., Scheckel, K., 2014. Remediation of heavy metal(loid)s contaminated soils – To mobilize or not to mobilize? *J. Hazar. Mater.* 266, 141–166.

Brady, N.C., Weil, R.R., 2002. *The Nature and Properties of Soils*, 13th edition. Prentice Hall, Upper Saddle River, NJ.

Carlon, C., 2007. Derivation Methods of Soil Screening Values in Europe: A review and Evaluation of National Procedures Towards Harmonization. Available at https://esdac.jrc.ec.europa.eu/content/derivation-methods-soil-screening-values-europe-review-and-evaluation-national-procedures (Last accessed on June 3, 2019).

CCME (Canadian Council of Ministers of the Environment), 2018. Soil Quality Guidelines for the Protection of Environmental and Human Health. Available at http://st-ts.ccme.ca/en/ (Last accessed on June 3, 2019).

CEC (Council of the European Communities), 1986. The Protection of the Environment, and in Particular of the Soil, When Sewage Sludge Is Used in Agriculture. Council Directive of June 12, 1986. *Official Journal of the European Communities* No L 181/6.

CF (Common Forum – Working Document), 2009. Compilation of Standards for Contamination of Surface Water, Groundwater, Sediments and Soil. Available at https://www.commonforum.eu/Documents/WorkingDocument/CFWD_Standards.pdf (Last accessed on June 3, 2019).

Chaney, R.L., 1980. Health risks associated with toxic metals in municipal sludge, In: G.L. Bitton, B.L. Damron, G.T. Edds, and J.M. Davidson (Eds.). *Sludge: Health Risks of Land Application*. Ann Arbor Science Publishers, Woburn, MA, p. 59.

Chernova, O.V., Beketskaya, O.V., 2011. Permissible and background concentrations of pollutants in environmental regulation (heavy metals and other chemical elements). *Eurasian Soil Sci.* 44, 1008–1017.

CLAIRE, 2018. Introduction to Soil Guideline Values: Heavy Metals and Other Inorganic Compounds. Available at https://www.claire.co.uk/information-centre/water-and-land-library-wall/44-risk-assessment/178-soil-guideline-values?showall=&start=1. (Last accessed on June 3, 2019).

CO (Coordination Officieuse), 2008. 5 décembre 2008 – Décret Relatif à la Gestion des Sols (1) (M.B. 18.02.2009 – add. 06.03.2009 – entrée en vigueur le 18.05.2009). Available at http://environnement.wallonie.be/legis/solsoussol/sol003.htm (Last accessed on June 3, 2019).

DEPE (Danish Environmental Protection Agency), 2002. Environmental Guidelines No. 7, 2002. Vejledning fra Miljøstyrelsen. Guidelines on Remediation of Contaminated Sites. Available at https://www2.mst.dk/udgiv/publications/2002/87-7972-280-6/pdf/87-7972-281-4.pdf. (Last accessed on September 16, 2018).

DoEC (Department of Environment and Conservation, Australia), 2010. Contaminated Sites Management Series Assessment Levels for Soil, Sediment and Water, Version 4, Revision 1, February 2010. Available at https://www.der.wa.gov.au/images/documents/your-environment/contaminated-sites/guidelines/2009641_-_assessment_levels_for_soil_sediment_and_water_-_web.pdf (Last accessed on June 3, 2019).

During, A., Rinklebe, J., Böhme, F., Wennrich, R., Stärk, H.-J., Mothes, S., Du Laing, G., Schulz, E., Neue, H.-U., 2009. Mercury volatilization from three floodplain soils at the Central Elbe River (Germany). *Soil Sediment Contam.* 18, 429–444.

El-Naggar, A., Shaheen, S.M., Ok, Y.S., Rinklebe, J., 2018. Biochar affects the dissolved and colloidal concentrations of Cd, Cu, Ni, and Zn and their phytoavailability and potential mobility in a mining soil under dynamic redox-conditions. *Sci. Total Environ.* 624, 1059–1071.

Elbana, T.A., Selim, H.M., Akrami, N., Newman, A., Shaheen, S.M., Rinklebe, J., 2018. Freundlich sorption parameters for cadmium, copper, nickel, lead, and zinc for different soils: Influence of kinetics. *Geoderma* 324, 80–88.

Evans, L.J., Barabash, S.J., 2010. Molybdenum, silver, thallium and vanadium. In: P.S. Hooda (Ed.), *Trace Elements in Soils*. John Wiley & Sons, Chichester, UK, pp. 515–549.

Frohne, T., Rinklebe, J., Langer, U., Du Laing, G., Mothes, S., Wennrich, R., 2012. Biogeochemical factors affecting mercury methylation rate in two contaminated floodplain soils. *Biogeosci.* 9, 493–507.

GGS (Government Gazette Staatskoerant, South Africa), 2012. National Environmental Management: Waste Act (59/2008): Draft National Norms and Standards for the Remediation of Contaminated Land and Soil Quality: For Public Comments No. 35160. Available at https://cer.org.za/wp-content/uploads/2010/03/national-environmental-management-waste-act-59-2008-national-norms-and-standards-for-the-remediation-of-contaminated-land-and-soil-quality_20140502-GGN-37603-00331.pdf (Last accessed on June 3, 2019).

Gupta, N., Yavad, K.K., Kumar, V., Kumar, S., Chadd, R.P., Kumar, A., 2019. Trace elements in sol-vegetables interface: Translocation, bioaccumulation, toxicity and amelioration – A review. Sci. Total Environ. 651, 2927–2942.

He, Z.L., Shentu, J., Yang, X.E., 2010. Manganese and selenium. In: P.S. Hooda (Ed.), *Trace Elements in Soils*. John Wiley & Sons, Chichester, UK, pp. 481–495.

Hodson, M.E., 2004. Heavy metals – geochemical bogey men? *Environ. Pollut.* 129, 341–343.

INSURE, 2017. EQS Limit and Guideline Values for Contaminated Sites – Report 2017. Available at https://www.meteo.lv/fs/CKFinderJava/userfiles/files/EQS_limit_and_guideline_values.pdf (Last accessed on June 3, 2019).

Jiang, Y., Chao, S., Liu, J., Yang, Y., Chen, Y., Zhang, A., Cao, H., 2017a. Source apportionment and health risk assessment of heavy metals in soil for a township of Jiangsu Province, China. *Chemosphere* 168, 1658–1668.

Kusinm, F.M., Azanim N.N.M., Hasan, S.N.M.S., Sulong, N.A., 2018. Distribution of heavy metals and metalloid in surface sediments of heavily-mined area for bauxite ore in Pengerang, Malaysia and associated risk assessment. *CATENA* 165, 454–464.

Liu, B., Ai, S., Zhang, W., Huang, D., Zhang, Y., 2017a. Assessment of the bioavailability, bioaccessibility and transfer of heavy metals in the soil-grain-human systems near a mining and smelting area in NW China. *Sci. Total Environ.* 609, 822–829.

Liu, G., Wang, J., Liu, X., Liu, X., Li, X., Ren, Y., Wang, J., Dong, L., 2018. Partitioning and geochemical fractions of heavy metals from geogenic and anthropogenic sources in various soil particle size fractions. *Geoderma* 312, 104–113.

Liu, J., Luo, X., Wang, J., Xiao, T., Chen, D., Sheng, G., Yin, M., Lippold, H., Wang, C., Chen, Y., 2017b. Thallium contamination in arable soils and vegetables around a steel plant – A newly-found significant source of Tl pollution in South China. *Environ. Pollut.* 224, 445–453.

McBride, M.B., 1995. Toxic metal accumulation from agricultural use of sludge: Are US EPA regulations protective? *J. Environ. Qual.* 24, 5–18.

MoE (Ministry of Environment, Korea), 2018. Soil Contaminants and Control Limits. Available at http://eng.me.go.kr/eng/web/index.do?menuId=311 (Last accessed on June 3, 2019).

MvV (Ministerie von Volkshuissvesting), 2000. Dutch Target and Intervention Values, 2000 (the New Dutch List). Annexes: Circular on target values and intervention values for soil remediation. Available at https://www.esdat.net/Environmental%20Standards/Dutch/annexS_I2000Dutch%20Environmental%20Standards.pdf. (Last accessed on September 16, 2018).

Nadal, M., Rovira, J., Diaz-Ferrero, J., Schuhmacher, M., Domingo, J.L., 2016. Human exposure to environmental pollutants after a tire landfill fire in Spain: Health risks. *Environ. Int.* 97, 37–44.

Natasha, Shahid, M., Niazi, N.K., Khalid, S., Murtaza, B., Bibi, I., Rashid, M.I., 2018. A critical review of selenium biogeochemical behavior in soil-plant system with an inference to human health. *Environ. Pollut.* 234, 915–934.

Niazi, N.K., Bibi, I., Shahid, M., Ok, Y.S., Burton, E.D., Wang, H., Shaheen, S.M., Rinklebe, J., Luttge, A., 2018. Arsenic removal by perilla leaf biochar in aqueous solutions and groundwater: An integrated spectroscopic and microscopic examination. *Environ. Pollut.* 232, 31–41.

Ontario, 2011. Soil, Groundwater and Sediment Standards for Use under Part XV.1 of the Environmental Protection Act. Available at https://www.ontario.ca/page/soil-ground-water-and-sediment-standards-use-under-part-xv1-environmental-protection-act (Last accessed on June 3, 2019).

Page, A.L., 1974. Fate and Effects of Trace Elements in Sewage Sludge when Applied to Agricultural Lands. EPA Document #670/2-74-005. Available at https://nepis.epa.gov (Last accessed on June 3, 2019).

Pan, L., Wang, Y., Ma, J., Hu, Y., Su, B., Fand, G., Wang, L., Xiang, B., 2018. A review of heavy metal pollution levels and health risk assessment of urban soils in Chinese cities. *Environ. Sci. Pollut. Res.* 25, 1055–1069.

Pourret, O., Bollinger, J.-C. 2018. "Heavy metal"–What to do now: To use or not to use? *Sci. Total Environ.* 419–420.

Pourrut, B., Shahid, M., Dumat, C., Winterton, P., Pinelli, E., 2011. Lead uptake, toxicity, and detoxification in plants. *Rev. Environ. Contam. Toxicol.* 213, 113–136.

Puschenreiter, M., Wieczorek, S., Horak, O., Wenzel, W.W., 2003. Chemical changes in the rhizosphere of metal hyper-accumulator and excluder *Thlaspi* species. *J. Plant Nutr. Soil Sci.* 166, 579–584.

Rinklebe, J., During, A., Overesch, M. Du Laing, G. Wennrich, R. Stärk, H.-J. Mothes, S., 2010. Dynamics of mercury fluxes and their controlling factors in large Hg-polluted floodplain areas. *Environ. Pollut.* 158, 308–318.

Rinklebe, J., Shaheen, S.M., 2017. Geochemical distribution of Co, Cu, Ni, and Zn in soil profiles of Fluvisols, Luvisols, Gelysols, and Calcisols originating from Germany and Egypt. *Geoderma* 307, 122–238.

Shaheen, S.M. 2009. Sorption and lability of cadmium and lead in different soils from Egypt and Greece. *Geoderma* 153, 61–68.

Shaheen, S.M., Balbaa, A.A., Khatab, A.M., Antoniadis, V., Wang, J., 2019. Biowastes alone and combined with sulfur affect the phytoavailability of Cu and Zn to barnyard grass and sorghum in a fluvial alkaline soil under dry and wet conditions. *J. Environ. Manage.* 234, 440–447.

Shaheen, S.M., Kwon, E.E., Biswas, J.K., Tack, F.M.G., Ok, Y.S., Rinklebe, J., 2017a. Arsenic, chromium, molybdenum, and selenium: Geochemical fractions and potential mobilization in riverine soil profiles originating from Germany and Egypt. *Chemosphere* 180, 553–463.

Shahid, M., Dumat, C., Khalid, S., Niazi, N.K., Antunes, P.M.C., 2016. Cadmium bioavailability, uptake, toxicity and detoxification in soil-plant system. In: *Reviews of Environmental Contamination and Toxicology.* Springer, New York, pp. 1–65.

Tang, Z., Chai, M., Cheng, J., Jin, J., Yang, Y., Nie, Z., Huang, Q., Li, Y., 2017. Contamination and health risks of heavy metals in street dust from a coal-mining city in eastern China. *Ecotox. Environ. Safe.* 138, 83–91.

Tepanosyan, G., Sahakyan, L., Belyaeva, O., Maghakyan, N., Saghatelyan, A., 2017. Human health risk assessment and riskiest heavy metal origin identification in urban soils of Yerevan, Armenia. *Chemosphere* 184, 1230–1240.

UoV (University of Vermont-Extension), 2011. Interpreting the Results of Soil Tests for Heavy Metals University of Vermont. Available at https://www.uvm.edu/vtvegandberry/factsheets/interpreting_heavy_metals_soil_tests.pdf (Last accessed on June 3, 2019).

US EPA (United States Environment Protection Agency), 2016. Regional Screening Levels (RSLs) – Generic Tables (May 2016) (TR=1E-06 THQ=0.1). Available at http://www.sviva.gov. (Last accessed on June 3, 2019).

Wang, G., Shan, Y., 2013. Soil Environmental Standards/Screening Values in China. Available at http://www.iccl.ch/download/durban_2013/Slides_ICCL_2013_Meeting/I_C2_ICCL_2013_GQWANG_10OCT2013@DURBAN_SA.pdf (Last accessed on June 3, 2019).

Wu, W., Wu, P., Yang, F., Sun, D.-L., Zhang, D.-X., Zhou, Y.-K., 2018. Assessment of heavy metal pollution and human health risks in urban soils around an electronics manufacturing facility. *Sci. Total Environ.* 630, 53–61.

Yang, Q., Li, Z., Lu, X., Duan, Q., Huang, L., Bi, J., 2018. A review of soil heavy metal pollution from industrial and agricultural regions in China: Pollution and risk assessment. *Sci. Total Environ.* 642, 690–700.

5 Emerging Contaminants in Soil and Groundwater

*Wentao Jiao, Xiaomin Dou, Jian Hu, Shiyu Wang, Longjie Ji,
Ziyu Han, Yinqing Fang, and Zhirui Qin*

CONTENTS

5.1 SOURCE AND OCCURRENCE OF EMERGING CONTAMINANTS

The term "emerging contaminants" (ECs) generally refers to compounds previously not considered or known to be significant about their concentration and toxicity. In recent years, they are frequently detected and found to induce ecological or human health risks (Stuart et al., 2012; Shores et al., 2017). The majority of the ECs have not been regulated. However, regulatory agencies such as World Health Organization (WHO), United States Geological Survey (USGS), and United States Environmental Protection Agency (US EPA) have paid attention to ECs in the environment. In regard to the category of ECs, there has been no explicit definition. Here, we address the source and occurrence of pesticides, pharmaceuticals and personal care products (PPCPs), endocrine disrupting chemicals (EDCs), industrial additives and by-products, food additives and flame/fire retardants, etc.

5.1.1 Pesticides

Pesticides refer to chemicals used in agriculture, gardening, and domestic activities to control pests and diseases and regulate plant growth. Pesticides usually come from agricultural sources. About 2×10^6 tons of pesticides in agriculture are consumed all over the world, and 75% of them are consumed by developed areas such as Japan, North America, and Western Europe. In India pesticides are generally divided into four categories, including insecticides, herbicides, fungicides, and bactericides, according to their usage. Their use has inevitably resulted in the contamination of soil and groundwater (Zhao and Pei, 2012). Pesticides have been detected frequently in soil and groundwater (Alfy and Faraj, 2016; Dowling et al., 2019). Diazinon is a typical pesticide, the concentration of which was 0.13 ng L^{-1} with a 100% detection in the soil of Klang River estuary, Malaysia (Omar et al., 2018). The concentrations of pesticides in groundwater in Italy were between 90 and 4.78×10^5 ng L^{-1} (Meffe and Bustamante, 2014). Because of their toxicity, atrazine and simazine have been listed as priority substances reported in the Directive 2008/105/EC. The concentration of atrazine ranged from 500 to 15000 ng L^{-1} in Hungary (Szekacs et al., 2015). And the total concentration of organochlorine pesticides ranged from 0.01 to 100.45 ng g^{-1} in the surface soil in central China (Gereslassie et al., 2019). Pesticide metabolites are also detected in soil and groundwater. Pesticides in the soil and groundwater environment can cause various adverse ecological and/or human health effects.

5.1.2 Pharmaceuticals and Personal Care Products

PPCPs broadly refer to products with healthcare, medical purposes, or personal care for humans and/or animals, mainly including pharmaceuticals and personal care products. Pharmaceuticals usually include antibiotics, blood lipid regulators, analgesics, and anti-inflammatory and cytostatic drugs. They are mainly from Western medicine and result in hundreds of tons of products being produced and consumed each year. After being ingested by human beings or animals, most of the medicines are excreted in the form of raw drugs and metabolites through their feces and urine, and then enter into WWTPs through sewage discharge pipelines. In addition, some unused or expired drugs are also imported into municipal domestic sewage through toilet disposal. These pollutants could be effectively removed through the conventional sewage treatment system, and the efficiency of treatment is generally 30%–80%. Pharmaceuticals are commonly transmitted to soil and groundwater when the reclaimed water is recharged. Sewage treatment plants have become another important source of antibiotics in the soil and groundwater environment. Animal manure and slurries have been extensively used as fertilizers in agriculture. The veterinary pharmaceuticals in the manure and slurries will be transmitted to soil and leached to groundwater. Therefore, animal manure is reused as fertilizers, which is another important route of pharmaceuticals'

entry in the soil and groundwater (Wojslawski et al., 2019; Yang et al., 2017b). Personal care products generally include shampoos, toothpastes, cosmetics, body washes, sunscreens, hand lotions, etc. With daily use, these personal care products were discharged into surface water (Gago-Ferrero et al., 2013; Serra–Roig et al., 2016). PPCPs been detected in soil and groundwater. The mean concentrations of five most commonly reported PPCPs including carbamazepine, sulfamethoxazole, caffeine, ibuprofen, and diclofenac in groundwater in North America, Europe, Asia, and the Middle East were 1.5 µg L^{-1}, 5 µg L^{-1}, 9.8 µg L^{-1}, and 252 ng L^{-1}, respectively (Lapworth et al., 2012). For carbamazepine and ibuprofen, which are two commonly used analgesic, the concentrations were 41 and 32000–128000 ng L^{-1}, respectively, in leachates from 19 landfills in the United States (Masoner et al., 2014). In groundwater samples, the concentration is reported of two samplings from wells. All six wells investigated in India contained ciprofloxacin and cetirizine with a high concentration up to 14 and 28 mg L^{-1}, respectively (Jechalke et al., 2014; Gani et al., 2017). The concentrations of propylparaben and oxybenzone, as two typical personal care products, were 5.5 and 70.4 µg L^{-1}, respectively, in groundwater in the United Kingdom (Stuart et al., 2012).

5.1.3 Endocrine Disrupting Chemicals

Endocrine disrupting chemicals (EDCs) refer to the exogenous chemicals that interfere with living organisms or the human body formation, secretion, transport, binding, reaction, and metabolism with the synthesis of natural hormones, which could affect the production of substances or with human reproductive, neurological, immune, and other functions (Zhou et al., 2010). Most EDCs enter into the groundwater and soil through the disposal of untreated sewage. At present, many countries have a huge gap between sewage generation and sewage treatment capacity. Their transfer to the aquatic environment occurs through the runoff from the application site, and based on the compound solubility these chemicals can bioaccumulate in living organisms, soil, or sediments (Gonzalez et al., 2012). These substances can lead to genital disorders, behavioral abnormalities, reduction of reproductive capacity, death of larva, and even extinction of animals (Oscar et al., 2017). EDCs mainly include hormones, pesticides, and some PPCPs. Several non-steroidal, and synthetic compounds are used as plasticizer, surfactants, and flame retardants. Natural estrogens such as estrone (E1), 17β-estradiol (E2) and estriol (E3), industrial compounds such as nonylphenol (NP), octylphenol (OP) and bisphenol A (BPA), and synthetic estrogens such as 17α-ethinylestradiol (EE2) are common EDCs, which are the most popularly studied and monitored in soil and groundwater (Li et al., 2013).

NP, as one of the priority pollutants, is the degradation product of alkyl phenol ethoxylates (APEs) and is used in the manufacture of surfactants. NP has been widely spread in soil and groundwater because of the reclaimed water irrigation in recent years. The mean concentration of NP in soil irrigated with reclaimed water in Southeast of China is 25 µg kg^{-1}. And the concentration of NP in 80 m depth groundwater samples in this area ranged from ND to 1407.9 ng L^{-1} (Wang et al., 2018a), which was much less than the standard of 6.6 µg L^{-1} in fresh water. E3 and E1 are used as estrogen and degraded estradiol, respectively, and the concentrations of them are 2 and 0.8 ng L^{-1} in leachates from 19 landfills in the United States (Masoner et al., 2014).

5.1.4 Industrial Additives and By-Products

Industrial additives are substances that can change the reaction efficiency but do not change the chemical properties of additives themselves. Industrial by-products are substances produced concurrently with the main industrial reaction products. Lots of hardly removable industrial additives and by-products permeate through soil by irrigation with industrial wastewater (Andrade et al., 2010; Hale et al., 2012). There are varieties of industrial compounds that are released to the soil and groundwater, and many of them could lead to environmental problems, such as the petroleum hydrocarbons and chlorinated solvents (Meffe and Bustamante, 2014; Moran et al., 2005; Verliefde et al., 2007). The industrial ECs usually include 1,4-dioxane, benzotriazole derivatives and dioxins. Some breakdown products may also be regarded as ECs.

Some industrial compounds such as phthalates (di(2-ethylhexyl) phthalate), and perfluorinated compounds perfluorooctane sulfonic acid (PFOS) are regarded as priority substances due to their toxicity, bioaccumulation ability, persistence, and high frequency of detection (at concentrations in the order of µg L^{-1}) (Li et al., 2013, 2015; Sakurai et al., 2016).

In industry, diethyl phthalate, methyl benzotriazole, naphthalene, and cresol are used for plasticizers, corrosion inhibitors, fumigants, surface active agent, respectively. The concentrations of them are 2000–80000, 1410, 100–4000, and 400–6000 ng L^{-1}, respectively, with an occurrence frequency of 60%, 50%, 79%, and 55% in leachates from 19 landfills in the United States (Masoner et al., 2014). Perfluorinated compounds have been detected with the frequency of 26% in groundwater monitoring sites in England and Wales in 2006. The highest detectable concentrations of PFOS was 78.8 ng L^{-1} in the groundwater of Taiwan in China (Lin et al., 2015).

5.1.5 Food Additives

Food additives are substances added to food during the processing or making of that food, which are used to preserve flavor or enhance its taste, appearance, or other qualities. There are a variety of ECs in food additives. Agriculture is an important source of food additives (Bernhardt et al., 2019). Triethyl citrate is one of the food additives, which is used to stabilize foams and is used in pharmaceutical coatings. Food additives usually include butylated hydroxytoluene (BHT), butylated hydroxyanisole (BHA), camphor, citral, citronellal, hexanoic acid, menthol, and heliotropin. And some of them may be implicated as oxidants or EDCs. With their wide use, food additives

are discharged into soil and groundwater, which can contribute contamination. In the United Kingdom, the max concentrations of BHT, BHT analogue, and 1(3H)-Isobenzofuranone (phthalide) in groundwater over the period 1992 to 2009 were 7, 4.2, and 9.3 µg L^{-1} (Stuart et al., 2012).

5.1.6 Flame/Fire Retardants

Flame retardants, functional additives that give flame retardancy to flammable polymers, are mainly designed for flame retardancy of polymer materials. They are commonly added to constituent polymers in electronics, furniture, and textiles. Flame retardants are mainly from the atmosphere (La Guardia et al., 2012). There are many types of flame retardants, which can be divided into additive flame retardants and reactive flame retardants according to their use methods. At present, the main additive of flame retardants are organic flame retardants and inorganic flame retardants, halogen (organic chlorides and organic bromides), and halogen-free flame retardants. Polybrominated diphenyl ethers, as one group of typical halogen flame retardants, are mostly used in resins for industrial and household use. They enter the environment via waste disposal to incineration and landfill. The concentration of 2-ethylhexyl diphenyl and phosphate TRCP, as usually used flame retardant plasticizers, is 2.7 and 4.9 µg L^{-1}, respectively (Stuart et al., 2012).

5.2 PATHWAY

The contamination of ECs entry into groundwater or soils was largely caused by anthropogenic activity such as industry, waste disposal, and agricultural practices (Sorensen et al., 2015). Point and non-point are two sources of ECs to soil and groundwater. The point source of contamination originated from discontinuous sources whose inputs into the environment can usually be defined as spatially explicit. In contrast, non-point-source of contamination originates from poorly defined, diffuse sources that occur over broad geographical scales, such as atmospheric deposition, urban and storm-water runoff, and agricultural runoff.

The transport of ECs in the environment can be set forth by a source-pathway-receptor model. Thus, a pathway is necessary to transport the contaminant between the source and the receptor. ECs can enter the environment through multiple pathways, including animal manure, crop/pest application, domestic wastewater discharges, landfill, industrial wastewater, and leakages and emergencies (Stuart et al., 2012; Wang et al., 2018a). ECs usually migrate from soil to the unsaturated zone, then finally to groundwater by crop/pest application and animal manure for pesticides and veterinary pharmaceuticals. PPCPs are discharged into soil and groundwater through domestic wastewater and hospital wastewater (Yang et al., 2017b). Compared with domestic wastewater, pharmaceuticals in hospital effluents are usually found to have a higher concentrations and detection frequency (Yang et al., 2017a). That is because drugs used for humans in hospitals are discharged into the environment directly or indirectly. And some non-metabolized or dissolved pharmaceutical ingredients are excreted from the human via urine and feces (Kim et al., 2011) and discharged into groundwater. There are some other ways for PPCP to be discharged into groundwater and soil, such as the disposal of unused drugs to landfills and the irrigation use of reclaimed water. However, the management of PPCPs is different in different countries. So the exposure pathways to PPCP also vary in different countries. Industrial wastewater is the main way of industrial compounds into soil and groundwater. Landfill and leakages and emergencies are two other ways for ECs to enter into soil and groundwater.

The unsaturated zone is an active zone where the hydrosphere, lithosphere, and biosphere interact. ECs spreading on the soil surface usually migrate through the soil profile, finally reaching groundwater. It is an important pathway for ECs' removal or entry into groundwater from reclaimed water reuse. ECs applied to the surface soil usually migrate through the soil zone, the unsaturated zone, and the saturated zone in a well-established way. This is usually the route for the components of sewage sludge and agricultural pesticides. Vertical distribution of ECs in the soils of the vadose zone is affected by the water sources of irrigation and conditions of geochemistry (Gaston et al., 2019).

Groundwater-surface water interaction is another important pathway. In many cases treated effluent from WWTPs and industrial plants are discharged to surface water, and then they infiltrate to groundwater. On-site sewage treatment facilities, especially septic systems combined with soil infiltration, are an important source of ECs in surface water and groundwater. For many ECs, the source pathway-receptor model is still very unclear. For the soil compartment, the receptors of ECs are the micro floras such as bacteria and fungi, plants, and soil fauna (Figure 5.1).

Many researchers have studied the pathway of ECs in the environment. However, there are still some critical knowledge gaps, which include lack of precise analytical methods and reference materials, and insufficient understanding on their fate and behavior.

5.3 THE ADVERSE EFFECTS AND ECOLOGICAL RISK ASSESSMENT ON SOIL AND GROUNDWATER

5.3.1 THE ADVERSE EFFECTS ON SOIL AND GROUNDWATER

ECs can enter the soil and groundwater environment through various channels, causing various adverse ecological and/or human health effects. Trace ECs in the environment often cause high hazards and risks. However, information about the toxicity, occurrence, and characteristics in the environment of most of the ECs is still less known than conventional contaminants, which requires extensive study (Li, 2014).

5.3.1.1 The Adverse Effects on Soil

Soil is one important kind of receptor for many contaminants. Some kinds of ECs can persist in the soil ecosystem for a long time and accumulate gradually, which have potential harmful effects on the health of soil ecosystems. Numerous studies have confirmed that ECs have adverse effects on soil fauna, plants, and microorganisms. The adverse effects of ECs in soil ecosystems include: (1) inhibition of soil microbial activity; (2) inhibition of seed germination and crop growth; (3) accumulation in plant leaves and fruits; (4) migrating into groundwater (Du and Liu, 2011).

Triclosan, a typical PPCP chemical, showed a significant inhibition effect on soil respiration with concentrations higher

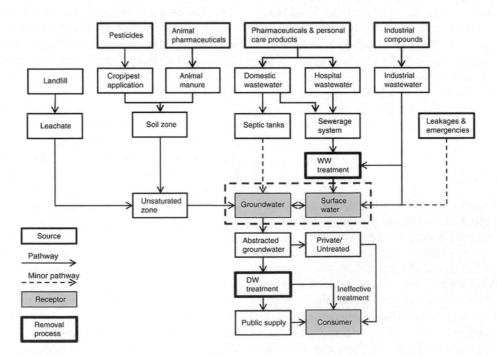

FIGURE 5.1   The pathway of ECs in the environment. (From Stuart, M. et al., *Sci. Total Environ.*, 416, 1–21, 2012.)

than 10 mg kg^{-1} dry weight. In addition, triclosan led to the evolution of microbial community structures in a 70-day incubation period. Within the range of 100-1000 mg kg^{-1} dw soil, with the increase of triclosan concentration in soil, the dominance of fungal community increased, among which the increase rate was 0.08–0.25 in loamy sand, 0.06–0.65 in clay soil, and 0.13–0.47 in sandy loam soil. The EC is complicated in the soil environment; the existing data is particularly important to take action toward keeping tolerance of soil and microbial communities with mixed existence of EC (Gomes et al., 2017).

The ECs in the soil ecosystems can inhibit of seed germination and crop growth. Barley and carrots were selected to examine the ability of growing and developing in contaminated soil with metformin, ciprofloxacin, and narasin. Metformin causes a decrease in carrot growth, while ciprofloxacin is more effective in reducing barley growth (Eggen et al., 2011). Concentrations that varied from 10 to 2000 mM of the food flavoring 1,8-cineole may decrease tobacco seeds' germination. When exposed to the concentrations higher than 200 mM, the growth of the hypocotyl was inhibited and root length decreased (Yoshimura et al., 2011).

ECs can accumulate in crop biomass. The uptake of sulfamethazine carbamazepine, diclofenac, fluoxetine, propranolol and sulfamethazinein soil by radish (*Raphanus sativus*), and ryegrass (*Lolium perenne*) was examined, and four chemicals accumulated in plant biomass (Carter et al., 2014). The pharmaceuticals, a typical chemical of ECs, can transfer from soil to plants, and can accumulate in crop biomass and bring about potential risks to the human body via the food supply chain. Exposure to pharmaceuticals may cause crop loss. When barley and carrots were exported in contaminated agricultural soil with metformin, ciprofloxacin, and the narration elected to, their ability to grow and develop were reduced. Metformin may cause a decrease in carrot growth, while ciprofloxacin is more effective in reducing barley growth (Eggen et al., 2011).

The ECs contained in the upper soil can migrate into groundwater through leaching/runoff diffusion. Types of ECs detected from groundwater sources around agricultural sources, factories, landfills, etc. are similar and have higher concentrations (Du and Liu, 2011; Gomes et al., 2017).

5.3.1.2 The Adverse Effects on Groundwater

ECs from agricultural waste, landfill, septic tanks, and wastewater can transport into the groundwater through application of organic fertilizer, sewage recharge, leaching, etc., and migrate into the groundwater in the horizontal and vertical directions. With the intensification of human activities and the advancement of analytical techniques, more and more trace amounts of ECs were detected continuously in groundwater (Husain, 2008; Lapworth et al., 2012). It is well known that the presence of EC in groundwater can bring about detrimental effects to both aquatic ecosystems and human health. The adverse impact of ECs in groundwater ecosystems mainly include: (1) threats to human health; (2) harm to aquatic organisms in groundwater; and (3) long-term presence in groundwater.

Groundwater is an important drinking water source in many countries and regions. At present, ECs have been detected in groundwater in many countries, with higher concentrations in some regions. Continuous and chronic exposure to these ECs may pose a threat to the human immunity, endocrine system, and nervous system, even with trace levels (Patel et al., 2019; Philip et al., 2018).

The presence of ECs in water environments can affect aquatic organisms and lead to changes that threaten the sustainability of aquatic ecosystems. The presence of 17-α-ethinyl estradiol in an aquatic environment at the ng L^{-1} level can modulate the activity of different physiological endpoints of aquatic organisms, such as controlling the activity of acetylcholinesterase and glutathione S-transferase enzymes in neurotransmitters and detoxification systems (Souza et al., 2013). Fish exposed to lipid-lowering agents, gemfibrozil and the non-steroidal anti-inflammatory drug diclofenac, GST inhibition, lipid peroxidation (LPO), and DNA damage were significantly affected (Sharma et al., 2019). In the past few decades, human estrogen has become a striking new type of water pollutant because it was found to cause endocrine disruption and negatively affects the sexual and reproductive systems of wildlife, fish, and humans (García–Galán et al., 2013; Gros et al., 2019; Hakk et al., 2016). Most of the studies on ECs has focused on surface waters, which are commonly found to contain higher concentrations and diversity of contaminants, and their monitoring is less difficult than groundwater (Stuart et al., 2012).

Because of the redox control of saturated regions and low microbial populations, ECs can stay long in groundwater, and then they may go further with the groundwater. In some cases, ECs may remain in groundwater for decades. Some ECs are frequently detected in groundwater, the footprints of which are global (Lapworth et al., 2018). Contaminant properties, concentration, and re-excitation conditions of ECs are key factors for the residence time and persistence of EC in groundwater.

5.3.2 Assessment of Environmental Risks to Soil and Groundwater Ecosystems

The ecological risk assessment consists of three main components: exposure concentration analysis, effect concentration analysis, and risk characterization; the main content and major components are shown in Figure 5.2.

Risk characterization requires the support from exposure concentration analysis and effect concentration analysis, exposure concentration analysis and effect concentration analysis. These analyses are the basis of risk assessment. The exposure concentration analysis and effect concentration analysis are relatively independent, and the supporting to risk characterization is parallel. The commonly methods used in ecological risk assessment include hazard quotients (HQs), probabilistic ecological risk assessment (PERA), and compound joint ecological risk assessment.

HQs are extensively used to determine whether a chemical contaminant at a concentration has a potentially harmful effect. The value of HQ is calculated by comparing the

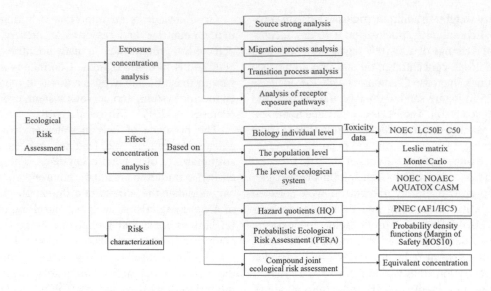

FIGURE 5.2 The main content and components of ecological risk assessment to ecosystems. (*Abbreviations:* LC50: Median lethal concentration, EC50: Half-maximal effective concentration, NOEC: No observed effect concentration, NOAEC: No observed adverse effect concentration, AQUATOX: AQUATOX model, CASM: CASM model, PENC: Predicted no effected concentrations, AFI: Evaluation factors, HQ: Hazard quotients, PERA: Probabilistic ecological risk assessment, MOS: Margin of safety.)

estimated environmental exposure concentration (ECE or PEC) to the model with the toxicity data (predicted non-effect concentration, PNEC) that characterizes the hazard level of the substance. When HQ is more than one, the risk potential is high, while when HQ is between 0.1 to 1, the level of risk is intermediate and when HQ is less than 0.1, risk is less or no eco-toxicological hazards. The required parameters are few and easy to obtain, so it has been widely used. However, HQ is usually conservative in determining the exposure and selecting the toxicity reference value. It is a qualitative or semi-quantitative method of determining risk.

PERA takes each exposure concentration and toxicity data as independent observation values, which considers their probability and statistical significance on this basis. PERA relies on statistical models and expresses risks in the form of probabilities, the results of which are closer to the actual situation. The method includes margin of safety (MOS10) and probability distribution curve. In the ecological risk assessment, the commonly used index is the PNEC. We can get PNEC by no observed effect concentration (NOEC). Due to the lack of NOEC for most compounds, the NOEC used in current ecological risk assessment needs to be extrapolated from LC_{50} or EC_{50}. The most successful models for the ecological risk analysis models currently used to obtain PNEC are AQUATOX, CASM, and so on.

Compound joint ecological risk assessment can be used to calculate the combined ecological risk of the mixture. In most cases, the ecological risk assessment only considers the ecological risks of monomeric compounds. In fact, receptor organisms are often exposed to mixed compounds. Therefore, this method integrates the HQ, the PERA, and other assessment methods, making full use of various methods and means for simple to complex risk assessment.

ECs have been widely used in human health and daily life for many years; they can be found in measurable concentrations or not in the environment. But there is very little available data to predict the environmental risks of the EC. More research is needed for ECs in the environment.

5.3.2.1 Assessment of Environmental Risks to Soil Ecosystems

The receptors of ECs in the soil compartment consist of plants, the microflora including actinomycetes, bacteria, fungi, and soil fauna such as protozoa and invertebrates. However, the ecotoxicological significance of some ECs remains largely unknown. It is an international challenge to conduct an effective risk assessment of ECs in the soil environment. More accurate endpoints and analytical methodologies in the complex soil environment are necessary to assess the ecotoxicity of the increasing number of ECs.

The ecotoxicological effects of ECs including emerged pesticides, pharmaceuticals, personal care products, food additives, and industrial compounds to soil organisms in soil environments are summarized in Table 5.1. Because the satisfying data to determine their risk often does not exist, the toxicology data was collected based on a variety of soil model organisms (Gomes et al., 2017). The test organism includes typical soil animals (earthworm, enchytraeids, springtail), plants (barley, carrot, tabacum), the endpoint were LC_{50}, EC_{50}, reduction in biomass, and DNA damage; the effect concentration ranged from 1.7 to 2000 mg kg^{-1}.

Since we cannot predict the mixed effect of a large number of compounds' impact on natural communities, the impact on the soil ecosystem function are poorly understood. In order to obtain more helpful information on the impact of emerging pollutants on natural communities, it is necessary to conduct

TABLE 5.1

Types of Emerging Contaminants and Studies to Assess Their Environmental Risks in Soil Organisms

Emerging Contaminants	Risk Assessment			Compound (class; PNECa)	Effect Concentration and 95% CI (w/w drysoil)	References
	Endpoint/Effect Evaluated	Results	Test Organism			
Pesticides	Mortality (LC_{50})	–	Earthworm (*Eisenia andrei*)	Abamectin (insecticide)	18 (15–22) mg kg⁻¹	Kolar et al. (2008)
	Reproduction (EC_{50})	–	Earthworm (*E. andrei*)	Abamectin (insecticide)	–	Marques et al. (2009)
	Avoidance behavior (EC_{50})	–	Earthworm (*E. andrei*)	Sulcotrione (herbicide)	1263 (ND) mg kg⁻¹	De Silva et al. (2010)
	Reproduction (EC_{50})	–	Earthworm (*Perionyx excavatus*)	Carbofuran (Pure)	1.7(1.6–2.0) mg kg⁻¹	
	Mortality (LC_{50})	–		Carbofuran (insecticide)	9 (8–10) mg kg⁻¹	
Pharmaceuticals	Reduction in biomass	Approx. 3 g in leaf and 3 g in seed Approx. 9 g in leaf and 25 g in root	Barley Carrot	Ciprofloxacin (26000 µg kg⁻¹)	6.5 ± 0.7 mg kg⁻¹	Eggen et al. (2011)
	Decrease in reproduction	Decrease of 45.6% of cocoons	Earthworm	Chlortetracycline (antibiotic)	100 mg kg⁻¹	Lin et al. (2012)
	DNA damage	13–48%	–		3–300 mg kg⁻¹	–
	Mortality (LC_{50})	–			96.1 mg kg⁻¹ for juvenile	
Personal care products	Fungal/bacterial ratio	Ratio of 0.08–0.25 in loamy sand soil, 0.13–0.47 in sandy loam soil, and 0.06–0.65 in clay soil	Microcosms (fungal/bacterial ratio)	Triclosan (2.1 µg kg⁻¹)	10–1000 mg kg⁻¹	Butler et al. (2012)
	Inhibition of plant growth (root length)	–	Cucumber		17 mg kg⁻¹	Liu et al. (2009)
	Inhibition of plant growth (root length, EC_{50})	–	Cucumber		108 (83–140) mg kg⁻¹	–
	DNA damage	1.7–22.0% after 7 days of exposure, 1.0–24.0% after 14 days of exposure	Earthworm (*E. fetida*)		1–300 mg kg⁻¹	Lin et al. (2010)
	Mortality (LC_{50})	–	Earthworm (*E. fetida*)	Triclocarban (antibacterial)	40 mg kg⁻¹	Snyder et al. (2011)
	Mortality (LC_{50})	No mortality	Earthworm (*E. fetida*)	Tetracycline (8800 µg kg⁻¹)	0.3–300 mg kg⁻¹	Dong et al. (2012)
	DNA damage	Superoxide dismutase and catalase activities	–			–
Life-style compounds product	Mortality (LC_{50})	–	Earthworm (*E. fetida*)	Caffeine (37 µg kg⁻¹)	19.3 µg cm⁻²	McKelvie et al. (2011)
Food additives	Decrease of seeds germination	50%	*Nicotium tabacum*	1,8-cineole	1200 µM	Yoshimura et al. (2011)
	Inhibition of hypocotyls growth (IC_{50})	–	–		200 µM	
	Inhibition of plant growth (root length, IC_{50})	–	–		2000 µM	

(Continued)

TABLE 5.1 (Continued)
Types of Emerging Contaminants and Studies to Assess Their Environmental Risks in Soil Organisms

Emerging Contaminants	Risk Assessment — Endpoint/Effect Evaluated	Results	Test Organism	Compound (class; PNECa)	Effect Concentration and 95% CI (w/w drysoil)	References
Industrial compounds	Decrease of biomass of root	Reduction of 28 g for Napoli	Carrots	TCEP and TCPP (Flame retardants)	0.6–1.0 mg kg^{-1}	Eggen et al. (2013)
	Decrease in female reproductive allocation	20.4–41.6%	Terrestrial isopod (*Porcellio scaber*)	Bisphenol A (Endocrine disruptor)	10–1000 mg kg^{-1}	Lemos et al. (2010)
	Increase of abortions	20%	–	–	–	
	Mortality	30%	–	–	1000 mg kg^{-1}	Lemos et al. (2009)
	Molting disturbances	25%	–	–	–	
	Mortality (LC$_{50}$)	–	–	–	910 (163–1658) mg kg^{-1}	Domene et al. (2009)
	Inhibition of plant germination (EC$_{50}$)	–	*Lolium perene* and *Brassica rapa*	Nonylphenol (0.3 mg kg^{-1})	1 g kg^{-1}	
	Reproduction inhibition (EC$_{50}$)	–	Enchytraeids and collembolans	–	64–226 mg kg^{-1}	
	Inhibition of plant germination (EC$_{50}$)	–	*Lolium perene* and *Brassica rapa*	Nonylphenol polyethoxylate	10 g kg^{-1}	
	Reproduction inhibition (EC$_{50}$)	–	Enchytraeids and collembolans	–	356–1876 mg kg^{-1}	
Nanomaterials	Decrease of microbial biomass	Approx. 13%	Microcosms	TiO$_2$	0.5–2.0 mg g^{-1}	Ge et al. (2011)
	Reproductive toxicity	Decrease of 3 cocoons	Earthworm	TiO$_2$	100 mg kg^{-1}	Canas et al. (2011)
	Reproductive inhibition	25.4% / 100%	–	TiO$_2$	1000 mg kg^{-1}	Heckmann et al. (2011)
	Decrease in cellulase activity	25%	–	MWCNT	1 g kg^{-1} of MWCNT and 5 mg kg^{-1} of nonylphenol	Hu et al. (2013)
	Decrease in Na$^+$/K$^+$-ATPase activity	75%	–	Nonylphenol	–	
	DNA damage	10%	–	MWCNT	1 g kg^{-1} of MWCNT	
	Growth inhibition	23%	–	AgNO$_3$	7.41 mg kg^{-1}	Shoults-Wilson et al. (2011)

Source: Gomes, A. R. et al., *J. Environ. Sci. Health A Tox. Hazard. Subst. Environ. Eng.*, 52, 992–1007, 2017.

Notes: ND, not determined; TCEP, tris-2-chloroethyl phosphate; TCPP, tris-2-chloroisopropyl phosphate; MWCNT, multi-walled carbon nanotubes; PNEC, predicted no effect concentration (these values are identified when available in the respective papers). PNEC values from Muñoz (Muñoz et al., 2009).

more ecotoxicological studies to get useful toxicity data for defining the PNEC values in the soil, by analyzing more sensitive endpoints.

5.3.2.2 Assessment of Environmental Risks to Groundwater Ecosystems

5.3.2.2.1 Hazard Quotients (HQs)

HQ value is the most extensively used method to assess of environmental risks in aquatic systems. Health and ecological risk assessment of ECs in groundwater were carried out in the Ganges River Basin, India. Occurrences of 15 personal care products (PPCPs) and 5 artificial sweeteners (ASWs) were confirmed, which ranged between 34–293 ng L^{-1} and 0.5–25 ng L^{-1}, respectively. The values of age-dependent RQ and age-dependent drinking water equivalent level (DWEL) were calculated for each detected PPCP. For all detected PPCPs, DWELs for all age groups ranged from 4.8 μg L^{-1} (for carbamazepine, 1–2 years age group) to 12.8 mg L^{-1} (for acetaminophen and 16–21 years age group) and RQs ranged from 1.5×10^{-7} (for acetaminophen, 16–21 years age group) to 0.0021 (for carbamazepine, 16–21 years age group). The PPCPs with higher RQs were carbamazepine, ciprofloxacin, ketoprofen, caffeine, ibuprofen, and triclosan (Sharma et al., 2019).

HQ was also used to evaluate the risk of PPCPs in the Pearl River, the Yellow River, Haihe River, and Liaohe River in China. These screening grades indicate that some PPCPs can cause relatively high environmental risks in some water bodies in China (Wang et al., 2010; Zhao et al., 2010). RQs were determined in treated landfill to leachate at three trophic levels (algae, invertebrates, and fish). The results showed that two kinds of PAHs, two kinds of PPCPs, BPA, and g-HCH were presented to be with high risks (Qi et al., 2018).

Based on the PNEC values, Frédéric classified 172 pharmaceuticals into 3 categories. Compounds of category 1 have a hazard quotient (HQ) < 1, with PNEC values higher than 1 μg L^{-1}; compounds in category 1 are less ecotoxic or no eco-toxicological hazards. Compounds of category 2 have a HQ values between 1 and 1000 with PNEC values higher than 100 ng L^{-1} but less than 1 μg L^{-1}; compounds in category 2 have intermediate ecotoxicity. Compounds of category 3 have HQs > 1000, with PNEC values less than 1 ng L^{-1}; these are hazardous (Fernández and Yves, 2014).

5.3.2.2.2 Probability Index

Worrall used a probability index to predict the risk and make a simple assessment of groundwater contamination caused by the pesticide metabolites for pesticides with UK usage in more than a 50,000-hectare area. This method had no regard for the activity or toxicity of these metabolites. Because of the lack of persistence data, many ECs cannot be assessed in the same way (Worrall et al., 2000).

5.3.2.2.3 Other Assessment Methods of Environmental Risks to Groundwater Ecosystems

A quantitative risk ranking model using a Monte Carlo simulation approach was developed for human exposure to 16 ECs

following treated municipal sewage sludge application to Irish agricultural land. Predicted environmental concentration (PEC) for ECs was calculated in soil and groundwater. Nonylphenols ranked the highest EC with mean concentration of 5.69 mg kg^{-1} and 2.22×10^{-1} μg L^{-1}, respectively, in soil and groundwater. The model emphasized that triclocarban and triclosan were ECs which required further investigation (Clarke et al., 2016).

Based on the DRASTIC model, the contamination risk of groundwater was evaluated by a multilevel fuzzy approach (Jafari and Nikoo, 2019). Assessment of groundwater contamination risk using hazard quantification, combined with a modified DRASTIC model and groundwater value was done in Beijing Plain, China. The study introduced a groundwater pollution risk assessment method which combines hazard, inherent vulnerability, and groundwater values together. It assessed the hazard by quantifying the nature of the contaminant and the infiltration of the contaminant (Wang et al., 2012; Zhang et al., 2013). The framework can be used to make decisions based on parameter uncertainty and the complexity of possible changes in groundwater contaminations.

With the continuous expansion of the scope of ecological risk assessment and the increasing complexity of evaluation content, the existing ecological risk assessment technology doesn't work well anymore. Ecological risk assessment techniques tend to be diversified and complex. Therefore, the research hotspots have been extended from traditional accidents and human health risk assessments gradually to large-scale regional ecological risk assessments, and multi-risk factors, multi-risk receptors, and multi-evaluation endpoints need to be considered when evaluation is taken.

5.3.2.2.4 Regulation

It is a huge challenge to regulate ECs effectively. So far there hasn't been sufficient information to fully understand the risks to human health and the environment, and regulatory science and policy decisions are still in progress. Industrial and technology breakthroughs have outpaced the regulatory practice (Naidu et al., 2016). There is only a small number of effective regulatory solutions and guidance regimes globally (Naidu et al., 2016).

The Environmental Protection Agency (EPA) is responsible for pushing forward on ECs regulation in United States based on several federal laws like the Clean Water Act, Safe Drinking Water Act, Toxic Substances Control Act, Food Quality Protection Act, Federal Insecticide, Fungicide, and Rodenticide Act, etc. Although some actions and measures are intended to regulate ECs in surface water and drinking water, the same catalog ECs in groundwater and soil are also subject to regulation and management. The EPA is moving forward with several major steps toward regulating contaminants by group/class. An obvious advantage for this measure is that a whole group of contaminants can be managed and the removal techniques can be developed through it. One of the disadvantages is that some chemicals in a group could be over- or under-regulated compared to the screening standard for the group. For example, a group of chemicals have low toxicities and risks, but there are exceptions, which may be neglected. Group chemicals occurred in surface and groundwater that appear most likely for potential regulations are volatile organic compounds (VOCs), nitrosamines, and perfluorinated compounds. The EPA published technical fact sheets for federal facilities, which provide brief summaries of ECs that present unique issues and challenges to the environment and community. The Minnesota Department of Health (MDH) drinking water contaminants of emerging concern program is considered as one of the best practices on ECs regulation. Although this program is targeting ECs in drinking water, the science-policy-advocacy integration advances are instructive for ECs guidance in groundwater and soil. The US Food and Drug Administration also gets involved in ECs supervision when these chemicals are discharged in the point of entry. The agency requires ecological testing and evaluation on new drug applications if these chemicals are equal to or exceed 1 ppb at the points of entry into the aquatic environment (Naidu et al., 2016).

In 2005, the European Commission initiated a project to promote a permanent network program (NORMAN) of reference laboratories, research centers, and related organizations for monitoring and prioritizing ECs (Dulio et al., 2018). After more than a decade, NORMAN has made progress on data collection, management, methods harmonization and validation, and evolution toward "big data" management from hundreds to tens of thousands of candidate substances. Figure 5.3 demonstrates a typical EC prioritization scheme in the NORMAN program. The program encourages the development of collaborative R&D strategies with a view to their integration into policy. After 10 years of activities, NORMAN has become an essential network in support of EU policies on ECs (Dulio et al., 2018). The recent adopted directive focusing on groundwater protection is Directive 2006/118/EC, which was

developed in response to the requirements of Article 17 of the Water Framework Directive (2000/60/EC). According to the common implementation strategy of the Water Framework Directive, there is a working group responsible for priority substances' monitoring and screening.

In EPAs, Australia, the Cooperative Research Centre for Contamination Assessment and Remediation of the Environment (CRC CARE) has reached agreement with the Australian government regarding the development of guidance for ECs. In 2012, CRC CARE had identified and prioritized ECs for contaminated site assessment, management, and remediation in Australia. The priority contaminants include the perfluorinated chemicals, PFOS and PFOA, MTBE, benzo[a]pyrene (B[a]P), PBDEs, etc. In August 2017, the heads of EPA, Australia and New Zealand released a consultation draft of the PFOS National Environmental Management Plan (NEMP). The NEMP was sought to provide a nationally consistent approach to environmental regulation of PFAS. The health screening levels (HSLs) of PFOS and PFOA in soil were suggested according to multiple land use possibilities in the NEMP draft. Recently, CRC CARE released a technical report on guidance to risk-based assessment, remediation, and management of PFAS site contamination. It suggested the health screening levels (HSLs) for PFOS and PFOA in groundwater and considered the groundwater used for different beneficial uses, such as concentration levels of 0.07 μg L^{-1} for PFOS + PFHxS and 0.56 μg L^{-1} for PFOA when using groundwater as the drinking water source (Table 5.2).

Regulation of ECs is ongoing. With more analytical methods developed, more environmental behaviors need to be understood and more eco-risks need to be considered, the concerned ECs may have more possibilities to go through the policy review process which may lead to progress on more guidance on ECs.

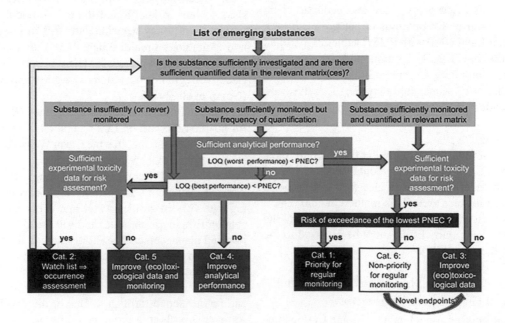

FIGURE 5.3   EC prioritization scheme of the NORMAN network. (From Dulio, V. et al., *Environ. Sci. Eur.*, 30, 5, 2018.)

TABLE 5.2

Groundwater HSLs for PFOS and PFOA and Considerations for Their Application[a]

Groundwater Beneficial Use	PFOS (+PFHxS)	PFOA	Considerations
Drinking water	0.07 µg L^{-1}	0.56 µg L^{-1}	Drinking water quality guidelines apply to all sources of drinking water, including extracted groundwater (DoH 2017).
Primary contact recreation	0.7 µg L^{-1}	5.6 µg L^{-1}	NHMRC 2008 specifies that it may be appropriate to apply a factor for 10 to drinking water standards to account for limited ingestion of recreational waters during swimming activities. The recreational water quality guideline presented by DoH (2017) has adopted this approach.
Stock water	Trigger points for food are presented in table 7 (FSANZ 2017).	Trigger points for food are presented in table 7 (FSANZ 2017).	PFOS and PFOA can biomagnify in the food chain to varying extents. Limited data and information on this issue makes the derivation of a screening level that will avoid contamination of food products difficult. Drinking water guideline values do not apply (an important consideration is that water ingestion rates differ for animals). Where PFAS concentrations are measured in soils or water used for stock water at residential properties higher than laboratory LOR. and domestic stock and animal products are identified at the property, it is recommended that direct sampling of animal products, especially poultry eggs, be undertaken, and compared to the FSANZ (2017b) trigger points. The practicalities of direct sampling need to be considered, especially where rapid management decisions are required.

Source: CRC CARE, Practitioner guide to risk-based assessment, remediation and management of PFAS site contamination CRC CARE Technical Report no. 43, CRC for Contamination Assessment and Remediation of the Environment, Newcastle, Australia, 2018.

[a] Note, the HSLs may change subject to more research evidence and more results in the future as reminded by the CRC CARE report team.

5.4 SOIL AND GROUNDWATER REMEDIATION

Some technologies have been developed to remove ECs in soils to warrant the safety of human health (Song et al., 2017). These technologies cover physical, chemical (Jinlan et al., 2011), thermal treatment, and biological methods (Agamuthu et al., 2010; Li et al., 2018; Ma and You, 2016). Because of the complexity of the soil/groundwater environment, it is not feasible to remove the contaminants only by a single remediation technique. Joint application of multiple technologies will become prevalent in the future (Lukić et al., 2018).

5.4.1 Soil Remediation

Remediation of polluted soil refers to the technical measures to reduce the concentration/activity of soil pollutants through physical, chemical, biological, and ecological principles, and to achieve the expected detoxification effect by adopting natural/artificial control measures, so as to realize the harmlessness and stabilization of pollutants. At present, the theoretically feasible remediation technologies for ECs include physical, chemical, thermal treatment, microbial, soil barrier, landfill, and other comprehensive methods.

5.4.1.1 Physical Methods

5.4.1.1.1 Soil Vapor Extraction (SVE)

Through the SVE well system, the pressure generated by vacuum or air injection is applied to accelerate the air flow in the unsaturated soil, so as to remove VOCs and eventually achieve the purpose of soil cleaning. SVE is an accepted, cost-effective technique for removing organic compounds with high free phase and volatility, especially for the contaminated soil with high permeability. It is not suitable for soil (like clay) with low permeability. The technical advantages of this technology are mainly reflected in relatively low cost and the significant "simplicity" of the project implementation.

A typical soil vapor extraction system consists of the following units: extraction equipment, vapor extraction well, gas distribution pipeline, monitoring device, controlling device, exhaust gas treatment device, etc. Vapor extraction equipment is the core component, which is used to generate negative pressure and pump volatile organic pollutants from underground to the exhaust gas treatment system. Nowadays, the treatment methods of gaseous pollutants can be divided into thermal oxidation, catalytic oxidation, adsorption, concentrating, biological filtration, and membrane filtration. The pollutants in the exhaust gas can reach the standard of discharge after being treated with the above methods, meanwhile, completing the remediation process of the contaminated sites (Figure 5.4).

5.4.1.1.2 Ambient Temperature Desorption (ATD)

ATD is implemented in a closed system by using a turner to conduct artificial disturbance on contaminated soil. Through increasing the contact area between soil and air, the pollutants adsorbed in the soil are driven to evaporate into the air under a concentration gradient and good permeability. This technology is suitable for ectopic treatment, especially for the low concentration and extremely VOCs. The engineering application of this remediation method has been proved to be economical and effective.

The technological process is shown in Figure 5.5. In order to keep the system temperature within a certain range, the enclosed carriers transport the polluted soil to a

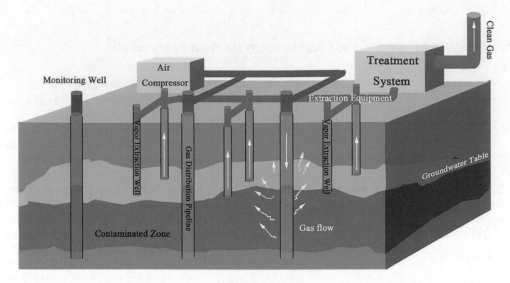

FIGURE 5.4   The schematic diagram of soil vapor extraction technology.

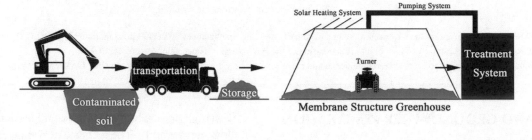

FIGURE 5.5   The schematic diagram of room temperature desorption technology.

room temperature desorption workshop, such as membrane structure greenhouse and auxiliary heating by solar energy. Thereafter, the turner is used to conduct artificial disturbance for reducing moisture content, increasing soil permeability, and accelerating the evaporation of pollutants. The volatilized exhaust gas is intensively collected by the pumping system and transported to the exhaust gas treatment system. After the adsorption and purification process, the pollutants are captured and destroyed so as to achieve the standard discharge.

5.4.1.2 Chemical Methods

5.4.1.2.1 Chemical Oxidation/Reduction

Due to the strong oxidative/reductive strength of chemical oxidants or reductants, chemical oxidation/reduction technology has been widely used in the implementation of soil remediation projects. They can oxidize plenty of contaminants including ECs at a high reaction rate. In the practical remediation projects, more oxidants will be consumed if humus acid, reducing metal, or other substances exist in the soil. Less permeable soil, such as clay, slows down the transfer rate. In addition, there may be adverse effects such as heat and gas production. Meanwhile, the pH value of soil will also influence the reaction more highly. Hydrogen peroxide, permanganate, and ozone are the most frequently used oxidants. Recently, persulfate ($Na_2S_2O_8$) has emerged as an alternative oxidant for chemical oxidation. Some typical reductive agents include ferrous sulfate, polysulfide calcium, divalent iron, zero valent iron, etc.

The process flow of this technology is shown in the Figure 5.6. The system consists of agent preparation/storage system, injection well, injection system, monitoring system, etc. The injection system consists of a storage tank, injection pump, mixing equipment, flowmeter and pressure gauge, etc. The quantity and depth of injection wells are designed according to the size and degree of the contaminated area. Monitoring wells should be designed around the injection wells and the polluted areas for monitoring the effects of the distribution and migration of pollutants and agents. Pumping wells can be set up to facilitate groundwater circulation to increase mixing, which contributes to the quick remediation of large areas of contamination.

5.4.1.2.2 Chemical Leaching

The mechanism of chemical leaching is to combine the eluent or chemical additives with the pollutants in the soil. This technology refers to the technology of dissolving, separating, and treating pollutants from contaminated soil by injecting chemical solvents which can promote the dissolution or migration of soil pollutants (Dermont et al., 2008). The soil is remediated under the function of desorption, chelation, dissolution, or fixation of the eluent. It can effectively remove the adsorbed ECs.

According to the location of treated soil, this technology can be grouped into *in situ* and *ex situ* soil leaching. *In situ* soil leaching (Figure 5.7) is applicable to porous and

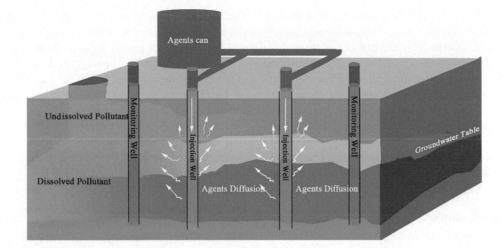

FIGURE 5.6   The schematic diagram of chemical oxidation/reduction technology.

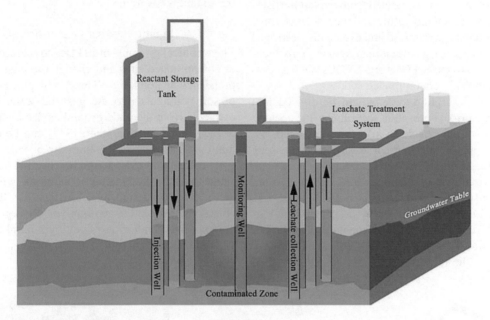

FIGURE 5.7   The schematic diagram of *in situ* soil leaching technology.

permeable soil with hydraulic conductivity greater than 10 cm s^{-1}, such as sand, gravel, alluvial, and coastal soil. *In situ* soil leaching requires the establishment of remediation facilities, including cleaning fluid feeding system, subsoil leachate collection system, and leachate treatment system. At the same time, it is necessary to seal off contaminated areas, usually using physical barriers or segmentation technique.

In *ex situ* soil leaching process (Figure 5.8), the contaminated soil is first dug up and sifted to remove its oversized

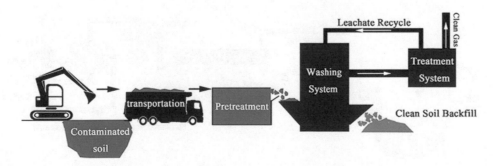

FIGURE 5.8   The schematic diagram of *ex situ* soil leaching.

components, separating the soil into coarse and fine materials. The polluted soil is decontaminated by removing contaminants with an eluent. Meanwhile, the effluent is treated for containing contaminants. *Ex situ* soil leaching is often used as a pretreatment to reduce the amount of contaminant in soil, mainly in combination with other remediation techniques. When the sand and gravel content of contaminated soil exceeds 50%, *ex situ* soil leaching is very effective. The separation and removal effect of *ex situ* soil leaching technology is very limited while contaminated soil with clay and powder content is over 30%. *Ex situ* soil leaching is suitable for soil with clay content less than 25%.

5.4.1.3 Thermal Treatment Methods

5.4.1.3.1 Ex Situ *Thermal Desorption*

Ex situ thermal desorption is the process of treating contaminated soil by direct or indirect heating at a certain temperature. The target pollutants are separated from the soil particles by controlling the system temperature and the residence time. These target pollutants are removed and eventually destroyed by running on the exhaust gas treatment system. This technique is applicable to dispose VOCs and SVOCs (such as pesticides, PBDEs, PFOS) in contaminated soil.

Ex situ thermal desorption systems can be divided into direct and indirect thermal desorption according to the contacting form between soil and flame. It also can be divided into the high and low temperature thermal desorption technology depended on the target remediation temperature.

The direct thermal desorption consists of feeding system, heating system, and exhaust gas treatment system (Figure 5.9). The contaminated soil is pre-treated by procedures such as screening, dewatering, crushing, and magnetic separation. After the contaminated soil is transported into the rotary kiln, it is heated to the target treatment temperature. When the target pollutants reach the gasification temperature, in the meantime, they are removed from the soil. The exhaust gas enriched with gasification pollutants can be completely removed and destroyed through cyclone dust removal, incineration, cooling, bag dust removal, alkali liquor leaching, etc.

The indirect thermal desorption also consists of a feeding system, desorption system, and exhaust gas treatment system (Figure 5.10). Differences with the direct thermal desorption lie in the heating system and the exhaust gas treatment system. The flame produced by the burner uniformly heats the outside of the rotary kiln. The pollutants are separated from soil while reaching their boiling points. The gas is discharged directly after combustion. The exhaust gas enriched with gasification pollutants passes through filters, condensers, and adsorption apparatus. After the gas passes through the condenser, the oil and water can be separated, then the organic pollutants can be concentrated and recovered.

5.4.1.3.2 In Situ *Thermal Desorption (ISTD)*

The technical principle of ISTD is to increase the temperature of contaminated sites and change the physical and chemical properties of pollutants. These will promote the desorption of soil pollutants into the gas and water phases, and then extracted from the underground environment and transferred to the ground for treatment. ISTD can be divided into thermal conductive heating, steam air injection, electrical resistive heating, and radio frequency heating according to heater elements. Thermodynamic conditions will be changed if thermal is transferred to subsurface. As the soil is heated, volatile, semi-volatile, and non-volatile organic contaminants are vaporized and/or destroyed by a number of mechanisms which include evaporation, steam distillation, boiling, oxidation, and pyrolysis.

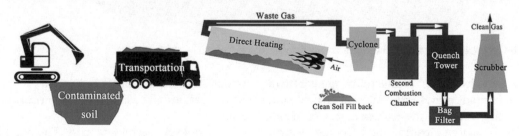

FIGURE 5.9   The schematic diagram of direct thermal desorption technology.

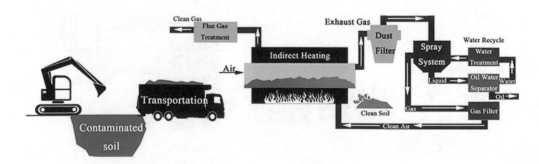

FIGURE 5.10   The schematic diagram of indirect thermal desorption technology.

This technology can be applied to remove a variety of organic compounds. Thermal and vacuum are applied simultaneously to the target media with a group of vertical or horizontal heater elements. *In situ* thermal technologies are known to have the following advantages compared to other remedial options: (1) shorter remediation period; (2) many pollutants can be treated at once; (3) these technologies are less sensitive to subsurface heterogeneities. The relative low temperatures speed the biodegradation of chlorinated solvents and petroleum hydrocarbons. Meanwhile, the higher temperatures cause contaminants to evaporate, even in clay and bedrock.

However, the potential disadvantages of ISTD lie in the following aspects: (1) more sophisticated design and operation are required; (2) pollutants may be enhanced to migrate to previously non-impacted areas; (3) higher soil temperature will exist for some time.

ISTD mainly includes an energy supply system, heating system, extraction system, monitoring system, and waste water and exhaust gas treatment system (Figure 5.11).

5.4.1.3.3 Co-Processing in Cement Kilns

The cement rotary kiln, with the characteristics of high temperature, long retention time, large heat capacity, good thermal stability, alkaline environment, and no waste residue discharge, attracts great attention on soil remediation. The organic contaminated soil is added into cement rotary kiln and the soil temperature could dramatically increase to 1450°C, with the gas phase temperature up to 1800°C. Under the high temperature of the cement kiln, pollutants are transformed into inorganic compounds. High temperature air fully contacts with the high fineness, concentration, adsorption, and high uniformity distribution of alkaline material (CaO, $CaCO_3$, etc.). Therefore, acid emissions are effectively inhibited, transforming sulfur and chlorine into inorganic salts.

Co-processing in cement kiln technology mainly consists of soil pretreatment system, feeding system, monitoring system, cement rotary kiln, and auxiliary system. The soil pretreatment system is carried out in a closed environment, including airtight storage facilities, screening facilities, and exhaust gas treatment system. The exhaust gas generated by the pretreatment system meets the discharge standard after passing through the exhaust gas treatment system. The feeding system mainly includes the storage hopper, plate feeder, belt scale, and elevator. The entire feeding process is conducted in a closed environment to avoid the emission of pollutants and dust into the air, resulting in secondary pollution. Cement rotary kiln and auxiliary systems mainly include preheater, rotary cement kiln, high temperature fan, burner, grate cooler, bag dust collector, screw conveyor, and trough conveyor. The monitoring system mainly includes online monitoring of oxygen, dust, nitrogen oxide, carbon dioxide, water and temperature, as well as regular monitoring of cement kiln exhaust gas and cement clinker for ensuring the treatment efficiency of contaminated soil and production safety.

5.4.1.4 Biological Methods

5.4.1.4.1 Microbial Remediation

Soil microbial remediation technology utilizes indigenous microorganisms or artificially domesticated microorganisms with specific functions to reduce the activity of harmful pollutants in soil or degrade them into harmless substances through their own metabolism under the appropriate environmental conditions.

This technology can be applied in the *in situ* treatment with relatively low costs. It also can deal with different kinds of ECs simultaneously, and has little disturbance to the environment and does not destroy the soil environment. Meanwhile, treatment forms are various and the operation process is relatively simple. However, microbial degradation technology has high specificity, and there is a limit on the maximum concentration of pollutants. Moreover, pollutants cannot be degraded by microorganisms if the solubility of pollutants is low or the combination of pollutants with soil humus and clay minerals is tight.

Soil microbial remediation technologies can be grouped into *in situ* and *ex situ* microbial remediation according to

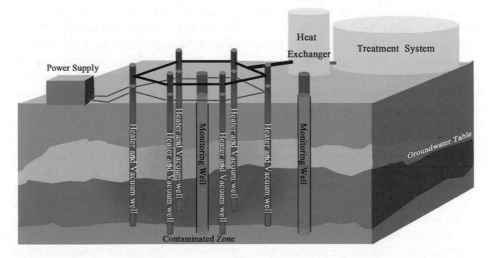

FIGURE 5.11 Typical *in situ* thermal desorption process.

the location of treated soil. *In situ* microbial remediation does not need to remove the contaminated soil from the site, but it can directly release nitrogen, phosphorus, and other nutrients and oxygen in the contaminated soil. These released elements will promote the metabolic activity of indigenous microorganisms or specific microorganisms in the soil and accelerate the degradation of pollutants. *In situ* bioremediation mainly consists of biological ventilation method, biological reinforcement, landfarming method, chemical activity gate remediation method, etc. Heterotopic microbial remediation is to dig out contaminated soil and conduct concentrated biodegradation. It mainly includes the precast bed method, stacking method, and mud bioreactor method.

5.4.1.4.2 Phytoremediation

Plants are used to remove, transform, and destroy the pollutants in soil by means of extraction, rhizosphere filtration, volatilization, and fixation. At present, most researches and applications of phytoremediation focus on heavy metal elements in contaminated soils. ECs including PBDEs, PFOS, pesticides, etc. can also be treated by phytoremediation. The remediation process can be summarized as follows: (1) investigating and evaluating contaminated soil including the content and distribution of ECs in contaminated soil, soil pH value, soil organic matter and nutrient content, soil moisture content, soil porosity, soil particle uniformity, etc.; (2) proposing remediation goals and making remediation plans; (3) selecting appropriate restoration plants and growing seedlings; (4) plant planting, management, and mowing of contaminated sites should be carried out according to the specific conditions of the soil, such as irrigation, fertilization, and addition of metal releasing agent; (5) harmless disposal of hyper accumulators.

5.4.1.5 Soil Barrier and Landfill

The contaminated soil or the treated soil will be placed in the impermeable barrier landfill site for inhibiting the migration and diffusion of pollutants by setting the barrier layer. Thus, the isolation between contaminated soil and the surrounding environment will avoid the direct interaction of pollutants with humans and migration with precipitation or groundwater. According to its implementation mode, it can be divided into *in situ* barrier coverage and *ex situ* barrier landfill.

The *in situ* barrier coverage is to build a barrier layer around and cover the isolation layer at the top of the contaminated area to completely separate the contaminated sites from the surrounding environment. According to the actual situation of the site and the risk assessment results, we can choose to build a barrier layer only near the site or build a cover layer only at the top. The *in situ* soil barrier system is mainly composed of soil barrier system, soil mulch system, and monitoring system. The strong water solubility or high permeability formation of soil and frequent geological activities and high groundwater level will influence its function.

The *ex situ* landfills are used to embed the contaminated soil or the treated soil barrier in an anti-seepage barrier landfill composed of high-density polyethylene film (HDPE) or other anti-seepage barrier materials. Contaminated soil is isolated from the surrounding environment, which can prohibit the migration of ECs with precipitation or groundwater. The exposure and migration of ECs on the surface will be reduced by this technology. However, the toxicity and volume of the pollutants in the soil cannot be reduced. The ectopic soil barrier landfill system is mainly composed of soil pretreatment system, landfill anti-seepage barrier system, leachate collection system, closure system, drainage system, and monitoring system. According to the geological conditions and contaminated soil conditions of the site, groundwater drainage system, gas extraction system, or ground ecological covering system is necessary.

5.4.2 Remediation Technologies for ECs in Groundwater

The remediation of polluted groundwater refers to the recovery of polluted groundwater to the original water quality by using pump-and-treat, permeable reactive barrier, multi-phase extraction, biological methods, chemical methods, groundwater monitored natural attenuation, etc. The control and remediation of polluted groundwater is significant to the sustainable utilization of groundwater resources. The methods which have been introduced at remediation of polluted soil are not listed here, including biological methods and chemical methods.

5.4.2.1 Pump and Treat (P&T)

The P&T technique is one of the most common methods for treatment of groundwater (Boal et al., 2015). Several suction pumps are implemented in the certain location for pumping contaminated groundwater, and then the surface treatment facilities are used to dispose the water. The operation and maintenance of this technology is relatively simple; only the pump well and pipeline valve need to be maintained during the operation. Maintenance of water and sewage treatment equipment should be adjusted according to different pollutants. The P&T can be used to remediate various types of heavily polluted groundwater include pesticides, PPCPs, EDCs, and so on. There are several limitations for the technique: (1) P&T is a time-consuming and high-cost method; (2) suction radii of pumping wells are restricted; (3) the technique cannot be applied to pollutants with high adsorption capacity and groundwater with non-aqueous phase liquid (NAPL) (Endres et al., 2007). Because of the limitation of hydrogeological conditions, the interaction between aquifer medium and pollutants will reduce the concentration of pollutants extracted with the progress of pumping project, resulting in trailing phenomenon.

P&T systems consist of groundwater pumps, treatment equipment, and a monitoring facility. The system and treatment process (Hashim et al., 2011) are shown in Figure 5.12. Main implementation of the P&T process includes: (a) pollution survey for location of source of pollution, characteristics of pollutants, soil properties, and so on.; (b) P&T system design, mathematic or computer model are used to analyze groundwater flow condition and evaluate the pumping process; (c) groundwater control system is built: first, pump wells are

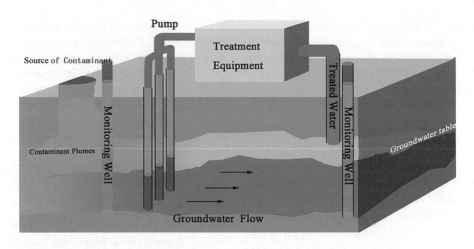

FIGURE 5.12 P&T system composition and treatment process.

dug in different stages for aims of obtaining the information of pollution sources and plume; second, the pumps are installed; finally, pulsed pump groundwater to achieve optimal pollutant removal efficiency by extracting the minimum amount of groundwater; (d) groundwater treatment; (e) remediation effects evaluation; (f) system shutdown and dismantling.

5.4.2.2 Permeable Reactive Barrier (PRB)

This technology has shown good potential for remediation of underground water since the twentieth century (Faisal et al., 2018). PRB refers to a passive in site reaction material area which can degrade and retard the pollutants in groundwater flow through the barrier. Practically, pollutants are delivered to the treatment media in the barrier with groundwater through natural or artificial hydraulic gradients. Disposal methods in the treatment medium include degradation, adsorption, oxido-reduction, leaching, and some bioremediation processes (Faisal et al., 2018). The function of the medium is different for various ECs. The oxidizing agent is used for oxidizing ECs. The chelating agent is applied for fixing ECs. The nutrient substance is employed to promote biodegradation process for the ECs.

The PRB technology is a cost-effective technique for removing ECs such as pesticides, PPCPs, PBDEs, and so on. (Faisal et al., 2018; Kong et al., 2012; Phillips et al., 2010;

Richmond et al., 2001). However, the barrier cannot be used as a load bearing structure. Therefore, the technique is limited in a confined aquifer. In addition, it is not suitable for contaminated groundwater deeper than 10 m because the barrier needs to be maintained regularly.

Generally, the PRB can be divided into single PRB process system and multi-unit PRB processing system (Wanner et al., 2013). The basic structure types of single process system include continuous type PRB and funnel-gate type PRB, as well as some improved configurations, such as curtain type, injection type, siphon type, *in situ* water-barrier reactor type, and so on., which are suitable for sites with relatively single pollutants, low pollution concentration, and small area. The multi-unit process system PRB is suitable for the sites with integrated pollutants and complex geological conditions (Krishna et al., 2010). Multi-unit processing systems can be divided into series structure and parallel structure. Series structure is implemented site with complex pollution composition, for the different pollutants series barriers can fill with corresponding treatment medium (Baldwin et al., 2009). Parallel structure is utilized when the pollution plume is relatively wide and pollution component is single.

An atypical PRB process system is shown in Figure 5.13. There are two main influence factors for the system: the

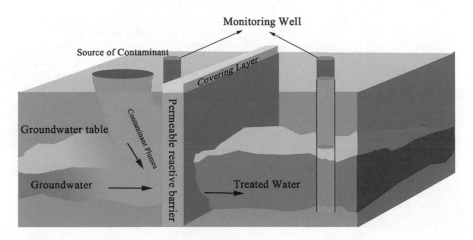

FIGURE 5.13 Typical PRB process system.

barrier must be completely embedded into the aquiclude or aquitard to avoid groundwater migrating under the barrier. Polluted water must have sufficient reaction time in the barrier. Monitoring wells are located both upstream and downstream of the contaminated zone to monitor the remediation process (Moraci et al., 2016).

5.4.2.3 Multi-Phase Extraction (MPE)

Through vacuum extraction, soil gas and groundwater in underground polluted areas are extracted for phase separation and treatment, and pollutants present in both soil and water are concentrated (Baldwin et al., 2009). This technology is suitable for non-aqueous phase liquid pollutants with high volatility, mobility, such as BTEX (Lari et al., 2016). It cannot be used in the site with poor permeability or large fluctuation of groundwater level (Suthersan et al., 2015).

The MPE system is usually a three-part system: polyphase extraction, polyphase separation, and pollutants treatment. A typical MPE process is shown in Figure 5.14. The main equipment includes MPE pump, gas-liquid separator, NAPL-liquid separator, drive pump, treatment facilities, etc. The MPE pump is the core facility to extract both gases and liquids from the contaminated area to treatment process. MPE can be divided into single pump system and double pump system. Extraction power of the single pump system is provided by vacuum equipment only. The double pump system is provided by vacuum equipment and pump together.

5.4.2.4 Groundwater Monitored Natural Attenuation (GMNA)

PRB refers to through implementing of planned monitoring strategy on the basis of naturally occurring physical, chemical, and biological reactions to reduce the environmental risk of pollutants in groundwater to an acceptable level. These reactions include adsorption, oxidation/reduction, precipitation, bio-degradation, and so on. to change the amount, toxicity, and mobility of pollutants (Azadpour-Keeley et al., 2010). The GMNA technology can be applied for degradation of

ECs such as pesticides (Richmond et al., 2001), PPCPs, EDCs (Krishna et al., 2010), and so on GMNA has a high requirement of long-term monitor and operation.

GMNA mainly includes net of monitor well system, monitoring strategy, natural attenuation evaluating system, and emergency plan. The implementation process follows five steps: (a) primary feasibility evaluation; (b) developing a monitoring strategy; (c) construction of the groundwater monitoring system; (d) evaluation of the efficiency of GMNA in detail; and (e) establishment of the emergency plan (Xiaoping and Xi, 2014).

5.4.3 Effects of Soil and Groundwater Remediation Methods on ECs

At present, the remediation technology for organic pollution in soil and groundwater mainly focuses on traditional pollutants such as halogenated VOCs, BTEX, SVOCs (others), VOCs (others), PAHs, halogenated SVOCs, etc. Not much literature discusses emerging contaminant sites remediation. ECs have occurred in soil and groundwater environments and caused various adverse ecological and/or human health effects. Therefore, studies and technologies to remove ECs in soil and groundwater should be given more concern. The existing remediation methods significantly affect the ECs.

5.4.3.1 Effects of Physical Disposal Methods on ECs

The majority of remediation technologies use the physical priciples, such as adsorption, desorption, leaching, volatilization to separate the pollutants with soil or groundwater.

Adsorption and desorption are the main mechanisms for SVE, ATD, CL, thermal remediation, PRB, etc. The pH condition has significant influence on adsorption and desorption processes. Higher concntrations of organic acid can promote the desorption of pestesides in soil particles (Mariana et al., 2010). Some soil conditioners are used to increase the adsorption stability of pesticides; for instance, the organic carbon and surfactant can obviously decrease desorption capacity

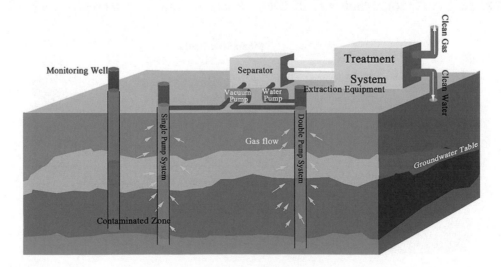

FIGURE 5.14   A typical MPE process.

of soil (Trinh et al., 2017). For some EDCs and PPCPs, the adsorption/desorption capacities are closely related to redox conditions. It has been proven that under sufficient oxidation conditions, a large amount of estrone cannot adsorb on the soil (Ying and Kookana, 2010). Waste recycling is an important source of flame/fire retardants in the environment (Wang et al., 2018b). TOC is a major factor that influences absorption/desorption and transfer ability of some flame retardants such as brominated flame retardants and tetrabromobisphenol-A (Wang et al., 2015).

5.4.3.2 Effects of Chemical Disposal Methods on ECs

Chemical methods are implemented among most of the soil and groundwater remediation technologies for disposition and purification of ECs, for example chemical oxidation/reduction, P&T, PRB, MPE, etc.

Most of the organic pollutants are sensitive to oxidizing agents. It has been reported that oxidizing agents like H_2O_2, sodium persulfate, and Fenton's reagent are used to remedy the contaminated soil by pesticides and flame/fire retardants (Han et al., 2008; Parker et al., 2017). Soil polluted by PPCPs can also be controlled through oxidation process, and the efficiency of removal is high (Klavarioti et al., 2009). Ozonation has been considered to be an effective way for the reduction of organic pollutants; the degradation of EDCs (estrone, estriol, 17α-ethynylestradiol, bisphenol A, and 4-nonylphenol) through the oxidation process is remarkable (Qiang et al., 2013).

5.4.3.3 Effects of Biological Disposal Methods on ECs

Most ECs are the carbon sources of microorganisms, thus biological disposal methods are effective for the degradation of ECs. It is important to take advantage and recognize the biodegradable capabilities of microorganisms.

Microorganisms can degrade the pesticide of clorfenvinphos with high toxicity to acceptable levels (Oliveira et al., 2015), as well as biolysis of pharmaceuticals and EDCs which have significant influence on biofunctionality (Baenanogueras et al., 2017). Biodegradation processes are included in most remediation technologies for soil and groundwater, especially the long-term technologies such as GMNA and *in situ* microbial remediation. The biological and protein activity are limited under some extreme circumstances like high temperature (e.g., thermal remediation) or strong electric strength (e.g., electrical resistance heating).

REFERENCES

Agamuthu, P., et al., 2010. Phytoremediation of soil contaminated with used lubricating oil using *Jatropha curcas*. *J. Hazard. Mater.* 179, 891–894.

Alfy, M. E., Faraj, T., 2016. Spatial distribution and health risk assessment for groundwater contamination from intensive pesticide use in arid areas. *Environ. Geochem Health.* 39, 1–23.

Andrade, N. A., et al., 2010. Persistence of polybrominated diphenyl ethers in agricultural soils after biosolids applications. *J. Agric. Food. Chem.* 58, 3077–3084.

Azadpour-Keeley, A., et al., 2010. Monitored natural attenuation of contaminants in the subsurface: Applications. *Ground Water Monit. Remediat.* 21, 136–143.

Baenanogueras, R. M., et al., 2017. Degradation kinetics of pharmaceuticals and personal care products in surface waters: Photolysis vs biodegradation. *Sci. Total Environ.* 590–591, 643–654.

Baldwin, B. R., et al., 2009. Enumeration of aromatic oxygenase genes to evaluate biodegradation during multi-phase extraction at a gasoline-contaminated site. *J. Hazard. Mater.* 163, 524–530.

Bernhardt, D. C., et al., 2019. Husks of *Zea mays* as a potential source of biopolymers for food additives and materials' development. *Heliyon.* 5, e01313.

Boal, A. K., et al., 2015. Pump-and-treat groundwater remediation using chlorine/ultraviolet advanced oxidation processes. *Ground Water Monit. Remediat.* 35, 93–100.

Butler, E., et al., 2012. The effect of triclosan on microbial community structure in three soils. *Chemosphere.* 89, 1–9.

Canas, J. E., et al., 2011. Acute and reproductive toxicity of nano-sized metal oxides (ZnO and TiO_2) to earthworms (*Eisenia fetida*). *J. Environ. Monit.* 13, 3351–3357.

Carter, L. J., et al., 2014. Fate and uptake of pharmaceuticals in soil-plant systems. *J. Agri. Food Chem.* 62, 816–825.

Clarke, R., et al., 2016. A quantitative risk ranking model to evaluate emerging organic contaminants in biosolid amended land and potential transport to drinking water. *Hum. Ecol. Risk Assess.* 22, 958-990.

CRC CARE, 2018. Practitioner guide to risk-based assessment, remediation and management of PFAS site contamination CRC CARE Technical Report No. 43, CRC for Contamination Assessment and Remediation of the Environment, Newcastle, Australia.

De Silva, P. M. C. S., et al., 2010. Toxicity of chlorpyrifos, carbofuran, mancozeb and their formulations to the tropical earthworm *Perionyx excavatus*. *Appl. Soil Ecol.* 44, 56–60.

Dermont, G., et al., 2008. Soil washing for metal removal: A review of physical/chemical technologies and field applications. *J. Hazard. Mater.* 152, 1–31.

Dodd, M., Addison, J. A., 2010. Toxicity of methyl tert butyl ether to soil invertebrates (springtails: *Folsomia candida*, *Proisotoma minuta*, and *Onychiurus folsomi*) and lettuce (*Lactuca sativa*). *Environ. Toxicol. Chem.* 29, 338–346.

DoH, 2017, Health based guidance values for PFAS for use in site investigations in Australia, Department of Health, available at www.health.gov.au/internet/main/publishing.nsf/Content/2200FE086D480353CA2580C900817CDC/$File/fs-HealthBased-Guidance-Values.pdf, accessed 12 April 2017.

Domene, X., et al., 2009. Soil pollution by nonylphenol and nonylphenol ethoxylates and their effects to plants and invertebrates. *J. Soils Sediments.* 9, 555–567.

Dong, L., et al., 2012. DNA damage and biochemical toxicity of antibiotics in soil on the earthworm *Eisenia fetida*. *Chemosphere.* 89, 44–51.

Dowling, T. P., et al., 2019. Spatial assessment of pesticide leaching risk to groundwater: Sub-national decision making and model output aggregation. *Pest Manag. Sci* 75(10), 2575–2591.

Du, L., Liu, W., 2011. Occurrence, fate, and ecotoxicity of antibiotics in agro-ecosystems. A review. *Agron. Sustain. Dev.* 32, 309–327.

Dulio, V., et al., 2018. Emerging pollutants in the EU: 10 years of NORMAN in support of environmental policies and regulations. *Environ. Sci. Eur.* 30, 5.

Eggen, T., et al., 2011. Uptake and translocation of metformin, ciprofloxacin and narasin in forage- and crop plants. *Chemosphere.* 85, 26–33.

Eggen, T., et al., 2013. Uptake and translocation of organophosphates and other emerging contaminants in food and forage crops. *Environ. Sci. Pollut. Res.* 20, 4520–4531.

Endres, K. L., et al., 2007. Equilibrium versus nonequilibrium treatment modeling in the optimal design of pump-and-treat groundwater remediation systems. *J. Environ. Eng.* 133, 809–818.

Faisal, A. A. H., et al., 2018. A review of permeable reactive barrier as passive sustainable technology for groundwater remediation. *Int. J. Environ. Sci. Technol.* 15, 1123–1138.

Fernández, C., Yves, P., 2014. Pharmaceuticals in hospital wastewater: Their ecotoxicity and contribution to the environmental hazard of the effluent. *Chemosphere.* 115, 31–39.

Gago-Ferrero, P., et al., 2013. Fully automated determination of nine ultraviolet filters and transformation products in natural waters and wastewaters by on-line solid phase extraction–liquid chromatography–tandem mass spectrometry. *J. Chromatogr. A.* 1294, 106–116.

García-Galán, M. J., et al., 2013. Fate and occurrence of PhACs in the terrestrial environment. In: M. Petrovic et al., Eds., *Analysis, Removal, Effects and Risk of Pharmaceuticals in the Water Cycle: Occurrence and Transformation in the Environment,* Vol. 62. Elsevier, pp. 559–592.

Gaston, L., et al., 2019. Prioritization approaches for substances of emerging concern in groundwater: A critical review. *Environ. Sci. Technol.* 53, 6107–6122.

Ge, Y., et al., 2011. Evidence for negative effects of TiO_2 and ZnO nanoparticles on soil bacterial communities. *Environ. Sci. Technol.* 45, 1659–1664.

Gereslassie, T., et al., 2019. Determination of occurrences, distribution, health impacts of organochlorine pesticides in soils of central China. *Int. J. Environ. Res. Public Health.* 16, 146.

Gomes, A. R., et al., 2017. Review of the ecotoxicological effects of emerging contaminants to soil biota. *J. Environ. Sci. Health A Tox. Hazard. Subst. Environ. Eng.* 52, 992–1007.

Gonzalez, S., et al., 2012. Presence and biological effects of emerging contaminants in Llobregat River basin: A review. *Environ. Pollut.* 161, 83–92.

Gros, M., et al., 2019. Veterinary pharmaceuticals and antibiotics in manure and slurry and their fate in amended agricultural soils: Findings from an experimental field site (Baix Emporda, NE Catalonia). *Sci. Total Environ.* 654, 1337–1349.

Hakk, H., et al., 2016. Fate and transport of the beta-adrenergic agonist ractopamine hydrochloride in soil-water systems. *J. Environ. Sci. (China).* 45, 40–48.

Hale, R. C., et al., 2012. Polybrominated diphenyl ethers in U.S. sewage sludges and biosolids: Temporal and geographical trends and uptake by corn following land application. *Environ. Sci. Technol.* 46, 2055–2063.

Han, S. K., et al., 2008. Oxidation of flame retardant tetrabromobisphenol A by singlet oxygen. *Environ. Sci. Technol.* 42, 166–172.

Hashim, M. A., et al., 2011. Remediation technologies for heavy metal contaminated groundwater. *J. Environ. Manage.* 92, 2355–2388.

Heckmann, L. H., et al., 2011. Limit-test toxicity screening of selected inorganic nanoparticles to the earthworm *Eisenia fetida. Ecotoxicology.* 20, 226–233.

Hu, C., et al., 2013. Toxicological effects of multi-walled carbon nanotubes adsorbed with nonylphenol on earthworm *Eisenia fetida. Environ. Sci. Process. Impacts.* 15, 2125–2130.

Husain, S., 2008. Literature overview: Emerging organic contaminants in water and their remediation. *Remed. J.* 18, 91–105.

Jafari, S. M., Nikoo, M. R., 2019. Developing a fuzzy optimization model for groundwater risk assessment based on improved DRASTIC method. *Environ. Earth Sci.* 78, 109.

Jechalke, S., et al., 2014. Fate and effects of veterinary antibiotics in soil. *Trends Microbiol.* 22, 536–545.

Jinlan, X., et al., 2011. Chemical oxidation of cable insulating oil contaminated soil. *Chemosphere.* 84, 272–277.

Kim, K. R., et al., 2011. Occurrence and environmental fate of veterinary antibiotics in the terrestrial environment. *Water Air Soil Pollut.* 214, 163–174.

Klavarioti, M., et al., 2009. Removal of residual pharmaceuticals from aqueous systems by advanced oxidation processes. *Environ. Int.* 35, 402–417.

Kolar, L., et al., 2008. Toxicity of abamectin and doramectin to soil invertebrates. *Environ. Pollut.* 151, 182–189.

Kong, X. K., et al., 2012. Laboratory column study for evaluating a bio-chemical permeable reactive barrier to remove ammonium from groundwater. *Environ. Sci. Technol.* 35, 1–5.

Krishna, A. K., et al., 2010. Monitored natural attenuation as a remediation tool for heavy metal contamination in soils in an abandoned gold mine area. *Curr. Sci.* 99, 628–635.

La Guardia, M. J., et al., 2012. In situ accumulation of HBCD, PBDEs, and several alternative flame-retardants in the bivalve (*Corbicula fluminea*) and gastropod (*Elimia proxima*). *Environ. Sci. Technol.* 46, 5798–5805.

Lapworth, D. J., et al., 2012. Emerging organic contaminants in groundwater: A review of sources, fate and occurrence. *Environ. Pollut.* 163, 287–303.

Lapworth, D. J., et al., 2018. Deep urban groundwater vulnerability in India revealed through the use of emerging organic contaminants and residence time tracers. *Environ. Pollut.* 240, 938–949.

Lari, K. S., et al., 2016. Incorporating hysteresis in a multi-phase multi-component NAPL modelling framework; a multi-component LNAPL gasoline example. *Adv. Water Resour.* 96, 190–201.

Lemos, M. F. L., et al., 2009. Endocrine disruption in a terrestrial isopod under exposure to bisphenol A and vinclozolin. *J. Soil. Sediment.* 9, 492–500.

Lemos, M. F. L., et al., 2010. Reproductive toxicity of the endocrine disrupters vinclozolin and bisphenol A in the terrestrial isopod *Porcellio scaber* (Latreille, 1804). *Chemosphere.* 78, 907–913.

Li, B., et al., 2015. Occurrence and distribution of phthalic acid esters and phenols in Hun River Watersheds. *Environ. Earth Sci.* 73, 5095–5106.

Li, S., et al., 2018. Ten-year review and prospect of industrial contaminated site remediation in China. In: Y. Luo, C. Tu, Eds., *Twenty Years of Research and Development on Soil Pollution and Remediation in China.* Springer, Singapore, pp. 105–123.

Li, W. C., 2014. Occurrence, sources, and fate of pharmaceuticals in aquatic environment and soil. *Environ. Pollut.* 187, 193–201.

Li, Z., et al., 2013. Seasonal variation of nonylphenol concentrations and fluxes with influence of flooding in the Daliao River Estuary, China. *Environ. Monit. Assess.* 185, 5221–5230.

Lin, D., et al., 2010. Potential biochemical and genetic toxicity of triclosan as an emerging pollutant on earthworms (*Eisenia fetida*). *Chemosphere.* 81, 1328–1333.

Lin, D., et al., 2012. Physiological and molecular responses of the earthworm (*Eisenia fetida*) to soil chlortetracycline contamination. *Environ. Pollut.* 171, 46–51.

Lin, Y.-C., et al., 2015. Occurrence of pharmaceuticals, hormones, and perfluorinated compounds in groundwater in Taiwan. *Environ. Monit. Assess.* 187, 256.

Liu, F., et al., 2009. Terrestrial ecotoxicological effects of the antimicrobial agent triclosan. *Ecotoxicol. Environ. Saf.* 72, 86–92.

Lukić, B., et al., 2018. A review on the efficiency of landfarming integrated with composting as a soil remediation treatment. *Environ. Toxicol. Rev.* 6, 94–116.

Ma, W., You, X. Y., 2016. Numerical simulation of plant–microbial remediation for petroleum-polluted soil. *J. Soil Contam.* 25, 727–738.

Mariana, G., et al., 2010. Assessing pesticide leaching and desorption in soils with different agricultural activities from Argentina (Pampa and Patagonia). *Chemosphere.* 81, 351–358.

Marques, C., et al., 2009. Using earthworm avoidance behaviour to assess the toxicity of formulated herbicides and their active ingredients on natural soils. *J. Soil. Sediment.* 9, 137–147.

Masoner, J. R., et al., 2014. Contaminants of emerging concern in fresh leachate from landfills in the conterminous United States. *Enrion. Sci. Proc. Impacts.* 16, 2335–2354.

McKelvie, J. R., et al., 2011. Metabolic responses of *Eisenia fetida* after sub-lethal exposure to organic contaminants with different toxic modes of action. *Environ. Pollut.* 159, 3620–3626.

Meffe, R., Bustamante, I. D., 2014. Emerging organic contaminants in surface water and groundwater: A first overview of the situation in Italy. *Sci. Total Environ.* 481, 280–295.

Moraci, N., et al., 2016. Modelling long term hydraulic conductivity behaviour of zero valent iron column tests for PRB design. *Can. Geotech. J.* 53, 946–961.

Moran, M. J., et al., 2005. MTBE and gasoline hydrocarbons in ground water of the United States. *Ground Water.* 43, 615–627.

Muñoz, I., et al., 2009. Chemical evaluation of contaminants in wastewater effluents and the environmental risk of reusing effluents in agriculture. *TrAC, Trends Anal. Chem.* 28, 676–694.

Naidu, R., et al., 2016. Emerging contaminant uncertainties and policy: The chicken or the egg conundrum. *Chemosphere.* 154, 385–390.

Oliveira, B. R., et al., 2015. Biodegradation of pesticides using fungi species found in the aquatic environment. *Environ. Sci. Pollut. Res. Int.* 22, 11781–11791.

Omar, T. F. T., et al., 2018. Occurrence, distribution, and sources of emerging organic contaminants in tropical coastal sediments of anthropogenically impacted Klang River estuary, Malaysia. *Mar. Pollut. Bull.* 131, 284–293.

Oscar, M., et al., 2017. Treatment technologies for emerging contaminants in water: a review. *Chem. Eng. J.* 323, 361–380.

Parker, A. M., et al., 2017. UV/H$_2$O$_2$ advanced oxidation for abatement of organophosphorous pesticides and the effects on various toxicity screening assays. *Chemosphere.* 182, 477.

Patel, M., et al., 2019. Pharmaceuticals of emerging concern in aquatic systems: Chemistry, occurrence, effects, and removal methods. *Chem. Rev.* 119, 3510–3673.

Philip, J. M., et al., 2018. Emerging contaminants in Indian environmental matrices – A review. *Chemosphere.* 190, 307–326.

Phillips, D. H., et al., 2010. Ten year performance evaluation of a field-scale zero-valent iron permeable reactive barrier installed to remediate trichloroethene contaminated groundwater. *Environ. Sci. Technol.* 44, 3861–3869.

Qi, C., et al., 2018. Contaminants of emerging concern in landfill leachate in China: A review. *Emerg. Contam.* 4, 1–10.

Qiang, Z., et al., 2013. Degradation of endocrine-disrupting chemicals during activated sludge reduction by ozone. *Chemosphere.* 91, 366–373.

Richmond, S. A., et al., 2001. Assessment of natural attenuation of chlorinated aliphatics and BTEX in subarctic groundwater. *Environ. Sci. Technol.* 35, 4038–4045.

Sakurai, T., et al., 2016. Temporal trends for inflow of perfluorooctanesulfonate (PFOS) and perfluorooctanoate (PFOA) to Tokyo Bay, Japan, estimated by a receptor-oriented approach. *Sci. Total Environ.* 539, 277–285.

Serra-Roig, M. P., et al., 2016. Occurrence, fate and risk assessment of personal care products in river–groundwater interface. *Sci. Total Environ.* 568, 829–837.

Sharma, B. M., et al., 2019. Health and ecological risk assessment of emerging contaminants (pharmaceuticals, personal care products, and artificial sweeteners) in surface and groundwater (drinking water) in the Ganges River Basin, India. *Sci. Total Environ.* 646, 1459–1467.

Shores, A., et al., 2017. Produced water surface spills and the risk for BTEX and naphthalene groundwater contamination. *Water Air Soil Pollut.* 228, 1–13.

Shoults-Wilson, W. A., et al., 2011. Role of particle size and soil type in toxicity of silver nanoparticles to earthworms. *Soil Sci. Soc. Am. J.* 75, 365–377.

Snyder, E. H., et al., 2011. Toxicity and bioaccumulation of biosolids-borne triclocarban (TCC) in terrestrial organisms. *Chemosphere.* 82, 460–467.

Song, B., et al., 2017. Evaluation methods for assessing effectiveness of in situ remediation of soil and sediment contaminated with organic pollutants and heavy metals. *Environ. Int.* 105, 43–55.

Sorensen J.P.R., et al., 2015. Emerging contaminants in urban groundwater sources in Africa. *Water Res.* 72, 51–63.

Souza, M.S., et al., 2013. Low concentrations, potential ecological consequences: synthetic estrogens alter life-history and demographic structures of aquatic invertebrates. *Environ. Pollut.* 178, 237–243.

Stuart, M., et al., 2012. Review of risk from potential emerging contaminants in UK groundwater. *Sci. Total Environ.* 416, 1–21.

Suthersan, S., et al., 2015. Contemporary management of sites with petroleum LNAPL presence. *Groundwater Monit. Remediat.* 35, 23–29.

Szekacs, A., et al., 2015. Monitoring pesticide residues in surface and ground water in Hungary: Surveys in 1990–2015. *J. Chem.* 4, 1–15.

Trinh, H. T., et al., 2017. Simultaneous effect of dissolved organic carbon, surfactant, and organic acid on the desorption of pesticides investigated by response surface methodology. *Environ. Sci. Pollut. Res.* 24, 19338–19346.

Verliefde, A., et al., 2007. Priority organic micropollutants in water sources in Flanders and the Netherlands and assessment of removal possibilities with nanofiltration. *Environ. Pollut.* 146, 281–289.

Wang, J., et al., 2012. Assessment of groundwater contamination risk using hazard quantification, a modified DRASTIC model and groundwater value, Beijing Plain, China. *Sci. Total Environ.* 432, 216–226.

Wang, J., et al., 2015. Distribution of metals and brominated flame retardants (BFRs) in sediments, soils and plants from an informal e-waste dismantling site, South China. *Environ. Sci. Pollut. Res.* 22, 1020–1033.

Wang, L., et al., 2010. Occurrence and risk assessment of acidic pharmaceuticals in the Yellow River, Hai River and Liao River of north China. *Sci. Total Environ.* 408, 3139–3147.

Wang, S., et al., 2018a. Migration and health risks of nonylphenol and bisphenol a in soil-winter wheat systems with long-term reclaimed water irrigation. *Ecotoxicol. Environ. Saf.* 158, 28–36.

Wang, Y., et al., 2018b. Occurrence and distribution of organophosphate flame retardants (OPFRs) in soil and outdoor settled dust from a multi-waste recycling area in China. *Sci. Total Environ.* 625, 1056.

Wanner, C., et al., 2013. Unraveling the partial failure of a permeable reactive barrier using a multi-tracer experiment and Cr isotope measurements. *Appl. Geochem.* 37, 125–133.

Wojslawski, J., et al., 2019. Leaching behavior of pharmaceuticals and their metabolites in the soil environment. *Chemosphere.* 231, 269–275.

Worrall, F., et al., 2000. A. New approaches to assessing the risk of groundwater contamination by pesticides. *J Geol. Soc. London.* 57, 877–884.

Xiaoping, L., Xi, C., 2014. Applications of monitored natural attenuation in contaminated soil and groundwater. *Meteorol. Environ. Res.* 5, 31–35.

Yang, L., et al., 2017a. Occurrence, distribution, and attenuation of pharmaceuticals and personal care products in the riverside groundwater of the Beiyun River of Beijing, China. *Environ. Sci. Pollut. Res.* 24, 15838–15851.

Yang, Y. Y., et al., 2017b. Pharmaceuticals and personal care products (PPCPs) and artificial sweeteners (ASs) in surface and ground waters and their application as indication of wastewater contamination. *Sci. Total Environ.* 616–617, 816–823.

Ying, G., Kookana, R. S., 2010. Sorption and degradation of estrogen-like endocrine disrupting chemicals in soil. *Environ. Toxicol. Chem.* 24, 2640–2645.

Yoshimura, H., et al., 2011. 1,8-cineole inhibits both proliferation and elongation of BY-2 cultured tobacco cells. *J. Chem. Ecol.* 37, 320–328.

Zhang, Q., et al., 2013. Risk assessment of groundwater contamination: A multilevel fuzzy comprehensive evaluation approach based on DRASTIC model. *Sci. World J.* 2013, 610390.

Zhao, J. L., et al., 2010. Occurrence and risks of triclosan and triclocarban in the Pearl River system, South China: From source to the receiving environment. *J. Hazard. Mater.* 179, 215–222.

Zhao, Y. Y., Pei, Y. S., 2012. Risk evaluation of groundwater pollution by pesticides in China: A short review. *Procedia Environ. Sci.* 13, 1739–1747.

Zhou, H., et al., 2010. Behaviour of selected endocrine-disrupting chemicals in three sewage treatment plants of Beijing, China. *Environ. Monit. Assess.* 161, 107–121.

6 Arsenic Removal from Water Using Biochar-Based Sorbents

Production, Characterization, and Sequestration Mechanisms

Rabia Amen, Hamna Bashir, Irshad Bibi, Muhammad Mahroz Hussain, Sabry M. Shaheen, Muhammad Shahid, Muhammad Bilal Shakoor, Kiran Hina, Hailong Wang, Jochen Bundschuh, Yong Sik Ok, Jörg Rinklebe, and Nabeel Khan Niazi

CONTENTS

6.1 INTRODUCTION

Arsenic (As) is a well-known carcinogenic metalloid which is classified as a Class-I human carcinogen (Lin et al., 2015). It is the 14th richest element in seawater and the 12th richest element in the human body (Mandal and Suzuki, 2002).

The toxicity of As is closely related to its speciation and oxidation state. In the aquatic environment, inorganic As species are mainly present as arsenite (As(III)) and arsenate (As(V)), and the organic species are dimethylarsenic acid (DMA), monomethylarsonic acid (MMA), and arsenobetaine (AsB). The chemical speciation of As depends mainly on the pH and redox potential (E_h) of the water. Arsenite predominates in alkaline water (pH > 7.5) while As(V) dominates in the acidic conditions (Niazi and Burton, 2016; LeMonte et al., 2017). Arsenite is about 60 times more toxic than As(V), while the organic species of As (MMA and DMA) are 100 times less toxic than inorganic arsenicals (Rahman et al., 2011; Shakoor et al., 2016).

Natural sources such as weathering, hot springs, and volcanic eruption increase the concentration of As in the aqueous environment. Arsenic is used as a wood preservative and insecticide because it is resistant to corrosion and is bactericidal. In addition, mining activities, the production of pharmaceuticals and manufacturing of electronics, as well as industrial uses are the main causes of As contamination in water (Chung et al., 2014).

The natural occurrence of As in surface water and groundwater represents a serious health risk to nearly 200 million people worldwide (Saha and Sahu, 2016). According to the latest scientific report, more than 47 million people in Pakistan are vulnerable to drinking water contaminated with As (Shahid et al., 2018). Arsenic-contaminated drinking water and food are the main sources of As accumulation in the human body (Eisler, 2004; Rahman et al., 2009). Arsenic causes several human diseases such as papillary and cortical necrosis, skin lesions, cirrhosis, and blackfoot disease (Abdul et al., 2015). Chronic As exposure can cause fatal carcinogenic effects such as lung, liver, kidney, and skin cancers (Shakoor et al., 2015). Considering the lethal health effects of As, the World Health Organization (WHO) decreased the safe limit of As in drinking water from 50 μg L^{-1} to 10 μg L^{-1} in 2001 (Mohan and Pittman 2007; Nickson et al., 2005).

Remediation of As-contaminated water has attracted worldwide attention due to its carcinogenic and toxicological effects (Niazi et al., 2011, 2012; Shaheen and Rinklebe, 2015; Niazi et al., 2018a). Many water treatment technologies have been used to remove As, such as nanofiltration, lime softening, frying, activated alumina, zeolite filtration, flocculation, electrochemical techniques, chemical precipitation, flotation, ion exchange, and membrane separation (Shakoor et al., 2016; Niazi et al., 2016; Lata and Samadder, 2016). However, these technologies have significant drawbacks, such as large waste generation, high cost and energy requirements, and partial removal of PTEs like As (Barakat, 2011). Unlike conventional methods, adsorption is a sustainable, cost-effective, user-friendly, and effective method for removing As from contaminated drinking water (Niazi et al., 2018a; Bibi et al., 2016; Shaheen et al., 2013; Shaheen et al., 2019).

Many adsorbents have been used to treat As in aqueous solutions, such as iron (Fe)-sulfide, Fe-coated particles, activated carbon, fly ash, and Fe-oxide-based nanosorbents (Bibi et al., 2017; Cope et al., 2014; Lata and Samadder, 2016; Niazi and Burton, 2016). However, these adsorbents have some disadvantages such as high cost, hazardous chemical use in the production of adsorbents, low adsorption capacity, and stability. Biochar is a low-cost, carbonaceous, stable product obtained by the pyrolysis of raw materials with little or no oxygen (Sohi, 2012; Niazi et al., 2016). The raw materials used to produce biochar are easily available and inexpensive, which mainly come from solid waste and agricultural biomass (Shen et al., 2012; Yao et al., 2012; Qian and Chen, 2013; Xu et al., 2013).

In addition, the conversion of agricultural and solid waste into biochar is effective for environmental protection and waste management (Dong et al., 2013; Cao et al., 2009; Zheng et al., 2010). According to previous research, biochar has proven to be an excellent adsorbent for the remediation of pollutants such as organic pollutants, heavy metals, and other pollutants in aqueous media (Karakoyun et al., 2011; Xue et al., 2012; Zhang et al., 2012; Yang et al., 2014). Various types of biochar had been used for the As removal from aqueous media such as Perilla leaf and Japanese Oakwood biochars (Niazi et al., 2018a, 2018b), sewage sludge biochar (Wongrod et al., 2018), rice husk, municipal solid waste (Agrafioti et al., 2014), and *Cassia fistula* biochar (Alam et al., 2018).

Attention is currently being focused on improving the adsorption capacity of various biochars by using different modifications/pretreatment methods. Biochar has been modified by the impregnation of organic functional groups, nanoparticles, minerals, and reducing agents (Mohan et al., 2014b). The modification of biochar improved the removal capacity of magnetite coated biochar (Zhang et al., 2016), manganese oxide-modified biochar (Wang et al., 2015a), Fe-impregnated biochar (Hu et al., 2015), and chemically modified sewage sludge derived biochar for As in water (Wongrod et al., 2018). Wang et al. (2015a) showed that modified manganese oxide pine biochar showed higher adsorption capacity for As(V) (0.59 and 4.91 mg kg^{-1}) compared to unmodified biochar (0.20 and 2.35 mg kg^{-1}).

Biochar impregnated with Fe had a high adsorption capacity for As(V), which was 2.16 mg g^{-1} at pH 6 (Hu et al., 2015). Oak wood biochar removed 70% As(III) from aqueous solution at pH 3 to 4; the maximum removal of As(III) and As(V) by oak wood biochar was 81% and 84%, respectively (Niazi et al., 2018b). Various mechanisms such as electrostatic interaction, precipitation, ion exchange (Shakoor et al., 2019), and surface complexation are involved in the removal of As from water (Zama et al., 2017; Zhang et al., 2016). Electrostatic interactions and complexation are the most important mechanisms of As adsorption compared to precipitation and reduction (Li et al., 2017).

At present, numerous researches have been published on As removal from water, but still there is a need to critically review the As removal by different forms of pristine and modified biochar. Thus, this chapter focuses on As removal through pristine and modified/engineered biochars. We summarize the (i) As sources; (ii) physical and chemical characterization of biochar; (iii) application of biochar for As removal from wastewater and drinking well water; (iv) factors affecting the removal of As with biochar from water; (v) removal mechanisms of As by biochar; and (vi) biochar modification to enhance As removal.

6.2 SOURCES OF ARSENIC IN AN AQUEOUS ENVIRONMENT

Arsenic is a trace element in the Earth's crust and mainly occurs in the form of arsenite (As(III) or arsenate(V) and As sulfide. It is part of more than 200 minerals worldwide, such as arsenopyrite (FeAsS), arsenolite (As_4O_6), orpiment (As_2S_3), and realgar (AsS). The desorption of these As minerals is a major cause of groundwater contamination (Bissen and Frimmel, 2003; Shakoor et al., 2015). The higher rate of groundwater pumping lowers the groundwater level and leads to the oxidation of As minerals such as arsenopyrite, FeAsS, thereby releasing As into the groundwater system (Mohan and Pittman, 2007). Inorganic As species such as As(III) and As(V) mainly contribute to the pollution of groundwater (Naidu et al., 2006; Shakoor et al., 2015). Hot springs, volcanic eruptions, and stone weathering are main natural sources of As (Basu et al., 2014).

Human activities release large amounts of As into water bodies. Arsenic-contaminated well water (up to 3500 µg L^{-1}) is used in many countries, which is the only source of drinking water for humans (Rasheed et al., 2016; Waqas et al., 2017). Major human sources of As include agrochemicals, pharmaceutical wood preservatives, dyes, cosmetics, paints, incineration, tanning, and smelting (Mohan and Pittman 2007; Niazi et al., 2018b).

Oxides used in the production of pesticides include calcium arsenate $Ca_3(AsO_4)_2$ for Colorado beetle and sodium arsenite ($NaAsO_2$) for aphids. The most important factors for As release into water are industrial waste (40%), steel industry (13%), coal ash (22%), and mining (16%) (Basu et al., 2014; Eisler, 2004). In smelting plants, As-containing gases and fine dust are released which seriously pollute the surface water. Arsenic contaminated drinking water exists in many countries around the world, particularly in South Asia (Shakoor et al., 2015; Alam et al., 2016; Hoover et al., 2017; Zeng et al., 2018). In India, Bangladesh, and Pakistan, As-contaminated groundwater is mainly used for drinking purposes (Malik et al., 2009; Mondal et al., 2013).

6.3 PRODUCTION OF BIOCHAR

Various transformation techniques have been established to improve the properties of biochar (Czernik and Bridgwater, 2004; Mohan et al., 2006). Biotechnology (fermentation, anaerobic digestion, and hydrolysis) and thermal processes (gasification, incineration, hydrothermal carbonation, and pyrolysis) are used to convert biomass into fuels and by-products. Thermochemical processes are used to produce biochar from municipal waste, invasive plants, animal litter, bones, and wood biomass. Biochar produced by thermal technology has a high energy density (typically > 28 kJ g^{-1}).

Temperature and residence time are the two major factors involved in the classification of these biochar production technologies (Czernik et al., 2004; Mohan et al., 2006). The different technologies for producing biochar are shown in Figure 6.1. Here is a brief overview of the most important technologies to produce biochar.

6.3.1 PYROLYSIS

Pyrolysis is a technique in which the biomass is thermally decomposed in the presence of less or no oxygen (Figure 6.1). The pyrolysis technology is divided into two different types, namely the ratio of temperature increase, slow pyrolysis and rapid pyrolysis. In thermal decomposition, lignin, cellulose, and hemicellulose-containing biomass undergo specific reaction pathways, including fragmentation, cross-linking, and depolymerization at certain temperatures, resulting in viable gaseous, solid, and liquid products. The liquid and solid products are known as bio-oil and biochar. The gaseous product contains C1-C2, CO, H_2, and CO_2 hydrocarbons, so-called synthesis gas. The yield of the resulting biochar depends on the pyrolysis process and the properties of the starting material.

Parameters such as temperature, residence time, and heating rate influence the product obtained by the pyrolysis process. In general, the biocarbon yield has an antagonistic effect on the pyrolysis temperature, which is proportional to the pyrolysis temperature (Inguanzo et al., 2002; Al-Wabel et al. 2013).

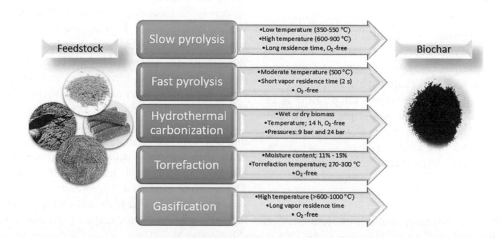

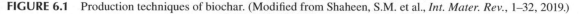

FIGURE 6.1 Production techniques of biochar. (Modified from Shaheen, S.M. et al., *Int. Mater. Rev.*, 1–32, 2019.)

As temperature increases, biochar production and acidic functional groups decrease, while pH, basic functional groups, ash content, and carbon stability increase with temperature (Mohammad et al., 2013; Zhang et al., 2015a). The increase in pH with rising temperature is due to a decrease in organic functional groups such as hydroxyl (–OH) and carboxyl (–COOH). Due to cracking at elevated pyrolysis temperatures, the bio-oil yield is also increased up to about 500°C (Chen et al., 2003).

6.3.1.1 Slow Pyrolysis

Slow pyrolysis is a conventional technique for the production of charcoal. Slow pyrolysis is a versatile technology for producing biochar that is commonly used to treat drinking water and wastewater (Tan et al., 2015). The production of biochar and its unique properties depend on the raw materials, thermochemical methods, and operational conditions. Biochar produced at low temperatures by slow pyrolysis has increased polarity and surface acidity, but low aromaticity and hydrophobicity. Most of the biomass is degraded in many steps over a temperature range of 200°C to 500°C, including partial lignin, whole cellulose, and partial and complete hemicellulose decomposition (Rutherford et al., 2012). The steam residence time varies between 5 and 30 minutes.

In conventional pyrolysis, vapors do not flow rapidly as in fast pyrolysis. This technique maintains the biochar production instead of generating the gaseous and liquid products (Tripathi et al., 2016; Lee et al., 2017). During slow pyrolysis, the increase in temperature (Manyà et al., 2012) could be a reason of intensified fixed carbon content in biochar which is apparent in the temperature range of 400°C–500°C. Decrease in temperature significantly lowers biochar yield in the slow pyrolysis technique (Lee et al., 2013). The surface area and pore size of biochar could also be influenced by high temperature, hence enhancing the sorption capacity of biochar for the pollutants removal from contaminated water (Kim et al., 2012; Abdel-Fattah et al., 2015; Awad et al., 2017; Niazi et al., 2018a). As compared to biochar yield, the fixed carbon yield is a superior index, as it shows chemical composition of the biochar. Biochar produced at high temperatures possess high porosity, aromaticity, and fixed carbon (Lee et al., 2017).

6.3.1.2 Fast Pyrolysis

Fast pyrolysis is very different compared to slow pyrolysis. Prerequisites for rapid pyrolysis are rapid heat transfer, dry feedstock (water < 10 wt%), steam residence time 1 s (up to 5 s), heating of small particle biomass (1–2 mm), and temperature range from 400°C to 500°C (Lima et al., 2010).

It has been observed that the pyrolysis temperature is an important factor that affects biochar properties. For example, biochar pyrolyzed at elevated temperatures showed relatively large pore volumes and surface area compared to biochar produced at low temperatures (Ahmad et al., 2012, 2013).

6.3.2 Gasification

Thermal gasification of biomass is an important method of bioenergy for biochar production (Hansen et al., 2015).

The gaseous fuel produced by the gasification technology is used to generate electricity and contribute in direct heat generation. Partial combustion of solid biomass occurs in gasification, and the resulting products (solids, liquids, and gases) are controlled by varying the gas composition, temperature, pressure, particle size, and residence time. Biochar produced by slow pyrolysis has high aromaticity compared to gasification and fast pyrolysis. In the biochar (7–8 ring) produced by slow and fast pyrolysis, the size of the bonded aromatic ring is similar, and the biochar produced by gasification has an extremely agglomerated ring (17 ring) (Brewer et al., 2009).

6.3.3 Torrefaction

Torrefaction usually implicates slow heating within temperature range of 250°C–300°C, liberate CO_2 and moisture both from the biomass, and subsequently produce solid fuel possessing lesser O/C ratio (Pimchuai et al., 2010). Torrefaction significantly decreases biomass weight, intensifies its hydrophobic nature and energy density, and increases commercial usage of wood to produce energy by decreasing transport expenditure.

6.3.4 Hydrothermal Carbonization

To distinguish biochar from hydrochar is very important. Hydrothermal carbonization produces hydrochar at elevated pressure and temperature in the water, generating char-water slurry, and the resultant solid char could be separated easily (Libra et al., 2011). Physical and chemical characteristics of the resultant hydrochar would be greatly varied from the feedstock biomass. Hydrochar produced by agricultural residue had been used for the treatment of soil and water (Wiedner et al., 2013).

Chars produced via hydrothermal carbonization are enriched with surface charges and O-containing functional groups, and are more acidic as compared to slow pyrolysis resultant chars (Tripathi et al., 2016). Conversely, hydrothermal carbonization involves higher energy consumption because of higher moisture content requirements (Malghani et al., 2013; Garlapalli et al., 2016).

It is also reported that the thermochemical techniques (time, temperature, and heating) and feedstock source significantly influence the biochar properties such as pore-size, yield, pH, surface area, porosity, hydrophobicity, composition, and water holding capacity. Hydrothermal carbonization and slow pyrolysis are greatly suggested for the biochar production as compared to gasification, torrefaction, and fast pyrolysis (Brewer et al., 2009). For the formation of alkaline biochar, slow pyrolysis is more suitable than hydrothermal carbonization.

The slow pyrolysis resultant biochars have higher aromaticity than the gasification or fast pyrolysis resultant biochars; therefore, most studies have reported biochar production through hydrothermal carbonization and slow pyrolysis. The biochar produced by hydrothermal carbonization and slow pyrolysis showed high surface functional groups and surface area which control sorption of contaminants on biochar (Shaheen et al., 2019).

6.4 PHYSICAL AND CHEMICAL CHARACTERIZATION OF BIOCHAR

The physical and chemical properties of biochar fluctuate with production techniques (Sun et al., 2012). As described before, thermochemical techniques, pyrolysis temperatures, feedstocks, residence times, and the properties of biochar affect the biochar production. Similarly, these factors significantly affect the absorption of carbon dioxide by biochar. From their point of view, temperature can be a key factor. Pyrolysis temperatures have a large effect on the isothermal shape and structural characteristics of biochar compared to raw biomass (Chen et al., 2012). Some important physical and chemical properties of different biochars are given in Table 6.1.

6.4.1 Physical Properties

6.4.1.1 Surface Area and Porosity

Surface area and porosity are the most important physical properties of biochar, affecting its ability to adsorb As and other PTEs. In the pyrolysis process of biomass, the loss of moisture due to the dehydration process causes micropore formation. The pore size of biochar is very erratic, including macroscopic (>50 nm), microscopic (<2 nm), and nanoporous (<0.9 nm) (Ahmedna et al., 2004). The pore size is necessary to remove As and other metals; for example, larger sorbates cannot be captured by biochar with a small pore size.

Brunauer–Emmett–Teller (BET) is a technique commonly used to estimate the surface area of biochar at a lower temperature (77 K) (Keiluweit et al., 2010; Wang et al., 2013; Igalavithana et al., 2017). The surface area of the biochar can therefore be calculated by a CO_2 adsorption technique at a relatively high temperature (273 K). This technology provides a more accurate surface area measurement for biochar and has been used widely (Kasozi et al., 2010; Wang et al., 2013).

Studies have shown that biochar produced at high temperatures has a larger pore size and therefore has a higher surface area. For example, straw biochar is made at three different temperatures (300°C, 500°C, and 700°C). The effects of production temperature on Pb removal mechanism and straw biochar adsorption characteristics were investigated by intermittent adsorption experiments, sequential metal extraction, and microstructure evaluation. A relatively high surface area and pH are observed in biochar produced at high temperatures.

The production temperature of rapeseed biochar was positively correlated with ash content, pH value, fixed-C, microporous structure, and surface area, and negatively correlated with density, functional groups, yield, O/H ratio, and volatile matter. The morphology and surface area of biochar are significantly influenced by resistance time, which is often overlooked in publications (Zhao et al., 2018).

The surface area of the biosolid biochar increased from 25.4 to 67.6 m^2 g^{-1} by raising the temperature from 500°C to 600°C, and the porosity increased from 0.056 to 0.099 cm^3 g^{-1} (Chen et al., 2014). In contrast, it has been observed that in some cases the porosity and surface area of biochar decrease with increasing temperature. The porous structure of biochar can be blocked by tar or removed due to elevated temperatures, resulting in a reduced surface area.

The surface of biochar also depends on the nature of the raw material. For example, oak bark biochar has a relatively large surface area (8.8 m^2 g^{-1}) compared to oak biochar (6.1 m^2 g^{-1}) (Mohan et al., 2014a). The surface area of biosolids and manure biochar (5.4–94.2 m^2 g^{-1}) is rather small compared to plant biochar (112–642 m^2 g^{-1}) such as wheat, pine needles, and oaks.

The modification or activation of biochar by bases such as KOH and NaOH may increase the surface of the biochar compared to the integration of the nanoparticles, resulting in increased As removal. For example, KOH-activated residential waste carbon fuel increased the surface area by 29 to 49 m^2 g^{-1}, thus increasing the As(V) sorption from water (Jin et al., 2014).

6.4.2 Chemical Characteristics

6.4.2.1 pH and Surface Charge

Surface charge is another important feature that influences As adsorption on the surface of biochar. The pH of the medium in which the biochar is added to remove the metalloid greatly affects the surface charge of the sorbent. Biochar with a zero-charge point (pH_{pzc}) means that the net charge on its surface is zero. If the pH of the solution is greater than pH_{pzc}, it means that the biochar is negatively charged and attracts the cationic metals such as Hg^{2+}, Pb^{2+}, and Ni^{2+}. If the pH of the solution is below than pzc, this indicates that the surface of the biochar is positive and has strong affinity to bind the anions such as $HCrO_4^-$ and $HAsO_4^{2-}$. The pH of the biosolid biochar increased from 8.58 to 10.2 by increasing the temperature from 400°C to 600°C (Chen et al., 2014).

The pH of biochar also depends on the feedstock and pyrolysis temperature. Acidic (4.84–4.91) biochar is produced when the oak is pyrolyzed in the temperature range of 350°C to 600°C (Nguyen et al., 2010). A low pH (4.60) was detected in pyrolyzed oak biochar at 200°C. On the other hand, biochar produced at 400°C and 600°C showed neutral to alkaline pH (6.90–9.50) (Zhang et al., 2015a). Rapeseed stalk biochar was pyrolyzed at various temperatures (200°C–700°C) and the effects of various parameters were evaluated. It was determined that the pyrolysis temperature is the critical parameter that positively correlates with the pH (Zhao et al., 2018).

Various materials such as switchgrass, peanut shell, poultry litter, and pecan shell were pyrolyzed at various temperatures ranging from 250°C to 700°C. The results demonstrated that the increased pyrolysis temperature increases the ash content, surface area, and pH while decreasing yield and total surface charge (Novak et al., 2009). Low temperature biochar may

TABLE 6.1
Physiochemical Properties of Different Biochars

Feedstock	Temperature (°C)	pH	Surface Area (m² g⁻¹)	Porosity	Yield	Ash Content (%)	C (%)	H (%)	O (%)	N (%)	References
Broiler	350	–	60.0	0.000	–	–	45.60	4.00	18.30	4.50	Uchimiya et al. (2010)
Litter	700	–	94.0	0.018	–	–	46.00	1.42	7.40	2.82	Tong et al. (2011)
Canola straw	400	–	–	–	27.4	–	45.70	–	–	0.19	Ro et al. (2010)
Chicken litter	620	–	–	–	43–49	53.2	41.50	1.20	0.70	2.77	Zama et al. (2017)
Corncobs	350	10.0	12.44	–	36.8	6.11	69.6	4.50	24.4	1.36	
	450	10.1	14.4	–	29.6	8.40	76.3	3.3	18.9	1.39	
Mulberry wood	350	10.2	16.56	–	37.5	7.52	67.9	4.53	25.3	2.16	
	450	11.1	31.45	–	32.7	7.72	70.8	3.32	23.8	1.92	
	550	10.6	58.03	–	26.2	9.82	77.0	2.41	18.8	1.68	
	650	10.6	24.46	–	22.8	9.77	80.1	1.63	16.6	1.58	
	500	–	0.0	–	28.9	7.9	87.50	2.82	7.60	1.50	
	800	–	322.0	–	24.2	9.2	90.0	0.60	7.00	1.90	
Feed lot	350	28.7	1.3	–	51.1	28.7	53.32	4.05	15.70	3.64	Cantrell et al. (2012)
	700	10.3	145.2	–	32.2	44.0	52.41	0.91	7.20	1.70	
Oak wood	450	–	2.7	0.410	–	2.9	82.83	2.70	8.05	0.31	Mohan et al. (2011)
Peanut shell	300	7.8	3.1	–	36.9	1.2	68.27	3.85	25.89	1.91	Ahmad et al. (2012)
	700	10.6	448.2	0.200	21.9	8.9	83.76	1.75	13.34	1.14	
Hickorywood	600	–	256	–	–	–	84.7	1.8	11.3	0.3	Ding et al. (2016)
Hickorywood (Alkali modified)	600	–	873	–	–	–	82.1	2.2	13.2	0.25	
Bagasse	500	9.3	–	–	43.7	8.57	85.59	–	–	–	Lee et al. (2013)
Conocarpus wastes	200	7.37	–	–	51.33	4.53	64.19	–	–	–	Al-Wabel et al. (2013)
	600	12.21	–	–	27.22	8.56	82.93	–	–	–	
	800	12.38	–	–	23.19	8.64	84.97	–	–	–	
Cotton stem	350	8.5	0.69	–	44	8.77	62.37	4.6	22.5	1.65	Samsuri et al. (2013)
	550	10.4	0.18	–	23	10.77	72.1	2.8	12.9	1.3	
Eucalyptus sawdust	350	7.6	1.08	–	42	2.48	68.4	4.6	23.9	0.52	
	550	9.2	23.1	–	21	2.87	83.8	2.7	9.7	0.82	
Perilla leaf	300	9.7	3.2	0.01	38.6	28.4	67.9	3.93	20.9	1.3	Niazi et al. (2017)
	700	10.6	473.4	0.1	23.4	41.9	71.8	0.9	15.3	1.5	
Municipal solid waste	400	–	2.0	0.007	–	50.1	80.2	7.4	8.5	2.8	Li et al. (2015)
	600	–	4.0	0.029	–	53.8	85.0	2.3	6.7	3.6	

have a higher acidic functional groups, which could result in a lower pH of the biochar (e.g., carboxyl and phenolic groups) (Zhang et al., 2015b).

6.4.3 Spectroscopic Characterization

Various spectral techniques such as Fourier transform infrared (FTIR) spectroscopy, X-ray photoelectron spectroscopy (XPS), Raman spectroscopy, and synchrotron-based near edge X-ray absorption fine structure spectroscopy (NEXAFS) are used to analyze various carbon surface materials such as biochar and other raw materials. These studies provide qualitative knowledge about the processes involved in the functionalization and aging of biochar.

6.4.3.1 FTIR Spectroscopy

The FTIR is used to examine the structural properties and characterization of functional groups on the surface of biochar. It can track the differences in the functional groups of raw materials and their biochar. It also analyzes the changes before and after As adsorption. For example, the change in the FTIR spectrum of Fe-impregnated biochar lies mainly in a wavenumber range between 1900 and 800 cm⁻¹. For pre- and post-adsorption analyses of Fe-impregnated biochar, a new peak at 1807–1523 cm⁻¹ was observed. After Fe-impregnation, the peak appeared at 1326 cm⁻¹, but it was significantly weakened after the adsorption of As (Figure 6.2). The peak was considerably discolored at 881 cm⁻¹ and led to a blue shift after impregnation of As and Fe, confirming that Fe has a significant role in the adsorption of As (Figure 6.2). In the Fe-impregnated biochar and the original biochar, the characteristics peaks of C=O was observed at about 1577 cm⁻¹ and remained unchanged after adsorption. Therefore, both adsorption and Fe impregnation on biochar change the positions of surface functional groups (Hu et al., 2015).

Niazi et al. (2018b) reported the FTIR spectrum of Japanese oak wood biochar (OW-BC) (Figure 6.3). In general, the main spectral structure of OW-BC consisted of aromatic surface groups. The levels of oxygen, carbon, nitrogen, and hydrogen of OW-BC determine the polarity, hydrophobicity, and carbonation of the biochar. The aromatic compound is increased by condensation upon dehydrogenation of the carbohydrate (Zama et al., 2017; Cope et al., 2014).

6.4.3.2 X-Ray Photoelectron Spectroscopy (XPS)

The XPS is used for surface analysis of various materials and provides valuable information on chemical states and quantification (Wu et al., 2012). The XPS is used to analyze the surface of the sample by photoelectron excitation. In this process, X-rays are emitted from the surface of the sample, and the energy emitted by the photoelectrons is calculated

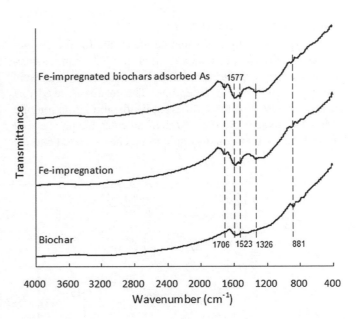

FIGURE 6.2 FTIR spectra of pristine biochar and pre- and post-sorption Fe-impregnated biochar. (Reproduced with permission from Hu, X. et al., *Water Res.*, 68, 206–216, 2015.)

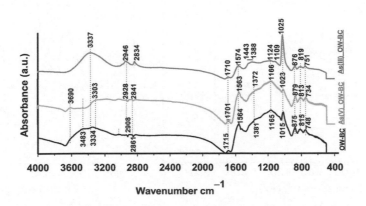

FIGURE 6.3 The FTIR absorbance spectra of Japanese oak wood-derived biochar prepared at 500°C (OW-BC); solid black line (——) shows OW-BC_As-unloaded, grey line (——) shows OWBC_As(V)-loaded, solid gray line (——) shows OW-BC_As(III)-loaded spectra. (Reproduced with permission from Niazi, N.K. et al., *Sci. Total Environ.*, 621, 1642–1651, 2018b.)

with an electron energy analyzer (Shaheen et al., 2019). XPS is an excellent tool for detecting significant changes in the percentage of aromatic carbon in aged and fresh biochar (Singh et al., 2014).

The XPS is used to identify functional groups on the surface of biochar (e.g., –OH, –CH, –CH). Zama et al. (2017) used XPS and found that functional groups such as C–O–C, C–O, and C–OH were present on the surface of mulberry biochar. They strongly believe that mineral-containing elements (P, Mg, K, and Ca) (e.g., $CaPO_4$ and $CaCO_3$) were sequestered on the surface of wood biochar. The XPS analysis showed that the Fe hydroxide particles present on the surface of biochar played an important role in the adsorption of Pb, and the adsorption process was mainly controlled by the chemical adsorption mechanism (Figure 6.4).

6.4.3.3 Raman Spectroscopy

Raman spectroscopy is a valuable technique for the characterization of carbon-based nanostructures. The Raman spectra of the original and treated multiwalled nanotubes were excited on a 514.5 nm laser line. The results showed that spectra consisted of three different bands, the formation of a D-band at about 1338 cm⁻¹, which indicates the presence of amorphous carbon in the sample to be characterized and the

formation of a G band in the graphene sheet by significant diffusion of CC bond (Datsyuk et al., 2008).

Raman spectroscopy has been used to evaluate the crystalline structure of carbon in biochar (see Figure 6.5) (Vithanage et al., 2017; Parikh et al., 2014; Wu et al., 2009). However, the efficiency of Raman spectroscopy to characterize biochar and other carbon-based sorbents is considered to be less because of the fluorescence effects that can occur in the study of polycyclic aromatic hydrocarbons and graphite compounds in the biochar, thus Raman spectroscopy is not widely used compared to FTIR spectroscopy (Shaheen et al., 2019).

6.4.3.4 Near Edge X-Ray Absorption Fine Structure (NEXAFS) Spectroscopy

The NEXAF has been used to distinguish extremely unpredictable and complex forms of carbon-based materials such as carbon elements (Lehmann et al., 2005; Liang et al., 2008). Black carbon plays an important role in soil biogeochemistry. In the environment, complexity is a challenge for scientists, mainly because it is not a total carbon capture method but is beneficial for the detection of different types of functional groups.

NEXAFS serves to illustrate the functionalities and C-chemistry of a wide range of raw materials (pine, oak, and

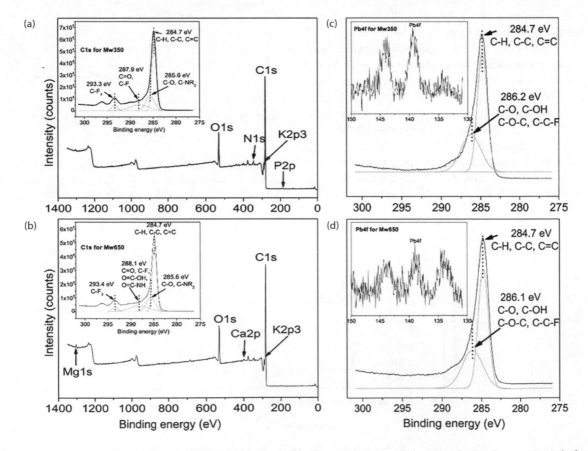

FIGURE 6.4 XPS spectra of MW350 and MW650 before and after Pb sorption (c and d, respectively). Survey scans before sorption (a and b), C1s scans before sorption (a and b insets), C1s scans after Pb sorption (c and d), and Pb4f scans after Pb sorption (c and d insets). (Reproduced with permission from Zama, E.F. et al., *J. Clean. Prod.*, 148, 127–136, 2017.)

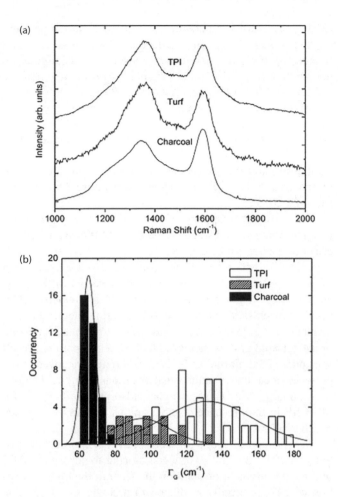

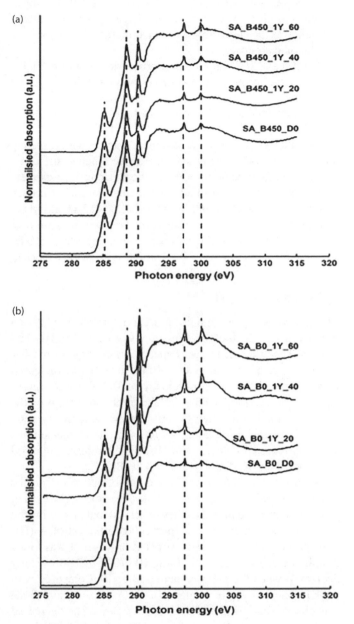

FIGURE 6.5 Statistical Raman spectroscopy study for crystalline dimensionality. (a) Raman spectra from TPI-carbon, turf (middle), and charcoal (bottom). (b) Statistical analysis of the GG from TPI-carbons, turf, and charcoal obtained with a 632.8 nm laser. (Reproduced with permission from Jorio, A. et al., *Soil Till. Res.*, 122, 61–66, 2012.)

FIGURE 6.6 -C-1s NEXAFS spectra of the light fraction of (a) the SA soil treated with B450 biochar (SA_B450) and (b) the control SA soil (SA_B0) without biochar after incubation at different times and temperatures. (Reproduced with permission from Singh, B. et al., *Org. Geochem.*, 77, 1–10, 2014.)

chestnut) and soot. The unique resonances in the NEXAFS spectrum provide potentially invasive reference materials for the direct formation of molecules in carbon chemistry and environmental conditions derived from a variety of raw material experiments and conventional biochar. The reference material for biochar is characterized by a C-based aromatic region, which accounts for about 40% of the total adsorption capacity but addresses certain areas of the process (Wiedemeier et al., 2015).

Singh et al. (2014) presented two peaks at ~288.5 and 285.1 eV in soils and controls modified with eucalyptus wood, followed by incubation in the laboratory, as shown in Figure 6.6. The authors associated these peaks with changes in carbonyl-C, aromatic C, carboxamide-C, and carboxyl-C. Future research will need to expand the NEXAFS application to characterize the surface groups and biochemical components of biochar related to the removal of contaminants such as As from aqueous media.

6.5 FACTORS AFFECTING THE REMOVAL EFFICACY OF ARSENIC IN WATER

6.5.1 Biochar Dose

The adsorption efficiency depends on the amount of adsorbent used. The use of biochar with an optimal dosage is crucial for a cost-effective technology. Increasing biochar concentration reduces the adsorption efficiency (Chen et al., 2011). The highest metalloid adsorption capacity obtained in maize straw and hardwood biochar was 1 g L⁻¹. However, a high biochar to water ratio also increases overall removal efficiency as the entire active site also increases with increasing the biochar dose.

Alam et al. (2018) varied the adsorbent dose from 1 to 8 g L⁻¹ in batch experiments with optimized initial As(III) and As(V) concentrations. First, the removal efficiency of biochar was increased by increasing the dose up to 4 g L⁻¹ (79.6%) for As(V) and 6 g L⁻¹ (74.5%) for As(III). As the sorbent dose increases, solidification occurs when heavy metals are removed, mainly due to the continued increase of the active binding surface on the adsorbent at specific As concentrations. However, regardless of the amount of adsorbent added, the final removal efficiency was not increased (Das and Mondal, 2011). At high adsorbent concentrations, the agglomerates form between the adsorbent particles which limits the number of binding sites, thereby reducing the surface area and the absorption capacity of biochar.

6.5.2 Solution pH

The pH value is an important parameter which plays an important role in the sorption of As species (As(III) and As(V)). The effect of the pH depends on the type of biochar and the targeted contaminants. When the pH was increased from 7 to 9, the removal efficiency of BC 700 was significantly improved for As (88%–90%). Functional groups on the surface of the biochar significantly contribute to the sorption of As. Compared to BC 300, BC 700 showed a larger surface area and more aromaticity, thus BC 300 had higher removal efficiency than BC 700 (Niazi et al., 2018a).

Niazi et al. (2018b) investigated the effect of pH on the adsorption and removal rates of As(III) and As(V) using Japanese oak wood biochar. They observed that increasing the pH from 3 to 7 increased the percentage removal of As(III) adsorption from 75% to 81%. At pH 7, however, it was found that the percentage removal of As(III) (76%–69%) drastically decreased when the pH was increased to 10. Compared with As(III), the highest percentage of As(V) adsorption (84%) was observed at pH 6, and the As(V) removal rate decreased significantly as the pH was increased from 7 to 10 (80%–59%).

Alam et al. (2018) considered the pH value as one of the most important parameters influencing the adsorption of metalloids by stimulating changes in non-metallic species. They used Cassia Seed biochar to analyze the effect of the pH from 2 to 10 by keeping other factors constant. The maximum percentage removal of As(V) (84.8%) was observed at pH 2, while As(III) at pH 6 reached the maximum removal level (78.1%). They claim that changing the pH significantly changes the total charge of biochar, especially the charge of the surface functional groups and transformations that cause ionization and speciation of the adsorbate.

6.5.3 Temperature and Background Electrolytes

Previous studies have shown that the adsorption of PTEs like As on biochar is an endothermic process. As the temperature increases, the adsorption capacity of biochar increases

(Zhang et al., 2013). In contrast, the influence of temperature on the removal of As from the aqueous solution was examined by keeping the other parameters constant and changing the temperature from 27°C to 80°C at pH 7. At room temperature (27°C), As(V) (78.8%) and As(III) (74.7%) have the highest adsorption efficiency. When the temperature rises to 80°C, the removal effect was decreased gradually (Alam et al., 2018). From the above observations, it can be concluded that the adsorption of As above a certain temperature is essentially exothermic and that As gradually desorbs from the surface of the adsorbent with increasing temperature (Ahmet and Mustafa, 2010).

The adsorption capacity of wood biochar in warm water is higher than in cold water. Zhou et al. (2017) evaluated the effect of background electrolytic cations on the use of As(V) oxyanions in MG-CSB. They estimated that the adsorption capacity of MG-CSB for the removal of As(V) in the presence of monovalent cations (K⁺ and Na⁺) is greater at electrolyte concentrations of 0.01 to 0.1 M and in the presence of divalent cations (Mn²⁺, Mg²⁺, and Ca²⁺). When the proportion of background anions increased to 1 M, a comparative result was found. In addition, As(V) is more strongly removed in the presence of cations than the negative anions (SO₄²⁻ and NO₃⁻). In summary, As(V) has a higher adsorption efficiency for MS-CSB in the presence of other anions than in the absence of other anions.

The concentration and presence of the background electrolyte can affect the ability of wood biochar to adsorb the PTE of As in the eradication solution. In general, the high adsorption of PTE is usually evaluated in a single metal solution (single adsorption system) and not in a multi-metal solution (competitive adsorption system) (Shaheen et al., 2013, 2015). The effect of common ions such as sulfate, carbonates, phosphate, and biocarbonate was evaluated having concentrations of 1, 5, and 10 mg. Compared with the other three anions, phosphate showed the greatest influence in reducing As adsorption, and its adsorption efficiency was significantly reduced. In summary, by increasing the co-occurring ions concentration, the As adsorption was decreased in the order of $PO_4^{3-} \rightarrow HCO_3^- \rightarrow SO_4^{2-} \rightarrow CO_3^{2-}$ (Alam et al., 2018).

Most of the published adsorption studies use synthetic wastewater, so there is still a gap in the actual state of the As-rich wastewater or groundwater. Biochar applications to treat real wastewater or drinking water are still a challenge. In addition, many ions and complex contaminants are normally present in real water. The effect of these complex ions on the equilibrium adsorption capacity of the adsorbent cannot be ignored. Careful evaluation is required and should be performed separately to allow valuable consideration of the mechanism of influence and concentration of background electrolytes (anions and cations) (Tan et al., 2015).

6.6 MODIFICATION OF BIOCHAR

The incorporation of nanoparticles into biochar can increase the efficacy of biochar for water treatment (Inyang et al., 2014). Although biochar can remove PTEs including As from aqueous solutions, its efficacy is generally lower compared to some common biosorbents, such as activated carbon. Therefore, the focus of modern research is on improving the adsorption capacity of biochar by modifying biochar.

For example, efforts have been made to improve the porosity, functional groups, surface area, and pH_{PZC} of biochar by different modification methods. Biochar modification strategies include activation by alkaline solutions, impregnation of nanoparticles, reducing agents, minerals, and organic functional groups. Mineral modification was carried out, including treatment of biochar with various materials such as birnessite, hematite (Fe_2O_3), calcium oxide, magnetite, aqueous Mn-oxide, and zero-valent Fe (Table 6.2). The modification can be carried out by loading minerals before and after pyrolysis of the feedstock (Li et al., 2017). The Fe-impregnated biochar is considered a low-cost adsorbent. Iron-impregnated biochar is made by an innovative technology that hydrolyzes Fe salts on raw biochar (pecan biochar). Iron impregnation leads to a significant reduction in the surface area of biochar and less As adsorption capacity compared to the original biochar (Hu et al., 2015).

Magnetite biochar is made by pyrolysis of a mixture of pine and hematite minerals. For example, Fe_2O_3 impregnation on the surface of the biochar provides an excess As(V) adsorption site due to the electrostatic interaction between Fe oxide (positively charged) and As(V) (negatively charged). Adsorption efficiency of magnetite coated biochar (4.29 mg g^{-1}) was higher compared to the adsorption capacity of crude biochar (2.65 mg g^{-1}) (Wang et al., 2015b) Wang et al., (2015a).

Two modification methods were used to increase the adsorption capacity of biochar for As and Pb. The pine material with the combination of $MnCl_2 \cdot 4H_2O$ (MPB) was pyrolyzed, and on the other hand birnessite was injected by precipitation on the pine material followed by pyrolysis (BPB). The adsorption capacity of As(V) and Pb by BPB (0.91 and 47.05 g kg^{-1}) and MPB (0.59 and 4.91 g kg^{-1}) was significantly higher than that of the original biochar (0.20 and 2.35 g kg^{-1}).

In addition, the adsorption capacity of biochar can also be increased by activation or modification with an alkali solution such as KOH and NaOH, which increases the surface of the biochar, thereby enhancing the metal adsorption capacity. The KOH activated municipal solid waste showed a high As(V) adsorption capacity (25–31 mg kg^{-1}), mainly due to the increased surface area (29–40 m^2 g^{-1}) (Jin et al., 2014).

6.7 REMOVAL MECHANISM AND FATE OF ARSENIC ON BIOCHAR

6.7.1 Removal Mechanisms

Possible removal mechanisms for PTEs typically include surface complexation, electrostatic attraction, physical adsorption, precipitation, and ion exchange. Many functional groups on the surface of biochar (mainly O groups such as hydroxyl groups, –OH, and carboxylates, –COOH) strongly attract As through electrostatic attraction, surface complexation, and ion exchange. Many of the mechanisms proposed for the interaction of As with biochar are shown in Figure 6.7. These interactions are evidenced by changes in the functional groups before and after metalloid adsorption on the surface of the biochar (Khare et al., 2013).

Reduction and precipitation are relatively simple mechanisms that affect the adsorption of As by biochar. The XRD analyses of pine wood biochar before and after As(V) adsorption showed that after adsorption no new peaks and minerals were formed and the precipitation mechanism had no apparent adsorption effect on As(V) of biochar (Wang et al., 2015b). The XPS analyses showed that magnetic biochar (prepared by precipitation of Fe^{2+}/Fe^{3+} on water hyacinths) sequestered 89% of As was present as As(V) on the surface of biochar which indicated less dominance of reduction mechanism (Zhang et al., 2016). Electrostatic interactions and complexation are key mechanisms for the adsorption of As in biochar compared to reduction and precipitation.

TABLE 6.2
Sorption Capacity of Modified Biochars for Removal of As(III)/As(V)

Biochar	Type of Biochar	Temperature (°C)	As(III)/As(V)	Maximum Removal (mg g^{-1})	References
Fe-coated rice husk biochar	Rice husk	700	As(V)	6.0	Cope et al. (2014)
Mn oxide modified biochar	Pine wood	600	As(V)	0.59	Wang et al. (2015a)
Magnetic biochar	Pine wood	600	As(V)	0.43	Wang et al. (2015b)
Magnetic biochar	Cotton wood	600	As(V)	3.1	Zhang et al. (2013)
Rice husk Fe-biochar	Rice husk	700	As(III)	30.7	Samsuri et al. (2013)
			As(V)	16.9	
Empty fruit bunch Fe-biochar	Empty fruit brunch	700	As(III)	31.4	
			As(V)	15.2	
Fe-biochar	Hickory chips	550	As(V)	2.16	Hu et al. (2015)
Birnessite-modified biochar	Pine wood	600	As(V)	0.91	Wang et al. (2015b)
KOH-activated biochar	Municipal solid waste	–	As(III)	2.16	Jin et al. (2014)
Hydrogel biochar	Rice husk	300	As(V)	28.0	Sanyang et al. (2016)

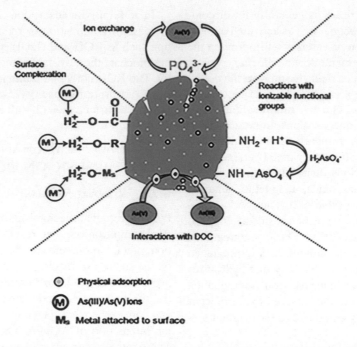

FIGURE 6.7 Mechanism of As adsorption on biochar surface. (Reproduced with permission from Vithanage, M. et al., *Carbon*, 113, 219–230, 2017.)

The adsorption capacity of two biochars (rice and empty fruit bunches) was evaluated for the removal of As(III) and As(V) (Samsuri et al., 2013). It was suggested that adsorption on biochar depends on the complexation of As with a functional group such as alcohol, ester, a hydroxyl group, and a carboxyl groups (Samsuri et al., 2013). In addition to complexation, electrostatic interaction is another important mechanism for the adsorption of As(V) by biochar.

Pine biochar showed maximum As(V) adsorption of 0.03 mg kg^{-1} at pH 7. At this pH, As(V) is mainly in the form of HAsO$_4^{-2}$ and the surface of the biochar is due to protonation of the functional group positively charged, since the pH of the solution is <pH$_{PZC}$. Arsenate interacts with a positively charged functional group through electrostatic interaction. Compared with the solutions high pH, biochar has more protonated functional groups at a lower pH of the solution and thus has a stronger As(V) adsorption capacity through electrostatic interaction (Wang et al., 2015b).

From the above discussion, it can be concluded that electrostatic interaction and complexation are important mechanisms for the adsorption of As by biochar and that functional groups mainly control the adsorption of As. Biochar produced at low temperature from a suitable raw material having a large number of functional groups can contribute to the adsorption of As from water.

6.7.2 Fate of Arsenic on Biochar

Recently, Niazi et al. (2018b) determined the fate of As(III) and As(V) sorbed on Japanese oak wood biochar using microscopic and spectroscopic techniques and macroscopic adsorption data. They found that the adsorption of As(III/V) involves a complex redox conversion from As(III) to As(V). Using XANES, they clearly demonstrated that Japanese oak wood biochar can seize the surface of the As species (As(III/V)), regardless of how the redox conditions of As change, As(V) is reduced to As(III) by biochar. It is a serious environmental risk because As(III) has higher bioavailability and toxicity than As(V) (Choppala et al., 2016). Appropriate management decisions are needed to safely treat spent biochar in order to reduced the risk of secondary pollution from sorbed As species.

6.8 APPLICATION OF BIOCHAR TO REMOVE ARSENIC FROM DRINKING WELL WATER

Arsenic must be removed from the drinking water due to its toxic effects. The adsorption capacity of various types of biochar used to remove As(III/V) is given in Table 6.3. Japanese oak wood biochar was used to remove As from groundwater, and results showed that 92% to 100% of total As was removed from As-contaminated groundwater (which is used for drinking). It was noticeable that the concentration of As in drinking water after treatment with Japanese oak wood biochar was lower than the WHO approved limit (10 μg L^{-1}), indicating that the presence of various ions (e.g., PO$_4^{3-}$, SO$_4^{2-}$, CO$_3^{2-}$) had no significant effect on the adsorption efficiency of biochar used in their study.

Moreover, the biochar pyrolysis temperature has a significant effect on the pH of the solution and thus significantly affects the removal of As. Perilla leaf biochar was produced at two different temperatures of 300°C and 700°C and was referred to as BC-700 and BC-300. Both biochars removed >93% of As from groundwater. However, compared to BC-300, BC-700 had a higher adsorption capacity and

TABLE 6.3

Sorption Capacity of Different Biochars for Removal of As(III)/As(V)

Biochar	Temperature °C	pH	As(III)/As(V)	Maximum Removal	References
Rice husk	300	6.7–7	As(V)	2.59 μg g^{-1}	Agrafioti et al. (2014)
Sewage sludge	300	6.7–7	As(V)	4.25 μg g^{-1}	Agrafioti et al. (2014)
	400	–	As(III)	6.04 mg g^{-1}	Zhang et al. (2015c)
	500	–	As(III)	5.60 mg g^{-1}	
	600	–	As(III)	1.21 mg g^{-1}	
Solid waste	300	6.7–7	As(V)	3.54 μg g^{-1}	Agrafioti et al. (2014)
Municipal solid waste	400	6.0	As(V)	24 mg g^{-1}	Agrafioti et al. (2014)
	500	6.0	As(V)	25 mg g^{-1}	Jin et al. (2014)
	600	6.0	As(V)	28 mg g^{-1}	
Empty fruit bunch	700	8.0	As(III)	18.9 mg g^{-1}	Samsuri et al. (2013)
		6.0	As(V)	5.1 mg g^{-1}	
Rice husk	700	8.0	As(III)	19.3 mg g^{-1}	
		6.0	As(V)	7.1 mg g^{-1}	
Oak wood char	400–450	5.0	As(III)	4.13 mg g^{-1}	Mohan et al. (2007)
	500	6	As(V)	3.89 mg g^{-1}	Niazi et al. (2018b)
	500	7	As(III)	3.16 mg g^{-1}	
Oak bark	400–450	3.5	As(III)	7.4 mg g^{-1}	Mohan et al. (2007)
Corn straw	400	–	As(V)	6.80 mg g^{-1}	He et al. (2018)
Pine wood char	400	5.0	As(III)	2.62 mg g^{-1}	Mohan et al. (2007)
Oak bark char	400	5.0	As(III)	3.00 mg g^{-1}	
Pine bark char	400	5.0	As(III)	13.1 mg g^{-1}	
Sewage sludge	550	3–3.5	As(III)	0.07 mg g^{-1}	Tavares et al. (2012)
Pine cone biochar	500	4.0	As(III)	0.006 mg g^{-1}	Wang et al. (2015a)
Rice husk biochar	500	9.5	As(V)	0.35 mg g^{-1}	Norazlina et al. (2014)
Empty fruit bunch biochar	300–350	10.2	As(V)	0.42 mg g^{-1}	

Source: Vithanage, M. et al., *Carbon*, 113, 219–230.

eliminated 97%–100% of total As, while BC-300 removed 94%–100% of As from drinking water/groundwater (Niazi et al., 2018a).

Many scientists focus on the elimination of As from single metal systems and only a few studies have investigated the As removal by natural and modified biochar in polymetallic systems (Samsuri et al., 2013). Furthermore, to our knowledge, no experimental studies have been conducted to investigate the use of biochar for removal of organic As species such as thioarsenate and methylated arsenate. Future research must also focus on the real water system rich in As contamination.

6.9 CONCLUSION AND FUTURE CHALLENGES

The biochar is a viable new adsorbent which is often used to adsorb As from water. This chapter summarized the use of biochar to treat synthetic as well as real water systems (groundwater/wastewater). To produce biochar, different raw materials have been used and the physical and chemical properties of biochar are strongly influenced by pyrolysis techniques, temperature, and residence time. The investigation of the adsorption mechanism shows that among adsorption

mechanisms (ion exchange, physical adsorption, and precipitation), complexation and electrostatic interaction are important mechanisms for As adsorption.

It has been evaluated that raw biochar is less effective in adsorbing As than the biochar modified by alkali solution, nanoparticles, organic functional groups, minerals, and reducing agents. The increased adsorption of modified biochar is due to an increase in porosity, surface area, functional groups, and pH$_{zpc}$. It is also concluded that the biochar produced at low pyrolysis temperature could be more effective for As sorption compared to the biochar produced at high pyrolysis. The spectral characterization provides knowledge on the effectiveness and mechanism of As adsorption on biochar, which is crucial for assessing the use of biochar in the adsorption of heavy metals and metalloids (As) in aqueous media. Although the use of biochar as an adsorbent is increasing, there are good research gaps:

1. Sorption experiments have usually been performed only at laboratory scale involving mono-metal sorption from synthetic solution. In real water systems (wastewater, ground/well water), various ions co-occur with the different contaminants, thus

competition occurs among As and other ions or organic contaminants for sorption sites on surface of biochar. Future research should focus on biochar usage to treat real water to ensure its pertinence in real water systems.

2. No experimental study focused on biochar interaction with organic As species so far, which is necessary for future research.

3. Although use of modified biochar for the sorption of As is increasing, more analysis is crucial to evaluate the use of nano-biochar composites. Biochar can also be activated via nitrogen, which results in the formation of amide functional groups on biochar surface that are useful in enhancing As sorption, but limited research has been focused on it. Thus, there is need to focus on evaluating interaction between nitrogen-activated biochars and various As species.

4. There is a necessity of future research on use of combined spectroscopic techniques, for example, XAFS and μ-XRF spectroscopy to examine the stability and mechanisms involved in As removal by biochar in water.

5. Limited data is available which focused on cost-benefit analyses of biochar, which could be proved as imperative research gap linked with As biochar interaction.

REFERENCES

Abdel-Fattah, T.M., Mahmoud, M.E., Ahmed, S.B., Huff, M.D., Lee, J.W. and Kumar, S. 2015. Biochar from woody biomass for removing metal contaminants and carbon sequestration. *Journal of Industrial and Engineering Chemistry*, 22, 103–109.

Abdul, K.S.M., Jayasinghe, S.S., Chandana, E.P., Jayasumana, C. and De Silva, P.M.C. 2015. Arsenic and human health effects: A review. *Environmental Toxicology and Pharmacology*, 40(3), 828–846.

Agrafioti, E., Kalderis, D. and Diamadopoulos, E. 2014. Ca and Fe modified biochars as adsorbents of arsenic and chromium in aqueous solutions. *Journal of Environmental Management*, 146, 444–450.

Ahmad, M., Lee, S.S., Dou, X., Mohan, D., Sung, J.K., Yang, J.E. and Ok, Y.S. 2012. Effects of pyrolysis temperature on soybean stover-and peanut shell-derived biochar properties and TCE adsorption in water. *Bioresource Technology, 118*, 536–544.

Ahmad, M., Lee, S.S., Rajapaksha, A.U., Vithanage, M., Zhang, M., Cho, J.S., Lee, S.E. and Ok, Y.S. 2013. Trichloroethylene adsorption by pine needle biochars produced at various pyrolysis temperatures. *Bioresource Technology, 143*, 615–622.

Ahmedna, M., Marshall, W.E., Husseiny, A.A., Rao, R.M. and Goktepe, I. 2004. The use of nutshell carbons in drinking water filters for removal of trace metals. *Water Research, 38*(4), 1062–1068.

Alam, M.A., Shaikh, W.A., Alam, M.O., Bhattacharya, T., Chakraborty, S., Show, B. and Saha, I. 2018. Adsorption of As (III) and As (V) from aqueous solution by modified *Cassia fistula* (golden shower) biochar. *Applied Water Science, 8*(7), 198.

Alam, M.O., Shaikh, W.A., Chakraborty, S., Avishek, K. and Bhattacharya, T. 2016. Groundwater arsenic contamination and potential health risk assessment of Gangetic Plains of Jharkhand, India. *Exposure and Health, 8*(1), 125–142.

Al-Wabel, M.I., Al-Omran A., El-Naggar A.H., Nadeem M. and Usman A.R. 2013. Pyrolysis temperature induced changes in characteristics and chemical composition of biochar produced from conocarpus wastes. *Bioresource technology.* 131:374–379.

Awad, Y.M., Vithanage, M., Niazi, N.K., Rizwan, M., Rinklebe, J., Yang, J.E., Ok, Y.S. and Lee, S.S. 2017. Potential toxicity of trace elements and nanomaterials to Chinese cabbage in arsenic- and lead-contaminated soil amended with biochars. *Environmental Geochemistry and Health*, 1–15.

Barakat, M.A. 2011. New trends in removing heavy metals from industrial wastewater. *Arabian Journal of Chemistry, 4*(4), 361–377.

Bibi, I., Icenhower, J., Niazi, N.K., Naz, T., Shahid, M. and Bashir, S. 2016. Clay minerals: Structure, chemistry, and significance in contaminated environments and geological CO_2 sequestration. *Environmental Materials and Waste* (pp. 543–567). Academic Press, Amsterdam, the Netherlands.

Bibi, S., A. Farooqi, Yasmin A., Kamran M.A. and Niazi N.K. 2017. Arsenic and fluoride removal by potato peel and rice husk (pprh) ash in aqueous environments. *International journal of phytoremediation.* 19:1029–1036.

Bissen, M. and Frimmel, F.H. 2003. Arsenic—a review. Part I: Occurrence, toxicity, speciation, mobility. *Acta Hydrochimica et Hydrobiologica, 31*(1), 9–18.

Brewer, C.E., Schmidt-Rohr, K., Satrio, J.A. and Brown, R.C. 2009. Characterization of biochar from fast pyrolysis and gasification systems. *Environmental Progress & Sustainable Energy: An Official Publication of the American Institute of Chemical Engineers, 28*(3), 386–396.

Cantrell, K.B., Hunt, P.G., Uchimiya, M., Novak, J.M. and Ro, K.S. 2012. Impact of pyrolysis temperature and manure source on physicochemical characteristics of biochar. *Bioresource Technology*, 107, 419–428.

Cao, X., Ma, L., Gao, B. and Harris, W. 2009. Dairy-manure derived biochar effectively sorbs lead and atrazine. *Environmental Science & Technology, 43*(9), 3285–3291.

Chen, G., Andries, J., Luo, Z. and Spliethoff, H. 2003. Biomass pyrolysis/gasification for product gas production: The overall investigation of parametric effects. *Energy Conversion and Management, 44*(11), 1875–1884.

Chen, T., Zhang, Y., Wang, H., Lu, W., Zhou, Z., Zhang, Y. and Ren, L. 2014. Influence of pyrolysis temperature on characteristics and heavy metal adsorptive performance of biochar derived from municipal sewage sludge. *Bioresource Technology, 164*, 47–54.

Chen, X., Chen, G., Chen, L., Chen, Y., Lehmann, J., McBride, M.B. and Hay, A.G. 2011. Adsorption of copper and zinc by biochars produced from pyrolysis of hardwood and corn straw in aqueous solution. *Bioresource Technology, 102*(19), 8877–8884.

Chen, Z., Chen, B., Zhou, D. and Chen, W. 2012. Bisolute sorption and thermodynamic behavior of organic pollutants to biomass-derived biochars at two pyrolytic temperatures. *Environmental Science & Technology, 46*(22), 12476–12483.

Choppala, G., Bolan, N., Kunhikrishnan, A. and Bush, R. 2016. Differential effect of biochar upon reduction-induced mobility and bioavailability of arsenate and chromate. *Chemosphere, 144*, 374–381.

Chung, J.Y., Yu, S.D. and Hong, Y.S. 2014. Environmental source of arsenic exposure. *Journal of Preventive Medicine and Public Health, 47*(5), 253.

Cope, C.O., Webster, D.S. and Sabatini, D.A. 2014. Arsenate adsorption onto iron oxide amended rice husk char. *Science of the Total Environment*, 488, 554–561.

Czernik, S. and Bridgwater, A.V. 2004. Overview of applications of biomass fast pyrolysis oil. *Energy & Fuels*, 18(2), 590–598.

Das, B. and Mondal, N.K. 2011. Calcareous soil as a new adsorbent to remove lead from aqueous solution: Equilibrium, kinetic and thermodynamic study. *Universal Journal of Environmental Research & Technology*, 1(4), 515–530.

Datsyuk, V., Kalyva, M., Papagelis, K., Parthenios, J., Tasis, D., Siokou, A., Kallitsis, I. and Galiotis, C. 2008. Chemical oxidation of multiwalled carbon nanotubes. *Carbon*, 46(6), 833–840.

Ding, Z., Hu, X., Wan, Y., Wang, S. and Gao, B. 2016. Removal of lead, copper, cadmium, zinc, and nickel from aqueous solutions by alkali-modified biochar: Batch and column tests. *Journal of Industrial and Engineering Chemistry*, 33, 239–245.

Dong, X., Ma, L.Q., Zhu, Y., Li, Y. and Gu, B. 2013. Mechanistic investigation of mercury sorption by Brazilian pepper biochars of different pyrolytic temperatures based on X-ray photoelectron spectroscopy and flow calorimetry. *Environmental Science & Technology*, 47(21), 12156–12164.

Eisler, R. 2004. Arsenic hazards to humans, plants, and animals from gold mining. *Reviews of Environmental Contamination and Toxicology* (pp. 133–165). Springer, New York.

Garlapalli, R.K., Wirth, B. and Reza, M.T. 2016. Pyrolysis of hydrochar from digestate: Effect of hydrothermal carbonization and pyrolysis temperatures on pyrochar formation. *Bioresource Technology*, 220, 168–174.

Hansen, V., Müller-Stöver, D., Ahrenfeldt, J., Holm, J.K., Henriksen, U.B. and Hauggaard-Nielsen, H. 2015. Gasification biochar as a valuable by-product for carbon sequestration and soil amendment. *Biomass and Bioenergy*, 72, 300–308.

He, R., Peng Z., Lyu H., Huang H., Nan Q. and Tang J. 2018. Synthesis and characterization of an iron-impregnated biochar for aqueous arsenic removal. *Science of the Total Environment*. 612:1177–1186.

Hoover, J., Gonzales, M., Shuey, C., Barney, Y. and Lewis, J. 2017. Elevated arsenic and uranium concentrations in unregulated water sources on the Navajo nation, USA. *Exposure and Health*, 9(2), 113–124.

Hu, X., Ding, Z., Zimmerman, A.R., Wang, S. and Gao, B. 2015. Batch and column sorption of arsenic onto iron-impregnated biochar synthesized through hydrolysis. *Water Research*, 68, 206–216.

Igalavithana, A.D., Mandal, S., Niazi, N.K., Vithanage, M., Parikh, S.J., Mukome, F.N., Rizwan, M. et al. 2017. Advances and future directions of biochar characterization methods and applications. *Critical Reviews in Environmental Science and Technology*, 47(23), 2275–2330.

Inguanzo, M., Domınguez, A., Menéndez, J.A., Blanco, C.G. and Pis, J.J. 2002. On the pyrolysis of sewage sludge: The influence of pyrolysis conditions on solid, liquid and gas fractions. *Journal of Analytical and Applied Pyrolysis*, 63(1), 209–222.

Inyang, M., Gao, B., Zimmerman, A., Zhang, M. and Chen, H. 2014. Synthesis, characterization, and dye sorption ability of carbon nanotube–biochar nanocomposites. *Chemical Engineering Journal*, 236, 39–46.

Jin, H., Capareda, S., Chang, Z., Gao, J., Xu, Y. and Zhang, J. 2014. Biochar pyrolytically produced from municipal solid wastes for aqueous As (V) removal: Adsorption property and its improvement with KOH activation. *Bioresource Technology*, 169, 622–629.

Jorio, A., Ribeiro-Soares, J., Cançado, L.G., Falcao, N.P.S., Dos Santos, H.F., Baptista, D.L., Ferreira, E.M., Archanjo, B.S. and Achete, C.A. 2012. Microscopy and spectroscopy analysis of carbon nanostructures in highly fertile Amazonian anthrosoils. *Soil and Tillage Research*, 122, 61–66.

Karakoyun, N., Kubilay, S., Aktas, N., Turhan, O., Kasimoglu, M., Yilmaz, S. and Sahiner, N. 2011. Hydrogel–biochar composites for effective organic contaminant removal from aqueous media. *Desalination*, 280(1–3), 319–325.

Kasozi, G.N., Zimmerman, A.R., Nkedi-Kizza, P. and Gao, B. 2010. Catechol and humic acid sorption onto a range of laboratory-produced black carbons (biochars). *Environmental Science & Technology*, 44(16), 6189–6195.

Keiluweit, M., Nico, P.S., Johnson, M.G. and Kleber, M. 2010. Dynamic molecular structure of plant biomass-derived black carbon (biochar). *Environmental Science & Technology*, 44(4), 1247–1253.

Khare, P., Dilshad U., Rout P., Yadav V. and Jain S. 2013. Plant refuses driven biochar: Application as metal adsorbent from acidic solutions. *Arab. J. Chem.* http://dx. doi. org/10.1016/j. arabjc. 47.

Kim, K.H., Kim, J.Y., Cho, T.S. and Choi, J.W. 2012. Influence of pyrolysis temperature on physicochemical properties of biochar obtained from the fast pyrolysis of pitch pine (*Pinus rigida*). *Bioresource Technology*, 118, 158–162.

Lata, S. and Samadder, S.R. 2016. Removal of arsenic from water using nano adsorbents and challenges: A review. *Journal of Environmental Management*, 166, 387–406.

Lee, J., Yang, X., Cho, S.H., Kim, J.K., Lee, S.S., Tsang, D.C., Ok, Y.S. and Kwon, E.E. 2017. Pyrolysis process of agricultural waste using CO_2 for waste management, energy recovery, and biochar fabrication. *Applied Energy*, 185, 214–222.

Lee, Y., Ryu, C., Park, Y.K., Jung, J.H. and Hyun, S. 2013. Characteristics of biochar produced from slow pyrolysis of Geodae-Uksae 1. *Bioresource Technology*, 130, 345–350.

Lehmann, J., Liang, B., Solomon, D., Lerotic, M., Luizão, F., Kinyangi, J., Schäfer, T., Wirick, S. and Jacobsen, C. 2005. Near-edge X-ray absorption fine structure (NEXAFS) spectroscopy for mapping nano-scale distribution of organic carbon forms in soil: Application to black carbon particles. *Global Biogeochemical Cycles*, 19(1), 1–12.

LeMonte, J.J., Stuckey, J.W., Sanchez, J.Z., Tappero, R., Rinklebe, J. and Sparks, D.L. 2017. Sea level rise induced arsenic release from historically contaminated coastal soils. *Environmental Science & Technology*, 51(11), 5913–5922.

Li, H., Dong, X., da Silva, E.B., de Oliveira, L.M., Chen, Y. and Ma, L.Q. 2017. Mechanisms of metal sorption by biochars: Biochar characteristics and modifications. *Chemosphere*, 178, 466–478.

Liang, B., Lehmann, J., Solomon, D., Sohi, S., Thies, J.E., Skjemstad, J.O., Luizao, F.J., Engelhard, M.H., Neves, E.G. and Wirick, S. 2008. Stability of biomass-derived black carbon in soils. *Geochimica et Cosmochimica Acta*, 72(24), 6069–6078.

Libra, J.A., Ro, K.S., Kammann, C., Funke, A., Berge, N.D., Neubauer, Y., Titirici, M.M. et al. 2011. Hydrothermal carbonization of biomass residuals: A comparative review of the chemistry, processes and applications of wet and dry pyrolysis. *Biofuels*, 2(1), 71–106.

Li, G., Shen B., Li F., Tian L., Singh S. and Wang F. 2015. Elemental mercury removal using biochar pyrolyzed from municipal solid waste. *Fuel Processing Technology*. 133:43–50.

Lima, I.M., Boateng, A.A. and Klasson, K.T. 2010. Physicochemical and adsorptive properties of fast-pyrolysis bio-chars and their steam activated counterparts. *Journal of Chemical Technology & Biotechnology*, 85(11), 1515–1521.

Lin, S.C., Chang, T.K., Huang, W.D., Lur, H.S. and Shyu, G.S. 2015. Accumulation of arsenic in rice plant: A study of an arsenic-contaminated site in Taiwan. *Paddy and Water Environment*, *13*(1), 11–18.

Malghani, S., Gleixner, G. and Trumbore, S.E. 2013. Chars produced by slow pyrolysis and hydrothermal carbonization vary in carbon sequestration potential and greenhouse gases emissions. *Soil Biology and Biochemistry*, *62*, 137–146.

Malik, A.H., Khan, Z.M., Mahmood, Q., Nasreen, S. and Bhatti, Z.A. 2009. Perspectives of low cost arsenic remediation of drinking water in Pakistan and other countries. *Journal of Hazardous Materials*, *168*(1), 1–12.

Mandal, B.K. and Suzuki, K.T. 2002. Arsenic round the world: A review. *Talanta*, *58*(1), 201–235.

Manyà, J.J. 2012. Pyrolysis for biochar purposes: A review to establish current knowledge gaps and research needs. *Environmental Science & Technology*, *46*(15), 7939–7954.

Mohan, D., Kumar, H., Sarswat, A., Alexandre-Franco, M. and Pittman Jr, C.U. 2014a. Cadmium and lead remediation using magnetic oak wood and oak bark fast pyrolysis bio-chars. *Chemical Engineering Journal*, *236*, 513–528.

Mohan, D. and Pittman Jr, C.U. 2007. Arsenic removal from water/wastewater using adsorbents—a critical review. *Journal of Hazardous Materials*, *142*(1–2), 1–53.

Mohan, D., Pittman, C.U. and Steele, P.H. 2006. Pyrolysis of wood/biomass for bio-oil: A critical review. *Energy & Fuels*, *20*(3), 848–889.

Mohan, D., Rajput, S., Singh, V.K., Steele, P.H. and Pittman Jr, C.U. 2011. Modeling and evaluation of chromium remediation from water using low cost bio-char, a green adsorbent. *Journal of Hazardous Materials*, *188*(1–3), 319–333.

Mohan, D., Sarswat, A., Ok, Y.S. and Pittman Jr, C.U. 2014b. Organic and inorganic contaminants removal from water with biochar, a renewable, low cost and sustainable adsorbent–a critical review. *Bioresource Technology*, *160*, 191–202.

Mondal, P., Bhowmick, S., Chatterjee, D., Figoli, A. and Van der Bruggen, B. 2013. Remediation of inorganic arsenic in groundwater for safe water supply: A critical assessment of technological solutions. *Chemosphere*, *92*(2), 157–170.

Naidu, R., Smith, E., Owens, G. and Bhattacharya, P. 2006. *Managing Arsenic in the Environment: From Soil to Human Health*. CSIRO Publishing.

Nguyen, B.T., Lehmann, J., Hockaday, W.C., Joseph, S. and Masiello, C.A. 2010. Temperature sensitivity of black carbon decomposition and oxidation. *Environmental Science & Technology*, *44*(9), 3324–3331.

Niazi, N.K., Bibi, I., Shahid, M., Ok, Y.S., Burton, E.D., Wang, H., Shaheen, S.M., Rinklebe, J. and Lüttge, A. 2018a. Arsenic removal by perilla leaf biochar in aqueous solutions and groundwater: An integrated spectroscopic and microscopic examination. *Environmental Pollution*, *232*, 31–41.

Niazi, N.K., Bibi, I., Shahid, M., Ok, Y.S., Shaheen, S.M., Rinklebe, J., Wang, H., Murtaza, B., Islam, E., Nawaz, M.F. and Lüttge, A. 2018b. Arsenic removal by Japanese oak wood biochar in aqueous solutions and well water: Investigating arsenic fate using integrated spectroscopic and microscopic techniques. *Science of the Total Environment*, *621*, 1642–1651.

Niazi, N.K. and Burton, E.D. 2016. Arsenic sorption to nanoparticulate mackinawite (FeS): An examination of phosphate competition. *Environmental Pollution*, *218*, 111–117.

Niazi, N.K., Murtaza, B., Bibi, I., Shahid, M., White, J.C., Nawaz, M.F., Bashir, S., Shakoor, M.B., Choppala, G., Murtaza, G., Wang, H. 2016. Removal and recovery of metals by biosorbents and biochars derived from biowastes. In: Prasad, M.N.V., Shih, K. (Eds.), *Environmental Materials and Waste: Resource Recovery and Pollution Prevention* (pp. 149–177). Academic Press, Amsterdam, the Netherlands.

Niazi, N.K., Singh, B. and Shah, P. 2011. Arsenic speciation and phytoavailability in contaminated soils using a sequential extraction procedure and XANES spectroscopy. *Environmental Science & Technology*, *45*(17), 7135–7142.

Niazi, N.K., Singh, B., Van Zwieten, L. and Kachenko, A.G. 2012. Phytoremediation of an arsenic-contaminated site using *Pteris vittata* L. and *Pityrogramma calomelanos* var. *austroamericana*: A long-term study. *Environmental Science and Pollution Research*, *19*(8), 3506–3515.

Nickson, R.T., McArthur, J.M., Shrestha, B., Kyaw-Myint, T.O. and Lowry, D. 2005. Arsenic and other drinking water quality issues, Muzaffargarh District, Pakistan. *Applied Geochemistry*, *20*(1), 55–68.

Norazlina, A.S., Che F.I. and Rosenani A.B. 2014. Characterization of oil palm empty fruit bunch and rice husk biochars and their potential to adsorb arsenic and cadmium. *American Journal of Agricultural and Biological Sciences*. 9:450–456.

Novak, J.M., Lima, I., Xing, B., Gaskin, J.W., Steiner, C., Das, K.C., Ahmedna, M. et al. 2009. Characterization of designer biochar produced at different temperatures and their effects on a loamy sand. *Annals of Environmental Science*, 3, 195–206

Oh, T.K., Choi, B., Shinogi, Y. and Chikushi, J. 2012. Effect of pH conditions on actual and apparent fluoride adsorption by biochar in aqueous phase. *Water, Air, & Soil Pollution*, *223*(7), 3729–3738.

Parikh, S.J., Goyne, K.W., Margenot, A.J., Mukome, F.N. and Calderón, F.J. 2014. Soil chemical insights provided through vibrational spectroscopy. *Advances in Agronomy*, *126*, 1–148.

Pimchuai, A., Dutta, A. and Basu, P. 2010. Torrefaction of agriculture residue to enhance combustible properties. *Energy & Fuels*, *24*(9), 4638–4645.

Qian, L. and Chen, B. 2013. Dual role of biochars as adsorbents for aluminum: The effects of oxygen-containing organic components and the scattering of silicate particles. *Environmental Science & Technology*, *47*(15), 8759–8768.

Rahman, M.M., Asaduzzaman, M. and Naidu, R. 2011. Arsenic exposure from rice and water sources in the Noakhali district of Bangladesh. *Water Quality, Exposure and Health*, *3*(1), 1–10.

Rahman, M.M., Owens, G. and Naidu, R. 2009. Arsenic levels in rice grain and assessment of daily dietary intake of arsenic from rice in arsenic-contaminated regions of Bangladesh—implications to groundwater irrigation. *Environmental Geochemistry and Health*, *31*(1), 179–187.

Rasheed, H., Slack, R. and Kay, P. 2016. Human health risk assessment for arsenic: A critical review. *Critical Reviews in Environmental Science and Technology*, *46*(19–20), 1529–1583.

Ro, K.S., Cantrell, K.B. and Hunt, P.G. 2010. High-temperature pyrolysis of blended animal manures for producing renewable energy and value-added biochar. *Industrial & Engineering Chemistry Research*, *49*(20), 10125–10131.

Rutherford, D.W., Wershaw, R.L., Rostad, C.E. and Kelly, C.N. 2012. Effect of formation conditions on biochars: Compositional and structural properties of cellulose, lignin, and pine biochars. *Biomass and Bioenergy*, *46*, 693–701.

Saha, D. and Sahu, S. 2016. A decade of investigations on groundwater arsenic contamination in Middle Ganga Plain, India. *Environmental Geochemistry and Health*, *38*(2), 315–337.

Samsuri, A.W., Sadegh-Zadeh, F. and Seh-Bardan, B.J. 2013. Adsorption of As (III) and As (V) by Fe coated biochars and biochars produced from empty fruit bunch and rice husk. *Journal of Environmental Chemical Engineering*, *1*(4), 981–988.

Sanyang, M., Ghani, W.A.W.A.K. Idris A. and Ahmad M.B. 2016. Hydrogel biochar composite for arsenic removal from wastewater. *Desalination and Water Treatment*. 57:3674–3688.

Shaheen, S.M., Eissa, F.I., Ghanem, K.M., El-Din, H.M.G. and Al Anany, F.S. 2013. Heavy metals removal from aqueous solutions and wastewaters by using various byproducts. *Journal of Environmental Management*, *128*, 514–521.

Shaheen, S.M. and Rinklebe, J. 2015. Phytoextraction of potentially toxic elements by Indian mustard, rapeseed, and sunflower from a contaminated riparian soil. *Environmental geochemistry and Health*, *37*(6), 953–967.

Shaheen, S.M., Niazi, N.K., Hassan, N.E., Bibi, I., Wang, H., Tsang, D.C., Ok, Y.S., Bolan, N. and Rinklebe, J. 2019. Wood-based biochar for the removal of potentially toxic elements in water and wastewater: A critical review. *International Materials Reviews*, 1–32.

Shaheen, S.M., Tsadilas, C.D., Rupp, H., Rinklebe, J. and Meissner, R. 2015. Distribution coefficients of cadmium and zinc in different soils in mono-metal and competitive sorption systems. *Journal of Plant Nutrition and Soil Science*, *178*(4), 671–681.

Shahid, M., Niazi, N.K., Dumat, C., Naidu, R., Khalid, S., Rahman, M.M. and Bibi, I. 2018. A meta-analysis of the distribution, sources and health risks of arsenic-contaminated groundwater in Pakistan. *Environmental Pollution*, 242, 307–319.

Shakoor, M., Niazi, N., Bibi, I., Rahman, M., Naidu, R., Dong, Z., Shahid, M. and Arshad, M. 2015. Unraveling health risk and speciation of arsenic from groundwater in rural areas of Punjab, Pakistan. *International Journal of Environmental Research and Public Health*, *12*(10), 12371–12390.

Shakoor, M.B., Niazi, N.K., Bibi, I., Murtaza, G., Kunhikrishnan, A., Seshadri, B., Shahid, M. et al. 2016. Remediation of arsenic-contaminated water using agricultural wastes as biosorbents. *Critical Reviews in Environmental Science and Technology*, *46*(5), 467–499.

Shakoor, M.B., Niazi, N.K., Bibi, I., Shahid, M., Saqib, Z.A., Nawaz, M.F., Shaheen, S.M. et al. 2019. Exploring the arsenic removal potential of various biosorbents from water. *Environment International*, *123*, 567–579.

Shen, Y.S., Wang, S.L., Tzou, Y.M., Yan, Y.Y. and Kuan, W.H. 2012. Removal of hexavalent Cr by coconut coir and derived chars–The effect of surface functionality. *Bioresource Technology*, *104*, 165–172.

Singh, B., Fang, Y., Cowie, B.C. and Thomsen, L. 2014. NEXAFS and XPS characterisation of carbon functional groups of fresh and aged biochars. *Organic Geochemistry*, *77*, 1–10.

Sohi, S.P. 2012. Carbon storage with benefits. *Science*, *338*(6110), 1034–1035.

Sun, H., Hockaday, W.C., Masiello, C.A. and Zygourakis, K. 2012. Multiple controls on the chemical and physical structure of biochars. *Industrial & Engineering Chemistry Research*, *51*, 3587–3597.

Tan, X., Liu, Y., Zeng, G., Wang, X., Hu, X., Gu, Y. and Yang, Z. 2015. Application of biochar for the removal of pollutants from aqueous solutions. *Chemosphere*, *125*, 70–85.

Tavares, D.S., Lopes C.B., Coelho J.P., Sánchez M.E., Garcia A.I., Duarte A.C., Otero M. and Pereira E. 2012. Removal of arsenic from aqueous solutions by sorption onto sewage sludge-based sorbent. *Water, Air, & Soil Pollution*. 223:2311–2321.

Tong, X.J., Li, J.Y., Yuan, J.H. and Xu, R.K. 2011. Adsorption of Cu (II) by biochars generated from three crop straws. *Chemical Engineering Journal*, *172*(2–3), 828–834.

Tripathi, M., Sahu, J.N. and Ganesan, P. 2016. Effect of process parameters on production of biochar from biomass waste through pyrolysis: A review. *Renewable and Sustainable Energy Reviews*, *55*, 467–481.

Uchimiya, M., Klasson, K.T., Wartelle, L.H. and Lima, I.M. 2011. Influence of soil properties on heavy metal sequestration by biochar amendment: 1. Copper sorption isotherms and the release of cations. *Chemosphere*, *82*(10), 1431–1437.

Uchimiya, M., Wartelle, L.H., Lima, I.M. and Klasson, K.T. 2010. Sorption of deisopropylatrazine on broiler litter biochars. *Journal of Agricultural and Food Chemistry*, *58*(23), 12350–12356.

Vithanage, M., Herath, I., Joseph, S., Bundschuh, J., Bolan, N., Ok, Y.S., Kirkham, M.B. and Rinklebe, J. 2017. Interaction of arsenic with biochar in soil and water: A critical review. *Carbon*, *113*, 219–230.

Wang, S., Gao, B., Li, Y., Mosa, A., Zimmerman, A.R., Ma, L.Q., Harris, W.G. and Migliaccio, K.W. 2015a. Manganese oxide-modified biochars: Preparation, characterization, and sorption of arsenate and lead. *Bioresource Technology*, *181*, 13–17.

Wang, S., Gao, B., Zimmerman, A.R., Li, Y., Ma, L., Harris, W.G. and Migliaccio, K.W. 2015b. Removal of arsenic by magnetic biochar prepared from pinewood and natural hematite. *Bioresource Technology*, *175*, 391–395.

Wang, Z., Zheng, H., Luo, Y., Deng, X., Herbert, S. and Xing, B. 2013. Characterization and influence of biochars on nitrous oxide emission from agricultural soil. *Environmental Pollution*, *174*, 289–296.

Waqas, H., Shan, A., Khan, Y.G., Nawaz, R., Rizwan, M., Rehman, M.S.U., Shakoor, M.B., Ahmed, W. and Jabeen, M. 2017. Human health risk assessment of arsenic in groundwater aquifers of Lahore, Pakistan. *Human and Ecological Risk Assessment: An International Journal*, *23*(4), 836–850.

Wiedemeier, D.B., Abiven, S., Hockaday, W.C., Keiluweit, M., Kleber, M., Masiello, C.A., McBeath, A.V. et al. 2015. Aromaticity and degree of aromatic condensation of char. *Organic Geochemistry*, *78*, 135–143.

Wiedner, K., Rumpel, C., Steiner, C., Pozzi, A., Maas, R. and Glaser, B. 2013. Chemical evaluation of chars produced by thermochemical conversion (gasification, pyrolysis and hydrothermal carbonization) of agro-industrial biomass on a commercial scale. *Biomass and Bioenergy*, *59*, 264–278.

Wongrod, S., Simon, S., van Hullebusch, E.D., Lens, P.N. and Guibaud, G. 2018. Changes of sewage sludge digestate-derived biochar properties after chemical treatments and influence on As (III and V) and Cd (II) sorption. *International Biodeterioration & Biodegradation*, *135*, 96–102.

Wu, H., Gao, G., Zhou, X., Zhang, Y. and Guo, S. 2012. Control on the formation of Fe_3O_4 nanoparticles on chemically reduced graphene oxide surfaces. *CrystEngComm*, *14*(2), 499–504.

Wu, H., Yip, K., Tian, F., Xie, Z. and Li, C.Z. 2009. Evolution of char structure during the steam gasification of biochars produced from the pyrolysis of various mallee biomass components. *Industrial & Engineering Chemistry Research*, *48*(23), 10431–10438.

Xu, X., Cao, X. and Zhao, L. 2013. Comparison of rice husk- and dairy manure-derived biochars for simultaneously removing heavy metals from aqueous solutions: Role of mineral components in biochars. *Chemosphere*, *92*(8), 955–961.

Xue, Y., Gao, B., Yao, Y., Inyang, M., Zhang, M., Zimmerman, A.R. and Ro, K.S. 2012. Hydrogen peroxide modification enhances the ability of biochar (hydrochar) produced from hydrothermal carbonization of peanut hull to remove aqueous heavy metals: Batch and column tests. *Chemical Engineering Journal*, *200*, 673–680.

Yang, Y., Wei, Z., Zhang, X., Chen, X., Yue, D., Yin, Q., Xiao, L. and Yang, L. 2014. Biochar from *Alternanthera philoxeroides* could remove Pb (II) efficiently. *Bioresource Technology*, *171*, 227–232.

Yao, Y., Gao, B., Chen, H., Jiang, L., Inyang, M., Zimmerman, A.R., Cao, X., Yang, L., Xue, Y. and Li, H. 2012. Adsorption of sulfamethoxazole on biochar and its impact on reclaimed water irrigation. *Journal of Hazardous Materials*, *209*, 408–413.

Zama, E.F., Zhu, Y.G., Reid, B.J. and Sun, G.X. 2017. The role of biochar properties in influencing the sorption and desorption of Pb (II), Cd (II) and As (III) in aqueous solution. *Journal of Cleaner Production*, *148*, 127–136.

Zeng, Y., Zhou, Y., Zhou, J., Jia, R. and Wu, J. 2018. Distribution and enrichment factors of high-arsenic groundwater in Inland Arid area of PR China: A case study of the Shihezi area, Xinjiang. *Exposure and Health*, *10*(1), 1–13.

Zhang, F., Wang, X., Xionghui, J. and Ma, L. 2016. Efficient arsenate removal by magnetite-modified water hyacinth biochar. *Environmental Pollution*, *216*, 575–583.

Zhang, H., Voroney, R.P. and Price, G.W. 2015a. Effects of temperature and processing conditions on biochar chemical properties and their influence on soil C and N transformations. *Soil Biology and Biochemistry*, *83*, 19–28.

Zhang, J., Liu, J. and Liu, R. 2015b. Effects of pyrolysis temperature and heating time on biochar obtained from the pyrolysis of straw and lignosulfonate. *Bioresource Technology*, *176*, 288–291.

Zhang, W., Zheng, J., Zheng, P., Tsang, D. C., & Qiu, R. 2015c. Sludge-derived biochar for arsenic (III) immobilization: effects of solution chemistry on sorption behavior. *Journal of environmental quality*, 44(4), 1119–1126.

Zhang, M., Gao, B., Yao, Y., Xue, Y. and Inyang, M. 2012. Synthesis of porous MgO-biochar nanocomposites for removal of phosphate and nitrate from aqueous solutions. *Chemical Engineering Journal*, *210*, 26–32.

Zhang, Z.B., Cao, X.H., Liang, P. and Liu, Y.H. 2013. Adsorption of uranium from aqueous solution using biochar produced by hydrothermal carbonization. *Journal of Radioanalytical and Nuclear Chemistry*, *295*(2), 1201–1208.

Zhao, B., O'Connor, D., Zhang, J., Peng, T., Shen, Z., Tsang, D.C. and Hou, D. 2018. Effects of pyrolysis temperature, heating rate, and residence time on rapeseed stem derived biochar. *Journal of Cleaner Production*, *174*, 977–987.

Zheng, W., Guo, M., Chow, T., Bennett, D.N. and Rajagopalan, N. 2010. Sorption properties of greenwaste biochar for two triazine pesticides. *Journal of Hazardous Materials*, *181*(1–3), 121–126.

Zhou, Z., Liu, Y.G., Liu, S.B., Liu, H.Y., Zeng, G.M., Tan, X.F., Yang, C.P., Ding, Y., Yan, Z.L. and Cai, X.X. 2017. Sorption performance and mechanisms of arsenic (V) removal by magnetic gelatin-modified biochar. *Chemical Engineering Journal*, *314*, 223–231.

7 Potential Value of Biowastes in the Remediation of Toxic Metal(loid)-Contaminated Soils

Jin Hee Park and Nanthi S. Bolan

CONTENTS

7.1 INTRODUCTION

Because of toxic effects of metal(loid)s on the environment and human health, metal(loid) contamination of soil is a global concern. Use of biowastes for remediation of metal(loid)-contaminated soils attracted research interest following increasing amounts of biowaste generation. Due to ever-increasing production of livestock and poultry products for human consumption, a large volume of organic wastes from these industries is generated. Biowastes including poultry manure compost and biosolids are used as not only a source of nutrients but also as a soil conditioner to improve the physicochemical properties of soils. Therefore, land treatment of biowastes is one of the important waste management practices.

Metal(loid)s have properties of not degrading, thereby persisting for a long time after introduction to soils. Therefore, remediation options for metal(loid) contaminated soil generally include amelioration of soils to reduce metal(loid) bioavailability. Bioavailability can be reduced by chemical and biological immobilization of metal(loid)s using a range of inorganic compounds and organic compounds. Inorganic amendments include phosphate compounds and lime while organic amendments include biowastes such as manure and organic waste-derived biochars (Cao et al., 2018; Jin et al., 2019; Seshadri et al., 2017; Tsang et al., 2016). Although the more localized and concentrated metal contamination found in urban environments is remediated by metal mobilization processes such as chemical washing, immobilization processes using biowastes are cost-effective and reduce the burden for waste disposal (Lee et al., 2013; Park et al., 2011).

This chapter focuses on the potential value of biowastes in the remediation of metal(loid) contaminated soils. Sources of biowastes and the reactions of metal(loid)s in soils are briefly introduced. The mechanisms for the enhanced remediation of metal(loid)s by biowastes are described. The practical implications of biowastes on remediation are discussed according to the sequestration and bioavailability of metal(loid)s in soils.

7.2 METAL(LOID) CONTAMINATION IN SOIL

Toxic metal(loid)s are introduced to the soil environment by both pedogenic and anthropogenic processes. Most metal(loid)s are naturally enriched in soil parent materials and exist mainly in the non-bioavailable forms. Because of their low bioavailability and solubility, the metal(loid)s present in the parent materials are often not available for plant uptake and cause minimum impact to soil organisms. Most of geogenic metal(loid)s except Se and As have limited impact on soil (Park et al., 2011; Shaheen et al., 2017). Metal(loid)s added through anthropogenic activities typically have high bioavailability compared to the pedogenic inputs (Yang et al., 2014). Anthropogenic metal(loid) sources are primarily industrial processes, urban disposal, agricultural, and industrial waste materials (Dickinson et al., 1996).

Phosphate fertilizers are a major source of Cd ranging from trace amounts up to 300 mg kg^{-1} (Grant, 2015). To prevent Cd contamination in soil, low Cd-containing P fertilizers should be used. Use of phosphate rocks (PRs) with low Cd or treatment of the PRs to remove Cd may help to reduce Cd contamination in soil (Nziguheba and Smolders, 2008; Mar and Okazaki, 2012). Although PRs with low Cd contents for manufacturing P fertilizers are available, P fertilizers with higher Cd contents are still used in many countries for practical and economic reasons.

Biowastes such as biosolid and poultry manure are also sources of metal(loid) contamination in soils. The toxic metal(loid)s in biosolid include Pb, Ni, Cd, Co, Cr, Cu, and Zn, which mainly originated from industrial waste water (Benitez et al., 2001; Kazi et al., 2005). Continuous application of animal manure is also major source of metal input in soils resulting in elevated metal(loid) concentrations in soil. For example, land application of pig manure for 5 years seriously contaminated arable soils with Zn (Ogiyama et al., 2005).

Livestock and poultry feedstuff are supplied with a number of metal(loid)s not only as essential nutrients but also as supplement to improve animal health and feed efficiency (Bolan et al., 2004). The metal(loid)s such as As, Co, Cu, Fe, Mn, Se, and Zn are added as feed additives to prevent diseases, enhance weight gains and feed conversion efficiency, and increase egg production for poultry (Moore et al., 1998; Bolan et al., 2010). Because major parts of the metal(loid)s ingested are excreted in feces or urine, metal(loid) concentrations in manure depend primarily on their concentrations in the feeding (Brugger and Windisch, 2015). Addition of organo-arsenical feed additives in poultry and long-term application of the poultry litter in pasture soils resulted in the increase of water soluble arsenic (Rutherford et al., 2003). Similarly, the excessive use of Cu compounds such as copper sulfate and copper lysine complex as a growth promoter and fungicide in swine and poultry might result in elevated Cu concentration in effluent and manure (Bolan et al., 2003a; Jondreville et al., 2003).

7.3 METAL(LOID) REACTIONS IN SOILS

Toxic metal(loids) undergo various reactions in soil, which include adsorption and complexation, precipitation, redox reactions, and methylation. Metal(loid)s are physicochemically adsorbed or form complexes on the surface of adsorbent. Adsorption of metal(loid)s by an adsorbent can be grouped into specific and non-specific adsorption. Non-specific adsorption refers to outer-sphere surface complexation while specific adsorption forms inner-sphere complexation. Specific adsorption is generally stronger and more selective than non-specific adsorption (Du et al., 2016; Wan et al., 2018).

Adsorption of metal(loid)s depends on soil pH and generally metal sorption increases with increasing pH due to increase in negative surface charges and precipitation of metal(loid)s as hydroxides (Jiang et al., 2016). Adsorption of toxic metal(loid)s is also affected by soil components including clay mineral, organic matter, and iron, aluminum, and manganese oxides (Diagboya et al., 2015; Uddin, 2017). Metals have a high affinity for ligands or functional groups of soil organic matter (Park et al., 2016). The adsorption of metal(loid)s such as As and Cr can be removed by adsorption on the iron or manganese oxide (Mishra and Mahato, 2016; Liu et al., 2015).

In the presence of anions such as carbonate, hydroxide, sulfide, and phosphate, precipitation is a significant process of metal(loid) immobilization (Blais et al., 2008). Co-precipitation also contributes to the immobilization of metal(loid)s. Metal(loid)s are co-precipitated with Fe and Al oxyhydroxides (Lee et al., 2002). Hydroxide precipitation is the most common precipitation technique for toxic metal(loid)s (Fu and Wang, 2011). Therefore, liming is an effective way of reducing metal(loid) mobility by the precipitation of metal(loid)s. Sulfide precipitation is also an effective process for the immobilization of metal(loid)s. The solubility of metal sulfides are very low under reducing conditions. Metal(loid) sulfide precipitation occurs at relatively low pH because of the low solubility of metal(loid) sulfide (Mokone et al., 2010). However, to precipitate metal(loid)s with sulfide and maintain low metal(loid) solubility, reducing conditions are required, which is not practical in surface soils.

Metal(loid)s undergo oxidation and reduction reactions, especially As, Cr, Hg, and Se are subjected to microbial oxidation/reduction reactions. The microbial oxidation/reduction reactions can be grouped into assimilatory and dissimilatory metabolism depending on the function. Metal(loid) substrate acts as a terminal electron acceptor in assimilatory metabolism while the metal(loid) substrate has no specific function in microorganisms in the dissimilatory reaction. Oxyanions such as As and Se can be used as terminal electron acceptors by anaerobic respiration of microbes. The oxidation of organic substrates such as acetate, lactate, pyruvate, and alcohol can be coupled to the reduction of As and Se (Stolz and Oremland, 1999). Metal(loid) reduction or oxidation changes mobility and toxicity of the metal(loid)s. For example, Cr(III) forms insoluble precipitates, which is less toxic while Cr(VI) is highly mobile and toxic in the environment (Barrera-Díaz et al., 2012).

Methylation is considered as a metal-dependent mechanism because some metal(loid)s can be methylated. Methylation is a major process of volatilization of As, Hg, and Se resulting in the removal of the metal(loid)s from soils and sediments (Roane et al., 2015). The methylation processes are mediated by microorganisms. Biogeochemical methylation and volatilization of As are important processes in As cycling in paddy soils (Huang et al., 2016). The methylation and volatilization of inorganic As is considered to be a detoxification process of As. Methylation of Se is also a major detoxification process of Se, which was biologically catalyzed (Okuno et al., 2001). However, methylation of Hg is known to increase Hg toxicity because methylated Hg easily penetrates lipid membranes of organisms (Bustamante et al., 2006).

7.4 USE OF BIOWASTES FOR THE REMEDIATION OF METAL(LOID)-CONTAMINATED SOIL

Large quantities of biowastes including crop residues, manure composts, and biosolids are produced over the world (Table 7.1). The biowastes have been beneficially utilized as soil amendments. Animal manures have been widely used as organic fertilizers to provide essential nutrients and improve physicochemical properties of soils. The food industry, including the wine and brewery industries, also produces large amounts of biowastes each year (Mahro and Timm, 2007; Ruggieri et al., 2009; Ryu et al., 2013). Municipalities generate several types

TABLE 7.1

Potential Quantity of Biowastes Produced ($\times 10^3$ Mg year^{-1}) in Selected Countries

Biowastes	Australia	United States	United Kingdom	New Zealand	China	Bangladesh	Japan	South Africa	India	Germany
Biosolids	407	5645	1120	80	24691	2940	2319	Not available	22086	1491
Chicken manure	893	22599	1765	145	51684	2454	3078	1722	8328	1228
Cattle manure	21953	77094	8304	8100	68814	18929	3593	11276	172613	10519
Pig manure	396	11233	772	58	82400	Not available	1696	276	1666	4586
Sheep manure	11272	930	5146	5391	22189	301	2	4056	12250	346
Plant residue	108056	454854	15512	680	918576	356850	49654	37840	1298080	26382

Source: Thangarajan, R. et al., *Sci. Total Environ.*, 465, 72–96, 2013.

of organic waste including municipal solid wastes (MSW) and biosolids, which is also referred to sewage sludge. Biosolid can be used as soil amendments because of its nutritional values (Hue and Sobieszczyk, 1999). Crop residues including stalks and stubble, leaves, and seed pods can be used as soil amendments to supply plant nutrients. Lal (2005) reported that the total global crop residue production is estimated to be 3.8 billion Mg comprised of 74% cereals, 8% legumes, 3% oil crops, 10% sugar crops, and 5% tubers.

Since most metal(loid)s do not undergo microbial or chemical degradation, immobilization and mobilization of metal(loid)s are employed in mitigating metal(loid) impacts and remediation of contaminated environments. Bioavailability of metal(loid)s plays a key role in the remediation of metal(loid) contaminated soils. The bioavailability of metal(loid)s in soil is determined by the amount of metal(loid)s in soil solution and exchangeable metal(loid)s adsorbed on soil particles. Therefore, reducing bioavailability can be a remedial goal of the metal(loid) contaminated soils.

Immobilization processes reduce the bioavailability of metal(loid)s by aging the soil or by adding soil amendments. The primary objective of the immobilization is to reduce the risk of metal(loid)s reaching the food chain through preventing plant uptake of metal(loid)s. Mobilization increases the bioavailability of metal(loid)s by transforming the metal(loid)s from the solid phase to the soil solution phase. The mobilized metal(loid)s are subsequently removed through soil washing or plant uptake. Soil washing removes metal(loid)s from soils, which is applicable to soils highly contaminated with metal(loid)s (Dermont et al., 2008). Removal of metal(loid)s through plant uptake is called phytoextraction. Metal(loid)s accumulated in the plants can be subsequently recovered or should be safely disposed through incineration and ashing (Sas-Nowosielska et al., 2004). However, removal of metal(loid)s is not always feasible and *in situ* immobilization can be considered as an option for the remediation of metal(loid) contaminated soils.

Biowastes can be used to immobilize metal(loid)s in soils through adsorption reactions, complexation, and precipitation. The immobilization of metal(loid)s by application of biowastes in soils is attributed to an increase in surface charges and the presence of complexation compounds (Bolan

et al., 2003b). Biosolid application as soil amendment reduced the bioavailability of Pb by 43% in soil (Brown et al., 2003). Sewage sludge and MSW compost also immobilized metals in soil resulting in increased biomass grown in the contaminated soils (Alvarenga et al., 2009). The biowastes corrected soil acidity and provided plant nutrients such as N, P, and K.

Animal manure has been used to reduce bioavailability of metal(loid)s and enhance plant growth. Liu et al. (2009) reported that application of chicken manure compost in Cd contaminated soil decreased soluble and exchangeable Cd in soil, thereby reducing phytotoxicity of Cd and Cd uptake by wheat. Chen et al. (2010) also reported that poultry manure compost reduced Cd uptake by pakchoi, which is related to the increased soil pH and Cd complexation with organic matter. The immobilization of Cd can be attributed to the increase of soil pH, Cd complexation by organic matter, and co-precipitation of Cd by P in the compost. Similarly, cattle manure increased maize height, root length, and dry biomass by reducing Ni concentration in the plant grown in Ni contaminated soil (Rehman et al., 2016).

Reduction of Cr(VI) is the detoxification process of Cr, which can be mediated by biowastes. Park et al. (2008) showed that pine bark reduced Cr(VI) to Cr(III), and the reducing capacity was significantly higher than that of chemical reductant ($FeSO_4$). Plant residues were reported to reduce bioavailability of metals when applied as green manure. Rapeseed residue increased soil organic matter content and microbial population in metal contaminated paddy soil. Rapeseed residue transformed easily accessible Cd and Pb fraction to less accessible fractions, thereby decreasing phytoavailability of the metals (Ok et al., 2011).

7.5 LIMITATION OF USING BIOWASTES FOR SOIL REMEDIATION

Land application of biowastes has shown to increase the crop productivity in agricultural soils and in metal contaminated soils. Although biowastes were successfully applied for remediation of metal(loid) contaminated soils in several researches, the long-term application of biowastes for metal(loid) immobilization is often questioned. Biowastes often contain environmental contaminants such as N, P, organic contaminants, and metal(loid)s. Long-term amendment of biowaste compost significantly increased total Zn and Pb concentrations in soil (Bartl et al., 2002).

Dissolved organic carbon mobilizes metal(loid)s such as Cu and Zn (Zhao et al., 2007). Release of dissolved organic carbon upon degradation of biowastes induces remobilization of metal(loid)s immobilized in soil. Merritt and Erich (2003) demonstrated that complexation of Cu with low molecular weight water-extractable carbon increased with humification of wheat straw.

Mobilization of toxic metal(loid)s by dissolved organic carbon and subsequent leaching to water bodies is of major concern. The possibility of metal leaching to groundwater affected by sewage sludge application in soil was tested by lysimeters. The result showed that NO_3-N, Cu, and Ni concentrations in leaching solution increased and trace metal concentrations in topsoil rapidly increased, which might result in toxicity to herbage (Keller et al., 2002). Therefore, application of biowastes for remediation of metal contaminated soils should be conducted with care, and long-term monitoring of metal(loid) leaching in biowaste applied soils should be accompanied.

7.6   CONCLUSIONS

Application of biowastes such as biosolid, animal manure, and plant residue to metal(loid) contaminated soils immobilizes metal(loid)s in the soils. However, biowastes may release contaminants such as N, P, and metal(loid)s upon the mineralization of the biowastes. Therefore, biowastes that are low in metal(loid)s can be effectively utilized to remediate soils contaminated with toxic metal(loid)s. Application of biowastes immobilizes metal(loid)s through adsorption, complexation, precipitation, and redox reactions, thereby reducing bioavailability of the metal(loid)s. The use of biowastes for the remediation of metal contaminated soils not only reduces burden for waste disposal, but also improves soil physicochemical properties. However, the immobilization of metal(loid)s using biowastes in the soils cannot be the ultimate solution for metal(loid) contaminated soils because metal(loid)s remain in soil and can be released when environmental conditions change.

REFERENCES

Alvarenga, P., Gonçalves, A.P., Fernandes, R.M., De Varennes, A., Vallini, G., Duarte, E. and Cunha-Queda, A.C., 2009. Organic residues as immobilizing agents in aided phytostabilization: (I) Effects on soil chemical characteristics. *Chemosphere*, 74(10), pp. 1292–1300.

Barrera-Díaz, C.E., Lugo-Lugo, V. and Bilyeu, B., 2012. A review of chemical, electrochemical and biological methods for aqueous Cr (VI) reduction. *Journal of Hazardous Materials*, 223, pp. 1–12.

Bartl, B., Hartl, W. and Horak, O., 2002. Long-term application of biowaste compost versus mineral fertilization: Effects on the nutrient and heavy metal contents of soil and plants. *Journal of Plant Nutrition and Soil Science*, 165(2), pp. 161–165.

Benitez, E., Romero, E., Gomez, M., Gallardo-Lara, F. and Nogales, R., 2001. Biosolids and biosolids-ash as sources of heavy metals in a plant-soil system. *Water, Air, and Soil Pollution*, 132(1–2), pp. 75–87.

Blais, J.F., Djedidi, Z., Cheikh, R.B., Tyagi, R.D. and Mercier, G., 2008. Metals precipitation from effluents. *Practice Periodical of Hazardous, Toxic, and Radioactive Waste Management*, 12(3), pp. 135–149.

Bolan, N., Adriano, D., Mani, S. and Khan, A., 2003a. Adsorption, complexation, and phytoavailability of copper as influenced by organic manure. *Environmental Toxicology and Chemistry: An International Journal*, 22(2), pp. 450–456.

Bolan, N.S., Khan, M.A., Donaldson, J., Adriano, D.C. and Matthew, C., 2003b. Distribution and bioavailability of copper in farm effluent. *Science of the Total Environment*, 309(1–3), pp. 225–236.

Bolan, N., Adriano, D. and Mahimairaja, S., 2004. Distribution and bioavailability of trace elements in livestock and poultry manure by-products. *Critical Reviews in Environmental Science and Technology*, 34(3), pp. 291–338.

Bolan, N.S., Szogi, A.A., Chuasavathi, T., Seshadri, B., Rothrock, M.J. and Panneerselvam, P., 2010. Uses and management of poultry litter. *World's Poultry Science Journal*, 66(4), pp. 673–698.

Brown, S., Chaney, R.L., Hallfrisch, J.G. and Xue, Q., 2003. Effect of biosolids processing on lead bioavailability in an urban soil. *Journal of Environmental Quality*, 32(1), pp. 100–108.

Brugger, D. and Windisch, W.M., 2015. Environmental responsibilities of livestock feeding using trace mineral supplements. *Animal Nutrition*, 1(3), pp. 113–118.

Bustamante, P., Lahaye, V., Durnez, C., Churlaud, C. and Caurant, F., 2006. Total and organic Hg concentrations in cephalopods from the North Eastern Atlantic waters: Influence of geographical origin and feeding ecology. *Science of the Total Environment*, 368(2–3), pp. 585–596.

Cao, X., Hu, P., Tan, C., Wu, L., Peng, B., Christie, P. and Luo, Y., 2018. Effects of a natural sepiolite bearing material and lime on the immobilization and persistence of cadmium in a contaminated acid agricultural soil. *Environmental Science and Pollution Research*, 25(22), pp. 22075–22084.

Chen. H.S., Huang, Q.Y., Li-Na, L.I.U., Peng, C.A.I., Liang, W. and Ming, L.I., 2010. Poultry manure compost alleviates the phytotoxicity of soil cadmium: Influence on growth of pakchoi (*Brassica chinensis* L.). *Pedosphere*, 20(1), pp. 63–70.

Dermont, G., Bergeron, M., Mercier, G. and Richer-Laflèche, M., 2008. Soil washing for metal removal: A review of physical/chemical technologies and field applications. *Journal of Hazardous Materials*, 152(1), pp. 1–31.

Diagboya, P.N., Olu-Owolabi, B.I. and Adebowale, K.O., 2015. Effects of time, soil organic matter, and iron oxides on the relative retention and redistribution of lead, cadmium, and copper on soils. *Environmental Science and Pollution Research*, 22(13), pp. 10331–10339.

Dickinson, W.W., Dunbar, G.B. and McLeod, H., 1996. Heavy metal history from cores in Wellington Harbour, New Zealand. *Environmental Geology*, 27(1), pp. 59–69.

Du, H., Chen, W., Cai, P., Rong, X., Feng, X. and Huang, Q., 2016. Competitive adsorption of Pb and Cd on bacteria–montmorillonite composite. *Environmental Pollution*, 218, pp. 168–175.

Fu, F. and Wang, Q., 2011. Removal of heavy metal ions from wastewaters: A review. *Journal of Environmental Management*, 92(3), pp. 407–418.

Grant, C.A., 2017. Influence of phosphate fertilizer on cadmium in agricultural soils and crops. In *Phosphate in Soils* (pp. 138–163). CRC Press H. Magdi Selim. Editor. 6000 Broken Sound Parkway NW, Suite 300.

H. Magdi Selim. Editor. 6000 Broken Sound Parkway NW, Suite 300.

Huang, K., Chen, C., Zhang, J., Tang, Z., Shen, Q., Rosen, B.P. and Zhao, F.J., 2016. Efficient arsenic methylation and volatilization mediated by a novel bacterium from an arsenic-contaminated paddy soil. *Environmental Science & Technology*, 50(12), pp. 6389–6396.

Hue, N.V. and Sobieszczyk, B.A., 1999. Nutritional values of some biowastes as soil amendments. *Compost Science & Utilization*, 7(1), pp. 34–41.

Jiang, S., Huang, L., Nguyen, T.A., Ok, Y.S., Rudolph, V., Yang, H. and Zhang, D., 2016. Copper and zinc adsorption by softwood and hardwood biochars under elevated sulphate-induced salinity and acidic pH conditions. *Chemosphere*, 142, pp. 64–71.

Jin, H., Yan, D., Zhu, N., Zhang, S. and Zheng, M., 2019. Immobilization of metal (loid) s in hydrochars produced from digested swine and dairy manures. *Waste Management*, 88, pp. 10–20.

Jondreville, C., Revy, P.S. and Dourmad, J.Y., 2003. Dietary means to better control the environmental impact of copper and zinc by pigs from weaning to slaughter. *Livestock Production Science*, 84(2), pp. 147–156.

Kazi, T.G., Jamali, M.K., Kazi, G.H., Arain, M.B., Afridi, H.I. and Siddiqui, A., 2005. Evaluating the mobility of toxic metals in untreated industrial wastewater sludge using a BCR sequential extraction procedure and a leaching test. *Analytical and Bioanalytical Chemistry*, 383(2), pp. 297–304.

Keller, C., McGrath, S.P. and Dunham, S.J., 2002. Trace metal leaching through a soil–grassland system after sewage sludge application. *Journal of Environmental Quality*, 31(5), pp. 1550–1560.

Lal, R., 2005. World crop residues production and implications of its use as a biofuel. *Environment International*, 31(4), pp. 575–584.

Lee, S.S., Lim, J.E., El-Azeem, S.A.A., Choi, B., Oh, S.E., Moon, D.H. and Ok, Y.S., 2013. Heavy metal immobilization in soil near abandoned mines using eggshell waste and rapeseed residue. *Environmental Science and Pollution Research*, 20(3), pp. 1719–1726.

Lee, G., Bigham, J. M. and Faure, G. 2002. Removal of trace metals by coprecipitation with Fe, Al and Mn from natural waters contaminated with acid mine drainage in the Ducktown Mining District, Tennessee. *Applied Geochemistry*, 17(5), 569–581.

Liu, L., Chen, H., Cai, P., Liang, W. and Huang, Q., 2009. Immobilization and phytotoxicity of Cd in contaminated soil amended with chicken manure compost. *Journal of Hazardous Materials*, 163(2–3), pp. 563–567.

Liu, Y., Luo, C., Cui, G. and Yan, S., 2015. Synthesis of manganese dioxide/iron oxide/graphene oxide magnetic nanocomposites for hexavalent chromium removal. *RSC Advances*, 5(67), pp. 54156–54164.

Mahro, B. and Timm, M., 2007. Potential of biowaste from the food industry as a biomass resource. *Engineering in Life Sciences*, 7(5), pp. 457–468.

Mar, S.S. and Okazaki, M., 2012. Investigation of Cd contents in several phosphate rocks used for the production of fertilizer. *Microchemical Journal*, 104, pp. 17–21.

Merritt, K.A. and Erich, M.S., 2003. Influence of organic matter decomposition on soluble carbon and its copper-binding capacity. *Journal of Environmental Quality*, 32(6), pp. 2122–2131.

Mishra, T. and Mahato, D.K., 2016. A comparative study on enhanced arsenic (V) and arsenic (III) removal by iron oxide and manganese oxide pillared clays from ground water. *Journal of Environmental Chemical Engineering*, 4(1), pp. 1224–1230.

Mokone, T.P., Van Hille, R.P. and Lewis, A.E., 2010. Effect of solution chemistry on particle characteristics during metal sulfide precipitation. *Journal of Colloid and Interface Science*, 351(1), pp. 10–18.

Moore, P.A., Daniel, T.C., Gilmour, J.T., Shreve, B.R., Edwards, D.R. and Wood, B.H., 1998. Decreasing metal runoff from poultry litter with aluminum sulfate. *Journal of Environmental Quality*, 27(1), pp. 92–99.

Nziguheba, G. and Smolders, E., 2008. Inputs of trace elements in agricultural soils via phosphate fertilizers in European countries. *Science of the Total Environment*, 390(1), pp. 53–57.

Ogiyama, S., Sakamoto, K., Suzuki, H., Ushio, S., Anzai, T. and Inubushi, K., 2005. Accumulation of zinc and copper in an arable field after animal manure application. *Soil Science & Plant Nutrition*, 51(6), pp. 801–808.

Ok, Y.S., Usman, A.R., Lee, S.S., El-Azeem, S.A.A., Choi, B., Hashimoto, Y. and Yang, J.E., 2011. Effects of rapeseed residue on lead and cadmium availability and uptake by rice plants in heavy metal contaminated paddy soil. *Chemosphere*, 85(4), pp. 677–682.

Okuno, T., Kubota, T., Kuroda, T., Ueno, H. and Nakamuro, K., 2001. Contribution of enzymic α, γ-elimination reaction in detoxification pathway of selenomethionine in mouse liver. *Toxicology and Applied Pharmacology*, 176(1), pp. 18–23.

Park, D., Ahn, C.K., Kim, Y.M., Yun, Y.S. and Park, J.M., 2008. Enhanced abiotic reduction of Cr (VI) in a soil slurry system by natural biomaterial addition. *Journal of Hazardous Materials*, 160(2–3), pp. 422–427.

Park, J.H., Lamb, D., Paneerselvam, P., Choppala, G., Bolan, N. and Chung, J.W., 2011. Role of organic amendments on enhanced bioremediation of heavy metal (loid) contaminated soils. *Journal of Hazardous Materials*, 185(2–3), pp. 549–574.

Park, J.H., Ok, Y.S., Kim, S.H., Cho, J.S., Heo, J.S., Delaune, R.D. and Seo, D.C., 2016. Competitive adsorption of heavy metals onto sesame straw biochar in aqueous solutions. *Chemosphere*, 142, pp. 77–83.

Rehman, M.Z.U., Rizwan, M., Ali, S., Fatima, N., Yousaf, B., Naeem, A., Sabir, M., Ahmad, H.R. and Ok, Y.S., 2016. Contrasting effects of biochar, compost and farm manure on alleviation of nickel toxicity in maize (*Zea mays* L.) in relation to plant growth, photosynthesis and metal uptake. *Ecotoxicology and Environmental Safety*, 133, pp. 218–225.

Roane, T.M., Pepper, I.L. and Gentry, T.J., 2015. Microorganisms and metal pollutants. In *Environmental Microbiology* (pp. 415–439) Ian Pepper, Charles Gerba, Terry Gentry. Editors. Academic Press, San Diego, CA.

Ruggieri, L., Cadena, E., Martínez-Blanco, J., Gasol, C.M., Rieradevall, J., Gabarrell, X., Gea, T., Sort, X. and Sánchez, A., 2009. Recovery of organic wastes in the Spanish wine industry. Technical, economic and environmental analyses of the composting process. *Journal of Cleaner Production*, 17(9), pp. 830–838.

Rutherford, D.W., Bednar, A.J., Garbarino, J.R., Needham, R., Staver, K.W. and Wershaw, R.L., 2003. Environmental fate of roxarsone in poultry litter. Part II. Mobility of arsenic in soils amended with poultry litter. *Environmental Science & Technology*, 37(8), pp. 1515–1520.

Ryu, B.G., Kim, J., Kim, K., Choi, Y.E., Han, J.I. and Yang, J.W., 2013. High-cell-density cultivation of oleaginous yeast *Cryptococcus curvatus* for biodiesel production using organic waste from the brewery industry. *Bioresource Technology*, 135, pp. 357–364.

Sas-Nowosielska, A., Kucharski, R., Małkowski, E., Pogrzeba, M., Kuperberg, J.M. and Kryński, K., 2004. Phytoextraction crop disposal – an unsolved problem. *Environmental Pollution*, 128(3), pp. 373–379.

Seshadri, B., Bolan, N.S., Choppala, G., Kunhikrishnan, A., Sanderson, P., Wang, H., Currie, L.D., Tsang, D.C., Ok, Y.S. and Kim, G., 2017. Potential value of phosphate compounds in enhancing immobilization and reducing bioavailability of mixed heavy metal contaminants in shooting range soil. *Chemosphere*, *184*, pp. 197–206.

Shaheen, S. M., Kwon, E. E., Biswas, J. K., Tack, F. M., Ok, Y. S., & Rinklebe, J. (2017). Arsenic, chromium, molybdenum, and selenium: Geochemical fractions and potential mobilization in riverine soil profiles originating from Germany and Egypt. *Chemosphere*, *180*, 553–563.

Stolz, J.F. and Oremland, R.S., 1999. Bacterial respiration of arsenic and selenium. *FEMS Microbiology Reviews*, *23*(5), pp. 615–627.

Thangarajan, R., Bolan, N.S., Tian, G., Naidu, R. and Kunhikrishnan, A., 2013. Role of organic amendment application on greenhouse gas emission from soil. *Science of the Total Environment*, *465*, pp. 72–96.

Tsang, D.C., Zhou, F., Zhang, W. and Qiu, R., 2016. Stabilization of cationic and anionic metal species in contaminated soils using sludge-derived biochar. *Chemosphere*, *149*, pp. 263–271.

Uddin, M.K., 2017. A review on the adsorption of heavy metals by clay minerals, with special focus on the past decade. *Chemical Engineering Journal*, *308*, pp. 438–462.

Wan, S., Wu, J., Zhou, S., Wang, R., Gao, B. and He, F., 2018. Enhanced lead and cadmium removal using biochar-supported hydrated manganese oxide (HMO) nanoparticles: Behavior and mechanism. *Science of the Total Environment*, *616*, pp. 1298–1306.

Yang, J., Chen, L., Liu, L.Z., Shi, W.L. and Meng, X.Z., 2014. Comprehensive risk assessment of heavy metals in lake sediment from public parks in Shanghai. *Ecotoxicology and Environmental Safety*, *102*, pp. 129–135.

Zhao, L.Y., Schulin, R., Weng, L. and Nowack, B., 2007. Coupled mobilization of dissolved organic matter and metals (Cu and Zn) in soil columns. *Geochimica et Cosmochimica Acta*, *71*(14), pp. 3407–3418.

8 Microbiological Interfacial Behaviors in the Removal of Polycyclic Aromatic Hydrocarbons from Soil and Water

Dong Zhang and Li Lyu

CONTENTS

8.1 INTRODUCTION

Polycyclic aromatic hydrocarbons (PAHs) are recognized as one of the most widespread and ubiquitous hydrophobic organic compounds (HOCs), and the pollution of PAHs in soil, surface water, and groundwater has become a major environmental and human concern (Li et al., 2014; Zhang and Zhu, 2014). Due to the toxic, carcinogenic, and mutagenic nature of PAHs, the contaminated sites pose ecological risks and a threat to human health, and effective remediation of PAH-polluted environment is needed. In the past three decades, numerous efforts including physical, chemical, and biological treatments have been conducted to remove PAHs from polluted environments (Lighty et al., 1990; Yeung and Gu, 2011; Siegrist et al., 1995; Pignatello et al., 2006; Jonsson et al., 2010; Singh and Borthakur, 2018; Pohren et al., 2019). Zhu et al. (2010) provided an extensive overview of remediation technologies for treatment of PAH-contaminated soils and generally suggested bioremediation/biodegradation as one of the most promising, relatively efficient, and cost-effective approaches. To date, more than 55% of published literatures (Web of Science®) for remediation of PAH-contaminated soil are related to bioremediation.

Although the direct degradation of solid PAHs or crystalline PAHs through extracellular enzyme is possible (Johnsen et al., 2005), the uptake and subsequent degradation via intercellular enzyme of dissolved PAHs is found in most cases. Actually, there usually is a physical barrier for microbes to contact the dissolved target molecules in contaminated soil. Owing to the high hydrophobicity, PAHs mainly are retarded within the soil matrix. Desorption of soil-sorbed PAHs can be considered to be the first step for microbes to uptake the molecules. Then, the biodegradation process for dissolved PAHs in soil system or simply in aquatic environment refers to several interfacial processes as suggested by Zhang and Zhu (2014): (1) Sorption of PAH molecules onto the cell surface of certain degraders (mainly bacteria) from aqueous phase (surface water or soil solutions). (2) Transmembrane transport from the outer surface of biological cells into the internal matrix through the lipid bilayer (Song et al., 2010; Zhang et al., 2013a). This transport is often recognized as the rate-limited process in the PAHs biodegradation, due to the hydrophilic area of cytoplasmic membrane lipids and lipopolysaccharide (LPS, Gram negative bacteria only) as a "barrier" for HOCs (Hearn et al., 2009; Wiener and Horanyi, 2011). (3) Biotransformation or mineralization of PAHs within microbial cells facilitated with intercellular enzymes (Cerniglia, 1992; Poulos, 1995; Mueller et al., 1997; Dean-Ross et al., 2001; Karlsson et al., 2003; Baboshin et al., 2008; Fanesi et al., 2018; Kadri et al., 2018). These interfacial interactions of PAH molecules with microbial cell surface, cytoplasmic membrane lipids, and proteins, as well as intercellular enzymes, determine the efficiency of biodegradation of PAHs.

Therefore, in this chapter, we mainly discuss the biological sorption of PAHs by microbial surface, transmembrane transport, and underlying mechanisms of PAHs, as well as the role of each biological process or interfacial behaviors in biotransformation of PAHs in aqueous environment (both aquatic environment and soil solution).

8.2 INTERFACIAL INTERACTION OF PAHS WITH BACTERIAL SURFACES

8.2.1 BIOSURFACE SORPTION/ACCUMULATION OF PAHS AND ITS ROLE IN PAH DEGRADATION

The bioremediation or biodegradation of PAHs by microbes such as bacteria and fungi contains a series of complex and consecutive interfacial processes, referring to cell surface sorption, transmembrane transport, and intercellular degradation (Zhang and Zhu, 2014). As the first step of bioremediation or biodegradation, the surface sorption of PAHs is considered to be the crucial step and plays a critical role in biodegradation of PAHs (Stringfellow and Alvarez-Cohen, 1999; Zhang and Zhu, 2012; Xu et al., 2013; Zhang et al., 2013a; Sanches et al., 2017). The sorption process not only acts as a remover of aqueous pollutants, but also provides more directly available substances for the microbial cells. Due to the poor water solubility and high hydrophobicity, very low concentrations of PAHs are usually found in aquatic environment and soil solutions. Biosorption, using the sorption capability of dead or inactive biomass, is one of the most efficient methods to concentrate and reduce organic pollutants from wastewater or other aqueous environments (Volesky, 2003; Aksu, 2005; Chen and Ding, 2012; Gu et al., 2015; Arias et al., 2017). Likewise, the surface sorption plays a similar role in live microbial systems. In addition, microorganisms with higher sorption affinity for PAHs may take the advantage of obtaining more "food," being beneficial to biodegradation of PAHs. Therefore, insight of surface sorption of PAHs by microbes and underlying mechanisms is useful to have a better understanding of bioremediation of PAH-contaminated environments.

8.2.1.1 Sorption of PAHs

In the period of 1920s to 1940s, the isolation and characteristic of naphthalene-degrading bacterium *Pseudomonas aeruginosa*, which could use naphthalene as its only source of carbon and energy, indicated the beginning of microbial degradation of organic pollutants (Gray and Thornton, 1928;

Strawinski and Stone, 1943). Numerous works had been done to illustrate the biodegradation pathways and affecting factors (Smith, 1948; Voerman and Besemer, 1970; Siddaramappa et al., 1973). However, in 1973, biosorption of HOCs (typically p, p'-DDT and methoxychlor) by inactive biomass of *Aerobacter aerogenes* and *Bacillus subtilis* was first conducted to evaluate the biomagnification (Johnson and Kennedy, 1973). At the end of last century, the relationship between the sorption of PAHs to bacterial surface and biodegradation was evaluated (Stringfellow and Alvarez-Cohen, 1999), and from then on, the significant role of (bio)sorption of PAHs in the biodegradation and bioremoval of PAHs has been gradually realized.

The sorption of organic pollutants by biosolids or biomass can be recognized as a branch of biotechnology that effectively reduces chemical concentrations in aqueous environment, as well as a crucial step for the subsequent transmembrane transport and intercellular biodegradation that uptake substance from surrounding aqueous environment (Bokbolet et al., 1999; Zhang et al., 2018a). Sorption of hydrophobic organic compounds and pseudo hydrophobic organic compounds including PAHs, polychlorinated biphenyls (PCBs), organic pesticides, and pharmaceutical and personal care products (PPCPs) has been extensively studied onto soil, sewage sludge, microorganisms, and nanoparticles (Steen and Karickhoff, 1984; Tsezos and Bell, 1989; Chung et al., 2007; Li et al., 2007; Carballa et al., 2008; Chen et al., 2010; Zhuang et al., 2011; Zhang and Zhu, 2012; Zhang et al., 2018a). Results indicated that the sorption played an essential role in both bioremoval of PAHs from aqueous environments and facilitation to subsequent intercellular degradation.

8.2.1.1.1 Biosorption and Its Relation to the Bioremoval of PAHs from Aqueous Solutions

Biosorption is considered to be a high-efficiency and low-cost method for the treatment of PAH-bearing wastewater and other aquatic environments (Loukidou et al., 2004; Ata et al., 2012). Many dead or inactive biomass of various microorganisms including bacteria (Giri et al., 2013; Xu et al., 2013), fungi (Chen et al., 2010; Chen and Ding, 2012; Gu et al., 2015, 2016), and algae (Ata et al., 2012; Zhang et al., 2013b, 2015; Arias et al., 2017) have been used as low-cost biosorbents and remarkable performance was conducted in removal of heavy metals (Park and Chon, 2016; Albert et al., 2018; Bano et al., 2018), pesticides (Tsezos and Bell, 1989; Gao et al., 2014; Li et al., 2018), PPCPs (Carballa et al., 2008; Blair et al., 2015; Ashfaq et al., 2017), PAHs (Dimitriou-Christidis et al., 2007; Zhang et al., 2018b), and dyes (Chandra et al., 2015; Zhao and Zhou, 2016; Cheng et al., 2018). As typical HOCs, PAHs are relatively hydrophobic and nonpolar organic pollutants, and theoretically PAHs are easily sorbed onto microbial biomass. Selected microbial biomass (derived from bacteria, fungi, and algae) as well as their sorption parameters for PAHs are listed in Table 8.1.

TABLE 8.1

Sorption Properties of PAHs onto Biosurfaces

Compounds	MW^a	$\log K_{ow}{}^b$	Biomass	Capacity, mg kg^{-1} ^{c}	$\log K_d$, L kg^{-1}	$\log K_{oc}$, L kg^{-1}	k_{biol}, mg L^{-1} d^{-1}	References
Bacteria								
Phenanthrene	178.3	4.46	Soil indigenous bacterial consortium	1363.8	3.84			Zhang et al. (2018b)
Phenanthrene	178.3	4.46	Microcystis aeruginosa 7820	6780	2.86^d			Bai et al. (2016)
Naphthalene	128.2	3.30	Bacterial consortium	2100	–			Xu et al. (2013)
Phenanthrene	178.3	4.46	Bacterial consortium	1900	–			
Pyrene	202.3	4.88	Bacterial consortium	3800	–			
Pyrene	202.3	4.88	Klebsiella oxytoca PYR-1		4.35		0. 0024	Zhang and Zhu (2012)
Naphthalene	128.2	3.30	Activated sludge	66.36	1.86			Meng et al. (2016)
Phenanthrene	178.3	4.46	Activated sludge	65.66	3.15			
Pyrene	202.3	4.88	Activated sludge	26.17	4.44			
Phenanthrene	178.3	4.46	Corynebacterium urealyticum				1.82–80.12	Kamil et al. (2016)
Fluoranthene	202.3	5.16	Sludge	100	3.51	3.72		Zolfaghari et al. (2017)
Phenanthrene	178.3	4.46	Escherichia coli	356.6	3.59			Xiao et al. (2007)
Phenanthrene	178.3	4.46	Gordona bronchialis RR2		4.56			String fellow and Alvarez-Cohen (1999)
Phenanthrene	178.3	4.46	Rhodococcus rhodochrous RR1		4.48			
Phenanthrene	178.3	4.46	Rhodococcus erythropolis GPMEX2		4.18			
Phenanthrene	178.3	4.46	Mycobacterium parafortuitum ATCC 19686		4.15			
Phenanthrene	178.3	4.46	Pseudomonas fluorescens		4.04			
Phenanthrene	178.3	4.46	Nostocodia JS8		4.04			
Phenanthrene	178.3	4.46	Municipal activated sludge		4.00			
Phenanthrene	178.3	4.46	Acinetobacer sp.		3.99			
Phenanthrene	178.3	4.46	Refinery activated sludge (lab)		3.98			
Phenanthrene	178.3	4.46	Refinery activated sludge (plant)		3.95			
Phenanthrene	178.3	4.46	Escherichia coli W3110		3.72			
Phenanthrene	178.3	4.46	Pseudomonas stutzeri P-16		3.76			
Phenanthrene	178.3	4.46	Pseudomonas stutzeri ATCC 11607		3.69			
Phenanthrene	178.3	4.46	Pseudomoans aeruginosa CV1		3.68			
Phenanthrene	178.3	4.46	Micrococcus luteus		3.40			
Phenanthrene	178.3	4.46	Pseudomonas sp. Ph6		4.37			Ma et al. (2018)
Phenanthrene	178.3	4.46	Pseudomonas synxantha LSH-7'	15000				Meng et al. (2019)
Pyrene	202.3	4.88	Pseudomonas synxantha LSH-7'	12500				

(Continued)

TABLE 8.1 (Continued)
Sorption Properties of PAHs onto Biosurfaces

Compounds	MW[a]	log K_{ow}[b]	Biomass	Capacity, mg kg^{-1} [c]	log K_d, L kg^{-1}	log K_{oc}, L kg^{-1}	k_{biol}, mg L^{-1} d^{-1}	References
Naphthalene	128.2	3.30	Klebsiella oxytoca PYR-1	6128	2.56			Zhang et al. (2018a)
Acenaphthene	154.2	3.92	Klebsiella oxytoca PYR-1	821	3.24			
Fluorene	166.2	4.18	Klebsiella oxytoca PYR-1	951	3.40			
Phenanthrene	178.3	4.46	Klebsiella oxytoca PYR-1	1418	3.87			
Anthracene	178.3	4.45	Klebsiella oxytoca PYR-1	168	3.80			
Pyrene	202.3	4.88	Klebsiella oxytoca PYR-1	1630	4.36			
Fluoranthene	202.3	5.16	Klebsiella oxytoca PYR-1	829	4.29			
Fungi								
Phenanthrene	178.3	4.46	Phanerochaete chrysosporium	1768.8	3.97	4.33[d]		Gu et al. (2015)
Phenanthrene	178.3	4.46	Methylation of amino groups of Phanerochaete chrysosporium	1552.9	3.85	4.23[d]		
Phenanthrene	178.3	4.46	Acetylation of hydroxyl groups of Phanerochaete chrysosporium	1570.9	3.84	4.21[d]		
Phenanthrene	178.3	4.46	Lipid removal of Phanerochaete chrysosporium	1346.4	3.80	4.17[d]		
Phenanthrene	178.3	4.46	Esterification of carboxyl groups of Phanerochaete chrysosporium	991.0	3.54	3.92[d]		
Phenanthrene	178.3	4.46	Base hydrolysis of Phanerochaete chrysosporium	2041.5	3.95	4.34[d]		
Phenanthrene	178.3	4.46	Phanerochaete chrysosporium		3.61		0.068[d]	Ding et al. (2013)
Pyrene	202.3	4.88	Phanerochaete chrysosporium		4.24		0.0015[d]	
Pyrene	202.3	4.88	Brevibacillus brevis	410[d]	2.84[d]		0.099[d]	Liao et al. (2015)
Phenanthrene	178.3	4.46	Aspergillus niger ATC 16404	45				Hamzah et al. (2018)
Phenanthrene	178.3	4.46	Fusarium solani	100				Fayeulle et al. (2014)
Benzo(a) pyrene	252.3	5.97	Fusarium solani	1200				
Anthracene	178.3	4.45	Acremonium sp. P0997 (live)		2.59			Ma et al. (2014)
Anthracene	178.3	4.45	Acremonium sp. P0997 (dead)		2.56			
Fluoranthene	202.3	5.16	Agaricus bisporus	53.48	1.80			Xie et al. (2015)
Fluoranthene	202.3	5.16	Agaricus bisporus + Lentinus edodes	71.94	2.10			
Naphthalene	128.2	3.30	Consortium of white-rot fungi		2.56	2.92		Chen et al. (2010)
Acenaphthene	154.2	3.92	Consortium of white-rot fungi		3.22	3.58		
Fluorene	166.2	4.18	Consortium of white-rot fungi		3.46	3.82		
Phenanthrene	178.3	4.46	Consortium of white-rot fungi		3.83	4.19		
Pyrene	202.3	4.88	Consortium of white-rot fungi		4.38	4.74		

(Continued)

TABLE 8.1 (*Continued*)
Sorption Properties of PAHs onto Biosurfaces

Compounds	MW[a]	log K_{ow}[b]	Biomass	Capacity, mg kg[-1][c]	log K_d, L kg[-1]	log K_{oc}, L kg[-1]	k_{biol}, mg L[-1] d[-1]	References
Anthracene	178.3	4.45	*Gomphidius viscidus* (dead)	1515				Huang et al. (2010)
Anthracene	178.3	4.45	*Gomphidius viscidus* (live)	1887				
Algae								
Phenanthrene	178.3	4.46	*Spirulina*		3.70	4.17		Zhang et al. (2013b)
Phenanthrene	178.3	4.46	*Seaweed*		3.37	3.91		
Phenanthrene	178.3	4.46	*Porphyra*		3.59	3.96		
Phenanthrene	178.3	4.46	*Chlorella*		3.58	4.17		
Phenanthrene	178.3	4.46	*Sphaerellopsis*		3.61	4.20		
Phenanthrene	178.3	4.46	Zooplankton LHH13		3.84	4.41		
Phenanthrene	178.3	4.46	Phytoplankton LHH25		3.77	4.47		
Naphthalene	128.2	3.30	*Chlorella vulgaris*	4540	3.51			Ashour et al. (2008)
Phenanthrene	178.3	4.46	*Microcystis aeruginosa*		4.68			Tao et al. (2014)
Pyrene	202.3	4.88	*Microcystis aeruginosa*		4.72			
Benzo(a)pyrene	252.3	5.97	*Microcystis aeruginosa*		5.50			
Phenanthrene	178.3	4.46	*Rhosomonas baltica*		3.67[d]	4.45		Arias et al. (2017)
Pyrene	202.3	4.88	*Rhosomonas baltica*		4.00[d]	4.78		
Fluoranthene	202.3	5.16	*Rhosomonas baltica*		3.95[d]	4.73		
Phenanthrene	178.3	4.46	*Sargassum hemiphyllum*	450.6	3.82	3.96		Chung et al. (2007)

a   Molecular weight.
b   Obtained from Zhang et al. (2018a).
c   Obtained from literatures based on maximal sorption capacity or stimulated equilibrium capacity.
d   Calculated or modified based on literature data.

Numerous studies reported high-efficiency removal of PAHs. For example, in a bioreactor of immobilized cells of *Mycoplana* sp. MVMB2, more than 99.43% of phenanthrene was removed within 3 hours hydraulic retention time, whereas about 73.3% of removed phenanthrene was attributed to biosorption (Lakshmi et al., 2012). Zhang et al. (2018b) found the biosorption efficiency escalated from 60.65% to 78.78% for phenanthrene by inactive soil indigenous bacterial biomass at biomass dosages of 0.25–1.25 g L^{-1} within 8 hours. Tam et al. (2010) reported removal and biodegradation of 3 PAHs by immobilized microalgal beads. Within 48 hours, 93% of phenanthrene, 100% of fluoranthene, and 100% of pyrene were removed at the dosage of 12 bead mL^{-1}, respectively. At the same condition, 52%, 65%, and 68% of these three PAHs were degraded, respectively. Furthermore, at less than 6 hours, the removal efficiencies for the three PAHs reached almost the same removal level, whereas the degraded percentages of 3 PAHs were less than 20%. The results indicated biosorption played an important role in the removal of PAHs from aqueous solutions especially in fast removal conditions. Sanches et al. (2017) also provided the similar conclusion that the bioremoval of PAHs was strongly governed by PAH biosorption. Over 88% of acenaphthene and 85% of phenanthrene were biosorptive removed in the inactive cells set after 1.5–2 hours of incubation.

In addition, microbial biomass usually exhibited relatively high sorption capacity for PAHs. Xu et al. (2013) reported high sorption capacities of 3 typical PAHs by microbial consortium, which were 2100, 1900, and 3800 mg kg^{-1} for naphthalene, phenanthrene, and pyrene, respectively. Comparable results were illustrated by several other studies. For example, 1418 mg kg^{-1} for phenanthrene and 1630 mg kg^{-1} for pyrene on bacterial cells of *Klebsiella oxytoca* were achieved (Zhang et al., 2018a). Sorption amount of 1768.8 mg kg^{-1} for phenanthrene on fungal biomass of *Phanerochaete chrysosporium* was obtained (Gu et al., 2015). Soil indigenous bacterial biomass presented a sorption capacity of 1363.8 mg kg^{-1} (Zhang et al., 2018b). However, pretty low and high sorption capacities were also reported. For example, sorption capacity of 6780 mg kg^{-1} for phenanthrene on *Microcystis aeruginosa* (Bai et al., 2016) and 15,000 mg kg^{-1} for phenanthrene on *Pseudomonas synxantha* (Meng et al., 2019) were obtained. No more than 100 mg kg^{-1} sorption capacity for phenanthrene (e.g., 100 mg kg^{-1} on *Fusarium solani*, 45 mg kg^{-1} on *Aspergillus niger*, and 65.66 mg kg^{-1} on activated sludge) were reported (Fayeulle et al., 2014; Meng et al., 2016; Hamzah et al., 2018). The high deviation of sorption capacity appears mainly owing to the various states of microbial surface properties, initial concentrations of target pollutants, and ratio of biosorbent to solution volume.

In addition to the high removal potential, biosorption of PAHs by inactive biomass can be used to predict the bioremoval or biodegradation of PAHs by underlying living cells of microorganisms. Meng et al. (2019) evaluated the relationship between biodegradation rates (representing bioremoval) of phenanthrene (or pyrene) by living *Pseudomonas synxantha* LSH-7' and biosorption of phenanthrene (or pyrene) on

inactive biomass of *P. synxantha* LSH-7', regardless of initial PAHs concentrations, addition of fertilizer, sediments, and surfactants. A highly positive correlation was observed and shown as following:

$$\text{For phenanthrene}: k_b = 0.3245Q + 1.5990$$

$$\left(n = 30, \ R^2 = 0.9098\right) \quad (8.1)$$

$$\text{For pyrene}: k_b = 0.2020Q + 0.4103 \left(n = 30, \ R^2 = 0.9241\right) \quad (8.2)$$

where k_b represents the biodegradation rate, mg L^{-1} d^{-1}, and Q represents the biosorption amount, mg g^{-1}.

The authors believed that as long as the ratio of bacterial biosorption could be improved, the biodegradation rate of PAHs would be enhanced. In addition, the slope values of the two correlations determined the biodegradation potential of sorbed PAHs. The value for phenanthrene (0.3245) is larger than that for pyrene (0.2020), and it indicated that phenanthrene had higher biodegradation potential than pyrene. In a previous study, Zhang and Zhu (2012) proposed a similar relationship between surfactant-induced biodegradation capability and sorption capability for pyrene (as shown in Figure 8.1). The highly positive correlation ($n = 9$, $R = 0.9927$, $P < 0.0001$) suggested that more sorbed pyrene induced by surfactant meant there would be more pyrene available for intracellular enzymes. The results also indicated that the promotion of biodegradation was derived from sorption-related bioavailability enhancement. The relatively higher slope value (0.8440, near 1.0) implied the bacterial strain (*Klebsiella oxytoca* PYR-1) could easily degrade sorbed pyrene.

Although sorption efficiency and sorption capacity are useful parameters to evaluate the removal of PAHs, these parameters are restricted to certain conditions and can't exhibit the sorption capability per unit biosorbent and the

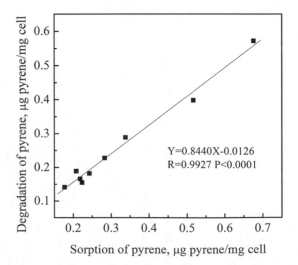

FIGURE 8.1 The relationship between biodegradation capability of pyrene and sorption capability of *Klebsiella oxytoca* PYR-1.

sorption affinity with PAHs. In addition, these parameters are not comparable through different studies with different microbial biomass under different conditions. Therefore, the distribution/partition coefficient (K_d) and carbon normalized distribution coefficient (K_{oc}, equals to K_d/f_{oc}, where f_{oc} is the carbon content of microbial biomass) are usually used to evaluate the distribution of PAHs between microbial surface and aqueous phases. These parameters are useful and exclude the influences of physicochemical properties such as initial PAHs concentrations, ratio of microbial biomass to solution volume, and so on. Basically, the distribution coefficient mainly depends on the property of microbial biomass and target pollutants rather than environmental conditions. Therefore, for a specific contaminant, the sorption affinity of different biosorbent/microbial biomass can be directly compared using this parameter. In general, the biomass with higher hydrophobic surface has higher distribution coefficient for PAHs with more aromatic rings and higher hydrophobicity. Theoretically, K_{oc} is more suitable than K_d to evaluate the relationship between distribution coefficient and PAH property, owing to excluding the influence of microbial carbon content. However, K_{oc} or carbon content of microbial biomass is usually not provided in published literatures. For example, as summered distribution coefficient shown in Table 8.1, 39 valid K_d for phenanthrene from 15 published papers were obtained, whereas only 16 valid K_{oc} for phenanthrene from 5 published papers were directly available or indirectly calculated. It will be appreciated for researchers and authors to provide K_d, K_{oc} as well as specific organic carbon content of microbial biomass in future published papers. As shown in Table 8.1, the log K_d values of PAHs cover a large range due to the difference of various biomasses. Taking phenanthrene for example, the obtained and calculated log K_d of phenanthrene valued with a wide range of 2.86–4.68 for biomass derived from bacteria, fungi, and algae. The average value and mean value of summarized log K_d of phenanthrene are 3.83 and 3.83, respectively. Among these data, 43.5% of reported log K_d values fall the range of 3.60–3.90. These values are much higher than those of soil-water system (1.48–2.30) (Yang et al., 2006; Zhou and Zhu, 2007) and are comparable or slightly higher than live tissues such as wood chips (3.40) (Chen et al., 2011), plant roots (3.32–3.70) (Zhu et al., 2007; Chen et al., 2011), and tender tea leaves (3.52–3.54) (Lin et al., 2007). As expected, they are lower than compositions of soil organic matter such as cuticle of plants (4.21–4.73) (Li and Chen, 2009) and humic acid (4.45) (Salloum et al., 2002). Therefore, microbial biosorption may play an important role in the fate and transport of PAHs in soils and surface water (Chen et al., 2010). As mentioned above, higher log K_d values are expected for PAHs with more aromatic rings and higher hydrophobicity (usually using log K_{ow} to represent the hydrophobicity of organic pollutants, where K_{ow} is the octanol-water distribution coefficient). For example, Zhang et al. (2018a) examined the biosorption of naphthalene, acenaphthene, and fluorene on bacterial biomass of *Klebsiella oxytoca* and the obtained log K_d followed the order of fluorene (3.40) > acenaphthene (3.24) > naphthalene (2.56), which is in accordance with their

log K_{ow} values (4.18, 3.92, and 3.30 for fluorene, acenaphthene, and naphthalene, respectively). Fungal biosorption of phenanthrene and pyrene using inactive biomass of *Phanerochaete chrysosporium* were conducted and distribution coefficient for pyrene was much larger than the one for phenanthrene (Ding et al., 2013). A similar result has been reported by Arias et al. (2017) using algal biomass of *Rhosomonas baltica*, with log K_d of 3.67 and 4.00 for phenanthrene and pyrene, respectively. Based on limited data, the values of log K_{oc} also expressed the same trend. As summarized in Table 8.1, the log K_{oc} values of five typical PAHs obtained by biosorption sets regardless of microbial types (bacteria, fungi, and algae) followed the order: pyrene (log K_{ow} = 4.88, log K_{oc}: 4.74–4.78) > phenanthrene (log K_{ow} = 4.46, log K_{oc}: 3.91–4.45) > fluorene (log K_{ow} = 4.18, log K_{oc} = 3.82) > acenaphthene (log K_{ow} = 3.92, log K_{oc} = 3.58) > naphthalene (log K_{ow} = 3.30, log K_{oc} = 2.92) (Chung et al., 2007; Chen et al., 2010; Zhang et al., 2013b; Gu et al., 2015; Arias et al., 2017). Actually, several studies have established correlations between log K_d (or log K_{oc}) and log K_{ow}. We will talk about this in Section 8.1.2.2.

8.2.1.1.2 As a Step and Relationship between Sorption and Degradation

Biosorption can be considered to effectively remove PAHs from aqueous environment with microbial biomass as low-cost biosorbents (Liu and Liu, 2008; Vijayaraghavan and Yun, 2008; Chojnacka, 2010; Bai et al., 2016). In addition, the sorption of PAHs by living microorganisms is widely considered a first step during the biodegradation process, followed by transmembrane process and intercellular metabolism (illustrated in Figure 8.2, modified based on Zhang and Zhu, 2014). In general, to uptake and utilize PAHs, there are at least three consecutive interfacial processes a microorganism needs to make progress, including cell surface (bio)sorption, transmembrane transport, and intercellular enzyme-related metabolism. Furthermore, transport of microbes and/or mass transport of target pollutants in environment may also be needed before the actual connect between organic compounds and microbes. As the first "real contact" to pollutant

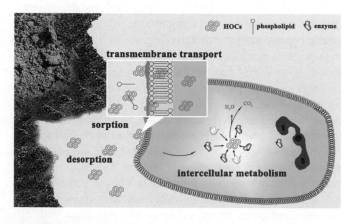

FIGURE 8.2 Schematic diagram of consecutive process during the biodegradation of PAHs in soil.

molecules, (bio)sorption is usually considered a crucial process during biodegradation and to provide readily acceptable "food" to the microbial cells. Stringfellow and Alvarez-Cohen (1999) suggested the sorption of PAHs on non-degrading bacterial strain would reduce the biodegradation rate by degrading strain. The rates of fluoranthene and pyrene degradation by degrading strain *Pseudomonas stutzeri* P-16 were reduced from 10.2 μM per g biomass per day and 4.2 μM per g biomass per day to 3.5 μM per g biomass per day and 1.7 μM per g biomass per day, respectively, in the presence of nondegrading strain *Rhodococcus rhodochrous* RR1. This result reminds us of the potential adverse effect of biosorption on biodegradation. However, it should not weaken the positive role of biosorption in biodegradation. First of all, in the study conducted by Stringfellow and Alvarez-Cohen (1999), the bioavailability of PAHs to degrading bacteria was reduced by the sequestration of PAHs by non-degraders, which actually acted as another biosolids phase similar to soil and/or sediment. Second, as mentioned above, the sorption affinity of bacterial biomass with PAHs is much higher than that of soil. With the dying and decomposing, large part of the non-degrader-sorbed PAHs could release into the aqueous phase. And then the released PAHs are probably easier to uptake by degraders, compared with direct uptake from soil-sorbed PAHs. Furthermore, for microorganisms or microbial community with high sorption abilities, the biosorption process will facilitate the intracellular biodegradation of PAHs confirmed by several recent studies (Chen et al., 2010; Zhang and Zhu, 2012; Zhang et al., 2013a; Hadibarata et al., 2014; Liao et al., 2015; Sanches et al., 2017; Ma et al., 2018; Meng et al., 2019).

During the biodegradation process, PAHs (such as phenanthrene and pyrene) are first adsorbed on the outer cell membrane or other cell surface part, which then facilitate the acceptability to and uptake by *Klebsiella oxytoca* (Zhang and Zhu, 2012; Zhang et al., 2018a), *Phanerochaete chrysosporium* (Ding et al., 2013; Gu et al., 2016), *Arthrobacter* sp. (Kallimanis et al., 2007), *Pseudomonas* sp. (Ma et al., 2018; Meng et al., 2019), and *Brevibacillus brevis* (Liao et al., 2015). The more PAHs sorbed on the cell surface, the more available PAHs for intracellular enzymes and thus the more degradation of PAHs occurs. Ma et al. (2018) analyzed the phenanthrene concentrations in cytochylema of a *Pseudomonas* strain, representing the directly degradable PAHs. A significant positive linear correlation ($C_{cyto} \propto 0.00267\ Q_{Ph6-envelops}$, $n = 5$, $R^2 = 0.962$) was exhibited between the phenanthrene contents in the cytoplasm (C_{cyto}) and phenanthrene contents on envelop (surface sorbed, $Q_{Ph6-envelops}$). The results validated that the increased phenanthrene biosorption did lead to the enhanced phenanthrene concentration in the cytoplasm. A similar relationship between biodegradation promotion and apparent partition coefficient of inactive biomass for pyrene was obtained (as shown in Figure 8.3). The parameters B^* and K_d^* represent biodegradation efficiency and apparent partitioning coefficients in the presence of surfactants with different types and concentrations, respectively. The positive correlation ($n = 41$, $P < 0.0001$) suggested surfactant-induced

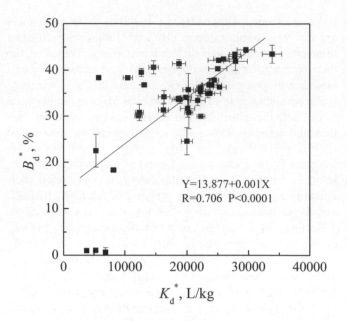

FIGURE 8.3 The relationship between apparent biodegradation efficiency (B^*) with apparent sorption affinity.

biosorption governed the surfactant-induced pyrene biodegradation. Biosorption and biotransformation mechanisms existed simultaneously during the removal of fluoranthene by the white-rot fungus *Pleurotus eryngii* F032 (Hadibarata et al., 2014). The major mechanism of fluoranthene elimination was the mineralization via enzymes such as laccase. However, a similar trend of fluoranthene biosorption versus incubation time was found with biotransformation kinetics and the presence of biosorption accelerated the elimination of fluoranthene.

8.2.1.2 Sorption Mechanisms

There are two main sorption mechanisms in the sorption of organic pollutants by microbial biomass, namely absorption and adsorption. (1) Absorption (or partition): hydrophobic interactions of the aliphatic and aromatic groups of a compound with the hydrophobic sites of biomass including biolipid chains, lipid fractions, and lipophilic cell membrane of the microorganisms. (2) Adsorption: interactions such as Van der Waals, dipole-dipole interactions, electrostatic forces, and weak intermolecular associations.

Partitioning was considered to be the primary sorption mechanism for PAHs by inactive biomass or live cells of microorganisms and the isotherm sorption curves usually fit well with the linear model (Ternes et al., 2004; Chen et al., 2010; Zhang and Zhu, 2012; Fernandez-Fontaina et al., 2013; Zhang et al., 2013a). Xu et al. (2013) investigated the sorption of petroleum hydrocarbons by a heat-killed microbial consortium and results inferred that the sorption of PAHs was a fast process and the sorption content was constant for each PAH (e.g., naphthalene, phenanthrene, and pyrene). They reproved that the biosorption mechanism for nonpolar organic compounds by dead biomass is partition (Xu et al., 2013). The partition/sorption sites on microbial surfaces mainly refer to hydrophobic groups. FTIR results indicated functional groups in bacterial surface provided effective binding sites for phenanthrene, including saturated aliphatic carbon (3000–2850 cm^{-1}, –CH$_3$, –CH$_2$–), unsaturated aliphatic carbon (near 980 cm^{-1}, C=C), and aromatic structure (1650 cm^{-1}) (Zhang et al., 2018b). In addition, bacterial cell envelop usually functions as a wrapped layer. The bacterial cell envelop mediates the adhesion of hydrophobic organic compounds to bacterial surface or the three-dimensional polymer network that interconnects intracellular systems by anchoring the hydrophobic groups on the cell surface (Ma et al., 2018).

In some cases, the isotherm sorption exhibits nonlinear and several interactions were proposed (Chan et al., 2006). A mechanism of cation-π interactions between the metal cations and PAHs was proposed by Zhu et al. (Xiao et al., 2007; Qu et al., 2008). In the interaction, PAHs function as electron acceptor and π donor, whereas the coexisting metal cations or amine groups function as electron donor. The binding of metal cation mainly attributes to the deprotonated carboxyl groups on the bacterial surface. The deprotonated carboxyl groups are imbedded within structures of hydrophobic subunits such as aliphatic chains, relieving the barrier of "desolvation penalty" (Qu et al., 2008) and facilitating the cation-π interactions. Similar results have been reported by several other groups (Tao et al., 2014; Chen et al., 2016). Tao et al. (2014) believed the cation-π interactions between the metal cations (e.g., Cu^{2+}, Cd^{2+}) and PAHs facilitated the sorption of PAHs by *Microcystics aeruginosa*. The metal cations were binding in the cell surface by carboxyl groups. Sorption improvement of the PAHs caused by the cation-π interactions increased with the softness order of the metals (Cd^{2+} < Cu^{2+} < Ag$^+$) and the π donor strength order of the PAHs (phenanthrene < pyrene < benzo(a)pyrene) (Tao et al., 2014).

The PAHs uptake process on porous carbon consists of three successive steps, and three main mechanisms have been proposed for PAH sorption: the H-bonding formation, the electron donor-acceptor interaction, and the π–π interaction (Yuan et al., 2010; Meng et al., 2016). The π–π and electron donor–acceptor interactions are thought to be major mechanisms for the sorption of planar structure PAHs (e.g., phenanthrene) on hyphae of fungi (e.g., *Phanerochaete chrysosporium*), and cell surface containing numerous chemical groups (e.g., –OH, –COO–, O–C=O, CO–NH) as well as high carbon content lipid vesicles acted as active sorption sites (Gu et al., 2015, 2016).

Physical entrapment is also a candidate mechanism for biosorption of PAHs. Entrapment of high molecular weight PAHs (HMW PAHs, e.g., benzo(a)pyrene, BaP for short) occurs via a "hole-filling" process formed by polysaccharides from cellulose materials (e.g., cell walls and chloroplast thylakoids) of the algal cells (Ke et al., 2010). Polysaccharides are reported to be weak domain for sorption of PAHs, being 400 times lower than octanol and up to 3 × 10^5 times lower than black carbon or coal (Jonker, 2008), due to low accessibility (Mackay and Gschwend, 2000). However, the presence of low molecular weight PAHs (LMW PAHs) can facilitate the swelling of the amorphous affinity for HMW PAHs (Ke et al., 2010).

In addition, some studies reported relatively high affinity was achieved even in the absence of LMW PAHs, with HMW PAHs physically trapped in the microfibrils of cellulose (Boki et al., 2007).

Specific electrostatic interactions usually were attributed to biosorption of many organic compounds such as PPCPs and dyes. These pollutants containing positively charged groups interacted with the negatively charged surfaces of the microorganisms caused by the presence of carboxyl group, hydroxyl group, and phosphate group (Golet et al., 2003; Siegrist et al., 2003; Zhang and Zhu, 2014). In addition, polar molecules highly depend on the chemical form(s) in particular solution conditions. Therefore, the pH of wastewater (usually between 7 and 9) or aqueous solutions will directly impact the ratio of non-ionized to ionized chemical form(s) of these compounds and thus biosorption property (Wells, 2006). However, since PAHs mainly contain negative or positive charges, specific electrostatic interactions are not a proper option for biosorption of PAHs.

8.2.2 Affecting Factors

In general, the sorption affinity of PAHs with microbial biomass depends on chemical nature of PAHs (e.g., log K_{ow}, S_w, and structure), co-existing substances, properties of microbial biomass (e.g., cell surface hydrophobicity), and practical conditions such as temperature, pH, ion strength, and contact time. These factors could also indirectly influence the subsequent biodegradation.

8.2.2.1 PAH Properties

As we mentioned above, chemical nature plays a crucial role in the sorption (including adsorption and absorption) of PAHs during biological treatment of PAHs-bearing solutions. Chemical properties such as hydrophobicity and hydrophilicity are considered to be a driving force in environmental transport (Wells, 2006). (1) Polar properties. Polar moieties are recognized as one of the key properties affecting the fate of organic pollutants in the environment. Parent PAHs are basically nonpolar organic pollutants, whereas some of the intermediates or derivatives (e.g., hydroxynapthalene, dihydroxyphenanthrene, 3,4-dimethyl phenanthrene, monohydroxylated BaP) are of high polarity and are easily dissolved in aqueous phase (Semple et al., 1999; Ke et al., 2010). However, parent PAHs are strongly tend to sorbed in biomass, as mentioned above, with high sorptive removal efficiency and large sorption affinity. (2) Hydrophobicity. Basically, PAHs with higher hydrophobicity are much easier to sorb on microbial biomass rather than in liquid phases. Octane-water distribution coefficient (K_{ow}) and the number of aromatic rings are commonly used parameters to represent hydrophobicity of PAHs. Fayeulle et al. (2014) reported a twelve times higher biosorption capacity for BaP (5-ring PAH, log K_{ow} = 5.97) on *Fusarium solani* than that for phenanthrene (3-ring PAH, log K_{ow} = 4.46). As we all know, the distribution coefficient (K_d) and carbon normalized distribution coefficient (K_{oc}) are considered as a good parameter to evaluate the sorption property of HOCs from aqueous solutions onto solid phase or matrix. The relationships between log K_d (or log K_{oc}) with log K_{ow} for tested organic pollutants on particular sorbents are often established to evaluate, illustrate, and/or even predict the biosorption behaviors of microbial biomass. For example, based on reasonable assumptions for nonspecific lipophilic interactions, several relationships between log K_d (or log K_{oc}) with log K_{ow} were obtained (the following equations), and could be used to predict sorption behaviors for various hydrophobic organic compounds and microorganism-derived biosorbents (Zhang et al., 2018a).

Bacterial biomass for PAHs: log K_d = 1.01 log K_{ow} − 0.74 (n = 7, R^2 = 0.9590) (Zhang et al., 2018a).

Bacterial suspensions for unsubstituted and methylated PAHs: log K_d = 0.98($\pm$0.36) log K_{ow} − 2.71($\pm$1.55) (n = 11, R^2 = 0.8100) (Dimitriou-Christidis et al., 2007).

White-rot fungi for PAHs: log K_{oc} = 1.13 log K_{ow} − 0.84 (n = 5, R^2 = 0.9960) (Chen et al., 2010).

Activated sludge for PAHs: log K_d = 1.52 log K_{ow} − 3.27 (n = 3, R^2 = 0.9201) (Meng et al., 2016).

Based on the established equations of log K_d − log K_{ow}, for example, log K_d = 1.01 log K_{ow} − 0.74 from biosorption of 7 PAHs on *Klebsiella oxytoca* (Zhang et al., 2018a), biosorption of HOCs to microbial biomass can be predicted. We compared several predicted K_d values for organic compounds to microorganisms-derived biosorbents/biomass with available sorption data from the literature (as illustrated in Figure 8.4).

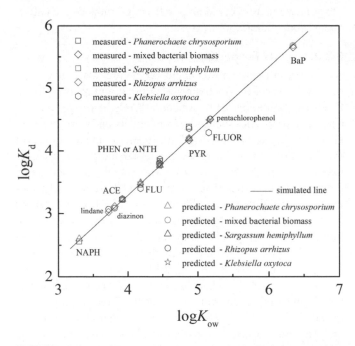

FIGURE 8.4 Predicted sorption parameter (K_d) onto various microorganisms using relationship (log K_d~log K_{ow}) established based on bacterial biomass. (Data obtained from Steen, W.C. and Karickhoff, S.W., *Chemosphere*, 10, 27–32, 1984; Tsezos, M. and Bell, J.P., *Water Res.*, 23, 561–568, 1989; Chung, M.K. et al., *Sep. Purif. Technol.*, 54, 355–362, 2007; Chen, B.L. et al., *J. Hazard. Mater.*, 179, 845–851, 2010; Zhang, D. et al., *Water*, 10, 675, 2018a. NAPH: naphthalene; ACE: acenaphthene; FLU: fluorene; PHEN: phenanthrene; ANTH: anthracene; PYR: pyrene; FLUOR: fluoranthene; BaP: benzo(a)pyrene. Simulated line was adapted from Zhang, D. et al., *Water*, 10, 675, 2018a.)

It showed that the predicted K_d values were pretty close to the measured ones, even for other types of HOCs such as lindane and pentachlorophenol. The results suggested that K_{ow} could be a useful parameter to predict the fate of PAHs for microorganisms.

8.2.2.2 Microbial Cell Surface Properties

Microbial cell surface is the first layer to contact and uptake organic substances. The cell surface is not a smooth capsule-like structure and consists of various hydrophobic and hydrophilic groups such as biolipid alkane chains and carboxyl and phosphate groups (Xiao et al., 2007; Qu et al., 2008; Zhang and Zhu, 2012). The biosorption of PAHs on microbial cells occurs related to cell surface components and their proportion, as well as the cell surface hydrophobicity (CSH). Simply, the more hydrophobic components the cell surface has, the larger CSH is, and the higher affinity between cell surface and PAHs is.

PAHs, as hydrophobic organic and nonpolar compounds, prefer to sorb on nonpolar areas (e.g., lipid) through hydrophobic interactions. Zhang et al. (2013b) established a multivariate correlation among the log K_{oc}, lipid contents, and C/N ratios. A positive correlation between log K_{oc} and lipid contents and negative correlations between log K_{oc} and C/N ratios as well as (N+O)/C ratios was observed, indicating the important role of lipid in the overall biosorption of PAHs by algal samples. The authors also believed that polar components, illustrated by (N+O)/C ratio, were responsible for the decreasing biosorption capacity.

Cell surface hydrophobic plays important roles in the microbial growth and biodegradation activity, and also can regulate the interfacial interactions between cell and HOC-molecules (Zhong et al., 2008; Zhang and Zhu, 2012). In general, strains with higher CSHs tended to higher sorption ability for PAHs, facilitating their ability of PAHs degradation (Stringfellow and Alvarez-Cohen, 1999). During the biodegradation of PAHs, the CSH measured by BATH (bacterial adherence to hydrocarbon) method became more hydrophilic, while the surface adsorption by living cells slightly decreased (Xu et al., 2013). In a recent study, significant linear relations were primarily found between CSH and $K_{c/w}$ and between CSH and $K_{e/w}$, where $K_{c/w}$ and $K_{e/w}$ were the sorption coefficient of phenanthrene by *Pseudomonas* sp. Ph6 cells and Ph6 envelops, respectively (Ma et al., 2018). The increased CSH facilitated the biosorption and acceptability of biodegraders for HOCs (e.g., phenanthrene). This can be explained that the CSH depends on the ratio of hydrophobic groups to hydrophilic groups on the cell surface. Therefore, increased CSH indicates more hydrophobic groups appear on the cell surface, which is benefit for the biosorption of hydrophobic PAHs.

8.2.2.3 Coexisting Substances

The sorption of PAHs by microbial biomass mainly depends on the interfacial interactions between microbial cell and target pollutants. Coexisting substances or other pollutants could act as competitors or inducers by affecting the cell surface properties, liquid phase conditions, and by interacting directly with PAHs. Herein, we mainly discuss two typical types of coexisting chemicals, namely heavy metals and surfactants. Heavy metals are of common co-occurrence with PAHs in contaminated environments such as manufactured gas plant sites, railway yards, and petrol stations (Thavamani et al., 2011, 2012; Bayen, 2012). Surfactants are often used to enhance the bioavailability and microbial biodegradation of PAHs, recognizing as surfactant-enhanced bioremediation (SEBR) (Kile and Chiou, 1989; Zhao et al., 2005; Yu et al., 2007; Zhu et al., 2010). Interactions among surfactant, PAHs, microbes, and environmental media (e.g., soil) play a crucial role in SEBR.

8.2.2.3.1 Heavy Metal Ions

The presence of heavy metals could have negative, positive, or no effect on the biosorption of PAHs mainly depending on the sorption mechanism and toxicity to microorganisms.

Zhu et al. proposed a mechanism of cation-π interactions between the metal cations and PAHs (Xiao et al., 2007; Qu et al., 2008). According to their hypothesis, the coexisting of heavy metals enhanced the sorption of organic compounds (e.g., phenanthrene), and the sorption amount of phenanthrene on *Escherichia coli* increased with the increasing of cation concentrations (Xiao et al., 2007). The presence of heavy metals (Ag^+, Cu^{2+}, and Fe^{3+}) increased the sorption distribution coefficient from 3900 L kg^{-1} to 5900, 5500, and 4600 L kg^{-1}, respectively. Complexation of transition metals with the deprotonated functional groups (mainly carboxyl) of bacterial cell walls neutralized the negative charge in cell surface, resulting in the bacterial surface being less hydrophilic and partition enhancement of phenanthrene (Xiao et al., 2007). These mechanisms were proposed: cation-π interactions with protonated amines and π H-bonding with pronated carboxyls. Other studies also reported similar results. The presence of heavy metals such as copper (Cu^{2+}), cadmium (Cd^{2+}), and silver (Ag^+) influenced the sorption of phenanthrene, pyrene, and BaP onto *Microcystis aeruginosa* (Tao et al., 2014). With the increasing of heavy metal concentrations, the sorption parameters increased initially and then decreased. The heavy metal facilitated sorption of PAHs on biomass was mainly attributed to the cation-π interactions between the metal cations and PAHs.

Adverse effects are also reported mainly due to competition and toxicity to microorganisms. The uptake of naphthalene by dead algal cells decreased significantly with the addition of metals such as nickel and copper ions, salts (e.g., NaCl), and chelating agents (e.g., citric acid), mainly due to competition for the adsorption sites on the surface (Ashour et al., 2008). Basically, the presence of Fe^{3+}, Zn^{2+}, Cu^{2+}, Pb^{2+}, Ni^{2+}, and Cd^{2+} resulted in negative effects on lipid accumulated (sorbed) and biodegraded anthracene, as well as the specific anthracene uptake rate and specific lipid accumulation rate (Goswami et al., 2017). In addition, the specific lipid accumulation rate correlated with the specific anthracene uptake rate in the absence and presence of heavy metals, and the result indicated sorption was essential for the biodegradation of PAHs.

Heavy metals can affect the partition uptake of PAHs resulting in different influences. Different heavy metals may show different influence on the biosorption and biodegradation (or removal) of PAHs by a heavy-metal resistance fungi *Acremonium* sp. P0997. The presence of Cu^{2+} led to an increase of biosorption of anthracene, whereas the co-existing Mn^{2+} resulted in negative effects on the biosorption and removal of anthracene (Ma et al., 2014). However, both positive and negative effects of heavy metal ions on the removal of PAHs were taken place through their effect on the partition process of PAHs between water and mycelia. Furthermore, these results indicated that the biosorption by microorganisms played an important role in the fate of PAHs even in the presence of heavy metals (Ma et al., 2014). Nevertheless, binding of copper with carboxyl groups on cell walls of *Stenotrophomonas maltophilia* did not obviously affect the uptake of BaP (Chen et al., 2016).

8.2.2.3.2 Surfactants

Surfactants are the most commonly used amphiphilic agents in industry, domestic, and environmental remediation. The surfactant can occupy, conceal, replace, and/or release the hydrophobic (or hydrophilic) sites in the cell surface, modifying the interactions between bacterial surface and pollutants. The interaction of bacterial surface with surfactant occurs via hydrophobic interaction, electrostatic interaction, π-electronic polarization, as well as H-bonding (Paria and Khilar, 2004; Yuan et al., 2007). The surfactant-induced change of cell surface property can modify the microbial biosorption of PAHs. In addition, the presence of surfactant micelles in aqueous phase will also change the sorption potential for PAHs onto biomass. The effects of surfactant on the biosorption of PAHs depend on the type and concentration of surfactants. For example, as illustrated in Figure 8.5, with the increasing of surfactant concentrations, the K_d^* tended to increase initially and then decrease. The most significant promotion appeared at their CMC, which were 1.27, 1.33, and 1.52 times higher than the surfactant-free control for Tween 20, Tween 40, and Tween 80, respectively. The promotion seems to positively relate to carbon chain length of surfactant (12, 16, and 18 for Tween 20, Tween 40, and Tween 80, respectively), and negatively relate to the hydrophile-lipophile balance (HLB) values (16.7, 15.6, and 15.0 for Tween 20, Tween 40, and Tween 80, respectively). Similar results were reported in other studies. Ma et al. (2018) demonstrated the same trends, where the sorption affinity for phenanthrene by *Pseudomonas* sp. Ph6 initially increased and then decreased at the rhamnolipid concentration range of 0–400 mg L^{-1}. Meng et al. (2019) investigated the effects of Tween 80 (ranging from 0 to 132 mg L^{-1}) on the sorption of phenanthrene and pyrene and showed increased sorption amount of both PAHs on *Pseudomonas synxantha* LSH-7'.

Modification of cell surface property (e.g., CSH) is responsible for the surfactant-mediated biosorption of organic pollutants on microbial surfaces. The presence of surfactants

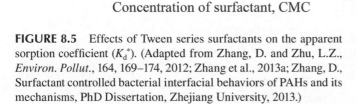

Concentration of surfactant, CMC

FIGURE 8.5 Effects of Tween series surfactants on the apparent sorption coefficient (K_d^*). (Adapted from Zhang, D. and Zhu, L.Z., *Environ. Pollut.*, 164, 169–174, 2012; Zhang et al., 2013a; Zhang, D., Surfactant controlled bacterial interfacial behaviors of PAHs and its mechanisms, PhD Dissertation, Zhejiang University, 2013.)

may either increase or reduce the CSH of microorganisms, depending on the initial surface hydrophobicity: increase for the hydrophilic cultures and decrease for the hydrophobic strains (Zhang and Zhu, 2014). Available evidence have been summarized and listed in Table 8.2. Owsianiak et al. (2009) analyzed the influence of surfactants (rhamnolipid and TX100, 150 mg L^{-1}) on modification of CSH. Generally, it could be stated that both surfactants decreased the CSH of strains with high hydrophobicity and increased for those with low CSH. It's pretty hard to identify a precise boundary for hydrophobic and hydrophilic cells. However, generally, strains with CSH less than 20% can be considered as hydrophilic cells, and hydrophobic strains can be identified with CSH higher than 40%. The hydrophilic strain *Citrobacter* sp. SA01 (CSH ranges from 10.5% to 11.4%) led to significant increase of CSH with the addition of both Tween 80 and SDBS at the concentration range of 0 to 100 mg L^{-1} (Li and Zhu, 2012). On the other hand, for strains with high CSH, for example, *Pseudomonas aeruginosa* ATCC 15442 (CSH 55%) and *P. aeruginosa* ATCC 27853 (CSH 74%) used by Zhang and Miller (1994), the presence of rhamnolipid (0–0.008 mmol L^{-1}) resulted in rapid decreases of CSH.

The modification of CSH caused by surfactants mainly occurs through two ways referring surfactant-cell interactions (Zhang and Zhu, 2014). The first one is the sorption of surfactant on the cell surface (Owsianiak et al., 2009; Zhang et al., 2013a). The orientation of sorbed surfactants determines the degree of CSH variation. For microorganisms with low CSH, the hydrophilic sites on cell surface can be replaced by surfactants exposing the surfactant hydrophobic tails outside. For microbial strains with high CSH, hydrophobic interactions between hydrophobic sites of cell surface and hydrophobic tails of surfactants easily take place, via monomer adsorption (Noda and Kanemasa, 1986). Thus, the exposing hydrophilic head of surfactants to aqueous phase leads to significant decrease of CSH.

The other interactions to change CSH by surfactants is that surfactant induces the release of outer membrane components (primarily lipopolysaccharides, LPS) for Gram negative bacteria, resulting in an increased CSH (Al-Tahhan et al., 2000; Li and Zhu, 2012). Three possible mechanisms for the removal of LPS have been summarized by Zhang and Zhu (2014): (1) Direct removal of LPS or O-antigen part of LPS through surfactant micellar capture, leaving hydrophobic lipid A of LPS or hydrophobic sites in outer membrane (Zhao et al., 2011; Caroff and Karibian, 2003). (2) Indirect release by formation of complex of surfactant with Mg^{2+} in the outer membrane. The absence of Mg^{2+} wrecks the strong LPS-LPS interaction (Al-Tahhan et al., 2000). (3) Modify the structure of outer membrane protein, which controls the synthetization and secretion of LPS (Bos et al., 2004).

8.2.2.4 Other Factors

Temperature increasing may lead to damage of active binding sites in the microbial biomass, therefore resulting in the decrease of sorption affinity of microorganisms for PAHs (Gu et al., 2015). Decreased distribution coefficients (K_d) of phenanthrene on *Klebsiella oxytoca* and natural and modified biomass of *Phanerochaete chrysosporium* were reported (Gu et al., 2015; Zhang et al., 2018a). In addition, the thermodynamic parameters (e.g., ΔG and ΔS) implied the biosorption process was spontaneous and exothermic.

Incubation time would influence the surface property (e.g., CSH) of microorganisms. Zhang and Miller (1994) reported a continuous increase of CSH of *Pseudomonas aeruginosa* ATCC 9027. During the biodegradation process, the contribution of biosorption may change due to the surface property changing. In the bioremoval of pyrene by *Klebsiella oxytoca* PYR-1, up to 89% of the removed pyrene was attributed to the sorption at day 5 (Zhang and Zhu, 2012). However, the sorption contribution decreased to 57.8% at day 18 (Zhang and Zhu, 2012), partly due to the decrease of CSH along with the incubation (Ma et al., 2018).

Different substrates (or carbon sources) may also modify the surface property of microorganisms, and subsequently influence the biosorption of PAHs. For example, the K_d for phenanthrene sorption by *Pseudomonas aeruginosa* CV1 was slightly higher when grown on phenol compared to peptone. In addition, even though grown in different concentrations of the same media (e.g., peptone), the K_d for phenanthrene sorption by *Pseudomonas stutzeri* P-16 was slightly different (Stringfellow and Alvarez-Cohen, 1999).

TABLE 8.2

Effects of Surfactants on Bacterial Cell Surface Hydrophobicity

Surfactants	Concentration	Microorganisms (CSH)[a]	Substrates	Effects[b]	References
Rhamnolipids	20–400 mg L^{-1}	*Bacillus subtilis* BUM (73.5%)	Phenanthrene	−	Zhao et al. (2011)
Rhamnolipids	20–400 mg L^{-1}	*Pseudomonas aeruginosa* P-CG3 (24.7%)	Phenanthrene	+	
Tween 80	0.24 mmol L^{-1}	Two *Mycobacterium* (both hydrophobic)[c]	Fluoranthene	−	Willumsen et al. (1998)
Tween 80	0.24 mmol L^{-1}	Two *Sphingomonas* (both hydrophilic)[c]	Fluoranthene	+	
TX100, Tween 80, two rhamnolipids	0.25–3 CMC[d]	*Bacillus* sp. 15UM (hydrophobic)[c]	Phenanthrene	−	Wong et al. (2004)
TX100, rhamnolipid, saponin	120 mg L^{-1}	*Aeromonas hydrophila* (CSH 7%)	Diesel oil	+	Kaczorek et al. (2010)
Tween 80	0–8 CMC	*Klebsiella oxytoca* PYR-1 (11.7%)	Pyrene	+	Zhang and Zhu (2012)
Tween 20, Tween 40	0–8 CMC	*Klebsiella oxytoca* PYR-1 (11.7%)	Pyrene	+	Zhang et al. (2013a)
Rhamnolipid	6 mmol L^{-1}	*Pseudomonas aeruginosa* ATCC 9027 (CSH 20%)	Glucose, hexadecane	+	Al-Tahhan et al. (2000)
Rhamnolipid	6 mmol L^{-1}	*Pseudomonas aeruginosa* ATCC 27853 (CSH 10%)	Glucose, hexadecane	+	
Rhamnolipid	0–1 mmol L^{-1}	*Pseudomonas aeruginosa* CCTCC AB93072 (CSH nearly 0)	Glucose	+	Zhong et al. (2008)
Rhamnolipid	0–1 mmol L^{-1}	*Pseudomonas aeruginosa* CCTCC AB93072 (CSH 30%)	Glucose	−	
Tween 80, SDBS	0–80 mg L^{-1}	Citrobacter sp. SA01[c]	Phenanthrene	+	Li and Zhu (2012)
Rhamnolipid	0–0.008 mmol L^{-1}	*Pseudomonas aeruginosa* ATCC 9027 (CSH 27%), *P. aeruginosa* NRRL 3198 (CSH 40%), *P. aeruginosa* ATCC 15442 (CSH 55%), *P. aeruginosa* ATCC 27853 (CSH 74%)	Octadecane	−	Zhang and Miller (1994)

[a] CSH, cell surface hydrophobicity.
[b] + represents positive effect; − represents negative effect.
[c] Didn't give CSH values.
[d] CMC, critical micelle concentration.

8.3 TRANSMEMBRANE MECHANISMS OF PAHs

The cell membrane consists of a phospholipid bilayer and membrane proteins (e.g., transport proteins, specific enzymes) along with oligosaccharides (Singer and Nicolson, 1972; Helenius and Simons, 1975; Yeagle, 2005). The cell membrane performs as a natural protective barrier and plays a number of essential roles in normal cellular activities such as nutrient transport, ion conduction, signal transduction, synthesis, and secretion of specific extracellular enzymes (Green et al., 1980; Carrière and Legrimellec, 1986; Rauchova et al., 1999; Pande et al., 2005; Song et al., 2010). One of the first physiological demonstrations of the existence of membrane is the semipermeable property of cell envelopes (Al-Awqati, 1995). This fact indicates that not all substances can cross the membrane easily. Moreover, in the case of Gram negative bacteria, hydrophilic LPS present out of the outer membrane (Sikkema et al., 1995). Chemical molecules, especially hydrophobic organic compounds (e.g., PAHs), transport across the cell membrane (as a "barrier") with different one-way flux through several mechanisms. These main hypotheses for HOCs transport into cells include the active transport mechanism (Fayeulle et al., 2014), the passive transport mechanism (Verdin et al., 2005; Gu et al., 2016), and the endocytotic internalization-like mechanism (Luo et al., 2013; Fayeulle et al., 2014). No matter the transmembrane transport of HOC molecule is driven by which mechanism, the transport follows two routes including through the lipid bilayer and various protein channels. Assuming that a given chemical contaminant has some degree of water solubility with a certain concentration in aqueous phase, the obstacle to biological degradation would be the accessibility for the pollutants to enzymatic machinery of the cell (might be the real bioavailability) (Bressleer and Gray, 2003). The rates at which organic pollutant molecules transport across membrane are a thousand to a million times smaller than the diffusion rates of the same molecules through a water layer of equal thickness (Stein, 1967). Thus, their transmembrane transport is the rate-limited process in the biodegradation kinetics, and practical evidence have been reported in several studies (Dimitriou-Christidis et al., 2007; Li and Zhu, 2014; Gu et al., 2016; Ma et al., 2018). Therefore, the membrane-crossing rate (flux, J) is one of the most important parameters we focus on. Generally, the flux depends on the target contaminant concentration outside the cell membrane (Vander et al., 2001), as well as the type and structure of protein channels.

8.3.1 Transmembrane Route

There mainly are two routes for organic compounds transport across the cell membrane into the cytoplasmic environment, regardless of transmembrane mechanism, namely through lipid bilayer and protein channels. The cytoplasmic membrane lipids of bacteria cells play an essential role in the function of transmembrane processes (Green et al., 1980; Carrière and Legrimellec, 1986). The permeation involved lipid bilayer allows nonpolar HOCs' diffusion through membrane to be possible. Integral membrane proteins can span the lipid bilayer, some of which can form channels allowing ions as well as organic compounds to diffuse across the membrane (Vander et al., 2001).

8.3.1.1 Diffusion through the Lipid Bilayer

Phospholipid, as the form of bilayer, is one of the most important components to form the cell membrane. The amphiphilic property of phospholipids, consisting of a polar hydrophilic head group and two nonpolar hydrophobic "tails," divides the lipid bilayer into three unique polarity domains. The polar heads orient toward the outside of both sides of the membrane (favorable to polar water) and form two considerable fairly polar domains. The nonpolar and hydrophobic central region of the bilayers spontaneously formed due to the fatty acid moieties of the phospholipids. The polar-nonpolar-polar structure of lipid bilayers makes most polar chemicals difficult or even impossible to enter through (Stein, 1967), whereas nonpolar organic compounds (e.g., parent PAHs) can diffuse across plasma membranes more rapidly. The reason is that nonpolar molecules can be dissolved in the central nonpolar regions of membrane, which is occupied by nonpolar fatty acid chains. In general, nonpolar organic compounds with high

octane-water distribution coefficient (log K_{ow}) readily partition into regular membrane lipid bilayers and easily enter into the internal matrix. For example, the release of lipid bilayer-captured pyrene (log K_{ow} = 4.88) was conducted using liposome (as experimental models for membranes due to simple and good mimic of biological membranes) prepared by lecithin (phosphatidylcholine) (Zhang et al., 2013a). The released concentration of pyrene could eventually reach around 0.1 mg L^{-1} (water solubility of pyrene is 0.12 mg L^{-1} at 25°C), though it seemed to be a time-consuming process. Some of polar contaminants could sorb on the membrane surface; however, a small part of sorbed pollutants can enter into the membrane. The transmembrane transport of two antibacterial drugs (kanamycin and chloramphenicol) with different hydrophobicity was compared using lecithin liposome (Song et al., 2010). Relatively strong sorption of both drugs onto liposome was found; however, it showed totally different permeability through the lipid bilayer. More than 89% of the adsorbed hydrophilic kanamycin (log K_{ow} = −6.70) was located on the outer surface of the liposome, but over 80% of the sorbed weak hydrophobic chloramphenicol (log K_{ow} = 1.14) entered the internal matrix.

8.3.1.2 Transport through the Transmembrane Protein Channels

Integral membrane proteins provide diffusion channels for chemicals unable to diffuse through the lipid bilayer, known as mediated-transport systems. Movement of substances across the membrane through the mediated-transport systems requires conformational changes in these transporters (basically the integral membrane proteins). Two possible transport mechanisms can be envisioned.

1. "Classical" integral protein transport. The hatch undergoes conformational changes to create a transient channel, where hydrophobic molecules can transport from the extracellular medium directly to the aqueous periplasm (van den Berg et al., 2004; Karuppiah et al., 2011). The transporter is open outwards, and the transported chemical molecules must first bind to a specific site on a transporter. The binding of solute triggers the conformational change, getting on the transporter at the intercellular side and off at the other side. Then, the dissociation of the substance from the transporter binding site completes the process to move the organic pollutant through membrane.

2. Alternative "lateral diffusion" transport (Hearn et al., 2009; Martínez et al., 2013). In the case of Gram-negative bacteria (as one of the most important pools of biodegraders for PAHs), the outer membrane associated with cytoplasmic membrane acts as an effective barrier for the passage of hydrophobic molecules, owing to the presence of the polar and hydrophilic LPS layer on the outside of the bacterial cells (Hearn et al., 2009; Martínez et al., 2013). The transport of hydrophobic

contaminants through LPS layer is energetically unfavorable. The release of outer membrane components (primarily LPS) for Gram negative bacteria led to an increased CSH (Al-Tahhan et al., 2000; Li and Zhu, 2012) and improvement of phenanthrene uptake by a *Citrobacter* strain (Li and Zhu, 2012). The results indirectly indicated LPS functions as a barrier for hydrophobic molecules. Fortunately, microorganisms (e.g., bacteria) capable of utilizing and degrading hydrophobic substrates evolve strategy to avoid the hydrophilic layer of LPS. This hydrophobic passageway for HOCs diffusion actually is an outer membrane protein with lateral opening, allowing hydrophobic molecules to get through the barrel wall (LPS layer) to move into the outer membrane and then diffuse into the periplasm. To open the passageway, a detergent-like molecule bounded lateral opening can be trigged. The lateral opening proteins are located in the region of the polar-nonpolar interface of the outer leaflet of the outer membrane. A brief schematic diagram is illustrated in Figure 8.6. To date, several outer membrane proteins (Omp) containing lateral opening for transport hydrophobic molecules have been observed and characterized, including *OprG* (Touw et al., 2010), *AlkL* (Julsing et al., 2012), *NapQ* (Eaton, 1994), *FhaC* (Karuppiah et al., 2011), *YrbC* (Malinverni and Silhavy, 2009), and *FadL* family (Hearn et al., 2009; Wiener and Horanyi, 2011; Martínez et al., 2013).

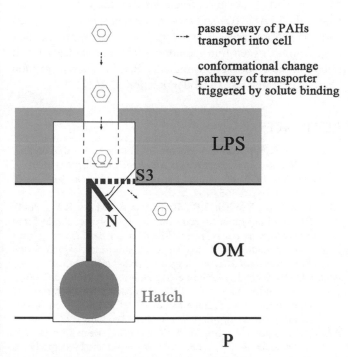

FIGURE 8.6 Schematic diagram of outer membrane protein with lateral opening. (Modified from Hearn, E.M. et al., *Nature*, 458, 367–370, 2009.)

8.3.2 THEORETICAL TRANSMEMBRANE FLUX

The amount of material (target chemical molecules) crossing a surface (cell membrane) in a unit of time is known as a flux, which can reflect the transmembrane rate. The membrane-crossing flux can be simply evaluated as following (Stein, 1967; Bressleer and Gray, 2003):

$$J = k_m \left(C_L^{out} - C_L^{in} \right) \tag{8.3}$$

where J is the flux of substrate (hydrophobic contaminant) across the membrane; C_L^{in} and C_L^{out} are intracellular and external concentration of target pollutant, respectively; k_m is the bilipid membrane permeability constant.

Considering the case where transmembrane transport is the rate-limited process in biodegradation, the active intracellular enzyme can significantly reduce the intracellular concentration of target compound to a relative low level. In some cases, part of the intercellular PAHs is not immediately degraded by active enzymes. For example, transport of BaP into fungus appeared first in hyphae apex and fungal branching areas (which are involved in fungal growth and among the most metabolically active areas) and then all the hyphae including lipid bodies and endocytosis vacuoles without being degraded (Fayeulle et al., 2014). However, most of these un-degraded PAHs are mainly trapped within lipid tissues such as vesicles (Verdin et al., 2005) and PHBs (Li et al., 2014), and act as "inactive" concentration. Therefore, we can propose the intercellular concentration of contaminant to be nearly zero ($C_L^{in} \approx 0$). The external concentration (C_L^{out}) represents the compound concentration in the aqueous phase outside of the membrane (C_{aq}). It is assumed that the aqueous concentration of pollutant remains essentially constant, because the extracellular fluid is much larger than the intracellular volume. The bilipid membrane permeability constant (k_m) has a positive correlation with oil-water partition coefficient ($K_{oil\text{-}water}$) and is negatively related with the square root of molecular weight of target compound. The $K_{oil\text{-}water}$ can be estimated from the octanol-water distribution coefficient (K_{ow}) (typically, $K_{oil\text{-}water} = 0.1\ K_{ow}$). The k_m can be written as the following:

$$k_m = 0.003 K_{ow}\ MW^{-0.5} \tag{8.4}$$

where K_{ow} is the octanol-water distribution coefficient of target compound, MW is the molecular weight of target chemical, and 0.003 is the empirical constant.

From Equations (8.3) and (8.4), the maximum membrane flux can be estimated as the following equation (Bressleer and Gray, 2003):

$$J_{max} = 0.003 K_{ow}\ C_{aq}\ MW^{-0.5} \tag{8.5}$$

Based on Equation (8.5), Bressleer and Gray (2003) demonstrated a pretty good correlation between maximum aerobic degradation rate and maximum membrane flux of 16 chemicals including aromatics, chlorine compounds, and oxygenated compounds. They suggested that a broad trend of maximum possible bioremediation rate could be correlated with an estimation of membrane flux, calculating from the basic physicochemical properties of contaminants. A similar trend was provided by Dimitriou-Christidis et al. (2007). A positive correlation was found between specific uptake rate and theoretical maximal transmembrane flux of 20 unsubstituted and methylated PAHs (excluded 2,3,5-trimethylnaphthalene). Results indicated membrane transport acted as the rate-determining process.

8.3.3 Relationship between Biosorption and Transmembrane Rate

Cell surface sorption, transmembrane transport, and intercellular enzyme-related degradation of PAHs act as interfacial behaviors during biodegradation, as well as consecutive physiological processes. The previous step (e.g., sorption) will definitely affect the follow-up process (e.g., transmembrane process). Theoretical correlations between sorption and transmembrane flux, as well as practical evidence, could provide support this hypothesis.

For hydrophobic PAHs, a considerable amount of compounds are accumulated (sorbed) on the surface of microorganisms (see Section 8.1.1). The direct available external concentration (C_L^{out}) in Equation (8.3) actually should be the sorbed concentration (C_{ad}) rather than the aqueous concentration (Zhang, 2013). Therefore, the C_L^{out} can be written as the following:

$$C_L^{out} = C_{ad} = K_d C_{aq} \qquad (8.6)$$

where K_d is the sorption affinity coefficient and C_{aq} is the aqueous concentration of target PAHs.

From Equations (8.3), (8.4), and (8.6), the apparent maximal membrane flux of PAHs can be estimated as the following equation (Stein, 1967; Bressleer and Gray, 2003; Dimitriou-Christis et al., 2007; Zhang, 2013):

$$J_{max}^* = 0.003 K_{ow} \, K_d \, C_{aq} \, MW^{-0.5} \qquad (8.7)$$

where K_{ow} is the octanol-water distribution coefficient of target compound, MW is the molecular weight of target chemical, and 0.003 is the empirical constant.

According to Equation (8.7), the apparent maximal membrane flux is positively related to surface sorption parameter (K_d), indicating sorption process plays an important role in the controlling of the transmembrane rate. Unfortunately, to date, limited data is available to validate the deduction by evaluating the relationship between examined degradation rate of PAHs (k_b) and apparent transmembrane flux (J_{max}^*) calculated from sorption data (K_d). It is strongly encouraged and appreciated for researcher to provide sorption and biodegradation parameters in their further studies.

On the other hand, practical evidence have been reported to approve the relationship between sorption and the transmembrane process. Ma et al. (2018) examined the phenanthrene content in the cytoplasm and to the envelope of *Pseudomonas* sp. Ph6, and a significant positive linear correlation was found.

Results indicated that the more phenanthrene transported into inside of cell with respect to higher sorbed amount of phenanthrene out of cell surface. The sorbed phenanthrene by the outer cell membrane acted as the reservoir for the diffusion of phenanthrene into the cytoplasm, which increased the possibility and rate of phenanthrene to transport across cell membrane. In another recent work, higher phenanthrene biosorption outside the cells of *Phanerochaete chrysosporium* further indicated the higher active efflux and passive transport of phenanthrene by *P. chrysosporium* (Gu et al., 2016). Moreover, inhibition and toxicological tests indirectly suggested the transmembrane transport related to biosorption process. For example, higher toxicity of two well-known inhibitors (colchicine and cytochalasin B) to mycelial pellets was conducted compared to spores of *P. chrysosporium*, which might contribute to higher biosorption and lower active efflux in mycelial pellets than those in spores.

8.4 SUMMARY OF THE ROLES OF SORPTION IN THE TRANSMEMBRANE TRANSPORT AND BIODEGRADATION OF PAHs

The sorption, transmembrane process, and intercellular enzymatic degradation of PAHs can be recognized as successive steps during biodegradation process containing several interfacial interactions. The cell surface sorption of PAHs will provide more available substance for the subsequent process. The biosorption of PAHs onto cell surface acts as a substrate reservoir and facilitates the transmembrane transport into intracellular cytoplasm. The sorption and accelerated transmembrane rate of PAHs are benefit for the intercellular enzyme-related degradation of PAHs due to sufficient substance providing. Positive correlations are found between surface sorbed PAHs (Q) and biodegradation rate (k_b), between sorption parameter (K_d) and apparent transmembrane flux (J_{max}^*), and between transmembrane flux and biodegradation rate (k_b). These correlations indicate that, in general, sorption plays an essential role in biodegradation of PAHs.

REFERENCES

Aksu Z., 2005, Application of biosorption for the removal of organic pollutants: A review, *Process Biochemistry* 40, 997–1026.

Al-Awqati Q., 1995, Membrane permeability, *Current Opinion in Cell Biology* 7, 463–464.

Al-Tahhan R.A., Sandrin T.R., Bodour A.A., Maier R.M., 2000, Rhamnolipid-induced removal of lipopolysaccharide from *Pseudomonas aeruginosa*: Effect on cell surface properties and interaction with hydrophobic substrates, *Applied and Environmental Microbiology* 66, 3262–3268.

Albert Q., Leleyter L., Lemoine M., Heutte N., Rioult J.P., Sage L., Baraud F., Garon D., 2018, Comparison of tolerance and biosorption of three trace metals (Cd, Cu, Pb) by soil fungus *Absidia cylindrospora*, *Chemosphere* 196, 386–392.

Arias A.H., Souissi A., Glippa O., Roussin M., Dumoulin D., Net S., Ouddane B., Souissi S., 2017, Removal and biodegradation of phenanthrene, fluoranthene and pyrene by the marine algae *Rhodomonas baltica* enriched from North Atlantic Coasts, *Bulletin of Environmental Contamination and Toxicology* 98, 392–399.

Ashfaq M., Li Y., Wang Y.W., Chen W.J., Wang H., Chen X.Q., Wu W., Huang Z.Y., Yu C.P., Sun Q., 2017, Occurrence, fate, and mass balance of different classes of pharmaceuticals and personal care products in an anaerobic-anoxic-oxic wastewater treatment plant in Xiamen, China, *Water Research* 123, 655–667.

Ashour I., Abu Al-Rub F.A., Sheikha D., Volsky B., 2008, Biosorption of naphthalene from refinery simulated waste-water on blank alginate beads and immobilized dead algal cells, *Separation Science and Technology* 43, 2208–2224.

Ata A., Nalcaci O.O., Ovez B., 2012, Macro algae *Gravilaria verrucosa* as a biosorbent: A study of sorption mechanisms, *Algal Research* 1, 194–204.

Baboshin M., Akimov V., Baskunov B., Born T.L., Khan S.U., Golovleva L., 2008, Conversion of polycyclic aromatic hydrocarbons by *Sphingomonas* sp VKM B-2434, *Biodegradation* 19(4), 567–576.

Bai L.L., Xu H.C., Wang C.H., Deng J.C., Jiang H.L., 2016, Extracellular polymeric substances facilitate the biosorption of phenanthrene on cyanobacteria *Microcystis aeruginosa*, *Chemosphere* 162, 172–180.

Bano A., Hussain J., Akbar A., Mehmood K., Anwar M, Hasni M.S., Ullah S., Sajid S., Ali I., 2018, Biosorption of heavy metals by obligate halophilic fungi, *Chemosphere* 199, 218–222.

Bayen S., 2012, Occurrence, bioavailability and toxic effects of trace metals and organic contaminants in mangrove ecosystems: A review, *Environmental International* 48, 84–101.

Blair B., Nikolaus A., Hedman C., Klaper R., Grundl T., 2015, Evaluating the degradation, sorption, and negative mass balances of pharmaceuticals and personal care products during wastewater treatment, *Chemosphere* 134, 395–401.

Bokbolet M., Yenigun O., Yucel I., 1999, Sorption studies of 2,4-D on selected soils, *Water, Air, & Soil Pollution* 111(1), 75–88.

Boki K., Kadota S., Takahashi M., Kitakouji M., 2007, Uptake of polycyclic aromatic hydrocarbons by insoluble dietary fiber, *Journal of Health Science* 53, 99–106.

Bos M.P., Tefsen B., Geurtsen J., Tommassen J., 2004, Identification of an outer membrane protein required for the transport of lipopolysaccharide to the bacterial cell surface, *Proceedings of National Academy of Sciences of the United States of America* 101, 9417–9422.

Bressleer D.C., Gray M.R., 2003, Transport and reaction processes in bioremediation of organic contaminnats. 1. Review of bacterial degradation and transport, *International Journal of Chemical Reactor Engineering* 1, R3.

Carballa M., Fink G., Omil F., Lema J.M., Ternes T., 2008, Determination of the solid-water distribution coefficient (K_d) for pharmaceuticals, estrogens and musk fragrances in digested sludge, *Water Research* 42, 287–295.

Carrière B., Legrimellec C., 1986, Effects of benzyl alcohol on enzyme activities and D-glucose transport in kidney brush border membranes, *Biochimica Et Biophysica Acta* 857, 131–138.

Caroff M., Karibian D., 2003, Structure of bacterial lipopolysaccharides, *Carbohydrate Research* 338, 2431–2447.

Cerniglia C.E., 1992, Biodegradation of polycyclic aromatic hydrocarbons, *Biodegradation* 3, 351–368.

Chan S.M.N., Luan T.G., Wong M.H., Tam N.F.Y., 2006, Removal and biodegradation of polycyclic aromatic hydrocarbons by *Selenastrum capricornutum*, *Environmental Toxicology and Chemistry* 25, 1772–1779.

Chandra T.S., Mudliar S.N., Vidyashankar S., Mukherji S., Sarada R., Krishnamurthi K., Chauhan V.S., 2015, Defatted algal biomass as a non-conventional low-cost adsorbent: Surface characterization and methylene blue adsorption characteristics, *Bioresource Technology* 184, 395–404.

Chen B.L., Ding J., 2012, Biosorption and biodegradation of phenanthrene and pyrene in sterilized and unsterilized soil slurry systems stimulated by *Phanerochaete chrysosporium*, *Journal of Hazardous Material* 229, 159–169.

Chen B.L., Wang Y.S., Hu D.F., 2010, Biosorption and biodegradation of polycyclic aromatic hydrocarbons in aqueous solutions by a consortium of white-rot fungi, *Journal of Hazardous Materials* 179, 845–851.

Chen B.L., Yuan M.X., Liu H., 2011, Removal of polycyclic aromatic hydrocarbons from aqueous solution using plant residue materials as a biosorbent, *Journal of Hazardous Materials* 188, 436–442.

Chen S.N., Yin H., Tang S.Y., Peng H., Liu Z.H., Dang Z., 2016, Metabolic biotransformation of copper-benzo[a]pyrene combined pollutant on the cellular interface of *Stenotrophomonas maltophilia*, *Bioresource Technology* 204, 26–31.

Cheng N., Li Q.Y., Tang A.X., Su W., Liu Y.Y., 2018, Decolorization of a variety of dyes by *Aspergillus flavus* A5p1, *Bioprocess and Biosystem Engineering* 41, 511–518.

Chojnacka K., 2010, Biosorption and bioaccumulation – The prospects for practical applications, *Environmental International* 36, 299–307.

Chung M.K., Tsui M.T.K., Cheung K.C., Tam N.F.Y., Wong M.H., 2007, Removal of aqueous phenanthrene by brown seaweed *Sargassum hemiphyllum*: Sorption-kinetic and equilibrium studies, *Separation and Purification Technology* 54, 355–362.

Ding J., Chen B.L., Zhu L.Z., 2013, Biosorption and biodegradation of polycyclic aromatic hydrocarbons by *Phanerochaete chrysosporium* in aqueous solution, *Chinese Science Bulletin* 58(6), 613–621.

Dimitriou-Christidis P., Autenrieth R.L., McDonald T.J., Desai A.M., 2007, Measurement of biodegradability parameters for single unsubstituted and methylated polycyclic aromatic hydrocarbons in liquid bacterial suspensions, *Biotechnology and Bioengineering* 97, 922–932.

Dean-Ross D., Moody J.D., Freeman J.P., Doerge D.R., Cerniglia C.E., 2001, Metabolism of anthracene by a *Rhodococcus* species, *FEMS Microbiology Letters* 204(1), 205–211.

Eaton R.W., 1994, Organization and evolution of naphthalene catabolic pathways: Sequence of the DNA encoding 2-hydroxychromene-2-carboxylate isomerase and *trans*-o-hydroxybenzylidenepyruvate hydratase-aldolase from the *Nah*7 plasmid, *Journal of Bacteriology* 176, 7757–7762.

Fanesi A., Zegeye A., Mustin C., Cébron A., 2018, Soil particles and phenanthrene interact in defining the metabolic profile of *Pseudomonas putida* G7: A vibrational spectroscopy approach, *Frontiers in Microbiology* 9, 2999.

Fayeulle A., Veignie E., Slomianny C., Dewailly E., Munch J.C., Rafin C., 2014, Energy-dependent uptake of benzo[a]pyrene and its cytoskeleton-dependent intracellular transport by the telluric fungus *Fusarium solani*, *Environmental Science and Pollution Research* 21, 3515–3523.

Fernandez-Fontaina E., Pinho I., Carballa M., Omil F., Lema J.M., 2013, Biodegradation kinetic constants and sorption coefficients of micropollutants in membrane bioreactors, *Biodegradation* 24, 165–177.

Gao J., Ye J.S., Ma J.W., Tang L.T., Huang J., 2014, Biosorption and biodegradation of triphenyltin by *Stenotrophomonas maltophilia* and their influence on cellular metabolism, *Journal of Hazardous Materials* 276, 112–119.

Giri A.K., Patel R.K., Mahapatra S.S., Mishra P.C., 2013, Biosorption of arsenic (III) from aqueous solution by living cells of *Bacillus cereus*, *Environmental Science and Pollution Research* 20, 1281–1291.

Gray P.H.H., Thornton H.G., 1928, Soil bacteria that decompose certain aromatic compounds, *Centr Bakt Parasitenk II Abt* 73, 74–96.

Golet E., Xifra I., Siegrist H., Alder A., Giger W., 2003, Environmental exposure assessment of fluoroquinolone antibacterial agents from sewage to soil, *Environmental Science & Technology* 37, 3243–3249.

Goswami L., Manikandan N.A., Pakshirajan K., Pugazhenthi G., 2017, Simultaneous heavy metal removal and anthracene biodegradation by the oleaginous bacteria *Rhodococcus opacus*, *3 Biotech* 7, 37.

Green D.E., Fry M., Blondin G.A., 1980, Phospholipids as the molecular instruments of ion and solute transport in biological membranes, *Proceedings of the National Academy of Sciences of the United States of America* 77, 257–261.

Gu H.P., Luo J., Wang H.Z., Yang Y., Wu L.S., Wu J.J., Xu J.M., 2016, Biodegradation, biosorption of phenanthrene and its transmembrane transport by *Massilia* sp WF1 and *Phanerochaete chrysosporium*, *Frontier in Microbiology* 7, 38.

Gu H.P., Luo X.Y., Wang H.Z., Wu L.S., Wu J.J., Xu J.M., 2015, The characteristics of phenanthrene biosorption by chemically modified biomass of *Phanerochaete chrysosporium*, *Environmental Science and Pollution Research* 22, 11850–11861.

Hadibarata T., Kristanti R.A., Hamdzah M., 2014, Biosorption and biotransformation of fluoranthene by the white-rot fungus *Pleurotus eryngii* F032, *Biotechnology and Applied Biochemistry* 61(2), 126–133.

Hamzah N., Kamil N.A.F.M., Singhal N., Padhye L., Swift S., 2018, Comparison of phenanthrene removal by *Aspergillus niger* ATC 16404 (filamentous fungi) and *Pseudomonas putida* KT2442 (bacteria) in enriched nutrient-liquid medium, *Earth and Environmental Science* 140, 012047.

Hearn E.M., Patel D.R., Lepore B.W., Indic M., van den Berg B., 2009, Transmembrane passage of hydrophobic compounds through a protein channel wall, *Nature* 458, 367–370.

Helenius A., Simons K., 1975, Solubilization of membranes by detergents, *Biochiminca Et Biophysica Acta* 415, 29–79.

Huang Y., Zhagn S.Y., Lv M.J., Xie S.G., 2010, Biosorption characteristics of ectomycorrhizal fungal mycelium for anthracene, *Biomedical and Environmental Sciences* 23, 378–383.

Johnsen A.R., Wick L.Y., Harms H., 2005, Principles of microbial PAH-degradation in soil, *Environmental Pollution* 133, 71–84.

Johnson B.T., Kennedy J.O., 1973, Biomagnification of *p, p'*-DDT and methoxychlor by bacteria, *Applied Microbiology* 26(1), 66–71.

Jonker M.T.O., 2008, Absorption of polycyclic aromatic hydrocarbons to cellulose, *Chemosphere* 70, 778–782.

Jonsson S., Lind H., Lundstedt S., Haglund P., Tysklind M., 2010, Dioxin removal from contaminated soils by ethanol washing, *Journal of Hazardous Materials* 179(1–3), 393–399.

Julsing M.K., Schrewe M., Cornelissen S., Hermann I., Schmid A., Bühler B., 2012, Outer membrane protein *AlkL* Boosts biocatalytic oxyfunctionalization of hydrophobic substrates in *Escherichia coli*, *Applied and Environmental Microbiology* 78, 5724–5733.

Kaczorek E., Urbanowicz M., Olszanowski A., 2010, The influence of surfactants on cell surface properties of *Aeromonas hydrophila* during diesel oil biodegradation, *Colloids Surface B-Biointerfaces* 81, 363–368.

Kadri T., Magdouli S., Rouissi T., Brar S.K., Daghrir R., Lauzon J., 2018, Bench-scale production of enzymes from the hydrocarbonoclastic bacteria *Alcanivorax borkumensis* and biodegradation tests, *Journal of Biotechnology* 283, 105–114.

Kallimanis A., Frillingos S., Drainas C., Koukkou A.I., 2007, Taxonomic identification, phenanthrene uptake activity, and membrane lipid alterations of the PAH degrading *Arthrobacter* sp. strain Sphe3, *Applied Microbiology and Biotechnology* 76, 709–717.

Kamil N.A.F.M., Hamzah N., Talib S.A., Hussain N., 2016, Improving mathematical model in biodegradation of PAHs contaminated soil using Gram-positive bacteria, *Soil and Sediment Contamination* 25, 443–458.

Karlsson A., Parales J.V., Parales R.E., Gibson D.T., Eklund H., Ramaswamy S., 2003, Crystal structure of naphthalene dioxygenase: Side-on binding of dioxygen to iron, *Science* 299(5609), 1039–1042.

Karuppiah V., Berry J.L., Derrick J.P., 2011. Outer membrane translocons: Structural insights into channel formation. *Trends in Microbiology* 19(1), 40–48.

Ke L., Luo L.J., Wang P., Luan T.G., Tam N.F.Y., 2010, Effects of metals on biosorption and biodegradation of mixed polycyclic aromatic hydrocarbons by a freshwater green alga *Selenastrum capricornutum*, *Bioresource Technology* 101, 6950–6961.

Kile D.E., Chiou C.T., 1989, Water solubility enhancements of DDT and trichlorobenzene by some surfactants below and above the critical micelle concentration, *Environmental Science & Technology* 23, 832–838.

Lakshmi M.B, Muthukumar K., Velan M., 2012, Immobilization of *Mycoplana* sp. MVMB2 isolated from petroleum contaminated soil onto papaya stem (*Carica papaya* L.) and its application on degradation of phenanthrene, *Clean Soil, Air, Water* 40(8), 870–877.

Lepore B.W., Indic M., Pham H., Hearn E.M., Patel D.R., van den Berg B., 2011, Ligand-gated diffusion across the bacterial outer membrane, *Proceedings of the National Academy Sciences of the United States of America* 108, 10121–10126.

Li A., Tai C., Zhao Z.S., Wang Y.W., Zhang Q.H., Jiang G.B., Hu J.T., 2007, Debromination of decabrominated diphenyl ether by resin-bound iron nanoparticles, *Environmental Science & Technology* 41(19), 6841–6846.

Li F., Zhu L.Z., 2012, Effect of surfactant-induced cell surface modifications on electron transport system and catechol 1,2-dioxygenase activities and phenanthrene biodegradation by *Citrobacter* sp. SA01, *Bioresource Technology* 123, 42–48.

Li F., Zhu L.Z., Zhang D., 2014, Effect of surfactant on phenanthrene metabolic kinetics by *Citrobacter* sp. SA01, *Journal of Environmental Sciences* 26, 2298–2306.

Li Y.G., Chen B.L., 2009, Phenanthrene sorption by fruit cuticles and potato periderm with different compositional characteristics, *Journal of Agricultural and Food Chemistry* 57, 637–644.

Li Y.R., Zhang J., Liu H., 2018, Removal of chloramphenicol from aqueous solution using low-cost activated carbon prepared from *Typha orientalis*, *Water* 10, 351.

Liao L.P., Chen S.N., Peng H., Yin H., Ye J.S., Liu Z.H., Dang Z., Liu Z.C., 2015, Biosorption and biodegradation of pyrene by *Brevibacillus brevis* and cellular responses to pyrene treatment, *Ecotoxicology and Environmental Safety* 115, 166–173.

Lighty J.S., Silcox G.D., Pershing D.W., Cundy V.A., Linz D.G., 1990, Fundamentals for the thermal remediation of contaminated soils – Particle and bed desorption models, *Environmental Science & Technology* 24(5), 750–757.

Lin D.H., Pan B., Zhu L.Z., Xing B.S., 2007, Characterization and phenanthrene sorption of tea leaf powders, *Journal of Agricultural and Food Chemistry* 55, 5718–5724.

Liu Y., Liu Y.J., 2008, Biosorption isotherms, kinetics and thermodynamics, *Separation and Purification Technology* 61, 22–242.

Loukidou M.X., Zouboulis A.I., Karapantsios T.D., Matis K.A., 2004, Equilibrium and kinetic modeling of chromium (VI) biosorption by *Aeromonsa caviae, Colloids and Surfaces A – Physicochemical and Engineering Aspects* 242, 93–104.

Luo Y.C., Teng Z., Wang T.T.Y., Wang Q., 2013, Cellular uptake and transport of zein nanoparticles: Effects of sodium caseinate, *Journal of Agricultural and Food Chemistry* 61, 7621–7629.

Ma X.K., Wu L.L., Fam H., 2014, Heavy metal ions affecting the removal of polycyclic aromatic hydrocarbons by fungi with heavy-metal resistance, *Applied Microbiology and Biotechnology* 98, 9817–9827.

Ma Z., Liu J., Dick R.P., Li H., Shen D., Gao Y.Z., Waigi M.G., Ling W.T., 2018, Rhamnolipid influences biosorption and biodegradation of phenanthrene by phenanthrene-degrading strain *Pseudomonas* sp. Ph6, *Environmental Pollution* 240, 359–367.

Mackay A.A., Gschwend P.M., 2000, Sorption of monoaromatic hydrocarbons to wood, *Environmental Science & Technology* 34, 839–845.

Malinverni J.C., Silhavy T.J., 2009, An ABC transport system that maintains lipid asymmetry in the Gram-negative outer membrane, *Proceedings of the National Academy Sciences of the United States of America* 106, 8009–8014.

Martínez E., Estupiñán M., Pastor F.I.J., Busquets M., Díaz P., Manresa A., 2013, Functional characterization of ExFadLO, an outer membrane protein required for exporting oxygenated long-chain fatty acids in *Pseudomonas aeruginosa, Biochimie* 95, 290–298.

Meng L., Li W., Bao M.T., Sun P.Y., 2019, Great correlation: Biodegradation and chemotactic adsorption of *Pseudomonas synxantha* LSH-7' for oil contaminated seawater bioremediation, *Water Research* 153, 160–168.

Meng X.J., Li H.B., Zhang Y.X., Cao H.B., Sheng Y.X., 2016, Analysis of polycyclic aromatic hydrocarbons (PAHs) and their adsorption characteristics on activated sludge during biological treatment of coking wastewater, *Desalination and Water Treatment* 57(50), 23633–23643.

Mueller J.G., Devereus R., Santavy D.L., Lantz S.E., Willis S.G., Pritchard P.H., 1997, Phylogenetic and physiological comparisons of PAH-degrading bacteria from geographically diverse soils, *Antonie Van Leeuwenhoek International Journal of General and Molecular Microbiology* 71(4), 329–343.

Noda Y., Kanemasa Y., 1986, Determination of hydrophobicity on bacterial surfaces by nonionic surfactants, *Journal of Bacteriology* 167, 1016–1019.

Owsianiak M., Szulc A., Chrzanowski L., Cyplik P., Bogacki M., Olejnik-Schmidt A.K., Heipieper H.J., 2009, Biodegradation and surfactant-mediated biodegradation of diesel fuel by 218 microbial consortia are not correlated to cell surface hydrophobicity, *Applied Microbiology and Biotechnology* 84, 545–553.

Pande A.H., Qin S., Tatulian S.A., 2005, Membrane fluidity is a key modulator of membrane binding, insertion, and activity of 5-lipoxygenase, *Biophysical Journal* 88, 4084–4094.

Paria S., Khilar K.C., 2004, A review on experimental studies of surfactant adsorption at the hydrophilic solid-water interface, *Advances in Colloid and Interface Science* 110, 75–95.

Park J.H., Chon H.T., 2016, Characterization of cadmium biosorption by *Exiguobacterium* sp. isolated form farmland soil near Cu-Pb-Zn mine, *Environmental Science and Pollution Research* 23, 11814–11822.

Pignatello J.J., Oliveros E., MacKay A., 2006, Advanced oxidation processes for organic contaminant destruction based on the Fenton reaction and related chemistry, *Critical Reviews in Environmental Science & Technology* 36(1), 1–84.

Pohren R.S., Rocha J.A.V., Horn K.A., Vargas V.M.F., 2019, Bioremediation of soils contaminated by PAHs: Mutagenicity as a tool to validate environmental quality, *Chemosphere* 214, 659–668.

Poulos T.L., 1995, Cytochrome P450, *Current Opinion in Structural Biology* 5(6), 767–774.

Qu X.L., Xiao L., Zhu D.Q., 2008, Site-specific adsorption of 1,3-dinitrobenzene to bacterial surfaces: A mechanism of *n*-π electron-donor-acceptor interactions, *Journal of Environmental Quality* 37, 824–829.

Rauchova H., Drahota Z., Koudelova J., 1999, The role of membrane fluidity changes and thiobarbituric acid-reactive substances production in the inhibition of cerebral cortex Na^+/K^+-ATPase activity, *Physiological Research* 48, 73–78.

Sanches S., Martins M., Silva A.F., Galinha C.F., Santos M.A., Pereira I.A.C., Crespo M.T.B., 2017, Bioremoval of priority polycyclic aromatic hydrocarbons by a microbial community with high sorption ability, *Environmental Science and Pollution Research* 24, 3550–3561.

Salloum M.J., Chefetz B., Hatcher P.G., 2002, Phenanthrene sorption by aliphatic-rich natural organic matter, *Environmental Science & Technology* 36, 1953–1958.

Semple K.T., Cain R.B., Schmidt S., 1999, Biodegradation of aromatic compounds by microalgae, *FEMS Microbiology Letters* 170, 291–300.

Siegrist H., Joss A., Alder A., McArdell-Bürgisser C., Göbel A., Keller E., Ternes T.A., 2003, Micropollutants – New challenge in wastewater disposal? *Eawag News* 57, 7–10.

Siegrist R.L., West O.R., Morris M.I., Pickering D.A., Greene D.W., Muhr C.A., Davenport D.D., Gierke J.S., 1995, In-situ mixed region vapor stripping in low-permeability media. 2. Full scale field experiments, *Environmental Science & Technology* 29(9), 2198–2207.

Siddaramappa R., Rajaram K.P., Sethunathan N., 1973, Degradation of parathion by bacteria isolated from flooded soil, *Applied Microbiology* 26, 846–849.

Sikkema J., Debont J.A.M., Poolman B., 1995, Mechanisms of membrane toxicity of hydrocarbons, *Microbiological Reviews* 59, 201–222.

Singer S.J., Nicolson G.L., 1972, Fluid mosaic model of structure of cell membranes, *Science* 175, 720–731.

Singh P., Borthakur A., 2018, A review on biodegradation and photocatalytic degradation of organic pollutants: A bibliometric and comparative analysis, *Journal of Cleaner Production* 196, 1669–1680.

Smith M.S., 1948, Persistence of DDT and benzene hexachloride in soil, *Nature* 161(4085), 246.

Song C., Gao N.Y., Gao H.W., 2010, Transmembrane distribution of kanamycin and chloramphenicol: insights into the cytotoxicity of antibacterial drugs, *Molecular Biosystems* 6, 1901–1910.

Steen W.C., Karickhoff S.W., 1984, Biosorption of hydrophobic organic pollutants by mixed microbial populations, *Chemosphere* 10, 27–32.

Stein W.D., 1967, *The Movement of Molecules across Cell Membranes*, New York: Academic Press.

Strawinski R.J., Stone R.W., 1943, Conditions governing the oxidation of naphthalene and the chemical analysis of its products, *Journal of Bacteriology* 45, 16.

Stringfellow W.T., Alvarez-Cohen L., 1999, Evaluating the relationship between the sorption of PAHs to bacterial biomass and biodegradation, *Water Research* 33, 2535–2544.

Tam N.F.Y., Chan M.N., Wong Y.S., 2010, Removal and biodegradation of polycyclic aromatic hydrocarbons by immobilized microalgal beads, *Waste Management and the Environment* 140, 391–402.

Tao Y.Q., Xue B., Yang Z., Yao S.C., Li S.Y., 2014, Effects of heavy metals on the sorption of polycyclic aromatic hydrocarbons by *Microcystis aeruginosa*, *Journal of Environmental Quality* 43, 1953–1962.

Ternes T.A., Herrmann N., Bonerz M., Knacker T., Siegrist H., Joss A., 2004, A rapid method to measure the solid-water distribution coefficient (K_d) for pharmaceuticals and musk fragrances in sewage sludge, *Water Research* 38, 4075–4084.

Thavamani P., Megharaj M., Krishnamurti G.S., McFarland R., Naidu R., 2011, Finger printing of mixed contaminants from former manufactured gas plant (MGP) site soils: Implications to bioremediation, *Environmental International* 37, 184–189.

Thavamani P., Malik S., Beer M., Megharaj M., Naidu R., 2012, Microbial activity and diversity in long-term mixed contaminated soils with respect to polyaromatic hydrocarbons and heavy metals, *Journal of Environmental Management* 99, 10–17.

Tsezos M., Bell J.P., 1989, Comparison of the biosorption and desorption of hazardous organic pollutants by live and dead biomass, *Water Research* 23, 561–568.

Touw D.S., Patel D.R., Berg B., 2010, The crystal structure of *OprG* from *Pseudomonas aeruginosa*, a potential channel for transport of hydrophobic molecules across the outer membrane, *Plos One* 5(11), e15016.

van den Berg B., Black P.N., Clemons Jr. W.M., Rapoport T.M., 2004, Crystal structure of the long-chain fatty acid transporter FadL, *Science* 304, 1506–1509.

Vander J., Sherman J., Luciano D., 2001, Movement of molecules across cell membranes. Chapter 6 of *Human physiology: the mechanisms of body function*, 8th ed., New York: McGraw-Hill Higher Eduction.

Verdin A., Sahraoui A.L.H., Newsam R., Robinson G., Durand R., 2005, Polycyclic aromatic hydrocarbon storage by *Fusarium solani* in intracellular lipid vesicles, *Environmental Pollutant* 133, 283–291.

Vijayaraghavan K., Yun Y.S., 2008, Bacterial biosorbents and biosorption, *Biotechnology Advances* 26, 266–291.

Voerman S., Besemer A.F.H., 1970, Residues of dieldrin, lindane, DDT, and parathion in a light sandy soil after repeated application throughout a period of 15 years, *Journal of Agricultural and Food Chemistry* 18(4), 717–719.

Volesky B., 2003, Sorption and biosorption, BV Sorbex: Quebec, Canada.

Wells M.J.M., 2006, Log D_{OW}: Key to understanding and regulating wastewater-derived contaminants, *Environmental Chemistry* 3, 439–449.

Wiener M.C., Horanyi P.S., 2011, How hydrophobic molecules traverse the outer membranes of Gram-negative bacteria, *Proceedings of the National Academy Sciences of the United States of America* 108(27), 10929–10930.

Wilhelm R.C., Hanson B.T., Chandra S., Madsen E., 2018, Community dynamics and functional characteristics of naphthalene-degrading populations in contaminated surface sediments and hypoxic anoxic groundwater, *Environmental Microbiology* 20(10), 3543–3559.

Willumsen P.A., Karlson U., Pritchard P.H., 1998, Response of fluoranthene-degrading bacteria to surfactants, *Applied Microbiology and Biotechnology* 50, 475–483.

Wong J.W.C., Fang M., Zhao Z.Y., Xing B.S., 2004, Effect of surfactants on solubilization and degradation of phenanthrene under thermophilic conditions, *Journal of Environmental Quality* 33, 2015–2025.

Xiao L., Qu X., Zhu D.Q., 2007, Biosorption of nonpolar hydrophobic organic compounds to *Escherichia coli* facilitated by metal and proton surface binding, *Environmental Science & Technology* 41(8), 2750–2755.

Xie H., Chen Y.J., Wang C., Shi W.J., Zuo L., Xu H., 2015, The removal of fluoranthene by *Agaricus bisporus* immobilized in Ca-alginate modified by *Lentinus edodes* nanoparticles, *RSC Advances* 5, 44812–44823.

Xu N.N., Bao M.T., Sun P.Y., Li Y.M., 2013, Study on bioadsorption and biodegradation of petroleum hydrocarbons by a microbial consortium, *Bioresources Technology* 149, 22–30.

Yang K., Zhu L.Z., Xing B.S., 2006, Enhanced soil washing of phenanthrene by mixed solutions of TX100 and SDBS, *Environmental Science & Technology* 40, 4274–4280.

Ye Q.H., Liang C.Y., Wang C.Y., Wang Y., Wang H., 2018, Characterization of a phenanthrene-degrading methanogenic community, *Frontiers in Environmental Science & Engineering* 12(5), 4.

Yeagle P.L., 2005, *The Structure of Biological Membranes*, 2nd edition. New York: CRC Press.

Yeung A.T., Gu Y.Y., 2011, A review on techniques to enhance electrochemical remediation of contaminated soils, *Journal of Hazardous Materials* 195, 11–29.

Yu H.S., Zhu L.Z., Zhou W.J., 2007, Enhanced desorption and biodegradation of phenanthrene in soil-water systems with the presence of anionic-nonionic mixed surfactants, *Journal of Hazardous Materials* 142, 354–361.

Yuan M.J., Tong S.T., Zhao S.Q., Jia C.Q., 2010, Adsorption of polycyclic aromatic hydrocarbons from water using petroleum coke-derived porous carbon, *Journal of Hazardous Materials* 181, 1115–1120.

Yuan X.Z., Ren F.Y., Zeng G.M., Zhong H.G., Fu H.Y., Liu J.X., Xu X.M., 2007, Adsorption of surfactants on a *Pseudomonas aeruginosa* strain and the effect on cell surface lypohydrophilic property, *Applied Microbiology and Biotechnology* 76, 1189–1198.

Zhang D., 2013, Surfactant controlled bacterial interfacial behaviors of PAHs and its mechanisms, PhD Dissertation, Zhejiang University (in Chinese).

Zhang D., Zhu L.Z., 2012, Effects of Tween 80 on the removal, sorption and biodegradation of pyrene by *Klebsiella oxytoca* PYR-1, *Environmental Pollution* 164, 169–174.

Zhang D., Zhu L.Z., 2014, Controlling microbiological interfacial behaviors of hydrophobic organic compounds by surfactants in biodegradation process, *Frontiers in Environmental Science & Engineering* 8(3), 305–315.

Zhang D., Lu L., Zhao H.T., Jin M.Q., Lü T., Lin J., 2018a, Application of *Klebsiella oxytoca* biomass in the biosorptive treatment of PAH-bearing wastewater: Effect of PAH hydrophobicity and implications for prediction, *Water* 10, 675.

Zhang D., Lu S.G., Song X.Q., Zhang J.F., Huo Z.M., Zhao H.T., 2018b, Synergistic and simultaneous biosorption of phenanthrene and iodine form aqueous solutions by soil indigenous bacterial biomass as a low-cost biosorbent, *RSC Advances* 8, 39274–39283.

Zhang D., Zhu L.Z., Li F., 2013a, Influences and mechanisms of surfactants on pyrene biodegradation based on interactions of surfactant with a *Klebsiella oxytoca* strain, *Bioresource Technology* 142, 454–461.

Zhang D.N., Ran C.Y., Yang Y., Ran Y., 2013b, Biosorption of phenanthrene by pure planktons and their fractions, *Chemosphere* 93, 61–68.

Zhang D.N., Ran Y., Cao X.Y., Mao J.D., Cui J.F., Schmidt-Rohr K., 2015, Biosorption of nonylphenol by pure algae, field-collected planktons and their fractions, *Environmental Pollution* 198 61–69.

Zhang Y., Miller R., 1994, Effect of a *Pseudomonas* rhamnolipid biosurfactant on cell hydrophobicity and biodegradation of octadecane, *Applied and Environmental Microbiology* 60, 2101–2116.

Zhao B.W., Zhu L.Z., Li W., Chen B.L., 2005, Solubilization and biodegradation of phenanthrene in mixed anionic-nonionic surfactant solutions, *Chemosphere* 58, 33–40

Zhao S.X., Zhou T.S., 2016, Biosorption of methylene blue from wastewater by an extraction residual of *Salvia miltiorrhiza* Bge, *Bioresource Technology* 219, 330–337

Zhao Z.Y., Selvam A., Wong J.W.C., 2011, Effects of rhamnolipids on cell surface hydrophobicity of PAH degrading bacteria and the biodegradation of phenanthrene, *Bioresource Technology* 102(5), 3999–4007.

Zhong H., Zeng G.M., Liu J.X., Xu X.M., Yuan X.Z., Fu H.Y., Huang G.H., Liu Z.F., Ding Y., 2008, Adsorption of monorhmnolipid and dirhamnolipid on two *Pseudomonas aeruginosa* strains and the effect on cell surface hydrophobicity, *Applied Microbiology and Biotechnology* 79, 671–677.

Zhou W.J., Zhu L.Z., 2007, Enhanced desorption of phenanthrene from contaminated soil using anionic/nonionic mixed surfactant, *Environmental Pollution* 147, 350–357.

Zhu L.Z., Lu L., Zhang D., 2010, Mitigation and remediation technologies for organic contaminated soils, *Frontiers in Environmental Science & Engineering* 4(4), 373–386.

Zhu Y., Zhang S., Zhu Y., Christie P., Shan X.X., 2007, Improved approaches for modeling the sorption of phenanthrene by a range of plant species, *Environmental Science & Technology* 41, 7818–7823.

Zhuang Y., Ahn S., Seyfferth A.L., Masue-Slowey Y., Fendorf S., Luthy R.G., 2011, Dehalogenation of polybrominated diphenyl ethers and polychlorinated biphenyl by bimetallic, impregnated, and nanoscale zerovalent iron, *Environment Science & Technology* 45(11), 4896–4903.

Zolfaghari M., Drogui P., Brar S.K., Buelna G., Dubé R., 2017, Insight into the adsorption mechanisms of trace organic carbon on biological treatment process, *Environmental Technology* 38, 2324–2334.

9 Remediation of Soil and Groundwater Contaminated with Per- and Poly-Fluoroalkyl Substances (PFASs)

S.T.M.L.D. Senevirathna and Reza Mahinroosta

CONTENTS

9.1 INTRODUCTION

Protecting our fresh water resources has become a real challenge due to the exponential growth of water usage, global warming and extreme climatic events, and various pollution loads being discharged into the water environment through a variety of anthropogenic activities. In particular, contamination of the water environment by various domestic and industrial activities is identified as an emerging environmental issue.

Organic pollutants that resist natural degradation and accumulate in living organisms, are toxic to human and environmental health, and are widely distributed or categorized as persistent organic pollutants (POPs) (Gupta and Ali, 2012; Ritter et al., 1995). Conventional strategies may not be sufficient to eliminate POPs from the environment. The list of POPs grows with time, and remediation is more challenging as the behavior of some chemicals has not been fully understood.

9.1.1 WHAT ARE PFASS?

Per- and polyfluoroalkyl substances are commonly known as PFASs. These substances were first synthesized during the late 1940s. PFASs are used in various industrial and consumer products such as fluorinated polymers, surfactants, insecticides, cosmetics, packing materials, kitchen

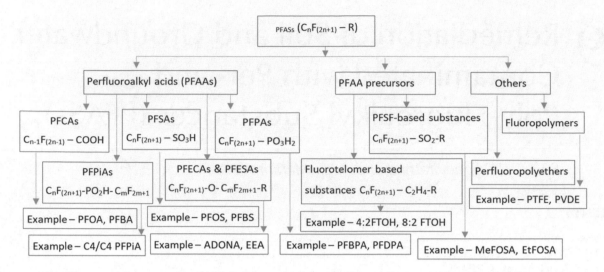

FIGURE 9.1   "Family tree" of PFASs suggested by Wang et al. (2017).

appliances, medical instruments, pesticides, textiles, leather, inks, mining, apparel, and aqueous firefighting foams. Figure 9.1 illustrates the PFAS "family tree" as suggested by Wang et al. (2017).

PFASs contain many fluorine atoms connected with carbon. Perfluorinated compounds (PFCs) are a category of PFAS in which all the carbon-hydrogen bonds are replaced with carbon-fluorine bonds, except one bonding site at the end of chain. The structural arrangements of man-made PFASs are significantly different from natural organofluorine molecules (about 30) in the environment, which contain only one fluorine atom per molecule. Table 9.1 shows the molecular

structures of frequently reported PFCs. Specific characteristics of fluorine, such as strong electronegativity and small atomic size, make perfluoroalkyl moiety ($C_nF_{2n+1}^-$) more stable and repels both oil and water.

More than 3000 PFASs have been produced and used for various domestic and industrial applications. However, the adverse effects of most PFASs on the environmental and public health have been poorly understood. Most PFAS research has been limited to a few well studied medium chain PFASs, particularly perfluorooctanesulfonate (PFOS), perfluorooctanoic acid (PFOA), and their precursors.

TABLE 9.1

Molecular Structures of Per- and Polyfluoroalkyl Substances

PFAS	Example
Perfluoroalkyl carboxylates	

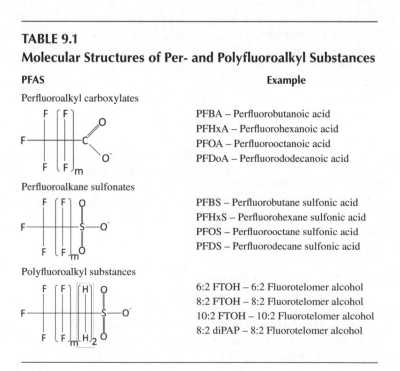

	PFBA – Perfluorobutanoic acid
	PFHxA – Perfluorohexanoic acid
	PFOA – Perfluorooctanoic acid
	PFDoA – Perfluorododecanoic acid
Perfluoroalkane sulfonates	
	PFBS – Perfluorobutane sulfonic acid
	PFHxS – Perfluorohexane sulfonic acid
	PFOS – Perfluorooctane sulfonic acid
	PFDS – Perfluorodecane sulfonic acid
Polyfluoroalkyl substances	
	6:2 FTOH – 6:2 Fluorotelomer alcohol
	8:2 FTOH – 8:2 Fluorotelomer alcohol
	10:2 FTOH – 10:2 Fluorotelomer alcohol
	8:2 diPAP – 8:2 Fluorotelomer alcohol

9.1.2 Applications of PFASs

Unique chemical and physical characteristics of PFASs make them an essential chemical for various commercial, industrial, and domestic applications including:

- Surface protectors (carpet, textiles, leather protection, paper, and boards)
- Firefighting foams and polymerization aids
- Speciality surfactants (cosmetics, electronics)
- Various applications in medical instruments
- Various industrial processes
- Wiring insulation

PFOA is a well-known PFAS that is used in the production of fluoropolymers such as PVF (polyvinylfluoride), PTFE (polytetrafluoroethylene), PFPE (perfluoropolyether), and PVDF (polyvinylidene fluoride). Manufacturing facilities for fluorochemicals and related products are identified as one of the largest point sources of emissions for PFCs.

9.1.3 POP Properties of Some PFASs

PFOS is one of the most studied PFASs and is known for its persistence in the environment and bioaccumulation in the food chain. Many researchers have detected PFASs, particularly PFOS and PFOA, in animal tissues. Higher concentrations have been observed in animals at higher levels in the food chain, such as fish-eating eagles, confirming the bio-accumulative properties of PFOS (Haukås et al., 2007). Several lab-scale toxicology investigations with rats and rabbits have provided strong evidence of PFOS toxicity (Wang et al., 2015a; Yin et al., 2018; Zhang et al., 2016). Even though some studies have suggested potential adverse effects of PFOS on human health (Wang et al., 2015a, 2015b), the PFOS risk to human health is not yet fully understood.

9.1.3.1 Persistence

Many studies have reported the stability of PFOS even in advanced oxidation processes including ozone, ozone/UV, ozone/H_2O_2, and Fenton reagent. Also, chromium potassium oxide ($Cr_2O_7{}^{2-}$) and potassium permanganate acid ($MnO_4{}^-$) have been observed to be unable to oxidize almost all PFCs – confirming the stability of PFCs in real environmental conditions.

9.1.3.2 Bioaccumulation

Usually the potential of bioaccumulation is estimated using the partition coefficient (K_{ow}), which is defined as the ratio of concentrations between octane and water phases. However, it is not available for most PFASs. For example, PFCs are surfactants and a third layer is formed during measurement, which reduces the accuracy of the partition coefficient estimate.

Bioaccumulation factors (BAFs) indicate the accumulation potential of organics from the environment to organisms. The following equation is used to calculate BAFs (Equation 9.1):

$$\text{Bioaccumulation factors (BAFs)} = \frac{\text{average concentration in organism}}{\text{concentration in water}} \tag{9.1}$$

Table 9.2 shows the BAFs for some PFASs. It is worth noting that the Log BAF has been reported as 6.38 for some fish, highlighting the risk of fish consumption near PFAS contaminated areas.

9.1.3.3 Toxicity

Although several research studies have been conducted, the toxicokinetics of PFASs is still not fully understood. More studies are needed to explain the pathological process and to confirm the adverse effects on human health. Available results generally suggest that short chain PFASs are less toxic than long chain and medium chain PFASs. Laboratory

---

TABLE 9.2

Log Bioaccumulation Factors (BAFs) of Some PFASs

PFAS	log BAF	Biota	References
PFBuA	1.11–2.56	Echinoderm (*Holuthuria tubulosa* intestine)	Martín et al. (2019)
PFPeA	6.38	Fish and eel	Campo et al. (2016)
PFHxA	1.4	Fish (*Cyprinus carpio*)	Pignotti et al. (2017)
PFHpA	1.81	Gastropod	Naile et al. (2013)
PFOA	2.57–4.21	*Holuthuria tubulosa* intestine	Martín et al. (2019)
	2.91	Fish (*Barbus graellsii, Cyprinus carpio, Micropterus salmoides*)	Campo et al. (2015)
	2.21	Aquatic snails (*Bithynia tentaculata*)	Wilkinson et al. (2018)
	2.6	Crab	Hong et al. (2015)
PFOS	3.5–5.2	Fish (*Leuciscus cephalus*)	Labadie and Chevreuil (2011)
	1.72–2.41	Crab	Naile et al. (2013)
	2.65	Prawn	Wang et al. (2013)
	3.82–4.66	Fish	Fujii et al. (2007)

experiments with animals show that several factors including species, gender, and age affect PFAS toxicity in animals. It is estimated that the average half-life of PFOA in the human body is 4.37 years. However, some researchers argue the natural elimination of PFOA from the human body is possible.

9.1.4 Physiochemical Properties of PFCs

Depending on the chain length, PFASs exist in either liquid or solid form. PFASs are heavier than water and have low solubility in water and high solubility in most organic solvents such as ethanol, acetonitrile, and methanol. The physical properties of PFASs are mainly determined by the length of the carbon chain (the number of carbon atoms in the molecule). For example, long chain PFASs with higher carbon atoms in the molecules show comparatively higher density, vapor pressure, viscosity, boiling point, surface tension, and refractive index. Some properties of selected PFASs are shown in Tables 9.3 and 9.4.

9.1.5 PFAS-Related Regulations

Although PFAS research was accelerated after the year 2000, PFASs were synthesized and used for several decades previously for various domestic and industrial applications. For example, Teflon was first introduced by DuPont during the 1950s. Little attention was given to health and environmental

concerns in the early application of PFASs (from 1960 to 2000). PFAS levels in blood and tap water were first determined in 1968 and 2000, respectively. Adverse environmental impacts of PFOSs were first investigated by the company 3M, which was a main producer of PFASs. In 2000, the company announced its plan to phase out PFOS use.

9.1.5.1 PFCs and the Stockholm Convention

The Stockholm Convention on Persistent Organic Pollutants (POPs) was adopted in 2001 and came into force in 2004. The convention is a global treaty to protect human health and the environment from chemicals that accumulate in the fatty tissue of humans and wildlife, become widely distributed geographically, and remain intact in the environment for long periods. The United Nations Environment Programme, which is based in Geneva, Switzerland, administers the Stockholm Convention. Relevant parties commit to eliminate or reduce the discharge of POPs into the environment by signing the Stockholm Convention.

Candidate chemicals to be categorized as POPs are continuously investigated and reviewed by an independent committee appointed as a subsidiary body to the Stockholm Convention. In 2009, the committee recommended amending the list of POPs and to categorize PFOS as a POP. This recommendation was accepted by the fourth meeting of the Conference of the Parties (COP4) to the Stockholm Convention, and PFOS and its salts were categorized as POPs (decision SC-4/17).

TABLE 9.3
Basic Information of Selected PFASs

Abbr. Name	Full Name	Molecular Structure	MW	CAS No.
PFBA	Perfluorobutyric acid	$CF_3(CF_2)_2COOH$	214	375-22-4
PFHxA	Perfluorohexanoic acid	$CF_3(CF_2)_4COOH$	314	307-24-4
PFHpA	Perfluoroheptanoic acid	$CF_3(CF_2)_5COOH$	364	375-85-9
PFOA	Perfluorooctane acid	$CF_3(CF_2)_6COOH$	414	335-67-1
PFDA	Perfluorodecanoic acid	$CF_3(CF_2)_8COOH$	514	335-76-2
PFOS	Perfluorooctane sulfonate	$CF_3(CF_2)_7SO_3$	500	2795-39-3

Abbreviations name – Name abbreviated, MW – molecular weight.

TABLE 9.4
Basic Physiochemical Properties of Selected PFASs

PFCs	pK_a [a]	Melting Point[b] °C	Boiling point[c] °C	Specific Gravity	Texture
PFBA			120.8–121.0	1.65	Yellow liquid
PFHxA			159–160	1.76	Clear liquid
PFHpA			175	1.79	Crystalline
PFOA	2.50	55–56	189	1.70	White powder
PFDA		83–85	218		White powder
PFOS	3.27	>400		2.05	White powder

Note: a = PFOA (OECD, 2002), PFOS (US EPA, 2002); b = melting point, data from material safety data sheet (MSDS) of Wako Company and ExFluor Company; c = boiling point, from ExFluor MSDS.

OECD report ENV/JM/RD(2002)17/FINAL, Nov. 21, 2002. Hazard assessment of perfluorooctanesulfonate (PFOS) and its salts.

US EPA (2002), Hazard assessment of perfluorooctanoic acid and its salts, 107 pp, Office of Pollution Prevention and Toxics—Risk Assessment Division, US.

9.1.5.2 Acceptable Purposes

Although PFOS is categorized as a POP, provision has given to use it for limited application. Some of the applications are listed below:

- Photo-imaging
- Photo-resistant and anti-reflective coatings for semiconductors
- Etching agents for compound semiconductors and ceramic filters
- Aviation hydraulic fluids
- Metal plating (hard metal plating) only in closed-loop systems
- Certain medical devices
- Firefighting foams

9.2 WATER TREATMENT

9.2.1 LEVEL OF PFAS CONTAMINATION

Several researchers have reported on PFAS contamination levels in various water environments. Unfortunately, most of the PFAS monitoring studies have focused on the northern hemisphere water environment, meaning there is limited information available on contamination levels for the southern hemisphere water environment. However, the limited literature available identifies some highly contaminated sites in the southern hemisphere (Taylor and Johnson, 2016).

Teflon manufacturers are commonly identified as PFAS polluters, and the PFOS and PFOA levels measured in tap water samples collected around Teflon industries have varied from 10 ng L^{-1} to 40 ng L^{-1}. Tap water PFAS levels reported from uncontaminated areas are normally less than 5 ng L^{-1}.

9.2.2 UNIT PROCESSES FOR PFAS TREATMENT

Several studies have reported a positive correlation between the PFOS concentration in raw water and tap water samples, suggesting possible inefficiencies in PFOS elimination at conventional unit processes for water purification.

9.2.2.1 Membrane Filtration

Laboratory scale investigations have suggested that reverse osmosis (RO) membrane filtration is a promising method to treat water that is highly contaminated with PFOS. Tang et al. (2007) observed a more than 99% PFOS removal efficiency, with inlet PFOS concentration ranging from 0.5 to 1500 mg L^{-1}. They have also reported a reduction of permeate flux when the PFOS concentration is increased.

The results of a laboratory scale treatment trial showed more than 95% rejection of heavier PFAS (molecular weight > 300 g mol^{-1}) in nano-filtration systems. pH plays an important role in nano-filtration systems, and better removal efficiency is observed at higher pH levels.

A combined treatment using powdered activated carbon and membrane filtration has been identified as a promising method to eliminate PFAS. Rattanaoudom (2011) argues that the combined method is the best of the limited options available to eliminate PFOS and PFOA. In addition to a higher removal efficiency, the combined method has several other advantages including a reduction of fouling and the formation of a hydrotalcite layer on the membrane, which enhances PFOS and PFOA rejection.

The findings of Tang et al. (2006) indicated the feasibility of RO membranes to eliminate PFOS from semiconductor wastewater. Their results showed more than 99% PFOS removal efficiency with the RO membrane system (feed concentrations 0.5–1500 ppm). Also, the researchers found that the rejection was better for tighter membranes but was not affected by membrane zeta potential. Flux decreased with increasing PFOS concentration. Baudequin et al. (2011) have suggested installing a suitable pretreatment unit before the membrane process to minimize the fouling issue.

9.2.2.2 PFAS Adsorption

Adsorption is identified as one of the most promising methods to eliminate PFAS in water, and many studies have been done to help understand the PFAS adsorption characteristics of various granular materials. Many researchers have reported that the PFAS adsorption isotherm of granular materials is best explained by the homogenous surface diffusion model (HSDM) and the Freundlich equation. The length and number of carbon atoms in the molecules control the isothermal and kinetic characteristics of PFASs on granular activated carbon.

The pH and natural organic matter (NOM) of the bulk solution affect the adsorption kinetics and isotherm. NOM in bulk water may reduce the PFAS adsorption capacity of granular activated carbon (GAC) through molecular competition and carbon fouling. A lower pH in the bulk solution may be helpful for the adsorption of PFAS. There are a number of GAC products commercially available with different PFAS adsorption characteristics. It is essential to optimize the adsorbent for the relevant wastewater characteristics, prior to designing the adsorption unit process to remove PFAS.

A number of resins have been developed to eliminate PFAS selectively from water. Comparative studies have proven these resins perform better than GAC. For example, we tested the adsorption characteristics of three non-ion-exchange polymers (DowV493, DowL493, and AmbXAD4), two ion-exchange polymers (DowMarathonA and AmbI-RA400), and one GAC (Filtersorb400) at low PFOS concentrations (100–1000 ng L^{-1} equilibrium concentrations). The sorption capacities at 1 µg L^{-1} equilibrium concentration decreased in the following order: Ion-exchange polymers > non-ion-exchange polymers > GAC, but at further low equilibrium concentration (100 ng L^{-1}), non-ion-exchange polymers showed higher adsorption capacity than other adsorbents. In the case of sorption kinetics, GAC and ion-exchange polymers reached equilibrium concentration within 4 h and AmbXAD4 within 10 h. DowV493 and DowL493 took more than 80 h to reach equilibrium concentration. The results of this study provide strong evidence that the PFAS adsorption characteristics of resins are better than those of GAC. Tables 9.5 and 9.6 summarize the studies done to evaluate different materials to eliminate PFCs, mainly PFOS and PFOA.

TABLE 9.5

Summary of Previous Studies on Materials to Adsorb PFOS

Material	Condition mg L^{-1} of PFOS	Freundich Constants		Langmuir Constants		References
		K_f	n^{-1}	q_m	b	
GAC	20–250	0.43 mmol$^{(1-1/n)}$ L$^{1/n}$ g^{-1}	0.18	0.37 mmol g^{-1}	39 L mol^{-1}	Yu et al. (2009)
PAC	20–250	1.27 mmol$^{(1-1/n)}$ L$^{1/n}$ g^{-1}	0.18	1.04 mmol g^{-1}	55 L mol^{-1}	Yu et al. (2009)
AI400 (Ion exchange resign)	20–250	0.52 mmol$^{(1-1/n)}$ L$^{1/n}$ g^{-1}	0.17	0.42 mmol g^{-1}	69 L mol^{-1}	Ochoa-Herrera and Sierra-Alvarez (2008), Yu et al. (2009)
GAC (Calgon F300)	15–150	960.90 mg$^{(1-1/n)}$ L$^{1/n}$ g^{-1}	0.332	196.20 mg g^{-1}	0.068 L mg^{-1}	Ochoa-Herrera and Sierra-Alvarez (2008)
NaY Zeolite	15–150	38.50 mg$^{(1-1/n)}$ L$^{1/n}$ g^{-1}	1.577	–	–	Ochoa-Herrera and Sierra-Alvarez (2008)
Amb IRA-400	0.01–5	108.9 (μg g^{-1}) (μg L^{-1})$^{-n}$	2.08	–	–	Senevirathna (2010)
DowMarathonA	0.01–5	95.9 (μg g^{-1}) (μg L^{-1})$^{-n}$	1.68	–	–	Senevirathna (2010)
DowV493	0.01–5	81.3 (μg g^{-1}) (μg L^{-1})$^{-n}$	0.94	–	–	Senevirathna (2010)

TABLE 9.6

Summary of Previous Studies on Materials to Adsorb PFOA

Material	Condition mg L^{-1} of PFOA L^{-1}	Freundich Constants		Langmuir Constants		References
		K_f	n^{-1}	q_m	b	
GAC	20–250	0.5 mmol$^{(1-1/n)}$ L$^{1/n}$ g^{-1}	0.28	0.39 mmol g^{-1}	18 L mol^{-1}	Yu et al. (2009)
PAC	20–250	0.8 mmol$^{(1-1/n)}$ L$^{1/n}$ g^{-1}	0.20	0.67 mmol g^{-1}	59 L mol^{-1}	Yu et al. (2009)
GAC (Calgon F400)	15–150	11.8 mg$^{(1-1/n)}$ L$^{1/n}$ g^{-1}	0.443	112.1 mg g^{-1}	0.038 L mg^{-1}	Ochoa-Herrera and Sierra-Alvarez (2008), Yu et al. (2009)
AI400	20–250	3.4 mmol$^{(1-1/n)}$ L$^{1/n}$ g^{-1}	0.13	2.92 mmol g^{-1}	69 L mol^{-1}	Ochoa-Herrera and Sierra-Alvarez (2008), Yu et al. (2009)
Dowexoptopore V493	0.01–5	998.5 (μg g^{-1}) (μg L^{-1})$^{-n}$	0.40	—	—	Senevirathna (2010)
Dowexoptopore L493	0.01–5	208.3 (μg g^{-1}) (μg L^{-1})$^{-n}$	0.50	—	—	Senevirathna (2010)
Amberlite XAD 4	0.01–5	554.5 (μg g^{-1}) (μg L^{-1})$^{-n}$	0.54	—	—	Senevirathna (2010)
Filtrasorb 400 (GAC)	0.01–5	602.4 (μg g^{-1}) (μg L^{-1})$^{-n}$	1.67	—	—	Senevirathna (2010)

9.2.2.3 PFC Oxidation

Fluorine is identified as a powerful inorganic oxidant with 3.6 V (Equation 9.2) reduction potential. In PFAS molecules, carbon-hydrogen bonds are completely replaced by a carbon-fluoride bond, which makes PFAS strong, stable, and reluctant to oxidize. Fluorine is nearly always found in the (–I) valence state with the only exception being F_2 where its oxidation state is (0).

$$F \cdot + e^- \rightarrow F^- \quad (E_0 = 3.6 \text{ V}) \quad (9.2)$$

Fluorination inductively reduces head group electron density, which reduces the oxidizability of the ionic head group (SO_3^- for PFOS and CO_2^- for PFOA).

A number of persistent organics are eliminated from water by advanced oxidation processes. The process utilizes a hydroxyl radical to oxidize the target pollutant. There are several processes to generate hydroxyl radicals including:

- Photo-Fenton
- Ozonation
- Hydrogen peroxide photolysis
- Peroxone chemistry

Normally, a H atom is extracted from saturated organic by the hydroxyl radical to form water (Equation 9.3). The hydroxyl radical reacts with most aliphatic and aromatic organics at near diffusion-controlled rates. However, PFOS and PFOA don't have any hydrogen at the real environmental conditions (pressure, temperature, and pH), and hydroxyl radicals act through a direct electron transfer to form the less thermodynamically favored hydroxyl ion.

$$\cdot OH + e^- \rightarrow H_2O \ (E_0 = 2.7 \text{ V}) \quad (9.3)$$

$$\cdot OH + e^- \rightarrow OH^- \ (E_0 = 1.9 \text{ V}) \quad (9.4)$$

Literature data suggest that conventional advanced oxidation methods utilizing oxygen-based radicals are not promising methods for the decomposition of PFAS. For example, replacement of all organic hydrogen with fluorine makes PFOA and PFOS molecules stronger and more stable, as well as inert to advanced oxidation techniques.

Thus, the direct addition of H_2O_2 is not favorable to the photolytic degradation of PFOA by competitive photons adsorption because the second-order rate constant of "$\cdot OH + PFOA$" ($k_{\cdot OH + PFOA} \leq 10^5$ L mol^{-1} s^{-1}) is significantly slower than the reaction of the hydroxyl radical with most hydrocarbons.

9.2.2.3.1 Persulfate Photolysis – Sulfate Radical Oxidation

Persulfate photolysis has shown its capability to oxidize a number of organic pollutants.

As shown in Equation (9.5), two sulfate radicals are generated by persulfate photolysis. Equation (9.6) shows the sulfate radical is an oxidizing radical that reacts by a direct one-electron transfer to form sulfate. The sulfate radical has a one-electron reduction potential of 2.3 V, making it a stronger direct electron transfer oxidant than the hydroxyl radical.

$$S_2O_8^{2-} + hv \ (270 \ nm)/\Delta \rightarrow 2SO_4^- \qquad (9.5)$$

$$SO_4^- + e^- \ \rightarrow \ SO_4^{2-} \qquad (9.6)$$

Kutsuna and Hori (2007) have suggested a mechanism for the degradation of perfluoroalkylcarboxylates with sulfate radicals.

Equation (9.7) shows the initial step of the degradation, in which the sulfate radical receives an electron from the carboxylate terminal group.

$$CF_3(CF_2)_6 COO^- + SO_4^{\cdot-} \rightarrow CF_3(CF_2)_6 COO\cdot + SO_4^{2-} \quad (9.7)$$

Subsequently, as shown in Equation (9.8), a perfluoroheptyl radical is formed by decarboxylating oxidized PFOA.

$$CF_3(CF_2)_6 COO\cdot \rightarrow CF_3(CF_2)_5 CF_2\cdot + CO_2 \qquad (9.8)$$

As shown in Equation (9.9), a perfluoroheptylperoxy radical is formed by molecular oxygen reacting with the perfluoroheptyl radical.

$$CF_3(CF_2)_5 CF_2\cdot + O_2 \rightarrow CF_3(CF_2)_5 CF_2OO\cdot \qquad (9.9)$$

Due to unavailability of reductants in the solution to yield two perfluoroalkoxy radicals and molecular oxygen, the only possible reaction is between two perfluoroheptylperoxy radicals as shown in Equation (9.10).

$$CF_3(CF_2)_5 CF_2OO\cdot + R_FOO\cdot \rightarrow CF_3(CF_2)_5 CF_2O\cdot + R_FO + O_2 \qquad (9.10)$$

Two branching pathways are identified for the reaction in this stage – the first pathway is shown in Equation (9.11), in which unimolecular decomposition yields carbonyl fluoride and a perfluorohexyl radical.

$$CF_3(CF_2)_5 CF_2O\cdot \rightarrow CF_3(CF_2)_4 CF_2\cdot + COF_2 \qquad (9.11)$$

$$COF_2 + H_2O \ \rightarrow \ CO_2 + 2 \ HF \qquad (9.12)$$

In the second pathway, perfluoroheptanol is yielded by H-atom abstraction from an acid such as HSO_4^- (Equation 9.13).

$$CF_3(CF_2)_5 CF_2O\cdot + HSO_4^- \rightarrow CF_3(CF_2)_5 CF_2OH + HSO_4^\cdot{}^- \qquad (9.13)$$

The perfluoroheptanol from Equation (9.13) will unimolecularily decompose to give the perfluoroheptylacyl fluoride and HF (Equation 9.14).

$$CF_3(CF_2)_5 CF_2OH \ \rightarrow \ CF_3(CF_2)_5 COF + HF \qquad (9.14)$$

Perfluoroheptylacyl fluoride will hydrolyze to yield perfluoroheptanoate (Equation 9.15).

$$CF_3(CF_2)_5 COF + H_2O \ \rightarrow \ CF_3(CF_2)_5 COO^- + HF + H^+ \qquad (9.15)$$

Production of HF during the photolysis reduces the pH of the solution (Equation 9.13). It is reported that the second-order rate constants of the sulfate radical with various chain-length perfluorocarboxylates is in the order of 10^4 L mol^{-1} s^{-1}, which is a relatively slow rate when compared to second-order rates of the sulfate radical with hydrocarbons; short-chain alcohols and carboxylic acids are at the lower end with reaction rates in the order of 10^6 L mol^{-1} s^{-1} and aromatic organics are at the upper end with reaction rates being diffusion-controlled, 10^9–10^{10} L mol^{-1} s^{-1}.

9.2.2.3.2 Direct Ultraviolet (UV) Photolysis

Photolysis is the application of light to break chemical bonds. UV light adsorption yields an electronically excited molecule, which has an electron (molecular or atomic) promoted to an anti-bonding orbital. In terms of treatment, an electronically excited molecule is more susceptible to a chemical reaction.

Terrestrial solar-driven photolytic processes require utilization of 290–600 nm photons due to the atmospheric absorption of higher energy light.

Although a number of organics are reported to be photolyzed directly with solar irradiation, PFOS and PFOA have proven their stability even after 30 days of direct exposure to simulated sunlight.

Ultraviolet-C (UV-C, $\lambda < 300$ nm) and vacuum ultraviolet (VUV, $\lambda < 200$ nm) are high energy intense advanced oxidation processes. VUV irradiation photodissociates water into a H-atom and HO· (Equation 9.16) (Getoff, 1996).

$$H_2O + hv \ (\lambda < 200 \ nm) \rightarrow H\cdot + \cdot OH \qquad (9.16)$$

However, in the real application, hyper oxidizing region is generated near the lamp surface limiting the penetration depth of VUV (<100 mm).

Equation (9.17) shows the VUV photolysis of trifluoroacetic acid (gas phase) (Osborne et al., 1999).

$$CF_3COOH + hv \ (\lambda = 172 \ nm) \rightarrow CF_3COOH^* \rightarrow CF_3\cdot + CO_2 + H\cdot \qquad (9.17)$$

Equations 9.18 and 9.19 show the first two steps in VUV photolysis of PFOA (liquid phase).

$$CF_3(CF_2)_6 COO^- + hv \ (\lambda < 172 \ nm) \ \rightarrow \ CF_3(CF_2)_6 COO^{-*} \tag{9.18}$$

$$CF_3(CF_2)_6 COO^{-*} + H^+ \rightarrow \ CF_3(CF_2)_6 \cdot + CO_2 + H \cdot \tag{9.19}$$

A high concentration of the hydroxyl radical near the VUV lamp may oxidize PFOA and form perfluoroalkyl radicals. The perfluoroalkyl radical reacts with HO at the rate of diffusion control (photolysis conditions are anoxic) (Equation 9.20).

$$CF_3(CF_2)_6 \cdot + HO \cdot \rightarrow \quad CF_3(CF_2)_6 OH \tag{9.20}$$

Photolytic degradation of PFOS shows shorter chain perfluorocarboxylates and perfluoroalkyl alcohols as reaction intermediates. However, the photolysis rate of PFOS is much slower than PFOA.

9.2.2.3.3 *Phosphotungstic Acid Photocatalysis*

Phosphotungstic acid ($H_3PW_{12}O_{40}$) is used in the photocatalytic degradation of contaminants, with its predominant form of $PW_{12}O_{40}^{-3}$ at lower pH of <2. When the UV light is adsorbed, $PW_{12}O_{40}^{-3}$ enhances its oxidation strength by entering a photo-excited state (Equation 9.21).

$$PW_{12}O_{40}^{3-} + hv \ (\lambda < 390 \ nm) \rightarrow PW_{12}O_{40}^{3-*} \tag{9.21}$$

Evidence is available for the $H_3PW_{12}O_{40}$ photocatalysis decomposition of PFOA and PFPA.

It has been reported that the amount of fluoride production in $H_3PW_{12}O_{40}$ photocatalysis is in a similar range to that observed during persulfate photolysis. A similar degradation mechanism discussed in persulfate photolysis is suggested for $H_3PW_{12}O_{40}$ photocatalysis, where a shorter-chain perfluoroalkylcarboxylate is formed after removing the carboxylate headgroup.

$PW_{12}O_{40}^{3-}$ photocatalytic PFOA decomposition involves a photo-Kolbe type mechanism. Equations 9.22 and 9.23 show the first and second reactions of $PW_{12}O_{40}^{3-}$ photocatalytic PFOA decomposition (Hori et al., 2004b). PFOA first complexes with $PW_{12}O_{40}^{3-}$ and then an electron is transferred from PFOA to $PW_{12}O_{40}^{3-}$.

$$CF_3(CF_2)_6 COO^- + PW_{12}O_{40}^{3-} \rightarrow CF_3(CF_2)_6 COO^- .. PW_{12}O_{40}^{3-} \tag{9.22}$$

$$CF_3(CF_2)_6 COO^- .. PW_{12}O_{40}^{3-} + hv \ (\lambda < 390 \ nm)$$
$$\rightarrow CF_3(CF_2)_5 CF_2 \cdot + CO_2 + PW_{12}O_{40}^{4-} \tag{9.23}$$

Similar to the sulfate radical mechanism, the perfluoroheptyl radical will be formed by decarboxylating PFOA.

Oxygen is essential to the photocatalytic cycle in that it accepts an electron from the reduced phosphotungstic acid, $PW_{12}O_{40}^{4-}$ (Equation 9.23), returning it to its photoactive state. $PW_{12}O_{40}^{4-}$ is returned to its photoactive state by releasing an electron, and oxygen is available to accept this electron (Equation 9.24).

$$PW_{12}O_{40}^{4-} + O_2 \rightarrow PW_{12}O_{40}^{3-} + O_2 \cdot^- \tag{9.24}$$

Some PFAS can be degraded by $UV_{254 + 185}$ photolysis, although the efficiency of the process may be questionable.

Feasibility and the efficiency of the process will depend on various factors, such as existence of NOM, the volume to be treated, the PFAS level in the contaminated water, the level of treatment expected, the treatment rate, and other physical and chemical characteristics of wastewater. Previous studies on PFOA and PFOS degradation are summarized in Tables 9.7 and 9.8, respectively.

9.3 SOIL TREATMENTS

One of the main pathways of groundwater contamination is the migration of contaminants from the soil via water flow. This usually happens due to the existing groundwater flow, rainfall, and human activities. Ground lithology, soil texture, soil type, and soil condition are the major factors affecting the contamination migration and causing contaminants to distribute evenly or over decades. Therefore, a thorough understanding of the soil condition and the level of contamination is required.

9.3.1 SOIL INVESTIGATION AND THE CONCEPTUAL SITE MODEL

Before treatment starts, the source of the contaminant, its pathway, and the end destination should be studied to gain a clear understanding of the contaminant transfer and possible ways to reduce or eliminate the contaminants. For instance, it is now common engineering practice to prepare a conceptual site model (CSM), which shows the nature and source of a contaminant, its pathway, and the final target considering environmental settings, soil hydrogeological condition, surface conditions, and any constraints. For instance, geoenvironmental investigation including *in situ* and laboratory testing are required to identify first, the nature of the contaminants and their concentration levels through chemical analysis and second, the stratigraphy of the soil layers and physical characteristics such as soil type and classifications, density, and permeability. For instance, soil sampling from both surface and depth of the soil layers is required using a procedure that will not add any new pollution to the soil. Soil sampling should be done in such a way that it helps to establish a reasonable picture of the concentration of contaminants as well as to detect any unknown contaminants.

TABLE 9.7
Summary of Previous Studies on PFOA Degradation

Technique	Condition	Power (W) & Volume (mL)	k	Product	Energy (kJ)	References
UV direct photolysis	9.6 mg L^{-1} of PFOA $\lambda = 220–460$ nm	200 22	0.69 d^{-1} $\tau^{1/2} = 1440$ min	33% F^- 38% CO_2 65% PF acids	792000	Hori et al. (2004a)
UV phosphotungstic photocatalysis	9.6 mg L^{-1} of PFOA, $\lambda = 220–460$ nm 0.48 MPa of O_2, 6.6 mmol L^{-1} of PTA	200 22	2.0 d^{-1} $\tau^{1/2} = 500$ min	30% F^- 25% CO_2 70% PF acids	276000	Hori et al. (2004a)
UV direct photolysis	20 g L^{-1} of PFOA $\lambda = 185$ nm	23 1000	0.017 min^{-1} $\tau^{1/2} = 41$ min	10% F^- 90% PF acids	49	Chen and Zhang (2006)
UV persulfate photolysis	20 g L^{-1} of PFOA, $\lambda = 254$ nm 1.5 mmol L^{-1} of $S_2O_8^{2-}$	23 1000	0.012 min^{-1} $\tau^{1/2} = 58$ min	5% F^- 95% PF acids	69	Chen and Zhang (2006)
UV persulfate photolysis	540 g L^{-1} of PFOA, $\lambda = 220–460$ nm, 0.48 MPa of O_2, 10 mmol L^{-1} of $S_2O_8^{2-}$, pH = 2–3, 10 mmol L^{-1} of $S_2O_8^{2-}$	200 22	0.69 h^{-1} $\tau_{1/2} = 58$ min	12% F^- 85% PF acids	33600	Hori et al. (2005)
Sonolysis	8.2 L^{-1} of PFOA $f = 354$ kHz	150 600	0.018 min^{-1} $\tau^{1/2} = 39$ min	95% F^-	670	Vecitis et al. (2008)
Sonolysis	82 µg L^{-1} of PFOA $f = 354$ kHz	150 600	0.047 min^{-1} $\tau^{1/2} = 15$ min	95% F^-	260	Vecitis et al. (2008)
UV-KI photolysis	8.2 g L^{-1} of PFOA $\lambda = 254$ nm	1.5 30	0.0014 min^{-1} $\tau^{1/2} = 500$ min	10% F^- Gaseous fluoroalkanes	1500	Park et al. (2009)
UV-KI photolysis	200 nmol L^{-1} of PFOA $l = 254$ nm	1.5 30	0.0025 min^{-1} $\tau^{1/2} = 280$ min	10% F^- Gaseous fluoroalkanes	820	Park et al. (2009)
Ferrophotolysis	14.3 g L^{-1} of PFBA 2.5 mmol L^{-1} of $Fe_2(SO_4)_3$ $\lambda = 220–460$ nm	200 105	0.028 h^{-1} $\tau^{1/2} = 1490$ min	45% F^- 55% short chains	89400	Hori et al. (2007)

TABLE 9.8
Summary of Previous Studies on PFOS Degradation

Technique	Condition	Power (W) & Volume (mL)	k	Product	Energy (kJ)	References
Sub-critical Fe(0)	185 mg L^{-1} of PFOS 0.5 g of Fe(0) 350°C, 20 MPa	0 10	0.013 min^{-1} $\tau^{1/2} = 53$ min	50% F^-	2000	Hori et al. (2006)
UV direct photolysis	20 mg L^{-1} of PFOS $\lambda = 254$ nm	32 750	0.13 d^{-1} $\tau^{1/2} = 7700$ min	71% F^- 90% SO_4^{2-}	17000	Yamamoto et al. (2007)
UV alkaline IPA photolysis	20 mg L^{-1} of PFOS $\lambda = 254$ nm	32 750	0.93 d^{-1} $\tau^{1/2} = 1070$ min	$NaF_{(s)}$	2500	Yamamoto et al. (2007)
Sonolysis	10 mg L^{-1} of PFOS $f = 354$ kHz	150 600	0.011 min^{-1} $\tau^{1/2} = 63$ min	95% F^- 100% SO_4^{2-}	945	Vecitis et al. (2008)
UV-KI photolysis	10 mg L^{-1} of PFOS $\lambda = 254$ nm, [KI] = 10 mmol L^{-1}	1.5 30	0.002 min^{-1} $\tau^{1/2} = 350$ min	50% F^- 50% fluoroalkanes	960	Park et al. (2009)
UV-KI photolysis	100 mg L^{-1} of PFOS $\lambda = 254$ nm, [KI] = 10 mmol L^{-1}	1.5 30	0.008 min^{-1} $\tau^{1/2} = 87$ min	50% F^- 50% fluoroalkanes	260	Park et al. (2009)
Sonolysis	100 mg L^{-1} of PFOS, $f = 354$ kHz	150 600	0.023 min^{-1} $\tau^{1/2} = 30$ min	95% F^- 100% SO_4^{2-}	450	Vecitis et al. (2008)

After thoroughly examining the site conditions and contamination levels, different soil treatments should be considered. In general, there are three methods for treatment of lands for PFAS contamination: containment, off-site disposal in landfills, and soil remediation. Soil remediation can be done either *ex situ* or *in situ*; in the former soil is excavated from the ground for off-site treatment while in the latter the soil remains in place during treatment. A combination of these methods may also be conducted in the field.

9.3.2 CONTAINMENT TECHNOLOGIES

Containment or encapsulation is a very traditional way of controlling the movement of the pollution. In this method, the contaminated soil is surrounded by impermeable material from the top (capping), around the pollution (vertical barriers), and beneath the pollution (horizontal base). In fact, based on the soil condition, one or more of these barriers may be required. The method is a common practice in geotechnical engineering and is effectively used in sealing areas from penetration of water. Vertical systems are applied to induce low permeable zones around the pollution using low permeable materials such as grouting curtains, pile walls, cut of walls, trench barriers, or soil-cement columns. Horizontal barriers can be blankets of clay, soil-cement, or impermeable geosynthetics (geomembranes) at the top of the pollution, and grouting/jet grouting at the bottom.

The method is useful to prevent water flow into the pollution – reducing PFAS migration to other areas. However, complete isolation is almost impossible and needs extensive investment and accurate application. In reality, it is very difficult to ensure that a complete containment is achieved; chemical attacks to the barriers and their weathering may reduce their long-term stability and leachate may increase through time, which requires continuous monitoring and long-term management. More importantly, the method is not a real treatment to the contaminants, and the pollution will remain in the environment for a long time considering the long biological half-life of PFASs.

9.3.2.1 Off-Site Disposal in Landfills

Another treatment technology involves the excavation of contaminated soil to be transferred and deposited into a specific landfill site. This can involve the whole contaminated soil or immobilized contamination, depending on the required threshold applied by the relevant agencies. In some countries, very restrict guidelines are applied. For example, the threshold applied in Australia for total concentration of PFAS for landfilling is 50 mg kg^{-1} (HEPA, 2018). Landfills with specific designs and construction procedures with continuous monitoring are required. For the existing landfills, specific assessment is required to determine whether it is appropriate for accepting PFAS contaminated soil.

The landfill itself needs to be completely sealed by impermeable materials and be located on impermeable soil layers. In modern landfills, a combination of geosynthetic materials is used for sealing the whole disposal materials and to prevent

filtration and drainage from rainfall. However, rainwater usually enters during the landfilling process and mixes all the contaminants, and the wastewater flows to the potential exit points or boundaries. The leachate from the landfill needs to be collected and treated and the resultant PFAS must be detected and destroyed. Leachate should be checked regularly with a clear reference to environmental guideline values. Immobilization methods may be used for PFAS control before dumping in the landfill to reduce the PFAS transport into the leachate.

Although PFAS is reduced in the contaminated site, it remains in another location, which needs continuous monitoring. In addition, other chemicals in the form of liquid in the landfill may interact with PFASs and may cause them to release in the leachate. Therefore, the long-term liability of this method may be in doubt, particularly if PFAS is deposited in traditional landfills with low monitoring facilities and maintenance.

9.3.3 SOIL REMEDIATION

In soil remediation, different physical and chemical activities are conducted on the soil mass to immobilize, destruct, or remove the PFAS contaminants. Soil remediation technologies are more reliable due to the reduction of PFAS in the soil mass or the leachate. Different strategies have been applied from small scale in the laboratory to field application. Some of these methods have been commercialized. Most of the methods involve *ex situ* treatment; in fact, contaminated soil is excavated and transferred to a remediation center for treatment, and the treated soil is taken back to the site and compacted based on the requirement of the land use. Options for *in situ* remediation of PFAS contaminated lands are very limited and need further study.

In this section soil remediation for PFAS contamination is divided to four categories: immobilization technologies, destruction technologies, separation technologies, and other methods. These methods are described and their pros and cons are discussed with reference to the recent literature.

9.3.3.1 Immobilization Technologies

Immobilization (fixation) is a technique to fix the contaminant in place and usually is used for treatment of contaminants at shallow depths. For instance, the soil particles are mixed with sorbent as a binding agent, making them insoluble or low soluble compounds. The sorbents may include activated carbon (AC; powdered, PAC, or granular, GAC), resins, and oregano-modified clay-based agents, carbon nanotubes, and biomaterials.

Commercially available products exist for remediation through immobilization. However, laboratory tests should be conducted on the soil samples from the contaminated site to evaluate the effectiveness of the product and identify the best sorbent before the field application. The products were used in laboratory scale, field trials, and field conditions. Some products with the clay-based material, such as MatCARE™ (Modified clay adsorbent) and RemBind™

(AC + Aluminum hydroxide + Kaolin clay), have been used to stabilize PFAS in soil mass with successful results on soil samples from Royal Australian Air Force (RAAF) bases (Das et al., 2013) and two other airport sites with more than 99% efficiency in PFOS sorption (Zilteck, 2017). In the study on soils from a firefighting training site at an airport in Norway, the addition of AC to contaminated soils resulted in almost complete removal of PFCs from the water phase (Kupryianchyk et al., 2016). AC and montmorillonite were studied on sandy soil samples from a Norwegian airport site and achieved a 99% and 35% reduction in PFOS in the leachate, respectively (Hale et al., 2017). However, the efficiency of GAC and PAC is reduced in the presence of organic co-contaminants (National Groundwater Association [NGWA], 2017). Other minerals such as silica, iron oxides, and zeolites were used to remove contaminants from soil and water (ITRC, 2018). Stabilizers such as cement, fly ash, and pozzolans can be added to the soil to further immobilize the particles. However, care should be taken when applying these stabilizers as cement increases pH in the soil matrix, which in turn increases PFAS mobility.

Factors such as clay content, organic content, and pH are effective in sorption of PFAS to the soil particles. An increase in clay content and organic content is associated with the higher attachment of PFOS to the soil particles, while pH value has an opposite effect (Das et al., 2013). In fact, hydrophobic interaction, ion exchange, surface complexing, and hydrogen bonding all play a role in the process of PFOS bonding onto soil particles (Wei et al., 2017).

Although the method has been commercialized and been effective in reducing PFAS concentrate in groundwater, the technique does not destroy or remove PFAS from soil media, and treated soil will potentially become toxic due to weathering and aging effects. In fact, the long-term efficiency of the method needs further study. In addition, in practice a large amount of these adsorbents may be required, which increases the cost of remediation. Also, the options for land reuse may be limited because of changes of the soil matrix due to the use of mostly clay-based material and potential toxicity in the future.

9.3.3.2 Destruction Technologies

In destruction technologies, the contaminants are destroyed in the soil mass mostly through three methods – chemical oxidation, thermal treatment, and biological remediation. However, due to a strong C-F bound and a high melting point of PFAS substances, options for remediation are very limited.

9.3.3.2.1 Thermal Treatment

In thermal treatment methods, the chemicals are restructured by applying heat directly to the contaminated soil as an *ex situ* technology. For instance, after excavation and initial processing and size reduction screening, the contaminated soil is heated in a rotary kiln, usually up to 500°C–600°C (more than the boiling points of PFASs) to vaporize organic contaminants including PFAS from the soil matrix. The resultant gas stream is converted to stream with carbon dioxide and heated at high temperature (>1200°C) to break the PFAS compounds

and capture fluoride with a scrubber to meet emission standards. The cleaned soil is reused as fill. The whole process is called thermal desorption (TD).

Although the boiling points of PFASs are high, which affects the feasibility of the remediation method due to extensive energy use, remediation of 20 types of PFASs was reported by (Enviropacific, 2017) with more than 99.9% efficiency in PFOS and PFOA removal. Endpoint (2017) introduced a Vapor Energy Generator (VEG) system to remediate PFAS contaminated soil. The efficacy of the method was checked on spiked soil samples with 99% efficiency with a temperature of 950°C for 30 minutes. The method needs further development and more study to determine the optimum treatment condition and to develop mass balance to be sure that PFASs are completely destroyed.

Thermal treatment is an energy intensive approach and requires a high capital cost, especially for large contaminated sites. Contaminated soils with high clay and moisture content require higher energy input than sand, silt, and peat (Sarsby, 2013). Although there are potential options for *in situ* thermal treatment in general (US EPA, 2012), no *in situ* treatment has yet been applied for PFAS remediation, probably due to the high temperature required for PFAS desorption and the long treatment time required when the concentration level is high and contaminated site is large and deep.

9.3.3.2.2 Chemical Oxidation

In permeable soils, where a better contact is possible between contaminants and oxidants, chemical oxidation can be used for soil treatment. However, the process needs to be combined with other methods for higher efficiency. For *in situ* chemical oxidation (ISCO), oxidant agents can be applied at one side of the contaminated land, and groundwater can be extracted from the other side. This method has been used in soil contaminated by chlorinated solvents and petroleum hydrocarbons (Amarante, 2000).

Due to high electronegativity of fluorine atoms around the carbon chain, the rate of oxidation of PFASs, including PFOS and PFOA, is very slow (Vecitis et al., 2009). Breaking C-F bound requires high energy, and even an advanced oxidation process was not successful in the treatment of PFOS using ozone, ozone and UV, and Fenton's reagent (Schröder and Meesters, 2005). The decomposition rate of PFOS was reported to be 46.7% and 71.7% using permanganate (MnO_4^-) after 18 days in reactors at 65°C and 85°C, respectively, with high acidity (Liu et al., 2012). A reduction in pH and the addition of MnO_2 helped to increase the decomposition rate. The efficiency of the ISCO treatment method was reported by Eberle et al. (2017) for remediation of chlorinated volatile organic compounds and PFAA co-contaminants at a former firefighting site using peroxone activated sodium persulfate (OxyZone) technology in a pilot scale field test. The PFAA concentration in samples from both groundwater and soil was reduced with the total PFAS concentration between 21% and 79% in the monitoring well.

A thorough literature review on the use of the oxidation technique was explained by Dombrowski et al. (2018), indicating an evolving chemical mechanism for oxidation destruction of PFAS. Their results on a diverse list of PFASs showed that heat-activated persulfate was the oxidation method with the best degradation of PFAS in a soil slurry. They suggested repeated injection of the reagent with a lower dosage for the efficient removal of PFOA. However, the efficiency for PFOS oxidation was less than 20%.

9.3.3.2.3 Biological Remediation

Another destruction method is biological remediation (bioremediation), although the success rate for PFAS remediation has been very low.

Biodegradation of PFAS is not an effective method for the mineralization of PFAS. To initiate an attack on the PFAS substances via bacteria, at least one hydrogen atom in the perfluroalkyl chain is required (Key and Criddle, 1998). In addition, fluorine atoms form a dense hydrophobic layer around C–C bonds, which causes PFASs not to be used by microorganisms as carbon and energy sources (Colosi et al., 2009).

9.3.3.3 Separation Technologies

In separation technologies, the contaminants are detached from soil particles either *ex situ* or *in situ* via the application of a combination of water and solvent. This method needs pre-processing for soil screening and post treatment for water. *Ex situ* remediation is done via soil washing technology, and *in situ* is possible through soil-flushing technology.

9.3.3.3.1 Soil Washing Technique

Because of the mineralogy and high surface of fine particles, contaminants have a high tendency to attach to them. Soil washing can be used to separate contaminants from the surface of course particles and remove contaminated fine particles from soil mass. The result is a reduction in the volume of the contaminated soil. In addition to water, solvents may be added to the soil mass to detach PFASs from soil particles and release them to the leachate. Amendments such as surfactants, acids, and chelating agents can be used to increase the solubility of contaminants or desorption of contaminants from the soil particles. A subsequent treatment method is required for PFAS removal from the extracted contaminant.

Soils with low contents of fluoride and sulfide, CEC between 50 and 100 meq kg^{-1}, coarse particle sizes (0.25–2 mm), and contaminant solubility in water of more than 1000 mg L^{-1} can most effectively be cleaned by soil washing (Mulligan et al., 2001). PFASs are soluble in water and have a tendency to release to aqueous phase, especially substances like PFOA.

Adsorption and desorption of PFAS to and from soil particles plays an important role in the soil washing remediation technology. PFOA has lesser adsorption to soil with high solubility in water, so it can be found mostly in the liquid phase, while PFOS is attached strongly to the soil particles (Higgins and Luthy, 2006), and an appropriate solvent is needed to detach it from soil to the water phase.

The efficiency of regular soil washing and advanced soil washing on PFAS contaminated soil samples was reported as 99.7% and 99.9%, respectively (Pancras et al., 2013). Bench scale trials were conducted by DEC Environmental Solutions for PFOS removal with soil washing technology, with efficiencies of more than 99% and 94% in sandy soils and sandy clay soils, respectively (De Bruecker, 2015). With little information about remediation of PFAS contaminated soils with soil washing, further studies are required to identify the solvent characteristics, its optimization, and the dilution factor.

9.3.3.3.2 In Situ *Soil Flushing Technique*

In this technique, the contaminated site is flooded with an appropriate solution to remove the contaminant from the soil. The flushing solvent enables contaminants to mobilize by solubilization, the formation of emulsions, or a chemical reaction with the solvent. Flushing solvent is collected after passing the contaminated zone and is pumped back to the surface for disposal, recirculation, or on-site treatment and reinjection. Traditional flushing techniques rely on the ability to deliver, control the flow, and recover the flushing fluid via a pump-and-treat system.

A number of flushing solutions are applied in the field including water, acidic aqueous solutions, basic solutions, chelating or complexing agents, reducing agents, co-solvents, and surfactants. Water is an effective solvent to extract water-soluble (hydrophilic) or water-mobile constituents. For organic contaminants, co-solvents and surfactants have given promising results. Heating the flushing solution also can help to mobilize organic contaminants (Technology and Council, 2009).

The selection of the *in situ* soil flushing technology for a given contaminated site is largely dependent on the site characterization, the extent of remediation required, the initial PFAS concentration in the soil, and the flushing solvent. Although the method is highly effective in highly permeable soil, a significant removal of contaminant is achievable in contaminated land with low permeability silty soil conditions in both shallow and deep aquifers (Mulligan et al., 2001).

Depending on the groundwater flow, the location of injection of the flushing solvent and the pumping well should be designed in such a way that the process does not contaminate other areas. Maximum attempt is needed to gather the applied solvent from the groundwater not to contaminate the site additionally. For instance, precise monitoring is required through different bore holes to check and control any increase in the PFAS contaminant in the soil mass and groundwater.

Even though soil flushing is considered a mature technology because of its use for decades for both organic and inorganic contaminants, the feasibility of this technique to remediate PFAS contaminated soil has not been reported. As with soil washing technology, there is a need to identify the flushing solvent and its dilution factor and to recover the solvent and remove the PFAS contaminants from the pumped water.

9.3.3.4 Other Methods

Due to growing concerns about PFAS contaminants, several other methods have been studied mostly at the laboratory scale to try to find new remediation strategies. Some methods are still under development due to a recent focus of governments on PFAS decontamination. The use of ball milling was reported by Zhang et al. (2013) for the destruction of PFOS and PFOA in the laboratory with high efficiency in 6 and 3 hours, respectively. The method is based on the impact of steel balls with the soil particles inside a cylindrical container. Electron beam irradiation was used by Wang et al. (2016) to defluorinate PFOA in synthetic treated wastewater. Ma et al. (2017) reported decomposition efficiencies of PFOA and PFOS as 95.7% and 85.9%, respectively, using an electron beam. Again the method was used at laboratory scale only, and further development of the method is required.

More recently, soil liquefractionation is suggested by Niven et al. (2018) as a potential new approach to remediate PFAS contaminated land. The method involves a combination of foam fractionation and soil liquefaction. Foam fractionation has been used by OPEC Systems (2017) to remove PFAS *ex situ* in water treatment facilities or *in situ* from bore holes by injecting micro bubbles into the water and extracting contaminants from the bubbles that rise to the surface. Soil liquefractionation involves liquefying the soil particles and the application of foam fractionation, which is likely to be more successful in cohesion-less soils.

REFERENCES

Amarante, D. (2000) Applying in situ chemical oxidation. *Pollution Engineering* 32(2), 40–42.

Baudequin, C., Couallier, E., Rakib, M., Deguerry, I., Severac, R. and Pabon, M. (2011) Purification of firefighting water containing a fluorinated surfactant by reverse osmosis coupled to electrocoagulation–filtration. *Separation and Purification Technology* 76(3), 275–282.

Campo, J., Lorenzo, M., Pérez, F., Picó, Y., la Farré, M. and Barceló, D. (2016) Analysis of the presence of perfluoroalkyl substances in water, sediment and biota of the Jucar River (E Spain). Sources, partitioning and relationships with water physical characteristics. *Environmental Research* 147, 503–512.

Campo, J., Pérez, F., Masiá, A., Picó, Y., la Farré, M. and Barceló, D. (2015) Perfluoroalkyl substance contamination of the Llobregat River ecosystem (Mediterranean area, NE Spain). *Science of the Total Environment* 503, 48–57.

Chen, J. and Zhang, P. (2006) Photodegradation of perfluorooctanoic acid in water under irradiation of 254 nm and 185 nm light by use of persulfate. *Water Science and Technology* 54(11–12), 317–325.

Colosi, L.M., Pinto, R.A., Huang, Q. and Weber, W.J., Jr. (2009) Peroxidase-mediated degradation of perfluorooctanoic acid. *Environmental Toxicology and Chemistry* 28(2), 264–271.

Das, P., Arias E, V., Kambala, V., Mallavarapu, M. and Naidu, R. (2013) Remediation of perfluorooctane sulfonate in contaminated soils by modified clay adsorbent – A risk-based approach. *An International Journal of Environmental Pollution* 224(12), 1–14.

De Bruecker, T. (2015) Status Report PFOS Remediation, DEC Environmental Solutions, p. 12.

Dombrowski, P.M., Kakarla, P., Caldicott, W., Chin, Y., Sadeghi, V., Bogdan, D., Barajas-Rodriguez, F. and Chiang, S.-Y. (2018) Technology review and evaluation of different chemical oxidation conditions on treatability of PFAS. (Report). *Remediation: The Journal of Environmental Cleanup Costs, Technologies & Techniques* 28(2), 135.

Eberle, D., Ball, R. and Boving, T.B. (2017) Impact of ISCO treatment on PFAA co-contaminants at a former fire training area. *Environmental Science & Technology* 51(9), 5127–5136.

Endpoint (2017) Bench-scale VEG research & development study: Implementation memorandum for ex-situ thermal desorption of perfluoroalkyl compounds (PFCs) in soils.

Enviropacific (2017) Treatment of PFAS in Soils, Sediments and Water.

Fujii, S., Polprasert, C., Tanaka, S., Lien, H., Pham, N. and Qiu, Y. (2007) New POPs in the water environment: Distribution, bioaccumulation and treatment of perfluorinated compounds–a review paper. *Journal of Water Supply: Research and Technology-AQUA* 56(5), 313–326.

Getoff, N. (1996) Radiation-induced degradation of water pollutants – state of the art. *Radiation Physics and Chemistry* 47(4), 581–593.

Gupta, V.K., Ali, I. 2012. Environmental water: advances in treatment, remediation and recycling. Elsevier, Amsterdam, the Netherlands, 7–8.

Hale, S.E., Arp, H.P.H., Slinde, G.A., Wade, E.J., Bjørseth, K., Breedveld, G.D., Straith, B.F., Moe, K.G., Jartun, M. and Høisæter, Å. (2017) Sorbent amendment as a remediation strategy to reduce PFAS mobility and leaching in a contaminated sandy soil from a Norwegian firefighting training facility. *Chemosphere* 171, 9–18.

Haukås, M., Berger, U., Hop, H., Gulliksen, B. and Gabrielsen, G.W. (2007) Bioaccumulation of per-and polyfluorinated alkyl substances (PFAS) in selected species from the Barents Sea food web. *Environmental Pollution* 148(1), 360–371.

HEPA (2018) PFAS National Environmental Management Plan, Heads of EPAs Australia and New Zealand (HEPA).

Higgins, C. and Luthy, R. (2006) Sorption of perfluorinated surfactants on sediments. *Environmental Science & Technology* 40(23), 7251–7256.

Hong, S., Khim, J.S., Wang, T., Naile, J.E., Park, J., Kwon, B.-O., Song, S.J., Ryu, J., Codling, G. and Jones, P.D. (2015) Bioaccumulation characteristics of perfluoroalkyl acids (PFAAs) in coastal organisms from the west coast of South Korea. *Chemosphere* 129, 157–163.

Hori, H., Hayakawa, E., Einaga, H., Kutsuna, S., Koike, K., Ibusuki, T., Kiatagawa, H. and Arakawa, R. (2004a) Decomposition of environmentally persistent perfluorooctanoic acid in water by photochemical approaches. *Environmental Science & Technology* 38(22), 6118–6124.

Hori, H., Hayakawa, E., Koike, K., Einaga, H. and Ibusuki, T. (2004b) Decomposition of nonafluoropentanoic acid by heteropolyacid photocatalyst $H_3PW_{12}O_{40}$ in aqueous solution. *Journal of Molecular Catalysis A: Chemical* 211(1–2), 35–41.

Hori, H., Nagaoka, Y., Yamamoto, A., Sano, T., Yamashita, N., Taniyasu, S., Kutsuna, S., Osaka, I. and Arakawa, R. (2006) Efficient decomposition of environmentally persistent perfluorooctanesulfonate and related fluorochemicals using zerovalent iron in subcritical water. *Environmental Science & Technology* 40(3), 1049–1054.

Hori, H., Yamamoto, A., Hayakawa, E., Taniyasu, S., Yamashita, N., Kutsuna, S., Kiatagawa, H. and Arakawa, R. (2005) Efficient decomposition of environmentally persistent perfluorocarboxylic acids by use of persulfate as a photochemical oxidant. *Environmental Science & Technology* 39(7), 2383–2388.

Hori, H., Yamamoto, A., Koike, K., Kutsuna, S., Osaka, I. and Arakawa, R. (2007) Photochemical decomposition of environmentally persistent short-chain perfluorocarboxylic acids in water mediated by iron (II)/(III) redox reactions. *Chemosphere* 68(3), 572–578.

ITRC (2018) Remediation Technologies and Methods for Per- and Polyfluoroalkyl Substances (PFAS), Interstate Technology Regulatory Council, Washington, DC.

Key, B.D. and Criddle, C.S. (1998) Defluorination of organofluorine sulfur compounds by *Pseudomonas* sp. strain D2. *Environmental Science and Technology* 32(15), 2283–2287.

Kupryianchyk, D., Hale, S.E., Breedveld, G.D. and Cornelissen, G. (2016) Treatment of sites contaminated with perfluorinated compounds using biochar amendment. *Chemosphere* 142, 35–40.

Kutsuna, S. and Hori, H. (2007) Rate constants for aqueous-phase reactions of SO_4^- with $C_2F_5C(O)O^-$ and $C_3F_7C(O)O^-$ at 298 K. *International Journal of Chemical Kinetics* 39(5), 276–288.

Labadie, P. and Chevreuil, M. (2011) Partitioning behaviour of perfluorinated alkyl contaminants between water, sediment and fish in the Orge River (nearby Paris, France). *Environmental Pollution* 159(2), 391–397.

Liu, C.S., Shih, K. and Wang, F. (2012) Oxidative decomposition of perfluorooctanesulfonate in water by permanganate. *Separation and Purification Technology* 87(C), 95–100.

Ma, S.-H., Wu, M.-H., Tang, L., Sun, R., Zang, C., Xiang, J.-J., Yang, X.-X., Li, X. and Xu, G. (2017) EB degradation of perfluorooctanoic acid and perfluorooctane sulfonate in aqueous solution. *Nuclear Science and Techniques* 28(9), 137.

Martín, J., Hidalgo, F., García-Corcoles, M.T., Ibáñez-Yuste, A.J., Alonso, E., Vilchez, J.L. and Zafra-Gómez, A. (2019) Bioaccumulation of perfluoroalkyl substances in marine echinoderms: Results of laboratory-scale experiments with *Holothuria tubulosa* Gmelin, 1791. *Chemosphere* 215, 261–271.

Mulligan, C.N., Yong, R.N. and Gibbs, B.F. (2001) Remediation technologies for metal-contaminated soils and groundwater: An evaluation. *Engineering Geology* 60(1), 193–207.

Naile, J.E., Khim, J.S., Hong, S., Park, J., Kwon, B.-O., Ryu, J.S., Hwang, J.H., Jones, P.D. and Giesy, J.P. (2013) Distributions and bioconcentration characteristics of perfluorinated compounds in environmental samples collected from the west coast of Korea. *Chemosphere* 90(2), 387–394.

National Ground Water Association. (2017) Groundwater and PFAS: State of knowledge and practice. National Groundwater Association Press. Available at: https://www.ngwa.org/publications-andnews/Newsroom/ (accessed on 28 January 2020).

Niven, R., Khalili, N., Pashley, R., Taylor, M., Strezov, V., Wilson, S., Murphy, P. and Phillips, S. (2018) PFAS source zone remediation by foam fractionation and in situ fluidisation, Australian Research Council (ARC), Scheme Round Statistics for Approved Proposals – Special Research Initiatives 2018 round 1.

Ochoa-Herrera, V. and Sierra-Alvarez, R. (2008) Removal of perfluorinated surfactants by sorption onto granular activated carbon, zeolite and sludge. *Chemosphere* 72(10), 1588–1593.

OPEC Systems (2017) Downhole Foam Fractionation (DFF) Solutions.

Osborne M.C., Li Q. and Smith I.W.M. (1999) Products of the ultraviolet photodissociation of trifluoroacetic acid and acrylic acid. *Physical Chemistry Chemical Physics* 1(7), 1447–1454.

Osborne, M.C., Li, Q. and Smith, I.W.M. (1999) Products of the ultraviolet photodissociation of trifluoroacetic acid and acrylic acid.

Pancras, T., Plaisier, W., Barbier, A., Ondreka, J., Burdick, J. and Hawley, E. (2013) *Challenges of PFOS Remediation*, Barcelona, Spain.

Park, H., Vecitis, C.D., Cheng, J., Choi, W., Mader, B.T. and Hoffmann, M.R. (2009) Reductive defluorination of aqueous perfluorinated alkyl surfactants: Effects of ionic headgroup and chain length. *The Journal of Physical Chemistry A* 113(4), 690–696.

Pignotti, E., Casas, G., Llorca, M., Tellbüscher, A., Almeida, D., Dinelli, E., Farré, M. and Barceló, D. (2017) Seasonal variations in the occurrence of perfluoroalkyl substances in water, sediment and fish samples from Ebro Delta (Catalonia, Spain). *Science of the Total Environment* 607, 933–943.

Rattanaoudom, R. (2011) Membrane hybrid system for removal of PFOS and PFOA in industrial waste water: Application of conventional adsorbents and nanoparticles, PhD dissertation Environmental Engineering and Management Inter-University.

Ritter, L., Solomon, K., Forget, J., Stemeroff, M. and O'leary, C. (1995) A review of selected persistent organic pollutants. International Programme on Chemical Safety (IPCS). PCS/95.39. Geneva: World Health Organization 65, 66.

Sarsby, R.W. (2013) *Environmental Geotechnics*, ICE Publishing, London, UK.

Schröder, H.F. and Meesters, R.J.W. (2005) Stability of fluorinated surfactants in advanced oxidation processes – A follow up of degradation products using flow injection–mass spectrometry, liquid chromatography–mass spectrometry and liquid chromatography–multiple stage mass spectrometry. *Journal of Chromatography A* 1082(1), 110–119.

Senevirathna, S.T.M.L.D., 2010. Development of effective removal methods of PFCs (perfluorinated compounds) in water by adsorption and coagulation, Doctoral Dissertation (Engineering), Kyoto University (http://hdl.handle.net/2433/126798)

Tang, C.Y., Fu, Q.S., Robertson, A.P., Criddle, C.S., Leckie, J.O., 2006. Use of reverse osmosis membranes to remove perfluorooctane sulfonate (PFOS) from semiconductor wastewater. *Environ. Sci. Technol.* 40, 7342–7349.

Tang, C.Y., Fu, Q.S., Criddle, C.S. and Leckie, J.O. (2007) Effect of flux (transmembrane pressure) and membrane properties on fouling and rejection of reverse osmosis and nanofiltration membranes treating perfluorooctane sulfonate containing wastewater. *Environmental Science & Technology* 41(6), 2008–2014.

Taylor, M.D. and Johnson, D.D. (2016) Preliminary investigation of perfluoroalkyl substances in exploited fishes of two contaminated estuaries. *Marine Pollution Bulletin* 111(1–2), 509–513.

Technology, I. and Council, R. (2009) Evaluating LNAPL remedial technologies for achieving project goals, Author, Washington, DC.

Tseng, N., Wang, N., Szostek, B. and Mahendra, S. (2014) Biotransformation of 6:2 fluorotelomer alcohol (6:2 FTOH) by a wood-rotting fungus. *Environmental Science & Technology* 48(7), 4012–4020.

US EPA (2012) A Citizen's Guide to In Situ Thermal Treatment, United States Environmental Protection Agency.

Vecitis, C., Park, H., Cheng, J., Mader, B. and Hoffmann, M. (2008) Enhancement of perfluorooctanoate and perfluorooctanesulfonate activity at acoustic cavitation bubble interfaces. *The Journal of Physical Chemistry C* 112(43), 16850–16857.

Vecitis, C.D., Park, H., Cheng, J., Mader, B.T. and Hoffmann, M.R. (2009) Treatment technologies for aqueous perfluorooctanesulfonate (PFOS) and perfluorooctanoate (PFOA). *Frontiers of Environmental Science & Engineering in China* 3(2), 129–151.

Wang, J., Zhang, Y., Zhang, F., Yeung, L.W., Taniyasu, S., Yamazaki, E., Wang, R., Lam, P.K., Yamashita, N. and Dai, J. (2013) Age-and gender-related accumulation of perfluoroalkyl substances in captive Chinese alligators (*Alligator sinensis*). *Environmental Pollution* 179, 61–67.

Wang, L., Batchelor, B., Pillai, S.D. and Botlaguduru, V.S.V. (2016) Electron beam treatment for potable water reuse: Removal of bromate and perfluorooctanoic acid. *Chemical Engineering Journal* 302, 58–68.

Wang, Y., Liu, W., Zhang, Q., Zhao, H. and Quan, X. (2015a) Effects of developmental perfluorooctane sulfonate exposure on spatial learning and memory ability of rats and mechanism associated with synaptic plasticity. *Food and Chemical Toxicology* 76, 70–76.

Wang, Y., Rogan, W.J., Chen, H.-Y., Chen, P.-C., Su, P.-H., Chen, H.-Y. and Wang, S.-L. (2015b) Prenatal exposure to perfluroalkyl substances and children's IQ: The Taiwan maternal and infant cohort study. *International Journal of Hygiene and Environmental Health* 218(7), 639–644.

Wang, Z., DeWitt, J.C., Higgins, C.P. and Cousins, I.T. (2017) A never-ending story of per- and polyfluoroalkyl substances (PFASs)? *Environmental Science & Technology* 51(5), 2508–2518.

Wei, C., Song, X., Wang, Q. and Hu, Z. (2017) Sorption kinetics, isotherms and mechanisms of PFOS on soils with different physicochemical properties. *Ecotoxicology and Environmental Safety* 142, 40–50.

Wilkinson, J.L., Hooda, P.S., Swinden, J., Barker, J. and Barton, S. (2018) Spatial (bio) accumulation of pharmaceuticals, illicit drugs, plasticisers, perfluorinated compounds and metabolites in river sediment, aquatic plants and benthic organisms. *Environmental Pollution* 234, 864–875.

Yamamoto, T., Noma, Y., Sakai, S.-i. and Shibata, Y. (2007) Photodegradation of perfluorooctane sulfonate by UV irradiation in water and alkaline 2-propanol. *Environmental Science & Technology* 41(16), 5660–5665.

Yin, N., Yang, R., Liang, S., Liang, S., Hu, B., Ruan, T. and Faiola, F. (2018) Evaluation of the early developmental neural toxicity of F-53B, as compared to PFOS, with an in vitro mouse stem cell differentiation model. *Chemosphere* 204, 109–118.

Yu, Q., Zhang, R., Deng, S., Huang, J. and Yu, G. (2009) Sorption of perfluorooctane sulfonate and perfluorooctanoate on activated carbons and resin: Kinetic and isotherm study. *Water Research* 43(4), 1150–1158.

Zhang, K., Huang, J., Yu, G., Zhang, Q., Deng, S.B. and Wang, B. (2013) Destruction of perfluorooctane sulfonate (PFOS) and perfluorooctanoic acid (PFOA) by ball milling. *Environmental Science & Technology* 47(12), 6471–6477.

Zhang, Q., Zhao, H., Liu, W., Zhang, Z., Qin, H., Luo, F. and Leng, S. (2016) Developmental perfluorooctane sulfonate exposure results in tau hyperphosphorylation and β-amyloid aggregation in adults rats: Incidence for link to Alzheimer's disease. *Toxicology* 347, 40–46.

Zilteck (2017) RemBind™ Product Overview.

10 Phytomanagement of Pollutants in Soil and Groundwater

Sofie Thijs, Andon Vasilev Andonov, Malgorzata Wojcik, and Jaco Vangronsveld

CONTENTS

10.1 INTRODUCTION

Past and present industrial activities, agricultural, and residential practices continue to result in the pollution of soils and groundwater by organic and inorganic pollutants. As a remedial action, experts chose more and more for greener and low environmental footprint remediation solutions, as alternative to the existing methods to tackle the pollution. *Phytoremediation* is such a green remediation option that involves the use of plants and their associated microorganisms to degrade, detoxify, or stabilize pollutants *on-site*, and thereby maximize sustainability, reduce energy usage and emissions, promote carbon neutrality, and protect and preserve soil structure and land resources. Because of these advantages, phytoremediation is gaining increased popularity compared to the traditional remediation methods. This chapter provides an overview of the state of the art of phytotechnologies that can be adopted for metal polluted soils and groundwater polluted with organics compounds.

Soil and groundwater pollution is of all times. Between 800,000 and 1 million potential brownfield sites exist in Europe (Megharaj and Naidu, 2017). In the United States alone, over 500,000 polluted sites have been identified (Doty, 2008), and a similar number in Asia. Pollutants get into soil from accidental leaks and spills, deliberate dumping, or accumulated polluted dust and aerial deposition. Once in the soil, pollutants may evaporate, bind to soil particles, or migrate downward. Groundwater flow through the soil can direct a plume of pollutants. Over time, pollutants will spread through the soil potentially polluting drinking water supplies.

About $6–8 billion are spent annually for managing and remediating polluted sites (Glass, 1999; Tsao, 2003). The overwhelming majority of that goes to managing, using dig and dump to take the pollutants elsewhere, or cap and contain to seal them off. These approaches do not solve the problem, not in the long term anyway, but they are fast. Of the million-dollar budget costs for remediation, only about $10 million is spent on bioremediation, a figure that could grow to $40 million in the next decade. Traditional remediation techniques used for brownfield rehabilitation usually involve the excavation of polluted soil and its disposal into an off-site landfill, or the capping of polluted soil with clay or concrete and then importing clean fill for planting over the top of the cap. Polluted groundwater left on-site is usually mechanically pumped, treated, and recharged.

Phytoremediation to treat polluted soils and groundwater has received increased attention over the last decades as a green alternative. It is a technology that refers to the direct utilization of living green plants and their associated microbiota for *in situ* removal (pump and treat), degradation, or containment of pollutants. It has been successfully applied to a variety of sites including pipelines, industrial and municipal landfills, agricultural fields, wood treating sites, military bases, fuel storage tank farms, gas stations, army ammunition plants, mining sites, and residential sites. Field studies have included the remediation of typical inorganic pollutants such as salts, heavy metals, metalloids, and radionuclides and organic pollutions such as chlorinated compounds, petroleum hydrocarbons, crude oil, polychlorinated biphenyls (PCBs), pesticides, and explosives compounds (Sleegers, 2010).

Phytoremediation gained in popularity because it is considered one of the cheapest and most cost-effective ways to remove pollutants from soils and groundwater. There is no need for classical pump-and-treat installations or lengthy physicochemical treatments, both of which are prohibitively expensive and have a high carbon footprint. In this sense, plants provide a solar-powered, low-cost alternative to reduce leaching of pollutants through groundwater and avoid dispersion; it is most effective, however, if the pollutants are present at a shallow depth, in non-phytotoxic concentrations, and over large areas. An often mentioned drawback of phytoremediation are the sometimes longer remediation times and the climatic dependency of the photosynthetic activity of the plants. However, the longer time span can be turned into an advantage because phytoremediation co-creates a greener landscape and increases local biodiversity (Sleegers, 2010). Moreover, in the broader context of phytomanagement of polluted soils, production of bioenergy, fibers and aromatics, watershed management, carbon sequestration, can also be considered as potential advantages (for more details see below).

10.2   DESIGN AND IMPLEMENTATION CONSIDERATIONS

Prior to the consideration of phytoremediation as a viable remediation strategy for polluted groundwater, it is important to pose several site-specific questions, which will lead to the selection of the most suitable phytoremediation mechanism. Some considerations are evaluated for any remediation strategy, whether classic or bio-based, such as the operational and maintenance costs/requirements, analytical costs/needs, public perspective, and regulatory and legislative rules set by the local authority for the protection of the environment. In addition, some more specific investigations are needed for the remediation, such as knowledge about the impacted medium and location, pollutant(s) of concern and exposure assessment, hydrogeology and geochemistry, climate conditions, and existing vegetation. Below, some criteria in respect to phytoremediation are discussed.

Can the pollutant be taken up by the plant? Typically, for phytoremediation of organic pollutants, it is important to consider that the octanol:water partition coefficient (log K_{ow}) of the chemical is between 1 and 3.5 for uptake by plants to occur (Burken and Schnoor, 1998). The plant root membrane consists of a lipid bilayer making it partially hydrophobic and nonpolar. Too hydrophobic chemicals (log $K_{ow} > 3.5$) therefore are not sufficiently soluble in the transpiration stream or bind so strongly to the surface of roots that they cannot be easily translocated into the plant xylem. Too water-soluble compounds, in contrast, are not sufficiently sorbed by the roots, or actively transported through plant membranes because of

their high polarity. Most chlorinated solvents, in addition to benzene, toluene, ethylbenzene, xylene (BTEX) chemicals, and short-chain aliphatics, have a log K_{ow} falling within the range that allows them to be up-taken by plants, and are thus amenable for groundwater phytoremediation. Plant uptake of inorganic pollutants generally is quite efficient. In case of essential elements like zinc (Zn), specific membrane transporters are present. Non-essential elements like cadmium (Cd) and lead (Pb) "sneak in" via transporters for essential elements.

Will the pollutant accumulate in the plant? This can be a concern mainly for the inorganic pollutants because they do not degrade (Kennen and Kirkwood, 2015). Typically, concentrations are not of that level that they represent a risk for trophic level transfer, also because insects or birds migrate and do not solely consume the polluted biomass in a confined area. Organics usually do not accumulate and are susceptible to phyto/biodegradation or volatilization.

Plants are equipped with unique metabolic and absorption capabilities, and with transport vessels that can take up nutrients or pollutants from the polluted soil or groundwater. Phytoremediation of polluted groundwater, for instance, involves growing plants on the polluted site for the required period, to remove pollutants via volatilization, or facilitate immobilization (binding/containment) or degradation and detoxification (Kennen and Kirkwood, 2015). Several plants are known to tolerate high levels of pollutants, such as sunflowers that "capture" Zn (Thijs et al., 2018), ferns that "thrive on" arsenic (Ma et al., 2001), herbs that hoard Zn, mustards that "lap up" Pb, clovers that "eat" oil, and poplar trees that "destroy" chlorinated solvents (Weyens et al., 2010). Plants, however, do not act alone. They are associated with millions of microorganisms (fungi, bacteria, archaea), with the highest density typically found in the soil layer just nearby the roots, the rhizosphere. The microbial communities become less diverse when moving inside plant tissues, from roots to aboveground organs. In the phyllosphere, however, on the photosynthetically active leaves exposed to the ambient air, a reported number of 106–107 cells cm^{-2} has been reported (Vorholt, 2012). Microbes are known to form relationships with plants that generate many benefits for them, including increased tolerance to heat, drought, pollutants, and other stresses; accelerated development of seedlings; and increased growth and yield (de Vrieze, 2015). A particular group of plant beneficial microorganisms, endophytes, which live inside plant tissues without causing obvious visible symptoms, have a highly intimate association with plants and are of high importance to phytoremediation. They can assist their host in the degradation of water-soluble organic pollutants that travel through the xylem or in coping with inorganic pollutants that are absorbed by the plant. The term used to consider all the biotic and abiotic factors that determine together the plant growth, health, and functioning is called the phytobiome. The interactions that determine and grow a phytobiome are complex, and a lot remains to be discovered in this area.

In bacteria-assisted phytoremediation, the symbiotic relationships among microbes and plants is targeted for enhancement of phytoremediation in a way that depends on the purpose of the phytoremediation actions. Once the microbial composition is characterized and the roles of individual microbes are known, new techniques can be developed to harness the signaling cascades between plants and microorganisms to improve the benefits that they generate for their host plant, e.g. improving metal uptake, metabolizing organic pollutants (Quiza et al., 2015), or alternately, decrease the association with microorganisms that negatively impact plant health and phytoremediation efficiency.

10.3 MICROORGANISM-AUGMENTED PHYTOREMEDIATION TO SOLVE GROUNDWATER POLLUTION WITH ORGANIC POLLUTANTS

In this section, first some design and implementation considerations for groundwater phytoremediation are reviewed. Subsequently the underlying plant-microbe interactions and physicochemical processes important to phytoremediation are discussed, providing evidence of its working. We also propose different research directions and contexts to enhance the predictability and efficiency of this green remediation technology, and end with an outlook on further integration of genomics into groundwater remediation.

Is target groundwater available or can the target depth be accessed by special installations? Depths within 4.5 m are generally accessible by deep-rooting trees (Kennen and Kirkwood, 2015). Depth of the root penetration is dictated by the nutrient availability, moisture, oxygen, and ease or resistance to grow in the matrix. Compact formations, such as clay, bedrock, or layers with low porosity and lower water transmissivity are more resistant to root penetration. Also, regardless of the ability of roots to penetrate soil layers, plants will extend their roots just enough to supply the necessary nutrients and moisture for themselves. If this is achieved prior to accessing the underlying impacted polluted groundwater, additional measures should be taken in order not to restrict the efficacy of the phytoremediation system. Boring, trenching, ploughing, and installation of "straw" TreeWell® (Figure 10.1) are methods currently employed to bring roots in contact with target impacted groundwater that is otherwise not accessible.

Is thickness of the impacted groundwater layer a problem? Except for wetland conditions, and some species of willow and alder, roots typically do not immerse into the water saturated zone nor tolerate extended saturated conditions. Thus, if the pollution resides below these depths, the impacted zones will not be directly affected by the vegetation unless a TreeWell system is used. Phytotoxicity by organics is rarely an issue with the TreeWell systems, including LNAPL and dissolved phase in the 100–1000 mg L^{-1} range, but in other cases first removing pure product, e.g. using low-cost passive sorption systems, will allow reaching the cleanup goal faster.

Is time a limitation? Phytoremediation treatments may take longer than traditional pump-and-treat methods to reach

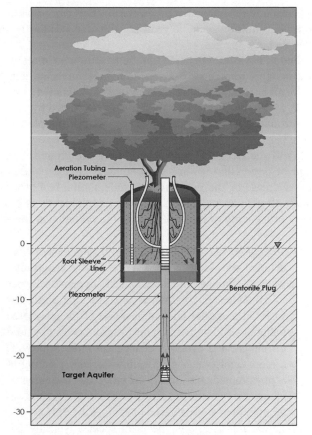

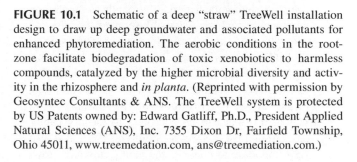

FIGURE 10.1 Schematic of a deep "straw" TreeWell installation design to draw up deep groundwater and associated pollutants for enhanced phytoremediation. The aerobic conditions in the root-zone facilitate biodegradation of toxic xenobiotics to harmless compounds, catalyzed by the higher microbial diversity and activity in the rhizosphere and *in planta*. (Reprinted with permission by Geosyntec Consultants & ANS. The TreeWell system is protected by US Patents owned by: Edward Gatliff, Ph.D., President Applied Natural Sciences (ANS), Inc. 7355 Dixon Dr, Fairfield Township, Ohio 45011, www.treemedation.com, ans@treemediation.com.)

final cleanup goals. That is mainly because the time of treatment must include the time needed to establish the plant community and reach optimal depth. However, if the phytoremediation can be started up as early as possible in the remediation plan, time is not necessarily a constraint, as the pump capacity of phreatophytic trees is high in summer, equaling traditional pump installations. In addition, pump and treat has diminishing returns as concentrations get lower, which makes it fairly impossible to get to the contaminants that are entrained in the capillary pores of the soil matrix, and this is not a problem with phytoremediation.

Is there sufficient area to plant or are obstructions present? Free space to put plants/trees is an obvious requirement for phytoremediation, though infrastructure can sometimes hinder the installation of a vegetation cover. A general rule of thumb is a planting density of 25 m^2 per tree, planting trees on an average 5 to 7 m centers using staggered rows (Kennen and Kirkwood, 2015). For many trees used in phytoremediation systems, a full canopy area of vegetation can remove between 10 and 20 million liters/ha/year, but this depends on the climate and tree species.

The removal rate per tree can be based on estimations or *in situ* measurements. Transpiration measurements can be performed using sap flow sensors or lysimeters (Mirck and Volk, 2009). Performance can also be determined by assessing groundwater and pollutant fluxes within the plantation using passive flux meters (Haluska et al., 2018; Verreydt et al., 2013). If the trees are actively pumping the groundwater, a cone of depression toward the boundary of the plantation, or a decrease of the groundwater table should be observed, which all give an indication whether the plantation is effectively influencing or interacting with the pollution at the site.

> *Hydrogeological conditions.* In groundwater remediation, detailed knowledge is needed about the static groundwater conditions but also about the dynamic changes in groundwater conditions (laterally and vertically, seasonal and tidal fluctuations) and of the associated pollutants and nutrients. This information can be used to mitigate the dynamic changes in order to provide a more stable water supply, and to guide planting of a phytoremediation stand covering the groundwater plume throughout the seasons based on the groundwater migration speed (Verreydt et al., 2010, 2015).

> *Plant species selection.* The plant species typically utilized in phytoremediation of groundwater pollution include phreatophytic species like poplars (*Populus* sp.) and willows (*Salix* sp.), and, for natural or constructed wetlands, *Phragmites* sp., *Typha* sp., and *Juncus* sp. (Coleman et al., 2002; Vymazal, 2011). Online phytoremediation-specific databases exist, for example, https://clu-in.org/products/phyto/search/phyto_search.cfm, that can be consulted to identify candidates from a list of species (McIntyre, 2003). It is also very important to consider the plants currently growing at the polluted site or in the region for performing phytoremediation as these are already adapted to the prevailing conditions (Bell et al., 2015). If there is no usable information to base the selection on, a feasibility study is typically conducted under controlled settings designed to mimic the site conditions as much as possible, which entails using water and soil from the site, and growing conditions representative for the climate of the area. The criteria for screening in these feasibility studies are biomass, root growth, transpiration rate, and signs of stress (chlorosis of leaves, curling leaves, wilting, etc.). When encountering toxic concentrations of pollutants, plants may respond by closing stomata and reducing sap flow in order to avoid the uptake of toxicants (Ferro et al., 2013). Toxic pollutant concentrations may eliminate phytoremediation as a viable option for site cleanup.

It is clear that for phytoremediation, detailed site information is crucial, as well as having the remediation goal(s) in mind. In the following paragraph several phytoremediation mechanisms are described, distinguished based on the distinct fate of pollutants in groundwater and the subsurface soil.

10.3.1 PHYTOREMEDIATION STRATEGIES FOR GROUNDWATER REMEDIATION

Once phytoremediation is considered for site cleanup, strategies are designed based on the pollutant characteristics but also on the remediation goals, which include containment of pollutants through stabilization or sequestration, volatilization and evapotranspiration, detoxification, transformation, and degradation, or a combination thereof. In the following sections, the most important phytoremediation mechanisms for treating organic pollutants are summarized.

> *Phytohydraulics* is a strategy whereby the plant pulls up water and dissolved pollutants can come with it (Ferro et al., 2003). Planting can change the direction of groundwater flows and thus can intercept plumes of polluted groundwater (Barac et al., 2009). The plant may utilize other processes including rhizodegradation, phytovolatilization, or phytodegradation to handle the pollutant.

> *Rhizodegradation* is an example in which the pollutants are degraded by plant exudates (e.g., secreted enzymes) and microorganisms outside the plant in the root zone (rhizosphere). The ideal scenario is when the organic pollutants are completely degraded by the plant and its associated microorganisms and accordingly are fully removed from the environment.

> *Phytovolatilization.* The plant takes up the pollutant or toxic degradation intermediates and transpires them to the atmosphere as a gas thus removing it from the site (Limmer and Burken, 2016). The gas is usually released sufficiently slowly such that the surrounding atmosphere is not significantly affected.

> *Phytodegradation.* Plants can break down and degrade organic pollutants while they usually contain and stabilize inorganics (Newman and Reynolds, 2004). The evapotranspiration makes them act as a pump attracting pollutants to the root zone. The pollutants are degraded in the rhizosphere and inside the plant after they are taken up in the plant. This degradation into smaller, usually non-toxic molecules is performed by plant enzymes and/or plant-associated microorganisms. The plant uses the by-products from the breakdown for its growth processes, leaving little pollution (Kennen and Kirkwood, 2015).

At all times, a specific plant often performs multiple processes and can deal with multiple organics or inorganics at the same time. The processes of phytoremediation which are described above ultimately result in the pollutant being transported into the plant or attracted in the root zone, leaving only residual levels of pollutants in the matrix that should be remediated.

10.3.2 EVIDENCE FOR ITS WORKING

Phytoremediation is not a very new science. It was first tested in the 1990s (Vangronsveld et al., 2009) and has since then become increasingly popular (Thijs et al., 2017). However, current field applications are still lagging behind a common obstacle of insufficient field trials (Schwitzguébel et al., 2002; Van der Lelie et al., 2001; Vangronsveld et al., 2009). Discussion points of phytoremediation include field evidence for its efficiency, reliability, and a better understanding of its working (Bell et al., 2014). Therefore, a demonstration of the benefits of microbe-assisted phytoremediation and evidence for its working can be recapitulated in this chapter. Understanding the basic plant physiology, hydrogeology, and the plant as a holobiont in its environment interacting with millions of microorganisms and the consequences of management are needed to explain and fully understand phytoremediation. This information can confirm remediation strategies, refine prediction models, and aid in optimization of the system and performance.

In the following paragraphs the benefits of phytoremediation systems and microbe-assisted phytoremediation of organic pollutants in groundwater over natural attenuation are discussed. In proposing phytoremediation, feasibility tests are sometimes needed if site-specific information is missing (addressed above), and future projects can use that experience for other polluted sites.

Evapotranspiration is the first mechanism that plants use for the evaporation and transpiration of water but is also relevant to *phytohydraulic control*, forming a hydraulic barrier. Poplar and willow trees provide hydraulic control of polluted groundwater and can prevent further horizontal and vertical dispersion and migration of the subsurface pollutant plume (Cook et al., 2010; Ferro et al., 2003, 2013). Estimated water use (per tree) for hybrid poplars is 0.31–6 L day^{-1} for one-year-old trees, 1.9–42 L day^{-1} for two years of age, and 23–59 L day^{-1} for four-year-old trees (Ferro et al., 2001; Vose et al., 2000). In a study in Raleigh, North Carolina, a mixed stand of four-year-old willow and poplar trees was shown to extract 493 L day^{-1} of groundwater, which was sufficient, relative to the calculated rate of groundwater flux beneath the stand, to achieve a high level of plume control (Ferro et al., 2013). Trees actively evapotranspiring can be observed also from a lowering of the groundwater table.

A welcome side-effect of the *upward movement of groundwater during evapotranspiration is that pollutants are drawn into the aerobic zone where they can be oxidized* (Karthikeyan and Kulakow, 2003). More oxygen enhances degradation. This is especially important for benzene, toluene, ethylbenzene, and xylene (collectively known as BTEX) and methyl tert-butyl ether (MTBE) because their half-life in aerobic environments is relatively short compared to saturated, anaerobic conditions (El-Naas et al., 2014). Both field and laboratory studies have shown increased concentrations of oxygen at deeper soil depths due to groundwater extraction by trees and the release of oxygen from tree roots, which enhances degradation (Weishaar et al., 2009). Greater oxygen input to soils resulted in increased abundance of BTEX degrading microorganisms and increased

degradation (Weishaar et al., 2009). Barac et al. (2004, 2009) observed increased abundance of toluene-degrading bacteria in rhizosphere soils and endophytic bacteria once poplar roots reached groundwater polluted with BTEX, but the abundance of these bacteria declined after BTEX concentrations declined (Barac et al., 2004, 2009). The latter suggests that degradative genes are plasmid encoded and can be easily gained and lost. Indeed, genomics analyses of a strain *Paraburkholderia aromaticivorans* BN5 isolated from petroleum polluted soil showed that the genes are acquired via horizontal gene transfers and/or gene duplication, resulting in enhanced ecological fitness by enabling strain BN5 to degrade all compounds in the polluted soil, including naphthalene, BTEX, and short aliphatic hydrocarbons (Lee et al., 2019). Horizontal gene transfer was also observed on a field site after inoculation of a *Pseudomonas putida* strain containing a degradative plasmid (Weyens et al., 2009).

Studies have shown that trees, especially phreatophytic species like poplars and willows, can effectively dissipate fuel pollutants such as BTEX, MTBE, and some polycyclic aromatic hydrocarbons (PAHs) in the root zone by *rhizodegradation* (Burken and Schnoor, 1998; Fernández et al., 2012; Maila et al., 2005). The roots of plants release sugars and phytochemicals which stimulate microbiological activity in the rhizosphere. The microorganisms in the rhizosphere break down the pollutant by utilizing it as a carbon source. The plants' roots and microorganisms in the rhizosphere also allow the transportation of chemicals from the soil into the plant when the pollutant is dissolved in water (Zhang et al., 2017). Thus water solubility of the pollutant is an important factor. The plants and their associated rhizosphere also increase the microbiological activity in the soil which results in improved soil structure, along with reintroducing organic matter into the soil through the deposition of leaves, branches, and root cells (Sleegers, 2010). In addition, plants can manipulate their microbiome because of the secretion of exudates to influence the local environment which creates conditions that support enhanced growth and aid in pollutant dissipation (Hassan et al., 2019).

Changing groundwater levels due to natural seasonal groundwater fluctuations, or by *tree interception helps in dispersing and smearing of pollutants in the aerobic zone.* Tree roots will interact with pollution as roots penetrate smear zones or if precipitation events elevate groundwater levels and displace light non-aqueous phase liquid (LNAPL) fuel product upward through the vadose zone to the tree roots (Figure 10.2). This is another mechanism showing the benefits of using plants in phytoremediation systems. Gasoline range organics (GROs), such as BTEX and two- to three-ringed PAHs, have been shown to be effectively remediated in this way (Ferro et al., 2013).

The relatively lower log K_{ow} ranges (1.5–2.9) of GROs can also facilitate their uptake by trees and subsequently *in planta* phytodegradation. GRO taken up via the roots can be metabolized in tree tissues or transpired to the atmosphere (Limmer and Burken, 2016), although the latter mechanism is not the primary route of removal. Usually GROs are found in woody stem tissue, particularly lower stem tissues as concentrations

FIGURE 10.2 Phytoforensics to find pollutants. Photograph showing Dr. Joel Burken (Missouri S&T) sampling at a phytoremediation site with TCE pollution near Antwerp (Belgium) (September 17, 2010). (Courtesy of Jaco Vangronsveld, UHasselt. Published with kind agreement of Dr. Burken.)

of pollutants have been shown to decline exponentially with height on the trunk. Burken and Schnoor (1998) have developed and described innovative techniques *for sampling and analyzing roots and hardwood tree trunks for detecting pollutants* (Figure 10.2), for example trichloroethylene (TCE), but also BTEX, and MTBE. The technique has been used in the context of phytoforensics to map pollution plumes in groundwater using trees (Burken et al., 2011; Wilson et al., 2018).

Bioaugmented phytoremediation can be additionally used to speed up and enhance the remediation of very difficult to degrade pollutants such as heavy oil fractions, diesel range organics (DRO), fuel oil, and coal tar (Agnello et al., 2016). The degradation of these compounds is more difficult due to the more hydrophobic character and higher molecular weight range compared to other hydrocarbon compounds. To enhance degradability of this class of compounds, contact between the pollutant and microbes and plant roots should be as high as

possible. Contact can be enhanced by bioaugmentation, i.e., inoculation of degradative bacteria which produce *biosurfactants and emulsifying compounds* (Borah and Yadav, 2017) (Figure 10.3). More field data are, however, needed to show the benefit of adding emulsifying compounds in groundwater and subsurface soil on the removal of heavier aliphatic (>C20) and aromatic hydrocarbons, fuel oil, coal tar, and creosote. Most published studies have focused primarily on grasses (Cook and Hesterberg, 2013; Mezzari et al., 2011) but not on deep-rooting trees.

10.3.3 MANAGING AND MODIFYING THE PLANT MICROBIOME IN PHYTOREMEDIATION OF GROUNDWATER POLLUTANTS

From the previous part, it is clear that the plant-associated microorganisms play an important role in the dissipation of groundwater pollutants. Although there has been significant progress in knowledge on plant-microbiome interactions, many gaps remain. For example, indirect effects and multispecies interactions, such as bacteria-bacteria and plant-bacteria effects on plant gene expression and pollutant uptake, still are poorly understood. For too long, these interactions have been ignored by studying processes in simple pot or hydroponics settings. It is crucial to go to the field and to use our -omics "backpack" to learn from natural systems to derive testable hypothesis for laboratory experiments. In the latter experiments, simplified approaches, for example, using synthetic communities, can be used and interaction studies can be performed to later translate the results to the multi-complexity in the field. This is not a dilemma in bacteria-assisted phytoremediation alone, but in crop growth and plant protection in general. Researchers have summarized and proposed future research avenues for phytoremediation (Bell et al., 2014; Thijs et al., 2016) and where phytobiomes research (Bell et al., 2019) should go to in the near future. We summarize below some aspects of these review papers and what can be done to better manage and modify the plant microbiome toward tailored endpoints: soil and groundwater phytoremediation.

FIGURE 10.3 Isolation, in vitro screening for diesel degradation, and inoculation of hydrocarbon degrading strains in the rhizosphere of poplar. On the right, poplar and willow plantation with inoculation tubes. (Courtesy of [left]: P. Gkorezis [UHasselt]; [middle]: W. Sillen [UHasselt]; [right]: S. Thijs [UHasselt], with permission of the owners.)

How do groundwater and subsurface soil properties constrain phytobiome manipulation? Practically speaking, inherent groundwater properties (pH, salts, redox, oxygen) and subsurface properties (e.g., depth to bedrock, slope, and texture) are determined, making them difficult or impossible to change, and imposing *a priori* constraints on which plant-microbiomes can be established (Afzal et al., 2011; Ulrich and Becker, 2006). Alterable groundwater and soil properties are strongly shaped by biotic factors, which are affected by vegetation, microbial activity, and human management. For plants, resource availability is the most important category of properties influencing the strength and direction of plant-soil feedbacks. Under low nutrient availability, plants allocate less photosynthates to roots than they do under high nutrient availability, leading to lower overall soil microbial biomass and weaker interactions between roots and microbial biotrophs (Johnson, 2010). Groundwater properties (pH, redox potential, and nutrient chemical speciation) are also dynamic and will affect functional and morphological microbial groups in different ways (Tedersoo et al., 2014). Microbial activity itself results in changes of the groundwater, aggregate formation, and stabilization of the soil structure (Six and Paustian, 2014), macropore creation, and altered hydraulic conductivity. These changes can have contrasting effects on phytoremediation outcomes.

Pollutant stress. Recent studies that compared the groundwater microbiome of groundwater polluted with chlorinated solvents versus pristine groundwater aquifers, demonstrated that under pollutant stress, changes occur in the microbial assemblages with a higher phylogenetic diversity in the pristine community encoding more genes for efficient geochemical and nutrient cycling than the stressed community (Hemme et al., 2015). It has also been shown with high throughput sequencing that before and after treatment of chlorinated solvent polluted aquifers with amendments, a stable population of organohalide-respiring bacteria was established which could efficiently convert toxic vinyl chloride to harmless ethene supported by the slow release of electron donors from the amendment (Matturro et al., 2018). A better understanding of how pollution, electron donors, and plant roots shape the groundwater microbiome may lead to using the discoveries of a pollutant induced enrichment in certain microbiome types and genera as a potential blueprint for manipulating plant and groundwater microbiomes for improved phytoremediation.

10.4 PHYTOMANAGEMENT OF HEAVY METALS IN POLLUTED SOILS

10.4.1 INTRODUCTION

Soil pollution by heavy metals (HM) represents a serious threat for both the environment and human health. Due to their elemental character HM are not degradable and may persist in soils for thousands of years. Conventional civil-engineering based technologies for the remediation of HM polluted soils are based on physical or chemical processes. These technologies are expensive and can hardly be applied at large scale. In addition, they are ecologically unacceptable as they affect soil structure and organic matter and often turn it into a biologically inactive "substrate." Therefore, scientists started to pay significant attention to alternative approaches for HM polluted environments, including phytoremediation.

The latter includes a set of plant-based technologies aiming to handle HM polluted soils by HM phytoaccumulation in the harvestable plant parts (*phytoextraction*) or to stabilize (inactivate) the pollutants through decreasing their mobility and bioavailability (*phytostabilization*). In addition, metalloid-polluted (such as Hg, As, and Se) soils may be approached by *phytovolatization* techniques, based on biological conversion of present metalloid forms into less toxic forms and, if acceptable from an environmental point of view, eventually their diffusion into the atmosphere. The main advantages of the phytoremediation technologies are their "gentle" character, relatively low cost, and high public acceptance (Ali et al., 2013; Chaney and Baklanov, 2017; Mench et al., 2010; Vangronsveld et al., 2009).

The choice of a suitable phytoremediation strategy depends on many factors, one of which is the level of the risk presented by the metal-polluted soil. In principle, there are two schemes for assessing the risks of toxicants, including HM: (a) human health risks and (b) environmental risks (Adriano et al., 2004). In the second scheme, along with standard chemical and physical methods, bioassays with animal species, microorganisms, and plants are recommended. This is due to the fact that the total soil HM concentrations in environmental media are not directly related to their mobility, accessibility, and biotoxicity. The HM contents in different extractable soil fractions are also not sufficiently indicative, and no generally accepted methodologies exist. In addition, in the majority of cases soils are polluted with a complex mixture of HM and therefore their biotoxicity can be the result of both a particular metal and the interactions between them – synergistic, antagonistic, or additive. Plant-based tests have been successfully used for monitoring of metal phytotoxicity (Van Assche and Clijsters, 1990; Vangronsveld et al., 1995a, 1995b, 2009; Vassilev et al., 2004) as well as efficiency of the applied phytoremediation projects over the time (Geebelen et al., 2003; Meers et al., 2005; Ruttens et al., 2006).

10.4.2 PHYTOSTABILIZATION OF METAL-POLLUTED SOILS

Phytostabilization of metal polluted soils is a technology for decreasing environmental risks by lowering the mobility/availability of HM using soil amendments and eventually adapted/tolerant plants (Adriano et al., 2004; Bolan et al., 2011; Vangronsveld and Cunningham, 1998; Vangronsveld et al., 1995a, 1995b). It involves the establishment of a plant cover on the surface of the polluted soils with the aim of reducing the mobility of pollutants within the vadose zone through accumulation by roots or immobilization within the rhizosphere, thereby reducing off-site pollution. The process includes transpiration and root growth that immobilizes pollutants by reducing leaching, wind, and water erosion; creates a more aerobic environment in the root zone; and adds organic matter to the soil that binds the pollutants. Moreover, the microbial activity associated with the rhizosphere may cause the degradation of organic pollutants.

Phytostabilization has been developed as a technology based on the experience of restoration of industrially polluted soils around old mines or metallurgical plants, where, due to the high phytotoxicity and poor soil physical and chemical characteristics, vegetation is virtually absent. Besides heavily polluted soils, it is suitable for soils with low levels of radionuclides if their half-life is not long, for example ^{137}Cs and ^{90}Sr. Phytostabilization is also of interest for slightly polluted agricultural soils, because the addition of suitable amendments may decrease the mobility of HM and their transfer to the food chains.

The amendments used in phytostabilization are often similar to those for soil fertility restoration. Among the most often used are (alone or in combination) different forms of lime, phosphate fertilizers, biosolid compost, cyclonic ashes, iron (Fe) oxides, zeolites, clays, etc. (Vangronsveld et al., 2009). The mechanisms by which they lower metal mobility in the soils are different – precipitation, sorption, complexation, redox transformations, etc. (Adriano et al., 2004). Phosphate fertilizers are most effective for immobilizing Pb, and the other amendments are suitable e.g. for arsenic (As), Cd, nickel (Ni), and Zn.

The role of plants in phytostabilization is to fulfill two main functions: (a) protecting the soil from wind and water erosion; and (b) reducing the water filtration in the underground horizons, thereby reducing the risk of HM percolation to the groundwater (Vangronsveld et al., 1995b). Furthermore, plants contribute to the stabilization of HM in the soil by: (c) adsorbing HM on the root surface; (d) changes in HM mobility patterns as a result of changes in pH and redox potential in the rhizosphere; and (e) the effects of associated soil microorganisms (bacteria and mycorrhiza) on the mobility of HM.

An early successful phytostabilization project was carried out in an area around an old metallurgical plant in Belgium (Vangronsveld et al., 1995a, 1995b). In brief, the surface "substrate" in the plot was completely devoid of vegetation, because of the extremely high content of HM (mg kg^{-1}: 18550 (Zn), 138 (Cd), 4900 (Pb) at pH 7.4 and low cation exchange capacity – 2.83 cmol$_c$ kg^{-1}). Compost (100 t ha^{-1}) and modified aluminosilicate (beringite, 120 t ha^{-1}) were introduced into the soil and after 4 weeks a mixture of tolerant ecotypes of *Festuca rubra* and *Agrostis capillaries* was sown. The plants developed normally, the formed plant roof reduced the total wind erosion, and the HM washing in the underground layers decreased by 85%. The stability of the rehabilitation effect over time has been confirmed by long-term monitoring.

Another interesting field case that is still running is a former wood preservation site (10 ha) located in Saint-Médard-d'Eyrans, SW France (Mench et al., 2018). Twenty-eight field plots with total topsoil copper (Cu) in the 198–1169 mg kg^{-1} range were assessed. Different amendments and plant species have been tested since 2008.

Although this technology may be very effective in the immobilization/inactivation of metal(loid)s, such a site needs regular monitoring to ensure that the stabilizing conditions are sustained (Vangronsveld et al., 2009). Cases are not always successful due to: (1) the inability to achieve, in some cases, a sufficient degree of immobilization of HM in the soil that would allow the development of stable plant associations; (2) too much immobilization of elements that are essential for plants (e.g., manganese (Mn), magnesium (Mg), …), leading to mineral deficiencies; (3) insufficient stability of the immobilization effect achieved over time; and others.

10.4.3 METAL PHYTOEXTRACTION

The idea of using plants to clean up HM-polluted environments is very old, but Chaney (1983) was the first who proposed the concept of metal phytoextraction through growing and harvesting crops on HM-polluted soils. Since then a lot of research efforts have been invested leading to novel and useful information concerning: (1) promising HM hyperaccumulating species; (2) different approaches for phytoextraction improvement; (3) unraveling of the mechanisms behind the extraordinary metal phytoaccumulation; etc. Most of the early studies were performed at growth chamber and greenhouse scale (hydroponics and soils polluted with HM) and were rightly criticized for reporting unrealistic estimations of both phytoaccumulation and phytoextraction potential (Chaney and Baklanov, 2017). During the last decades, several trials were performed at field scale, providing reliable data and important lessons for properly coping with the practical problems of metal phytoextraction (Chaney et al., 2007; Hammer and Keller, 2003; Michels et al., 2018; Thijs et al., 2018; Vangronsveld et al., 2009).

In principle, metal phytoextraction is simple and includes several steps: (1) plant cultivation on HM-polluted soil; (2) harvesting HM-enriched plant biomass; (3) processing this biomass; and (4) storing the residual product as toxic waste. Follow-up treatments aim to significantly reduce the mass of the residual product and to obtain "added value" by harvested plant biomass. Possibilities in this regard are direct combustion, gasification or pyrolysis for the production of thermal and electrical energy and useful chemicals, as well as production of liquid fuels (Cundy et al., 2016; Meers et al., 2010; Ruttens et al., 2006; Sas-Nowosielska et al., 2004; Vassilev et al., 2004). Among them, pyrolysis seems to be a promising option. Stals et al. (2010) mentioned that three valuable products can be obtained from HM polluted wood of willow, with up to 52% oil, 25% of biogas, and 23% of char. The char that contains most of the HM can be used as an adsorbent for water fine purification. The economics of the willow pyrolysis have been positively assessed by Thewys and Kuppens (2008).

The phytoextraction concept was initially developed based on the use of metal hyperaccumulating plants. The term "hyperaccumulator" was introduced by Brooks et al. (1977) and the definition was further detailed by Baker (1981), Reeves and Brooks (1983), and Van der Ent et al. (2013). Now it is widely accepted that plant species are considered as hyperaccumulator if they can accumulate one or more elements in concentrations higher than following thresholds in mg per kg of dry shoots, provided that the plant grows in a natural habitat (Pollard et al., 2014; Van der Ent et al., 2013): 100 for Cd, selenium (Se), or thallium (Tl); 300 for cobalt (Co), chromium (Cr), or Cu; 1000 for Ni or Pb; 3000 for Zn; and 10,000 mg for Mn. They developed these remarkable traits growing on HM-mineralized or highly HM-polluted soils (Wójcik et al., 2017). Among the most known hyperaccumulators are *Noccaea caerulescens* (formerly *Thlaspi caerulescens*, Brassicaceae) for Zn (Reeves and Brooks, 1983) and Cd (Reeves et al., 2001), *Alyssum bertolonii* for Ni (Gambi et al., 1977; Li et al., 2003), and *Pteris vittata* for the metalloid As (Ma et al., 2001). More than 720 hyperaccumulating species have been identified so far with a majority (over 90%) being Ni hyperaccumulators (Reeves et al., 2018).

The phytoextraction technology based on HM hyperaccumulators has been named *natural* or *continuous phytoextraction* due to the estimated long-lasting phytoremediation process (McGrath et al., 1993). Later, new directions were developed, namely (1) *chemically assisted phytoextraction* based on the use of chelates to increase HM mobility in the soil and subsequently their uptake in the harvestable plant biomass (Blaylock et al., 1997), as well as (2) *high biomass plant-based phytoextraction*, where high biomass crops and trees are used and agronomic techniques are applied to further increase their yields (Thijs et al., 2018).

The two approaches have both advantages and disadvantages and need optimization. One of the biggest disadvantages is the time needed for the realization of cleanup by any of these phytotechnologies. Soil depollution by phytoextraction is not a fast process. Several authors proposed that the duration of practical phytoextraction should not exceed 10 years (Robinson et al., 1998; Vangronsveld et al., 2009). In fact, recent calculations based on field experiments with high biomass crops and trees showed that long time periods are needed (at least several decades) to decrease the pseudo-total contents of HM to the legal threshold values as for instance for Cd on agricultural land in northeast Belgium (Thijs et al., 2018).

Based on the accumulated knowledge, there exists some agreement concerning the real phytoextraction potential for the most problematic HM (Pb, Zn and Cd) in polluted soils. Phytoextraction of HM, such as Pb and Zn, seems to be unrealistic due to their high contents and often low plant-availability (especially for Pb) in polluted soils and the very long time needed for its removal (Ernst, 2005; Robinson et al., 2015). The case of Cd in agricultural soils is considered differently. It is often present in lower levels due to long-time use of phosphate fertilizers containing Cd impurities. These levels might be extracted by suitable plants (McGrath et al., 1993; Robinson et al., 1998).

Accordingly, an opinion developed that phytoextraction should mainly concentrate on the reduction of the plant available fractions of HM in the soil (Greger, 1999; Herzig et al., 2014). When the phytoremediation project aims at removing only the most labile fractions of HM, its duration may be significantly shorted. This approach has been called bioavailable contaminant stripping (BCS) (Hamon and McLaughlin, 1999). Herzig et al. (2014) have demonstrated that the phytoextraction of the "bioavailable" Zn pool in the soils in northeastern Switzerland can be lowered by 45%–70% within a 5-year period. An important criticism for this approach is that a new equilibrium will develop between more stable and labile fractions and thus the labile fraction will be replenished (Herzig et al., 2014).

10.4.3.1 Phytoextraction Using Hyperaccumulator Plants

The potential of any plant species to be used in phytoextraction results from two important variables – HM concentration in the harvested biomass and the biomass yield. The continuous metal phytoextraction in most cases is hampered by the second variable – biomass yield. For example, many authors showed that the most studied hyperaccumulator of Zn and Cd, *Noccaea caerulescens*, is characterized by low biomass yield, rosette growth form, and high stress sensitivity, making it unsuitable for harvesting (Chaney and Baklanov, 2017; Ernst, 2005; Hammer and Keller, 2003). Li et al. (2012) pointed out that the insufficient biomass yield of this species is also due to its high sensitivity to fungal pathogens and herbivores, as well as insufficient knowledge for adequate cultivation.

A completely different case is the Ni phytoextraction by the hyperaccumulators of the genus *Alyssum* (Baker and Brooks, 1989). These species are highly productive (up to 9 tons DM ha^{-1}) with Ni concentrations in the shoots around 2%–3% and a phytoextraction potential over 100 kg Ni ha^{-1} year^{-1} (Bani et al., 2015). In fact, this case is not related to polluted soils, but rather to serpentine soils, where phytomining seems to be a promising option.

Significant variations in metal phytoaccumulation have been found among *N. caerulescens* populations and ecotypes, but it was reported that the variability in that and other hyperaccumulating species cannot be further enhanced (Gonneau et al., 2014). In an attempt to initiate the creation of "phytoextraction cultivars," Sterckeman et al. (2019) performed breeding experiments on *N. caerulescens*. For this purpose, plants were selected from 60 populations for high shoot biomass, and Cd, Ni, and Zn concentrations accumulated in the shoots (Sterckeman et al., 2019). Plants were self-pollinated, and the selection and fixation were continued for three generations. Resulting plants demonstrated a dry matter production of 5–10 t ha^{-1}. However, the high biomass genotypes could not be fixed, possibly because of both their complexity and the sensitivity of this trait to environmental conditions, and particularly plant density. Enhancements of the Cd and Zn accumulation capacities were obtained, yet the Ni accumulation of these plants decreased. Metal accumulation seemed to be more heritable than biomass production.

Obviously, as hyperaccumulators in general already accumulate high HM concentrations in the harvestable parts, further improvement of natural phytoextraction should mainly focus on increasing their biomass production. Therefore, significant efforts have been made to improve the phytoextraction potential of several commonly known hyperaccumulators by different agronomic means.

The effects of pesticide use, fertilization, establishment strategy (transplantation vs direct sowing), planting density, cultural cycles, etc. have been studied in the case of *N. caerulescens* (Jacobs et al., 2017, 2018). Chaney and Baklanov (2017) proposed the use of monocot herbicides to limit competition with other species and increase the density of *N. caerulescens*, its survival, and phytoextraction potential.

Hammer and Keller (2003) found that growing transplants is a better strategy than direct sowing, as the former approach resulted in twice the biomass yield (1.5 t ha^{-1} vs 0.8 t ha^{-1}). Jacobs et al. (2018) confirmed that the use of transplants was more efficient for Zn extraction, and also found higher removal after a longer growing season (12 rather than 7 months). The biomass yield of *N. caerulescens* may reach of 5 t ha^{-1} in one growing season through suitable selection and cultivation (Sterckeman et al., 2019). Phytoextraction capacity also increased with planting densities (Jacobs et al., 2018). Although higher densities decreased the biomass of the individuals, a higher total biomass allowed obtaining the highest metal phytoextraction efficiency.

Mineral fertilization also influences metal phytoextraction (Jacobs et al., 2018). Nitrogen fertilization increased plant biomass, but decreased to some extent the concentrations of Zn and Cd in the leaves. The expected integral result could be the enhancement of phytoextraction potential due to higher biomass per unit area, but surprisingly the fertilization did not lead to really clear-cut results. The phytoextraction capacity of *N. caerulescens* plants was enhanced when the fertilization did not provoke collateral effects as fungal pathogen development and retarded growth. The authors calculated that using *N. caerulescens* (Ganges population) and the optimal treatments, it is possible to reduce soil exchangeable concentrations by about 25% for Cd and by 9% for Zn (with a non-metallicolous population) during a 6-month period.

As has been mentioned above, the case of Cd is different from that of Zn, Pb, and other problematic HM in polluted soils. Cadmium has a high mobility in the plant–soil system, and its contents in the harvestable plant parts often exceed the established threshold value for edible crops (0.05–0.2 mg Cd kg^{-1} fresh weight). On the other hand, due to its lower soil content as compared with other metals, it seems that the risks presented by Cd could be entirely eliminated by phytoextraction. Estimations, based on field trials, support this conclusion. For example, according to Robinson et al. (1998) a single cropping of *N. caerulescens* would reduce 10 mg Cd kg^{-1} soil by nearly a half after 1–2 croppings only. More recent data confirming this opinion were presented by Simmons et al. (2015). They evaluated the efficiency of a complex agronomic approach, including soil pH adjustment (addition of sulfur and marl), mineral fertilization, as well as fungicide applications in order to improve effective Cd phytoextraction by *N. caerulescens* from rice paddy soils in Thailand. One of the tested French populations (1613) showed the capacity to reduce soil Cd from 10 mg kg^{-1} to 3.0 mg kg^{-1} in less than

10 croppings. Also promising results for Cd/Zn phytoextraction were reported in field experiments using *Sedum plumbizincicola* (Deng et al., 2016). Different agronomic strategies were tested. The results indicated that *S. plumbizincicola* at an appropriate planting density and intercropped with maize (*Zea mays* L.) could achieve a high extraction efficiency without affecting the productivity of the cereal. An important remark that should be made with the mentioned experiments is the labor-intensive character of the cropping systems that were used. The feasibility of their large-scale use therefore is questionable.

10.4.3.2 Phytoextraction Using High Biomass Crops and Trees

The efficiency of HM phytoextraction could be greatly increased if high-yielding plants would be able to accumulate significant amounts of HM in their above-ground mass. Unfortunately, these plants are non-metal accumulators and are not tolerant to HM. According to Blaylock and Huang (2000), the practical phytoextraction by high-yielding plants may be realistic if they produce more than 20 t ha^{-1} year^{-1} dry biomass and can reach HM concentrations in the harvestable plant parts of not less than 1% (w/w). The effective dry mass production of several crops and trees is even higher than these proposed yields; therefore, they could be good candidates for cleanup of moderately polluted soils, if it would be possible to enhance their metal uptake and accumulation. Among the species studied for that purpose are maize (Meers et al., 2010; Wuana and Okieimen, 2010), sunflower (*Helianthus annuus* L.) (Nehnevajova et al., 2009; Thijs et al., 2018), tobacco (*Nicotiana tabacum* ssp.) (Thijs et al., 2018), Indian mustard (*Brassica juncea* (L.) Czern.) (Tlustoš et al., 2006) as well as willow (Courchesne et al., 2017; Janssen et al., 2015; Thijs et al., 2018), poplar (Ruttens et al., 2011; Thijs et al., 2018), and others.

The main focus in the biomass phytoextraction approach is primarily on searching possibilities to achieve high shoot metal concentrations. It is well known that key factors determining the phytoaccumulation of HM are HM solubility in the soil, uptake of metal ions by roots, as well as their translocation to harvestable plant parts (mostly shoots). The available forms of the targeted HM in the soil are different, so their uptake by plant roots is not similar. For example, Zn and Cd occur more in easily assessable exchangeable forms, whereas the plant availability of Pb is generally very low. Therefore, one of the initial ideas for enhancing metal uptake value was to enhance the concentration of soluble metals in the soil using chemical agents to manipulate metal availability.

Blaylock et al. (1997), Huang et al. (1997), and Wang et al. (2007) reported that application of the chelating agent EDTA (ethylenediamine-tetra acetic acid) at 2 g kg^{-1} soil increased the Pb concentration in the shoots of Indian mustard, maize, pea (*Pisum sativum* L.), and *Bidens maximowicziana*. Positive, but smaller effects on both mobility and plant accumulation of Zn, Cd, and Cu were also found using nitrile triacetate (NTA) and diethylene triaminepentaacetate

(DTPA), organic acids (oxalic, salicylic), etc. (Puschenreiter et al., 2001). Kayser et al. (2000) demonstrated that the application of NTA could increase the solubility of Zn, Cd, and Cu 21, 58, and 9 times, respectively, but that accumulation in plants increased only by a factor 2 to 3 (Kayser et al., 2000). Meers et al. (2008) presented an overview of potential soil amendments for enhancing HM uptake by phytoextraction crops, with a special focus on the more easily degradable alternatives.

A high Pb accumulation in plants, however, was typically accompanied by severe phytotoxicity, retarded growth, and very often plant mortality. These effects limit the phytoextraction process to a short time after the addition of the chelates (Barocsi et al., 2003). Trying to avoid the problem, Barocsi et al. (2003) studied multiple applications of EDTA at lower doses, thus monitoring and controlling the EDTA-induced metal accumulation and phytotoxicity.

Chelate-assisted phytoextraction is not accepted unambiguously and is criticized by a number of scientists (Chaney and Baklanov, 2017; Meers et al., 2008; Nowack et al., 2006; Vangronsveld et al., 2009; Wenzel et al., 2003). They believe that the use of high doses of EDTA, in addition to the phytotoxic effects, leads to uncontrolled mobility and percolation of HM to the groundwater. As EDTA is poorly photo-, chemo-, and biodegradable (Nörtemann, 1999), it has a long persistence in soil environment. It inhibits the development of arbuscular mycorrhiza (Geebelen et al., 2002) and has negative effects on soil fauna (Bouwman et al., 2005). Therefore, it is now acknowledged that the use of chelating agents to promote phytoextraction of HM from polluted soils is environmentally unacceptable (Nowack et al., 2006). Moreover, Chaney and Mahoney (2014) calculated that the price of EDTA, applied at the most effective concentration (10 mmol kg^{-1} soil), would represent a cost of about \$23,500 ha^{-1} year^{-1} (Chaney and Mahoney, 2014).

Another approach to achieve higher metal accumulation in plants is lowering the pH of the soil. Puschenreiter et al. (2001) and Kayser et al. (2000) reported that it is possible to obtain similar, although weaker, effects as for chelators by applying elemental sulfur or acid fertilizers, such as NH$_4$SO$_4$. While the chelates could be used to solubilize Pb in the soil, the lowering of pH approach is more suitable to enhance plant uptake of Cd and Zn from neutral or slightly alkaline soils. It has been shown (Kayser et al., 2000) that the application of elemental sulfur (S) on carbonate-rich soils would be a valuable, but long-term approach for Zn, Cd, and Cu phytoextraction, creating minimal risks as this amendment is oxidized gradually by sulfur-oxidizing bacteria. However, Chaney et al. (1999) warned that there might be some negative effects associated with soil acidification. If soil pH decreases below 5.4, soil microbes which normally oxidize Mn^{2+} to Mn^{4+} will be inactivated, so Mn^{2+} can accumulate in the exchangeable ion pool to phytotoxic levels (Chaney et al., 1999). Moreover, a stronger decrease of pH will lead to Al^{3+} dissolving in pore water at concentrations that are toxic to most plant species. Accordingly, the soil pH management, together with other agronomic practices, may include two steps. First, pH is

lowered to facilitate Cd phytoextraction, and then after harvest, pH is adjusted to >6.5 to minimize continuing Cd uptake or potential Zn phytotoxicity to next crop species (Chaney and Baklanov, 2017).

Another approach for improving HM phytoextraction based on biomass crops is focusing on the optimization of the plant's characteristics related to enhanced HM tolerance and phytoextraction efficiency. Two main options have been explored: (1) improving phytoextraction by non-GM fast tracking breeding (Herzig et al., 2014; Nehnevajova et al., 2009) and (2) improving phytoextraction by use of GM plants (Kärenlampi et al., 2000). The most reliable data have been obtained using the non-GM plants grown in field conditions, while experiments with transgenic plants are limited to the laboratory scale in many European countries.

Using conventional in vitro breeding and selection techniques, Guadagnini (2000) developed somaclonal variants of tobacco with increased metal tolerance, accumulation, and extraction properties. They found that the best ones are distinguished by the mother lines with an improved annual shoot HM removal that was 5–7 times greater for Cu, 2–5 times for Cd, and 0.5 for Zn in a hydroponic system (Guadagnini, 2000). The best tobacco clones confirmed these properties when grown on spiked potting soil with realistic labile Cd, Zn, and Cu pools. Also on polluted sites, the best clones showed enhanced HM removal: 1.8 times for Cd, 3.2 times for Zn, and 2.0 times for Pb at the Rafz site (Switzerland) and by a factor 12.4 for Cd, 13.7 for Zn, and 13.5 for Pb at the Lommel site (Belgium) (Vangronsveld et al., 2009).

Sunflower is another crop practiced for metal phytoextraction optimization. Nehnevajova et al. (2007, 2009) used chemical mutagenesis to enhance both metal tolerance and extraction properties of a sunflower inbred line IBL04. The mutants in the second generation confirmed stability of these traits. When grown at a sewage sludge polluted site, the best of these mutants showed increased removals for Cd, Zn, and Pb up to 7.5-, 9.2-, and 8.2-fold, respectively (Nehnevajova et al., 2007, 2009).

The ideal plants for metal phytoextraction should be highly productive and able to assimilate and translocate a significant portion of the problematical HM in their harvested parts. Additional favorable features are rapid growth, a deep and extended root system, etc. Some tree species, mainly from the genera *Salix* and *Populus*, comply to a great extent with these features, making them candidates for phytoextraction purposes (Pulford and Watson, 2003). Importantly, Unterbrunner et al. (2007) reported that Zn and Cd are preferentially accumulated in the leaves and not in the wood of these species. This means that the accumulated HM can be continuously removed from the site by harvesting leaf biomass (Unterbrunner et al., 2007).

In order to find out which high biomass crop possessed the highest and most constant (in time) phytoextraction potential, a large-scale (8 ha) field experiment was performed on a HM polluted former maize field near a Zn smelter in

Lommel (Belgium) (Thijs et al., 2018). Biomass production and metal uptake by pre-selected tobacco somaclonal variants and pre-selected sunflower mutants were investigated during four consecutive years. Phytoextraction potentials of poplar and willow in short rotation coppice (SRC) were evaluated at the end of the first cutting cycle (after four growing seasons). Tobacco and sunflower revealed to be efficient extractors of Cd and Zn, respectively; the highest simultaneous extraction potential for Cd and Zn was obtained with some SRC clones. However, high annual variations in biomass production and metal removal capacity were observed for all crops. To reach legal threshold values for the pseudo-total content of Cd in this specific soil, a treatment duration of at least 60 years was estimated. Combining estimated phytoextraction potential and economic and environmental aspects, SRC appeared as the best approach for this specific case. Due to the fact that the highest metal concentrations were found in the leaves, very significant improvements of the metal removal capacities could be obtained in case leaves of SCR plants would be harvested.

Of course, when adopting "bioavailable" metal concentrations, much shorter remediation periods would be estimated. However, as discussed before, this way of thinking is not generally accepted due to the uncertainties regarding equilibria between the various metal species in the soil and the eventual replenishment of the "bioavailable" metal pool in the longer term. In any case, an important conclusion was again that economic revenues through biomass conversion and a rewarding for environmental benefits of a phytoextraction crop plantation are crucial for large-scale, commercial implementation of metal phytoextraction.

A general conclusion that can be made from several field studies is that the focus of growing crops on polluted land should be shifted toward valorization of the produced biomass. In this way, an income is generated for the farmer and the phytoremediation time is no longer an issue. Remediation of the soil thus becomes a secondary, but of course not unimportant, longer-term objective.

In any case, there is a need for enhancement of natural metal phytoextraction potential, and several studies have addressed this problem (Thijs et al., 2018; Vassilev et al., 2004). The plant-rhizosphere interactions controlling metal uptake by roots are of primary interest. Indeed, also for phytoremediation of metals, plant-associated microbes might be of high importance (Sessitsch et al., 2013). To what extent root exudates can mobilize metals (as was shown for Fe and possibly Zn) or if microbial rhizosphere communities "shaped" by these root exudates (Baetz and Martinoia, 2014) can contribute to trace element phytoavailability remains to be further elucidated. As certain plants can use microbial siderophores to improve their Fe uptake, it was hypothesized 20 years ago that bacterial trace element chelators, such as siderophores, can eventually improve the uptake of metals by plants (Van der Lelie et al., 2001). In the meantime, several authors reported increases of metal uptake after inoculation of specific strains (Benizri and Kidd, 2018; Kong and Glick, 2017).

10.4.4 Adaptable Agriculture: Case Studies on Metal-Polluted Soils Near the City of Plovdiv, Bulgaria

As mentioned above, phytoremediation of metal polluted soils would be more economically feasible if, in addition to the risk reduction, the used crops produce biomass with added value (Vassilev et al., 2004). An example of such an approach, also termed "adaptable agriculture," has been proposed and implemented in the agricultural region near the city of Plovdiv, Bulgaria. More than 2100 ha agricultural soils in that area have been polluted by HM through atmospheric deposition from the KCM-Plovdiv smelter. The smelter is surrounded by fertile land that was traditionally used by the local farmers for production of grains, corn, peanuts, grapes, fruits, and vegetables. The climate conditions (continental, annual rainfall 600–650 mm) and soil properties (calcareous drained with a pH 7.1–7.2, moderate organic matter content) are suitable for agriculture, except the soil concentrations of Pb, Zn, Cd, and Cu, which exceed the threshold values for food and fodder production (Angelova et al., 2010). Intensive studies have indicated that leafy vegetables and some other crops may accumulate high amounts of HM, making them unsuitable for human consumption. Nowadays, the atmospheric pollution of the region has been reduced to a minimum as a result of the new technologies that were implemented in the middle of 1990s. Nevertheless, certain recommendations for proper use of land resources that have been proposed in the past are still valid (Benderev et al., 2016).

The main goal of adaptable agriculture is risk prevention together with maintaining a sustainable economic income for the farmers. In addition, phytoextraction of some quantity of HM could be achieved in the long-term perspective. Therefore, it has been suggested to replace the cultivation of food by non-food crops. In fact, a relatively large choice of alternative crops for cultivation in that region exists: fiber, aromatic, medicinal, and so on. The ideal crop for adaptable agriculture should fulfill several requirements, namely (1) to be relatively tolerant to excess HM; (2) to prevent HM accumulation in the valorized part of the biomass; (3) to allow obtaining end products of high quality; as well as (4) to possess some metal phytoextraction potential. In this respect, a significant number of crops have been studied since the beginning of 1990s: flax (*Linum usitatissimum* L.), cotton (*Gossypium arboreum* L.), and hemp (*Cannabis* sp.) (Angelova et al., 2016; Yankov et al., 2000), mint (*Mentha piperita* L.) (Zheljazkov and Nielsen, 1996a,b), lavender (*Lavandula angustifolia* Mill.) (Zheljazkov and Nielsen, 1996a,b), basil (*Ocimum basilicum* L.) (Zheljazkov et al., 2006), sunflower (Angelova et al., 2016; Yankov et al., 2000), bur marigold (*Bidens tripartita* L.), motherwort (*Leonurus cardiaca* L.), horehound (*Marrubium vulgare* L.), lemon balm (*Melissa officinalis* L.), winter marjoram (*Origanum heracleoticum* L.) (Zheljazkov et al., 2008), safflower (*Carthamus tinctorius* L.), clary sage (*Salvia sclarea* L.), rapeseed (*Brassica napus* L.), and milk thistle (*Silybum marianum* L.).

Most of these studies have been performed as field trials around the KCM-Plovdiv. As the metal concentrations in the soil decrease with increasing distance from the smelter, the different plots were arranged at different distances from the smelter (from 0.1 up to 15 km) (Angelova et al., 2004).

The first trials on the metal polluted soils around KCM-Plovdiv have been carried out with some aromatic and medicinal plant species, such as mint and lavender (Zheljazkov and Nielsen 1996a, 1996b). The results showed that the biomass yield decreased by 14% to 17% only in the plot located in the immediate vicinity of the smelter. Despite excessive metal accumulation (Pb, Zn, Cd, and Cu) in the aerial plant parts, oils as final product were not polluted, and only some decrease of the menthol content in mint was detected. Therefore, plantations of lavender (about 21 ha) have been established in 2001–2002 in the vicinity of the smelter, which still function well.

Yankov et al. (2000) proposed cotton as a suitable crop for adaptable agriculture around the smelter. They studied growth, development, productivity, and fiber quality of plants grown on different polluted sites in the region. The authors concluded that although cotton accumulates high concentrations of Cu, Zn, Cd, and Pb, the processing of fiber with boiling water significantly reduced their contents lowering them to levels found for the plants grown in non-polluted soils. In addition to cotton, other fiber crops such as flax and hemp have been studied (Angelova et al., 2004; Yanchev et al., 2000). Increased concentrations of HM were found in the harvested biomass of these crops, but they did not have any negative impact on both yields and fiber qualities. This fact together with the possibility for complete utilization of the harvested biomass for different industrial products (pulp and paper, furniture, etc.) made that flax and hemp were considered as suitable in the frame of adaptable agriculture. The authors suggested to pay attention if the fiber is processed for clothing as the higher HM contents may represent some health risk.

Also the suitability of sunflower for cultivation in that metal polluted region has been investigated (Angelova et al., 2016; Yankov et al., 2000). Sunflower is one of the most important crops in Bulgaria occupying a significant fraction of the arable land. One of the findings was that the HM present in the seed do not pass into oil during their processing or at least are below the threshold limits. In addition, the authors concluded that sunflower tolerated moderate HM pollution, accumulated Pb, Zn, and Cd in its aerial parts, and therefore possessed some phytoextraction capacity.

The fact that these alternative crops showed a relatively good growth performance when grown on these HM polluted soils is a partly due to the local soil properties limiting the plant availability of the HM, but also to the tolerance of the plants. A recent study confirmed the high quality of the lavender oil and also estimated its phytoremediation potential (Angelova et al., 2015, 2016) (Table 10.1). In brief, the quality of oil meets the requirements of ISO 3515:2002 for Bulgarian lavender oil and/or has values close to the threshold limits. The phytoremediation potential was evaluated based on HM content in the plants and calculated factors (bioconcentration factor [BCF], translocation factor [TF], enrichment factor [EF]).

The results indicated that both EF and TF values for lavender are higher than 1, so it could be categorized as an accumulator crop in terms of Pb, Zn, and Cd with significant phytoremediation potential. However, this should be taken with caution since no differentiation was made between HM concentrations before and after washing, meaning that it is not sure that all HM were taken up from the soil through the roots and translocated to the aerial parts and part of the total "HM load" may originate from HM polluted (soil?) particles that were deposed on the aerial parts. Moreover, previous studies showed some yield reductions for medicinal crops when grown in the polluted area. In fact, Zheljazkov et al. (2008) found that the biomass yields of several medicinal species (bur marigold, motherwort, horehound, lemon balm, and winter marjoram) growing in the immediate vicinity of the smelter (0.5 km) were lower than those of the respective reference locations. Negative effects on crop yields were not detected at higher distance. The calculated TF values were less than 1; therefore, the authors concluded that these species did not possess significant phytoextraction potential.

An attempt to increase the number of crops suitable for cultivation in that region has been recently done by testing other emerging crops, such as safflower, clary sage, rapeseed, and milk thistle (Angelova et al., 2015, 2016, 2017). Safflower is grown for its flowers, which are used for painting textiles and preparing food for birds, ruminants, and other animals. Milk thistle and clary sage are cultivated due to their broad use in different industries, but the main demand is from pharmaceutics. Rapeseed has become an important and profitable crop in Bulgaria due to the use of its oil as biofuel. In addition, it belongs to the well-known Brassicaceae family, representatives of which have been shown to possess good capacities to accumulate some HM in their above-ground parts. The authors concluded that all tested species showed significant tolerance to HM and had some metal phytoextraction capacity (BSF for Pb, Cd, and Zn > 1). Furthermore, the HM concentrations in their marketable products are within the standards. For example, the concentrations of Pb, Cd, and Zn in the essential oil of clary sage were lower than the accepted maximum permissible limits and the main odor-determining ingredients of the oil meet the requirements of the European Pharmacopoeia and BS ISO 7609.

In order to improve the productivity, quality, and safe use of the alternative crops Angelova et al. (2013) investigated the influence of amendments such as compost and vermicompost on the characteristics of the metal polluted soil near the KCM–Plovdiv smelter. Different amounts of vermicompost and compost (5 and 10 g kg^{-1}) were incorporated into the top 20 cm of soil. The results indicated that the application of these amendments decreased the DTPA-extractable levels of HM in the soil (Angelova et al., 2013).

The application of compost and vermicompost led to lower metal availability on one hand, and an improved soil fertility on the other hand. As a result, higher yield and quality of the marketable end products were obtained, including a lowering of the HM concentrations (Angelova et al., 2017).

TABLE 10.1
HM Bioconcentration (BCF) and Translocation (TF) Factors and Biological Accumulation Coefficients (BACs) = Enrichment Factors (EFs) of Different Alternative Crops Grown Near the KCM Smelter

Element	BCF		TF		BAC (EF)	
	Polluted Plot	Reference Plot	Polluted Plot	Reference Plot	Polluted Plot	Reference Plot
Lavender						
Distance from the smelter	0.3 km	1 km Less polluted	0.3 km	1 km Less polluted	0.3 km	1 km Less polluted
Pb	1.92	0.42	3.69	4.38	9.72	3.27
Cd	5.22	2.35	0.70	1.24	4.56	5.34
Zn	5.16	1.24	1.64	1.71	10.49	4.11
Milk thistle						
Distance from the smelter	0.5 km	15 km Non-polluted	0.5 km	15 km Non-polluted	0.5 km	15 km
Pb	0.47	0.23	4.9	2.8	–	–
Cd	1.96	0.23	2.3	0.67	–	–
Zn	1.1	0.43	4.8	7.4	–	–
Clary sage						
Distance from the smelter			0.5 km	15 km Non-polluted	0.5 km	15 km
Pb	–		6.2	3.9	15.0	1.51
Cd	–		5.89	2.0	13.3	0.48
Zn	–		4.6	2.7	13.0	11.2
Sunflower						
Compost amendment (200 t per ha)			Polluted without compost	Polluted plus compost	Polluted without compost	Polluted plus compost
Pb	–	–	2.04	1.38	1.0	0.9
Cd	–	–	8.47	14.3	8.6	8.9
Zn	–	–	2.61	2.5	4.6	2.5

Source: Angelova, V.R. et al., *Sci. World J.*, 10, 273–285, 2010; Angelova, V.R. et al., *Int. J. Biol. Biomol. Agric. Food Biotechnol. Eng.*, 9, 522–529, 2015; Angelova, V.R. et al., *Int. J. Agr. Biosyst. Eng.*, 10, 690–700, 2016.

Abbreviations: BCF = [Metal]roots/[Available metal]soil; TF = [Metal]shoots/[Metal]roots; BAC (EF) = [Metal]shoots/[Available Metal]roots.

10.5 MANIPULATING THE PLANT HOLOBIONT FOR IMPROVED PHYTOREMEDIATION AND STRESS TOLERANCE

Indirectly, improving plant stress tolerance will affect the plants' performance and phytoremediation efficiency. Inoculation with bacteria that can influence the plant stress status through the biosynthesis of 1-aminocyclopropane-1-carboxylic acid (ACC)-deaminase has been shown to contribute to enhanced phytoremediation (Glick 2005, 2010). But the question remains how to successfully inoculate plants. For plants grown from seeds, the priming of seeds with single microorganisms is one approach that has often been used to supplement the degradation potential in the rhizosphere microbiome (Parnell et al., 2016). In the case of groundwater phytohydraulics, installation of drainage tubes in tree stands and infiltration with the microorganisms can be applied (Weyens et al., 2009) as in Figure 10.3, which allows the direct targeted localization of the microbes. As shown for soil microbiome transplantation studies, complex communities that target multiple niche spaces and exhibit functional redundancy might be better at entering and surviving a new environment than single species (Yergeau et al., 2015). Knowledge of how microbiomes coalesce is still minimal, but we know that established inhabitants can prevent the success of newcomers, a phenomenon referred to as priority effects (Vannette and Fukami, 2014). Environmental stressors can weaken strong interactions between primary colonizers, and it gets even more complex under the influence of a plant, which can partly determine which microorganisms stay, and which do not (Mueller and Sachs, 2015). Therefore, translating the theoretical and ecological framework into applied solutions for rapidly adapting plants to enhance phytohydraulics and biodegradation remains an unmet challenge.

Can we manage endophytes? The advantage of working with plant endophytes is that they usually have a tight association with the host plant, and when water soluble, volatile compounds are taken up, the contact time in the transport vessels between endophyte and pollutant is much longer (hours to days) than in the rhizosphere or groundwater, which is favorable for degradation. There is still little evidence that these endophyte communities and functionalities may be easily managed, however. It has been shown, at laboratory and field scale, that horizontal gene transfer can take place between an inoculated strain and the indigenous endophytic community, which resulted in enhanced toluene and TCE phytodegradation and significantly less volatilization of toxic intermediates (Taghavi et al., 2009; Weyens et al., 2009, 2010). To better understand the dynamics of endophytes, community connectivity can be studied in the endosphere and rhizosphere over a growing season, as performed by Shi et al. (2016). They found an increase in rhizosphere microbiomes over the course of a growing season, implying an increasing role for microbiome community organization and consistent assembly dynamics as the season progresses. How groundwater microbiomes alter in function of time is not known (Shi et al., 2016).

Determining the ideal microbe-host association? There is still much room for improvement in constructing microbial consortia for bioaugmentation and the application in the field. A common strategy for bioaugmentation starts with the isolation of bacteria or fungi from the polluted site by selective enrichments, testing of the isolates in the laboratory for degradation, and at the end selecting a consortium or a single microorganism that shows best performance concerning plant growth and pollutant dissipation in feasibility studies. Prior to planting in the field, poplar and willow cuttings can be pre-inoculated with the consortium using a "root-dip" (Figure 10.3). Slow-release inorganic nutrients in tablet form or powder can also be added to an excavation hole in the field. Later, more inoculation doses of the microbes can be added via drainage tubing around the roots (Figure 10.3). Determining the most ideal combinations between microbes and host plant also requires more research and understanding of diversity and composition of endophytic and rhizospheric microbial communities across plant host species, considering that the most suitable plant-microbe combination may vary depending on the environment, plant host (Bell et al., 2015), and stress (Fitzpatrick et al., 2018). More focus should be placed on the extent of plant endophytes and their host range, by screening a diversity of plant species used in phytoremediation to explore application of bioaugmentation in phytoremediation. One meta-analyses of 94 endophytes, 42 host plant species, and 3 stresses (drought, nitrogen deficiency, and excess salinity) found little evidence of plant-endophyte specificity (Rho et al., 2018). In addition, to aid in inoculation formulation, the development of efficient screening tools to explore inoculant impacts on plants and pollution dissipation is required.

Plant genetics. In practice, natural selection and breeding (artificial selection) succeed in improving mean growth and productivity of crops by shaping matching factors, such as composition and function of the microbial community with each plant's genotypes (Bell et al., 2019). Still many interactions of genotype and environment should be deciphered at a more comprehensive systems level. Experimentally, this will require studies that compare replicates of multiple genotypes at multiple sites over multiple generations. Quantitative geneticists (breeders) are endeavoring this and make various experimental clones available, for example for enhanced tolerance to heavy metals and different root-shoot metal translocation (Thijs et al., 2018). To advance in this field, it will be necessary to further characterize and refine plant genotype-by-environment-by-microbiome-by-management interactions (Busby et al., 2017; Glassman and Casper, 2012). In addition, it is key to establish causality in complex microbial networks to cross the next hurdles of our ability to make informed manipulations that lead to predictable outcomes in natural systems. Without systems to predict or pre-empt outcomes of plant and microbiome disturbance or manipulation, capabilities to understand the impacts of this new knowledge on remediation will be limited. For example, genotype-microbiome interactions were important to explain differences in Zn accumulation in willows in phytostabilization (Bell et al., 2015).

How microbial (gen)-omics will enhance site remediation. Genomics have revolutionized the microbiology research fields with emerging impacts on agriculture, forestry, and remediation (Gonzalez et al., 2018; Plewniak et al., 2018; Terrón-González et al., 2016). For example, since the publication of the first complete bacterial genomes almost a quarter of a century ago, the new computational developments and accompanied reductions in cost have markedly increased the number of available microbial genomes today. These genomic analyses are yielding unprecedented insights into microbial evolution and microbial diversity and are elucidating the extent and complexity of the genetic variation in microorganisms involved in phytoremediation (Imperato et al., 2019). Next-generation sequencing technologies, particularly portable sequencers, are showing great potential in the field, enabling rapid *in situ* identification.

10.6   GENERAL CONCLUSION

Phytoremediation undoubtedly has a high potential for the degradation and/or removal of organic pollutants from soils, sediments, and (undeep ground)waters. However, based on often unrealistic extrapolations of data obtained from pot and hydroponical experiments, too enthusiastic interpretations and promises have been made concerning the possibilities of metal phytoextraction. Only for Ni are promising data available that are also realistic and achievable from an economic perspective. At this moment, risk-managed phytostabilization and monitored natural attenuation seem to be realistic scenarios for brownfields, urban and industrial areas polluted with metals (Dickinson et al., 2009), and recalcitrant organics.

It is clear that, although there is a growing public and commercial interest and several successful pilot studies and field scale applications are available, more fundamental and applied research still is required to better exploit the genetic and metabolic diversity of the plants themselves, but also to better comprehend the multifaceted interactions between pollutants, soil, plant roots, and microorganisms (bacteria and mycorrhiza) in the rhizosphere.

Given the non-renewable nature of clean soil and groundwater ecosystems, the understanding of polluted soil and groundwater microbial communities is a basic need, because the remediation and treatment of currently polluted environments is a priority to safeguard future generations (Blaser et al., 2016). As shown in this chapter, there are great opportunities to commence further research in the field of phytoremediation.

REFERENCES

Adriano DC, Wenzel W, Vangronsveld J, Bolan N (2004) Role of assisted natural remediation in environmental cleanup. *Geoderma* 122: 121–142.

Afzal M, Yousaf S, Reichenauer TG, Kuffner M, Sessitsch A (2011) Soil type affects plant colonization, activity and catabolic gene expression of inoculated bacterial strains during phytoremediation of diesel. *J Hazard Mater* 186: 1568–1575. doi:10.1016/j.jhazmat.2010.12.040.

Agnello AC, Bagard M, van Hullebusch ED, Esposito G, Huguenot D (2016) Comparative bioremediation of heavy metals and petroleum hydrocarbons co-contaminated soil by natural attenuation, phytoremediation, bioaugmentation and bioaugmentation-assisted phytoremediation. *Sci Total Environ* 563–564: 693–703. doi:10.1016/j.scitotenv.2015.10.061.

Ali H, Khan E, Sajad MA (2013) Phytoremediation of heavy metals—concepts and applications. *Chemosphere* 91: 869–881.

Angelova S, Nikolova V, Molla N, Dudev T (2017) Factors Governing the host–guest interactions between IIA/IIB group metal cations and α-cyclodextrin: A DFT/CDM study. *Inorg Chem* 56: 1981–1987.

Angelova V, Akova V, Artinova N, Ivanov K (2013) The effect of organic amendments on soil chemical characteristics. *Bulg J Agric Sci* 19: 958–971.

Angelova V, Ivanova R, Delibaltova V, Ivanov K (2004) Bioaccumulation and distribution of heavy metals in fibre crops (flax, cotton and hemp). *Ind Crops Prod* 19: 197–205.

Angelova VR, Grekov DF, Kisyov VK, Ivanov KI (2015) Potential of lavender (*Lavandula vera* L.) for phytoremediation of soils contaminated with heavy metals. *Int J Biol Biomol Agric Food Biotechnol Eng* 9: 522–529.

Angelova VR, Ivanova RV, Todorov GM, Ivanov KI (2016) Potential of *Salvia sclarea* L. for phytoremediation of soils contaminated with heavy metals. *Int J Agric Biosyst Eng* 10: 690–700.

Angelova VR, Ivanova RV, Todorov JM, Ivanov KI (2010) Lead, cadmium, zinc, and copper bioavailability in the soil-plant-animal system in a polluted area. *Sci World J* 10: 273–285.

Baetz U, Martinoia E (2014) Root exudates: The hidden part of plant defense. *Trends Plant Sci* 19: 90–98. doi:10.1016/j.tplants.2013.11.006.

Baker AJM (1981) Accumulators and excluders - strategies in the response of plants to heavy metals. *Journal of Plant Nutrition* 3: 643–654.

Baker A, Brooks R (1989) Terrestrial higher plants which hyperaccumulate metallic elements. A review of their distribution, ecology and phytochemistry. *Biorecovery* 1: 81–126.

Bani A, Echevarria G, Sulçe S, Morel JL (2015) Improving the agronomy of *Alyssum murale* for extensive phytomining: A five-year field study. *Int J Phytoremediat* 17: 117–127.

Barac T, Taghavi S, Borremans B, Provoost A, Oeyen L, Colpaert JV, Vangronsveld J, van der Lelie D (2004) Engineered endophytic bacteria improve phytoremediation of water-soluble, volatile, organic pollutants. *Nat Biotechnol* 22: 583–588. doi:10.1038/nbt960.

Barac T, Weyens N, Oeyen L, Taghavi S, van der Lelie D, Dubin D, Spliet M, Vangronsveld J (2009) Field note: Hydraulic containment of a BTEX plume using poplar trees. *Int J Phytoremediat* 11: 416–424. doi:10.1080/15226510802655880.

Barocsi A, Csintalan Z, Kocsanyi L, Dushenkov S, Kuperberg JM, Kucharski R, Richter PI (2003) Optimizing phytoremediation of heavy metal-contaminated soil by exploiting plants' stress adaptation. *Int J Phytoremediat* 5: 13–23.

Bell TH, Cloutier-Hurteau B, Al-Otaibi F, Turmel MC, Yergeau E, Courchesne F, St-Arnaud M (2015) Early rhizosphere microbiome composition is related to the growth and Zn uptake of willows introduced to a former landfill. *Environ Microbiol* 17: 3025–3038. doi:10.1111/1462-2920.12900.

Bell TH, Hockett KL, Alcalá-Briseño RI, Barbercheck M, Beattie GA, Bruns MA, Carlson JE et al. (2019) Manipulating wild and tamed phytobiomes: Challenges and opportunities. *Phytobiomes J* 3: 3–21. doi:10.1094/pbiomes-01-19-0006-w.

Bell TH, Joly S, Pitre FE, Yergeau E (2014) Increasing phytoremediation efficiency and reliability using novel omics approaches. *Trends Biotechnol* 32: 271–280. doi:10.1016/j.tibtech.2014.02.008.

Benderev A, Kerestedjian T, Atanassova R, Mihaylova B, Singh V (2016) Dynamics and evolution of water and soil pollution with heavy metals in the vicinity of the KCM smelter, Plovdiv area, Bulgaria. *Bulg Chem Commun* 48: 92–104.

Benizri E, Kidd PS (2018) The role of the rhizosphere and microbes associated with hyperaccumulator plants in metal accumulation. *Agromining: Farming for Metals*. Springer, Cham, Switzerland.

Blaser MJ, Cardon ZG, Cho MK, Dangl JL, Donohue TJ, Green JL, Knight R et al. (2016) Toward a predictive understanding of Earth's microbiomes to address 21st century challenges. *mBio* 7: e00714–00716. doi:10.1128/mBio.00714-16.

Blaylock MJ, Huang JW (2000) Phytoextraction of metals. In: Raskin I, Ensley BD (eds) *Phytoremediation of Toxic Metals: Using Plants to Clean up the Environment*. Wiley, New York, pp. 53–70.

Blaylock MJ, Salt DE, Dushenkov S, Zakharova O, Gussman C, Kapulnik Y, Ensley BD, Raskin I (1997) Enhanced accumulation of Pb in Indian mustard by soil-applied chelating agents. *Environ Sci Technol* 31: 860–865.

Bolan NS, Park JH, Robinson B, Naidu R, Huh KY (2011) Phytostabilization: A green approach to contaminant containment. *Adv Agron*. 112: 145–204.

Borah D, Yadav RNS (2017) Bioremediation of petroleum based contaminants with biosurfactant produced by a newly isolated petroleum oil degrading bacterial strain. *Egypt J Pet* 26: 181–188. doi:10.1016/j.ejpe.2016.02.005.

Bouwman L, Bloem J, Römkens P, Japenga J (2005) EDGA amendment of slightly heavy metal loaded soil affects heavy metal solubility, crop growth and microbivorous nematodes but not bacteria and herbivorous nematodes. *Soil Biol Biochem* 37: 271–278.

Brooks RR, Lee J, Reeves RD, Jaffré T (1977) Detection of nickeliferous rocks by analysis of herbarium specimens of indicator plants. *Journal of Geochemical Exploration* 7: 49–57.

Burken JG, Schnoor JL (1998) Predictive Relationships for Uptake of Organic Contaminants by Hybrid Poplar Trees. *Environ Sci Technol* 32: 3379–3385. doi:10.1021/es9706817.

Burken JG, Vroblesky DA, Balouet JC (2011) Phytoforensics, dendrochemistry, and phytoscreening: New green tools for delineating contaminants from past and present. *Environ Sci Technol* 45: 6218–6226.

Busby PE, Soman C, Wagner MR, Friesen ML, Kremer J, Bennett A, Morsy M, Eisen JA, Leach JE, Dangl JL (2017) Research priorities for harnessing plant microbiomes in sustainable agriculture. *PLoS Biol* 15: e2001793. doi:10.1371/journal.pbio.2001793.

Chaney RL, Mahoney M (2014) Phytostabilization and phytomining: principles and successes. Paper 104. In: *Proceedings of life of mines conference*, Brisbane, Australia, 15–17 July 2014. Australian Institute of Mining and Metallurgy, Brisbane.

Chaney RL, Angle JS, Broadhurst CL, Peters CA, Tappero RV, Sparks DL (2007) Improved understanding of hyperaccumulation yields commercial phytoextraction and phytomining technologies. *J Environ Qual* 36: 1429–1443.

Chaney RL, Baklanov IA (2017) Phytoremediation and phytomining: Status and promise. *Adv Bot Res*. 83: 189–221.

Chaney RL, Li YM, Angle JS, Baker AJM, Reeves RD, Brown SL, Homer FA, Malik M, Chin M (1999) Improving metal hyperaccumulator wild plants to develop commercial phytoextraction systems: approaches and progress. In: Terry N., Bañuelos GS (eds), *Phytoremediation of contaminated soil and water*. CRC Press, Boca Raton, FL, 131–160.

Coleman JOD, Frova C, Schröder P, Tissut M (2002) Exploiting plant metabolism for the phytoremediation of persistent herbicides. Environmental Science and Pollution Research 9: 18.

Cook RL, Hesterberg D (2013) Comparison of trees and grasses for rhizoremediation of petroleum hydrocarbons. *Int J Phytoremediat* 15: 844–860.

Cook RL, Landmeyer JE, Atkinson B, Messier JP, Nichols EG (2010) Field note: successful establishment of a phytoremediation system at a petroleum hydrocarbon contaminated shallow aquifer: Trends, trials, and tribulations. *Int J Phytoremediat* 12: 716–732. doi:10.1080/15226510903390395.

Courchesne F, Turmel M-C, Cloutier-Hurteau B, Tremblay G, Munro L, Masse J, Labrecque M (2017) Soil trace element changes during a phytoremediation trial with willows in southern Québec, Canada. *Int J Phytoremediat* 19: 632–642.

Cundy A, Bardos R, Puschenreiter M, Mench M, Bert V, Friesl-Hanl W, Müller I, Li X, Weyens N, Witters N (2016) Brownfields to green fields: Realising wider benefits from practical contaminant phytomanagement strategies. *J Environ Manage* 184: 67–77.

de Vrieze J (2015) The littlest farmhands. *Science* 349(6249), 680. doi: 10.1126/science.349.6249.680.

Deng L, Li Z, Wang J, Liu H, Li N, Wu L, Hu P, Luo Y, Christie P (2016) Long-term field phytoextraction of zinc/cadmium contaminated soil by *Sedum plumbizincicola* under different agronomic strategies. *Int J Phytoremediat* 18: 134–140. doi:10.1080/15226514.2015.1058328.

Dickinson NM, Baker AJ, Doronila A, Laidlaw S, Reeves RD (2009) Phytoremediation of inorganics: Realism and synergies. *Int J Phytoremediat* 11: 97–114.

Doty SL (2008) Enhancing phytoremediation through the use of transgenics and endophytes. *New Phytol* 179: 318–333. doi:10.1111/j.1469-8137.2008.02446.x.

El-Naas MH, Acio JA, El Telib AE (2014) Aerobic biodegradation of BTEX: Progresses and prospects. *J Environ Chem Eng* 2: 1104–1122.

Ernst WH (2005) Phytoextraction of mine wastes–options and impossibilities. *Chem Erde-Geochem* 65: 29–42.

Fernández M, Niqui-Arroyo J, Conde S, Ramos J, Duque E (2012) Enhanced tolerance to naphthalene and enhanced rhizoremediation performance for Pseudomonas putida KT2440 via the NAH7 catabolic plasmid. *Appl Environ Microbiol* 78: 5104–5110. doi:10.1128/aem.00619-12.

Ferro A, Chard J, Kjelgren R, Chard B, Turner D, Montague T (2001) Groundwater capture using hybrid poplar trees: Evaluation of a system in Ogden, Utah. *Int J Phytoremediat* 3: 87–104.

Ferro A, Gefell M, Kjelgren R, Lipson DS, Zollinger N, Jackson S (2003) Maintaining hydraulic control using deep rooted tree systems. *Adv Biochem Eng Biotechnol* 78: 125–156.

Ferro AM, Adham T, Berra B, Tsao D (2013) Performance of deep-rooted phreatophytic trees at a site containing total petroleum hydrocarbons. *Int J Phytoremediat* 15: 232–244. doi:10.1080/15226514.2012.687195.

Fitzpatrick CR, Copeland J, Wang PW, Guttman DS, Kotanen PM, Johnson MTJ (2018) Assembly and ecological function of the root microbiome across angiosperm plant species. *Proceedings of the National Academy of Sciences* 115: E1157–E1165. doi:10.1073/pnas.1717617115.

Gambi OV, Pancaro L, Formica C (1977) Investigations on a nickel accumulating plant: *Alyssum bertolonii* Desv. I. Nickel, calcium and magnesium content and distribution during growth. *Webbia* 32: 175–188.

Geebelen W, Adriano D, van der Lelie D, Mench M, Carleer R, Clijsters H, Vangronsveld J (2003) Selected bioavailability assays to test the efficacy of amendment-induced immobilization of lead in soils. *Plant Soil* 249: 217–228.

Geebelen W, Vangronsveld J, Adriano DC, Van Poucke LC, Clijsters H (2002) Effects of Pb-EDTA and EDTA on oxidative stress reactions and mineral uptake in *Phaseolus vulgaris*. *Physiol Plant* 115: 377–384.

Glass DJ (1999) US and international markets for phytoremediation, 1999–2000. *D Glass Associates.* Accessed at: http://www.dgl-assassociates.com/INFO/phytrept.htm.

Glassman SI, Casper BB (2012) Biotic contexts alter metal sequestration and AMF effects on plant growth in soils polluted with heavy metals. *Ecology* 93: 1550–1559. doi:10.1890/10-2135.1.

Glick BR (2005) Modulation of plant ethylene levels by the bacterial enzyme ACC deaminase. *FEMS Microbiol Lett* 251: 1–7. doi:10.1016/j.femsle.2005.07.030.

Glick BR (2010) Using soil bacteria to facilitate phytoremediation. *Biotechnol Adv* 28: 367–374. doi:10.1016/j.biotec hadv.2010.02.001.

Gonneau C, Genevois N, Frérot H, Sirguey C, Sterckeman T (2014) Variation of trace metal accumulation, major nutrient uptake and growth parameters and their correlations in 22 populations of *Noccaea caerulescens*. *Plant Soil* 384: 271–287.

Gonzalez E, Pitre FE, Pagé AP, Marleau J, Nissim WG, St-Arnaud M, Labrecque M, Joly S, Yergeau E, Brereton NJ (2018) Trees, fungi and bacteria: Tripartite metatranscriptomics of a root microbiome responding to soil contamination. *Microbiome* 6: 53.

Greger M (1999) *Metal Availability and Bioconcentration in Plants. Heavy Metal Stress in Plants.* Springer, Berlin, Germany.

Guadagnini M (2000) In vitro-breeding for metal-accumulation in two tobacco (*Nicotiana tabacum*) cultivars. Mathematisch Naturwissenschftlichen Fakultät der Universität Freiburg, Breisgau, 109.

Haluska AA, Thiemann SM, Evans JP, Cho J, Annable DM (2018) Expanded application of the passive flux meter: In-situ measurements of 1,4-dioxane, sulfate, cr(VI) and RDX. *Water* 10. doi:10.3390/w10101335.

Hammer D, Keller C (2003) Phytoextraction of Cd and Zn with *Thlaspi caerulescens* in field trials. *Soil Use Manag* 19: 144–149.

Hamon R, McLaughlin M (1999) Use of the hyperaccumulator *Thlaspi caerulescens* for bioavailable contaminant stripping. *Extended Abstracts of the Fifth International Conference on the Biogeochemistry of Trace Elements (ICOBTE)*, Vienna, Austria.

Hassan MK, McInroy JA, Kloepper JW (2019) The interactions of rhizodeposits with plant growth-promoting rhizobacteria in the rhizosphere: A review. *Agriculture* 9: 142.

Hemme CL, Tu Q, Shi Z, Qin Y, Gao W, Deng Y, Nostrand JDV et al. (2015) Comparative metagenomics reveals impact of contaminants on groundwater microbiomes. *Front Microbiol* 6. doi:10.3389/fmicb.2015.01205.

Herzig R, Nehnevajova E, Pfistner C, Schwitzguebel J-P, Ricci A, Keller C (2014) Feasibility of labile Zn phytoextraction using enhanced tobacco and sunflower: Results of five-and one-year field-scale experiments in Switzerland. *Int J Phytoremediat* 16: 735–754.

Huang JW, Chen JJ, Berti WR, Cunningham SD (1997) Phytoremediation of lead-contaminated soils: role of synthetic chelates in lead phytoextraction. Environmental Science & Technology, 31: 800–805.

Imperato V, Portillo-Estrada M, McAmmond BM, Douwen Y, Van Hamme JD, Gawronski SW, Vangronsveld J, Thijs S (2019) Genomic diversity of two hydrocarbon-degrading and plant growth-promoting *Pseudomonas* species isolated from the oil field of bobrka (Poland). *Genes* 10. doi:10.3390/genes10060443.

Jacobs A, Drouet T, Noret N (2018) Field evaluation of cultural cycles for improved cadmium and zinc phytoextraction with *Noccaea caerulescens*. *Plant Soil* 430: 381–394.

Jacobs A, Drouet T, Sterckeman T, Noret N (2017) Phytoremediation of urban soils contaminated with trace metals using Noccaea caerulescens: Comparing non-metallicolous populations to the metallicolous 'Ganges' in field trials. *Environ Sci Pollut Res* 24: 8176–8188.

Janssen J, Weyens N, Croes S, Beckers B, Meiresonne L, Van Peteghem P, Carleer R, Vangronsveld J (2015) Phytoremediation of metal contaminated soil using willow: Exploiting plant-associated bacteria to improve biomass production and metal uptake. *Int J Phytoremediat* 17: 1123–1136.

Johnson NC (2010) Resource stoichiometry elucidates the structure and function of arbuscular mycorrhizas across scales. *New Phytol* 185: 631–647.

Kärenlampi S, Schat H, Vangronsveld J, Verkleij JAC, van der Lelie D, Mergeay M, Tervahauta AI (2000) Genetic engineering in the improvement of plants for phytoremediation of metal polluted soils. *Environ Pollut* 107: 225–231. doi:10.1016/s0269-7491(99)00141-4.

Karthikeyan R, Kulakow PA (2003) Soil plant microbe interactions in phytoremediation. In: Tsao DT (ed.) *Phytoremediation. Advances in Biochemical Engineering/Biotechnology*, Vol. 78. Springer, Berlin, Heidelberg, pp. 52–74.

Kayser A, Wenger K, Keller A, Attinger W, Felix HR, Gupta SK, Schulin R (2000) Enhancement of phytoextraction of Zn, Cd, and Cu from calcareous soil: The use of NTA and sulfur amendments. *Environ Sci Technol* 34: 1778–1783. doi:10.1021/es990697s.

Kennen K, Kirkwood N (2015) *Phyto: Principles and Resources for Site Remediation and Landscape Design.* Routledge, New York.

Kong Z, Glick BR (2017) The role of plant growth-promoting bacteria in metal phytoremediation. *Advances in Microbial Physiology* 71: 97–132.

Lee Y, Lee Y, Jeon CO (2019) Biodegradation of naphthalene, BTEX, and aliphatic hydrocarbons by *Paraburkholderia aromaticivorans* BN5 isolated from petroleum-contaminated soil. *Sci Rep* 9: 860. doi:10.1038/s41598-018-36165-x.

Li J-T, Baker AJM, Ye Z-H, Wang H-B, Shu W-S (2012) Phytoextraction of Cd-contaminated soils: current status and future challenges. *Critical Reviews in Environmental Science and Technology* 42: 2113–2152.

Li Y-M, Chaney RL, Brewer EP, Angle JS, Nelkin J (2003) Phytoextraction of nickel and cobalt by hyperaccumulator *Alyssum* species grown on nickel-contaminated soils. *Environ Sci Technol* 37: 1463–1468. doi:10.1021/es0208963.

Limmer M, Burken J (2016) Phytovolatilization of organic contaminants. *Environ Sci Technol* 50: 6632–6643. doi:10.1021/acs.est.5b04113.

Ma LQ, Komar KM, Tu C, Zhang W, Cai Y, Kennelley ED (2001) A fern that hyperaccumulates arsenic. *Nature* 409: 579.

Maila MP, Randima P, Cloete TE (2005) Multispecies and mono-culture rhizoremediation of polycyclic aromatic hydrocarbons (PAHs) from the soil. *Int J Phytoremediat* 7: 87–98. doi:10.1080/16226510890950397.

Matturro B, Pierro L, Frascadore E, Petrangeli Papini M, Rossetti S (2018) Microbial community changes in a chlorinated solvents polluted aquifer over the field scale treatment with poly-3-hydroxybutyrate as amendment. *Front Microbiol.* 9. doi:10.3389/fmicb.2018.01664.

McGrath S, Sidoli C, Baker A, Reeves R (1993) The potential for the use of metal-accumulating plants for the in situ decontamination of metal-polluted soils. *Integrated Soil and Sediment Research: A Basis for Proper Protection.* Springer, Dordrecht, the Netherlands.

McIntyre TC (2003) Databases and protocol for plant and microorganism selection: Hydrocarbons and metals. In: McCutcheon SC, Schnoor JL (eds) *Phytoremediation: Transformation and Control of Contaminants*, John Wiley & Sons, Inc. Hoboken, New Jersey, pp. 887–904.

Meers E, Ruttens A, Hopgood M, Lesage E, Tack FM (2005) Potential of *Brassic rapa*, *Cannabis sativa*, *Helianthus annuus* and *Zea mays* for phytoextraction of heavy metals from calcareous dredged sediment derived soils. *Chemosphere* 61: 561–572. doi:10.1016/j.chemosphere.2005.02.026.

Meers E, Tack FM, Van Slycken S, Ruttens A, Du Laing G, Vangronsveld J, Verloo MG (2008) Chemically assisted phytoextraction: A review of potential soil amendments for increasing plant uptake of heavy metals. *Int J Phytoremediat* 10: 390–414.

Meers E, Van Slycken S, Adriaensen K, Ruttens A, Vangronsveld J, Du Laing G, Witters N, Thewys T, Tack FM (2010) The use of bio-energy crops (*Zea mays*) for "phytoattenuation" of heavy metals on moderately contaminated soils: A field experiment. *Chemosphere* 78: 35–41. doi:10.1016/j.chemosphere.2009.08.015.

Megharaj M, Naidu R (2017) Soil and brownfield bioremediation. *Microbial Biotechnol* 10: 1244–1249. doi:10.1111/1751-7915.12840.

Mench M, Lepp N, Bert V, Schwitzguébel J-P, Gawronski S, Schröder P, Vangronsveld J (2010) Successes and limitations of phytotechnologies at field scale: Outcomes, assessment and outlook from COST Action 859. *J Soil Sediment* 10: 1039–1070. doi:10.1007/s11368-010-0190-x.

Mench MJ, Dellise M, Bes CM, Marchand L, Kolbas A, Le Coustumer P, Oustrière N (2018) Phytomanagement and remediation of Cu-Contaminated soils by high yielding crops at a former wood preservation site: Sunflower biomass and ionome. *Front Ecol Evol* 6. doi:10.3389/fevo.2018.00123.

Mezzari MP, Zimermann DMH, Corseuil HX, Nogueira AV (2011) Potential of grasses and rhizosphere bacteria for bioremediation of diesel-contaminated soils. *Rev Bras Ciênc Solo* 35: 2227–2236.

Michels E, Annicaerta B, De Moor S, Van Nevel L, De Fraeye M, Meiresonne L, Vangronsveld J, Tack FM, Ok YS, Meers E (2018) Limitations for phytoextraction management on metal-polluted soils with poplar short rotation coppice – Evidence from a 6-year field trial. *Int J Phytoremediat* 20: 8–15.

Mirck J, Volk TA (2009) Seasonal sap flow of four *Salix* varieties growing on the Solvay Wastebeds in Syracuse, NY, USA. *Int J Phytoremediat* 12: 1–23. doi:10.1080/15226510902767098.

Mueller UG, Sachs JL (2015) Engineering microbiomes to improve plant and animal health. *Trends Microbiol* 23: 606–617.

Nehnevajova E, Herzig R, Bourigault C, Bangerter S, Schwitzguébel J-P (2009) Stability of enhanced yield and metal uptake by sunflower mutants for improved phytoremediation. *Int J Phytoremediat* 11: 329–346.

Nehnevajova E, Herzig R, Federer G, Erismann K-H, Schwitzguébel J-P (2007) Chemical mutagenesis—a promising technique to increase metal concentration and extraction in sunflowers. *Int J Phytoremediat* 9: 149–165.

Newman LA, Reynolds CM (2004) Phytodegradation of organic compounds. *Curr Opin Biotechnol* 15: 225–230. doi:10.1016/j.copbio.2004.04.006.

Nörtemann B (1999) Biodegradation of EDTA. *Appl Microbiol Biotechnol* 51: 751–759.

Nowack B, Schulin R, Robinson BH (2006) Critical assessment of chelant-enhanced metal phytoextraction. *Environ Sci Technol* 40: 5225–5232.

Parnell JJ, Berka R, Young HA, Sturino JM, Kang Y, Barnhart DM, DiLeo MV (2016) From the lab to the farm: An industrial perspective of plant beneficial microorganisms. *Front Plant Sci* 7. doi:10.3389/fpls.2016.01110.

Plewniak F, Crognale S, Rossetti S, Bertin PN (2018) A genomic outlook on bioremediation: The case of arsenic removal. *Front Microbiol* 9: 820.

Pollard AJ, Reeves RD, Baker AJ (2014) Facultative hyperaccumulation of heavy metals and metalloids. *Plant Sci* 217: 8–17.

Pulford I, Watson C (2003) Phytoremediation of heavy metal-contaminated land by trees—a review. *Environ Int* 29: 529–540.

Puschenreiter M, Stöger G, Lombi E, Horak O, Wenzel WW (2001) Phytoextraction of heavy metal contaminated soils with *Thlaspi goesingense* and *Amaranthus hybridus*: Rhizosphere manipulation using EDTA and ammonium sulfate. *J Plant Nutr Soil Sci* 164: 615–621.

Quiza L, St-Arnaud M, Yergeau E (2015) Harnessing phytomicrobiome signaling for rhizosphere microbiome engineering. *Front Plant Sci* 6: 507. doi:10.3389/fpls.2015.00507.

Reeves RD, Baker AJ, Jaffré T, Erskine PD, Echevarria G, van der Ent A (2018) A global database for plants that hyperaccumulate metal and metalloid trace elements. *New Phytol* 218:407–411.

Reeves RD, Brooks RR (1983) Hyperaccumulation of lead and zinc by two metallophytes from mining areas of Central Europe. *Environ Pollut Series A Ecol Biol* 31: 277–285.

Reeves RD, Schwartz C, Morel JL, Edmondson J (2001) Distribution and metal-accumulating behavior of *Thlaspi caerulescens* and associated metallophytes in France. *Int J Phytoremediat* 3: 145–172.

Rho H, Hsieh M, Kandel SL, Cantillo J, Doty SL, Kim SH (2018) Do endophytes promote growth of host plants under stress? A meta-analysis on plant stress mitigation by endophytes. *Microb Ecol* 75: 407–418. doi:10.1007/s00248-017-1054-3.

Robinson B, Anderson C, Dickinson N (2015) Phytoextraction: Where's the action? *J Geochem Explor* 151: 34–40.

Robinson BH, Leblanc M, Petit D, Brooks RR, Kirkman JH, Gregg PE (1998) The potential of *Thlaspi caerulescens* for phytoremediation of contaminated soils. *Plant Soil* 203: 47–56.

Ruttens A, Boulet J, Weyens N, Smeets K, Adriaensen K, Meers E, Van Slycken S, Tack F, Meiresonne L, Thewys T (2011) Short rotation coppice culture of willows and poplars as energy crops on metal contaminated agricultural soils. *Int J Phytoremediat* 13: 194–207.

Ruttens A, Colpaert JV, Mench M, Boisson J, Carleer R, Vangronsveld J (2006) Phytostabilization of a metal contaminated sandy soil. II: Influence of compost and/or inorganic metal immobilizing soil amendments on metal leaching. *Environ Pollut* 144: 533–539. doi:10.1016/j.envpol.2006.01.021.

Sas-Nowosielska A, Kucharski R, Malkowski E, Pogrzeba M, Kuperberg JM, Krynski K (2004) Phytoextraction crop disposal – An unsolved problem. *Environ Pollut* 128: 373–379. doi:10.1016/j.envpol.2003.09.012.

Schwitzguébel J-P, van der Lelie D, Baker A, Glass DJ, Vangronsveld J (2002) Phytoremediation: European and American trends successes, obstacles and needs. *J Soil Sediment* 2: 91–99. doi:10.1007/bf02987877.

Sessitsch A, Kuffner M, Kidd P, Vangronsveld J, Wenzel WW, Fallmann K, Puschenreiter M (2013) The role of plant-associated bacteria in the mobilization and phytoextraction of trace elements in contaminated soils. *Soil Biol Biochem* 60: 182–194. doi:10.1016/j.soilbio.2013.01.012.

Shi S, Nuccio EE, Shi ZJ, He Z, Zhou J, Firestone MK (2016) The interconnected rhizosphere: High network complexity dominates rhizosphere assemblages. *Ecol Lett* 19: 926–936. doi:10.1111/ele.12630.

Simmons RW, Pongsakul P, Saiyasitpanich D, Klinphoklap S (2005) Elevated levels of cadmium and zinc in paddy soils and elevated levels of cadmium in rice grain downstream of a zinc mineralized area in Thailand: implications for public health. *Environmental Geochemistry and Health* 27: 501–511.

Six J, Paustian K (2014) Aggregate-associated soil organic matter as an ecosystem property and a measurement tool. *Soil Biol Biochem* 68: A4–A9.

Sleegers F (2010) Phytoremediation as green infrastructure and a landscape of experiences. *Proceedings of the Annual International Conference on Soils, Sediments, Water and Energy.* Vol. 15, Article 13, available at: https://scholarworks.umass.edu/soilproceedings/vol15/iss1/13.

Stals M, Thijssen E, Vangronsveld J, Carleer R, Schreurs S, Yperman J (2010) Flash pyrolysis of heavy metal contaminated biomass from phytoremediation: influence of temperature, entrained flow and wood/leaves blended pyrolysis on the behaviour of heavy metals. *Journal of Analytical and Applied Pyrolysis* 87: 1–7.

Sterckeman T, Cazes Y, Sirguey C (2019) Breeding the hyperaccumulator *Noccaea caerulescens* for trace metal phytoextraction: First results of a pure-line selection. *Int J Phytoremediat* 21: 448–455.

Taghavi S, Garafola C, Monchy S, Newman L, Hoffman A, Weyens N, Barac T, Vangronsveld J, van der Lelie D (2009) Genome survey and characterization of endophytic bacteria exhibiting a beneficial effect on growth and development of poplar trees. *Appl Environ Microbiol* 75: 748–757. doi:10.1128/AEM.02239-08.

Tedersoo L, Bahram M, Põlme S, Kõljalg U, Yorou NS, Wijesundera R, Ruiz LV, Vasco-Palacios AM, Thu PQ, Suija A (2014) Global diversity and geography of soil fungi. *Science* 346: 1256688.

Terrón-González L, Martín-Cabello G, Ferrer M, Santero E (2016) Functional metagenomics of a biostimulated petroleum-contaminated soil reveals an extraordinary diversity of extradiol dioxygenases. *Appl Environ Microbiol* 82: 2467–2478.

Thewys T, Kuppens T (2008) Economics of willow pyrolysis after phytoextraction. *International Journal of Phytoremediation* 10: 561–583.

Thijs S, Sillen W, Rineau F, Weyens N, Vangronsveld J (2016) Towards an enhanced understanding of plant-microbiome interactions to improve phytoremediation: Engineering the metaorganism. *Front Microbiol* 7: 341. doi:10.3389/fmicb.2016.00341.

Thijs S, Sillen W, Weyens N, Vangronsveld J (2017) Phytoremediation: State-of-the-art and a key role for the plant microbiome in future trends and research prospects. *Int J Phytoremediat* 19: 23–38. doi:10.1080/15226514.2016.1216076.

Thijs S, Witters N, Janssen J, Ruttens A, Weyens N, Herzig R, Mench M, Van Slycken S, Meers E, Meiresonne L, Vangronsveld J (2018) Tobacco, sunflower and high biomass SRC clones show potential for trace metal phytoextraction on a moderately contaminated field site in Belgium. *Front Plant Sci* 9: 1879.

Tlustoš P, Száková J, Hrubý J, Hartman I, Najmanová J, Nedělník J, Pavlíková D, Batysta M (2006) Removal of As, Cd, Pb, and Zn from contaminated soil by high biomass producing plants. *Plant Soil Environ* 52: 413–423.

Tsao DT (2003) Overview of phytotechnologies. *Adv Biochem Eng/Biotechnol* 78: 1–50.

Ulrich A, Becker R (2006) Soil parent material is a key determinant of the bacterial community structure in arable soils. *FEMS Microbiol Ecol* 56: 430–443.

Unterbrunner R, Puschenreiter M, Sommer P, Wieshammer G, Tlustoš P, Zupan M, Wenzel W (2007) Heavy metal accumulation in trees growing on contaminated sites in Central Europe. *Environ Pollut* 148: 107–114.

Van Assche F, Clijsters H (1990) A biological test system for the evaluation of the phytotoxicity of metal-contaminated soils. *Environ Pollut* 66: 157–172.

Van der Ent A, Baker AJ, Reeves RD, Pollard AJ, Schat H (2013) Hyperaccumulators of metal and metalloid trace elements: Facts and fiction. *Plant Soil* 362: 319–334.

Van der Lelie D, Schwitzguébel J-P, Glass DJ, Vangronsveld J, Baker A (2001) Peer reviewed: Assessing phytoremediation's progress in the United States and Europe. *Environ Sci Technol* 35: 446–452.

Vangronsveld J, Cunningham SD (eds) (1998) Metal-contaminated soils: In-situ inactivation and phytorestoration. Springer, Berlin.

Vangronsveld J, Herzig R, Weyens N, Boulet J, Adriaensen K, Ruttens A, Thewys T et al. (2009) Phytoremediation of contaminated soils and groundwater: Lessons from the field. *Environ Sci Pollut Res Int* 16: 765–794. doi:10.1007/s11356-009-0213-6.

Vangronsveld J, Sterckx J, Van Assche F, Clijsters H (1995a) Rehabilitation studies on an old non-ferrous waste dumping ground: Effects of revegetation and metal immobilization by beringite. *J Geochem Explor* 52: 221–229.

Vangronsveld J, Van Assche F, Clijsters H (1995b) Reclamation of a bare industrial area contaminated by non-ferrous metals: In situ metal immobilization and revegetation. *Environ Pollut* 87: 51–59. doi:10.1016/s0269-7491(99)80007-4.

Vannette RL, Fukami T (2014) Historical contingency in species interactions: Towards niche-based predictions. *Ecol Lett* 17: 115–124. doi:10.1111/ele.12204.

Vassilev A, Schwitzguébel J-P, Thewys T, Van Der Lelie D, Vangronsveld J (2004) The use of plants for remediation of metal-contaminated soils. *Sci World J* 4: 9–34.

Verreydt G, Annable M, Kaskassian S, Van Keer I, Bronders J, Diels L, Vanderauwera P (2013) Field demonstration and evaluation of the passive flux meter on a CAH groundwater plume. *Environ Sci Pollut Res* 20: 4621–4634.

Verreydt G, Bronders J, Van Keer I, Diels L, Vanderauwera P (2010) Passive samplers for monitoring VOCs in groundwater and the prospects related to mass flux measurements. *Ground Water Monit Remediat* 30: 114–126.

Verreydt G, Bronders J, Van Keer I, Diels L, Vanderauwera P (2015) Groundwater flow field distortion by monitoring wells and passive flux meters. *Groundwater* 53: 933–942.

Vorholt JA (2012) Microbial life in the phyllosphere. *Nat Rev Microbiol* 10: 828–840. doi:10.1038/nrmicro2910.

Vose JM, Swank WT, Harvey GJ, Clinton BD, Sobek C (2000) Leaf water relations and sapflow in eastern cottonwood (*Populus deltoides* Bartr.) trees planted for phytoremediation of a groundwater pollutant. *Int J Phytoremediat* 2: 53–73.

Vymazal J (2011) Constructed wetlands for wastewater treatment: five decades of experience. Environmental Science & Technology 45: 61–69.

Wang G, Koopmans GF, Song J, Temminghoff EJM, Luo Y, Zhao Q, Japenga J (2007) Mobilization of heavy metals from contaminated paddy soil by EDDS, EDTA, and elemental sulfur. Environmental Geochemistry and Health 29: 221-235.

Weishaar JA, Tsao D, Burken JG (2009) Phytoremediation of BTEX hydrocarbons: Potential impacts of diurnal groundwater fluctuation on microbial degradation. *Int J Phytoremediat* 11: 509–523. doi:10.1080/15226510802656326.

Wenzel WW, Unterbrunner R, Sommer P, Sacco P (2003) Chelate-assisted phytoextraction using canola (*Brassica napus* L.) in outdoors pot and lysimeter experiments. *Plant Soil* 249: 83–96.

Weyens N, Truyens S, Dupae J, Newman L, Taghavi S, van der Lelie D, Carleer R, Vangronsveld J (2010) Potential of the TCE-degrading endophyte Pseudomonas putida W619-TCE to improve plant growth and reduce TCE phytotoxicity and evapotranspiration in poplar cuttings. *Environ Pollut* 158: 2915–2919. doi:10.1016/j.envpol.2010.06.004.

Weyens N, Van der Lelie D, Artois T, Smeets K, Taghavi S, Newman L, Carleer R, Vangronsveld J (2009) Bioaugmentation with engineered endophytic bacteria improves contaminant fate in phytoremediation. *Environ Sci Technol* 43: 9413–9418. doi:10.1021/es901997z.

Wilson JL, Samaranayake VA, Limmer MA, Burken JG (2018) Phytoforensics: Trees as bioindicators of potential indoor exposure via vapor intrusion. *PLoS One* 13: e0193247. doi:10.1371/journal.pone.0193247.

Wójcik M, Gonnelli C, Selvi F, Dresler S, Rostański A, Vangronsveld J (2017) Metallophytes of serpentine and calamine soils–their unique ecophysiology and potential for phytoremediation. *Adv Bot Res.* 83, 1–42.

Wuana R, Okieimen F (2010) Phytoremediation potential of maize (*Zea mays* L.). A review. *Afr J Gen Agric* 6: 275–287.

Yanchev I, Jalnov I, Terziev J (2000) Hemp's (*Canabis sativa* L.) capacities for restricting the heavy metals soil pollution. *Rasteniev" dni Nauki* 37: 532–537.

Yankov B, Delibaltova V, Bojinov M (2000) Contents of Cu, Zn, Cd and Pb in the vegetative organs of cotton cultivars from industrially polluted region. *Rasteniev" dni Nauki* 37: 525–531.

Yergeau E, Bell TH, Champagne J, Maynard C, Tardif S, Tremblay J, Greer CW (2015) Transplanting soil microbiomes leads to lasting effects on willow growth, but not on the rhizosphere microbiome. *Front Microbiol* 6: 1436. doi:10.3389/fmicb.2015.01436.

Zhang C, Yao F, Liu Y-W, Chang H-Q, Li Z-J, Xue J-M (2017) Uptake and translocation of organic pollutants in plants: A review. *J Integr Agric* 16: 1659–1668.

Zheljazkov VD, Craker LE, Xing B (2006) Effects of Cd, Pb, and Cu on growth and essential oil contents in dill, peppermint, and basil. *Environ Exp Bot* 58: 9–16.

Zheljazkov VD, Jeliazkova EA, Kovacheva N, Dzhurmanski A (2008) Metal uptake by medicinal plant species grown in soils contaminated by a smelter. *Environ Exp Bot* 64: 207–216.

Zheljazkov VD, Nielsen NE (1996a) Effect of heavy metals on peppermint and cornmint. *Plant Soil* 178: 59–66.

Zheljazkov VD, Nielsen NE (1996b) Studies on the effect of heavy metals (Cd, Pb, Cu, Mn, Zn and Fe) upon the growth, productivity and quality of lavender (Lavandula angustifolia Mill.) production. *J Essent Oil Res* 8: 259–274.

11 Layered Double Hydroxides for Soil and Groundwater Remediation

Zhengtao Shen, Yiyun Zhang, and Deyi Hou

CONTENTS

11.1 INTRODUCTION

Various contaminants exist in soil and groundwater, threatening human health through different exposure pathways (Hou et al., 2014, 2018). To remove or immobilize these contaminants through adsorption is regarded as a green and sustainable approach due to less energy consumption, less secondary pollution, and lower life cycle carbon footprint compared with other conventional remediation methods. It is important to find out low-cost and high-efficient adsorptive materials to aid this technology for soil and groundwater remediation.

Layered double hydroxides (LDHs) are a class of natural or artificial anionic clays typically with two-dimensional structure (Goh et al., 2008). The general formula of LDH is $[M_{1-x}^{2+}M_x^{3+}(OH)_2]^{x+}(A^{n-})_{x/n} \cdot mH_2O$, where M^{2+} (e.g., Mg^{2+}, Ni^{2+}, or Zn^{2+}) and M^{3+} (e.g., Al^{3+}, Fe^{3+}, or Mn^{3+}) are divalent and trivalent cations, and A^{n-} is interlayer anion (e.g., NO_3^-, Cl^-, or CO_3^{2-}). As shown in Figure 11.1, due to isomorphous substitution, the M^{3+} can partially replace M^{2+} in the hydroxide lattice, resulting in the formation of hydrotalcite-like layer and the positive charge of the layer. Therefore, negatively charged anions intercalated in between the positively charged layers to compensate the positive charge. These abundant intercalated anions render LDHs superior anion exchange capacity. LDH is regarded as an eco-friendly adsorbent and good anion exchanger to remove anionic

LDH Structure

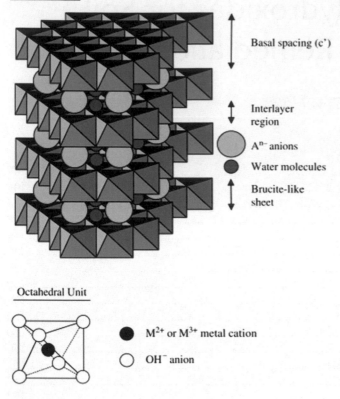

FIGURE 11.1 Schematic representation of LDH structure. (From Goh, K.H. et al., *Water Res.*, 42, 1343–1368, 2008.)

contaminants (e.g., oxyanions, organic dyes, and halogen anions) from soil and water (He et al., 2018a; Zubair et al., 2017). LDHs also reveal excellent ability to immobilize heavy metal cations through precipitation due to its alkaline nature and other mechanisms (Liang et al., 2013). It is suggested that LDH can be regenerated and reused through "memory effect" (detailed in Section 11.2.2.1) for many times. The starting materials to synthesize LDHs can be obtained from wastes (e.g., brine water and red mud). These give LDHs lower costs and high feasibility for large-scale usage.

A range of studies have investigated the removal of contaminants by LDHs in wastewater; however, the large-scale application of LDHs in soil and groundwater remediation is limited. Soil and groundwater pollution is a rising problem especially for developing countries (Hou and Ok, 2019; Shen et al., 2016, 2018). LDHs are of huge potential to be used in the upcoming projects to control and remediate contaminated soil and groundwater, for example, China's ambitious soil and water projects (Hou and Li, 2017). In view of these considerations, this chapter introduces the interactions between LDH and contaminants, and the current progress and future perspectives of the applications of LDHs in soil and groundwater. This chapter aims to give validation for scientific researchers and industrial participators a broad view of LDHs and their application in soil and groundwater remediation.

11.2 LDH PRODUCTION AND CHARACTERISTICS

The constituents of the interlayer cations and anions can be altered for different purposes. A number of transition metal-bearing LDHs (such as Fe-Al, Co-Al, and Ni-Al) are active in technological applications due to their special catalytic, electronic, optical, and magnetic properties (Othman et al., 2009). Different divalent and trivalent cationic substitutions of Mg and Al are able to affect the physicochemical properties, such as surface area, pore structure, and reducibility (Chmielarz et al., 2002). Likewise, the anion basicity has an impact on the reactivity of the LDH materials. For instance, the carbonate and hydroxide intercalated materials are more active than the chloride derivative (Constantino and Pinnavaia, 1995). The calcination of LDH at moderate temperatures (400°C–550°C) could lead to the formation of metastable mixed oxides that cannot be achieved by mechanical means.

11.2.1 PRODUCTION

Over the last decade, significant progress related to the synthesis of LDH with new compositions and morphologies has been made. There mainly exist five routes to synthesize LDHs, i.e., co-precipitation, urea-based method, salt-oxide method, ion-exchange (applicable when divalent or trivalent metal cations or the anions are unstable in mixed solution), and calcination-rehydration (reconstruction, based on the "memory effect" property) (Rives, 2001). Apart from the nature of host layer cations and the nature of interlayer species, different production methods and synthesis conditions will have determinant effects on the surface properties of LDHs as well. The ion-exchange method usually aims at synthesizing LDHs with poly anions or organic anions in the interlayer region, whereas the other ways generally only support inorganic anions (Duan and Evans, 2006). In addition, alternative methods such as sol-gel synthesis using ethanol and acetone solutions, electrochemical synthesis, and a fast nucleation process followed by a separate aging step, have also been reported (Othman et al., 2009). Figure 11.2 describes the relationship between different synthetic methods and different forms of LDH.

11.2.1.1 Co-Precipitation Method

Among the variety of methods available, co-precipitation remains the most widely spread and simplest way of preparing LDH (Palmer et al., 2009). In this "one-pot" direct method, the aqueous solution of chosen M^{2+} and M^{3+} metal precursors is mixed in alkaline solution that contains potential interlayer anions to generate LDH (He et al., 2006; Liu et al., 2006). It allows the precise control of a number of synthesis parameters independently, e.g., temperature in the precipitation process, concentration of metallic salts, anion species, pH of the reaction medium, addition rate of reactants, etc.

To ensure the simultaneous precipitation of two or more cations, it is recommended that synthesis be conducted under supersaturation conditions (Othman et al., 2009). The supersaturation can be adjusted by controlling the pH or the M^{2+}/M^{3+}

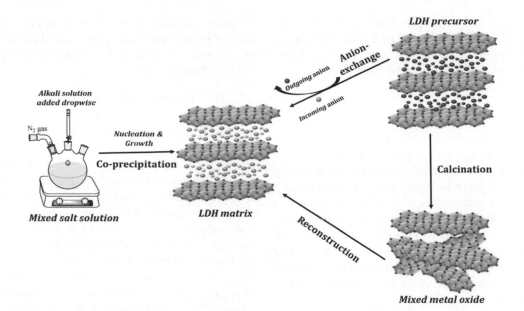

FIGURE 11.2 Schematic diagram of synthesis methods and post-synthesis modification of LDH. (From Mishra, G. et al., *Appl. Clay Sci.*, 153, 172–186, 2018.)

salt ratio in the reactor. In order to improve the crystallinity, a subsequent aging process at a higher temperature (over a few hours or several days) or a hydrothermal treatment is proposed (Paikaray and Hendry, 2014). The subsequent drying temperature also affects the properties of the LDH formed (Kuwahara et al., 2010).

11.2.1.2 Urea-Based Method

Although there are many similarities with co-precipitation method, urea method is generally classified as a separated method. Urea is a very weak Brønsted base ($pK_b = 13.8$), which is highly soluble in water, and its hydrolysis rate is highly sensitive to the temperature of the reaction (He et al., 2006). Therefore, control of the pH could be achieved by adjusting the temperature. Urea decomposition enables hydroxide formation from metal cations to be a gentler and slower process (Chubar et al., 2017). Homogeneous precipitation using urea usually leads to LDHs with higher crystallinity (Theiss et al., 2016).

In theory, this process results in more chemical phases in the final LDH product. Hydrothermal treatment has been adapted to improve the crystallinity of the LDH (Budhysutanto et al., 2010), which is generally performed in an autoclave in the presence of water vapor and under autogenous pressure. In addition, Lonkar et al. (2015) optimized the synthesis process by incorporating a microwave hydrothermal step.

11.2.1.3 Salt-Oxide Method

Solid compounds of divalent and trivalent metals, such Mg and Al hydrous oxides, can result in the formation of LDHs. This method has advantages in producing LDH in large quantities, without limited to large amounts of metallic salts. In 2004, Fogg et al. successfully produced a novel class of $[MAl_4(OH)_{12}](NO_3)_2 \cdot xH_2O$ (where M = Zn, Cu, Ni, and Co) using grinding-activated gibbsite (γ-Al(OH)$_3$) and metal

nitrates ($M(NO_3)_2$) solution. This method has been further developed to incorporate different M(II) and M(III) metals into the prepared LDHs. Chitrakar et al. (2011) conducted various studies on the synthesis of Mg-Al-Cl$^-$ LDH by mixing crystalline gibbsite γ-Al(OH)$_3$ and solid MgCl$_2 \cdot$6H$_2$O with subsequent hydrothermal treatment. The author also illustrated that the calcined Mg-Al LDH and Mg-Fe regained the layered structure through recovery in a low concentration Br$^-$ solution. Salomao et al. (2011) reported a novel co-precipitation method to synthesize Mg-Al LDH using various Mg, Al precursors, including MgO, Mg(OH)$_2$, Al$_2$O$_3$, and Al(OH)$_3$. The common theoretical expressions of salt-oxide process in different pH conditions are as below:

In the acidic/neutral condition:

$$aMg(OH)_2(s) + Al(OH)_4^- + xH_2O + A^-$$
$$\rightleftarrows Mg_aAl(OH)_{2+2a} A \cdot H_2O(s) + 2OH^-$$

In the basic condition:

$$aMg^{2+} + Al(OH)_3(s) + (2a-1)OH^- + xH_2O + A^-$$
$$\rightleftarrows Mg_aAl(OH)_{2+2a} A \cdot H_2O(s)$$

11.2.1.4 Ion-Exchange Method

A key feature of LDH is their anionic exchange capacity, which allows the prepared LDH to be modified by introducing new species. This method can be classified as a "post-synthesis" method. The LDH precursor (e.g., weak electrostatic interaction Cl$^-$, NO$_3^-$, ClO$_4^-$ anions bearing) is suspended in an aqueous solution containing a large excess (10–20 times excess) of the anions to be intercalated (Rives, 2001). pH plays a determining role in the exchange process. At higher pH values (10–12), the intercalation of CO_3^{2-} is strongly encouraged. Therefore, to complete the exchange of CO_3^{2-} LDH by other anions, the reaction must be performed under an acidic pH (4.5–6) (Rives, 2001). The chemical expression is as follows:

$$\left[M^{2+}-M^{3+}-X\right]+Y \rightarrow \left[M^{2+}-M^{3+}-Y\right]+X$$

Thermodynamically, ion exchange in LDH mainly relies on the electrostatic interactions between positive-charged hydroxylated sheets and the ready-to-exchange anions. The selectivity for divalent anions is generally higher than that of monovalent anions. Based on the calculations of ion-exchange isotherms, Miyata (1983) put forward a sequence of ion selectivity for various anions: for monovalent $OH^- > F^- > Cl^- > Br^- > NO_3^- > I^-$ and for divalent $CO_3^{2-} > HPO_4^{2-} > SO_4^{2-}$. Using the ion-exchange approach, organic anions with long chains (but dependent on the anion size) can directly replace the inorganic anions inside the LDH precursors (Meyn et al., 1990).

11.2.1.5 Calcination-Rehydration Method
Reconstruction is another important method to prepare LDH intercalated with various desirable anions, such as inorganic, organic, and biomedical anions. The rehydrated materials have been applied as highly active solid base catalysts (Nishimura et al., 2013). LDHs generally remain stable up to 400°C, but at 200°C, water molecules in the interlayer run away and anhydrous LDH is obtained (Lopez et al., 1996). Along with increasing temperature, LDHs dehydroxylate and subsequently de-anionize to form mixed metal oxides that cannot be obtained by mechanical means.

Although the periodic layer-layer structure collapses, the local cations remain evenly distributed in the mixed oxides (Xu and Lu, 2005). When back in contact with water and carbonate solution, the 2-D layer structure starts taking anions from the aqueous solution and the original layered structure is re-formed. Usually 24 h is required for rehydration for the sample calcined at or below 550°C. This is the so called structure memory effect. The procedure is relatively more complicated than co-precipitation or the ion-exchange method. The generated phases are mostly amorphous (He et al., 2006). LDHs produced via the calcination-rehydration method generally experience a decrease in crystallinity (Lv et al., 2006).

It is important to note that both the chemical composition of LDH sheets and the calcination temperature have a considerable effect on the reconstruction process. "Memory effect" decreases with an increase in the calcination temperature, since excessive calcination temperatures cause the solid-state diffusion of divalent cations into tetrahedral positions, which results in the formation of stable spinel. For instance, when the calcination temperature is above 450°C, Mg-Al CLDH transforms to the spinel $MgAl_2O_4$ that does not possess a reconstruction ability (Lopez et al., 1996).

11.2.2 Characteristics
This section focuses on the physical and chemical properties of the LDH materials. Modifying the synthetic process or the

chemical composition of LDHs could give rise to considerable differences on these properties. Therefore, the potential of LDH and their hybrids for environmental applications lies in their plasticity and variability.

11.2.2.1 Thermal Stability and Memory Effect
A valuable adsorbent should meet these key features: high adsorption capacity, rapid adsorption kinetics, and excellent recycle properties. According to Figure 11.2, the "memory effect" of LDH is a critical feature and an important advantage for wastewater treatment. The reconstruction process normally depends on the calcination temperature and chemical composition of the layered structure. The calcination temperature is very important and sensitive because it is the deciding factor as to whether or not the reconstruction of the layered structure is successful. The chosen temperature should ensure layer collapse but not exceed the formation of spinel phases. In general, the calcination temperature for Mg-Al LDH is set between 400°C and 600°C. Constantino and Pinnavaia (1995) noted that the thermally dehydrated LDH samples should be transferred quickly to characterization measurements or stored in a vacuum to minimize the rehydration effect.

11.2.2.2 Impacts of Cation Composition
It is possible to tune the basicity and acidity of layered double hydroxides, through incorporation of reducible cations into the layers (Valente et al., 2010). The inclusion of transition metals in LDHs such as Ni, Fe, Ti, Co, and Cu, has attracted a lot of interest (Wang et al., 2012). For instance, the incorporation of Ti^{4+} in LDHs enhanced anion exchange capacity but exhibited smaller surface areas (Das et al., 2009). A novel colloidal TiO_2/LDH was prepared, with a high surface area which accelerated the photodegradation of AO7 (Bauer et al., 1999). The inclusion of Fe (III) in the LDH structure enhances the efficiency of the adsorbent (Abou-El-Sherbini et al., 2015). In addition, magnetic LDH nanoparticles could be synthesized from various Fe minerals like spinel ferrite MFe_2O_4 (M = Mg, Zn, Co, Ni, etc.), magnesium-ferrite aluminate, iron oxides Fe_3O_4, and maghemite (γ-Fe_2O_3). Magnetic separation is a smart solution for water treatment as well as the recycling of adsorbents. Abou-El-Sherbini et al. (2015) successfully synthesized a series of Fe-containing LDHs that exhibited magnetic behavior and could be magnetically separated after being loaded with Isolan Dark Blue (IDB) or chromate.

11.2.2.3 Anion Exchange Properties
A key feature of LDHs is their anionic exchange capacity, which makes them attractive in wastewater treatment and soil remediation. This property makes LDHs controllable by introducing or removing various species, herein the reactivity of the interlayer region could be altered. Miyata (1983) pointed out that CO_3^{2-}-LDH has a theoretical anion exchange capacity of 3.6 meq g^{-1} if all the CO_3^{2-} in the general formula is exchanged. It is noteworthy that different synthesis conditions also have a considerable influence on the

TABLE 11.1

Nitrate Adsorption Capacity and Exchangeability by Different Counter Anions of LDHs Synthesized at Lab (L) or Pilot Plant (PP) Scale Using Either Aqueous Ammonia Solution or Potassium Hydroxide during Precipitation at Mg: Al Ratios of 2:1 and 5:1

Product	NO_3^--N Exchanged (mg g^{-1} Material)				Fraction of NO_3^- -N Exchanged (Percentage of HCO_3^-)		
	HCO_3^-	Cl^-	SO_4^{2-}	H_2O	Cl^-	SO_4^{2-}	H_2O
L(NH$_4$OH)$_{2:1}$	37	37	37	8	100	100	22
L(KOH)$_{2:1}$	46	46	46	12	100	100	26
L(NH$_4$OH)$_{5:1}$	28	27	18	10	96	64	36
L(KOH)$_{5:1}$	30	26	14	8	86	47	27
PP(KOH)$_{5:1}$	30	27	14	10	90	47	33

Source: Olfs, H.-W. et al., *Appl. Clay Sci.*, 43, 459–464, 2009.

nitrate exchange properties. Olfs et al. (2009) measured the exchangeable amounts and fractions from different nitrate-bearing LDHs by carbonate, chloride, sulfate, H$_2$O, and chloride, which is presented in Table 11.1. Torres-Dorante et al. (2008) focused on the selectively and capacity of the LDHs for nitrate adsorption in a simulated soil solution containing counter anions.

11.2.2.4 Morphology, Surface Area and Pore Size

The morphology and textural properties of LDH have determinant effects on their adsorption and desorption performances. In general, the size scale of lamellar texture and hexagonal crystals of LDH are several decades' nanometer. Figure 11.3 shows the surface morphology of Co-Al LDHs (Arai and Ogawa, 2009). Due to different synthesis conditions,

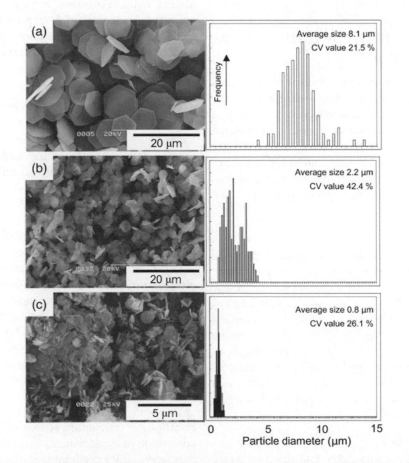

FIGURE 11.3 SEM images of Co-Al LDHs at different particle sizes: (a) average size 8.1 μm, (b) average size 2.2 μm, (c) average size 0.8 μm. (From Arai, Y. and Ogawa, M., *Appl. Clay Sci.*, 42, 601–604, 2009.)

the produced Co-Al LDHs have different particle sizes, resulting in different surface morphology. Budhysutanto et al. (2011) prepared a Mg-Al LDH with donut-like crystals via a microwave-assisted hydrothermal method, which provides a higher specific surface area. Vulić et al. (2008) studied how to control the morphology of the LDHs by varying the extent of trivalent cation substitutions. Zhao et al. (2002) proposed a colloid mill method to produce Mg-Al LDH with uniform nanoscale crystallites. Kuroda et al. (2018) achieved precise size control on the nanoparticles of reconstructed LDH using tripodal ligands. It is reported that LDH nanoparticles have excellent anion exchangeability under ambient conditions (Chubar et al., 2017).

Most of the as-synthesized LDHs originally have a surface area between 70 and 100 m^2 g^{-1} (Vulic et al., 2012). Thermal treatment significantly enlarged the surface area, attributed to the expulsion of CO_2 and H_2O from the LDH (Heredia et al., 2013). As Lv et al. (2006) report, thermally treated Mg-Al-CO_3^{2-} LDHs (2:1, Mg/Al, 400°C) and (3:1, Mg/Al, 500°C) has a surface area of 122.2 and 223.3 m^2/g, respectively. The specific surface area of the Mg–Fe LDH increases from 35.4 to 131 m^2 g^{-1} and the pore volume from 0.545 to 0.767 mL g^{-1} through heating at 500°C (Guo et al., 2013). Abou-El-Sherbini et al. (2015) pointed out that the surface area of calcined LDH varies with the kind and loading of transition metal (Cu and/or Co). A low content of transition metal maximizes the surface area.

11.3 INTERACTIONS BETWEEN LDH AND CONTAMINANTS

LDH can actively interact with environmental contaminants through a range of mechanisms including adsorption, intercalation, precipitation, degradation, and reduction depending on the type of contaminants and characteristics of LDH-based material. Typical contaminants that interact with LDH are organic dyes, oxyanions, halide anions, and heavy metal cations. Organic dyes may be the most frequently seen contaminants that LDH dealt with from existing literature.

11.3.1 Adsorption

Adsorption refers to "an accumulation of matter at the interface between an aqueous solution phase and a solid adsorbent without the development of a three–dimensional molecular arrangement" (Sposito, 1986). It distinguishes from absorption which refers to "the diffusion of an aqueous chemical species into a solid phase" (Sposito, 1986). The adsorption ability of LDH for a range of contaminants lays the foundation of the application of its application in environmental remediation. The typical adsorption behaviors include anion exchange, intercalation, electrostatic interaction, hydrogen bonding, and physical adsorption.

11.3.1.1 Anion Exchange

LDH is also well known as anionic clay, and anion exchange is its most famous function. Nearly all the negatively charged contaminants can be adsorbed by LDH through anion exchange, e.g., anionic organic dyes, oxyanions, and halide anions. As mentioned in Section 11.2.2, hydrated anions (e.g., CO_3^{2-}, HCO_3^{-}) are typically located in the interlamellar area of LDH lattice to maintain electroneutrality. These abundant anions give LDH relatively high anion exchange capacity (AEC). As reported by Abou-El-Sherbini et al. (2015), the AEC values of LDHs varies from 202 to 663 meq per 100 g. Even if sometimes the original interlayer anions are difficult to be exchanged from, the anion exchange may still occur on the surface of layers. Figure 11.4 illustrates the CO_3^{2-} on the outer planner surface were readily replaced by anionic dye anions. Therefore, when negatively charged dyes exist in the environment, they can easily be exchanged with the anions from LDH. To the best of the author's knowledge, anion exchange is the most important mechanism determining the adsorption of organic dyes to LDH. Chubar et al. (2017) believes that "LDH is the most promising candidate for the next-generation inorganic anion exchanger."

Anion exchange is a diffusion process (Hu et al., 2017); therefore, its kinetics in aqueous solutions is controlled by either film or intraparticle diffusion. It can be imagined that surface area and pore structure can affect the kinetics for the adsorption of contaminants to LDH through anion exchange as it is diffusion controlled. In the view of bonding energy, anion exchange is generally regarded as a weak bond. Therefore, in practical water treatment, the adsorbed anions on LDH can be easily desorbed and the LDHs can be reused for many cycles. This made LDH a cost-effective material for the removal of contaminants in wastewater.

11.3.1.2 Electrostatic Interaction

When the surface of LDH is positively charged, it can electrostatically attract negatively charged anions (e.g., anionic organic dyes, oxyanions, and halide anions). Likewise, if the surface of LDH is negatively charged, it can attract positively charged cations (e.g., heavy metal cations). Therefore, electrostatic interaction is dependent on both the charge of LDH and organic dyes. The surface charge of LDH is related to its pH at point of zero charge (pH_{pzc}). If the ambient pH is lower than the pH_{pzc} of LDH, its surface will be positively charged, and therefore it can adsorb anionic dyes through electrostatic interaction. If the ambient pH is higher than the pH_{pzc} of LDH, the negatively charged LDH can adsorb cationic dyes through electrostatic interaction. The stability of adsorbed contaminants on LDH is highly dependent on the ambient pH. Changes in environmental pH may lead to desorption of electrostatically bonded dyes on LDH. Therefore, in practical wastewater treatment, this can be used to discharge the adsorbed dyes on LDH and subsequently aid the reuse of it. A range of studies claim that electrostatic interaction is one of the mechanisms for the adsorption of contaminants (e.g., dyes) to LDH, although it may never play a predominant role (Abdelkader et al., 2011; Hu et al., 2017).

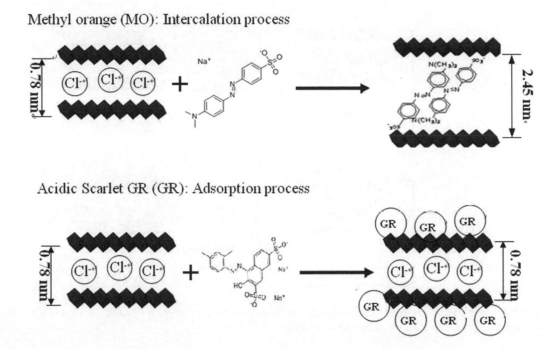

Methyl orange (MO): Intercalation process

Acidic Scarlet GR (GR): Adsorption process

FIGURE 11.4 Schematic diagram of MO and GR interacted with Ca-Al LDH. (From Zhang, P. et al., *J. Colloid Interface Sci.*, 365, 110–116, 2012.)

11.3.1.3 Hydrogen Bonding

The hydrogen from LDH can adsorb negatively charged dyes through hydrogen bonding. Hydrogen bonding has been extensively observed in the adsorption of contaminants to LDHs (Chakraborty and Nagarajan, 2015; Zhang et al., 2017). Hydrogen bonding has a weak binding energy. The adsorbed contaminants on LDH through hydrogen bonding can be easily desorbed, making it easier for the reuse of LDH in wastewater treatment. Similar to electrostatic interaction, hydrogen may not play a major role in the adsorption of contaminants to LDH.

11.3.1.4 Physical Adsorption

Physical adsorption refers to the interaction between contaminants and LDH through van der Waals force. It occurs in nearly all adsorption processes. However, physical adsorption has a very week binding energy and is reversible (Shen et al., 2017). It is highly related to the surface area and pore structure of LDH, as more abundant adsorption sites can aid physical adsorption. Although physical adsorption itself may not play a significant role in contaminant adsorption onto LDH, it is very important for other adsorption mechanisms (e.g., anion exchange and electrostatic interaction). The contaminants need to be seated on LDH through physical adsorption before it can have chemical interaction with LDH.

11.3.2 Intercalation

Intercalation refers to the entrance of contaminants (typically anions) to the interlayer space of LDH structure. It is perhaps the most unique mechanism for LDH to interact with contaminants and plays a very important role in the interaction between LDH and contaminants. It is distinguished

from adsorption, as adsorption is basically a surface process. Intercalation has been extensively reported as a mechanism for the removal of dyes, anions, and oxyanions by LDH in aqueous solutions (Goh et al., 2008; Theiss et al., 2014; Zubair et al., 2017). Abou-El-Sherbini et al. (2015) observed that the mechanisms of dye removal in aqueous solutions turn from anion exchange to intercalation after the calcination of LDH due to the loss of strongly held CO_3^{2-}. Alexandrica et al. (2015) noted that both adsorption and intercalation contribute to the removal of anionic dye by LDH in aqueous solutions.

The calcination (heated at 400°C and above) of LDH turns it to mixed metal oxides (CLDH). When place the CLDH into anionic solutions, it can reconstruct its layered structure. The interlayer of the newly formed double-layer structure can therefore hold more anionic contaminants through intercalation. Guo et al. (2013) noted the Brilliant Blue sorption capacity of CLDHs was more than 10 times bigger compared to its parent LDHs. The authors argued that reconstruction is the predominant adsorption mechanism for the majority of dye removal by CLDH. The calcination-rehydration process was described as follows, in which dye AB 14 was intercalated by chemisorption:

$$\left[Mg_{1-x}Fe_x(OH)_2 \right](CO_3)_{x/2}$$
$$\rightarrow Mg_{1-x}Fe_xO_{1+x/2} + (x/2)CO_2 + H_2O \quad (11.1)$$

$$Mg_{1-x}Fe_xO_{1+x/2} + (x/n)A^{n-} + (x)AB14 + (1+x/2)H_2O$$
$$\rightarrow Mg_{1-x}Fe_x(OH)_2(AB14)_x + xOH^- \quad (11.2)$$

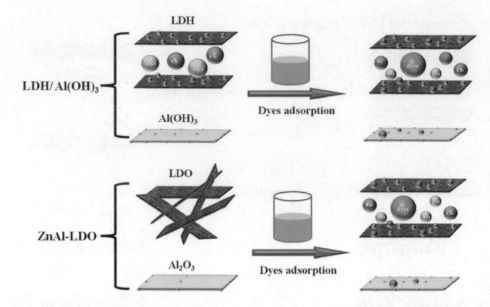

FIGURE 11.5 Intercalation of Methyl Orange to layered double metal oxides LDO (e.g., after Zn-Al LDH calcined at 500°C for 2 h). (From Guo, X. et al., *Microporous Mesoporous Mater.*, 259, 123–133, 2018.)

In another adsorption study, Guo et al. (2018) provided a schematic drawing of the reconstruction process of Zn-Al CLDH in Methyl Orange (MO) aqueous solution (Figure 11.5). For practical application, the dye-loaded LDH can be heated at 400°C or above to discharge the dyes. Then they can be reused relying on the memory effect. This provides LDH higher sustainability and lower cost as a sorbent.

11.3.3 PRECIPITATION

Precipitation is the one of the main mechanisms for LDH to immobilize heavy metals (e.g., Pb^{2+}, Cu^{2+}, Ni^{2+}, Zn^{2+}, and Cd^{2+}). The LDHs synthesized from co-precipitation typically have a pH value ranging from 10–12 (Salomao et al., 2011). This pH range is favored for the precipitation of most heavy metals on the surface of LDH. Precipitation is a fast process and can result in rapid removal of heavy metals from water. However, the binding between heavy metals and LDHs is relatively strong, which may bring difficulties in contaminant desorption and LDH reuse in water treatment. In the context of soil environment, the precipitated heavy metals are relatively stable and their long-term stability depends on the combined effect of acidic rain dissolution and LDH's buffering capacity. It is of note that in addition to the interactions between LDH and contaminants mentioned above, LDH can immobilize heavy metals through the bonding by its surface hydroxyl groups, isomorphic substitution, and chelation (Figure 11.6).

11.3.4 CATALYSIS

One of the most important effects of LDH-based materials on contaminants is their role as catalyst. When applied in photocatalysis, LDH is capable to degrade organic dyes and pesticides, which cannot be treated biologically. The detailed

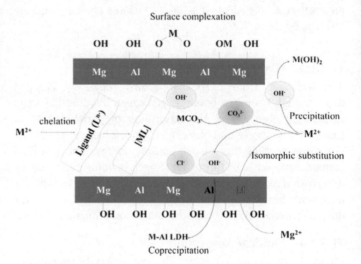

FIGURE 11.6 Schematic representation of the interaction between heavy metals and LDH. (From Liang, X.F. et al., *Colloid Surface A*, 433, 122–131, 2013.)

mechanisms for the photocatalysis process can be found in recent review works (Mohapatra and Parida, 2016). Briefly, the semiconductor surfaces under photoexcitation with ultraviolet-visible radiation can generate photoactivated electron–hole pairs. Therefore, electrons (e^-) migrate to the conduction band and meanwhile holes (h^+) are generated in the valance band. The oxidation reaction is mediated by the holes, and hydroxyl radicals are produced. Meanwhile, the reduction of the dissolved O_2 is mediated by the electrons, and the superoxide radicals form. The superoxide radicals are protonated and the hydroperoxyl radical $HO_2^·$ and subsequently H_2O_2 are created. The further decomposition of H_2O_2 results in the formation of $OH^·$ radicals. These radicals can oxidize the organic molecules that produce intermediates, H_2O, and end products.

Therefore, the resulting hydroxyl radicals (OH·) have stronger oxidization capacity than other conventional oxidants such as ozone, hydrogen peroxide, chlorine dioxide, and chlorine, which are capable to decompose non-biodegradable contaminants. Although LDH (e.g., Mg/Al) is not semiconductor, it could act as an induced semiconductor and termed as a photoassisted system instead of photocatalyst. In addition to degradation through photocatalysis, LDH-based materials were also observed to aid the oxidation of urea through electrocatalysis (Zeng et al., 2019), to aid the degradation of antibiotics through sonophotocatalysis (Abazari et al., 2019), to aid the reduction of Cr(VI) through photocatalysis (Yuan et al., 2017), and to remediate phenol and phenolic derivatives by catalytic wet peroxide oxidation (Hosseini et al., 2017).

11.4 LDH FOR SOIL AND GROUNDWATER REMEDIATION

LDH is a potential effective material for soil and groundwater remediation. Current studies about the environmental application of LDH mainly focus on wastewater treatment. Only a few have involved soil and groundwater remediation. However, the effective performance of LDH-based materials in wastewater treatment suggests a promising future for their applications in soil and groundwater remediation. Therefore, understanding the current applications and performances of LDH-based materials in environmental remediation can aid its future applications. The applications of LDH-based materials in environmental remediation include the removal of contaminants from water and groundwater mainly through adsorption, the remediation of contaminants through catalysis, and the potential usage in soil remediation.

11.4.1 REMOVAL OF CONTAMINANTS FROM WATER AND GROUNDWATER

Wastewater treatment may be the most typically environmental application of LDH-based materials. Such investigations can date back to the 2000s (Chuang et al., 2008; Prasanna and Kamath, 2009) and rapidly developed recently (He et al., 2018a; Wang et al., 2018). The most important mechanisms for LDH-based materials to remove contaminants from wastewater are the immobilization of the anionic contaminants through adsorption and intercalation, and of the cationic heavy metals through precipitation, isomorphic substitution, and chelation. For instance, Mg/Al LDH and sodium dodecyl sulfate intercalated Mg/Al LDH effectively removed both molecular and anionic 2-chlorophenol from aqueous solutions through adsorption and intercalation (Chuang et al., 2008). Likewise, Mg/Al LDH exhibited superior adsorption capacity (2.25 mmol g^{-1}) for phosphate, and removal mechanisms were attributed to anion exchange, electrostatic interaction, and surface complexation (Luengo et al., 2017). Ca/Al LDHs were observed to successfully treat highly contaminated wastewater by Cu^{2+}, Cd^{2+} and Pb^{2+} through precipitation (Rojas, 2014). Despite the extensive investigations on the removal

of contaminants from wastewater, the research focusing on groundwater remediation using LDH is limited. Kovacevic et al. (2013) investigated the potential of using Mg/Al LDH to remediate arsenic (in the form of arsenate) contaminated groundwater by conducting laboratory adsorption studies. They concluded that Mg/Al LDH is a suitable substrate for arsenic adsorption, and observed that the presence of phosphate did not significantly affect the arsenic adsorption. Likewise, Chao et al. (2018) conducted laboratory adsorption studies to reveal the potential of Mg/Al LDHs in chromium (Cr)(VI) removal from groundwater. They observed that the Mg/Al LDHs can effectively remove (di)chromate (maximum of 399 mg g^{-1} for Cr(VI)) mainly through anion exchange. Sheng et al. (2016) explored the potential application of nanoscale zerovalent iron (NZVI) supported on LDH (NZVI/LDH) in TcO$_4^-$ removal from groundwater via X-ray absorption fine structure (XAFS) approach and suggested NZVI/LDH is a promising strategy for the decontamination of Tc(VII) from groundwater. It is of note that all of these investigations are based on laboratory studies using synthetic contaminated solution. To the best of the author's knowledge, no practical applications of LDH-based materials in groundwater remediation have been reported.

Although LDHs themselves showed excellent performance of contaminated removal in water, they may not be universally effective depending on contaminant type and contamination level. Therefore, extensive studies have focused on the modification of LDH to improve its performance in the removal of contaminants from water. For instance, Yu et al. (2019) synthesized NZVI and LDH compost and observed that the removal capacity of uranium (VI) was enhanced by ~50% due to the NZVI modification. Ma et al. (2017) synthesized MoS$_4^{2-}$ intercalated LDH (MOS$_4$-LDH), and it revealed highly selective and exceptionally efficient and rapid capture of oxoanions of As(III)/As(V) (HAsO$_3^{2-}$/HAsO$_4^{2-}$) and Cr(VI) (CrO$_4^{2-}$). Tan et al. (2016) pyrolyzed biochar from Mg/Al LDH pre-coated ramie biomass, and the produced biochar revealed excellent ability to remove crystal violet from actual industrial wastewater and groundwater.

The removal ability of LDH-based materials for contaminants from water is typically revealed by laboratory batch adsorption tests. Isotherm, kinetics, and pH-dependence are three of the most important characteristics obtained from batch adsorption tests, and are mostly typically used to indicate the removal ability.

Adsorption isotherms are typically obtained through equilibrium study. It indicates how the molecules distribute between the liquid and solid phase when reaching the equilibrium state. A certain amount of adsorbent is mixed with solutions containing different sorbate concentrations. The mixtures are typically shaken at a designated time to reach equilibrium. The adsorbed amount of sorbate by adsorbent at equilibrium is plotted against the equilibrium concentration of sorbate. Empirical models are typically used to fit the results to obtain the isotherm curves and other parameters. Table 11.2 lists the empirical models that are typically used for isotherm fitting. It can be observed that both the nonlinear

TABLE 11.2

Adsorption Isotherm Models

Isotherm	Nonlinear Form	Linear Form	Plot
Langmuir	$q_e = \dfrac{Q_0 b C_e}{1 + b C_e}$	$\dfrac{C_e}{q_e} = \dfrac{1}{b Q_0} + \dfrac{C_e}{Q_0}$	$\dfrac{C_e}{q_e}$ vs C_e
		$\dfrac{1}{q_e} = \dfrac{1}{Q_0} + \dfrac{1}{b Q_0 C_e}$	$\dfrac{1}{q_e}$ vs $\dfrac{1}{C_e}$
		$q_e = Q_0 - \dfrac{q_e}{b C_e}$	q_e vs $\dfrac{q_e}{b C_e}$
		$\dfrac{q_e}{C_e} = b Q_0 - b q_e$	$\dfrac{q_e}{C_e}$ vs q_e
Freundlich	$q_e = K_F C_e^{1/n}$	$\log q_e = \log K_F + \dfrac{1}{n} \log C_e$	$\log q_e$ vs $\log C_e$
Dubinin-Radushkevich	$q_e = (q_s) \exp(-k_{ad} \varepsilon^2)$	$\ln q_e = \ln(q_s) - k_{ad} \varepsilon^2$	$\ln(q_e)$ vs ε^2
Tempkin	$q_e = \dfrac{RT}{b_T} \ln A_T C_e$	$q_e = \dfrac{RT}{b_T} \ln A_T + \left(\dfrac{RT}{b_T} \right) \ln C_e$	q_e vs $\ln C_e$
Flory-Huggins	$\dfrac{\theta}{C_0} = K_{FH} (1 - \theta)^{n_{FH}}$	$\log\left(\dfrac{\theta}{C_0} \right) = \log(K_{FH}) + n_{FH} \log(1 - \theta)$	$\log\left(\dfrac{\theta}{C_0} \right)$ vs $\log(1 - \theta)$
Hill	$q_e = \dfrac{q_{s_H} C_e^{n_H}}{K_D + C_e^{n_H}}$	$\log\left(\dfrac{q_e}{q_{s_H} - q_e} \right) = n_H \log(C_e) - \log(K_D)$	$\log\left(\dfrac{q_e}{q_{s_H} - q_e} \right)$ vs $\log(C_e)$
Redlich-Peterson	$q_e = \dfrac{K_R C_e}{1 + a_R C_e^g}$	$\ln\left(K_R \dfrac{C_e}{q_e} - 1 \right) = g \ln(C_e) + \ln(a_R)$	$\ln\left(K_R \dfrac{C_e}{q_e} - 1 \right)$ vs $\ln(C_e)$
Sips	$q_e = \dfrac{K_s C_e^\beta s}{1 + a_S C_e^\beta s}$	$\beta_S \ln(C_e) = -\ln\left(\dfrac{K_s}{q_e} \right) + \ln(a_S)$	$\ln\left(\dfrac{K_s}{q_e} \right)$ vs $\ln(C_e)$
Toth	$q_e = \dfrac{K_T C_e}{\left(a_T + C_e \right)^{1/t}}$	$\ln\left(\dfrac{q_e}{K_T} \right) = \ln(C_e) - \dfrac{1}{t} \ln(a_T + C_e)$	$\ln\left(\dfrac{q_e}{K_T} \right)$ vs $\ln(C_e)$
Koble-Corrigan	$q_e = \dfrac{A C_e^n}{1 + B C_e^n}$	$\dfrac{1}{q_e} = \dfrac{1}{A C_e^n} + \dfrac{B}{A}$	—
Khan	$q_e = \dfrac{q_s b_K C_e}{\left(1 + b_K C_e \right)^{a_K}}$	—	—
Radke-Prausnitz	$q_e = \dfrac{a_{RP} r_R C_e^\beta R}{a_{RP} + r_R C_e^\beta R^{-1}}$	—	—
BET	$q_e = \dfrac{q_s C_{BET} C_e}{(C_s - C_e)\left[1 + (C_{BET} - 1)(C_e / C_s)\right]}$	$\dfrac{C_e}{q_e (C_s - C_e)} = \dfrac{1}{q_s C_{BET}} + \dfrac{(C_{BET} - 1)}{q_s C_{BET}} \dfrac{C_e}{C_s}$	$\dfrac{C_e}{q_e (C_s - C_e)}$ vs $\dfrac{C_e}{C_s}$
FHH	$\ln\left(\dfrac{C_e}{C_s} \right) = -\dfrac{\alpha}{RT} \left(\dfrac{q_s}{q_e d} \right)^r$	—	—
MET	$q_e = q_s \left(\dfrac{k}{\ln(C_s / C_e)} \right)^{1/3}$	—	—

Source: Foo, K.Y. and Hameed, B.H., *Chem. Eng. J.*, 156, 2–10, 2010.

and linear forms of the models can be used. Among all the models listed in Table 11.2, Langmuir and Freundlich models are most frequently used in existing adsorption studies.

The nonlinear form of Langmuir model is:

$$q_e = \frac{Q_{max} b C_e}{1 + b C_e} \tag{11.3}$$

where Q_{max} (mg g^{-1}) is the maximum monolayer adsorption capacity, b is the Langmuir isotherm constant (L mg^{-1}), q_e is the adsorbed amount of sorbate on adsorbent at equilibrium, and C_e (mg L^{-1}) is the equilibrium concentration of sorbate in solution. Q_{max} is typically used to compare the adsorption capacities among different adsorbents. The equilibrium parameter R_L, as shown in Equation (11.4), can be used to express the essential characteristics of a Langmuir isotherm.

$$R_L = \frac{1}{1 + bC_{max}} \qquad (11.4)$$

where b is defined as per Equation (11.3), C_{max} (mM) is the highest initial sorbate concentration. The value of R_L indicates the type of the isotherm to be either unfavorable ($R_L > 1$), linear ($R_L = 1$), favorable ($0 < R_L < 1$) or irreversible ($R_L = 0$).

The nonlinear Freundlich model is:

$$q_e = K_f C_e^{1/n} \qquad (11.5)$$

where K_f (mg g^{-1}) is the Freundlich isotherm constant, n represents the adsorption intensity, q_e and C_e are the same as per Equation (11.3). "$1/n$" ranges between 0 and 1 and is a measure of adsorption intensity or surface heterogeneity. A lower $1/n$ value indicates a greater degree of heterogeneity on the surface of adsorbent.

Figure 11.7 and Table 11.3 show the adsorption isotherm of acid brown 14 dye on LDH and calcined LDH. The plots are the results of an equilibrium study. The solid line is the Langmuir isotherm fitting curve of the results. The solid-dash line is the Freundlich isotherm curve of the results. It can be observed

from the fitting results (Table 11.3) that the Langmuir model, with higher R^2 values, was better fitted for the results than the Freundlich model. This suggests the adsorption of the acid brown 14 dye to the LDH and CLDH was monolayer adsorption. The Q_{max} values of the LDH and CLDH calculated from Langmuir fitting were 41.7 and 370 mg g^{-1}, respectively.

Table 11.4 shows the selected maximum adsorption capacity of contaminants onto LDHs. It can be observed that Mg/Al LDHs were the most frequently investigated. Pb^{2+} and organic dye (RBB-150) showed the highest maximum adsorption capacities. The proposed mechanisms varied among different studies. The table also suggests although LDH is famous as an anion exchanger; its removal ability for heavy metal cations is excellent. LDH-based materials may also have the potential for heavy metal removal in water and groundwater.

Kinetics reveal the time dependence of an adsorption process. For a kinetics study, a certain amount of adsorbent is mixed with a solution containing fixed concentration of sorbate. The adsorbed sorbate amount on the adsorbent is measured and plotted at different designated times. Similar to isotherm, the kinetics results are typically fitted by empirical models to indicate the kinetics in more depth. Pseudo-first-order, pseudo-second-order, and intraparticle diffusion models are three of the most commonly used kinetics models.

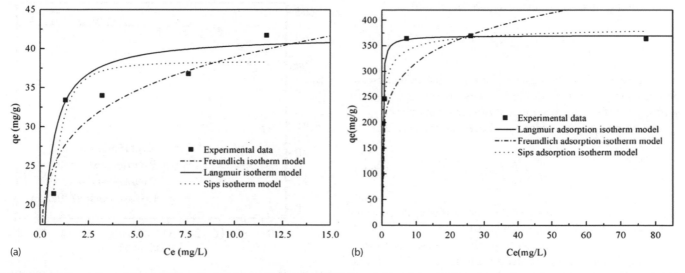

FIGURE 11.7 Adsorption isotherm of acid brown 14 dye on (a) LDH and (b) calcined LDH. (From Guo, Y. et al., *Chem. Eng. J.*, 219, 69–77, 2013.)

TABLE 11.3

Parameters of Adsorption Isotherm of Acid Brown 14 Dye on LDH and Calcined LDH

Material	Langmuir Isotherm Model			Freundlich Isotherm Model		
	q_m (mg g^{-1})	K_L (L mg^{-1})	R^2	K_F (mg$^{1-1/n}$ L$^{1/n}$ g^{-1})	$1/n$	R^2
LDH	41.7	3.00	0.991	26.5	0.188	0.771
CLDHs	370.0	6.76	0.999	215.7	0.167	0.877

Source: Guo, Y. et al., *Chem. Eng. J.*, 219, 69–77, 2013.

TABLE 11.4

Selected Maximum Adsorption Capacity of Contaminants onto LDHs

LDH Type	Contaminants	Maximum Adsorption Capacity (mg g⁻¹ Unless Noted)	Proposed Removal Mechanisms	References
Mg/Al LDH	2-chlorophenol	21.4	Adsorption and intercalation	Chuang et al. (2008)
Sodium dodecyl sulfate intercalated Mg/Al LDH	2-chlorophenol	18.1	Adsorption and intercalation	Chuang et al. (2008)
Mg/Al LDH	Phosphate	2.25 mmol g⁻¹	Anion exchange, electrostatic interaction, and surface complexation	Luengo et al. (2017)
Ca/Al LDH	Cu^{2+}	381	Precipitation	Rojas (2014)
Ca/Al LDH	Pb^{2+}	974	Precipitation	Rojas (2014)
Ca/Al LDH	Cd^{2+}	360	Precipitation	Rojas (2014)
Mg/Al LDH	Cr(VI)	399	Anion exchange	Chao et al. (2018)
Nanoscale zero-valent iron supported on LDH	U(VI)	176	Adsorption and reduction	Yu et al. (2019)
MoS_4^{2-} intercalated Mg/Al LDH	As(III)	99	As-S interactions	Ma et al. (2017)
MoS_4^{2-} intercalated Mg/Al LDH	As(V)	56	As-S interactions	Ma et al. (2017)
MoS_4^{2-} intercalated Mg/Al LDH	Cr(VI)	130	Reduction	Ma et al. (2017)
Mg/Al LDH	RR-120	830	Nitrate interlayer anions play a minor role	Boubakri et al. (2018)
Mg/Al LDH	RBB-150	960	Nitrate interlayer anions play a minor role	Boubakri et al. (2018)

The pseudo-first-order model can be expressed as:

$$q_t = q_e \left(1 - e^{-k_1 t}\right) \tag{11.6}$$

where k_1 (h⁻¹) is the rate constant of pseudo-first-order adsorption, q_e (mg g⁻¹) is the equilibrium adsorption capacity, and q_t (mg g⁻¹) is the adsorbed amount of sorbate at time t.

The pseudo-second-order model can be expressed as:

$$q_t = \frac{k_2 q_e^2 t}{1 + k_2 q_e t} \tag{11.7}$$

where k_2 (g mg⁻¹ h⁻¹) is the pseudo-second-order rate constant and q_e and q_t are as per Equation (11.6).

In practical application, both nonlinear and linear forms of pseudo first and second order models are used. Figure 11.8 and Table 11.5 show typical kinetics data of dye adsorption on LDH and the fitting using the two pseudo models. The regression coefficient and other parameters can be observed through data fitting. These parameters can be used to compare the kinetics of dye adsorption among different LDHs.

In addition to the pseudo first and second order models, the intraparticle diffusion model is typically used to fit the kinetics data and reveal the diffusion dynamics. It can be expressed as

$$q_t = k_i t^{0.5} + C \tag{11.8}$$

where k_i (mg g⁻¹ h⁻¹) is the coefficient of intraparticle diffusion and q_t is as per Equation (11.6).

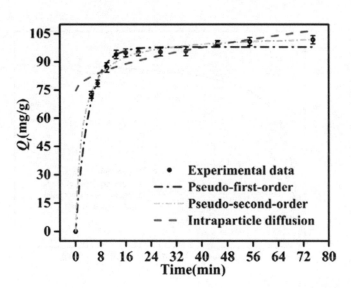

FIGURE 11.8 Kinetics of methylene blue adsorption on PDOPA-LDH. (From Zhao, J. et al., *J. Colloid Interface Sci.*, 505, 168–177, 2017.)

From a kinetic view, there are four stages in an adsorption process (Choy et al., 2004):

1. Transport of sorbate from the bulk solution to the exterior film surrounding of the adsorbent (remain in liquid phase)
2. Movement of sorbate across the external liquid film boundary layer to external surface sites of adsorbent (defined as film diffusion)

TABLE 11.5
Kinetic Parameters for the Adsorption of Methylene Blue on PDOPA-LDH

Models	Parameters	Initial Concentration (mg L^{-1}) 50
Pseudo-first-order equation	Q_e (cal) (mg g^{-1})	98
	K_1 (min^{-1})	0.2446
	R^2	0.9918
Pseudo-second-order equation	Q_e (cal) (mg g^{-1})	105
	k_2 (g mg^{-1} min^{-1})	0.004546
	R^2	0.9952
Intraparticle diffusion	k_p (mg g^{-1} min$^{-0.5}$)	3.814
	C	73.63
	R_2	0.6821

Source: Zhao, J. et al., *J. Colloid Interface Sci.*, 505, 168–177, 2017.

3. Migration of sorbate within the pores of the adsorbent by intraparticle diffusion (defined as intraparticle diffusion)
4. Adsorption of sorbate at internal surface sites

All the four processes control the kinetics of the adsorption. However, in a shaking-based laboratory adsorption study, the transport of sorbate from the bulk solution to the exterior film (Step 1) is very fast, and the time costed can be neglected. The adsorption of sorbate at the surface sites (Step 4) is also rapid. Therefore, in most adsorption processes, film diffusion (Step 2) and intraparticle diffusion (Step 3) are the rate-controlling steps. The intraparticle diffusion model (Equation 11.8) can be used to identify the film and diffusion steps of an adsorption.

Figure 11.9 shows the kinetics data of acid yellow 42 adsorption to a calcined LDH fitted by intraparticle diffusion

model. The first stage (k_{D1}) is regarded as the film diffusion step. The second stage (k_{D2}) is attributed to intraparticle diffusion step, while the third stage (k_{D2}) was regarded as a slow-down of intraparticle diffusion to equilibrium. It can be observed from Figure 11.9 that, although both film and intraparticle diffusion control the kinetics of acid yellow 42 adsorption to the calcined LDH, intraparticle diffusion is the main rate-limiting step.

In many cases, the adsorption of contaminants to LDH is a pH-dependent process. The influence of solution pH on the adsorption is typically investigated for a better understanding of the adsorption characteristics and mechanisms. As mentioned in Section 11.3.1.2, the ambient pH determined whether LDH is positively or negatively charged, and subsequently its electrostatic interaction with contaminants with charges. In addition, the anion exchange between anionic dyes and the anions in LDH can also be affected by solution pH. De Sa et al. (2013) investigated the influence of initial solution pH on the adsorption of sunset yellow FCF dye (in form of anion) to a Ca-Al-NO$_3$ LDH and identified a significant influence of initial solution pH on the dye adsorption to the LDH (Figure 11.10). They proposed the mechanisms for both pH below and above the pH$_{pzc}$ (7.29) of the LDH: for the solution pH values below pH$_{pzc}$, both anion exchange and electrostatic interaction contributes to the adsorption of the dye to the LDH; for the solution pH values above pH$_{pzc}$, deprotonation of surface hydroxyls groups occurred and the positive change of LDH deceases, and the increased OH$^-$ in the solution competed with the negatively charged dye for anion exchange with the LDH.

It should be noted that the adsorption test cannot distinguish the different interactions (e.g., adsorption, intercalation, precipitation). All the removed contaminants from water are the combination of these different interactions. If the understanding of clearer mechanisms are needed, spectroscopy analyses are suggested.

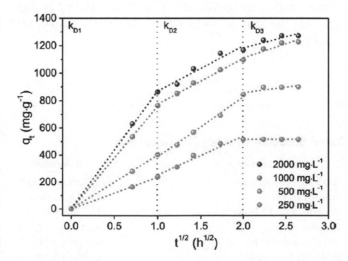

FIGURE 11.9 Intraparticle diffusion modeling of acid yellow 42 adsorption to calcined LDH. (From dos Santos, R.M.M. et al., *Appl. Clay Sci.*, 140, 132–139, 2017.)

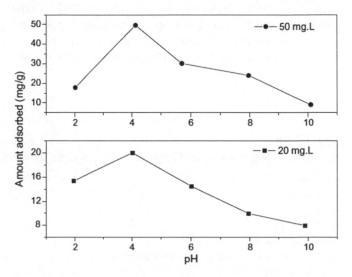

FIGURE 11.10 Effect of initial solution pH on the adsorption of sunset yellow FCF dye to CaAl-LDH-NO$_3$. (From De Sa, F.P. et al., *Chem. Eng. J.*, 215, 122–127, 2013.)

11.4.2 Remediation of Contaminants through Catalysis

Because the remediation of contaminants through catalysis is a totally different mechanism from the removal of them through adsorption, intercalation, and precipitation, it is discussed separately. As mentioned in Section 11.3.4, the degradation of organic pollutants by LDH-based materials through photocatalysis is an important application of LDHs in environmental remediation. In general, LDH-based materials showed excellent ability to aid the degradation of the organics and may be more efficient than biodegradation. Valente et al. (2009) observed that Mg-Zn-Al LDHs can effectively degrade phenol and 2,4 dichlorophenoxiacetic acid through photocatalysis, and the degradation only takes several hours. Khan et al. (2016) reported the successful decoloration and mineralization of five organic dyes by Cd-Al/C LDH nanocatalyst in both solar and visible light exposure. More investigations on the photocatalytic degradation of organic contaminants by LDH-based materials are summarized in Table 11.6. It can be observed that organic dyes are the most typical contaminants that can be effectively degraded through photocatalysis by LDH-based materials. In addition of photocatalysis, Abazari et al. (2019) observed the nanocomposites of g-C_3N_4@Ni–Ti LDH revealed high and ultrafast sonophotocatalytic performance of the degradation of antibiotics.

LDH-based materials were also observed to aid the reduction of inorganics through catalysis. Wong et al. (2019) investigated the sonocatalytic reduction of NO_3^- using magnetic Mg-Cu-Al LDH. They observed NO_3^- was firstly reduced to NO_2^- by Cu^0, and then to N_2/NH_4^+ by Mg^0, while Cu^0 served as reducing sites. Therefore, they conclude that sonocatalytic reduction using magnetic LDH has the potential to remediate NO_3^- contaminated water. Yuan et al. (2017) observed that calcined Zn/Al LDH was highly effective in photocatalytic reduction of Cr(VI).

It should be noted that LDH was observed to significantly aid the degradation of pesticides under normal conditions. Park et al. (2004) found that the presence of Ca/Fe LDHs resulted in quick decomposition of chlorinated pesticides (endosulfan sulfate) and the decomposition rate reached 45% within 15 min, which is comparable to those by photocatalysts.

11.4.3 LDH for Soil Remediation

The excellent ability for LDH to immobilize or degrade contaminants suggests its potential for soil remediation. However, such investigations are very limited to date. Zhou et al. (2017) carried out laboratory incubation experiments investigating the effect of arsenite immobilization by a ternary CaMgFe-LDH in a paddy soil. They achieved a removal efficiency for As(III) of 47% and a total As concentration of 346 μg L^{-1} in capillary water after 40-days incubation, and suggested the precipitation of As with Ca was a predominant immobilization mechanism. He et al. (2018b) used Fe(II)-Al LDH to remediate a Cr(VI) contaminated soil, and observed that the pre-adsorbed Cr(VI) in the soil (2080 mg kg^{-1}) was completely immobilized. Liu et al. (2018) used heterogeneous Fe-based LDH as persulfate (PS) activators to degrade phenylurea herbicide residual (isoproturon) in soil. They observed that 1% (w/w) LDH and 70 mM PS can completely degrade 500 mg kg^{-1} isoproturon in soil within 10 h. Despite the real investigations that applied LDHs to soil, several studies investigated the potential of LDH application in soil remediation by aqueous tests. Park et al. (2004) investigated the decomposition of enodsulfan substances by LDHs in suspension and found that the presence of LDH resulted in the 45% decomposition of contaminants within 15 min. They therefore suggest the potential of LDH for beneficial chemical remediation of the soils contaminated with chlorinated pesticides. Cordova et al. (2009) found that Mg/Al LDH can effectively immobilize enzyme laccase and suggested LDH was a suitable support for use in soil remediation.

In addition to contaminant remediation, LDH was also shown to enhance the soil fertility as an amendment. Berber et al. (2014) found that nitrate-layered double hydroxide nanoparticles can act as a controlled nitrogen release source when added into soil and become a sustainable nitrogen fertilizer. Therefore, the potential of LDH-based materials in soil remediation and amendment should be addressed for future studies considering their multiple benefits in contaminant immobilization and degradation and soil nutrient supply.

TABLE 11.6
Successful Degradation of Organic Contaminants by LDH-Based Materials through Photocatalysis

LDH Type	Contaminants	References
Mg-Zn-Al LDHs	2,4 dichlorophenoxiacetic acid	Valente et al. (2009)
Cd-Al/CLDH	Five organic dyes	Khan et al. (2016)
Mg-Fe-Ti LDH	Herbicide 2,4,5 trichlorophenoxyacetic acid	Phuong et al. (2016)
Co-, Cu-substituted ZnAl ternary LDHs	Orange II dye	Kim et al. (2017)
Composite of Cu-doped TiO_2 nanoparticles and NiAl LDH	Methyl orange and isoniazid	Jo et al. (2018)
Bi_2S_3/Zn-Al LDH	Methylene blue dye	Li et al. (2018)
Ag nanoparticles decorated NiAl-layered double hydroxide/graphitic carbon nitride (Ag/LDH/g-C_3N_4) nanocomposites	Rhodamine B and 4-chlorophenol	Tonda and Jo (2018)

11.5 CONCLUSIONS AND RECOMMENDATIONS FOR FUTURE WORK

This chapter introduces the synthesis and characteristics of LDHs, the interactions between LDHs and contaminants, and the application of LDHs for soil and groundwater remediation. The synthesis methods mainly include co-precipitation, urea-based method, salt-oxide method, ion exchange, and calcination. The unique characteristics of LDHs include memory effect and high anion exchange capacity. LDHs can remove/immobilize anionic contaminants mainly through adsorption and intercalation, and cationic heavy metals mainly through precipitation. It can also aid the degradation of contaminants through serving as a catalyst. LDHs have been applied in wastewater treatment; however, they show huge potential for soil and groundwater remediation. Therefore, future work is recommended to investigate the performance of LDHs in soil and groundwater remediation at both laboratory and field scales.

REFERENCES

Abazari, R., Mahjoub, A.R., Sanati, S., Rezvani, Z., Hou, Z.Q. and Dai, H.X. (2019) Ni-Ti layered double hydroxide@graphitic carbon nitride nanosheet: A novel nanocomposite with high and ultrafast sonophotocatalytic performance for degradation of antibiotics. *Inorganic Chemistry* 58(3), 1834–1849.

Abdelkader, N.B.-H., Bentouami, A., Derriche, Z., Bettahar, N. and De Menorval, L.-C. (2011) Synthesis and characterization of Mg–Fe layer double hydroxides and its application on adsorption of Orange G from aqueous solution. *Chemical Engineering Journal* 169(1–3), 231–238.

Abou-El-Sherbini, K.S., Kenawy, I.M., Hafez, M.A., Lotfy, H.R. and AbdElbary, Z.M. (2015) Synthesis of novel CO_3^{2-}/Cl^--bearing 3 (Mg+ Zn)/(Al+ Fe) layered double hydroxides for the removal of anionic hazards. *Journal of Environmental Chemical Engineering* 3(4), 2707–2721.

Alexandrica, M.C., Silion, M., Hritcu, D. and Popa, M.I. (2015) Layered double hydroxides as adsorbents for anionic dye removal from aqueous solutions. *Environmental Engineering and Management Journal* 14(2), 381–388.

Arai, Y. and Ogawa, M. (2009) Preparation of Co-Al layered double hydroxides by the hydrothermal urea method for controlled particle size. *Applied Clay Science* 42(3–4), 601–604.

Bauer, C., Jacques, P. and Kalt, A. (1999) Investigation of the interaction between a sulfonated azo dye (AO7) and a TiO_2 surface. *Chemical Physics Letters* 307, 397–406.

Berber, M.R., Hafez, I.H., Minagawa, K. and Mori, T. (2014) A sustained controlled release formulation of soil nitrogen based on nitrate-layered double hydroxide nanoparticle material. *Journal of Soils and Sediments* 14(1), 60–66.

Boubakri, S., Djebbi, M.A., Bouaziz, Z., Namour, P., Jaffrezic-Renault, N., Amara, A.B., Trabelsi-Ayadi, M., Ghorbel-Abid, I. and Kalfat, R. (2018) Removal of two anionic reactive textile dyes by adsorption into MgAl-layered double hydroxide in aqueous solutions. *Environmental Science and Pollution Research* 25(24), 23817–23832.

Budhysutanto, W.N., Kramer, H.J.M., van Agterveld, D., Talma, A.G. and Jansens, P.J. (2010) Pre-treatment of raw materials for the hydrothermal synthesis of hydrotalcite-like compounds. *Chemical Engineering Research and Design* 88, 1445–1449.

Budhysutanto, W.N., Van Den Bruele, F.J., Rossenaar, B.D., Van Agterveld, D., Van Enckevort, W.J.P. and Kramer, H.J.M. (2011) A unique growth mechanism of donut-shaped Mg–Al layered double hydroxides crystals revealed by AFM and STEM–EDX. *Journal of Crystal Growth* 318, 110–116.

Chakraborty, P. and Nagarajan, R. (2015) Efficient adsorption of malachite green and Congo red dyes by the surfactant (DS) intercalated layered hydroxide containing Zn^{2+} and Y^{3+}-ions. *Applied Clay Science* 118, 308–315.

Chao, H.P., Wang, Y.C. and Tran, H.N. (2018) Removal of hexavalent chromium from groundwater by Mg/Al-layered double hydroxides using characteristics of in-situ synthesis. *Environmental Pollution* 243, 620–629.

Chitrakar, R., Sonoda, A., Makita, Y. and Hirotsu, T. (2011) Calcined Mg–Al layered double hydroxides for uptake of trace levels of bromate from aqueous solution. *Industrial & Engineering Chemistry Research* 50, 9280–9285.

Chmielarz, L., Kuśtrowski, P., Rafalska-Łasocha, A., Majda, D. and Dziembaj, R. (2002) Catalytic activity of Co-Mg-Al, Cu-Mg-Al and Cu-Co-Mg-Al mixed oxides derived from hydrotalcites in SCR of NO with ammonia. *Applied Catalysis B: Environmental* 35, 195–210.

Choy, K.K.H., Ko, D.C.K., Cheung, C.W., Porter, J.F. and McKay, G. (2004) Film and intraparticle mass transfer during the adsorption of metal ions onto bone char. *Journal of Colloid and Interface Science* 271(2), 284–295.

Chuang, Y.H., Tzou, Y.M., Wang, M.K., Liu, C.H. and Chiang, P.N. (2008) Removal of 2-chlorophenol from aqueous solution by Mg/Al layered double hydroxide (LDH) and modified LDH. *Industrial & Engineering Chemistry Research* 47(11), 3813–3819.

Chubar, N., Gilmour, R., Gerda, V., Mičušík, M., Omastova, M., Heister, K., Man, P., Fraissard, J. and Zaitsev, V. (2017) Layered double hydroxides as the next generation inorganic anion exchangers: Synthetic methods versus applicability. *Advances in Colloid and Interface Science* 245, 62–80.

Constantino, V.R.L. and Pinnavaia, T.J. (1995) Basic properties of $Mg^{2+}_{1-x}Al^{3+}_x$ layered double hydroxides intercalated by carbonate, hydroxide, chloride, and sulfate anions. *Inorganic Chemistry* 34, 883–892.

Cordova, D.I.C., Borges, R.M., Arizaga, G.G.C., Wypych, F. and Krieger, N. (2009) Immobilization of Laccase on hybrid layered double hydroxide. *Quimica Nova* 32(6), 1495–1499.

Das, N.N., Konar, J., Mohanta, M.K. and Upadhaya, A.K. (2009) Synthesis, characterization and adsorptive properties of hydrotalcite-like compounds derived from titanium rich bauxite. *Reaction Kinetics, Mechanisms and Catalysis* 99, 167–176.

De Sa, F.P., Cunha, B.N. and Nunes, L.M. (2013) Effect of pH on the adsorption of Sunset Yellow FCF food dye into a layered double hydroxide (CaAl-LDH-NO_3). *Chemical Engineering Journal* 215, 122–127.

dos Santos, R.M.M., Gonsalves, R.G.L., Constantino, V.R.L., Santilli, C.V., Borges, P.D., Tronto, J. and Pinto, F.G. (2017) Adsorption of Acid Yellow 42 dye on calcined layered double hydroxide: Effect of time, concentration, pH and temperature. *Applied Clay Science* 140, 132–139.

Duan, X., & Evans, D. G. (eds.). (2006) Layered double hydroxides (Vol. 119). Springer Science & Business Media.

Foo, K.Y. and Hameed, B.H. (2010) Insights into the modeling of adsorption isotherm systems. *Chemical Engineering Journal* 156(1), 2–10.

Goh, K.H., Lim, T.T. and Dong, Z. (2008) Application of layered double hydroxides for removal of oxyanions: A review. *Water Research* 42(6–7), 1343–1368.

Guo, X., Yin, P. and Yang, H. (2018) Superb adsorption of organic dyes from aqueous solution on hierarchically porous composites constructed by ZnAl-LDH/Al(OH)(3) nanosheets. *Microporous and Mesoporous Materials* 259, 123–133.

Guo, Y., Zhu, Z., Qiu, Y. and Zhao, J. (2013) Enhanced adsorption of acid brown 14 dye on calcined Mg/Fe layered double hydroxide with memory effect. *Chemical Engineering Journal* 219, 69–77.

He, J., Wei, M., Li, B., Kang, Y., Evans, D.G. and Duan, X. (2006) Preparation of layered double hydroxides. *Layered Double Hydroxides* 119, 89–119.

He, X., Qiu, X.H., Hu, C.Y. and Liu, Y.W. (2018a) Treatment of heavy metal ions in wastewater using layered double hydroxides: A review. *Journal of Dispersion Science and Technology* 39(6), 792–801.

He, X., Zhong, P. and Qiu, X.H. (2018b) Remediation of hexavalent chromium in contaminated soil by Fe(II)-Al layered double hydroxide. *Chemosphere* 210, 1157–1166.

Heredia, A.C., Oliva, M.I., Agú, U., Zandalazini, C.I., Marchetti, S.G., Herrero, E.R. and Crivello, M.E. (2013) Synthesis, characterization and magnetic behavior of Mg–Fe–Al mixed oxides based on layered double hydroxide. *Journal of Magnetism and Magnetic Materials* 342, 38–46.

Hosseini, S.A., Davodian, M. and Abbasian, A.R. (2017) Remediation of phenol and phenolic derivatives by catalytic wet peroxide oxidation over Co-Ni layered double nano hydroxides. *Journal of the Taiwan Institute of Chemical Engineers* 75, 97–104.

Hou, D. and Li, F. (2017) Complexities surrounding China's soil action plan. *Land Degradation & Development* 28(7), 2315–2320.

Hou, D. and Ok, Y.S. (2019) Soil pollution: Speed up global mapping. *Nature* 566, 455.

Hou, D., Ding, Z., Li, G., Wu, L., Hu, P., Guo, G., Wang, X., Ma, Y., O'Connor, D. and Wang, X. (2018) A sustainability assessment framework for agricultural land remediation in China. *Land Degradation & Development* 29(4), 1005–1018.

Hou, D., O'Connor, D. and Al-Tabbaa, A. (2014) Modeling the diffusion of contaminated site remediation technologies. *Water, Air, & Soil Pollution* 225(9).

Hu, P., Zhang, Y., Lv, F., Tong, W., Xin, H., Meng, Z., Wang, X. and Chu, P.K. (2017) Preparation of layered double hydroxides using boron mud and red mud industrial wastes and adsorption mechanism to phosphate. *Water and Environment Journal* 31(2), 145–157.

Jo, W.K., Kim, Y.G. and Tonda, S. (2018) Hierarchical flower-like NiAl-layered double hydroxide microspheres encapsulated with black Cu-doped TiO$_2$ nanoparticles: Highly efficient visible-light-driven composite photocatalysts for environmental remediation. *Journal of Hazardous Materials* 357, 19–29.

Khan, S. A., Khan, S. B., & Asiri, A. M. (2016). Layered double hydroxide of Cd-Al/C for the mineralization and de-coloration of dyes in solar and visible light exposure. *Scientific reports*, 6(1), 1–15.

Kim, S., Fahel, J., Durand, P., Andre, E. and Carteret, C. (2017) Ternary layered double hydroxides (LDHs) based on Co-, Cu-Substituted ZnAl for the design of efficient photocatalysts. *European Journal of Inorganic Chemistry* (3), 669–678.

Kovacevic, D., Dzakula, B.N., Hasenay, D., Nemet, I., Roncevic, S., Dekany, I. and Petridis, D. (2013) Adsorption of arsenic on MgAl layered double hydroxide. *Croatica Chemica Acta* 86(3), 273–279.

Kuroda, Y., Oka, Y., Yasuda, T., Koichi, T., Muramatsu, K., Wada, H., Shimojima, A. and Kuroda, K. (2018) Precise size control of layered double hydroxide nanoparticles through reconstruction using tripodal ligands. *Dalton Transactions* 47, 12884–12892.

Kuwahara, Y., Ohmichi, T., Kamegawa, T., Mori, K. and Yamashita, H. (2010) A novel conversion process for waste slag: Synthesis of a hydrotalcite-like compound and zeolite from blast furnace slag and evaluation of adsorption capacities. *Journal of Materials Chemistry* 20, 5052.

Li, Z., Zhang, Q.W., Liu, X.Z., Chen, M., Wu, L. and Ai, Z.Q. (2018) Mechanochemical synthesis of novel heterostructured Bi$_2$S$_3$/Zn-Al layered double hydroxide nano-particles as efficient visible light reactive Z-scheme photocatalysts. *Applied Surface Science* 452, 123–133.

Liang, X.F., Zang, Y.B., Xu, Y.M., Tan, X., Hou, W.G., Wang, L. and Sun, Y.B. (2013) Sorption of metal cations on layered double hydroxides. *Colloids and Surfaces A: Physicochemical and Engineering Aspects* 433, 122–131.

Liu, Y., Lang, J., Wang, T., Jawad, A., Wang, H.B., Khan, A., Chen, Z.L. and Chen, Z.Q. (2018) Enhanced degradation of isoproturon in soil through persulfate activation by Fe-based layered double hydroxide: Different reactive species comparing with activation by homogenous Fe(II). *Environmental Science and Pollution Research* 25(26), 26394–26404.

Liu, Z., Ma, R., Osada, M., Iyi, N., Ebina, Y., Takada, K. and Sasaki, T. (2006) Synthesis, anion exchange, and delamination of Co-Al layered double hydroxide: Assembly of the exfoliated nanosheet/polyanion composite films and magneto-optical studies. *Journal of the American Chemical Society* 128, 4872–4880.

Lonkar, S.P., Raquez, J.-M. and Dubois, P. (2015) One-pot microwave-assisted synthesis of graphene/layered double hydroxide (LDH) nanohybrids. *Nano-Micro Letters* 7, 332–340.

Lopez, T., Bosch, P., Ramos, E., Gomez, R., Novaro, O., Acosta, D. and Figueras, F. (1996) Synthesis and characterization of sol-gel hydrotalcites. Structure and texture. *Langmuir* 12, 189–192.

Luengo, C.V., Volpe, M.A. and Avena, M.J. (2017) High sorption of phosphate on Mg-Al layered double hydroxides: Kinetics and equilibrium. *Journal of Environmental Chemical Engineering* 5(5), 4656–4662.

Lv, L., He, J., Wei, M., Evans, D.G. and Duan, X. (2006) Factors influencing the removal of fluoride from aqueous solution by calcined Mg–Al–CO$_3$ layered double hydroxide. *Journal of Hazardous Materials* 133, 119–128.

Ma, L.J., Islam, S.M., Liu, H.Y., Zhao, J., Sun, G.B., Li, H.F., Ma, S.L. and Kanatzidis, M.G. (2017) Selective and efficient removal of toxic oxoanions of As(III), As(V), and Cr(VI) by layered double hydroxide intercalated with MoS$_4^{2-}$. *Chemistry of Materials* 29(7), 3274–3284.

Meyn, M., Beneke, K. and Lagaly, G. (1990) Anion-exchange reactions of layered double hydroxides. *Inorganic Chemistry* 29, 5201–5207.

Mishra, G., Dash, B. and Pandey, S. (2018) Layered double hydroxides: A brief review from fundamentals to application as evolving biomaterials. *Applied Clay Science* 153, 172–186.

Miyata, S. (1983) Anion-exchange properties of hydrotalcite-like compounds. *Clays Clay Miner* 31, 305–311.

Mohapatra, L. and Parida, K. (2016) A review on the recent progress, challenges and perspective of layered double hydroxides as promising photocatalysts. *Journal of Materials Chemistry A* 4(28), 10744–10766.

Nishimura, S., Takagaki, A. and Ebitani, K. (2013) Characterization, synthesis and catalysis of hydrotalcite-related materials for highly efficient materials transformations. *Green Chemistry* 15, 2026–2042.

Olfs, H.-W., Torres-Dorante, L.O., Eckelt, R. and Kosslick, H. (2009) Comparison of different synthesis routes for Mg–Al layered double hydroxides (LDH): Characterization of the structural phases and anion exchange properties. *Applied Clay Science* 43, 459–464.

Othman, M.R., Helwani, Z., Martunus and Fernando, W.J.N. (2009) Synthetic hydrotalcites from different routes and their application as catalysts and gas adsorbents: A review. *Applied Organometallic Chemistry* 23, 235–246.

Paikaray, S. and Hendry, M.J. (2014) Formation and crystallization of Mg^{2+}–Fe^{3+}–SO_4^{2-}–CO_3^{2-}-type anionic clays. *Applied Clay Science* 88–89, 111–122.

Palmer, S.J., Frost, R.L. and Nguyen, T. (2009) Hydrotalcites and their role in coordination of anions in Bayer liquors: Anion binding in layered double hydroxides. *Coordination Chemistry Reviews* 253, 250–267.

Park, M., Lee, C.I., Lee, E.J., Choy, J.H., Kim, J.E. and Choi, J. (2004) Layered double hydroxides as potential solid base for beneficial remediation of endosulfan-contaminated soils. *Journal of Physics and Chemistry of Solids* 65(2–3), 513–516.

Phuong, N.T.K., Beak, M.W., Huy, B.T. and Lee, Y.I. (2016) Adsorption and photodegradation kinetics of herbicide 2,4,5-trichlorophenoxyacetic acid with MgFeTi layered double hydroxides. *Chemosphere* 146, 51–59.

Prasanna, S.V. and Kamath, P.V. (2009) Anion-exchange reactions of layered double hydroxides: Interplay between Coulombic and H-bonding interactions. *Industrial & Engineering Chemistry Research* 48(13), 6315–6320.

Rives, V. (2001). Layered double hydroxides: present and future. Nova Publishers.

Rojas, R. (2014) Copper, lead and cadmium removal by Ca Al layered double hydroxides. *Applied Clay Science* 87, 254–259.

Salomao, R., Milena, L.M., Wakamatsu, M.H. and Pandolfelli, V.C. (2011) Hydrotalcite synthesis via co-precipitation reactions using MgO and Al(OH)(3) precursors. *Ceramics International* 37(8), 3063–3070.

Shen, Z., Hou, D., Xu, W., Zhang, J., Jin, F., Zhao, B., Pan, S., Peng, T. and Alessi, D.S. (2018) Assessing long-term stability of cadmium and lead in a soil washing residue amended with MgO-based binders using quantitative accelerated ageing. *Science of the Total Environment* 643, 1571–1578.

Shen, Z., Som, A.M., Wang, F., Jin, F., McMillan, O. and Al-Tabbaa, A. (2016) Long-term impact of biochar on the immobilisation of nickel (II) and zinc (II) and the revegetation of a contaminated site. *Science of the Total Environment* 542(Pt A), 771–776.

Shen, Z., Zhang, Y., Jin, F., McMillan, O. and Al-Tabbaa, A. (2017) Qualitative and quantitative characterisation of adsorption mechanisms of lead on four biochars. *Science of the Total Environment* 609, 1401–1410.

Sheng, G.D., Tang, Y.N., Linghu, W.S., Wang, L.J., Li, J.X., Li, H., Wang, X.K. and Huang, Y.Y. (2016) Enhanced immobilization of ReO_4^- by nanoscale zerovalent iron supported on layered double hydroxide via an advanced XAFS approach: Implications for TcO_4^- sequestration. *Applied Catalysis B: Environmental* 192, 268–276.

Sposito, G. (1986) Distinguishing adsorption from surface precipitation. *Geochemical Processes at Mineral Surfaces* 323(1986), 217–228.

Tan, X.F., Liu, Y.G., Gu, Y.L., Liu, S.B., Zeng, G.M., Cai, X., Hu, X.J., Wang, H., Liu, S.M. and Jiang, L.H. (2016) Biochar pyrolyzed from MgAl-layered double hydroxides pre-coated ramie biomass (*Boehmeria nivea* (L.) Gaud.): Characterization and application for crystal violet removal. *Journal of Environmental Management* 184, 85–93.

Theiss, F.L., Ayoko, G.A. and Frost, R.L. (2016) Synthesis of layered double hydroxides containing Mg^{2+}, Zn^{2+}, Ca^{2+} and Al^{3+} layer cations by co-precipitation methods: A review. *Applied Surface Science* 383, 200–213.

Theiss, F.L., Couperthwaite, S.J., Ayoko, G.A. and Frost, R.L. (2014) A review of the removal of anions and oxyanions of the halogen elements from aqueous solution by layered double hydroxides. *Journal of Colloid and Interface Science* 417, 356–368.

Tonda, S. and Jo, W.K. (2018) Plasmonic Ag nanoparticles decorated NiAl-layered double hydroxide/graphitic carbon nitride nanocomposites for efficient visible-light-driven photocatalytic removal of aqueous organic pollutants. *Catalysis Today* 315, 213–222.

Torres-Dorante, L.O., Lammel, J., Kuhlmann, H., Witzke, T. and Olfs, H.-W. (2008) Capacity, selectivity, and reversibility for nitrate exchange of a layered double-hydroxide (LDH) mineral in simulated soil solutions and in soil. *Journal of Plant Nutrition and Soil Science* 171, 777–784.

Valente, J.S., Hernandez-Cortez, J., Cantu, M.S., Ferrat, G. and López-Salinas, E. (2010) Calcined layered double hydroxides Mg–Me–Al (Me: Cu, Fe, Ni, Zn) as bifunctional catalysts. *Catalysis Today* 150, 340–345.

Valente, J.S., Tzompantzi, F., Prince, J., Cortez, J.G.H. and Gomez, R. (2009) Adsorption and photocatalytic degradation of phenol and 2,4 dichlorophenoxiacetic acid by Mg-Zn-Al layered double hydroxides. *Applied Catalysis B-Environmental* 90(3–4), 330–338.

Vulić, T., Hadnadjev, M. and Marinković-Neducin, R. (2008) Structure and morphology of Mg-Al-Fe-mixed oxides derived from layered double hydroxides. *Journal of Microscopy* 232, 634–638.

Vulic, T.J., Reitzmann, A.F.K. and Lázár, K. (2012) Thermally activated iron containing layered double hydroxides as potential catalyst for N_2O abatement. *Chemical Engineering Journal* 207–208, 913–922.

Wang, J.Y., Zhang, T.P., Li, M., Yang, Y., Lu, P., Ning, P. and Wang, Q. (2018) Arsenic removal from water/wastewater using layered double hydroxide derived adsorbents, a critical review. *RSC Advances* 8(40), 22694–22709.

Wang, S.-H., Wang, Y.-B., Dai, Y.-M. and Jehng, J.-M. (2012) Preparation and characterization of hydrotalcite-like compounds containing transition metal as a solid base catalyst for the transesterification. *Applied Catalysis A: General* 439–440, 135–141.

Wong, K.T., Saravanan, P., Nah, I.W., Choi, J., Park, C., Kim, N., Yoon, Y. and Jang, M. (2019) Sonocatalytic reduction of nitrate using magnetic layered double hydroxide: Implications for removal mechanism. *Chemosphere* 218, 799–809.

Xu, Z.P. and Lu, G.Q. (2005) Hydrothermal synthesis of layered double hydroxides (LDHs) from mixed MgO and Al_2O_3: LDH formation mechanism. *Chemistry of Materials* 17, 1055–1062.

Yu, S.J., Wang, X.X., Liu, Y.F., Chen, Z.S., Wu, Y.H., Liu, Y., Pang, H.W., Song, G., Chen, J.R. and Wang, X.K. (2019) Efficient removal of uranium(VI) by layered double hydroxides supported nanoscale zero-valent iron: A combined experimental and spectroscopic studies. *Chemical Engineering Journal* 365, 51–59.

Yuan, X.Y., Jing, Q.Y., Chen, J.T. and Li, L. (2017) Photocatalytic Cr(VI) reduction by mixed metal oxide derived from ZnAl layered double hydroxide. *Applied Clay Science* 143, 168–174.

Zeng, M., Wu, J.H., Li, Z.Y., Wu, H.H., Wang, J.L., Wang, H.L., He, L. and Yang, X.J. (2019) Interlayer effect in NiCo layered double hydroxide for promoted electrocatalytic urea oxidation. *ACS Sustainable Chemistry & Engineering* 7(5), 4777–4783.

Zhang, B., Dong, Z.H., Sun, D.J., Wu, T. and Li, Y.J. (2017) Enhanced adsorption capacity of dyes by surfactant-modified layered double hydroxides from aqueous solution. *Journal of Industrial and Engineering Chemistry* 49, 208–218.

Zhang, P., Qian, G., Shi, H., Ruan, X., Yang, J. and Frost, R.L. (2012) Mechanism of interaction of hydrocalumites (Ca/Al-LDH) with methyl orange and acidic scarlet GR. *Journal of Colloid and Interface Science* 365(1), 110–116.

Zhao, J., Huang, Q., Liu, M.Y., Dai, Y.F., Chen, J.Y., Huang, H.Y., Wen, Y.Q., Zhu, X.L., Zhang, X.Y. and Wei, Y. (2017) Synthesis of functionalized MgAl-layered double hydroxides via modified mussel inspired chemistry and their application in organic dye adsorption. *Journal of Colloid and Interface Science* 505, 168–177.

Zhao, Y., Li, F., Zhang, R., David G. Evans, A. and Duan, X. (2002) Preparation of layered double-hydroxide nanomaterials with a uniform crystallite size using a new method involving separate nucleation and aging steps. *Chemistry of Materials* 14, 4286–4291.

Zhou, J.Z., Shu, W.K., Gao, Y., Cao, Z.B., Zhang, J., Hou, H., Zhao, J., Chen, X.P., Pan, Y. and Qian, G.R. (2017) Enhanced arsenite immobilization via ternary layered double hydroxides and application to paddy soil remediation. *RSC Advances* 7(33), 20320–20326.

Zubair, M., Daud, M., McKay, G., Shehzad, F. and Al-Harthi, M.A. (2017) Recent progress in layered double hydroxides (LDH)-containing hybrids as adsorbents for water remediation. *Applied Clay Science* 143, 279–292.

12 Nanoremediation
The Next Generation of In Situ *Groundwater Remediation Technologies*

Elio Brunetti, Christian Schemel, Alina Gawel, and Julian Bosch

CONTENTS

12.1 INTRODUCTION

Nanoremediation is an innovative and promising technology for the remediation of contaminated aquifers, which has received great attention over the last 15 years. It is based on the injection of aqueous suspensions of nano- and microparticles (NMPs) into the subsurface. Due to their small particle size and correlated high surface area, NMPs offer fast mass transfer kinetics and high reactivity in contaminant degradation. Therefore, in comparison with conventional technologies, nanoremediation has the potential to significantly reduce the time and costs required to clean up a contaminated site (Lien et al., 2006; Macé et al., 2006; Karn et al., 2009; Müller and Nowack, 2010; Comba et al., 2011; Liu et al., 2015). The first field application of nanoremediation was performed in the early 2000s in the United States, and several pilot- and large-scale field applications have been conducted to date in the United States and Europe, mainly for the remediation of aquifers contaminated with chlorinated aliphatic hydrocarbons (CAHs) using nanoscale zero-valent iron (nZVI) (Lien and Zhang, 2001; Quinn et al., 2005; Gavaskar et al., 2005; O'Hara et al., 2006; Henn and Waddill, 2006; Wei et al., 2010; Truex et al., 2011; Su et al., 2013; Elliott and Zhang, 2001).

In the last years, big research efforts were made to improve the properties of traditional nZVI-based nanoremediation and innovative techniques have been developed to limit the particle aggregation, and thus enhance the suspension injectability and optimize the NMPs mobility in the contaminated aquifer, for example by modification with biopolymers like carboxymethyl cellulose (CMC) or guar gum (Mackenzie et al., 2012; Busch et al., 2014b; Georgi et al., 2015; Busch, 2015; Velimirovic et al., 2014; Gastone et al., 2014a, 2014b; Luna et al., 2015). Also, the remediation of source areas for the first time came to the perspectives of nanoremediation, as emulsified nZVI and composite materials with higher mobility and hydrophobicity might offer the possibility to directly access non-aqueous phase liquids (NAPLs) (Quinn et al., 2005; Su et al., 2013; Mackenzie and Georgi, 2019). To overcome the intrinsic disadvantages of the nZVI-based materials, like the formation of the toxic by-products vinyl chloride (VC) and 1,2-dichloroethene (DCE), the development of novel particle types with unique properties and increased reactivity against the target CAHs was necessary (Bosch et al., 2018; Mackenzie et al., 2016). To treat a wider spectrum of pollutants, such as VOCs (volatile organic compounds), BTEX (benzene, toluene, ethylbenzene, and xylenes), PAHs (polycyclic aromatic hydrocarbons), heavy metals, and even emerging recalcitrant pollutants like pharmaceuticals and pesticides, new advanced nanoparticles, and composite materials have also been recently developed and licensed (NanoRem and CL:AIRE).

Beginning with a short description of the process fundamentals, the most recent developments in conceptual design and field application of nanoremediation are highlighted in this chapter. Thereby, the emphasis is put on completely novel particle types especially engineered for nanoremediation application by the combination of contaminant adsorption and degradation.

In detail, Section 12.2 includes a summary of the main contaminant removal mechanisms provided by nanoremediation (Section 12.2.1), addresses the factors for controlling the suspension stability and particle mobility in the porous media (Section 12.2.2), and finally provides an overview of the most advanced nanomaterials for groundwater remediation, with a focus on their reactivity, physicochemical properties, and potential impacts on ecosystems (Sections 12.2.3 and 12.2.4). Section 12.3 presents the techniques adopted for their practical application. High-resolution site investigations are preliminarily employed to characterize local subsurface heterogeneities (Section 12.3.1). Laboratory experiments on-site samples are then carried out to assess the particle reactivity against the target contaminant and their mobility in the contaminated aquifer (Section 12.3.2). Finally, field injections are designed, executed (Sections 12.3.3 and 12.3.4), and evaluated including a groundwater monitoring plan. Section 12.3.5 gives a case study for a successful nanoremediation application where a site impacted by CAHs was treated with the composite material Carbo-Iron. Section 12.3.6 discusses possible future technology developments and challenges.

12.2 NANOREMEDIATION FOR GROUNDWATER TREATMENT: AN OVERVIEW

12.2.1 MECHANISMS FOR CONTAMINANT REMOVAL AND PARTICLE OVERVIEW

The main benefit of particle-based *in situ* groundwater remediation techniques toward conventional approaches based on solution chemistry is the retention of the active particles in the contaminated zone instead of being transported away by groundwater flow. As every remedy released into environment, NMPs have to be based on environmentally compatible materials, like carbon, iron, minerals, and clays.

The nanoparticle-based removal of contaminants from groundwater relies either on the transformation to non-hazardous products by a chemical reaction or stimulated biological process, or immobilization by adsorption processes or phase transformation. The chemical reactions can be grouped into chemical reduction and oxidation. Below, these processes are briefly presented.

12.2.1.1 Reduction

A chemical reduction mechanism is applied primarily to the contaminant class of CAHs, which are among the most frequently detected groundwater contaminants. The application of nZVI for the reduction of CAHs is today the most prominent application of nanoremediation. Assuming a generalized chlorinated hydrocarbon (RCl), the transformation to the corresponding hydrocarbon (RH) can be represented by the simplified reaction given in Equation (12.1) (Bartke et al., 2018; Mueller et al., 2012).

$$Fe^0 + RCl + H_2O \rightarrow Fe^{2+} + OH^- + Cl^- + RH \qquad (12.1)$$

Due to requirements for environmental safety, *in situ* applicable chemical reductants are mostly based on reduced iron species, namely metallic, zero-valent iron or ferrous iron Fe(II). As an example, for a particulate ferrous reductant to be applied in nanoremediation, iron(II) sulfide (FeS) can be mentioned (Gong et al., 2016). The application of chemical reductants in groundwater is limited to anaerobic conditions with concentrations of dissolved oxygen below 2 mg L^{-1} (Braun et al., 2017). Also other oxidizing matrix ingredients like nitrate might lead to significant undesired reductant consumption and have to be considered in barrier planning.

12.2.1.2 Oxidation

Chemical oxidation is capable to attack a broad range of, especially organic, contaminants. Halogenated organics like VC, DCE, and those with aromatically bound chlorine, which are not treatable by the reduction agents mentioned before, can be attacked and defanged by chemical oxidation (Georgi et al., 2012). Radical-providing substances like hydrogen peroxide (H_2O_2) and permanganate (MnO_4^-) are examples for environmentally compatible oxidants. Nevertheless, no oxidative nanoremediation approach has been commercialized hitherto. As an approach for an oxidizing and *in situ* applicable particle, $KMnO_4$ covered with a protective layer of manganese dioxide (MnO_2) was reported as alternative to slow-release permanganate candles (Christenson et al., 2016). The benefit of this particle-based technique, when compared to conventional, dissolved $KMnO_4$, is a more prolonged oxidative effect in the aquifer, which is achieved by slow release of MnO_4^- through the protective layer (Rusevova et al., 2012). However, one limitation of *in situ* chemical oxidation, no matter if particle- or solution-based methods, is the lack of selectivity toward the target compounds. Oxidizing compounds are undesirably consumed by subsurface constituents (von Guten, 2018). Furthermore, there is the possibility of producing equally or even more harmful intermediate products, especially in case of halogenated compounds and in the areas of deficient oxidant delivery (Wang et al., 2018). Thus, monitoring of reaction products is strongly recommended.

12.2.1.3 Adsorption

For some contaminants, neither a chemical oxidation nor reduction can achieve destruction and thus, full risk elimination. On the one hand, this applies to heavy metals, which can be transformed to different more or less toxic redox states, but still remain in the aquifer. On the other hand, polar recalcitrant pollutants, for which no *in situ* applicable techniques for degradation are available at current state-of-the-art, are included in that category (Kucharzyk et al., 2017). In those cases, nanoremediation can achieve the immobilization of the pollutants to cutoff contaminant plumes and prevent their further spread. For this purpose, colloidal activated carbon can be injected into the aquifer to form a sorption barrier (Georgi et al., 2015; Regenesis, 2019). Depending on the interaction between the adsorbent and the contaminant, adequate barrier design can lead to lifetimes of several decades. In some cases, also immobilization of principally degradable compounds can be regarded as a treatment option. As an example, PAHs have a sufficiently high affinity to activated carbon to consider adsorption as a stand-alone remediation strategy, and barrier lifetimes of several hundred years can be achieved (Hale et al., 2011). Also for heavy metals there is a broad range of suitable adsorbents reported (Fu and Wang, 2011; Hashim et al., 2011), e.g. various commercial or modified activated carbons (Han et al., 2000; Acharya et al., 2009; Dwivedi et al., 2008), zeolites (Hong et al., 2019), iron-based adsorbents (Chowdhury and Yanful, 2010; Buerge-Weirich et al., 2003; Hajji et al., 2019; Hu et al., 2005), and other minerals (Lazaridis, 2003). In any case, when using a non-destructive immobilizing approach, observation of the long-term stability of the technique is especially important, as changes in the physicochemical properties of the aquifer, e.g., redox potential or pH value, can lead to the release of the immobilized contaminants. In case of technologies only based on physical adsorption, also the constant equilibrium between adsorption and desorption makes the process reversible in principle and thus, contaminant release over time possible.

12.2.1.4 Bioremediation

Some nanoremediation materials do not interact directly with the contaminant molecules, but improve the conditions for their microbial degradation. As an example, goethite nanoparticles (FeOOH with NOM coating) can act as electron acceptors for metal reducing bacteria, like *Geobacter*, which are thus able to oxidize contaminants like BTEX and other hydrocarbons (Braun et al., 2017; Bosch et al., 2010; Fuchs et al., 2011; Mackenzie et al., 2016). Hydrogen emerging from the corrosion of ZVI-based particles can stimulate the growth of sulfate-reducing bacteria producing sulfide, which in turn immobilizes various heavy metals by precipitation as insoluble metal sulfides (Muyzer and Stams, 2008). On the other hand, those bacteria are reported to be capable of metabolizing hydrocarbons, like alkanes, BTEX, but also explosives (So and Young, 1999; Edwards et al., 1992; Boopathy et al., 1998). Technologies based on stimulation of biodegradation are often characterized by relatively high cost efficiency.

12.2.1.5 Combined Removal Approaches

There are further nanoremediation techniques, which work with a combination of different treatment approaches. On the example of BTEX, adsorption by activated carbon is insufficient for a long-lasting retardation in the aquifer. However, it provides time for microbial degradation (Mason et al., 2000), which can be enhanced by the addition of goethite nanoparticles to the injected particle suspension (Bosch et al., 2010; Mackenzie et al., 2016). The presence of oxidants from *in situ* chemical oxidation (ISCO) approaches is, under controlled pH, also capable to stimulate the growth of bacteria in the aquifer, for example in case of diesel and PCE remediation (Chen et al., 2016; Bou-Nasr et al., 2006). Chemical reduction of a range of heavy metals transforms them into less mobile species, but cannot be termed as destructive removal. Often, a complex combination of reduction and adsorption or (co-)precipitation takes place. In the application of ZVI on contaminations with arsenic, As(III) was reported to be oxidized to As(V) by the iron corrosion products and simultaneously co-precipitated with them (Manning et al., 2002). Another example is the application of iron sulfide on chromate, which led to the formation of mixed solid Fe(III)/Cr(III) hydroxides (Patterson et al., 1997).

Two types of NMPs (Carbo-Iron and Trap-Ox® Fe-zeolites), which were especially designed for nanoremediation application and are based on a combination of adsorption and chemical degradation, will exemplarily be described in detail. These NMPs were chosen to point out the benefits of combined removal approaches and to illustrate different mechanisms of both suspension stabilization on the one hand, and degradation on the other hand. In Section 12.2.2, their suspension stability and mobility are shortly described. A detailed description of their reactivity and adsorptive properties are presented in Section 12.2.3.

Table 12.1 summarizes a selection of examples for contaminant classes, which perspectively can be treated by nanoremediation approaches. Thereby, different removal mechanisms and therefore different NMP types might be worth considering.

12.2.2 Particle Stability and Mobility

Besides the particles' reactivity, a successful remediation strategy requires that during field injection the suspension is stable enough to be readily injected into the subsurface and that the particles are sufficiently mobile to ensure a uniform and controlled distribution in the target zone. After the injection, the particles adsorb on the aquifer sediments providing a long-term removal capacity toward the target contaminants.

On the contrary, a poorly stable suspension results in very limited travel distances of the particles from the injection point, due to their strong tendency to aggregate, resulting in a progressive clogging of the media (Phenrat et al., 2007, 2008).

12.2.2.1 Particle Stability and Mobility
Colloidal particle transport in saturated porous media has been widely studied in the last decades (Yao et al., 1971; Elimelech and O'Melia, 1990; Logan et al., 1995; Grolimund et al., 2001; Bradford et al., 2003; Ryan and Elimelech, 1996; Bradford et al., 2002; Elimelech et al., 1995). The tendency of a particle suspension to resist aggregation and sedimentation for a specified time is described as the particle or suspension stability. It is determined by measuring the sedimentation rate in static sedimentation experiments carried out in closed containers (Phenrat et al., 2007). Particle mobility refers to the ability of the particles to move through the porous media during field injection. Higher suspension stability results in prolonged particle mobility, which, in turn, optimizes the particle distribution in the contaminated aquifer and thus assures the best technology performance.

12.2.2.2 Retention Mechanisms
As the particles are retained by the soil matrix, their deposition rates may decrease, increase, or remain constant, depending on the nature of the particle-particle and particle-soil matrix interactions. Indeed, besides advection and dispersion processes, particles are subject to various more complex dynamic retention mechanisms, such as blocking and ripening (Figure 12.1) (Kretzschmar et al., 1999; Sen and Khilar, 2006; Tosco et al., 2014, 2018; Tosco and Sethi, 2010; Molnar et al., 2015; Wiesner and Bottero, 2017).

TABLE 12.1

Summary of Selected Contaminant Classes with the Perspective to Be Treated by Nanoremediation and Examples of Particle Types Suitable for That Application

Contaminant Class	Mechanism	Particle Examples
Chlorinated aliphatic compounds (CAHs)[a]	Reductive hydrodechlorination	ZVI-based particles/composites (nZVI, Carbo-Iron) (Mackenzie et al., 2012; Vogel et al., 2018)
Heavy metals	Reduction, adsorption, (co) precipitation	ZVI-based particles (Rabbani and Park, 2016; Zou et al., 2016), Iron(II) sulfide (Patterson et al., 1997; Wilkin and Beak, 2017; Park et al., 2018) Nano magnetite (Chowdhury and Yanful, 2010; Ferreira et al., 2017)
	Adsorption	Activated carbons (Deliyanni Eleni et al., 2015) Nano goethite (Buerge-Weirich et al., 2003; Montalvo et al., 2018; Hajji et al., 2019; Zhang et al., 2019)
Nitroaromatic explosives	Reduction	ZVI-based particles (Agrawal and Tratnyek, 1996), Nano magnetite (Gorski et al., 2010)
BTEX	Biostimulation	Nano goethite (Mackenzie et al., 2016)
Oxidizable organic compounds[b]	Oxidation/mineralization	Trap-Ox Fe-zeolites (Georgi et al., 2012; Gonzalez-Olmos et al., 2009, 2013) MnO₂-coated KMnO₄ (Rusevova et al., 2012)
PAHs	Adsorption	Activated carbons (Brändli et al., 2008; Jakob et al., 2012)
PFCs	Adsorption	Activated carbons (Du et al., 2014; McGregor, 2018) Aluminum hydroxide (Ziltek, 2019)

[a] Except VC, DCE.

[b] For example, explosives, MTBE, chlorinated compounds.

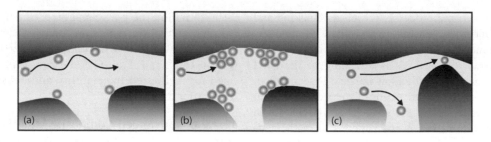

FIGURE 12.1 Pore scale schematic view of blocking (a), ripening (b), straining and mechanical filtration (c) effects. (Adapted from Tosco, T., et al., *J. Clean. Prod.*, 77, 10–21, 2014.)

Blocking is a surface exclusion process. When repulsive particle-particle interactions prevail, the particle deposition is progressively limited by the particles already deposited on the soil matrix, resulting in declining deposition rates. On the other hand, when attractive particle-particle interactions dominate, ripening may occur, as in the case of elevated salt concentration. In this case, the particle deposition rates on the solid matrix increases further with the increasing amount of already deposited particles, until the sediments surrounding the injection are completely clogged. With similar particle-particle and particle-soil matrix interactions, the deposition rate remains constant (Kretzschmar et al., 1999; Tosco et al., 2014, 2018; Tosco and Sethi, 2010). For large particles and aggregates or for fine-grained sediments, physical processes such as straining may prevail, with particles being retained in the smaller pores. Straining is relevant when the ratio of the suspended particle diameter to the soil grain diameter is higher than approximately 0.2 (Elimelech et al., 1995; Tosco et al., 2014; Tosco and Sethi, 2010; Tiraferri et al., 2011).

Along with the particle and soil properties, the retention mechanisms are strongly dependent on hydro-chemical properties, such as the solution pH and ionic strength, and also physical properties, such as the pore flow velocity, injection pressure, and particle diameter (Grolimund et al., 2001; Tiraferri et al., 2011; Hahn et al., 2004; Tosco et al., 2009, 2012; Li, et al., 2005). Below, the influence of ionic strength is briefly presented, since it is one of the parameters strongly controlling the particle transport during field injections.

12.2.2.3 The Influence of Ionic Strength

It has been largely documented that high ionic strength results in the reduction of the repulsive electrostatic double-layer forces between particles and leads to ripening, as the interaction is dominated by the attractive van der Waals forces. This effect is more pronounced in suspensions dominated by divalent cations (Ca^{2+}, Mg^{2+}) compared to monovalent cations (Na^+). On the other hand, water having low ionic strength causes the blocking phenomenon, that is, the aggregation rate decreases as the surface charge is balanced by a thicker repulsive double layer, extending from the surface to the bulk solution, resulting in higher particle mobility (Grolimund et al., 2001; Tiraferri et al., 2011; Tosco et al., 2009, 2012).

12.2.2.4 Particle Transport

The transport of NMPs through a homogeneous porous media during column tests is generally represented using the classical colloid filtration theory (CFT), where a particle deposition term is added to the 1D advection-dispersion equation under transient and fully saturated conditions (Yao et al., 1971; Logan et al., 1995):

$$\varepsilon \frac{\partial c}{\partial t} = -q \frac{\partial c}{\partial x} + \varepsilon D \frac{\partial^2 c}{\partial x^2} - \rho_b \frac{\partial s}{\partial t} \tag{12.2}$$

$$\rho_b \frac{\partial s}{\partial t} = \varepsilon k_a c \tag{12.3}$$

where ε is the porosity of the packed bed, c is the concentration of the particle in the solution, D is the dispersion coefficient, q is the Darcy velocity, ρ_b is the solid bulk density, and k_a is the particle attachment rate coefficient. Integrating the transport equation and neglecting the particle dispersion and accumulation terms in the water phase (i.e. steady state regime), the travel distance (L_T) for uniform particles can be calculated according to the following equation:

$$L_T = -\ln\left(\frac{C}{C_0}\right)\frac{1}{\lambda} = -\ln\left(\frac{C}{C_0}\right)\frac{2d_c}{3(1-\varepsilon_{eff})\alpha\eta_0} \tag{12.4}$$

where C/C_0 is the removed particle percentage at the travel distance L_T, λ is the filter coefficient, ε_{eff} is the effective porosity, d_c the average grain size, α the attachment coefficient, and η_0 the single collector efficiency. Based on Equation (12.4), the travel distance at which a certain degree of particles is retained in the porous media can be estimated (e.g., 50% particle removal at $C/C_0 = 0.5$) (Elimelech et al., 1995). The calculated travel distance cannot be applied to determine the absolute particle mobility in the field but acts as a rough indicator and can also be used for comparison of transport properties of different particles.

The CFT model neglects the particle detachment mechanisms and represents the deposition rate with a first-order attachment term. This linear, non-reversible interaction between particle and solid matrix is represented by Equation (12.3). Consequently, the transfer of particles from the solid to the water phase (i.e. reversible process), as well as blocking and ripening, are not included. For this reason, the CFT model is generally used at the early stage of deposition kinetics, when only a small amount of particles is attaching to the solid matrix, and for suspensions with a low particle concentration. A modified version of the 1D advection-dispersion governing equation for homogeneous and fully saturated porous media under transient conditions has been proposed to mathematically describe the most relevant transport and retention processes occurring at the porous scale (Tosco et al., 2014, 2018; Tosco and Sethi, 2010; Tiraferri et al., 2011; Hu et al., 2017).

$$\frac{\partial}{\partial t}\left(\varepsilon c\right)+\sum_{i}\left(\rho_{b}\frac{\partial s_{i}}{\partial t}\right)+\frac{\partial}{\partial x}\left(qc\right)+\frac{\partial^{2}}{\partial x^{2}}\left(\varepsilon Dc\right)=0 \qquad (12.5)$$

$$\rho_{b}\frac{\partial s_{i}}{\partial t}=\varepsilon k_{a,i}f_{att,i}c-\rho_{b}k_{d,i}s_{i} \qquad (12.6)$$

where $k_{a,i}$ and $k_{d,i}$ are, respectively, the particle attachment and detachment rate coefficients and $f_{att,i}$ is a dimensionless particle attachment function to be selected based on the ongoing i-th retention processes (e.g., blocking, ripening, linear reversible attachment). In more detail:

- The linear, reversible process is mathematically represented with $f_{att,i}=1$
- Blocking is described with a value of $f_{att,i}$ lower than 1 and expressing the attachment function as $f_{att,i}=1-s/s_{max}$, with s_{max} being the maximum concentration of particles that can be deposited on the solid matrix
- Ripening is employed with a value of $f_{att,i}$ higher than 1. The attachment function has been calculated as $f_{att,i}=1+As^{B}$, with A > 0 and B > 0, which are, respectively, a multiplier and an exponent coefficient used to define the interaction dynamics

Additional equations have also been formulated to represent the dependence of the retention mechanisms on the physico- and hydro-chemical parameters, such as flow velocity and water salinity (Tosco et al., 2009, 2014, 2018; Tiraferri et al., 2011). A formulation has also been proposed to simulate the progressive reduction of porosity and permeability when particles retain in soil pores, resulting in a progressive clogging of the media (Tosco and Sethi, 2010; Tosco et al., 2018).

12.2.2.5 Stabilization Techniques

Several techniques have been developed to stabilize the particle dispersions against aggregation and sedimentation and thus obtain the requested mobility under site-specific conditions. In many cases, suspension stabilization is achieved by adding commercially available colloid stabilizers, such as carboxymethyl cellulose (CMC) (Phenrat et al., 2008; Li et al., 2016; Raychoudhury et al., 2012; Kocur et al., 2013; Hotze et al., 2010; He et al., 2010; Johnson et al., 2013), guar gum (Velimirovic et al., 2014; Gastone et al., 2014a, 2014b; Luna et al., 2015), or polyacrylic acids (Laumann et al.,

2013) to inhibit the particle aggregation by promoting electrostatic and steric repulsion forces. Below, two advanced composite materials with unique properties are presented (Carbo-Iron® and Trap-Ox® Fe-zeolites), recently synthetized for facilitated injectability and enhanced mobility in the aquifer media.

12.2.2.6 Advanced Composite Materials

Different carrier colloids with a size around 1 μm, which is the particle size associated to optimized mobility in soil, have been designed to improve the transport of reactive particles in the subsurface (Busch et al., 2014a, 2015). Carbo-Iron is an advanced composite material with an average particle size of 0.8–1.2 μm, which contains metallic iron nano-structures, embedded in the pores of colloidal activated carbon (Figure 12.2).

The overall particle suspension stability of Carbo-Iron is influenced by particle size and density, agglomeration tendency, and surface charge. Compared to conventional nZVI particles, its lower density, weakened magnetic properties, and more negative surface charge inhibit the aggregation and enhance the mobility (Mackenzie et al., 2012). The size and density dependence of the sedimentation velocity can be seen in the Stokes equation and were already discussed in previous chapters. The Zeta potential for Carbon-Iron was measured at $\zeta = -23$ mV at pH 7 (Mackenzie et al., 2012). In general, particle suspensions with a Zeta potential of $\zeta \geq \pm 30$ mV are considered stable, since the particle-particle repulsion is strong enough to prevent agglomeration (Mahl, 2011). Electrostatic repulsion is one of the key parameters for high stability and high mobility. Negatively charged particles are expected to have the best mobility in aquifers, due to the repulsion with silica minerals, which show net-negative surface charge under aquifer conditions (Wang et al., 1997). The suspension stability can be enhanced further by the addition of polyanionic stabilizers. Carbo-Iron suspensions are stabilized by CMC, whose influence on Carbo-Iron's mobility has been widely investigated during column tests (Figure 12.3). CMC concentrations of up to 20 wt.% improve the long-term ($t > 3$ h) stability of the suspension (for 70%–80% of the particle mass). A maximum travel distance (L_T 99%) ranging from approximately 3–16 m was estimated in saturated porous

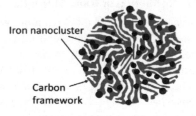

FIGURE 12.2 Sketch of the composite material Carbo-Iron. (Adapted from Mackenzie, K.G. et al., DL 3.2 Assessment of nanoparticle performance for the removal of contaminants: Non-ZVI and composite nanoparticles, in *NanoRem WP3: Design, Improvement and Optimized Production of Nanoparticles: Non-ZVI and Composite Nanoparticles*, 2016.)

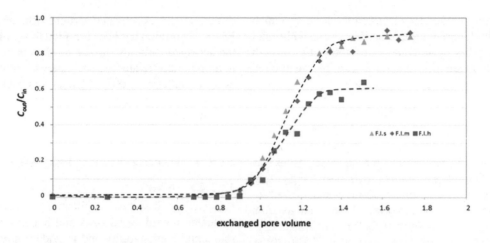

FIGURE 12.3 Column test breakthrough curves for CMC-stabilized Carbo-Iron prepared in different types of standard water, ranging from low (F.I.s.) to very hard water (CF.I.h.). The concentrations are 5.7 g L^{-1} for Carbo-Iron and 1.1 g L^{-1} for CMC. (From Micić Batka, V. and Hofmann, T. *DL4.2: Stability, Mobility, Delivery and Fate of optimized NPs under Field Relevant Conditions*, NanoRem, 2016.)

media for a Carbo-Iron suspension stabilized with CMC (Mackenzie et al., 2012, 2016a, 2016b; Busch et al., 2014b, 2015; Georgi et al., 2015).

Higher CMC content does not improve the stability, since CMC only occupies the outer surface of the Carbo-Iron particles (Mackenzie et al., 2012; Micić Batka and Hofmann, 2016). These stable suspensions show longer travel distances during the injection, which makes them suitable for "plume" treatment. When the concentration of CMC decreases (down to 5–7 wt.%), a "meta-stable" suspension is formed. These suspensions are only stable at $t < 1$ h and show a significant decrease in stability after this time period. Thus, an emplacement of particle closer to the injection point makes these suspensions suitable for "source" treatment.

Trap-Ox Fe-zeolites are a group of innovative particles especially synthetized for *in situ* trapping of organic contaminants by adsorption and catalytic oxidation in combination with hydrogen peroxide. With adjusted slightly alkaline conditions (pH 8–8.5), close to the values commonly detected in groundwater (approximately 6–8.5), the suspension presented a very high stability ($t > 24$ h), even with high ionic strength and high particle concentrations of up to 10 g L^{-1}. Under these conditions, results of column tests evidenced a high mobility in saturated media, also with low water flow velocities. When pH is decreased to slightly acidic conditions (5.5), the suspensions become highly unstable, as most of the particles (70%–80%) are lost from the observed area of the experimental setup within 0.3–1 h. The observed instability at pH 5.5 cannot be ascribed to an inversion of net-surface charge, as an overall negative Zeta potential was observed over the entire pH range (2.8–9.7) (Gillies et al., 2017). Instead, the pH dependence of the suspension stability can be traced to the Fe-loading. Without this loading, the zeolites show no difference of stability in dependence of pH. However, the Zeta potential is a net parameter, which does not take into account charge heterogeneities at the surface. The Fe-loading leads to the formation of small iron oxide clusters at the outer surface

of the zeolites, which can be readily observed via UV/Vis spectroscopy. These heterogeneities show significant relevance, as most iron oxides have a point of zero charge (PZC) in the range of pH 7–8 and thus, the positive charge of iron oxide clusters at pH 5.5 leads to attractive interaction with the negative surface charge of another particle, which results in agglomeration and sedimentation (Schwidder et al., 2005; Kosmulski, 2004). Based on these results, minor amounts of NaOH for pH adjustment are considered as sufficient additive for further field injections (Gillies et al., 2017).

12.2.3 Advances on Particle Performance: Particle Reactivity

Big efforts were made in order to engineer particles, which at least partially overcame the disadvantages and limitations of conventional nanoremediation materials. Following, examples for advanced NMPs especially engineered for nanoremediation application are presented.

12.2.3.1 Advanced Reductive Particles

Even though ZVI nano- and microparticles are regularly applied and researched for nanoremediation, they have various undesired properties inseparably combined to their physicochemical nature. In addition to their limited mobility caused by high density and magnetism (Section 12.2.2), depending on their fabrication conditions, some types of ZVI particles induce the formation of undesired by-products, like VC and *cis*-DCE from tetrachloroethylene (PCE) (Arnold and Roberts, 2000; Liu et al., 2005). However, those by-products show even increased toxicity compared to their parent compound and are not sufficiently fast degraded by ZVI, as the rate of this reaction is too low. Activated carbon-iron nanocomposite materials, such as Carbo-Iron, were shown to avoid the formation of the toxic by-products mentioned above. It is suspected that adsorption to the activated carbon matrix increases the contact between the contaminant molecules and the embedded iron nano-clusters, which leads to more complete reduction reactions. As an additional benefit, the probability of activated carbon to act as growth medium for dehalogenating bacteria can be mentioned (Vogel et al., 2018).

12.2.3.2 Advanced Oxidative Particles

To overcome the lack in selectivity of *in situ* oxidation processes mutual to the injection of liquids, a similar sorptive-reactive approach as for the activated carbon-iron-nanocomposites was chosen. So called Trap-Ox Fe-zeolites are iron-exchanged zeolites, which act both as a contaminant adsorber and, in a subsequent regeneration step, as a catalyst for a Fenton-like reaction to produce hydroxyl radicals from H_2O_2 for contaminant degradation (Figure 12.4) (Gonzalez-Olmos et al., 2009, 2013). From the broad variety of reactions involved into Fenton-like processes, the initial generation of hydroxyl radicals and the closure of the catalytic cycle are given in Eqs (12.7) and (12.8).

$$Fe^{2+} + H_2O_2 \rightarrow Fe^{3+} + \cdot OH + OH^- \qquad (12.7)$$

$$Fe^{3+} + H_2O_2 \rightarrow Fe^{2+} + HOO \cdot + H^+ \qquad (12.8)$$

As there are different types of zeolites, the adsorptive properties can be widely adjusted to the contamination by choosing best fit pore sizes and hydrophobicity (Shahbazi et al., 2014).

12.2.3.3 Advanced Adsorptive Particles

Significant progress was made in the field of the optimization and specialization of activated carbons for contaminant adsorption. On the example of various polar micropollutants, adsorption on bituminous coal-based re-agglomerated activated carbon was reported to be superior than on coconut-based direct activated carbon (McNamara et al., 2018), and that surface basicity-modification by ammonia gas or heat treatment significantly influenced the adsorption properties of

different types of activated carbons (Zhi and Liu, 2016). Also PAHs' adsorption has been reported to significantly depend on the activated carbon's surface properties, whereby preferably hydrophobic activated carbons showed best adsorption properties (Ania et al., 2007). Against this background, choosing a specialized activated carbon seems reasonable to enhance adsorptive barrier lifetimes.

12.2.4 Ecotoxicological Risk Assessment

12.2.4.1 NMPs

Prior to the first field application, a new particle type is evaluated regarding its risks and benefits to assure its environmental compatibility and provide the required regulatory evidence. Nanoparticles brought into an aquifer, in most cases, remain there for an indefinite time, which makes it necessary to clearly identify their fate and *in situ* effects. In addition to this consideration linked to possible exposure scenarios, ecotoxic effects of the particles must be generally precluded below concentrations of 100 mg L^{-1} for potentially affected aquatic species, bacteria, and soil species (Bardos et al., 2018; Gillett and Nathanail, 2017). In this context, ageing processes of the particles must also be considered. For a broad range of nanoremediation materials, like nZVI, Carbo-Iron, iron oxide nano-particles, and Trap-Ox Fe-zeolites, ecotoxicological harmlessness was already proven (Gillett and Nathanail, 2017; Hjorth et al., 2017).

12.2.4.2 Degradation Products

In case of a yet unfamiliar contaminant class, first batch experiments are done to show the principal functionality of the chosen approach. For remedial success, it is not enough to just undercut contaminants' threshold values. To also assure ecological benefit of a nanoremediation application, degradation pathways should be identified and the products must be tested for their toxic potential. If reactive pathways cannot be identified, at least the product mixtures at different points of the reaction progress should be subjected to ecotoxicological tests. In case of an incomplete oxidation, especially halogenated contaminants have the potential to form even more toxic products from the contaminants. Under non-optimized oxidation conditions, chloroethanes and –ethenes can, for instance, form toxic chloroacetic acids (Georgi et al., 2012). Also, there is a broad range of micro pollutants whose early stage oxidation products show similar or even increased toxicity, e.g., diclofenac, diuron, or carbamazepine (Wang et al., 2018; Yu et al., 2013). As default model species applied in the ecotoxicological assays, luminescent bacteria are often used to evaluate the toxicity development of a contamination (Parvez et al., 2006; De Zwart and Slooff, 1983). In special cases, the survey on other species is also reasonable, for example green algae for herbicides. Ecotoxicity assessment for product mixtures of compounds with low acute, but chronic toxic effects is not recommended. In those cases, identification and individual evaluation of the degradation products is inevitable.

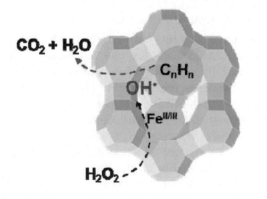

FIGURE 12.4 Scheme of the working principle of the Trap-Ox Fe-zeolites. (Adapted from Mackenzie, K.G. et al., DL 3.2 Assessment of nanoparticle performance for the removal of contaminants: Non-ZVI and composite nanoparticles, in *NanoRem WP3: Design, Improvement and Optimized Production of Nanoparticles: Non-ZVI and Composite Nanoparticles*, 2016.)

12.3 FIELD APPLICATION OF NANOREMEDIATION FOR GROUNDWATER REMEDIATION

In advance of any field application of nanoremediation, a pre-screening should be performed to ascertain the applicability of the NMPs to decontaminate a given site and remedial action objectives (RAOs) clearly identified. At this stage, the contamination type and distribution in the subsurface are evaluated, as well as the hydrogeological and hydrochemical site conditions. The optimum particle, which is potentially effective against the target contaminant, is selected. Based on the particle-specific properties and activity spectrum, the favorable site conditions and potential constraints are identified and evaluated. An environmental site assessment is also required, including the evaluation of potential risks to human and ecological receptors (Bruns et al., 2019). If, based on the outcomes of the pre-screening, the site appears suitable for the application of the selected NMP, a series of investigations are performed to design the field application.

12.3.1 HIGH-RESOLUTION SITE INVESTIGATION

For the nanoremediation technology, whose field performance is strongly dependent on the site hydrogeological conditions, such as groundwater flow direction and velocity, field investigations are needed. Indeed, low permeability layers and subsurface heterogeneities may strongly control the NMPs spatial distribution in the target zone. Therefore, high-resolution hydrogeological tests should be performed to determine the extension of the hydrostratigraphic units and their properties, such as hydraulic conductivity and porosity (Braun et al., 2017; Bruns et al., 2019).

The extent and mass of the contaminant in the target zone should also be estimated focusing on the horizontal and vertical distribution of the contaminants, both in soil and groundwater media. For this purpose, direct push technologies (DPTs) provide an extensive number of tools especially suited for high-resolution hydrostratigraphic characterization and contaminant profiling, such as DP injection logging (DPIL) and membrane interface probe (MIP) (Kaestner et al., 2012).

Depending on the particle-specific properties, the aquifer hydrochemistry also needs to be assessed, especially the type and quantity of redox-active species, which may potentially affect the NMPs performance. In highly heterogeneous aquifer systems, a three-dimensional assessment of the redox conditions should be performed, as this may result in a non-uniform distribution of the groundwater geochemistry. In the case of reductive particles, competing species such as dissolved oxygen, nitrates, and sulfate should be identified and quantified. Indeed, the particle may not only reduce the organic contaminants but also react with other electron acceptors. Other potential electron donors, such as natural organic matter, should be considered especially if oxidizing techniques are applied. Composition of the soil matrix, pH, and ionic strength of groundwater should also be under

investigation, as they may strongly influence NMPs distribution in the surrounding of the injection point (Bruns et al., 2019).

Recently, new innovative approaches have been developed for a high-resolution physicochemical characterization of groundwater flow and aquifer heterogeneities. IntraSense® consists of an integrated package of multiple field probes able to collect a wide range of 3D high-resolution groundwater data in existing monitoring points, including real-time measurements of several geochemical indicators, groundwater flow direction, and velocity. This system is especially suited for highly heterogeneous environments with zones of preferential groundwater flow (Intrapore, 2019).

Based on the results of the site investigations, a 3D multiscale high-resolution conceptual site model (CSM) should be constructed (Figure 12.5), ranging from the injection area (local scale) up to the groundwater body (site scale). The CSM should represent a three-dimensional description of the geological, hydrogeological, hydrochemical, and microbiological conditions. The extent of the source area and contaminant plume in groundwater should also be delineated and the natural processes controlling contaminant transport and degradation quantified. Uncertainties must always be clearly defined and listed, as well as the assumptions adopted. The high-resolution CSM is then employed to drive the design of the field injection and monitoring plan (Braun et al., 2017; Bruns et al., 2019).

12.3.2 LAB STUDIES FOR FIELD PREPARATION

Even if the real site conditions can hardly be reproduced in the laboratory, lab experiments are indispensable to support the design of the field injection. Several laboratory investigations on-site materials could be carried out, based on the site-specific factors and operation windows of the selected NMPs. Batch experiments with groundwater samples from the site are performed to look at the contaminant degradation rate, the adsorption kinetics, and the expected degradation products/intermediates. Column experiments with soil samples from the site are also carried out to estimate the suspension injectability and particle mobility in the aquifer media (Braun et al., 2017; Bruns et al., 2019).

12.3.2.1 Batch Experiments

Batch experiments with site groundwater and sediment are, in comparison with laboratory reference samples and stoichiometric calculations, used to determine the extent of site-specific matrix effects on the performance of the chosen NMPs. The kinetics of the reaction or adsorption process is evaluated to assess the necessary residence time of the contaminant in the reactive barrier for complete removal (Section 12.3.3). For adsorptive processes, adsorption isotherms are measured to gain information about the affinity of contaminants toward their adsorbents and the capacity of the latter to take up the contaminants. Therefore, different loadings of the target compound on the adsorbent are adjusted and the free concentrations are correlated by plotting the decadic

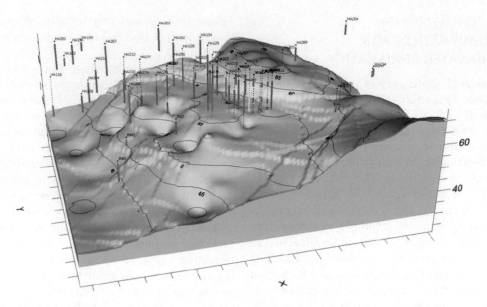

FIGURE 12.5   Example of a 3D CSM showing the morphology of the aquifer bottom layer.

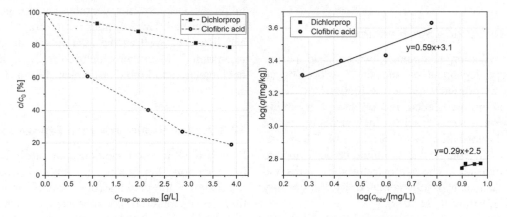

FIGURE 12.6   Change in the free concentration of the herbicides chlofibric acid and dichlorprop in dependence of the particle concentration of Trap-Ox zeolite FeBEA35 (left) and resulting adsorption isotherms according to Freundlich (right).

logarithm of loading q versus the decadic logarithm of the free concentration c_{free} as shown in Figure 12.6.

12.3.2.2   Column Experiments

The particle transport during column tests is predominantly one-dimensional and follows the general rules of the colloid transport theory and retention mechanisms (Section 12.2.2). Column experiments are conducted with saturated soil material from the site. The tests commonly include a pre-flushing phase with site water, an injection of the particle suspension, followed by flushing with site water and afterward with deionized or high pH water for the release of the retained particles (Figure 12.7). Conservative tracers are also employed to preliminarily define the effective porosity of the porous media. Particle transport is assessed by recording colloid effluent concentrations over time. Different scenarios may be explored to evaluate the impact of the parameters, which are assumed to strongly control the mobility of the selected NMP in the contaminated aquifer, during and after the injection,

such as ionic strength, pH, and injection velocities. Different NMP or stabilizer concentrations may be also investigated to optimize the suspension injectability and particle mobility (Georgi et al., 2015; Tosco et al., 2012; Hu et al., 2017; Gillies et al., 2017).

Based on the batch and column test outcomes, the main transport-controlling mechanisms are preliminary identified and computer codes may also be employed to numerically simulate the experimental results and to quantify the particle transport parameters (Tosco and Sethi, 2010; Tosco et al., 2014, 2016, 2018; Tiraferri et al., 2011; Šimůnek et al., 2008; Tosco and Sethi, 2009; Tian et al., 2010; Waghmare and Seshaiyer, 2015; Bianco et al., 2016, 2018). The attachment and detachment rate coefficients (k_a and k_d) are defined by inverse-fitting the colloid effluent concentration curves (Tosco et al., 2018), and the particle travel distance are also predicted (L_T) (Tosco et al., 2012; Gillies et al., 2017). Finally, the main retention mechanisms controlling the particle transport in the porous media are quantified.

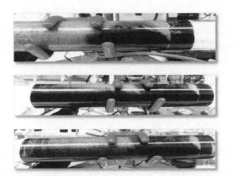

0.6 PV (injection period)

1.8 PV (injection period)

7 PV (flushing period)

FIGURE 12.7 Column experiment performed with site material presenting the Carbo-Iron migration ($c_{particle}$: 15 g L^{-1} and c_{CMC}: 1.5 g L^{-1}) during injection (0.6 PV and 1.8 PV) and flushing (7 PV). Groundwater flow from right to left. (Adapted from Micić Batka, V. and Hofmann, T. *DL4.2: Stability, Mobility, Delivery and Fate of optimized NPs under Field Relevant Conditions*, NanoRem, 2016.)

In case the particle travel distance largely differs from the expected outcomes, the particle suspension needs to be optimized by adding stabilizing agents like CMC, respectively, adjust the suspension's pH or ionic strength.

12.3.3 INJECTION AND MONITORING DESIGN

Based on the results of the site investigations and laboratory experiments, the particle mass dosage is calculated, and the injection layout is configured. Before any field application starts, regulatory approval is also required.

12.3.3.1 Particle Mass Dosage

The mass of NMPs to be injected is calculated considering the target mass of contaminant to be remediated, the extension of the contamination in groundwater, the calculated contaminant degradation rates, and adsorption kinetics. This estimation should also include a detailed analysis of the site geochemistry and redox conditions. Indeed, a wide range of naturally occurring species, other than the target contaminant, may potentially react with the selected NMPs, resulting in an excessive and costly dosage. A safety factor may be also considered to account for uncertainties, such as the native electron acceptor demands (e.g., oxygen, nitrate and sulfate) for reductive particles. Stabilizing agents may also be considered, based on the outcomes of the stability and mobility assessment.

12.3.3.2 Injection Configuration

If the NMPs are injected to remove the contamination source, the required mass and distribution of the particle in the subsurface are dependent on the mass of the contaminant in the target zone. For plume containment, the total mass of contaminant to be treated accounts for the dissolved phase in the saturated pore volume and the flux through the reactive zone over a certain period of time, depending on the desirable barrier lifetime (Braun et al., 2017; Bruns et al., 2019). In this case, key parameters are the natural groundwater flow velocity and the contact time between the contaminant and the NMPs, which is required for completing the treatment.

The hydraulic conductivity (K), hydraulic gradient (i), and the effective porosity need to be also calculated to define the linear groundwater velocity v at the site, based on the Darcy's law:

$$v = \frac{q}{\varepsilon_{eff}} = \frac{K\,i}{\varepsilon_{eff}} \tag{12.9}$$

The necessary residence time of the contaminant in the reactive barrier for its complete removal is assessed based on the kinetics of the reaction or adsorption process (Section 12.3.2). For adsorptive processes, the obtained partition coefficients, sorption capacities and affinities are used to approximate barrier breakthrough times and adjust the barrier length for the required retention according to the following equations:

$$t_{bt} = \frac{l}{v} \cdot R_i \tag{12.10}$$

$$R_i = \frac{v}{v_i} = 1 + \frac{\rho}{\varepsilon_{eff}} \cdot f \cdot K_d \tag{12.11}$$

$$K_d = c_i^n \cdot K_F \tag{12.12}$$

where t_{bt} is the approximated breakthrough time, l is the barrier length, R_i is the retardation coefficient of the compound i flowing through the barrier, v_i is the linear velocity of the compound i affected by adsorption to the sediment after adsorbent immobilization, ρ is the bulk density of the aquifer sediment, f is the adsorbent mass fraction on the sediment, K_d is the sorption coefficient of i on the adsorbent, c_i is the incoming concentration of i, n is the Freundlich exponent, and K_F the Freundlich coefficient (Georgi et al., 2015).

12.3.3.3 Monitoring Design

The effectiveness of the remediation needs to be verified and documented. Thus, a specific monitoring plan must be designed, based on the modeling results, including pre-injection, injection, and post injection monitoring campaigns. The pre-injection sampling event will be used to define the baseline and then continued after the injection with defined interval. Samples should be collected both on a short term (e.g., daily-weekly) and long term (e.g., monthly-yearly) basis to properly assess the remediation performance. An integrative approach is also recommended to assess changes in soil, groundwater, and soil vapor.

High-resolution measurements are suggested to record changes over space and time of groundwater chemistry and microbiological indicators in monitoring wells (Bruns et al., 2019). Physicochemical parameters should be recorded with a high frequency using a multiparameter probe. As the properties of the injected particle suspension mostly differ from the values commonly measured in groundwater, iron content, pH, redox, electric conductivity, temperature, dissolved oxygen, and turbidity may be considered as high-resolution and inexpensive tracers to monitor the NMP distribution in the aquifer. Geochemical analysis may also be combined with compound specific isotope analysis (CSIA) to identify whether natural attenuation is occurring at the site and to properly evaluate the performance of the remedial action. In any case, the list of parameters to be monitored is site-specific and strictly depended on the properties of the selected particle. Some NMPs may also present electrical and magnetic properties significantly different than the natural soil properties, so magnetic susceptibility measurements in field boreholes and electrical geophysical methods may be additional monitoring tools to detect the distribution of NMPs during subsurface delivery (Flores Orozco et al., 2015; Brunetti et al., 2013).

12.3.4   Suspension Preparation and Delivery

12.3.4.1   Preparation of the NMP Suspension

In accordance with the previous mass calculation criteria, the NMPs are manufactured by the producer and adequately stored in containers to preserve the original properties of the particles. In case of some particles, for example, Carbo-Iron, the contact with oxygen must be avoided both during transport and suspension preparation. Therefore, water has to be deoxygenated by purging with nitrogen before suspending the particles (i.e. a dissolved oxygen concentration < 0.1 mg L^{-1}) (Bruns et al., 2019).

Based on the total NMP mass, and assuming a desirable particle concentration in the suspension, the total volume of the suspension to be injected is calculated. Particle concentrations in the injected suspensions generally range from 1 to 20 g L^{-1}, depending on the agglomeration tendency of the particles. Biodegradable polymers, such as carboxymethylcellulose, guar gum, or polyacrylic acids, may be added to promote the suspension stability and particle mobility, based on the outcomes of the lab experiments. In general, the viscosity of the suspension increases with the concentration of the stabilizer and is higher for stabilizers with high molecular weight. If necessary, pH of the particle suspension can be adjusted with NaOH or H$_2$SO$_4$.

12.3.4.2   Injection Methods

Various injection techniques may be employed to target the NMPs suspension in the contaminated aquifer, such as gravity injection, direct push, pressure pulse technology, pneumatic injection, and hydraulic fracturing (Comba et al., 2011; Braun et al., 2017; Bruns et al., 2019; Christiansen et al., 2008). Recently, also new injection techniques have been developed, such as PIM™ (Carsico, Italy) (Carsico, 2019), which allows multiple injections over depth and time, and SPIN® injection technology (Injectis, Belgium) (Injectis, 2019). The injection method is usually site specific and is very dependent on the soil characteristic and properties of the suspension to be injected. Injections are commonly performed in unconsolidated saturated soils, with more favorable conditions represented by gravel and sandy aquifers. With increasing percentage of fine-grained sediments, the injection becomes progressively more difficult (Bruns et al., 2019).

Two different injection regimes may be identified: permeation and fracturing. Permeation is preferable to fracturing, as it is characterized by low injection pressure, usually less than 2 bar, resulting in a uniform distribution of the NMPs in the surrounding of the injection point. It is employed for medium to coarse sediments. Fracturing injection requires higher pressures, and is applied for medium to fine-grained formation (Luna et al., 2015; Bruns et al., 2019).

12.3.5   Field Application of Carbon-Iron®: A Case Study

The post injection monitoring results are finally analyzed and compared to the pre-injection scenario, to assess the efficiency of the remediation in terms of reduction of mass and concentration of the contaminant.

Below, a case study is presented where Carbo-Iron has been applied for the remediation of groundwater highly contaminated by CAHs. The aim was to enhance the ongoing remedial works at a former industrial facility where a pump-and-treat system has been applied for more than a decade.

12.3.5.1   Site Description

The selected site was used as a storage facility for chlorinated solvents. Despite the tank removal, elevated concentrations of tetrachloroethene (PCE) and trichloroethene (TCE) have been detected in groundwater. The aquifer setting includes an upper unconfined sandy unit, a middle fine-grained silty unit, and a lower sandy unit. The upper unit extends up to a depth of 8.5–15.5 m below ground level (bgl) and consists of fine to medium sand with a hydraulic conductivity from 7×10^{-6} to 9×10^{-5} m s^{-1}. The average depth to water table is 8 m and groundwater moves from NE to SW with a hydraulic gradient of 0.03 (Figure 12.8). Elevated concentrations of CAHs were detected in the upper unit of the aquifer, with concentrations of PCE equal to 33 mg L^{-1} from 8 to 12 m bgl, measured before injection.

12.3.5.2   Monitoring System Design

An innovative integrated package of multiple probes was preliminarily employed to collect real-time measurements in the existing monitoring wells, such as groundwater velocity, direction, and other relevant geochemical indicators (H$_2$, pH, ORP, dissolved oxygen, EC, temperature, and turbidity).

A new monitoring well (MW101) was placed approximately 10 m downgradient of the target contaminated zone, where the injections were planned (Figure 12.8). This well consisted of 2-inch diameter PVC screened from 6 to 10 m bgl. The monitoring points located upgradient and downgradient of the source area were also equipped with automatic dataloggers to identify groundwater table fluctuation, before, during, and after the injection. Groundwater sampling was carried out at MW101 to determine the baseline hydrochemistry and continued after the injections with defined intervals, approximately ranging from 2 days to 3 weeks, for the following 3 months. All the groundwater samples were analyzed for redox parameters, chlorinated ethenes, and dechlorination products.

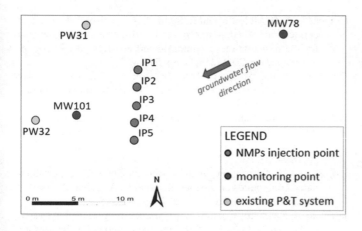

FIGURE 12.8 Injection layout.

12.3.5.3 Field Injection

Considering the contaminant vertical distribution, direct push injection technique was employed to enhance the particle permeation around the injection point. The injection took place in five points (IP1/5) to maximize the distribution of the particle in the contaminated area (Figure 12.8). The injection was carried out in the saturated zone at a depth of 9.5 m bgl. A total of 40 kg of Carbo-Iron was injected at low pressure and with a flow rate of approximately 1–1.5 m³ h⁻¹.

12.3.5.4 Results

Three months after, the injection the contaminant concentration in the source area was reduced by approximately 90% (Figure 12.9). The injection promoted a rapid PCE and TCE reduction due to adsorption processes and abiotic dechlorination, and no additional unwanted by-products, such as 1,2-DCE or VC, were detected. As evidence for the abiotic reductive dechlorination of the target compounds, non-chlorinated hydrocarbons, such as ethene and ethane, appeared in groundwater after the injection.

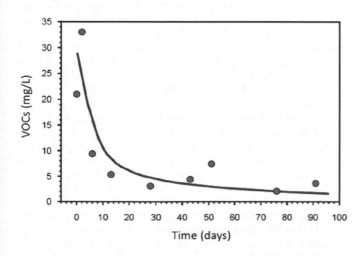

FIGURE 12.9 VOCs concentration in groundwater at monitoring point MW101 after Carbo-Iron injection.

12.3.6 FUTURE TECHNOLOGY DEVELOPMENTS

Nanoremediation is an innovative technology for the treatment of contaminated aquifers. Although there was significant progress made during the past years, this research area still offers various challenges. Even though particle-based technologies for the treatment of a broad range of groundwater contaminants do exist, numerous of them are still in the laboratory or pilot phase and still have to be brought to market maturity, which is hitherto hampered by the small number of validated case studies and, connected to that, limited access to operation windows based on application experience (Bartke et al., 2018). In addition, the application of nanoremediation to contaminations in the unsaturated subsurface as well as in contamination source areas is still hardly researched. For a series of traditional and emerging contaminations, no particle-based, or even no *in situ*-applicable technology, was developed until now. Especially hydrophilic compounds, which cannot be efficiently enriched on common adsorbents, offer a scientific challenge. The *in situ* degradation of those contaminants is still an open task for nanoremediation research.

REFERENCES

Acharya, J., et al., Removal of chromium(VI) from wastewater by activated carbon developed from Tamarind wood activated with zinc chloride. *Chemical Engineering Journal*, 2009. **150**(1): pp. 25–39.

Agrawal, A. and P.G. Tratnyek, Reduction of nitro aromatic compounds by zero-valent iron metal. *Environmental Science & Technology*, 1996. **30**(1): pp. 153–160.

Ania, C.O., et al., Effects of activated carbon properties on the adsorption of naphthalene from aqueous solutions. *Applied Surface Science*, 2007. **253**(13): pp. 5741–5746.

Arnold, W.A. and A.L. Roberts, Pathways and kinetics of chlorinated ethylene and chlorinated acetylene reaction with Fe(0) particles. *Environmental Science & Technology*, 2000. **34**(9): pp. 1794–1805.

Bardos, P., et al., Status of nanoremediation and its potential for future deployment: Risk-benefit and benchmarking appraisals. *Remediation Journal*, 2018. **28**(3): pp. 43–56.

Bartke, S., et al., Market potential of nanoremediation in Europe: Market drivers and interventions identified in a deliberative scenario approach. *Science of The Total Environment*, 2018. **619–620**: pp. 1040–1048.

Bianco, C., T. Tosco, and R. Sethi, A 3-dimensional micro- and nanoparticle transport and filtration model (MNM3D) applied to the migration of carbon-based nanomaterials in porous media. *Journal of Contaminant Hydrology*, 2016. **193**: pp. 10–20.

Bianco, C., T. Tosco, and R. Sethi, *MNMs 2018: Micro- and Nanoparticles Transport, Filtration and Clogging Model-Suite: A Comprehensive Tool for Design and Interpretation of Colloidal Particle Transport in 1D Cartesian and 1D Radial Systems*. 2018. DIATI, Politecnico di Torino: Torino, Italy.

Boopathy, R., et al., Metabolism of explosive compounds by sulfate-reducing bacteria. *Current Microbiology*, 1998. **37**(2): pp. 127–131.

Bosch, J., et al., *Application of Nanoremediation for Groundwater Decontamination: A Case Study at a Former Industrial Facility Contaminated by Chlorinated Solvents*, in REMTECH. 2018. Ferrara, Italy.

Bosch, J., et al., Nanosized iron oxide colloids strongly enhance microbial iron reduction. *Applied and Environmental Microbiology*, 2010. **76**(1): pp. 184.

Bou-Nasr, J., D. Cassidy, and D. Hampton. Comparative study of the effect of four ISCO oxidants on PCE oxidation and aerobic microbial activity, in *Remediation of Chlorinated and Recalcitrant Pollutants. Proceedings of the Fifth International Conference on Remediation of Chlorinated and Recalcitrant Compounds*. 2006. Monterey, CA: Battelle Press, Columbus, OH.

Bradford, S.A., et al., Modeling colloid attachment, straining, and exclusion in saturated porous media. *Environmental Science & Technology*, 2003. **37**(10): pp. 2242–2250.

Bradford, S.A., et al., Physical factors affecting the transport and fate of colloids in saturated porous media. *Water Resources Research*, 2002. **38**(12): pp. 63-1-63-12.

Brändli, R.C., et al., Sorption of native polyaromatic hydrocarbons (PAH) to black carbon and amended activated carbon in soil. *Chemosphere*, 2008. **73**(11): pp. 1805–1810.

Braun, J., et al., Generalized Guideline for Application of Nano-remediation. *NanoRem*, 2017.

Brunetti, E., et al., *Treatment of Chromium VI in the Unsaturated Zone Coupled with Geophysical Monitoring: A Field-Scale Pilot Test at the Industrial Site of Spinetta M. (AL), Italy*, in REMTECH. 2013. Ferrara, Italy.

Bruns, J., et al., Experiences from pilot- and large-scale demonstration sites from across the globe including combined remedies with NZVI, in *Nanoscale Zerovalent Iron Particles for Environmental Restoration*, T. Phenrat and G.V. Lowry, Editors. 2019. Springer, Switzerland.

Buerge-Weirich, D., PP. Behra, and L. Sigg, Adsorption of copper, nickel, and cadmium on goethite in the presence of organic ligands. *Aquatic Geochemistry*, 2003. **9**(2): pp. 65–85.

Busch, J., et al., A field investigation on transport of carbon-supported nanoscale zero-valent iron (nZVI) in groundwater. *Journal of Contaminant Hydrology*, 2015. **181**: pp. 59–68.

Busch, J., et al., Investigations on mobility of carbon colloid supported nanoscale zero-valent iron (nZVI) in a column experiment and a laboratory 2D-aquifer test system. *Environmental Science and Pollution Research*, 2014a. **21**(18): pp. 10908–10916.

Busch, J., et al., Transport of carbon colloid supported nanoscale zero-valent iron in saturated porous media. *Journal of Contaminant Hydrology*, 2014b. **164**: pp. 25–34.

Busch, J.P., Investigations on mobility of carbon colloid supported nanoscale zero-valent iron (nZVI) for groundwater remediation, in *Institut für Erd- und Umweltwissenschaften/Geoökologie*. 2015, Universität Potsdam. pp. 171.

Carsico. www.carsico.it. [cited March 19, 2019].

Chen, K.-F., Y.-C. Chang, and W.-T. Chiou, Remediation of diesel-contaminated soil using in situ chemical oxidation (ISCO) and the effects of common oxidants on the indigenous microbial community: A comparison study. *Journal of Chemical Technology & Biotechnology*, 2016. **91**(6): pp. 1877–1888.

Chowdhury, S.R. and E.K. Yanful, Arsenic and chromium removal by mixed magnetite-maghemite nanoparticles and the effect of phosphate on removal. *Journal of Environmental Management*, 2010. **91**(11): pp. 2238–2247.

Christenson, M., et al., A five-year performance review of field-scale, slow-release permanganate candles with recommendations for second-generation improvements. *Chemosphere*, 2016. **150**: pp. 239–247.

Christiansen, C.M., et al., Characterization and quantification of pneumatic fracturing effects at a clay till site. *Environmental Science & Technology*, 2008. **42**(2): pp. 570–576.

Comba, S., A. Di Molfetta, and R. Sethi, A comparison between field applications of nano-, micro-, and millimetric zero-valent iron for the remediation of contaminated aquifers. *Water, Air, & Soil Pollution*, 2011. **215**(1): pp. 595–607.

De Zwart, D. and W. Slooff, The Microtox as an alternative assay in the acute toxicity assessment of water pollutants. *Aquatic Toxicology*, 1983. **4**(2): pp. 129–138.

Deliyanni, E. A., et al., Activated carbons for the removal of heavy metal ions: A systematic review of recent literature focused on lead and arsenic ions. *Open Chemistry*, 2015. **13**(1): pp. 699–708.

Du, Z., et al., Adsorption behavior and mechanism of perfluorinated compounds on various adsorbents: A review. *Journal of Hazardous Materials*, 2014. **274**: pp. 443–454.

Dwivedi, C.P., et al., Column performance of granular activated carbon packed bed for Pb(II) removal. *Journal of Hazardous Materials*, 2008. **156**(1): pp. 596–603.

Edwards, E.A., et al., Anaerobic degradation of toluene and xylene by aquifer microorganisms under sulfate-reducing conditions. *Applied and Environmental Microbiology*, 1992. **58**(3): pp. 794.

Elimelech, M. and C.R. O'Melia, Kinetics of deposition of colloidal particles in porous media. *Environmental Science & Technology*, 1990. **24**(10): pp. 1528–1536.

Elimelech, M., et al., *Particle Deposition and Aggregation*, Butterworth-Heinemann. 1995. Boston.

Elliott, D.W. and W.-X. Zhang, Field assessment of nanoscale bimetallic particles for groundwater treatment. *Environmental Science & Technology*, 2001. **35**(24): pp. 4922–4926.

Ferreira, T.A., et al., Chromium(VI) removal from aqueous solution by magnetite coated by a polymeric ionic liquid-based adsorbent. *Materials* (Basel, Switzerland), 2017. **10**(5): pp. 502.

Flores Orozco, A., et al., Monitoring the injection of microscale zerovalent iron particles for groundwater remediation by means of complex electrical conductivity imaging. *Environmental Science & Technology*, 2015. **49**(9): pp. 5593–5600.

Fu, F. and Q. Wang, Removal of heavy metal ions from wastewaters: A review. *Journal of Environmental Management*, 2011. **92**(3): pp. 407–418.

Fuchs, G., M. Boll, and J. Heider, Microbial degradation of aromatic compounds—From one strategy to four. *Nature Reviews Microbiology*, 2011. **9**: pp. 803.

Gastone, F., T. Tosco, and R. Sethi, Green stabilization of microscale iron particles using guar gum: Bulk rheology, sedimentation rate and enzymatic degradation. *Journal of Colloid and Interface Science*, 2014a. **421**: pp. 33–43.

Gastone, F., T. Tosco, and R. Sethi, Guar gum solutions for improved delivery of iron particles in porous media (Part 1): Porous medium rheology and guar gum-induced clogging. *Journal of Contaminant Hydrology*, 2014b. **166**: pp. 23–33.

Gavaskar, A., L. Tatar, and W. Condit, *Cost and Performance Report Nanoscale Zero-Valent Iron Technologies for Source Remediation*. NAVTAC report, 2005.

Georgi, A., et al., Colloidal activated carbon for in-situ groundwater remediation—Transport characteristics and adsorption of organic compounds in water-saturated sediment columns. *Journal of Contaminant Hydrology*, 2015. **179**: pp. 76–88.

Georgi, A.R., K. Gonzales-Olmos, and K. Mackenzie, Nanostructured Fe-zeolite and orthoferrite nanoparticles: Fenton-like heterogeneous catalysts for oxidation of water contaminants, in *Nanotechnology for Water Purification*, T. Dey, Editor. 2012 BrownWalker Press, Boca Raton (Florida, USA).

Gillett, A.N., J. Nathanail, and PP. Nathanail, Risk screening model (RSM) for application of NanoRem nanoparticles to groundwater remediation, in Taking Nanotechnological Remediation Processes from Lab Scale to End User Applications for the Restoration of a Clean Environment. *Nanorem*, 2017.

Gillies, G., et al., Suspension stability and mobility of Trap-Ox Fe-zeolites for in-situ nanoremediation. *Journal of Colloid and Interface Science*, 2017. **501**: pp. 311–320.

Gong, Y., J. Tang, and D. Zhao, Application of iron sulfide particles for groundwater and soil remediation: A review. *Water Research*, 2016. **89**: pp. 309–320.

Gonzalez-Olmos, R., et al., Fe-zeolites as catalysts for chemical oxidation of MTBE in water with H_2O_2. *Applied Catalysis B: Environmental*, 2009. **89**(3): pp. 356–364.

Gonzalez-Olmos, R., et al., Hydrophobic Fe-zeolites for removal of MTBE from water by combination of adsorption and oxidation. *Environmental Science & Technology*, 2013. **47**(5): pp. 2353–2360.

Gorski, C.A., et al., Redox behavior of magnetite: Implications for contaminant reduction. *Environmental Science & Technology*, 2010. **44**(1): pp. 55–60.

Grolimund, D., M. Elimelech, and M. Borkovec, Aggregation and deposition kinetics of mobile colloidal particles in natural porous media. *Colloids and Surfaces A: Physicochemical and Engineering Aspects*, 2001. **191**(1): pp. 179–188.

Hahn, M.W., D. Abadzic, and C.R. O'Melia, Aquasols: On the role of secondary minima. *Environmental Science & Technology*, 2004. **38**(22): pp. 5915–5924.

Hajji, S., et al., Arsenite and chromate sequestration onto ferrihydrite, siderite and goethite nanostructured minerals: Isotherms from flow-through reactor experiments and XAS measurements. *Journal of Hazardous Materials*, 2019. **362**: pp. 358–367.

Hale, S., et al., Effects of chemical, biological, and physical aging as well as soil addition on the sorption of pyrene to activated carbon and biochar. *Environmental Science & Technology*, 2011. **45**(24): pp. 10445–10453.

Han, I., M.A. Schlautman, and B. Batchelor, Removal of hexavalent chromium from groundwater by granular activated carbon. *Water Environment Research*, 2000. **72**(1): pp. 29–39.

Hashim, M.A., et al., Remediation technologies for heavy metal contaminated groundwater. *Journal of Environmental Management*, 2011. **92**(10): pp. 2355–2388.

He, F., D. Zhao, and C. Paul, Field assessment of carboxymethyl cellulose stabilized iron nanoparticles for in situ destruction of chlorinated solvents in source zones. *Water Research*, 2010. **44**(7): pp. 2360–2370.

Henn, K.W. and D.W. Waddill, Utilization of nanoscale zero-valent iron for source remediation—A case study. *Remediation Journal*, 2006. **16**(2): pp. 57–77.

Hjorth, R., et al., Ecotoxicity testing and environmental risk assessment of iron nanomaterials for sub-surface remediation: Recommendations from the FP7 project NanoRem. *Chemosphere*, 2017. **182**: pp. 525–531.

Hong, M., et al., Heavy metal adsorption with zeolites: The role of hierarchical pore architecture. *Chemical Engineering Journal*, 2019. **359**: pp. 363–372.

Hotze, E.M., T. Phenrat, and G.V. Lowry, Nanoparticle aggregation: challenges to understanding transport and reactivity in the environment. *Journal of Environmental Quality*, 2010. **39**(6): pp. 1909–1924.

Hu, J., G. Chen, and I.M.C. Lo, Removal and recovery of Cr(VI) from wastewater by maghemite nanoparticles. *Water Research*, 2005. **39**(18): pp. 4528–4536.

Hu, Z., et al., Transport and deposition of carbon nanoparticles in saturated porous media. *Energies*, 2017. **10**(8).

Injectis. www.injectis.com. [cited March 19, 2019.].

Intrapore. November 11, 2019]; Available from: https://intrapore.com/en/intrasense/.

Jakob, L., et al., PAH-sequestration capacity of granular and powder activated carbon amendments in soil, and their effects on earthworms and plants. *Chemosphere*, 2012. **88**(6): pp. 699–705.

Johnson, R.L., et al., Field-scale transport and transformation of carboxymethylcellulose-stabilized nano zero-valent iron. *Environmental Science & Technology*, 2013. **47**(3): pp. 1573–1580.

Kaestner, M., et al., *Model-Driven Soil Probing, Site Assessment and Evaluation: Guidance on Technologies*. 2012: Sapienza Universita Editrice.

Karn, B., T. Kuiken, and M. Otto, Nanotechnology and in situ remediation: A review of the benefits and potential risks. *Environmental Health Perspectives*, 2009. **117**(12): pp. 1813–1831.

Kocur, C.M., D.M. O'Carroll, and B.E. Sleep, Impact of nZVI stability on mobility in porous media. *Journal of Contaminant Hydrology*, 2013. **145**: pp. 17–25.

Kosmulski, M., pH-dependent surface charging and points of zero charge II. Update. *Journal of Colloid and Interface Science*, 2004. **275**(1): pp. 214–224.

Kretzschmar, R., et al., Mobile subsurface colloids and their role in contaminant transport. *Advances in Agronomy*, 1999. **66**: pp. 121–193.

Kucharzyk, K.H., et al., Novel treatment technologies for PFAS compounds: A critical review. *Journal of Environmental Management*, 2017. **204**: pp. 757–764.

Laumann, S., et al., Carbonate minerals in porous media decrease mobility of polyacrylic acid modified zero-valent iron nanoparticles used for groundwater remediation. *Environmental Pollution*, 2013. **179**: pp. 53–60.

Lazaridis, N.K., Sorption removal of anions and cations in single batch systems by uncalcined and calcined Mg-Al-CO_3 hydrotalcite. *Water, Air & Soil Pollution*, 2003. **146**(1): pp. 127–139.

Li, J., S.R.C. Rajajayavel, and S. Ghoshal, Transport of carboxymethyl cellulose-coated zerovalent iron nanoparticles in a sand tank: Effects of sand grain size, nanoparticle concentration and injection velocity. *Chemosphere*, 2016. **150**: pp. 8–16.

Li, X., et al., Role of hydrodynamic drag on microsphere deposition and re-entrainment in porous media under unfavorable conditions. *Environmental Science & Technology*, 2005. **39**(11): pp. 4012–4020.

Lien, H. and W. Zhang, Nanoscale iron particles for complete reduction of chlorinated ethenes. *Colloids and Surfaces*, 2001. **191**: pp. 97–105.

Lien, H.-L., et al., Recent progress in zero-valent iron nanoparticles for groundwater remediation. *Journal of Environmental Engineering and Management*, 2006. **16**(6): pp. 371–380.

Liu, W., et al., Application of stabilized nanoparticles for in situ remediation of metal-contaminated soil and groundwater: A critical review. *Current Pollution Reports*, 2015. **1**(4): pp. 280–291.

Liu, Y., et al., TCE dechlorination rates, pathways, and efficiency of nanoscale iron particles with different properties. *Environmental Science & Technology*, 2005. **39**(5): pp. 1338–1345.

Logan, B.E., et al., Clarification of clean-bed filtration models. *Journal of Environmental Engineering*, 1995. **121**(12): pp. 869–873.

Luna, M., et al., Pressure-controlled injection of guar gum stabilized microscale zerovalent iron for groundwater remediation. *Journal of Contaminant Hydrology*, 2015. **181**: pp. 46–58.

Macé, C., et al., Nanotechnology and groundwater remediation: A step forward in technology understanding. *Remediation Journal*, 2006. **16**(2): pp. 23–33.

Mackenzie, K. and A. Georgi, NZVI synthesis and characterization, in *Nanoscale Zerovalent Iron Particles for Environmental Restoration*, T. Phenrat and G.V. Lowry, Editors. 2019, Springer.

Mackenzie, K., et al., Carbo-Iron: An Fe/AC composite : As alternative to nano-iron for groundwater treatment. *Water Research*, 2012. **46**(12): pp. 3817–3826.

Mackenzie, K., et al., Carbo-Iron as improvement of the nanoiron technology: From laboratory design to the field test. *Science of the Total Environment*, 2016. **563–564**: pp. 641–648.

Mackenzie, K., Georgi, A., Bleyl, S., Lloyd, J., Joshi, N., Meckenstock, R., Krok, B.A., Klaas, N., Herrmann, C., DL 3.2 assessment of nanoparticle performance for the removal of contaminants: Non-ZVI and composite nanoparticles, in *NanoRem WP3: Design, Improvement and Optimized Production of Nanoparticles: Non-ZVI and Composite Nanoparticles*. 2016 Nanorem.

Mahl, D., Synthese, Löslichkeit und Stabilität von Gold-Nanopartikeln in biologischen Medien, doctoral thesis, Universität Duisburg-Essen, 2011.

Manning, B.A., et al., Arsenic(III) and Arsenic(V) reactions with zerovalent iron corrosion products. *Environmental Science & Technology*, 2002. **36**(24): pp. 5455–5461.

Mason, C.A., et al., Biodegradation of BTEX by bacteria on powdered activated carbon. *Bioprocess Engineering*, 2000. **23**(4): pp. 331–336.

McGregor, R., In situ treatment of PFAS-impacted groundwater using colloidal activated Carbon. *Remediation Journal*, 2018. **28**(3): pp. 33–41.

McNamara, J.D., et al., Comparison of activated carbons for removal of perfluorinated compounds from drinking water. *Journal: American Water Works Association*, 2018. **110**(1): pp. E2–E14.

Micić Batka, V. and T. Hofmann, *DL4.2: Stability, Mobility, Delivery and Fate of optimized NPs under Field Relevant Conditions*. 2016, NanoRem.

Molnar, I.L., et al., Predicting colloid transport through saturated porous media: A critical review. *Water Resources Research*, 2015. **51**(9): pp. 6804–6845.

Montalvo, D., et al., Efficient removal of arsenate from oxic contaminated water by colloidal humic acid-coated goethite: Batch and column experiments. *Journal of Cleaner Production*, 2018. **189**: pp. 510–518.

Mueller, N.C., et al., Application of nanoscale zero valent iron (NZVI) for groundwater remediation in Europe. *Environmental Science and Pollution Research*, 2012. **19**(2): pp. 550–558.

Müller, N.C. and B. Nowack, *Nano Zero Valent Iron: The Solution for Water and Soil Remediation?* Observatory NANO focus report, 2010.

Muyzer, G. and A.J.M. Stams, The ecology and biotechnology of sulphate-reducing bacteria. *Nature Reviews Microbiology*, 2008. **6**: pp. 441.

NanoRem and CL:AIRE, *NanoRem Bulletin 1-12*. 2016–2017.

O'Hara, S., et al., Field and laboratory evaluation of the treatment of DNAPL source zones using emulsified zero-valent iron. *Remediation Journal*, 2006. **16**(2): pp. 35–56.

Park, M., et al., Removal of hexavalent chromium using mackinawite (FeS)-coated sand. *Journal of Hazardous Materials*, 2018. **360**: pp. 17–23.

Parvez, S., C. Venkataraman, and S. Mukherji, A review on advantages of implementing luminescence inhibition test (*Vibrio fischeri*) for acute toxicity prediction of chemicals. *Environment International*, 2006. **32**(2): pp. 265–268.

Patterson, R.R., S. Fendorf, and M. Fendorf, Reduction of hexavalent chromium by amorphous iron sulfide. *Environmental Science & Technology*, 1997. **31**(7): pp. 2039–2044.

Phenrat, T., et al., Aggregation and sedimentation of aqueous nanoscale zerovalent iron dispersions. *Environmental Science & Technology*, 2007. **41**(1): pp. 284–290.

Phenrat, T., et al., Stabilization of aqueous nanoscale zerovalent iron dispersions by anionic polyelectrolytes: Adsorbed anionic polyelectrolyte layer properties and their effect on aggregation and sedimentation. *Journal of Nanoparticle Research*, 2008. **10**: pp. 795–814.

Quinn, J., et al., Field demonstration of DNAPL dehalogenation using emulsified zero-valent iron. *Environmental Science & Technology*, 2005. **39**(5): pp. 1309–1318.

Rabbani, M.M.A., I., S.-J. Park, Application of Nanotechnology to Remediate Contaminated Soils, in *Environmental Remediation Technologies for Metal-Contaminated Soils*, H.R. Hasegawa, I. M. M.; Rahman, M. A., Editor. 2016 Springer, Japan.

Raychoudhury, T., N. Tufenkji, and S. Ghoshal, Aggregation and deposition kinetics of carboxymethyl cellulose-modified zerovalent iron nanoparticles in porous media. *Water Research*, 2012. **46**(6): pp. 1735–1744.

Regenesis. https://regenesis.com/en/remediation-products/plume-stop-liquid-activated-carbon/ [cited March 25, 2019].

Rusevova, K., F.D. Kopinke, and A. Georgi, Stabilization of potassium permanganate particles with manganese dioxide. *Chemosphere*, 2012. **86**(8): pp. 783–788.

Ryan, J.N. and M. Elimelech, Colloid mobilization and transport in groundwater. *Colloids and Surfaces A: Physicochemical and Engineering Aspects*, 1996. **107**: pp. 1–56.

Schwidder, M., et al., Selective reduction of NO with Fe-ZSM-5 catalysts of low Fe content: I. Relations between active site structure and catalytic performance. *Journal of Catalysis*, 2005. **231**(2): pp. 314–330.

Sen, T.K. and K.C. Khilar, Review on subsurface colloids and colloid-associated contaminant transport in saturated porous media. *Advances in Colloid and Interface Science*, 2006. **119**(2–3): pp. 71–96.

Shahbazi, A., et al., Natural and synthetic zeolites in adsorption/oxidation processes to remove surfactant molecules from water. *Separation and Purification Technology*, 2014. **127**: pp. 1–9.

Šimůnek, J., M.T. van Genuchten, and M. Šejna, Development and applications of the HYDRUS and STANMOD software packages and related codes. *Vadose Zone Journal*, 2008. **7**(2): pp. 587–600.

So, C.M. and L.Y. Young, Isolation and characterization of a sulfate-reducing bacterium that anaerobically degrades alkanes. *Applied and Environmental Microbiology*, 1999. **65**(7): pp. 2969.

Su, C., et al., Travel distance and transformation of injected emulsified zerovalent iron nanoparticles in the subsurface during two and half years. *Water Research*, 2013. **47**(12): pp. 4095–4106.

Tian, Y., et al., Transport of engineered nanoparticles in saturated porous media. *Journal of Nanoparticle Research*, 2010. **12**(7): pp. 2371–2380.

Tiraferri, A., T. Tosco, and R. Sethi, Transport and retention of microparticles in packed sand columns at low and intermediate ionic strengths: Experiments and mathematical modeling. *Environmental Earth Sciences*, 2011. **63**(4): pp. 847–859.

Tosco, T. and R. Sethi, MNM1D: A numerical code for colloid transport in porous media: Implementation and validation. *American Journal of Environmental Sciences*, 2009. **5**(4): pp. 517–525.

Tosco, T. and R. Sethi, Transport of non-Newtonian suspensions of highly concentrated micro- and nanoscale iron particles in porous media: A modeling approach. *Environmental Science & Technology*, 2010. **44**(23): pp. 9062–9068.

Tosco, T., A. Tiraferri, and R. Sethi, Ionic strength dependent transport of microparticles in saturated porous media: Modeling mobilization and immobilization phenomena under transient chemical conditions. *Environmental Science & Technology*, 2009. **43**(12): pp. 4425–4431.

Tosco, T., C. Bianco, and R. Sethi, An integrated experimental and modeling approach to assess the mobility of iron-based nanoparticles in groundwater systems, in *Iron Nanomaterials for Water and Soil Treatment*, M.I. Litter, N. Quici, and M. Meichtry, Editors. 2018: New York Pan Stanford Publishing, Singapore.

Tosco, T., et al., *D7.2 Simulation Module for Predicting Transport of NPs in Groundwater*. 2016, NanoRem.

Tosco, T., et al., Nanoscale zerovalent iron particles for groundwater remediation: A review. *Journal of Cleaner Production*, 2014. **77**: pp. 10–21.

Tosco, T., et al., Transport of ferrihydrite nanoparticles in saturated porous media: Role of ionic strength and flow rate. *Environmental Science & Technology*, 2012. **46**(7): pp. 4008–4015.

Tosco, T., F. Gastone, and R. Sethi, Guar gum solutions for improved delivery of iron particles in porous media (Part 2): Iron transport tests and modeling in radial geometry. *Journal of Contaminant Hydrology*, 2014. **166**: pp. 34–51.

Truex, M.J., et al., Injection of zero-valent iron into an unconfined aquifer using shear-thinning fluids. *Ground Water Monitoring & Remediation*, 2011. **31**(1): pp. 50–58.

Velimirovic, M., et al., Field assessment of guar gum stabilized microscale zerovalent iron particles for in-situ remediation of 1,1,1-trichloroethane. *Journal of Contaminant Hydrology*, 2014. **164**: pp. 88–99.

Vogel, M., et al., Combined chemical and microbiological degradation of tetrachloroethene during the application of Carbo-Iron at a contaminated field site. *Science of The Total Environment*, 2018. **628–629**: pp. 1027–1036.

von Gunten, U., Oxidation processes in water treatment: Are we on track? *Environmental Science & Technology*, 2018. **52**(9): pp. 5062–5075.

Waghmare, A. and P.P. Seshaiyer, Enhancing groundwater quality through computational modeling and simulation to optimize transport and interaction parameters in porous media. *Journal of Water Resource and Protection*, 2015. **7**(5): pp. 398–409.

Wang, F., et al., Surface properties of natural aquatic sediments. *Water Research*, 1997. **31**(7): pp. 1796–1800.

Wang, W.-L., et al., Potential risks from UV/H_2O_2 oxidation and UV photocatalysis: A review of toxic, assimilable, and sensory-unpleasant transformation products. *Water Research*, 2018. **141**: pp. 109–125.

Wei, Y.T., et al., Influence of nanoscale zero-valent iron on geochemical properties of groundwater and vinyl chloride degradation: A field case study. *Water Research*, 2010. **44**(1): pp. 131–140.

Wiesner, M.R. and J. Bottero, *Environmental Nanotechnology: Applications and Impacts of Nanomaterials*. 2017: Mc Graw Hill education.

Wilkin, R.T. and D.G. Beak, Uptake of nickel by synthetic mackinawite. *Chemical Geology*, 2017. **462**: pp. 15–29.

Yao, K.-M., M.T. Habibian, and C.R. O'Melia, Water and waste water filtration. Concepts and applications. *Environmental Science & Technology*, 1971. **5**(11): pp. 1105–1112.

Yu, H., et al., Degradation of diclofenac by advanced oxidation and reduction processes: Kinetic studies, degradation pathways and toxicity assessments. *Water Research*, 2013. **47**(5): pp. 1909–1918.

Zhang, L., F. Fu, and B. Tang, Adsorption and redox conversion behaviors of Cr(VI) on goethite/carbon microspheres and akaganeite/carbon microspheres composites. *Chemical Engineering Journal*, 2019. **356**: pp. 151–160.

Zhi, Y. and J. Liu, Surface modification of activated carbon for enhanced adsorption of perfluoroalkyl acids from aqueous solutions. *Chemosphere*, 2016. **144**: pp. 1224–1232.

Ziltek. https://ziltek.com/rembind/. [cited March 25, 2019].

Zou, Y., et al., Environmental remediation and application of nanoscale zero-valent iron and its composites for the removal of heavy metal ions: A review. *Environmental Science & Technology*, 2016. **50**(14): pp. 7290–7304.

13 Reclamation of Salt-Affected Soils

Viraj Gunarathne, J.A.I. Senadheera, Udaya Gunarathne,
Yaser A. Almaroai, and Meththika Vithanage

CONTENTS

13.1   INTRODUCTION

Since ancient times, soil degradation due to salinization is one of the significant environmental concerns, threatening the sustainability of the world's agricultural production, and it is prevalent in arid and semi-arid regions. Soil salinization is referred to as the accumulation of salts in the soil root zone to the extent that depresses plant growth (Rengasamy, 2006). Salt-affected soils are characterized by the high concentration of dissolved mineral salts which primarily composed with chlorides (Cl^-), sulfates (SO_4^{2-}), carbonates (CO_3^{2-}) and bicarbonates (HCO_3^-) of sodium, calcium, and magnesium (Manchanda and Garg, 2008). Globally, salt-affected soils are distributed across all continents, and about 100 countries all over the world face this menace. According to a recent estimate, more than 1128 million hectares of land has been affected with salinity all over the world (Wicke et al., 2011).

Salt-affected soils are generally formed due to either or both primary and secondary soil salinization processes. Primary soil salinization includes those soils which naturally have high inherent salts due to rock weathering, seawater intrusions, etc. (Rengasamy, 2006). In contrast, the secondary salinization process is human-induced and predominantly caused by prolonged irrigation with salt-rich water without adequate leaching, which involves concentrating salts in the root zone (Maggio et al., 2011). However, not all salinity problems are confined to the semi-arid regions of the world, and the salinization process may occur in the coastal zone under tropical conditions as well. The shallow and saline groundwater table in the coastal region allows upward movement of salt to surface through rapid evaporation and capillary action. Therefore, some of the potentially exploitable saline soils of the world are found in South and Southeast Asia, and about half of these are coastal saline soils (Ray et al., 2014).

Coastal soils are encountered with various abiotic stresses *viz.*, salinity, acidity, and sandy texture. Most of the coastal areas have problematic soils, such as saline, alkaline, and acid sulfate soils which are situated in low-lying areas, mainly along the estuaries. However, salinity is the main factor responsible for the reduced productivity of crops (Ray et al., 2014). Salinity in soil not only affects plant growth but also deteriorates soil health by altering the soil physical, chemical, and biological properties (Lakhdar et al., 2009; Manchanda and Garg, 2008). Low productivity of this ecosystem is attributed to the unfavorable agro-climatic conditions. In coastal zones, sodium (Na^+) is the primarily occurring soluble cation which is associated with saline soils. Clark et al. (2007) emphasize that the underdeveloped soil structure of salt-affected soils is attributed to unfavorable soil physical processes including swelling, slaking, and dispersion of clay. The study of Tejada and Gonzalez (2005) specified that the increases of soil electrical conductivity have adverse impacts on soil physical properties, including total porosity, bulk density, and structural stability. Furthermore, the salinity alters soil chemical properties such as pH, exchangeable sodium percentage (ESP), cation exchange capacity, plant available nutrients, and soil organic carbon content (Sumner, 2000).

Salinity generates adverse effects for soil microbial communities and their activity by changing the osmotic and matric potential of the soil solution (Rietz and Haynes, 2003). The raised salt concentration in saline soils affects negatively for soil microbial activity as well as soil physical and chemical properties, causing declines in plant productivity. Alteration of the composition of exchangeable ions in the soil solution induces ion-specific and osmotic effects which involve for imbalances of plant nutrients, such as deficiencies of macro- and micronutrients or elevated levels of Na^+ (Kaya et al., 2001). Further deterioration of physical and chemical properties of salt-affected soil is caused by reduced plant growth and lower carbon input into the soil due to salt toxicity and unfavorable osmotic potential (Wong et al., 2009).

13.2 ORIGIN AND CLASSIFICATION OF SALT-AFFECTED SOILS

13.2.1 FACTORS AFFECTING THE DEVELOPMENT OF SALINITY

The primary sources of salts in soils of arid and semi-arid regions are rainfall, mineral weathering, "fossil" salts, and many surface waters and groundwaters which redistribute accumulated salts, often as a result of human activities. Salinity can also result from seawater intrusion into soils in coastal areas where the water table has been declined by groundwater mining (Rengasamy, 2010).

13.2.1.1 Natural Factors

13.2.1.1.1 Precipitation (Rainfall)

The deposition of oceanic salts carried by wind and rain when droplets of water from oceanic sprays and turbulence produce atmospheric aerosols of suspended salt crystals or highly saline droplets. The salts that are thus brought to an area through precipitation have been termed "cyclic" salts (Hutton, 1958; Munns and Tester, 2008).

13.2.1.1.2 Mineral Weathering

The major concern which has attributed with the soils in arid and semi-arid regions is their relatively un-weathered nature, except for extremely ancient areas. Primary minerals constitute an excellent plant nutrient source but also a renewable source of salinity. This is because the parent rocks of primary

minerals contain salts, mainly chlorides of Na, Ca, and Mg, and to some extent, sulfates and carbonates (Rengasamy, 2002; Sharma et al., 2016).

13.2.1.1.3 Fossil Salts

Despite the substantial amounts of salt accumulated in many areas from atmospheric salt accretions and mineral weathering, the most noticeable occurrences of salt accumulation in arid regions are those involving "fossil" salts derived from prior salt deposits or connate (entrapped) solutions present in former marine sediments. Such release can occur either naturally or as a result of anthropogenic activities (Bresler et al., 1982).

13.2.1.1.4 Seawater Intrusion

Coastal soils are rich in salts, mainly due to the presence of saline groundwater table at shallow depth and frequent brackish water inundation in the low lying areas. The groundwater that is influenced by the sea and brackish water estuaries reaches the soil surface through capillary rise during dry seasons (Ray et al., 2014).

13.2.1.2 Anthropogenic Factors

13.2.1.2.1 Industrial Activities

Increased concentrations of atmospheric nitrogen and sulfur components are often found near industrial areas. These can increase the quantity of salts added annually from the atmosphere to soils of such regions. During water transportation processes such as irrigation, a human can also introduce salts to an area from deep within the earth's mantle or can impound water over heavily salinized geologic strata (Bresler et al., 1982).

13.2.1.2.2 Irrigation Activities

Furthermore, irrigation with poor quality water, shallow groundwater tables with poor drainage, and excessive evaporation than precipitation exacerbate salt accumulation in the surface horizon (Brinck and Frost, 2009; Smedema and Shiati, 2002).

Overgrazing, indiscriminate chemical fertilizer use, and deforestation are some of the other minor contributors to soil salinization (Pessarakli, 2016). Figure 13.1 summarizes the factors involved in the origination of salt-affected soils.

13.2.2 CLASSIFICATION OF SALT-AFFECTED SOIL

The general classification of salt-affected soils is based on their electrical conductivity of saturated paste extracts (EC_e), soil sodium adsorption ratio (SAR), and exchangeable sodium percentage (ESP) (Richards, 1954). Based on these properties, soils are classified as saline, sodic, or saline-sodic (Table 13.1).

$SAR_{1:5}$ sodium adsorption ratios are measured in 1:5 soil/water extract; SAR_e sodium adsorption ratio is measured in saturated extract (Data from Richards [1954] and Rengasamy [2002]).

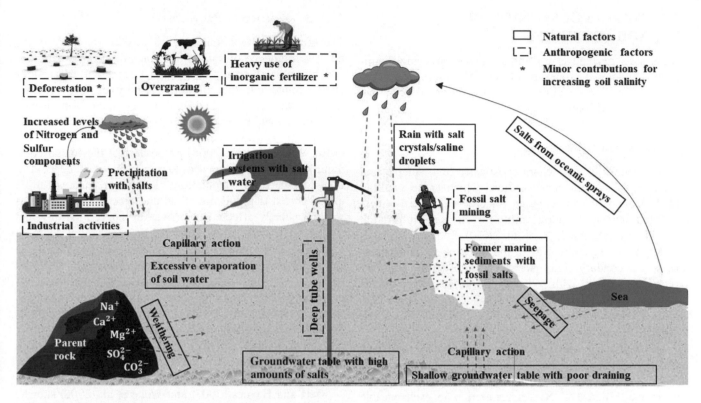

FIGURE 13.1 Factors influencing the origin of salt-affected soil.

TABLE 13.1
Chemical Characteristics and the Main Problems Associated with Salt-Affected Soils

Soil Salinity Class	EC_e (dS m^{-1})	pH	ESP (%)	SAR_e	$SAR_{1:5}$	Problems
Normal soil	<4.0	6–8	<5	<3	<5	• May have problems other than salts
Saline soil	>4.0	6–8	<15	<13	<5	• Osmotic effects
						• Possible toxicity of dominant anion or cation at high EC
Saline-sodic soil	>4.0	>8	>15	>13	>5	• Osmotic effects
						• HCO_3^- and CO_3^{2-} toxicity
						• Nutrient deficiency
						• High pH
						• Organic matter losses
						• The decline in plant growth
						• Slaking, swelling, dispersion, surface crust, and hard setting
						• Increased carbon loss
Sodic soil	<4.0	>8	>15	>13	>5	• HCO_3^- and CO_3^{2-} toxicity
						• Nutrient deficiency
						• High pH
						• Organic matter losses
						• The decline in plant growth
						• Slaking, swelling, dispersion, surface crust, and hard setting
						• Increased carbon loss
						• Seasonal waterlogging
						• Na^+ toxicity

13.3 IMPACTS OF SALINITY ON SOIL PROPERTIES

Excess salts may adversely influence the physical, chemical, and biological properties of soil. These property deteriorations have been well reported in the literature for saline, saline-sodic, and sodic soils.

13.3.1 PHYSICAL PROPERTIES

Salt-affected soils have characteristically high soluble and exchangeable sodium concentrations which cause swelling and dispersion of soils creating underdeveloped soil structure, thus limiting the permeability for soil-water infiltration (Gharaibeh et al., 2009; Yu et al., 2010). Sodicity, which is referred by an excess of Na^+ in the rhizosphere, is accountable as a secondary consequence of salinity. Clay soils are typically associated with sodicity, and it extensively affects numbers of physical properties of soil. The excessive Na^+ associated with exchangeable sites of this soil causes the deflocculation of clay particles.

Similarly, Läuchli and Epstein (1990) highlighted the occurrence of swelling and dispersion of clay particles and breaking of soil aggregates due to high concentrations of exchangeable Na^+ in clay. Increased Na^+ concentration in exchangeable sites involves generating higher repulsive forces among soil particles and widening the inter-particular spaces resulting in the breakdown of soil aggregates (Oster and Shainberg, 2001). However, the clay swelling and dispersion also result from high total electrolyte concentration (TEC) of irrigation water supply and in soil solution (Quirk, 2001; Dikinya et al., 2006).

13.3.2 CHEMICAL PROPERTIES

In addition to structural losses as a soil physical property, chemical and biological properties are also adversely influenced by high salinity and sodicity (Ganjegunte et al., 2008; Rietz and Haynes, 2003; Wong et al., 2008). Salinity changes soil chemical properties including pH, exchangeable sodium percentage, cation exchange capacity, soil organic carbon, and the osmotic and matric potential of the soil solution (Wang et al., 2014). Generally, salt-affected soils suffer from nutrient deficiencies, thus the intense application of fertilizers might be required. Macronutrients including nitrogen, potassium, and phosphorus in salt-affected soils occur at depleted concentrations. The deficiencies of micronutrients, which resulted as indirect soil complication of salinization, develop from both ion competition and alkalinization (Grattan and Grieve, 1998). Moreover, the high pH associated, especially with saline-sodic soils, also attributes with low availability of certain micronutrients including Zn, Mn, Cu, Fe, and Al (Pessarakli and Szabolcs, 1999; Lakhdar et al., 2009). Furthermore, the low carbon input into the soil due to less vegetation cover which resulted from unfavorable edaphic conditions including salt toxicity, unstable soil structure, and increased osmotic suction involve further degradation of chemical and physical properties of salt-affected soils (Wong et al., 2009).

13.3.3 BIOLOGICAL PROPERTIES

The study of Rietz and Haynes (2003) emphasized the negative influence of soil salinity on soil enzyme activities and microbial biomass. Similarly, Garcia and Hernandez (1996) found out the hindered activity of soil enzymes β-glucosidase and alkaline phosphatase, induced by excess soil salinity. Moreover, soil fungal biomass was extensively depleted in salt-affected soils (Walpola and Arunakumara, 2010). Wichern et al. (2006) proposed that the fungal communities are more exposed to increasing salt concentration than bacterial populations. Rietz and Haynes (2003) also reported that soil microbial and biochemical activities were negatively affected by irrigation-induced salinity and sodicity. The same study observed an exponential decrease in microbial biomass carbon with increasing soil EC_e and a linear decrease in biomass carbon with increasing ESP and SAR. The previous study also evaluated various biochemical enzyme activities and reported that enzyme activities were linearly decreased with increasing EC_e, ESP, and SAR. Several other studies have also been reported that the salinity and sodicity significantly decrease the soil microbial biomass and related enzyme activities (Garcia and Hernandez, 1996; Tripathi et al., 2006).

Rietz and Haynes (2003) and Wong et al. (2008) showed that the metabolic quotient (i.e., respiration per unit biomass) is increased with increasing salinity and sodicity, indicating the presence of more stressed microbial community. Similarly, Ghollarata and Raiesi (2007) also found an increase in metabolic quotient with an increment in soil salinity. Furthermore, Yuan et al. (2007) observed a shift in soil microbial communities with increasing salinity, which can be an adaptive mechanism that takes place to reduce salt stress with a lower metabolism. Table 13.2 summarizes the changes in physicochemical and biological properties of soil influenced by excessive salinity.

13.4 RECLAMATION OF SALT-AFFECTED SOIL BY AMENDMENTS

A variety of techniques are available for the reclamation of salt-affected soils (Figure 13.2). However, amendment application has been considered as one of the best methods for reclamation of salt-affected soils as it requires less technical knowledge and ease of application. Soil amendments usually modify both physical and chemical properties of affected soils, and the significant effects are on soil structure, water holding capacity, and cation exchange capacity (Walker and Bernal, 2008). The two major types of amendments used to mitigate the negative impacts of soil salinity are inorganic and organic amendments. Inorganic amendments such as vermiculite, perlite, gravel, and sand are extracted, mined, or synthesized by humans. On the other hand, organic amendments such as sphagnum peat, wood chips, grass clippings, straw, manure, compost, solid waste, wood ashes, and sludge arose from living materials, or their dead bodies, or their secretions (Bulluck Iii et al., 2002).

TABLE 13.2

Summary of Salinity Impacts on Soil Physicochemical and Biological Properties

Type of Property	Effect on Soil	References
Physical	High in both soluble sodium and exchangeable sodium, which cause soil swelling and dispersion that lead to poor structure	Gharaibeh et al. (2009)
	Limiting soil-water infiltration and permeability	Yu et al. (2010)
	High exchangeable Na+ percentage lead for swelling and dispersion of clays, as well as the breaking of soil aggregates	Läuchli and Epstein (1990)
	High Na+ concentration on the exchangeable sites of soil particles, increase repulsive forces which in turn enlarges the inter-particulate distances	Oster and Shainberg (2001)
Chemical	Soil chemical and biological properties are negatively affected by high salinity and sodicity	Ganjegunte et al. (2008)
	Affect for soil pH, cation exchange capacity (CEC), exchangeable sodium percentage (ESP), soil organic carbon, and alters the osmotic and matric potential of the soil solution	Wang et al. (2014)
	Micronutrient deficiency appears to be a side effect of salinization	Grattan and Grieve (1998)
	Salt-affected soils generally suffer from deficiencies of nitrogen, phosphorus, and potassium	Pessarakli and Szabolcs (1999)
	High pH also adversely affects the availability of micronutrients such as Fe, Al, Zn, Mn, and Cu	Lakhdar et al. (2009)
	High osmotic suction and degraded soil structure of salt-affected soils as a result of lower carbon inputs	Wong et al. (2009)
Biological	Salinity affect, both on soil microbial biomass carbon and enzyme activities	Rietz and Haynes (2003)
	Fungal part of the microbial biomass is strongly reduced in saline soils	Walpola and Arunakumara (2010)
	Fungal communities are more exposed to increasing salt concentration than bacterial populations	Wichern et al. (2006)
	Enzyme activities are linearly decreased with increasing EC$_e$, ESP, and SAR	Tripathi et al. (2006)
	Increase in metabolic quotient with an increase in soil salinity	Ghollarata and Raiesi (2007)
	A shift in soil microbial communities, and lower metabolisms takes place with increasing salinity	Yuan et al. (2007)

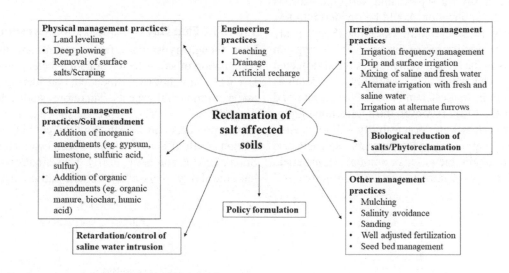

FIGURE 13.2 Reclamation strategies for salt-affected soils.

Gypsum ($CaSO_4 \cdot 2H_2O$), calcite ($CaCO_3$), calcium chloride ($CaCl_2 \cdot 2H_2O$), and organic matter (farmyard manure, green manure, organic amendment, and municipal solid waste) are some of the successful productive, low cost, and simple approaches that have been implemented worldwide to restore salt-affected soils (Sharma and Minhas, 2005; Tejada et al., 2006).

13.4.1 Inorganic Amendments

Reclamation of salt-affected soils including saline-sodic soils requires a soluble source of Ca^{2+} in order to replace high concentrations of Na+ ions associated with cation exchange sites of soil particles. Introduction of Ca^{2+} into salt-affected soil can be done in two ways, as the direct supply of soluble Ca^{2+} from a source of Ca^{2+} or solubilize the soil-associated insoluble calcium as a form of lime (Ahmad et al., 1990). In this regard, the most commonly used inorganic amendments to introduce Ca^{2+} into the salt-affected soils include gypsum, calcium chloride, sulfuric acid, and sulfur (Siyal et al., 2002). Inorganic amendments like gypsum and calcium chloride can supply Ca^{2+} into the soil, acting as a direct source. However, the amendments including sulfuric acid and sulfur are the other commonly used inorganic amendments which upon application to calcareous soil increase soil Ca^{2+} levels by dissolving native calcium carbonate pools in these soils (Sadiq et al., 2007; Vance et al., 2008).

13.4.1.1 Gypsum

Calcium-based gypsum ($CaSO_4 \cdot 2H_2O$), which is used to supply Ca^{2+} into soil, has been considered the most common, cost-effective, and effectual inorganic amendment for the reclamation of salt-affected soils (Keren and Miyamoto, 1990; Chen et al., 2015). Gypsum is cheap, easy to handle, slightly soluble in water, and generates no toxic effects for plants and the environment (Sakai et al., 2004). It involves reducing Na^+ concentration of salt-affected soil by Na^+-Ca^{2+} exchange mechanisms by inducing high ionic strength around soil colloids that enhance the physical properties of soil. Further, the dissolution of gypsum results in ionic complexes with soil SO_4^{2-} (Ahmad et al., 2006; Rengasamy and Olsson, 1991). Further, gypsum amendment enhances soil structure and porosity, improves soil-water permeability and leaching, prevents surface crusting, and reduces soil bulk density (Baumhardt et al., 1992; Greene et al., 1988; Southard et al., 1988; Gal et al., 1984; Lebron et al., 2002).

The required quantity of gypsum to mitigate the negative impact of salt-affected soils varies with the soil texture and exchangeable sodium percentage of soil (Figure 13.3). If the required gypsum quantity for a particular soil type exceeds five tons per acre, the application should be divided into few parts and applied with several time intervals (Siyal et al., 2002). Moreover, the application of fine particles of gypsum shows better results in the reclamation process of salt-affected soils since fine particles increase the solubility of gypsum as it is partially soluble in water (Chun et al., 2001; Kumar and Singh, 2003; Yu et al., 2010). However, the application of gypsum into the soil with less water permeability will reduce the reclamation ability of gypsum as impermeable soil is cause to remove Na^+ with negligible or fewer amounts (Ilyas et al., 1997). Therefore, the use of implementations to improve soil

permeability is necessary to take the maximum reclamation efficiency of gypsum as high permeability increase vertical leaching and removal of Na^+ ions from the root zone.

13.4.1.2 Sulfuric Acid

The potential of sulfuric acid for reclamation of calcareous salt-affected soil has long been studied (Ahmad et al., 2014; Miyamoto et al., 1975). Sulfuric acid supplies Ca^{2+} to the soil by an indirect way as it dissolves native insoluble lime deposits of calcareous salt-affected soils. Some previous studies revealed that the treatment with sulfuric acid gave better reclamation efficiencies than gypsum for calcareous saline-sodic soils (Amezketa et al., 2005; Ghafoor et al., 1986). Bower (1951) revealed the higher efficiency of sulfuric acid to decrease soil pH, particularly for soil with higher exchangeable sodium percentage than that of gypsum and sulfur.

Moreover, Vadyanina and Roi (1974) pointed out that the sulfuric acid has more effect on the mitigation of exchangeable sodium percentage than the application of gypsum. Water infiltration abilities of sodic soil also enhance with the treatment of sulfuric acid (Hussain et al., 2001). However, the study of Ahmad et al. (2013) indicated the better performance of gypsum than sulfuric acid when considering the biomass production of the crop since sulfuric acid induces toxicity for plants.

13.4.1.3 Flue Gas Desulfurization Gypsum

Even though gypsum's effectiveness as an amendment for reclamation of salt-affected soils has been proven, the extensive use of it is not recommended, considering the economic and environmental aspects. Therefore, studies to find out the capability of industrial by-products such as flue gas desulfurization products and fly ash from the coal industry for reclamation of salt-affected soils have been conducted (Kumar and Singh, 2003; Chun et al., 2001). These industrial by-products have the ability to dissolve insoluble native calcium carbonate

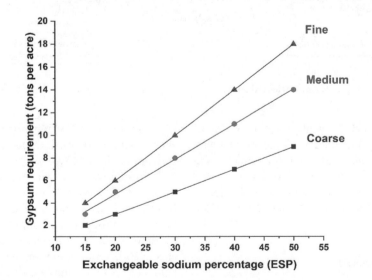

FIGURE 13.3 The requirement of gypsum for reclamation of salt-affected soils with different soil textures. (Data from Siyal, A. et al., *Pak. J. Appl. Sci.*, 2, 537–540, 2002.)

present in salt-affected soils which lead to replace exchangeable Na^+ with released Ca^{2+} ions and improve the physical properties of soil (Yu et al., 2010).

A majority of coal power plants are fitted with wet flue gas desulfurization scrubbers to maintain the quality of the flue gas that generates large quantities of flue gas desulfurization gypsum (Yonggan et al., 2017). Flue gas desulfurization gypsum is a fine particulate compound with a high percentage of gypsum, ranging from 70% to 90% (Yu et al., 2014). Coal power plants in China alone produced approximately 40 million tons of waste flue gas desulfurization gypsum in the year 2010 (Baligar et al., 2011). Studies found that it has a similar effect as pure gypsum for the reclamation of salt-affected soils and it is involved in improvement of the soil structure, enhanced soil-water infiltration, and reduced level of soil erosion (Amezketa et al., 2005; Rhoton and McChesney, 2011). Its powder-like nature particularly increases the dissolution with water releasing Ca^{2+} at high rates that accelerate the reclamation process of salt-affected soils.

Moreover, the nontoxic nature and applicability of flue gas desulfurization gypsum as an amendment for agricultural lands have been proven (Clark et al., 2001). Therefore, it is widely used to ameliorate salt-affected soils in countries such as China, which extensively generate electricity through coal power plants. The recent studies indicated that practically 120 km^2 of salt-affected land area in China had been amended by flue gas desulfurization gypsum (Yu et al., 2014; Wang et al., 2017).

Table 13.3 explains the changes in edaphic properties of salt-affected soils after treatment with different types of inorganic amendments.

13.4.2 ORGANIC AMENDMENTS

Over a long period of soil salinization, carbon storage decreases at a significant rate (Wong et al., 2010). Application of organic matter such as green manures, compost, and sewage sludge can both ameliorate and increase the carbon stocks and fertility of saline soils. As organic matter helps binding soil particles into aggregates, it is the most sustainable and persistent method used to ameliorate salt-affected soils (Nelson and Oades, 1998). De-protonation of humic and fulvic acids leads to the formation of large organic polyanions. These ions can bind clay particles into micro-aggregates by forming $[(A-B-C)_x]_y$ complexes (i.e., A, B, and C are clay particles), polyvalent cations, and organic matter, respectively (Tisdall and Oades, 1982). For instance, the addition of an organic amendment to a salt-affected soil can alter its chemical properties, enhancing both CEC and the soluble and exchangeable K^+. Potassium ions compete with Na^+ in terms of being adsorbed in saline soils and will limit the entry of Na^+ onto the exchange sites (Walker and Bernal, 2008). Chorom and Rengasamy (1997) added green manure to alkaline sodic soil and observed that decomposition and microbial respiration decreased pH and increased the solubility of $CaCO_3$ due to an increase in the partial pressure of CO_2.

Similarly, Liang et al. (2003) reported significant stimulation of soil respiration and enzymes such as urease and alkaline phosphatase activity after incorporation of organic manure to a soil derived from alluvial and marine deposits with 3.3 g kg^{-1} total salts. Moreover, Barzegar et al. (1997) observed that the dispersible clay decreases after the addition of pea straw irrespective of sodium adsorption ratio (SAR). Furthermore, the application of organic amendments is considered an effective management strategy for enhancement of salt-affected soils for improved plant growth. Therefore, organic amendments provide both soil nutrients and organic matter for lasting improvements in soil fertility. Soil salinity effects on the organic matter turnover and mineralization of C and N depend on the type of organic material incorporated (Walpola and Arunakumara, 2010). For example, the combined application of rice straw and pig manure affected enzymatic and microbial activity more significantly than applying these amendments alone (Liang et al., 2005). Therefore, the application of organic materials could provide an effective approach to mitigate toxic saline conditions. Hereafter, this chapter will discuss extensively the effectiveness of three potential organic amendments, namely biochar, solid waste compost, and municipal sewage sludge, on reclamation of salt-affected soils.

13.4.2.1 Biochar

The properties of biochar are more convenient to explain when dividing them into two separate fractions, namely carbon fraction (organic) and ash fraction (inorganic). The carbon fraction that consists of oxygen, hydrogen, and other elements is the most affected fraction by a biochar manufacturing process called pyrolysis. Pyrolysis conditions including temperature, heating time, and heating rate directly affect for the conversion of biomass consisting of organic carbohydrate compounds into char, which is a condensed aromatic structure of carbon. Conversely, the properties of feedstock ultimately influence the characteristics of inorganic ash fraction as mineral elements in feedstock biomass are accumulated in the ash fraction. However, pyrolysis conditions also affect the properties of ash and ash to carbon ratio of resulting char (Brewer, 2012).

Pyrolysis with low temperature produces an excellent yield of biochar and high volatile-matter composition than the high temperature pyrolysis. The yields of biochar and volatile-matter contents gradually reduced with the increasing pyrolysis temperature. High temperature pyrolysis involves producing biochars with large surface area, high carbon content, and high adsorption properties.

Furthermore, the feedstock characteristics also affect the yield of biochar and content of volatile matter. Among different types of biochar, the woody biochar exhibited a drastic change of volatile-matter content with increasing pyrolysis temperature from 400 to 800 than biochar from non-woody feedstocks. The high content of volatile matter in woody biochar resulted at low temperatures attributes with the high lignin content of woody feedstocks. Lignin is partially resistant

TABLE 13.3

Effect of Different Types of Inorganic Amendments on Reclamation of Salt-Affected Soil

Inorganic Amendment Used	Soil Type	Other Conditions Applied	Effect on Salt-Affected Soil	References
Gypsum	Calcareous saline-sodic soil	Crop rotation with wheat (*Triticum aestivm* L.) and rice (*Oryza sativa* L.)	Involved for removal of root zone Na$^+$ in greater concentrations than the control. Resulted in a minimal level of SAR and highest EC$_e$: SAR ratio	Ahmad et al. (2006)
Gypsum at 50% of gypsum required	Sodic soil	1% farmyard manure	Decreased ESP and significant increment of crop yield than other treatments	Dubey and Mondal (1993)
Gypsum (12.5 t ha^{-1})	Sodic soil	Farmyard manure (25 t ha^{-1})	Significantly enhanced pH, EC$_e$, ESP, available water capacity, infiltration rate, and osmotic potential	Makoi and Ndakidemi (2007)
Gypsum (25 mg ha^{-1})	Saline-sodic soil	Cropping with alfalfa	Significantly reduced SAR, pH, EC$_e$, and Cl$^-$. With alfalfa, the same parameters were improved up to 80 cm below the topsoil	Ilyas et al. (1997)
Gypsum	Saline sandy loam soil	—	Significant improvement of SAR, ESP, EC$_e$, exchangeable Na$^+$, and available water capacity	Makoi and Verplancke (2010)
Gypsum	Saline-sodic soil	—	No significant effect for microbial respiration	Yazdanpanah et al. (2013)
Sulfuric acid	Saline-sodic soil	—	Sulfuric acid is highly effective for the prevention of crusting than other inorganic amendments studied	Amezketa et al. (2005)
Sulfuric acid	Saline-sodic soil	—	Positive influence observed for the availability of K and P but no significant effect observed on microbial respiration	Yazdanpanah et al. (2013)
Sulfuric acid (95%) at 50% soil gypsum requirement	Saline-sodic soil	*Sesbania* sp. crop rotation	Significantly facilitated the leaching of Na$^+$ and other ions underwater with SAR $\geq$ 40	Ahmad et al. (2013)
Flue gas desulfurization gypsum	Saline-alkali soil	A five cm thick layer of straw at a soil depth of 30 cm	EC$_e$ of soil increased with the rate of gypsum application and the concentration of Na$^+$, Cl$^-$, and HCO$_3^-$ were reduced above the soil depth of 55 cm than control. The pH and ESP became reduced significantly, and further decrease occurred at an increasing rate of gypsum application	Yonggan et al. (2017)
Flue gas desulfurization gypsum (24.3 t ha^{-1})	Surface sodic soil	Four times of water leaching	Significantly reduced the pH and ESP	Wang et al. (2004)
Flue gas desulfurization gypsum	Sodic soil	Leaching with water	Increased soil porosity and reduced bulk density	Yu et al. (2014)
Flue gas desulfurization gypsum (23.1 Mg ha^{-1})	Salt-affected soil in the semiarid region	—	Significantly increased soluble and exchangeable Ca^{2+} and SO$_4^{2-}$ in soil, reduced soluble and exchangeable Na$^+$ and CO$_3^{2-}$. Significant increase in EC$_e$	Chun et al. (2007)
Polymeric aluminum ferric sulfate	Saline-sodic soils	—	The pH, ESP, EC$_e$, bulk density, soil CaCO$_3$, and silt-plus-clay contents were decreased while saturated hydraulic conductivity was increased	Luo et al. (2015)

Abbreviations: SAR = Sodium adsorption ratio, ECe = Electric conductivity, and ESP = Exchangeable sodium percentage.

to pyrolytic decomposition at relatively low temperatures like 400°C but readily decomposes at high temperatures such as 950°C (Jindo et al., 2014).

13.4.2.1.1 Physical Properties of Biochar

The porosity and surface area contribute to the most critical physical properties of biochar in order to improve soil characteristics such as soil adsorption capacity and water retention ability (Kalderis et al., 2014). Generally, the particle sizes of biochars produced at lower heating rates are similar to the particle sizes of the feedstock before pyrolysis. As the volatile matter is slowly removed during the pyrolysis process, the char becomes more porous but still holds its overall shape and size. The fine particles generated during pyrolysis, such as those one would find at the bottom of charcoal kilns, are the result of the partial feedstock combustion (high-ash chars) and the generation of dust from rubbing of now-friable char particles together. The rapid escape of volatiles at higher heating rates is believed to play an additional role in fine particles generation as fracturing of large particles (explode) that result from generated internal pressure (Brewer, 2012).

Usually, the particle density of biochar occurs between 1.5 to 1.7 g cm^{-3} and increases with the production temperature as the solid carbon condenses into dense structures containing aromatic rings. Further, the particle density increases with increasing mineral ash content and can exceed 2.0 g cm^{-3} for high-ash chars. The pores of biochar can be categorized into three groups considering the size of them. Biochar pores are divided into micropores, mesopores, and macropores, based on internal diameters of <2 nm, 2–50 nm, and >200 nm, respectively. However, soil scientists may use a different system of classification as pores with diameters <200 nm categorizing into micropores. The pores in each size range contribute to different properties of biochar. The surface area is another important physical property of biochar which can influence the magnitude of interactions in the soil environment. Biochar can involve for more chemical interactions in the soil when the surface area of it is high. Depending on the feedstock type, pyrolysis process, and pyrolyzing temperatures, some biochars can have surface areas in the hundreds, and even thousands of meters square per gram, potentially making them suitable for activated carbon applications (Brewer, 2012).

13.4.2.1.2 Chemical Properties of Biochar

The majority of chemical properties of biochar is related to two "carbon fraction" concepts, namely aromaticity and surface functionality. Aromaticity is defined as the fraction in biochar that participates for aromatic structures. Lignocellulosic feedstocks that have sugar polymers (all aliphatic carbons) and lignin (some aromatic rings) have relatively low aromaticity. As the pyrolysis reaction progresses, oxygen and hydrogen are removed, leaving the remaining carbons to form new aromatic carbon-carbon bonds.

The degree of aromatic condensation in biochar is believed to be related to recalcitrance in the environment since carbons in dense, aromatic structures are more resistant to oxidation and few microorganisms have enzymes capable of breaking down such bonds. This high stability of biochar comes from the fact that electrons are shared over more than one bond in aromatic molecules. Aromatic molecules can exist at lower energy states than non-aromatic molecules by "spreading out" electrons over the molecule. Such sharing of electrons is so efficient in graphite and some highly condensed chars that these materials can even conduct electricity (Brewer, 2012). Moreover, the pH of biochar increases with temperature, probably as a consequence of the relative concentration of non-pyrolyzed inorganic elements, already present in the original feedstocks (Novak et al., 2009a). Further, the analytical elements and both H/C and O/C ratios are considered as useful indicators for characterization of biochar (Nguyen and Lehmann, 2009). Increases of pyrolysis temperature result in higher losses of oxygen and hydrogen from feedstock materials compared to that of carbon. Additionally, the dehydrogenation of CH$_3$ as a result of thermal induction indicates a change in the biochar recalcitrance (Harvey et al., 2012).

13.4.2.1.3 Biochar as an Organic Amendment for Salt-Affected Soil

In recent years, research interest in biochar production and application has shown a dramatic increase. Biochar is highly recalcitrant due to its high aromaticity and can sequestrate carbon for very long periods (Fang et al., 2014; Kuzyakov et al., 2014). Therefore, biochar addition can be a potential pathway in order to improve and enhance native soil organic carbon and black carbon stabilization in cropping systems (Zhang et al., 2015). There is higher carbon accumulation in large macro-aggregates, suggesting that biochar-derived carbon can be physically protected within these aggregate sizes. Many studies have been undertaken to find out the effect of biochar application on soil organic carbon decomposition to discover solutions to enhance soil carbon sequestration (Jones et al., 2011; Luo et al., 2011). However, previous studies have shown that biochar may both suppress and stimulate native soil organic carbon decomposition (Liang et al., 2010; Luo et al., 2011). Differences in the nature of soil and biochar and incubation conditions used in different studies have led to these inconsistent results (Jones et al., 2011).

Moreover, biochar has received great interest in the remediation of organic and inorganic contaminants due to its high adsorption capacity (Ahmad et al., 2014; Mohan et al., 2014). The adsorption capacity of biochar is determined by the feedstock and production conditions (Paz-Ferreiro et al., 2014; Uchimiya et al., 2011). The high adsorption capacities of biochar are attributed to its characteristically high cation exchange capacity and surface area. Biochar exhibits an alkaline pH when the pyrolysis temperature is over 400°C (Gaskin et al., 2008; Novak et al., 2009b). However, the biochar usually has basic properties thus, can increase the pH of acid soils and decrease soil exchangeable Al^{3+} (Chan et al., 2008; Novak et al., 2009a). Furthermore, Table 13.4 shows numbers of positive effects generated by different types of biochar on reclamation of salt-affected soils.

TABLE 13.4

Summary of Biochar as an Organic Amendment for Salt-Affected Soil

Production Condition and Specifications of Biochar	Effect on Salt-Affected Soil	References
Not specified	Apart from being a source of carbon, biochar as an organic soil amendment has been shown to alter and improve the physical, chemical, and biological properties of soils	Rondon et al. (2007), Asai et al. (2009)
Not specified	Indirect mechanisms are involved in improving soil physical, chemical, and biological characteristics	Sohi et al. (2010), Xu et al. (2012)
Not specified	Significantly promote soil biological activity after biochar addition to soils, although this effect could be temporary	Lehmann et al. (2011)
Not specified	The availability of Ca^{2+} and Mg^{2+} increased after the addition of biochar	Major et al. (2010)
Not specified	The direct growth promotion under biochar amendment due to supplying of mineral nutrients (i.e., Ca, Mg, P, K, and S etc.), to the plants	Enders et al. (2012)
American beech (*Fagus grandifolia*) (SWM, MT)	Strong sorption of NaCl by biochar was indicated by EC measurements of weathered biochar collected from soil	Thomas et al. (2013b)
American beech (*Fagus grandifolia*) (SWM, MT)	The main mechanism for salt stress mitigation was sorption of NaCl resulting in reduced exposure	Thomas et al., (2013b)
American beech (*Fagus grandifolia*) (SWM, MT)	Addition of biochar derived from lingo-cellulosic material to soils can mitigate, or even eliminate, negative effects of salt on plant performance	Thomas et al. (2013b)
Not specified	Ameliorate salinity stress by reducing Na^+ uptake in plants	Lashari et al. (2013), Lashari et al. (2015), Thomas et al. (2013a)
Not specified	The enrichment of the exchange complex in Ca^{2+} and Mg^{2+} ions could decrease the proportion of Na^+ in the exchange complex and improves soil physical properties	Walker and Bernal (2008)
(250°C–400°C) (SWM, LT)	Biochars produced at lower temperatures have higher yield recoveries and contain more C=O and C-H functional groups that can serve as nutrient exchange sites after oxidation	Glaser et al. (2002)
(400°C–700°C) (SWM, HT)	Biochar produced at high temperatures have fewer ion-exchange functional groups due to dehydration and decarboxylation that potentially limit its usefulness in retaining soil nutrients	Glaser et al. (2002), Baldock and Smernik (2002), Hammes et al. (2006)
Peanut hulls (*Archis hypogaea*) (400°C) (NWM, LT)	Biochar produced at lower pyrolysis temperature significantly increased the water retention	Novak et al. (2009b)
Switchgrass (*Panicum virgatum*) (250°C) (NWM, LT)	Biochar produced at the lower pyrolysis temperature significantly involved for increased water retention	Novak et al. (2009b)
Pecan shells (*Carya illinoensis*) (350°C) (NWM, HT)	Significantly improve the fertility of a sandy coastal plain soil	Novak et al. (2009a)
Hardwood materials 400°C (HW, MT)	Biochar has an alkaline pH when the pyrolysis temperature is over 400°C	Gaskin et al. (2008), Novak et al. (2009b)

Abbreviations: HT, High temperature (T ≥ 500°C); MT, Moderate temperature (300°C–500°C); LT-Low temperature (T ≤ 300°C); HWM, Hard wood materials; SWM, Soft wood materials; NWM, None wood materials.

13.4.2.2 Solid Waste Compost

Compost is defined as aerobically decomposed, stable organic matter and result from a controlled decomposition process called "composting." Composting is a biological process occurring in favorable conditions and allowed to transform the initial raw material into a stable and mature end product (Amir et al., 2005). Chemical, physical, and biological properties of compost are varied with raw materials used and production method.

13.4.2.2.1 Properties of Solid Waste Compost

Nutrient contents of compost are often moderate but vary among sources due to differences in characteristics of waste materials and processing methods (Smith and Das, 1998).

The decomposition of organic matter gradually releases plant-available phosphorus (Ayaga et al., 2006). Compost contains magnesium and calcium which serve as bases when they exist as carbonates, oxides, and hydroxides when applied to the soil, and may counteract soil acidification and vary pH levels making soil nutrients more available to plants (Fricke and Vogtmann, 1994).

As a consequence of increased Ca^{2+} concentration in soil solution, Na^+ and Ca^{2+} exchange at the cation exchange sites of soil, resulting in leaching of exchanged Na^+ with percolating water causing a subsequent reduction in soil sodicity (Qadir and Oster, 2004). However, humic substances, which constitute a significant part of the organic matter of compost, can reduce metal solubility by forming stable metal chelates (Ross, 1994). Moreover, other edaphic properties such as mineral fractions (salt content, pH) and cation exchange capacity, as well as changes in the redox conditions of the soil also contribute for the metal availability of soil other than the effect from organic matters (humification) (Walker et al., 2004).

13.4.2.2.2 Compost as a Soil Amendment

Compost has a distinctive capability to enhance the chemical, physical, and biological properties of soils. It improves the soil structure and water retention by increasing the stability of soil aggregates (Cooperband, 2002). Solid waste compost is a humus-like, biologically stable substance that can be used for the amelioration of salt-affected soils (Wang et al., 2014). Moreover, it is widely used for agriculture and horticulture, especially for sandy soils because it improves soil quality and crop productivity by providing organic matters and macro- and micronutrients to the soil.

The beneficial effects of composts on affected soil properties depend on soil texture and moisture conditions, as well as on the origin of organic matter (De Leon-Gonzalez et al., 2000; Drozd, 2003). In saline soil, Na^+ constitutes a highly dispersive agent which directly affects the breakup of aggregates (Pernes-Debuyser and Tessier, 2004; Bronick and Lal, 2005). Exchangeable Na^+ in the soil solution and at the exchangeable sites contributes to repulsive charges that disperse clays particles. The application of organic matter in salt-affected soil promotes flocculation of clay minerals, which is an essential condition for the aggregation of soil particles and plays an important role in the control of erosion. The added organic matters aid to glue the tiny soil particles together into larger water-stable aggregates, increasing bio-pores spaces which increase soil air circulation necessary for the growth of plants and microorganisms (Rasool et al., 2007).

It is well known that the salinity in soil limits soil fertility, and most of the salt-affected soils contain depleted concentrations of nitrogen, potassium, and phosphorus. Addition of compost into such soils involves improving the rhizosphere, providing macro- and micronutrient elements, and counteracts with nutrient depletion (Lakhdar et al., 2008).

Furthermore, compost involves buffering the pH of saline and alkaline soils (Özenc and Caliskan, 2000). Soil pH often decreases with compost amended due to the effects of nitrification (Harrison et al., 1994). Conversely, compost has the potential to raise the pH of acidic soils or mitigate the acidification process of soils receiving N fertilizer (Walker et al., 2004).

13.4.2.3 Municipal Sewage Sludge

Sewage sludge is a residual waste product which results from sewage treatment of wastewater treatment facilities. These facilities are involved in treating high amounts of wastewater every day in order to eradicate bacteria, viruses, and other pollutants. Treated water is the main output of these facilities while the sewage sludge is the primary waste product (Bonfiglioli et al., 2014).

13.4.2.3.1 Properties of Sewage Sludge

Sewage sludge, which constitutes approximately 1% of wastewater, is treated by anaerobic digestion during the wastewater treatment process and is subjected to dehydrate. The dehydration step done by mechanical drying process involves reducing the amount of sewage sludge by 80%. The use of this sewage sludge as an organic amendment for agricultural lands has been considered for a long time because of its high fertilizer value. However, the nutrient content is strongly variable depending on the origin of sewage sludge (Bonfiglioli et al., 2014).

Sewage sludge is rich with organic matter and plant nutrients, mainly of nitrogen, potassium, and phosphorus which are considered essential plant nutrients and can substitute a large part of plant nutrient demand (Shafieepour et al., 2011). However, sewage sludge might contain toxic trace metals and pathogens; thus, the usage of it for agricultural lands is risky. Therefore, there is a need for eco-toxicological evaluation before its application to saline soils. In this aspect, it is well known that urban sludge is usually non-toxic because it originates from suspended matters of domestic wastewater and night soils, whereas sludge from industrial zones may be toxic (Natal-da-Luz et al., 2009). Referring back to the above-mentioned beneficial effects of sludge as a soil amendment, various studies addressed this approach and showed clearly the potential positive effects. However, positive effects may vary depending on the properties and characteristics of the added sludge. Tejada et al. (2006) reported that the soil microbial biomass and some enzymatic activities (e.g., alkaline and acid phosphatase, urease, and β-glucosidase, which are linked to carbon, nitrogen, phosphorus, and sulfur cycles) increased upon the addition of sludge.

13.4.2.3.2 Sludge as a Soil Amendment

Organic matter that is added to soil in the form of sewage sludge improves several soil properties, including bulk density, porosity, water-holding capacity, and cation exchange capacity (Sree Ramulu, 2001). Increases of cation exchange capacity of soils result in additional binding sites that may retain more essential plant nutrients in the rooting zone (Soon, 1981). Moreover, an increase in soil pH has been reported to occur in soils in which municipal sewage sludge was applied (Tsadilas et al., 1995). The chemical properties of sludge–soil mixtures do not only depend on the characteristics of soil or sludge or sludge application rates but also on soil pH and different interactions among these components (Parkpain et al., 1998).

The macronutrients present in sewage sludge serve as an excellent source of plant nutrients, and organic constituents impart beneficially for soil conditioning properties (Logan and Harrison, 1995; Singh and Agrawal, 2008). Furthermore, the digested or secondary sludge has higher nutrient content than primary sludge since most of the volatile organic matters were escaped as CH_4 or CO_2 during the digestion process (Hue and Ranjith, 1994).

Concerning the impact of amending sewage sludge to soils on plant growth and development, Morera et al. (2002) reported an increase in the dry weight of sunflower plants (*Helianthus annus* L.) grown in sludge-amended soil. Moreover, the yield of maize and barley was reported to be enhanced as a result of sludge application (Hernández et al., 1991). Conversely, sewage sludge has the potential to raise the pH of acidic soils or mitigate the acidification process of soils receiving nitrogen fertilizers (Walker et al., 2004). Koenig et al. (1998) reported that incorporation of sludge produced yields similar to those obtained with inorganic fertilizer. The same study found that the sludge application to grass hay gave yields that were intermediate between nitrogen fertilizer applied at 16.7 kg and an untreated (unfertilized) control. With grass hay, sludge addition resulted in significantly higher levels of calcium, magnesium, phosphorus, iron, copper, manganese, and zinc compared to ammonium nitrate fertilizer. The study of Togay et al. (2008) found that the dry grain yield of beans (*Phaseolus vulgaris* L.) increased significantly following the applications of sludge. In this study, the increase of dry bean yield is believed to be a result of minerals enrichment of saline fine-textured soils by the application of sludge. Furthermore, Samara (2009) reported a positive impact on the growth and development of Egyptian clover, followed by the addition of stabilized sewage sludge to saline soil.

13.5 CONCLUSIONS

Salinization of soil is one of the major global concerns, especially in arid and semi-arid regions, and thus, it is important to take reclamation priorities. Among the various methods available, application of soil amendments is one of the most effective methods which leads to improving soil physical, chemical, as well as biological properties. Gypsum is a well-known inorganic amendment used extensively for salt-affected soils, and it has proven its effectiveness on mitigation of salt-induced problems for a long period of time. However, the use of gypsum for a vast area of salt-affected soil is not an environmentally or economically viable option. Therefore, studies have been directed to find out low-cost but effective amendments for reclamation of salt-affected soils and revealed the capability of some of the by-products from leading industries such as flue gas desulfurization gypsum from wet flue gas desulfurization scrubbers of coal power plants. On the other hand, organic amendments such as compost, municipal sewage sludge, and biochar can be used not only as a cost-effective method for reclamation of salt-affected soils but also to ameliorate and increase the carbon stocks

and fertility of such soils. Although the suitability of biochar produced from the disposal garbage and sewage sludge for the reclamation of saline soil has been extensively addressed and the mechanisms have been well understood, there are some drawbacks and limitations. Therefore, further attention should be paid to fill the research gaps in the following areas before implementing large scale programs for remediation of salt-affected soils.

1. More detailed studies should be conducted under the field conditions in order to evaluate how soil amendments are effective in natural conditions and to monitor the long-term effects of amendments on treated salt-affected soil.
2. Further research studies are necessary to investigate the effects of biochar, compost, and sludge for alleviating the stress on soil properties, including flora, fauna, and soil microbial communities.
3. Studies should be conducted to understand the efficiency of biochar produced using different feedstock types for reclamation of different types of salt-affected soils and to find out an efficient way to remove heavy metals from the biochar produced using municipal solid wastes.
4. Synergistic effect of different amendments or different reclamation strategies should be investigated rather than the application of a single amendment alone.

REFERENCES

Ahmad, M., Rajapaksha, A. U., Lim, J. E., Zhang, M., Bolan, N., Mohan, D., Vithanage, M., Lee, S. S. and Ok, Y. S. 2014. Biochar as a sorbent for contaminant management in soil and water: A review. *Chemosphere* 99, 19–33. doi:https://doi.org/10.1016/j.chemosphere.2013.10.071.

Ahmad, N., Qureshi, R. and Qadir, M. 1990. Amelioration of a calcareous saline-sodic soil by gypsum and forage plants. *Land Degradation & Development* 2, 277–284. doi:https://doi.org/10.1002/ldr.3400020404.

Ahmad, S., Ghafoor, A., Akhtar, M. and Khan, M. 2013. Ionic displacement and reclamation of saline-sodic soils using chemical amendments and crop rotation. *Land Degradation & Development* 24, 170–178. doi:https://doi.org/10.1002/ldr.1117.

Ahmad, S., Ghafoor, A., Qadir, M. and Aziz, M. A. 2006. Amelioration of a calcareous saline-sodic soil by gypsum application and different crop rotations. *International Journal of Agriculture and Biology* 8, 142–146.

Amezketa, E., Aragüés, R. and Gazol, R. 2005. Efficiency of sulfuric acid, mined gypsum, and two gypsum by-products in soil crusting prevention and sodic soil reclamation. *Agronomy Journal* 97, 983–989. doi:https://doi.org/10.2134/agronj2004.0236.

Amir, S., Hafidi, M., Merlina, G., Hamdi, H. and Revel, J.-C. 2005. Fate of polycyclic aromatic hydrocarbons during composting of lagooning sewage sludge. *Chemosphere* 58, 449–458. doi:https://doi.org/10.1016/j.chemosphere.2004.09.039.

Asai, H., Samson, B. K., Stephan, H. M., Songyikhangsuthor, K., Homma, K., Kiyono, Y., Inoue, Y., Shiraiwa, T. and Horie, T. 2009. Biochar amendment techniques for upland rice production in Northern Laos: 1. Soil physical properties, leaf SPAD and grain yield. *Field Crops Research* 111, 81–84. doi:https://doi.org/10.1016/j.fcr.2008.10.008.

Ayaga, G., Todd, A. and Brookes, P. 2006. Enhanced biological cycling of phosphorus increases its availability to crops in low-input sub-Saharan farming systems. *Soil Biology and Biochemistry* 38, 81–90. doi:https://doi.org/10.1016/j.soilbio.2005.04.019.

Baldock, J. A. and Smernik, R. J. 2002. Chemical composition and bioavailability of thermally altered *Pinus resinosa* (Red pine) wood. *Organic Geochemistry* 33, 1093–1109. doi:https://doi.org/10.1016/S0146-6380(02)00062-1.

Baligar, V. C., Clark, R. B., Korcak, R. F. and Wright, R. J. 2011. Flue gas desulfurization product use on agricultural land. In: Donald L. S. (Ed.) *Advances in agronomy*. Burlington: Academic Press, 111, pp. 51–86. doi:https://doi.org/10.1016/B978-0-12-387689-8.00005-9.

Barzegar, A. R., Nelson, P. N., Oades, J. M. and Rengasamy, P. 1997. Organic matter, sodicity, and clay type: Influence on soil aggregation. *Soil Science Society of America Journal* 61, 1131–1137. doi:https://doi.org/10.2136/sssaj1997.03615995006100040020x.

Baumhardt, R., Wendt, C. and Moore, J. 1992. Infiltration in response to water quality, tillage, and gypsum. *Soil Science Society of America Journal* 56, 261–266. doi:https://doi.org/10.2136/sssaj1992.03615995005600010040x.

Bonfiglioli, L., Bianchini, A., Pellegrini, M. and Saccani, C. 2014. Sewage sludge: Characteristics and recovery options. doi:https://doi.org/10.6092/unibo/amsacta/4027.

Bower, C. A. 1951. The Improvement of an alkali soil by treatment with manure and chemical amendments: Owyhee Irrigation Project, Oregon.

Bresler, E., Mcneal, B. L. and Carter, D. L. 1982. Saline and sodic soils: Principles-dynamics-modeling. *In:* Yaron, B. (Ed.) *Advanced Series in Agricultural Sciences*, 1 ed. Springer Science & Business Media, Berlin, Germany.

Brewer, C. E. 2012. Biochar characterization and engineering. Graduate Theses and Dissertations. 12284. doi:https://doi.org/10.31274/etd-180810-2233.

Brinck, E. and Frost, C. 2009. Evaluation of amendments used to prevent sodification of irrigated fields. *Applied Geochemistry* 24, 2113–2122. doi:https://doi.org/10.1016/j.apgeochem.2009.09.001.

Bronick, C. J. and Lal, R. 2005. Soil structure and management: A review. *Geoderma* 124, 3–22. doi:https://doi.org/10.1016/j.geoderma.2004.03.005.

Bulluck Iii, L. R., Brosius, M., Evanylo, G. K. and Ristaino, J. B. 2002. Organic and synthetic fertility amendments influence soil microbial, physical and chemical properties on organic and conventional farms. *Applied Soil Ecology* 19, 147–160. doi:https://doi.org/10.1016/S0929-1393(01)00187-1.

Chan, K., Van Zwieten, L., Meszaros, I., Downie, A. and Joseph, S. 2008. Using poultry litter biochars as soil amendments. *Soil Research* 46, 437–444. doi:https://doi.org/10.1071/SR08036.

Chen, Q., Wang, S., Li, Y., Zhang, N., Zhao, B., Zhuo, Y. and Chen, C. 2015. Influence of flue gas desulfurization gypsum amendments on heavy metal distribution in reclaimed sodic soils. *Environmental Engineering Science* 32, 470–478. doi:https://doi.org/10.1089/ees.2014.0129.

Chorom, M. and Rengasamy, P. 1997. Carbonate chemistry, pH, and physical properties of an alkaline sodic soil as affected by various amendments. *Soil Research* 35, 149–162. doi:https://doi.org/10.1071/S96034.

Chun, S., Nishiyama, M. and Matsmoto, S. 2007. Response of corn growth in salt-affected soils of Northeast China to flue-gas desulfurization by-product. *Communications in Soil Science and Plant Analysis* 38, 813–825. doi:https://doi.org/10.1080/00103620701220833.

Chun, S., Nishiyama, M. and Matsumoto, S. 2001. Sodic soils reclaimed with by-product from flue gas desulfurization: Corn production and soil quality. *Environmental Pollution* 114, 453–459. doi:https://doi.org/10.1016/S0269-7491(00)00226-8.

Clark, G., Dodgshun, N., Sale, P. and Tang, C. 2007. Changes in chemical and biological properties of a sodic clay subsoil with addition of organic amendments. *Soil Biology and Biochemistry* 39, 2806–2817. doi:https://doi.org/10.1016/j.soilbio.2007.06.003.

Clark, R., Ritchey, K. and Baligar, V. 2001. Benefits and constraints for use of FGD products on agricultural land. *Fuel* 80, 821–828. doi:https://doi.org/10.1016/S0016-2361(00)00162-9.

Cooperband, L. 2002. Building soil organic matter with organic amendments. Centre for Integrated Agricultural Systems, College of Agricultural and Life Sciences, University of Wisconsin, Madison, WI.

De Leon-Gonzalez, F., Hernandez-Serrano, M., Etchevers, J., Payan-Zelaya, F. and Ordaz-Chaparro, V. 2000. Short-term compost effect on macro-aggregation in sandy soil under low rainfall in the valley of Mexico. *Soil and Tillage Research Journal* 56, 213–217.

Dikinya, O., Hinz, C. and Aylmore, G. 2006. Dispersion and re-deposition of fine particles and their effects on saturated hydraulic conductivity. *Soil Research* 44, 47–56. doi:https://doi.org/10.1071/SR05067.

Drozd, J. 2003. The risk and benefits associated with utilizing composts from municipal solid waste (MSW) in Agriculture. *In: Innovative Soilplant Systems for Sustainable Agricultural Practices*. Organisation for Economic Co-operation and Development (OECD), Paris, France, pp. 211–226.

Dubey, S. and Mondal, R. 1993. Sodic soil reclamation with saline water in conjunction with organic and inorganic amendments. *Arid Land Research and Management* 7, 219–231. doi:https://doi.org/10.1080/15324989309381352.

Enders, A., Hanley, K., Whitman, T., Joseph, S. and Lehmann, J. 2012. Characterization of biochars to evaluate recalcitrance and agronomic performance. *Bioresource Technology* 114, 644–653. doi:https://doi.org/10.1016/j.biortech.2012.03.022.

Fang, Y., Singh, B., Singh, B. and Krull, E. 2014. Biochar carbon stability in four contrasting soils. *European Journal of Soil Science* 65, 60–71. doi:https://doi.org/10.1111/ejss.12094.

Fricke, K. and Vogtmann, H. 1994. Compost quality: Physical characteristics, nutrient content, heavy metals and organic chemicals. *Toxicological & Environmental Chemistry* 43, 95–114. doi:https://doi.org/10.1080/02772249409358021.

Gal, M., Arcan, L., Shainberg, I. and Keren, R. 1984. Effect of exchangeable sodium and phosphogypsum on crust structure—Scanning electron microscope observations 1. *Soil Science Society of America Journal* 48, 872–878. doi:https://doi.org/10.2136/sssaj1984.03615995004800040035x.

Ganjegunte, G. K., King, L. A. and Vance, G. F. 2008. Cumulative soil chemistry changes from land application of saline–sodic waters. *Journal of Environmental Quality* 37, S-128–S-138. doi:https://doi.org/10.2134/jeq2007.0424.

Garcia, C. and Hernandez, T. 1996. Influence of salinity on the biological and biochemical activity of a calciorthird soil. *Plant and Soil* 178, 255–263. doi:https://doi.org/10.1007/BF00011591.

Gaskin, J. W., Steiner, C., Harris, K., Das, K. and Bibens, B. 2008. Effect of low-temperature pyrolysis conditions on biochar for agricultural use. *Transactions of the ASABE* 51, 2061–2069.

Ghafoor, A., Muhammad, S. and Mujtaba, G. 1986. Comparison on gypsum, sulphuric acid, hydrochloric acid and calcium chloride for reclaiming the subsoiled calcareous saline-sodic Khurrianwala soil series [Pakistan]. *Journal of Agricultural Research (Pakistan)* 24(3), 179–183. https://lib.dr.iastate.edu/etd/12284.

Gharaibeh, M. A., Eltaif, N. I. and Shunnar, O. F. 2009. Leaching and reclamation of calcareous saline-sodic soil by moderately saline and moderate-SAR water using gypsum and calcium chloride. *Journal of Plant Nutrition and Soil Science* 172, 713–719. doi:https://doi.org/10.1002/jpln.200700327.

Ghollarata, M. and Raiesi, F. 2007. The adverse effects of soil salinization on the growth of *Trifolium alexandrinum* L. and associated microbial and biochemical properties in a soil from Iran. *Soil Biology and Biochemistry* 39, 1699–1702. doi:https://doi.org/10.1016/j.soilbio.2007.01.024.

Glaser, B., Lehmann, J. and Zech, W. 2002. Ameliorating physical and chemical properties of highly weathered soils in the tropics with charcoal: A review. *Biology and Fertility of Soils* 35, 219–230. doi:https://doi.org/10.1007/s00374-002-0466-4.

Grattan, S. and Grieve, C. 1998. Salinity–mineral nutrient relations in horticultural crops. *Scientia Horticulturae* 78, 127–157. doi:https://doi.org/10.1016/S0304-4238(98)00192-7.

Greene, R., Rengasamy, P., Ford, G., Chartres, C. and Millar, J. 1988. The effect of sodium and calcium on physical properties and micromorphology of two red-brown earth soils. *Journal of Soil Science* 39, 639–648. doi:https://doi.org/10.1111/j.1365-2389.1988.tb01246.x.

Hammes, K., Smernik, R. J., Skjemstad, J. O., Herzog, A., Vogt, U. F. and Schmidt, M. W. 2006. Synthesis and characterisation of laboratory-charred grass straw (*Oryza sativa*) and chestnut wood (*Castanea sativa*) as reference materials for black carbon quantification. *Organic Geochemistry* 37, 1629–1633. doi:https://doi.org/10.1016/j.orggeochem.2006.07.003.

Harrison, R., Xue, D., Henry, C. and Cole, D. W. 1994. Long-term effects of heavy applications of biosolids on organic matter and nutrient content of a coarse-textured forest soil. *Forest Ecology and Management* 66, 165–177. doi:https://doi.org/10.1016/0378-1127(94)90155-4.

Harvey, O. R., Herbert, B. E., Kuo, L.-J. and Louchouarn, P. 2012. Generalized two-dimensional perturbation correlation infrared spectroscopy reveals mechanisms for the development of surface charge and recalcitrance in plant-derived biochars. *Environmental Science & Technology* 46, 10641–10650. doi:https://doi.org/10.1021/es302971d.

Hernández, T., Moreno, J. I. and Costa, F. 1991. Influence of sewage sludge application on crop yields and heavy metal availability. *Soil Science and Plant Nutrition* 37, 201–210. doi:https://doi.org/10.1080/00380768.1991.10415030.

Hue, N. and Ranjith, S. A. 1994. Sewage sludges in Hawaii: Chemical composition and reactions with soils and plants. *Water, Air, and Soil Pollution* 72, 265–283. doi:https://doi.org/10.1007/BF01257129.

Hussain, N., Hassan, G., Arshadullah, M. and Mujeeb, F. 2001. Evaluation of amendments for the improvement of physical properties of sodic soil. *International Journal of Agriculture and Biology* 3, 319–322.

Hutton, J. The chemistry of rainwater with particular reference to conditions in south eastern Australia. *Climatology and Microclimatology. Proceedings of the Canberra Symposium*, 1958. UNESCO Paris, France, pp. 285–290.

Ilyas, M., Qureshi, R. and Qadir, M. 1997. Chemical changes in a saline-sodic soil after gypsum application and cropping. *Soil Technology* 10, 247–260. doi:https://doi.org/10.1016/S0933-3630(96)00121-3.

Jindo, K., Mizumoto, H., Sawada, Y., Sanchez-Monedero, M. A. and Sonoki, T. 2014. Physical and chemical characterization of biochars derived from different agricultural residues. *Biogeosciences* 11, 6613–6621. doi:https://doi.org/10.5194/bg-11-6613-2014.

Jones, D., Murphy, D., Khalid, M., Ahmad, W., Edwards-Jones, G. and Deluca, T. 2011. Short-term biochar-induced increase in soil CO_2 release is both biotically and abiotically mediated. *Soil Biology and Biochemistry* 43, 1723–1731. doi:https://doi.org/10.1016/j.soilbio.2011.04.018.

Kalderis, D., Kotti, M., Méndez, A. and Gascó, G. 2014. Characterization of hydrochars produced by hydrothermal carbonization of rice husk. *Solid Earth* 5, 477–483. doi:https://doi.org/10.5194/se-5-477-2014.

Kaya, C., Kirnak, H. and Higgs, D. 2001. Enhancement of growth and normal growth parameters by foliar application of potassium and phosphorus in tomato cultivars grown at high (NaCl) salinity. *Journal of Plant Nutrition* 24, 357–367. doi:https://doi.org/10.1081/PLN-100001394.

Keren, R. and Miyamoto, S. 1990. Reclamation of saline, sodic, and boron-affected soils. Chapter 19, *In:* Tanji, K.K. (Ed.) *Agricultural Salinity Assessment and Management*, ASCE Manual No. 71. American Society of Civil Engineers, New York.

Koenig, R., Miner, D. and Goodrich, K. 1998. *Land Application of Biosolids a Guide for Farmers*. Utah State University Extension, Utah.

Kumar, D. and Singh, B. 2003. The use of coal fly ash in sodic soil reclamation. *Land Degradation & Development* 14, 285–299. doi:https://doi.org/10.1002/ldr.557.

Kuzyakov, Y., Bogomolova, I. and Glaser, B. 2014. Biochar stability in soil: Decomposition during eight years and transformation as assessed by compound-specific 14C analysis. *Soil Biology and Biochemistry* 70, 229–236. doi:https://doi.org/10.1016/j.soilbio.2013.12.021.

Lakhdar, A., Hafsi, C., Rabhi, M., Debez, A., Montemurro, F., Abdelly, C., Jedidi, N. and Ouerghi, Z. 2008. Application of municipal solid waste compost reduces the negative effects of saline water in *Hordeum maritimum* L. *Bioresource Technology* 99, 7160–7167. doi:https://doi.org/10.1016/j.biortech.2007.12.071.

Lakhdar, A., Rabhi, M., Ghnaya, T., Montemurro, F., Jedidi, N. and Abdelly, C. 2009. Effectiveness of compost use in salt-affected soil. *Journal of Hazardous Materials* 171, 29–37. doi:https://doi.org/10.1016/j.jhazmat.2009.05.132.

Lashari, M. S., Liu, Y., Li, L., Pan, W., Fu, J., Pan, G., Zheng, J., Zheng, J., Zhang, X. and Yu, X. 2013. Effects of amendment of biochar-manure compost in conjunction with pyroligneous solution on soil quality and wheat yield of a salt-stressed cropland from Central China Great Plain. *Field Crops Research* 144, 113–118. doi:https://doi.org/10.1016/j.fcr.2012.11.015.

Lashari, M. S., Ye, Y., Ji, H., Li, L., Kibue, G. W., Lu, H., Zheng, J. and Pan, G. 2015. Biochar–manure compost in conjunction with pyroligneous solution alleviated salt stress and improved leaf bioactivity of maize in a saline soil from central China: A 2-year field experiment. *Journal of the Science of Food and Agriculture* 95, 1321–1327. doi:https://doi.org/10.1002/jsfa.6825.

Läuchli, A. and Epstein, E. 1990. Plant responses to saline and sodic conditions. *Agricultural Salinity Assessment and Management* 71, 113–137.

Lebron, I., Suarez, D. and Yoshida, T. 2002. Gypsum effect on the aggregate size and geometry of three sodic soils under reclamation. *Soil Science Society of America Journal* 66, 92–98. doi:https://doi.org/10.2136/sssaj2002.9200.

Lehmann, J., Rillig, M. C., Thies, J., Masiello, C. A., Hockaday, W. C. and Crowley, D. 2011. Biochar effects on soil biota–a review. *Soil Biology and Biochemistry* 43, 1812–1836. doi:https://doi.org/10.1016/j.soilbio.2011.04.022.

Liang, B., Lehmann, J., Sohi, S. P., Thies, J. E., O'neill, B., Trujillo, L., Gaunt, J., Solomon, D., Grossman, J. and Neves, E. G. 2010. Black carbon affects the cycling of non-black carbon in soil. *Organic Geochemistry* 41, 206–213. doi:https://doi.org/10.1016/j.orggeochem.2009.09.007.

Liang, Y., Si, J., Nikolic, M., Peng, Y., Chen, W. and Jiang, Y. 2005. Organic manure stimulates biological activity and barley growth in soil subject to secondary salinization. *Soil Biology and Biochemistry* 37, 1185–1195. doi:https://doi.org/10.1016/j.soilbio.2004.11.017.

Liang, Y., Yang, Y., Yang, C., Shen, Q., Zhou, J. and Yang, L. 2003. Soil enzymatic activity and growth of rice and barley as influenced by organic manure in an anthropogenic soil. *Geoderma* 115, 149–160. doi:https://doi.org/10.1016/S0016-7061(03)00084-3.

Logan, T. and Harrison, B. 1995. Physical characteristics of alkaline stabilized sewage sludge (N-Viro Soil) and their effects on soil physical properties. *Journal of Environmental Quality* 24, 153–164. doi:https://doi.org/10.2134/jeq1995.00472425002400010022x.

Luo, J. Q., Wang, L. L., Li, Q. S., Zhang, Q. K., He, B. Y., Wang, Y., Qin, L. P. and Li, S. S. 2015. Improvement of hard saline–sodic soils using polymeric aluminum ferric sulfate (PAFS). *Soil and Tillage Research* 149, 12–20. doi:https://doi.org/10.1016/j.still.2014.12.014.

Luo, Y., Durenkamp, M., De Nobili, M., Lin, Q. and Brookes, P. 2011. Short term soil priming effects and the mineralisation of biochar following its incorporation to soils of different pH. *Soil Biology and Biochemistry* 43, 2304–2314. doi:https://doi.org/10.1016/j.soilbio.2011.07.020.

Maggio, A., De Pascale, S., Fagnano, M. and Barbieri, G. 2011. Saline agriculture in Mediterranean environments. *Italian Journal of Agronomy*, e7–e7. doi:https://doi.org/10.4081/ija.2011.e7.

Major, J., Rondon, M., Molina, D., Riha, S. J. and Lehmann, J. 2010. Maize yield and nutrition during 4 years after biochar application to a Colombian savanna oxisol. *Plant and Soil* 333, 117–128. doi:https://doi.org/10.1007/s11104-010-0327-0.

Makoi, J. H. and Ndakidemi, P. A. 2007. Reclamation of sodic soils in northern Tanzania, using locally available organic and inorganic resources. *African Journal of Biotechnology* 6. doi:http://dx.doi.org/10.5897/AJB2007.000-2292.

Makoi, J. H. and Verplancke, H. 2010. Effect of gypsum placement on the physical chemical properties of a saline sandy loam soil. *Australian Journal of Crop Science* 4, 556.

Manchanda, G. and Garg, N. 2008. Salinity and its effects on the functional biology of legumes. *Acta Physiologiae Plantarum* 30, 595–618. doi:https://doi.org/10.1007/s11738-008-0173-3.

Miyamoto, S., Prather, R. and Stroehlein, J. 1975. Sulfuric acid and leaching requirements for reclaiming sodium-affected calcareous soils. *Plant and Soil* 43, 573–585. doi:https://doi.org/10.1007/BF01928520.

Mohan, D., Sarswat, A., Ok, Y. S. and Pittman Jr, C. U. 2014. Organic and inorganic contaminants removal from water with biochar, a renewable, low cost and sustainable adsorbent: A critical review. *Bioresource Technology* 160, 191–202. doi:https://doi.org/10.1016/j.biortech.2014.01.120.

Morera, M., Echeverria, J. and Garrido, J. 2002. Bioavailability of heavy metals in soils amended with sewage sludge. *Canadian Journal of Soil Science* 82, 433–438. doi:https://doi.org/10.4141/S01-072.

Munns, R. and Tester, M. 2008. Mechanisms of salinity tolerance. *Annual Review of Plant Biology* 59, 651–681. doi:https://doi.org/10.1146/annurev.arplant.59.032607.092911.

Natal-Da-Luz, T., Tidona, S., Jesus, B., Morais, P. V. and Sousa, J. P. 2009. The use of sewage sludge as soil amendment. The need for an ecotoxicological evaluation. *Journal of Soils and Sediments* 9, 246. doi:https://doi.org/10.1007/s11368-009-0077-x.

Nelson, P. and Oades, J. 1998. Organic matter, sodicity and soil structure. *In*: Summer, M.E. and Naidu, R. (Eds.) *Sodic Soils: Distribution, Properties, Management and Environmental Consequences*. Oxford University Press, New York, pp. 51–75.

Nguyen, B. T. and Lehmann, J. 2009. Black carbon decomposition under varying water regimes. *Organic Geochemistry* 40, 846–853. doi:https://doi.org/10.1016/j.orggeochem.2009.05.004.

Novak, J. M., Busscher, W. J., Laird, D. L., Ahmedna, M., Watts, D. W. and Niandou, M. A. 2009a. Impact of biochar amendment on fertility of a southeastern coastal plain soil. *Soil Science* 174, 105–112. doi:https://doi.org/10.1097/SS.0b013e3181981d9a.

Novak, J. M., Lima, I., Xing, B., Gaskin, J. W., Steiner, C., Das, K. C., Ahmedna, M., Rehrah, D., Watts, D. W., Busscher, W. J. and Schomberg, H. 2009b. Characterization of designer biochar produced at different temperatures and their effects on a loamy sand. *Annals of Environmental Science* 3, 195–206.

Oster, J. and Shainberg, I. 2001. Soil responses to sodicity and salinity: Challenges and opportunities. *Soil Research* 39, 1219–1224. doi:https://doi.org/10.1071/SR00051.

Özenc, N. and Caliskan, N. Effects of husk compost on hazelnut yield and quality. *V International Congress on Hazelnut* 556, 559–566.

Parkpain, P., Sirisukhodom, S. and Carbonell-Barrachina, A. 1998. Heavy metals and nutrients chemistry in sewage sludge amended Thai soils. *Journal of Environmental Science & Health Part A* 33, 573–597. doi:https://doi.org/10.1080/10934529809376749.

Paz-Ferreiro, J., Lu, H., Fu, S., Méndez, A. and Gascó, G. 2014. Use of phytoremediation and biochar to remediate heavy metal polluted soils: A review. *Solid Earth* 5, 65–75. doi:https://doi.org/10.5194/se-5-65-2014.

Pernes-Debuyser, A. and Tessier, D. 2004. Soil physical properties affected by long-term fertilization. *European Journal of Soil Science* 55, 505–512. doi:https://doi.org/10.1111/j.1365-2389.2004.00614.x.

Pessarakli, M. 2016. *Handbook of Plant and Crop Stress*. CRC Press, Boca Raton, FL.

Pessarakli, M. and Szabolcs, I. 1999. Soil salinity and sodicity as particular plant/crop stress factors. *Handbook of Plant and Crop Stress* 2. Marcel Dekker, Inc., New York.

Qadir, M. and Oster, J. 2004. Crop and irrigation management strategies for saline-sodic soils and waters aimed at environmentally sustainable agriculture. *Science of the Total Environment* 323, 1–19. doi:https://doi.org/10.1016/j.scitotenv.2003.10.012.

Quirk, J. 2001. The significance of the threshold and turbidity concentrations in relation to sodicity and microstructure. *Soil Research* 39, 1185–1217. doi:https://doi.org/10.1071/SR00050.

Rasool, R., Kukal, S. and Hira, G. 2007. Soil physical fertility and crop performance as affected by long term application of FYM and inorganic fertilizers in rice: Wheat system. *Soil and Tillage Research* 96, 64–72. doi:https://doi.org/10.1016/j.still.2007.02.011.

Ray, P., Meena, B. L. and Nath, C. 2014. Management of coastal soils for improving soil quality and productivity. *Popular Kheti* 2, 95–99.

Rengasamy, P. 2002. Transient salinity and subsoil constraints to dryland farming in Australian sodic soils: An overview. *Australian Journal of Experimental Agriculture* 42, 351–361. doi:https://doi.org/10.1071/EA01111.

Rengasamy, P. 2006. World salinization with emphasis on Australia. *Journal of Experimental Botany* 57, 1017–1023. doi:https://doi.org/10.1093/jxb/erj108.

Rengasamy, P. 2010. Soil processes affecting crop production in salt-affected soils. *Functional Plant Biology* 37, 613–620. doi:https://doi.org/10.1071/FP09249.

Rengasamy, P. and Olsson, K. 1991. Sodicity and soil structure. *Soil Research* 29, 935–952. doi:https://doi.org/10.1071/SR9910935.

Rhoton, F. E. and Mcchesney, D. S. 2011. Erodibility of a sodic soil amended with flue gas desulfurization gypsum. *Soil Science* 176, 190–195. doi:https://doi.org/10.1097/SS.0b013e318212143d.

Richards, L. 1954. *Diagnosis and Improvement of Saline and Alkali Soils*, Handbook No. 60. US Department of Agriculture, Washington, DC, pp. 98–105.

Rietz, D. and Haynes, R. 2003. Effects of irrigation-induced salinity and sodicity on soil microbial activity. *Soil Biology and Biochemistry* 35, 845–854. doi:https://doi.org/10.1016/S0038-0717(03)00125-1.

Rondon, M. A., Lehmann, J., Ramírez, J. and Hurtado, M. 2007. Biological nitrogen fixation by common beans (*Phaseolus vulgaris* L.) increases with bio-char additions. *Biology and Fertility of Soils* 43, 699–708. doi:https://doi.org/10.1007/s00374-006-0152-z.

Ross, S. M. 1994. Retention, transformation and mobility of toxic metals in soils. *Toxic Metals in Soil-Plant Systems*, 63–152.

Sadiq, M., Hassan, G., Mehdi, S., Hussain, N. and Jamil, M. 2007. Amelioration of saline-sodic soils with tillage implements and sulfuric acid application. *Pedosphere* 17, 182–190. doi:https://doi.org/10.1016/S1002-0160(07)60024-1.

Sakai, Y., Matsumoto, S. and Sadakata, M. 2004. Alkali soil reclamation with flue gas desulfurization gypsum in China and assessment of metal content in corn grains. *Soil & Sediment Contamination* 13, 65–80. doi:https://doi.org/10.1080/10588330490269840.

Samara, N. M. 2009. Heavy metals concentrations in biosolids of Al-Bireh sewage treatment plant and assessment of biosolids application impacts on crop growth and productivity. Birzeit University.

Shafieepour, S., Ayati, B. and Ganjidoust, H. 2011. Reuse of Sewage Sludge for Agricultural Soil Improvement (Case Study: Kish Island).

Sharma, A., Rana, C., Singh, S. and Katoch, V. 2016. Soil salinity causes, effects and management in cucurbits. *In:* Pessarakli, M. (Ed.) *Handbook of Cucurbits: Growth, Cultural Practices, and Physiology*. CRC Press, Boca Raton, FL.

Sharma, B. R. and Minhas, P. S. 2005. Strategies for managing saline/alkali waters for sustainable agricultural production in South Asia. *Agricultural Water Management* 78, 136–151. doi:https://doi.org/10.1016/j.agwat.2005.04.019.

Singh, R. and Agrawal, M. 2008. Potential benefits and risks of land application of sewage sludge. *Waste Management* 28, 347–358. doi:https://doi.org/10.1016/j.wasman.2006.12.010.

Siyal, A., Siyal, A. and Abro, Z. 2002. Salt affected soils their identification and reclamation. *Pakistan Journal of Applied Sciences* 2, 537–540. doi:10.3923/jas.2002.537.540.

Smedema, L. K. and Shiati, K. 2002. Irrigation and salinity: A perspective review of the salinity hazards of irrigation development in the arid zone. *Irrigation and Drainage Systems* 16, 161–174. doi:https://doi.org/10.1023/A:1016008417327.

Smith, M. C., Das, K. C. and Tollner, E. W. 1998. Characterization of landfilled municipal solid waste following in situ aerobic bioreduction. *Proceedings of Composting in the Southeast*, University of Georgia, pp. 138–143.

Sohi, S. P., Krull, E., Lopez-Capel, E. and Bol, R. 2010. A review of biochar and its use and function in soil. In: Donald L. S. (Ed.) *Advances in Agronomy*. Burlington: Academic Press, 105, pp. 47–82. doi:https://doi.org/10.1016/S0065-2113(10)05002-9.

Soon, Y. 1981. Solubility and sorption of cadmium in soils amended with sewage sludge. *Journal of Soil Science* 32, 85–95. doi:https://doi.org/10.1111/j.1365-2389.1981.tb01688.x.

Southard, R., Shainberg, I. and Singer, M. 1988. Influence of electrolyte concentration on the micromorphology of artificial depositional crust1. *Soil Science* 145, 278–288.

Sree Ramulu, U. S. 2001. *Reuse of Municipal Sewage and Sludge in Agriculture*. Scientific Publishers, Jodhpur, India.

Sumner, M. E. 2000. *Handbook of Soil Science*. CRC Press, Boca Raton, FL.

Tejada, M. and Gonzalez, J. 2005. Beet vinasse applied to wheat under dryland conditions affects soil properties and yield. *European Journal of Agronomy* 23, 336–347. doi:https://doi.org/10.1016/j.eja.2005.02.005.

Tejada, M., Garcia, C., Gonzalez, J. L. and Hernandez, M. T. 2006. Use of organic amendment as a strategy for saline soil remediation: Influence on the physical, chemical and biological properties of soil. *Soil Biology and Biochemistry* 38, 1413–1421. doi:https://doi.org/10.1016/j.soilbio.2005.10.017.

Thomas, S., Anand, A., Chinnusamy, V., Dahuja, A. and Basu, S. 2013a. Magnetopriming circumvents the effect of salinity stress on germination in chickpea seeds. *Acta Physiologiae Plantarum* 35, 3401–3411. doi:https://doi.org/10.1007/s11738-013-1375-x.

Thomas, S. C., Frye, S., Gale, N., Garmon, M., Launchbury, R., Machado, N., Melamed, S., Murray, J., Petroff, A. and Winsborough, C. 2013b. Biochar mitigates negative effects of salt additions on two herbaceous plant species. *Journal of Environmental Management* 129, 62–68. doi:https://doi.org/10.1016/j.jenvman.2013.05.057.

Tisdall, J. M. and Oades, J. M. 1982. Organic matter and water-stable aggregates in soils. *Journal of Soil Science* 33, 141–163. doi:https://doi.org/10.1111/j.1365-2389.1982.tb01755.x.

Togay, N., Togay, Y. and Dogan, Y. 2008. Effects of municipal sewage sludge doses on the yield, some yield components and heavy metal concentration of dry bean (*Phaseolus vulgaris* L.). *African Journal of Biotechnology* 7(17), 3026–3030.

Tripathi, S., Kumari, S., Chakraborty, A., Gupta, A., Chakrabarti, K. and Bandyapadhyay, B. K. 2006. Microbial biomass and its activities in salt-affected coastal soils. *Biology and Fertility of Soils* 42, 273–277. doi:https://doi.org/10.1007/s00374-005-0037-6.

Tsadilas, C., Matsi, T., Barbayiannis, N. and Dimoyiannis, D. 1995. Influence of sewage sludge application on soil properties and on the distribution and availability of heavy metal fractions. *Communications in Soil Science and Plant Analysis* 26, 2603–2619. doi:https://doi.org/10.1080/00103629509369471.

Uchimiya, M., Wartelle, L. H., Klasson, K. T., Fortier, C. A. and Lima, I. M. 2011. Influence of pyrolysis temperature on biochar property and function as a heavy metal sorbent in soil. *Journal of Agricultural and Food Chemistry* 59, 2501–2510. doi:https://doi.org/10.1021/jf104206c.

Vadyanina, A. F. and Roi, P. K. 1974. Changes in aggregates status of saline sodic soil after their reclamation by different methods. *Vest Mask. Univ. Ser 6 Boil. Pochroned* 29: 111–117.

Vance, G. F., King, L. A. and Ganjegunte, G. K. 2008. Soil and plant responses from land application of saline–sodic waters: Implications of management. *Journal of Environmental Quality* 37, S-139–S-148. doi:https://doi.org/10.2134/jeq2007.0442.

Walker, D. J. and Bernal, M. P. 2008. The effects of olive mill waste compost and poultry manure on the availability and plant uptake of nutrients in a highly saline soil. *Bioresource Technology* 99, 396–403. doi:https://doi.org/10.1016/j.biortech.2006.12.006.

Walker, D. J., Clemente, R. and Bernal, M. P. 2004. Contrasting effects of manure and compost on soil pH, heavy metal availability and growth of *Chenopodium album* L. in a soil contaminated by pyritic mine waste. *Chemosphere* 57, 215–224. doi:https://doi.org/10.1016/j.chemosphere.2004.05.020.

Walpola, B. and Arunakumara, K. 2010. Effect of salt stress on decomposition of organic matter and nitrogen mineralization in animal manure amended soils. *Journal of Agricultural Sciences–Sri Lanka* 5. doi:https://doi.org/10.4038/jas.v5i1.2319.

Wang, J., Yang, P., Huang, G. and Pereria, S. 2004. The effect on physical and chemical properties of saline and sodic soils reclaimed with byproduct from flue gas desulphurization. *Land and Water Management Decision Tools and Practices* 2, 1015–1021.

Wang, L., Sun, X., Li, S., Zhang, T., Zhang, W. and Zhai, P. 2014. Application of organic amendments to a coastal saline soil in north China: Effects on soil physical and chemical properties and tree growth. *PLoS One* 9, e89185. doi:https://doi.org/10.1371/journal.pone.0089185.

Wang, S., Chen, Q., Li, Y., Zhuo, Y. and Xu, L. 2017. Research on saline-alkali soil amelioration with FGD gypsum. *Resources, Conservation and Recycling* 121, 82–92. doi:https://doi.org/10.1016/j.resconrec.2016.04.005.

Wichern, J., Wichern, F. and Joergensen, R. G. 2006. Impact of salinity on soil microbial communities and the decomposition of maize in acidic soils. *Geoderma* 137, 100–108. doi:https://doi.org/10.1016/j.geoderma.2006.08.001.

Wicke, B., Smeets, E., Dornburg, V., Vashev, B., Gaiser, T., Turkenburg, W. and Faaij, A. 2011. The global technical and economic potential of bioenergy from salt-affected soils. *Energy & Environmental Science* 4, 2669–2681.

Wong, V. N., Dalal, R. C. and Greene, R. S. 2008. Salinity and sodicity effects on respiration and microbial biomass of soil. *Biology and Fertility of Soils* 44, 943–953. doi:https://doi.org/10.1007/s00374-008-0279-1.

Wong, V. N., Dalal, R. C. and Greene, R. S. 2009. Carbon dynamics of sodic and saline soils following gypsum and organic material additions: A laboratory incubation. *Applied Soil Ecology* 41, 29–40. doi:https://doi.org/10.1016/j.apsoil.2008.08.006.

Wong, V. N., Greene, R., Dalal, R. and Murphy, B. W. 2010. Soil carbon dynamics in saline and sodic soils: A review. *Soil Use and Management* 26, 2–11. doi:https://doi.org/10.1111/j.1475-2743.2009.00251.x.

Xu, R. K., Zhao, A. Z., Yuan, J. H. and Jiang, J. 2012. pH buffering capacity of acid soils from tropical and subtropical regions of China as influenced by incorporation of crop straw biochars. *Journal of Soils and Sediments* 12, 494–502. doi:https://doi.org/10.1007/s11368-012-0483-3.

Yazdanpanah, N., Pazira, E., Neshat, A., Mahmoodabadi, M. and Sinobas, L. R. 2013. Reclamation of calcareous saline sodic soil with different amendments (II): Impact on nitrogen, phosphorous and potassium redistribution and on microbial respiration. *Agricultural Water Management* 120, 39–45. doi:https://doi.org/10.1016/j.agwat.2012.08.017.

Yonggan, Z., Yan, L., Shujuan, W., Jing, W. and Lizhen, X. 2017. Combined application of a straw layer and flue gas desulfurization gypsum to reduce soil salinity and alkalinity. *Pedosphere*. doi:https://doi.org/10.1016/S1002-0160(17)60480-6.

Yu, H., Yang, P., Lin, H., Rén, S. and He, X. 2014. Effects of sodic soil reclamation using flue gas desulphurization gypsum on soil pore characteristics, bulk density, and saturated hydraulic conductivity. *Soil Science Society of America Journal* 78, 1201–1213. doi:https://doi.org/10.2136/sssaj2013.08.0352.

Yu, J., Wang, Z., Meixner, F. X., Yang, F., Wu, H. and Chen, X. 2010. Biogeochemical characterizations and reclamation strategies of saline sodic soil in northeastern China. *CLEAN–Soil, Air, Water* 38, 1010–1016. doi:https://doi.org/10.1002/clen.201000276.

Yuan, B.-C., Li, Z.-Z., Liu, H., Gao, M. and Zhang, Y.-Y. 2007. Microbial biomass and activity in salt affected soils under arid conditions. *Applied Soil Ecology* 35, 319–328. doi:https://doi.org/10.1016/j.apsoil.2006.07.004.

Zhang, Q., Du, Z., Lou, Y. and He, X. 2015. A one-year short-term biochar application improved carbon accumulation in large macroaggregate fractions. *CATENA* 127, 26–31. doi:https://doi.org/10.1016/j.catena.2014.12.009.

14 Integrated Soil Remediation by Chemical-Enhanced Extraction and Biochar Immobilization for Potentially Toxic Elements

Jingzi Beiyuan, Daniel C.W. Tsang, Jörg Rinklebe, and Hailong Wang

CONTENTS

14.1 INTRODUCTION

14.1.1 MOBILIZATION BY CHELANT-ENHANCED EXTRACTION

Potentially toxic elements (PTEs), such as Cd, As, Cu, Pb, Cr, and Zn, can be released into soils through pedogenic and anthropogenic processes, including industrial activities, agricultural activities, and improper disposal of domestic or industrial waste (Ok et al., 2011; Peters, 1999). Studies have shown that PTEs released by anthropogenic processes can be more bioaccessible and/or bioavailable than pedogenic inputs (Bolan et al., 2014; Lamb et al., 2009).

Simply put, technology to control the PTE-induced soil contamination can be divided into two groups based on its mechanism: to mobilize (remove them from soils) or to immobilize (stabilize them in soils) (Bolan et al., 2014). Many chemical reagents can be used as extractants or reagents to increase the mobility of PTEs in soils, such as chelants, acidic and alkaline solvents, reducing and oxidizing agents, electrolytes, and surfactants (Hartley et al., 2014; Polettini et al., 2009; Tsang and Hartley, 2014; Zou et al., 2009). Chelants, also referred to as chelating agents, have been widely used in metal extraction techniques of soil remediation (e.g., soil washing, soil leaching, and phytoextraction) due to their strong complexation capacities with different metals (Hartley et al., 2014; Lestan et al., 2008; Tsang et al., 2013a). Ethylenediaminetetraacetic acid (EDTA), nitrilotriacetic

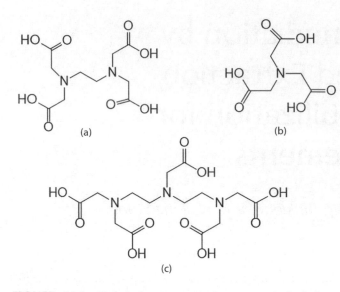

FIGURE 14.1 Chemical structures of some typical chelants: (a) EDTA, (b) NTA, and (c) DTPA.

acid (NTA), and diethylenetriamine pentaacetic acid (DTPA) are typical chelants that are commonly used in the industry because of their strong complexation capacities (Figure 14.1). However, according to the International Agency for Research on Cancer, NTA is classified as a Class II carcinogen and toxin, while DTPA is classified as a potential carcinogen (Peters, 1999; Zou et al., 2009). Later studies found that EDTA is difficult to be biodegraded or photodegraded without further interruption, which might cause adverse effects on soils, sediment, groundwater, and plants (Bucheli-Witschel and Egli, 2001; Duo et al., 2019; Jelusic and Lestan, 2014). The non-degraded EDTA can transport and mobilize PTEs in soils and sediments downstream, contaminate groundwater or drinking water, and increase human health risks (Tsang et al., 2009; Yip et al., 2009; Zhang et al., 2010a). A guideline value of 0.6 mg L^{-1} for EDTA in drinking water was recently set by the World Health Organization (WHO, 2017).

Biodegradable chelants, such as (*S,S*)-ethylenediaminedisuccinic acid (EDDS), methylglycinediacetic acid (MGDA), iminodisuccinic acid (IDSA), N,N-Bis(carboxymethyl)-L-glutamic acid (GLDA), polyaspartic acid (PASP), and glucomonocarbonic acid (GCA), etc. (Figure 14.2), which also have strong metal complexation capacities, were proposed to replace the role of toxic and/or non-degradable chelants in soil remediation (Arwidsson et al., 2010; Beiyuan et al., 2018b; Guo et al., 2017; Tandy et al., 2006; Wang et al., 2018). As a structural isomer of EDTA, EDDS is a promising substitute because of its strong metal-complexation capacity, lower toxicity to biota, and higher biodegradable capacity in the natural environment (with a biodegradation half-life of 7 to 32 days under natural conditions) (Grčman et al., 2003; Nowack, 2002; Yang et al., 2013). Successful cases of EDDS in metal extractions of soils have been frequently reported in the last decades (Luo et al., 2015; Wang et al., 2012; Yan et al., 2010). Nevertheless, the application of EDDS in metal extraction in soils is still hindered by its

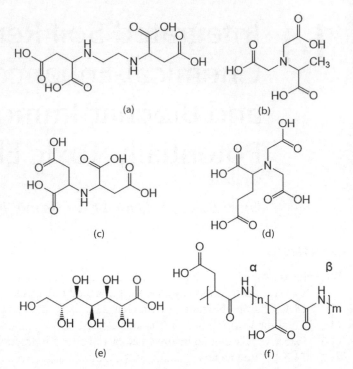

FIGURE 14.2 Chemical structures of some biodegradable chelants: (a) EDDS, (b) MGDA, (c) IDSA, (d) GLDA, (e) PASP, and (f) GCA.

properties. For example, the affinities of EDDS to various metals (for example, Cu, Zn, and Ni) are comparable to EDTA, but not for Pb ((log K(Pb-EDDS) = 12.7 and log K(Pb-EDTA) = 17.9) (Tandy et al., 2004). This probably contributes to a lower extraction capacity on Pb for EDDS compared with EDTA. Additionally, most of the chelants can hardly form stable chelant complexes with metalloids, such as arsenic (Tokunaga and Hakuta, 2002); therefore, it is difficult to extract them by EDDS. However, metals and metalloids commonly co-exist in the field. In addition, the residual metal-chelant complexes present in the soil before degradation could also be dangerous for treated soil.

14.1.2 IMMOBILIZATION BY BIOCHAR

Immobilization, or stabilization, by applying amendments, such as industrial by-products, natural minerals, alkaline compounds, agricultural waste, and bio-products, in soils can reduce the mobility, bioavailability, leachability, or transportation capacity of PTEs (Ahmad et al., 2014; Komarek et al., 2013; Rajapaksha et al., 2015; Tica et al., 2011). As one of them, biochar is a promising amendment because it not only immobilizes the PTEs but also reduces the amount of wastes added to landfills, contributing to carbon sequestration and improving the physicochemical and biological properties of soils (e.g., enhancing soil-water-holding capacity, soil microbial activities, biomass, and crop yields of plants) (El-Naggar et al., 2019a; Hussain et al., 2016; Lehmann et al., 2011; Park et al., 2011a;

Rajapaksha et al., 2016). The immobilization mechanisms of biochar are largely attributed to its porous surface, rich functional groups, slightly enhanced alkalinity, as well as its high cation exchange capacity (CEC), which result from the high contents of silicates and mineral oxide (Ahmad et al., 2016; Park et al., 2011b; Uchimiya et al., 2011). Ahmad et al. (2014) summarized the mechanisms for immobilization of PTEs by biochar: (i) electrostatic outer-sphere complexation of cations; (ii) electrostatic outer-sphere complexation of anions; (iii) surface ion change; (iv) co-precipitation and inner-sphere complexations; and (v) surface complexation with oxygen-containing functional groups.

Biochar is stable and difficult to be degraded or mineralized in soils because of its fused structure and the fact that large proportions of its carbons are in amorphous (low temperature pyrolysis) and turbostratic (high temperature pyrolysis) phases, making it suitable for soil remediation (Lehmann et al., 2011; Nguyen et al., 2010). However, its immobilization effects on various PTEs under changed environmental conditions are still a concern (Rinklebe et al., 2016). Recently, studies found that under varied redox potential conditions, the immobilization effects on Pb and Hg can be affected in different ranges (Beckers et al., 2019; Beiyuan et al., 2017a; El-Naggar et al., 2019b). In fact, soil contamination is complicated and rarely involves one contaminant. Multiple PTEs, including both metal and metalloids, are common. Thus, one single chelant or an amendment might be not efficient enough for complicated soil contamination cases, especially for those contaminated by both cationic and anionic PTEs. Therefore, methods using two or more techniques together, such as combining mobilization and immobilization techniques together, have been investigated recently.

Studies have supported that metals can be shifted to the weakly bound fraction and can increase the metal mobility after washing by chelants (Tsang and Hartley, 2014; Yang et al., 2015; Zhang et al., 2010b). Therefore, the key remedial goal of soil remediation should not only consider reducing the total contents of PTEs in soils but also the bioavailable/bioaccessible, mobilized, and leachable parts of PTEs. Additionally, the residual PTEs in soil, the variation of soil physicochemical and biological properties after remediation, and how to safely reuse the treated soils should also be considered in the current market (Tsang et al., 2013a; Udovic and Lestan, 2009). For example, soil enzyme activity, as one of key soil biological properties, can be a quick endpoint for evaluation of the remediation performance and can be affected by residual PTEs, washing reagents, and soil physicochemical properties (Im et al., 2015; Tica et al., 2011). In particular, activities of dehydrogenases (intracellular enzymes), β-glucosidase, urease, phosphatase, esterase, lipase, and protease are commonly evaluated in soils for being indicative of the average microbial population, reflecting the biogeochemical cycling of nitrogen and phosphate, etc. (Caldwell, 2005; Jelusic and Lestan, 2014). In addition, available and/or total nutrients (nitrogen, phosphorus, and potassium) are often evaluated for soil fertility.

14.2 COMBINED APPLICATION OF CHELANTS

As mentioned in Section 14.1.1, a reduced dosage of EDTA in metal extraction techniques in soils was suggested in the past ten years because of its non-biodegradable property, relatively high toxicity, and strong dissolution capacity, which might cause adverse effects on soil physicochemical and biological properties (Tsang et al., 2007; Zhang et al., 2010a). Luo et al. (2006) used a mixture usage of EDDS and EDTA at a 1:1 molecular ratio in a phytoextraction study to remove Pb from a soil, although previous studies found that EDTA is capable of higher Pb removal than EDDS because the stability of the Pb-EDTA complex is higher than that of Pb-EDDS (Tandy et al., 2004). A mixture usage of EDDS and EDTA (at a molar ratio of 1:2) caused a higher level of total phytoextracted Pb (2.1 and 6.1 times, respectively) by *Zea mays* L. compared with separate extractions using the same dosage of EDTA or EDDS alone. The authors concluded that EDDS increased the uptake and translocation of Pb from the roots to shoots of *Zea mays* L.

Yip et al. (2010) applied an equimolar mixture of EDDS and EDTA to remove Cu, Pb, and Zn in an artificially contaminated soil under deficient usage of chelant in a continuous extraction of over 30 h. The combined application of EDDS and EDTA extracted more Pb without noticeable Pb re-adsorption compared with using EDDS and EDTA separately at the same dosage. The authors suggested that the combination of chelants changed the metal speciation and contributed to less competition of chelants. In short, the mixture of chelants can make one chelant mainly focused on extraction to one or a less-specific metal, thus reducing the competition of metals to the chelant.

Beiyuan et al. (2018b) investigated the metal/chelant/metal-chelant complexes speciation in an equimolar mixture of EDDS and EDTA through the modeling results of Visual MINTEQ. The extraction results of a three-step washing within 3 h allowed the mixture of EDDS and EDTA to achieve comparable or even higher extraction efficiencies of Pb compared with using EDTA alone. The modeling results supported that an optimized chelant usage occurred, which allowed EDTA to focus on Pb and Zn extraction while EDDS focused on Cu removal. The modeling results also support the hypothesis by Yip et al. (2010). However, after a strong removal of Pb by EDTA in the first washing step, more and more Cu and Zn were complexed with EDTA than EDDS in the second and third washing steps. In contrast, more and more EDDS existed in an uncomplexed form in the system, which can lead to EDDS causing more mineral dissolution (mainly for Al oxides) before its complete degradation. Additionally, the mixed usage of chelants can reduce leachabilities (evaluated by TCLP, toxicity characteristic leaching procedure, US EPA Method 1311) of Cu, Zn, and Pb, mainly caused by EDTA, especially for those washing for over 24 h. The bioavailability of the PTEs evaluated by *in vitro* simplified bioaccessibility extraction test (SBET) was also reduced in the case of using mixed chelants.

Guo et al. (2017, 2019) used a mixture of EDTA, GLDA, and citric acid at a molecular ratio of 1:1:3 to remove Cd, Zn,

Pb, and Cu in soils. The pH condition was optimized to acidic condition for GLDA extraction by adding citric acid. The mixture of chelants successfully reduced the metal bioavailability. Slightly higher removal extraction efficiencies of all the targeted PTEs were achieved, although the usage of EDTA, which is non-biodegradable and toxic, was markedly reduced by 80%.

14.3 SELECTIVE DISSOLUTION BEFORE CHELANT-ENHANCED EXTRACTION

Insufficient extraction efficiencies of PTEs in soils by chelants have been reported (Beiyuan et al., 2016; Tsang et al., 2013a). In addition, field application of chelants is still a challenge, as chelants are relatively mild compared with some strong acid and alkaline reagents and the lack of selectivity of complexation during the extraction, especially for the mineral cations in soils (e.g., Fe, Al, Ca, etc.) (Komarek et al., 2009, 2010; Wang et al., 2012; Zhang and Tsang, 2013). Serious mineral dissolution in the soils can lead to unfavorable changes in soil physiochemical properties. Furthermore, the chelants have poor extraction capacities on anions, such as As, because they can hardly form stable complexes with chelants (Tokunaga and Hakuta, 2002). However, in field conditions, multiple PTEs, in both cationic and anionic forms, normally exist in contaminated soils. Therefore, washing reagents only suitable for cationic metals should be avoided. Additionally, chelant-enhanced extraction might be hindered by the physiochemical properties, contamination time, and type of contamination of soils. For example, chelants can extract the labile metals in soils, e.g., exchangeable and carbonate fractions, while it is difficult (or time consuming) to extract PTEs with affinity to Fe/Mn oxides, organic matter, and residue parts (Tsang and Hartley, 2014; Udovic and Lestan, 2010). The capacity of extracting PTEs from Fe/Mn oxides is varied, which means the capacities of chelants to dissolve Fe/Mn-containing minerals are different.

To overcome the mentioned drawbacks of chelant-enhanced extraction and release of PTEs bound to Fe/Mn oxides and/or organic matter, methods of selective dissolution have been proposed to use several types of reagents as pretreatments to selectively dissolve different parts in soils: (i) reducing agents to dissolve oxides, such as oxalate, ascorbic acid, hydroxylamine hydrochloride, and dithionite-enhanced methods (Im et al., 2015; Kim and Baek, 2015; Kim et al., 2015); (ii) oxidizing agents to dissolve organic matter, such as hydrogen peroxide and sulfate (He et al., 2013; Pham et al., 2012; Yan and Lo, 2012); (iii) alkaline acids, such as hydroxides and carbonates; (iv) organic ligands, such as oxalate and citrate, which can complex with PTEs and dissolve oxides (Drahota et al., 2014); and (v) moderate inorganic acids, such as phosphoric acid. Many attempts have been reported to use these reagents to selectively dissolve in multiple-step extractions. There is also evidence to support the idea that multiple-step washing by chelants, water, or other rinsing reagents can enhance the removal efficiencies of PTEs (Beiyuan et al., 2018b; Gusiatin and Klimiuk, 2012).

14.3.1 REDUCTANTS

As mentioned above, the use of EDDS to extract PTEs bound to Fe/Mn oxides and to form stable complexes with As is difficult (Tokunaga and Hakuta, 2002; Tsang and Yip, 2014; Tsang et al., 2014). However, As is expected to have a strong affinity to iron oxides, such as goethite and hematite; thus, the reductants can enhance its removal efficiency by dissolving the oxides (Smedley and Kinniburgh, 2002). Therefore, reducing agents, such as dithionite of the DCB method (sodium dithionite buffered with citrate and bicarbonate), acidified hydroxylamine hydrochloride (NH_2OH-HCl), and ascorbic acid, can selectively dissolve the Fe/Mn oxide before a chelant-enhanced extraction and are expected to enhance the removal of PTEs (Beiyuan et al., 2018a).

14.3.1.1 Dithionites

The DCB methods released large amounts of dissolved Fe and Mn in the solution (Im et al., 2015; Kim et al., 2015). The key player of the DCB method, dithionite ($S_2O_4^{2-}$), is a strong reductant (standard reduction potential at pH 7.3 (E^0) = 0.7 V) and follows Reactions 14.1 and 14.2 (Shang and Zelazny, 2008):

$$\text{Basic system: } S_2O_4^{2-} + 4OH^- \leftrightarrow 2SO_3^{2-} + 2H_2O + 2e^-$$
$$(14.1)$$

$$\text{Acidic system: } HS_2O_4^- + H_2O \leftrightarrow 2H_2SO_3 + H^+ + 2e^- \quad (14.2)$$

Citrate of the DCB method has a strong complex capacity with different metals by forming metal-citrate complexes to be extracted in solution. In addition, it can assist in the dissolution of Fe oxides by forming Fe-citrate, it is required after the reduction/dissolution of Fe/Mn oxides, and it can help to reduce the amount of metal re-adsorption. Therefore, an additional washing step with chelants can help to remove the metal-citrate complexes. Beiyuan et al. (2018a) found that only marginal amounts of Cu were removed by the DCB method. This might be because Cu-citrate complexes can be easily entrapped/re-adsorbed on the soil surface. In contrast, Cr(III)-citrate complexes, which are expected to generate at pH 7.3, showed little sorption onto soil surfaces (Puzon et al., 2008; Sarkar et al., 2013). Arsenite (As(III)) was expected as the dominant species in the strong reducing condition, and the neutral pH condition (7.3) is expected to be unfavorable for As adsorption (Gorny et al., 2015; Hartley et al., 2009). The strong reduction condition caused by dithionite can greatly dissolve Fe/Mn oxides, yet the decrease of valence of As might increase its mobility and toxicity because As(III) has higher mobility and toxicity than As(V). Furthermore, the DCB method also enhanced the leachability of Cr, Cu, and As, as evaluated by the TCLP method, yet their bioaccessibilities, studied by SBET, were all reduced.

Kim et al. (2016) also investigated the effects of a combination of dithionite and EDTA after 24-h extraction. Dithionite can dissolve a large amount of iron oxides and enhanced the removal of As, especially for the PTEs strongly bound to crystalline oxides. This further supports the potential of using dithionite before chelant-enhanced extraction. However, the combination of dithionite and EDTA showed limited removal efficiency of Cu, which might be caused by the formation of precipitation of Cu sulphides due to the self-decomposition of dithionite.

14.3.1.2 Hydroxylamine Hydrochloride

Acidic NH_2OH-HCl can contribute to a strong dissolution of amorphous iron and manganese oxides/hydroxides, and the dissolution reaction of Mn oxides by NH_2OH is presented in Reaction 14.3 (Chao and Zhou, 1983):

$$2NH_2OH + MnO_2 \rightleftharpoons Mn + N_2O(g) + 3H_2O \quad (14.3)$$

The acidified NH_2OH-HCl can only act as a reducing reagent and release the PTEs at the same time, while the DCB method includes steps of dissolution and complexation at the same time. This is further supported by the fact that the second step of EDDS washing in Beiyuan et al. (2018a) significantly enhanced PTE extraction by the DCB method, but not for the acidified NH_2OH-HCl extraction. Similarly, the leachability of PTEs after washing by acidified NH_2OH-HCl and EDDS reached a similar level as EDDS-enhanced extraction. However, cytotoxicity verified by luminescent bacteria (*V. fischeri*) was enhanced before and after washing by the acidified NH_2OH-HCl. Available nutrients (nitrogen and phosphorus) and relative enzyme activities (urease and acid phosphatase) were inhibited by the additional washing of acidified NH_2OH-HCl.

14.3.1.3 Ascorbic Acid

Ascorbic acid ($C_6H_8O_6$) is a mild reducing agent, with a standard reduction potential of 0.4 V compared to dithionite and NH_2OH-HCl. It can dissolve amorphous iron oxides, such as ferrihydrite and poorly crystallized lepidocrocite (Kim et al., 2015). A combination of ascorbic acid and EDTA enhanced the extraction of PTEs with a promoted amount of Fe dissolution. This indicates a possibility of using ascorbic acid before chelant-enhanced extraction.

14.3.2 Oxidants

Similarly, oxidants can be used to promote the dissolution of organic matter in soils, which can be beneficial for improving the extraction of PTEs by chelant-enhanced extraction.

14.3.2.1 Hydrogen Peroxide

Hydrogen peroxide (H_2O_2) and sodium hypochlorite ($NaClO$) are common and effective oxidizers for selectively removing soil organic matter, and they are commonly used in soil analysis methods (Shang and Zelazny, 2008). The major process is shown in Reaction 14.4:

$$C^0 + 2H_2O_2 \rightarrow CO_2 + 2H_2O \quad (14.4)$$

In addition to the decomposition of soil organic matter, Mn oxides can also be deconstructed by H_2O_2 as in Reaction 14.5, which might contribute to enhancing the removal of PTEs associated with Mn oxides:

$$MnO_2 + H_2O_2 + 2H^+ \rightarrow Mn^{2+} + 2H_2O + O_2 \quad (14.5)$$

Furthermore, hydrogen peroxide can react with amorphous Fe oxides via a catalyst-like (Fenton-like) reaction and generate OH· radicals. The Fenton-like reaction can oxidize organic matter in soils and release Cu and Pb (Mikutta et al., 2005). However, the usage of strong oxidants such as H_2O_2 can cause 60% to 80% loss of nutrients (carbon and nitrogen) and degradation of the water retention capacity (Sirguey et al., 2008).

14.3.2.2 Sodium Hypochlorite

Soil organic matter can be dissolved by $NaClO$ as in Reaction 14.6 (Shang and Zelazny, 2008):

$$C^0 + 2NaOCl \rightarrow CO_2 + 2NaCl \quad (14.6)$$

The $NaClO$ is relatively mild compared with H_2O_2. It cannot dissolve Mn oxides, enhance acidity, which might contribute to dissolution of mineral components, or perform a Fenton-like reaction. Consistently, H_2O_2 before EDDS-enhanced extraction improved the removal of Pb and Cu, while $NaClO$ showed limited improvement on the extraction efficiencies of the PTEs (Beiyuan et al., 2017b).

14.3.2.3 Oxone

Oxone (potassium peroxymonosulfate, $2KHSO_5 \cdot KHSO_4 \cdot K_2SO_4$) is commonly used as a cleanser, bleach, and other similar applications in industry. It is also applied as a transition metal activator for the degradation of some persistent organic pollutants in wastewater. It is regarded as an eco-friendly reagent because its main by-products, potassium salts, are harmless to the environment. The mechanisms of oxidation can be described as below (Nalliah, 2015), in Reactions 14.7 and 14.8:

$$Fe^{2+} + HSO_5^- \rightarrow Fe^{3+} + SO_4^{\bullet-} + OH^- \quad (14.7)$$

$$SO_4^{\bullet-} + organics \rightarrow [nSteps] \rightarrow CO_2 + H_2O + SO_4^{2-} \quad (14.8)$$

In the soil solution system, the co-presence of transition metals, e.g., Fe, can contribute to the generation of SO_4^{2-} radicals, which is the key component to the oxidation of soil organic matter (Liu et al., 2014; Nalliah, 2015). The combination of oxone and EDDS-enhanced extraction additionally removed 30% of Cu and Zn in soils due to the oxidation of soil organic matter by the SO_4^{2-} radicals of oxone (Beiyuan et al., 2017b). Similar to H_2O_2, dissolved transition metals, such as Fe^{2+}, coupled with oxone can speed up the oxidation by forming a catalyst-like reaction (Anipsitakis and Dionysiou, 2003).

14.3.3 ALKALINE SOLVENTS

14.3.3.1 Sodium Carbonate and Sodium Hydroxide

Strong alkaline conditions caused by alkaline solvents can dissolve aluminosilicates and release the As bound to these aluminum-containing minerals (Wang et al., 1981), desorb arsenate ions with via ion exchanges (Jang et al., 2007), and prevent re-adsorption due to electrostatic repulsion (Kim et al., 2015). Sodium carbonate ($NaCO_3$) and sodium hydroxide (NaOH) were used to enhance EDDS-enhanced extraction in the studies of Beiyuan et al. (2017b) and Beiyuan et al. (2018a). The carbonate and hydroxide extracted the highest amount of As in Beiyuan et al. (2018a) and improved Cu removal efficiency in both studies. However, the strong alkaline condition blocked the dissolution of Fe/Mn oxides, which led to marginal amounts of dissolved Fe and Mn. This effect contributed to the decreased removal of PTEs. This is probably because free Fe oxides generated under alkaline conditions can lead to blocking effects of Fe oxides on dissolution. In addition, low extraction efficiencies of Zn, Pb, and Cr were found. This might be because the metal carbonates and hydroxides formed have low solubility ($K_{sp(Zn(OH)_2)} = 3.0 \times 10^{-16}$, $K_{sp(ZnCO_3)} = 1.4 \times 10^{-11}$, $K_{sp(Pb(OH)_2)} = 1.4 \times 10^{-20}$, $K_{sp(PbCO_3)} = 1.4 \times 10^{-14}$, $K_{sp(Cr(OH)_2)} = 1.6 \times 10^{-16}$, and $K_{sp(Cr(OH)_3)}) = 6.3 \times 10^{-31}$ (Haynes, 2015). However, soluble $[Cu(OH)_4]^{2-}$ and $[Cr(OH)_4]^-$ can be formed via hydrolysis of $Cu(OH)_2$ and $Cr(OH)_3$ under strongly alkaline conditions. Additionally, the leachability of Cu sharply increased after washing by sodium carbonate and sodium hydroxide, which might be related to the dissolution of the precipitations generated by Cu.

14.3.4 ORGANIC LIGANDS

Similar to LMWOAs, organic ligands can be applied to enhance the extraction of PTEs in soils, such as oxalate ($C_2O_4^{2-}$) and citrate ($C_6H_5O_7^{3-}$). The two ligands are adopted in the selective methods on the removal of Fe oxides: acidified ammonium oxalate under darkness (AOD) and the DCB method, respectively (Shang and Zelazny, 2008). Oxalate and citrate can promote soil surface protonation and destabilization of Fe-O bonds in the solid lattice (Blesa et al., 1994).

14.3.4.1 Oxalate

In addition to organic ligands, oxalate is a mild reducing agent with a standard reduction potential (E^0) of -0.18 V and can selectively dissolve non-crystalline and poorly crystalline (hydr)oxides and aluminosilicates of Al, Fe, and Mn (Hodges and Zelazny, 1980; Shang and Zelazny, 2008). Consequently, oxalate can enhance the extraction of PTEs of non-crystalline and poorly crystalline (hydr)oxides and aluminosilicates but showed limited effects on the extraction of crystalline (hydr)oxides (Beiyuan et al., 2018a). However, a high dosage of oxalate might not be suitable for removing Pb before the EDDS extraction. Beiyuan et al. (2017b) found

that a high dosage (0.2 M) of oxalate inhibited the formation of Pb-EDDS complex but that a low dosage (2.2 mM) of oxalate could help to dissolve mineral components and promoted the removal of Pb. This might have contributed to the formation of Pb-oxalate (PbC_2O_4) precipitation, which has a low solubility product ($K_{sp(Pb-oxalate)} = 2.74 \times 10^{-11}$). In addition, oxalate significantly affected dehydrogenase activity, acid phosphatase activity, and urease activity after EDDS-enhanced extraction (Beiyuan et al., 2017b, 2018a), which is positive for soil remediation.

Because of the strong ligand property of oxalate, it can also be used as a chelant in extraction. Kim and Baek (2015) studied a combination of dithionite and oxalate to enhance the removal of As in a contaminated soil. The release of As(III) from the dissolution of Fe oxides might be easily re-adsorbed or incorporated onto the newly formed Fe oxides again (Kim et al., 2015). Oxalate complexed with Fe prevented the formation of new Fe oxides and therefore enhanced the iron oxide dissolution through a non-reductive dissolution. Furthermore, Wei et al. (2016) found that a combination of phosphoric acid (H_3PO_4), oxalic acid, and EDTA can facilitate the dissolution of minerals and residual fraction and enhance the extraction of As and Cd. The labile fractions of the As and Cd in soils were also reduced by the combination of PO_4^{3-}, $C_2O_4^{2-}$, and EDTA.

14.3.4.2 Citrate

Similar to oxalate, citrate can be used for removing the PTEs bound to organic matter in soil (Hwang et al., 2015). Citrate has been applied in soil and sediments extraction because it can also form stable complexes with metals, such as $[M–(citrate)_2]^{4+}$, $[M–(citrate)]^+$, $[MH–citrate(aq)]$, and $[MH_2–citrate]^+$ (Yoo et al., 2013). Among these forms, the $[M–(citrate)_2]^{4+}$ form was the most important one for metal complexation in the extractant. However, when citrate is the only extractant, it can greatly promote the dissolution of Fe oxides. This can increase the competition with target PTEs and inhibit the removal efficiencies of targeted PTEs. To avoid the undesired mineral dissolution, a coupled extraction of citrate and chelant might enhance the PTE extraction. Beiyuan et al. (2017b) investigated a pretreatment of citrate before EDDS extraction that elevated the extraction efficiencies of Cu, Zn, and Pb. However, slightly lower efficiencies of PTEs were observed because the removal efficiencies of PTEs bound to crystalline oxides were lower compared with oxalate (Beiyuan et al., 2018a). However, the variations of enzyme activities, such as dehydrogenase activity and urease activity, and the loss of total nitrogen and phosphorus after washing were mild compared with oxalate.

14.4 STABILIZATION BY BIOCHAR AND LOW-COST AMENDMENTS COMBINED WITH CHELANT-ENHANCED EXTRACTION

Chelant-enhanced extraction may cause residual metal-chelant complexes, which can disturb the metal distribution, shifting the metals to weakly bound fractions in soil (Tsang and Hartley, 2014; Yang et al., 2015). It can also inhibit (co-)

precipitation of PTEs and increase the long-term risk of release of PTEs after extraction. Unlike chelant-enhanced extraction, stabilization by low-cost amendment faces other challenges, such as long-term stabilities of the stabilized PTEs, especially facing varied environment conditions (Bolan et al., 2014; Komarek et al., 2013). Some studies suggested that even with stabilization by biochar or dewatered sludge, significant amounts of leaching PTEs were still observed under continuous leaching and acidic conditions (Houben et al., 2013; Tsang et al., 2013b). Therefore, a combination of chelant-enhanced extraction and stabilization by low-cost amendments such as biochar can be beneficial. The chelant-enhanced extraction leads to a pre-extraction of labile and weakly bound PTEs, while the amendments can help to reduce the potential release of metal-chelant complexes and/or re-adsorbed PTEs. In addition, the low-cost amendments could be good for recovering physicochemical properties of the treated soils.

14.4.1 BIOCHAR STABILIZATION COMBINED WITH CHELANT-ENHANCED EXTRACTION

Beiyuan et al. (2016) showed that the leachabilities (examined by TCLP method) of Cu, Zn, and Pb were successfully reduced more than 50% after 2-h EDDS-enhanced extraction compared to the untreated soil. Interestingly, the mobility (examined by the SPLP method, the synthetic precipitation leaching procedure US EPA Method 1312) of Cu was significantly elevated by EDDS extraction. The possible reason is that more EDDS tends to form Cu-EDDS complexes, which can be re-adsorbed on the treated soil particles, in contrast to complexes with Zn and Pb (Tsang et al., 2009). These metal-chelant complexes can be easily washed out and transported in soil, which is considered as the exchangeable/labile fraction (Tsang et al., 2013a; Tsang et al., 2013b). Furthermore, the extractant of TCLP is more aggressive than that of SPLP. More strongly bound metals were leached out by TCLP besides the increased leachability caused by residual Cu species. With a 2-month stabilization by soybean stover biochars produced at 300°C and 700°C, respectively, labile Cu was increased, and leachable PTEs by EDDS-enhanced extraction were dramatically reduced. The potential mechanisms mainly included the promoted alkalinity of soil and the rich surface area and functional groups. In contrast, the bioaccessible (examined by *in vitro* simple bioaccessibility extraction test, [Luo et al., 2012]) and plant-available (examined by DTPA extraction, [Luo et al., 2012]) PTEs were markedly decreased by the EDDS-enhanced extraction. The bioaccessible and plant-available PTEs were further reduced by the additional 2-month stabilization by biochars; yet, the biochar can only slightly reduce the bioaccessible and plant-available PTEs without the extraction by EDDS.

Guo et al. (2019) evaluated a mixture of chelants (EDTA, GLDA, and citric acid, at a molar ratio of 1:1:3) with organic amendments of biochar (produced at 600°C) and chicken manure. The mixed chelant washing significantly reduced the bioavailability (evaluated by $MgCl_2$ extraction) of Cd,

Pb, and Zn; yet soil pH, soil organic matter, and total and available/exchangeable nutrients were altered by the chelant-enhanced washing. With an additional amendment by biochar and chicken manure, the bioavailability of Pb was noticeably reduced (36%–65% and 64%–80%, respectively). The bioavailability of Cd was reduced by –3%–7% and 9%–28%, respectively. In contrast, the bioavailable Zn was only slightly changed or even increased, which might be due to the high content of Zn in the chicken manure (670.9 mg kg^{-1}) compared with biochar (57.6 mg kg^{-1}). A pot experiment showed that the application of biochar after the mixed chelant-enhanced washing elevated the total nutrients and available nutrients, therefore causing increments of shoot growth and seed germination.

14.4.2 BIOCHAR STABILIZATION COMBINED WITH FERRIC/CHELANT EXTRACTION

Yoo et al. (2018) studied the efficacy of biochar stabilization of a shooting range soil and a railway site soil after ferric nitrate ($Fe(NO_3)_3$) extraction and compared it with EDDS-enhanced extraction. Similar to EDDS extraction, a sludge biochar stabilization can reduce the mobilized metals caused by $Fe(NO_3)_3$ extraction. The mobilities of Cu and Pb (evaluated by European Council Waste Acceptance Criteria (ECWAC, prCEN/TS 12457-3) leaching test) were enhanced by the $Fe(NO_3)_3$ in the shooting range soil, which has more metals bound to exchangeable and soluble fractions, especially for Pb (95.6%). In contrast, the mobilities of both Cu and Pb were not increased by the $Fe(NO_3)_3$ extraction in the railway site soil, which has a higher amount of metals in the Fe-Mn oxy-hydroxides fraction; in addition, the extraction efficiencies of Cu and Pb were lower. The $Fe(NO_3)_3$ produces hydrogen ions that can improve the extraction of metals via ion exchange from soils by ferric iron hydrolysis, yet the ion exchange can be restricted between metals bound to Fe/Mn oxides and protons (Yoo et al., 2017a, 2017b). Furthermore, the study showed that the additional stabilization by biochars after $Fe(NO_3)_3$ extraction had marginal effects on controlling the bioaccessibility of Cu and Pb; in some cases, the bioaccessibility was promoted after stabilization. Additionally, elevated bioaccessibilities (evaluated by SBET method) of both Pb and Cu after a 2-month sludge biochar stabilization were found in both the shooting range soil and the railway site soil, which might be associated with the higher metal content in the sludge biochar. This further supports the idea that the combination of biochar stabilization and chemically enhanced extraction can be further tested in the field.

14.5 THE WAY FORWARD

Chelant-enhanced extraction is commonly used to remove PTEs in soils. Some of the latest findings were summarized in this chapter to improve chelant-enhanced extraction, especially for the biodegradable chelant EDDS, including the low extraction capacity to metals bound to Fe/Mn oxides and organic matter, the increased leachability and mobility of PTEs after extraction, and the low removal efficiency on metalloids, in view of its limitations. The combination of different chelants at various molecular ratios can reduce the amount of EDTA, redistribute and optimize the usage of chelants, and improve the removal of targeted PTEs. Selective dissolution by reductants (dithionites, NH_2OH-HCl, and ascorbic acid), oxidants (H_2O_2, NaClO, and oxone), alkaline solvents (sodium carbonate and sodium hydroxide), and organic ligands (oxalate and citrate) can promote the dissolution of Fe/Mn/Al oxides and/or organic matter, thus enhancing the removal of PTEs. A cost-effective biochar stabilization after chelant-enhanced extraction mitigated leachability and mobility of PTEs elevated by chelants and further controlled the bioaccessibility.

In view of the booming industrial market of soil remediation with higher standards, research in PTE-contaminated soil remediation requires an overall evaluation of soil properties rather than a mitigated total content of PTEs. The new findings reviewed in this chapter propose a future trend that one typical technique might not be sufficient for PTE-contaminated soils. Higher removal efficiencies of both metals and metalloids without greatly altering the soil properties in both physicochemical and biological are expected in the future. The combined application of traditional and new techniques reviewed in this study can optimize the removal results and reach remediation goals in a more environmentally friendly way. Therefore, further exploration of the combined techniques can be studied in field conditions, and more reasonable combinations of the current techniques besides those mentioned in this study can be investigated to meet the future requirements of soil remediation.

REFERENCES

Ahmad, M.; Lee, S.S.; Lee, S.E.; Al-Wabel, M.I.; Tsang, D.C.W.; Ok, Y.S. Biochar-induced changes in soil properties affected immobilization/mobilization of metals/metalloids in contaminated soils. *J Soil Sediment* 2016 17(3), 717–730. doi:10.1007/s11368-015-1339-4.

Ahmad, M.; Rajapaksha, A.U.; Lim, J.E.; Zhang, M.; Bolan, N.; Mohan, D.; Vithanage, M.; Lee, S.S.; Ok, Y.S. Biochar as a sorbent for contaminant management in soil and water: A review. *Chemosphere* 2014;99:19–33.

Anipsitakis, G.P.; Dionysiou, D.D. Degradation of organic contaminants in water with sulfate radicals generated by the conjunction of peroxymonosulfate with cobalt. *Environmental Science & Technology* 2003;37:4790–4797.

Arwidsson, Z.; Elgh-Dalgren, K.; von Kronhelm, T.; Sjoberg, R.; Allard, B.; van Hees, P. Remediation of heavy metal contaminated soil washing residues with amino polycarboxylic acids. *J Hazard Mater* 2010;173:697–704.

Beckers, F.; Awad, Y.M.; Beiyuan, J.; Abrigata, J.; Mothes, S.; Tsang, D.C.W.; Ok, Y.S.; Rinklebe, J. Impact of biochar on mobilization, methylation, and ethylation of mercury under dynamic redox conditions in a contaminated floodplain soil. *Environ Int* 2019;127:276–290.

Beiyuan, J.; Awad, Y.M.; Beckers, F.; Tsang, D.C.W.; Ok, Y.S.; Rinklebe, J. Mobility and phytoavailability of As and Pb in a contaminated soil using pine sawdust biochar under systematic change of redox conditions. *Chemosphere* 2017a;178:110–118.

Beiyuan, J.; Lau, A.Y.T.; Tsang, D.C.W.; Zhang, W.; Kao, C.-M.; Baek, K.; Ok, Y.S.; Li, X.-D. Chelant-enhanced washing of CCA-contaminated soil: Coupled with selective dissolution or soil stabilization. *Sci Total Environ* 2018a;612C:1463–1472.

Beiyuan, J.; Tsang, D.C.W.; Ok, Y.S.; Zhang, W.; Yang, X.; Baek, K.; Li, X.D. Integrating EDDS-enhanced washing with low-cost stabilization of metal-contaminated soil from an e-waste recycling site. *Chemosphere* 2016;159:426–432.

Beiyuan, J.; Tsang, D.C.W.; Valix, M.; Baek, K.; Ok, Y.S.; Zhang, W.; Bolan, N.S.; Rinklebe, J.; Li, X.-D. Combined application of EDDS and EDTA for removal of potentially toxic elements under multiple soil washing schemes. *Chemosphere* 2018b;205:178–187.

Beiyuan, J.; Tsang, D.C.W.; Valix, M.; Zhang, W.; Yang, X.; Ok, Y.S.; Li, X.D. Selective dissolution followed by EDDS washing of an e-waste contaminated soil: Extraction efficiency, fate of residual metals, and impact on soil environment. *Chemosphere* 2017b;166:489–496.

Blesa, M.A.; Morando, P.J.; Regazzoni, A.E. *Chemical Dissolution of Metal Oxides.* CRC Press, Boca Raton, FL; 1994.

Bolan, N.; Kunhikrishnan, A.; Thangarajan, R.; Kumpiene, J.; Park, J.; Makino, T.; Kirkham, M.B.; Scheckel, K. Remediation of heavy metal(loid)s contaminated soils–to mobilize or to immobilize? *J Hazard Mater* 2014;266:141–166.

Bucheli-Witschel, M.; Egli, T. Environmental fate and microbial degradation of aminopolycarboxylic acids. *FEMS Microbiol Rev* 2001;25:69–106.

Caldwell, B.A. Enzyme activities as a component of soil biodiversity: A review. *Pedobiologia* 2005;49:637–644.

Chao, T.T.; Zhou, L. Extraction techniques for selective dissolution of amorphous iron oxides from soils and sediments. *Soil Sci Soc Am J* 1983;47:225–232.

Drahota, P.; Grosslova, Z.; Kindlova, H. Selectivity assessment of an arsenic sequential extraction procedure for evaluating mobility in mine wastes. *Anal Chim Acta* 2014;839:34–43.

Duo, L.; Yin, L.; Zhang, C.; Zhao, S. Ecotoxicological responses of the earthworm Eisenia fetida to EDTA addition under turfgrass growing conditions. *Chemosphere* 2019;220:56–60.

El-Naggar, A.; Lee, S.S.; Rinklebe, J.; Farooq, M.; Song, H.; Sarmah, A.K.; Zimmerman, A.R.; Ahmad, M.; Shaheen, S.M.; Ok, Y.S. Biochar application to low fertility soils: A review of current status, and future prospects. *Geoderma* 2019a;337:536–554.

El-Naggar, A.; Shaheen, S.M.; Hseu, Z.Y.; Wang, S.L.; Ok, Y.S.; Rinklebe, J. Release dynamics of As, Co, and Mo in a biochar treated soil under pre-definite redox conditions. *Sci Total Environ* 2019b;657:686–695.

Gorny, J.; Billon, G.; Lesven, L.; Dumoulin, D.; Made, B.; Noiriel, C. Arsenic behavior in river sediments under redox gradient: A review. *Sci Total Environ* 2015;505:423–434.

Grčman, H.; Vodnik, D.; Velikonja-Bolta, Š.; Leštan, D. Ethylenediaminedissuccinate as a new chelate for environmentally safe enhanced lead phytoextraction. *J Environ Qual* 2003;32:500–506.

Guo, X.; Yang, Y.; Ji, L.; Zhang, G.; He, Q.; Wei, Z.; Qian, T.; Wu, Q. Revitalization of mixed chelator–washed soil by adding of inorganic and organic amendments. *Water Air Soil Poll* 2019;230.

Guo, X.; Zhang, G.; Wei, Z.; Zhang, L.; He, Q.; Wu, Q.; Qian, T. Mixed chelators of EDTA, GLDA, and citric acid as washing agent effectively remove Cd, Zn, Pb, and Cu from soils. *J Soil Sediment* 2017;

Gusiatin, Z.M.; Klimiuk, E. Metal (Cu, Cd and Zn) removal and stabilization during multiple soil washing by saponin. *Chemosphere* 2012;86:383–391.

Hartley, N.R.; Tsang, D.C.W.; Olds, W.E.; Weber, P.A. Soil washing enhanced by humic substances and biodegradable chelating agents. *Soil Sediment Contam* 2014;23:599–613.

Hartley, W.; Dickinson, N.M.; Riby, P.; Lepp, N.W. Arsenic mobility in brownfield soils amended with green waste compost or biochar and planted with Miscanthus. *Environ Pollut* 2009;157:2654–2662.

Haynes, W.M. *CRC Handbook of Chemistry and Physics*, 96th edition. CRC Press, Boca Raton, FL; 2015.

He, X.X.; de la Cruz, A.A.; Dionysiou, D.D. Destruction of cyanobacterial toxin cylindrospermopsin by hydroxyl radicals and sulfate radicals using UV-254 nm activation of hydrogen peroxide, persulfate and peroxymonosulfate. *J Photochem Photobio A* 2013;251:160–166.

Hodges, S.C.; Zelazny, L.W. Determination of noncrystalline soil components by weight difference after selective dissolution. *Clays Clay Miner* 1980;28:35–42.

Houben, D.; Evrard, L.; Sonnet, P. Mobility, bioavailability and pH-dependent leaching of cadmium, zinc and lead in a contaminated soil amended with biochar. *Chemosphere* 2013;92:1450–1457.

Hussain, M.; Farooq, M.; Nawaz, A.; Al-Sadi, A.M.; Solaiman, Z.M.; Alghamdi, S.S.; Ammara, U.; Ok, Y.S.; Siddique, K.H.M. Biochar for crop production: Potential benefits and risks. *J Soil Sediment* 2016 17(3), 685–716. doi:10.1007/s11368-016-1360-2.

Hwang, B.R.; Kim, E.J.; Yang, J.S.; Baek, K. Extractive and oxidative removal of copper bound to humic acid in soil. *Environ Sci Pollut Res Int* 2015;22:6077–6085.

Im, J.; Yang, K.; Jho, E.H.; Nam, K. Effect of different soil washing solutions on bioavailability of residual arsenic in soils and soil properties. *Chemosphere* 2015;138:253–258.

Jang, M.; Hwang, J.S.; Choi, S.I. Sequential soil washing techniques using hydrochloric acid and sodium hydroxide for remediating arsenic-contaminated soils in abandoned iron-ore mines. *Chemosphere* 2007;66:8–17.

Jelusic, M.; Lestan, D. Effect of EDTA washing of metal polluted garden soils. Part I: Toxicity hazards and impact on soil properties. *Sci Total Environ* 2014;475:132–141.

Kim, E.J.; Baek, K. Enhanced reductive extraction of arsenic from contaminated soils by a combination of dithionite and oxalate. *J Hazard Mater* 2015;284:19–26.

Kim, E.J.; Jeon, E.K.; Baek, K. Role of reducing agent in extraction of arsenic and heavy metals from soils by use of EDTA. *Chemosphere* 2016;152:274–283.

Kim, E.J.; Lee, J.C.; Baek, K. Abiotic reductive extraction of arsenic from contaminated soils enhanced by complexation: Arsenic extraction by reducing agents and combination of reducing and chelating agents. *J Hazard Mater* 2015;283:454–461.

Komarek, M.; Vanek, A.; Ettler, V. Chemical stabilization of metals and arsenic in contaminated soils using oxides–A review. *Environ Pollut* 2013;172:9–22.

Komarek, M.; Vanek, A.; Mrnka, L.; Sudova, R.; Szakova, J.; Tejnecky, V.; Chrastny, V. Potential and drawbacks of EDDS-enhanced phytoextraction of copper from contaminated soils. *Environ Pollut* 2010;158:2428–2438.

Komarek, M.; Vanek, A.; Szakova, J.; Balik, J.; Chrastny, V. Interactions of EDDS with Fe- and Al-(hydr)oxides. *Chemosphere* 2009;77:87–93.

Lamb, D.T.; Ming, H.; Megharaj, M.; Naidu, R. Heavy metal (Cu, Zn, Cd and Pb) partitioning and bioaccessibility in uncontaminated and long-term contaminated soils. *J Hazard Mater* 2009;171:1150–1158.

Lehmann, J.; Rillig, M.C.; Thies, J.; Masiello, C.A.; Hockaday, W.C.; Crowley, D. Biochar effects on soil biota: A review. *Soil Biol Biochem* 2011;43:1812–1836.

Lestan, D.; Luo, C.L.; Li, X.D. The use of chelating agents in the remediation of metal-contaminated soils: A review. *Environ Pollut* 2008;153:3–13.

Liu, H.; Bruton, T.A.; Doyle, F.M.; Sedlak, D.L. In situ chemical oxidation of contaminated groundwater by persulfate: Decomposition by Fe(III)- and Mn(IV)-containing oxides and aquifer materials. *Environ Sci Technol* 2014;48:10330–10336.

Luo, C.; Shen, Z.; Li, X.; Baker, A.J. Enhanced phytoextraction of Pb and other metals from artificially contaminated soils through the combined application of EDTA and EDDS. *Chemosphere* 2006;63:1773–1784.

Luo, C.; Wang, S.; Wang, Y.; Yang, R.; Zhang, G.; Shen, Z. Effects of EDDS and plant-growth-promoting bacteria on plant uptake of trace metals and PCBs from e-waste-contaminated soil. *J Hazard Mater* 2015;286:379–385.

Luo, X.S.; Yu, S.; Li, X.D. The mobility, bioavailability, and human bioaccessibility of trace metals in urban soils of Hong Kong. *Appl Geochem* 2012;27:995–1004.

Mikutta, R.; Kleber, M.; Kaiser, K.; Jahn, R. Review: Organic matter removal from soils using hydrogen peroxide, sodium hypochlorite, and disodium peroxodisulfate. *Soil Sci Soc Am J* 2005;69:120–135.

Nalliah, R.E. Oxone/Fe^{2+} degradation of food dyes: Demonstration of catalyst-like behavior and kinetic separation of color. *J Chem Educ* 2015;92:1681–1683.

Nguyen, B.T.; Lehmann, J.; Hockaday, W.C.; Joseph, S.; Masiello, C.A. Temperature sensitivity of black carbon decomposition and oxidation. *Environ Sci Technol* 2010;44:3324–3331.

Nowack, B. Environmental chemistry of aminopolycarboxylate chelating agents. *Environ Sci Technol* 2002;36:4009–4016.

Ok, Y.S.; Usman, A.R.; Lee, S.S.; Abd El-Azeem, S.A.; Choi, B.; Hashimoto, Y.; Yang, J.E. Effects of rapeseed residue on lead and cadmium availability and uptake by rice plants in heavy metal contaminated paddy soil. *Chemosphere* 2011;85:677–682.

Park, J.H.; Choppala, G.K.; Bolan, N.S.; Chung, J.W.; Chuasavathi, T. Biochar reduces the bioavailability and phytotoxicity of heavy metals. *Plant Soil* 2011a;348:439–451.

Park, J.H.; Lamb, D.; Paneerselvam, P.; Choppala, G.; Bolan, N.; Chung, J.W. Role of organic amendments on enhanced bioremediation of heavy metal(loid) contaminated soils. *J Hazard Mater* 2011b;185:549–574.

Peters, R.W. Chelant extraction of heavy metals from contaminated soils. *J Hazard Mater* 1999;66:151–210.

Pham, A.L.; Doyle, F.M.; Sedlak, D.L. Inhibitory effect of dissolved silica on H$_2$O$_2$ decomposition by iron(III) and manganese(IV) oxides: Implications for H$_2$O$_2$-based in situ chemical oxidation. *Environ Sci Technol* 2012;46:1055–1062.

Polettini, A.; Pomi, R.; Calcagnoli, G. Assisted washing for heavy metal and metalloid removal from contaminated dredged materials. *Water Air Soil Poll* 2009;196:183–198.

Puzon, G.J.; Tokala, R.K.; Zhang, H.; Yonge, D.; Peyton, B.M.; Xun, L. Mobility and recalcitrance of organo-chromium(III) complexes. *Chemosphere* 2008;70:2054–2059.

Rajapaksha, A.U.; Ahmad, M.; Vithanage, M.; Kim, K.R.; Chang, J.Y.; Lee, S.S.; Ok, Y.S. The role of biochar, natural iron oxides, and nanomaterials as soil amendments for immobilizing metals in shooting range soil. *Environ Geochem Health* 2015;37:931–942.

Rajapaksha, A.U.; Chen, S.S.; Tsang, D.C.W.; Zhang, M.; Vithanage, M.; Mandal, S.; Gao, B.; Bolan, N.S.; Ok, Y.S. Engineered/designer biochar for contaminant removal/immobilization from soil and water: Potential and implication of biochar modification. *Chemosphere* 2016;148:276–291.

Rinklebe, J.; Shaheen, S.M.; Frohne, T. Amendment of biochar reduces the release of toxic elements under dynamic redox conditions in a contaminated floodplain soil. *Chemosphere* 2016;142:41–47.

Sarkar, B.; Naidu, R.; Krishnamurti, G.S.; Megharaj, M. Manganese(II)-catalyzed and clay-minerals-mediated reduction of chromium(VI) by citrate. *Environ Sci Technol* 2013;47:13629–13636.

Shang, C.; Zelazny, L.W. Selective dissolution techniques for mineral analysis of soils and sediments. In: Ulery A.L., Richard Drees L., Eds., *Methods of Soil Analysis Part 5— Mineralogical Methods*. Soil Science Society of America, Madison, WI; 2008.

Sirguey, C.; Tereza de Souza e Silva, P.; Schwartz, C.; Simonnot, M.-O. Impact of chemical oxidation on soil quality. *Chemosphere* 2008;72:282–289.

Smedley, P.L.; Kinniburgh, D.G. A review of the source, behaviour and distribution of arsenic in natural waters. *Appl Geochem* 2002;17:517–568.

Tandy, S.; Ammann, A.; Schulin, R.; Nowack, B. Biodegradation and speciation of residual SS-ethylenediaminedisuccinic acid (EDDS) in soil solution left after soil washing. *Environ Pollut* 2006;142:191–199.

Tandy, S.; Bossart, K.; Mueller, R.; Ritschel, J.; Hauser, L.; Schulin, R.; Nowack, B. Extraction of heavy metals from soils using biodegradable chelating agents. *Environ Sci Technol* 2004;38:937–944.

Tica, D.; Udovic, M.; Lestan, D. Immobilization of potentially toxic metals using different soil amendments. *Chemosphere* 2011;85:577–583.

Tokunaga, S.; Hakuta, T. Acid washing and stabilization of an artificial arsenic-contaminated soil. *Chemosphere* 2002;46:31–38.

Tsang, D.C.W.; Hartley, N.R. Metal distribution and spectroscopic analysis after soil washing with chelating agents and humic substances. *Environ Sci Pollut Res Int* 2014;21:3987–3995.

Tsang, D.C.W.; Olds, W.E.; Weber, P. Residual leachability of CCA-contaminated soil after treatment with biodegradable chelating agents and lignite-derived humic substances. *J Soil Sediment* 2013a;13:895–905.

Tsang, D.C.W.; Olds, W.E.; Weber, P.A.; Yip, A.C.K. Soil stabilisation using AMD sludge, compost and lignite: TCLP leachability and continuous acid leaching. *Chemosphere* 2013b;93:2839–2847.

Tsang, D.C.W.; Yip, A.C.K. Comparing chemical-enhanced washing and waste-based stabilisation approach for soil remediation. *J Soil Sediment* 2014;14:936–947.

Tsang, D.C.W.; Yip, A.C.K.; Olds, W.E.; Weber, P.A. Arsenic and copper stabilisation in a contaminated soil by coal fly ash and green waste compost. *Environ Sci Pollut Res Int* 2014;21:10194–10204.

Tsang, D.C.W.; Yip, T.C.M.; Lo, I.M.C. Kinetic interactions of EDDS with soils. 2. metal–EDDS complexes in uncontaminated and metal-contaminated soils. *Environ Sci Technol* 2009;43:837–842.

Tsang, D.C.W.; Zhang, W.H.; Lo, I.M.C. Copper extraction effectiveness and soil dissolution issues of EDTA-flushing of artificially contaminated soils. *Chemosphere* 2007;68:234–243.

Uchimiya, M.; Chang, S.; Klasson, K.T. Screening biochars for heavy metal retention in soil: Role of oxygen functional groups. *J Hazard Mater* 2011;190:432–441.

Udovic, M.; Lestan, D. Pb, Zn and Cd mobility, availability and fractionation in aged soil remediated by EDTA leaching. *Chemosphere* 2009;74:1367–1373.

Udovic, M.; Lestan, D. Fractionation and bioavailability of Cu in soil remediated by EDTA leaching and processed by earthworms (*Lumbricus terrestris* L.). *Environ Sci Pollut Res Int* 2010;17:561–570.

Wang, A.; Luo, C.; Yang, R.; Chen, Y.; Shen, Z.; Li, X. Metal leaching along soil profiles after the EDDS application–A field study. *Environ Pollut* 2012;164:204–210.

Wang, C.; Kodama, H.; Miles, N.M. Effect of various pretreatments on x-ray diffraction patterns of clay fractions of podzolic B horizons. *Can J Soil Sci* 1981;61:311–316.

Wang, G.; Zhang, S.; Zhong, Q.; Xu, X.; Li, T.; Jia, Y.; Zhang, Y.; Peijnenburg, W.; Vijver, M.G. Effect of soil washing with biodegradable chelators on the toxicity of residual metals and soil biological properties. *Sci Total Environ* 2018;625:1021–1029.

Wei, M.; Chen, J.; Wang, X. Removal of arsenic and cadmium with sequential soil washing techniques using Na2EDTA, oxalic and phosphoric acid: Optimization conditions, removal effectiveness and ecological risks. *Chemosphere* 2016;156:252–261.

WHO. *Guidelines for Drinking-Water Quality*, 4th edition. World Health Organization, Geneva, Switzerland; 2017.

Yan, D.Y.S.; Lo, I.M.C. Pyrophosphate coupling with chelant-enhanced soil flushing of field contaminated soils for heavy metal extraction. *J Hazard Mater* 2012;199–200:51–57.

Yan, D.Y.S.; Yip, T.C.M.; Yui, M.M.; Tsang, D.C.W.; Lo, I.M.C. Influence of EDDS-to-metal molar ratio, solution pH, and soil-to-solution ratio on metal extraction under EDDS deficiency. *J Hazard Mater* 2010;178:890–894.

Yang, L.; Jiang, L.; Wang, G.; Chen, Y.; Shen, Z.; Luo, C. Assessment of amendments for the immobilization of Cu in soils containing EDDS leachates. *Environ Sci Pollut Res Int* 2015;22:16525–16534.

Yang, L.; Wang, G.; Cheng, Z.; Liu, Y.; Shen, Z.; Luo, C. Influence of the application of chelant EDDS on soil enzymatic activity and microbial community structure. *J Hazard Mater* 2013;262:561–570.

Yip, T.C.M.; Tsang, D.C.W.; Ng, K.T.W.; Lo, I.M.C. Kinetic interactions of EDDS with soils. 1. metal resorption and competition under EDDS deficiency. *Environ Sci Technol* 2009;43:831–836.

Yip, T.C.M.; Yan, D.Y.S.; Yui, M.M.; Tsang, D.C.W.; Lo, I.M.C. Heavy metal extraction from an artificially contaminated sandy soil under EDDS deficiency: Significance of humic acid and chelant mixture. *Chemosphere* 2010;80:416–421.

Yoo, J.-C.; Lee, C.-D.; Yang, J.-S.; Baek, K. Extraction characteristics of heavy metals from marine sediments. *Cheml Eng J* 2013;228:688–699.

Yoo, J.-C.; Lee, C.; Lee, J.-S.; Baek, K. Simultaneous application of chemical oxidation and extraction processes is effective at remediating soil Co-contaminated with petroleum and heavy metals. *J Environ Manage* 2017a;186:314–319.

Yoo, J.-C.; Park, S.-M.; Yoon, G.-S.; Tsang, D.C.W.; Baek, K. Effects of lead mineralogy on soil washing enhanced by ferric salts as extracting and oxidizing agents. *Chemosphere* 2017b;185:501–508.

Yoo, J.C.; Beiyuan, J.; Wang, L.; Tsang, D.C.W.; Baek, K.; Bolan, N.S.; Ok, Y.S.; Li, X.D. A combination of ferric nitrate/EDDS-enhanced washing and sludge-derived biochar stabilization of metal-contaminated soils. *Sci Total Environ* 2018;616–617:572–582.

Zhang, W.; Huang, H.; Tan, F.; Wang, H.; Qiu, R. Influence of EDTA washing on the species and mobility of heavy metals residual in soils. *J Hazard Mater* 2010a;173:369–376.

Zhang, W.; Tong, L.; Yuan, Y.; Liu, Z.; Huang, H.; Tan, F.; Qiu, R. Influence of soil washing with a chelator on subsequent chemical immobilization of heavy metals in a contaminated soil. *J Hazard Mater* 2010b;178:578–587.

Zhang, W.; Tsang, D.C.W. Conceptual framework and mathematical model for the transport of metal-chelant complexes during in situ soil remediation. *Chemosphere* 2013;91:1281–1288.

Zou, Z.; Qiu, R.; Zhang, W.; Dong, H.; Zhao, Z.; Zhang, T.; Wei, X.; Cai, X. The study of operating variables in soil washing with EDTA. *Environ Pollut* 2009;157:229–236.

15 Green and Sustainable Stabilization/Solidification

Lei Wang, Liang Chen, and Daniel C.W. Tsang

CONTENTS

15.1 INTRODUCTION

15.1.1 SOLIDIFICATION/STABILIZATION TECHNOLOGY

Solidification/stabilization (S/S) is a physical and chemical treatment method for waste/hazardous materials. It is aimed at entrapping the contaminants within a solid matrix (solidification) or binding/complexing toxic compounds into insoluble and stable materials (stabilization). The S/S method has been regarded as one of the best-demonstrated available methods for treating waste/hazardous materials because of the many advantages of S/S. These include its low cost, convenience of handling, high efficiency, high compatibility with wastes/impurities, relatively high durability, and good mechanical strength for further application. According to the United States Environmental Protection Agency (US EPA) statistics, S/S technology occupied approximately 24% share of the remediation market in terms of practical application. Moreover, S/S has been the final or only option for certain special wastes such as radioactive wastes.

In practice, ordinary Portland cement (OPC) is the predominant material for S/S. It is an inexpensive and robust material. After cement hydration, the hydrates (i.e., calcium silicate hydrate (C–S–H), calcium aluminum hydrate (C–A–H), and calcium hydroxide (CH)) could effectively immobilize the contaminants by precipitation, adsorption, and fixation (Ref.). However, the production of OPC is associated with high CO_2 emissions (0.7–1.0 tons CO_2 per ton of OPC) owing to the decomposition of limestone ($CaCO_3$) as well as the energy-intensive calcination

process at 1450°C (Mo et al., 2014). The CO_2 emissions from the cement industry contribute to approximately 7%–8% of global CO_2 emissions (Dung and Unluer, 2017). Therefore, researchers are exploring the use of low-carbon cementitious materials rather than OPC as green agents for S/S, such as alkali-activated cements (AACs), MgO-based cements, and other novel cementitious materials (Unluer and Al-Tabbaa, 2014).

15.1.2 SOLIDIFICATION/STABILIZATION FOR SOIL/SEDIMENT

For S/S of contaminated soil or sediment, it is important to understand the interaction between cement and the components of soil or sediment. Soil is a mixture of minerals, organic matter, organisms, liquids, and gases. It is formed by different physical, chemical, and biological processes. The major minerals in soil are quartz, orthoclase, muscovite, biotite, pyroxenes, amphiboles, and olivines, among others. These minerals are inert or relatively stable and exhibit low reactivity in cement/concrete systems. However, these minerals exhibit various microhardness and different particle-size distributions. Therefore, these parameters influence the hardness, pore size distribution, mechanical strength, and water absorption of the final products. Similarly, the particle size distribution of sediment also determines the final performance of S/S products. A recent study (Wang et al., 2019a) reported that silt and clay in marine sediments exhibit large surface area and micropores. This results in large porosity, particularly large volumes of capillary pores. The capillary pores generally cause microcracks and further destruction when pressed. Therefore, coarse aggregates are generally incorporated to reinforce the strength of S/S products.

In general, soils contain organic matter (5%–10%). Almost all these organic compounds exert retardant effects on cement setting and hardening. The mechanisms of inhibitory effect include the generation of insoluble Ca-compounds, surface adsorption, and complexation. The organic compounds easily form complexes with Ca and form a protective membrane around individual cement particles, which hinder further cement hydration. Certain organic components (methanol and phenol) delay the hydration process as well as form irregular structures, resulting in a low compressive strength of the S/S products. The organic matter can be removed by thermal pretreatment. With a 400°C pretreatment for 1 h, there was a 3.5-times increase in the mechanical strength compared to the untreated samples. This is because 90% of organic matter can be removed after similar treatment (Wang et al., 2015). At higher temperature (650°C or above), nearly 100% of organic matter decomposes. Moreover, a certain amount of clay minerals can be activated for engaging the cement reaction. Thus, although thermal treatment is a feasible pretreatment method, it is an energy-intensive approach. Alkaline extraction also can be used for organic matter removal, although the efficiency is relatively low. In addition, the use of strong alkalis (NaOH and KOH) is highly expensive, and certain remaining cations (Na^+ and K^+) interfere with the cement hydration. The addition of accelerators is another potential option. Chloride-based accelerators could boost the precipitation of oxychloride and promote hydration. Moreover, sulfate-based accelerators could promote the formation of suphoaluminate, accelerating early strength development. However, chloride and sulfate may cause a potential risk of low durability. Recently, accelerated carbonation was used to promote the cement hydration and early strength development. Highly concentrated CO_2 gas could carbonate the cement clinker (tricalcium silicate and dicalcium silicate) directly to generate calcium silicate hydrate (C–S–H) gel and calcium carbonates (CC). Furthermore, the calcium hydrates (CH) formed owing to hydration could be transformed into CC. Thus, accelerated carbonation could evade the Ca-complexing effect and enhance the reaction process directly. This carbonation process could enhance the setting and hardening rate, reduce porosity, densify microstructure, increase mechanical strength, and generate durable carbonates (Wang et al., 2017, 2018a). However, the commercial application of accelerated carbonated requires further investigation.

Although trace elements present in marginal amounts in soil minerals (<5%), they influence the cement hydration and performance of the final products significantly. Certain cations and anions exert an accelerating effect on the cement hydration. The accelerating efficiency of these cations can be ranked from high to low as follows: Ca, Mg, Sr, Ba, Li, K, Rb, Cs, Na, and NR_4. Moreover, the order of effectiveness of anions is Br, Cl, SCN, I, NO_3, ClO_4. It should be indicated that the efficiency depends highly on the size and charge of the ion. Small and highly charged ions are more effective. However, certain toxic elements such as Zn, As, and Pb retard the hydration process significantly. This is discussed in the following sections (Paria and Yuet, 2006).

15.1.3 TESTING METHODS

The properties of soil/sediment S/S blocks can be characterized by some standard test methods, including physical tests, chemical tests, and durability tests (Table 15.1). For the physical properties of S/S products, the compressive strength is the most important factor. The resistance to mechanical stresses of the product can be indicated by the compressive strength, which determines the quality and application of final products. Cohesive soils or cylindrical cement-based S/S blocks can be used for UCS tests. The test is conducted under controlled-strain condition, where the samples are only subjected to axial loading. The unconfined compressive strength (UCS) of 1 MPa is one basic requirement of backfill materials (HK EPD, 2011). If the UCS of blocks exceeds 7 MPa, the blocks conform with the requirement of non-load-bearing unit for partition blocks (BS EN 6073, 1981), whereas if UCS of blocks requires 30 MPa, it can be used as pedestrian paving blocks (HK ETWB, 2004).

The basic chemical tests include toxicity characteristic leaching procedure (TCLP), synthetic precipitation leaching procedure (SPLP), and semi-dynamic leaching tests. The TCLP is a single extraction batch test, which was designed to detect the leachability of waste when the wastes are disposed of in a landfill with municipal solid wastes. Nowadays, TCLP is widely used to distinguish hazardous waste and general waste. Two extraction solutions (i.e., pH of 4.93 ± 0.05 fluid and

TABLE 15.1
Performance Indicators of S/S Performance

Categories	Indicators	Standards
Physical tests	Density	BS EN 12390-7, 2009
	Water absorption	BS EN 772-11, 2011
	Water absorption rate	BS EN 772-11, 2011
	Thickness swelling	BS EN 772-11, 2011
	Compressive strength	BS EN 6717, 2001
	Tensile splitting strength	BS EN 6717, 2001
	Skid resistance	BS EN 6717, 2001
	Abrasion resistance	BS EN 6717, 2001
	Elastic modulus	ASTM C1585, 2013
Chemical tests	Toxicity characteristic leaching procedure	EPA Method 1311, 1992
	Synthetic precipitation leaching procedure	EPA Method 1312, 1992
	Semi-dynamic leaching test	EPA Method 1315, 1992
Durability tests	Drying shrinkage	ISO 1920-8, 2009
	Freeze-thaw resistance	ASTM D560, 2016
	Wet-dry cycles	ASTM C1185, 2008
	Water immersion	ASTM C1185, 2008
	Sulfate resistance	ASTM C452, 2015

pH of 2.88 ± 0.05 fluid) can be selected based on the alkalinity and buffering capacity of the materials. S/S samples should be crushed into particles with a size less than 9.5 mm. The liquid to solid ratio is 20:1 and samples should be agitated by a rotary machine for 18 h. If the leaching concentrations of toxic elements or components fulfill specified requirements, the waste can be disposal of in landfill; otherwise, the waste requires further treatment or especial disposal method. The SPLP is very similar to TCLP, except that the wastes are leached with different extraction solution, such as pH of 4.20 ± 0.05 fluid and pH of 5.00 ± 0.05 fluid. Organic acid (acetic acid) is used for the TCLP test, whereas inorganic acid is used for SPLP test, because SPLP is used to simulate the leachability of waste in the condition of acid rain. Semi-dynamic leaching tests also can be employed to reflect the leaching behavior of toxic components in S/S products. Different from TCLP or SPLP, intact samples are examined in the semi-dynamic leaching test. Usually, the leaching tests were conducted at room temperature and solution at different intervals would be detected. The cumulative fraction leached (CFL) of toxic elements in semi-dynamic leaching tests could be calculated, and its plot against time square root ($t^{1/2}$) enabled the computation of the effective diffusion coefficients (D_e) of toxic elements (ANS 16.1, 1986).

The durability of S/S product also is a very important factor. There are many tests to indicate durability, such as drying shrinkage, freeze-thaw resistance, wet-dry cycles, water immersion, sulfate resistance, depending on different conditions. In summary, S/S products should be examined and confirm with specified criteria before practical application.

15.2 S/S OF TOXIC METAL(LOID)S

Cement-based S/S usually can immobilize toxic metal or metalloids effectively through physical encapsulation and chemical fixation. The mechanisms and applications of S/S for specific toxic elements are introduced in this section.

15.2.1 S/S OF CHROMIUM (Cr)

Cr is a common toxic metal that is present mainly in two stable forms (i.e., trivalent chromium and hexavalent chromium). The toxicity of Cr varies substantially depending on the valence state. Trivalent Cr is an essential trace element in the human body and plays an important role in glucose and cholesterol metabolism. However, excessive inhalation can cause severe diseases. Hexavalent Cr compounds can contaminate groundwater, soil, and sediments through rainwater leaching and sewage discharge, resulting in severe environmental impacts (Rajapaksha et al., 2018). The toxicity of hexavalent Cr is a hundred times higher than that of trivalent Cr (Xia et al., 2019). Thus, Cr contaminated soil or sediment should be remediated with caution, particularly when Cr (VI) is the contaminant. Owing to the advancement of industry and economy, the demand for metallic Cr and Cr salts is increasing. Metallic Cr and Cr salts are considered as important strategic resources, and are used widely in electroplating, leather processing, printing and dyeing, as advanced alloy materials, chemicals, and in ceramic, spice, pigment, anti-corrosion, medicine, and other industries (Tsang et al., 2014; Xiong et al., 2019).

OPC is the primary binder applied for the S/S of Cr contaminated soil/sediment. During cement hydration, Cr can react with Ca to form an insoluble complex of calcium chromate ($CaCrO_4$). However, Cr exhibits a vigorous retardant effect on OPC hydration, resulting in delayed setting and hardening. Finally, the products presented a high leachability of contaminants and low mechanical strength owing to the low degree of cement hydration (Ivanov et al., 2016).

Certain alternative cements were used to immobilize Cr waste. Calcium aluminate cements (CACs) are special cements produced from limestone and bauxite. Therefore, CACs contain a high proportion of alumina (40%–80%) including monocalcium aluminate (CA), mayenite ($C_{12}A_7$), and tetrafalcium aluminate (C_4AF) (Claramunt et al., 2019). In the CAC system, Cr(VI) can be reduced to Cr(III). Furthermore, Cr may replace Al in the structure of CAC hydrates owing to the high content of Al(III) in C–A–H gel. The Cr-incorporated structures are similar to the components $Ca_2Cr_2O_5 \cdot 6H_2O$, $Ca_2Cr(OH)_7 \cdot 3H_2O$, and $Ca_2Cr_2O_5 \cdot 8H_2O$. Compared to OPC, CAC exhibits low pH; this is favorable for the precipitation of Cr(VI) because it is more soluble than Cr(III) at a high pH. Moreover, high Fe content in CACs also reduces Cr(VI) to Cr(III) (Ivanov et al., 2016).

Geopolymers also can be employed for immobilizing Cr. Although the hydrate of sodium aluminosilicate hydrate (N–S–A–H) does not immobilize Cr(VI) very effectively, Cr is sequestrated early in a geopolymer matrix (Zhang et al., 2008). Addition of sulphide in a geopolymer system can reduce the transformation from Cr(VI) to Cr(III). Moreover, sulphide is widely present in ground granulated blast-furnace slag (GGBS), an important precursor for geopolymers. Thus, the modified geopolymer exhibits remarkable performance regarding Cr immobilization (Duxson et al., 2007).

Research has revealed that GGBS can be incorporated into the OPC system to reduce the leachability of Cr(VI). The binary mixture can provide a relatively high content of hydrates (mainly C–S–H) and dense structure for Cr immobilization. The reduction effect on Cr(VI) is another reason for the low leachability. Furthermore, the addition of reactive MgO can improve the S/S performance owing to the generation of stable Mg–Cr spinel (Zhao et al., 2018).

15.2.2 S/S OF ZINC (Zn)

Zn is an essential trace element for humans, animals, plants, and microorganisms. However, an intake of high content of Zn influences health. Excess Zn can cause toxicity resulting in medical conditions such as severe hemolytic anemia, liver or kidney damage, vomiting, and diarrhea. Zn is widely used in the anti-corrosion, batteries, alloys, and chemical industries. Soil and sediment can be contaminated by Zn owing to the mining, refining, or discharge of zinc-bearing sewage. In large-scale hydrometallurgical Zn production plants, over 100 metric tons of Zn sludge/extraction residues can be generated per day (Erdem and Özverdi, 2010).

OPC is an effective reagent for S/S of Zn-contaminated materials. CH from cement hydration can immobilize Zn in cement matrices by generating insoluble calcium zincate ($CaZn_2(OH)_6 \cdot 2H_2O$). However, a dense membrane of calcium zincate can also prevent further cement hydration. Therefore, Ca-rich material is favorable for immobilizing Zn-waste. Zn contaminants may also promote the formation of ettringite (AFt), which can result in the expansion and cracking of cement-based blocks (Vinter et al., 2016). A few researchers stated that the process would benefit the immobilization of Zn because Zn^{2+} replaces the Ca^{2+} in the stable structure of AFt (Karamalidis and Voudrias, 2007). The pH value of the matrix is a significant factor affecting the S/S performance. The solubility of Zn in the pH range between 8.5 and 11.5 is relatively low owing to the formation of zinc hydroxide. Meanwhile, zincate anions can dissolve and easily leach out in a high pH condition (pH > 11.5). The long-term control of pH is highly challenging. The durability of S/S products with a high content of Zn is still problematic.

Addition of reactive MgO also can alleviate the inhibitory effect because MgO can provide abundant Mg^{2+} for Zn complexation. In general, the addition of GGBS to a cement system can densify the microstructure of the soil S/S matrix. However, for Zn-contaminated soil, the addition of GGBS up to 30% did not favor the immobilization of Zn. MgO-GGBS binary cement may be a potential binder for the S/S of Zn-contaminated soil/sediment (Goodarzi and Movahedrad, 2017). In addition, alternative cementitious materials may exhibit good compatibility with Zn-contaminated waste. In calcium sulphoaluminate (CSA) cements, ye'elimite (or tetracalcium trialuminate sulfate) are predominant components over belite (C_2S), which exhibit significantly higher compatibility with Zn than OPC. The setting and hardening of CSA are not hindered, even at a high concentration of Zn^{2+} (0.5 mol/L). Incorporation of gypsum (20%) can improve the S/S performance further. This can prevent an excessive temperature increase and cumulation during hydration. In addition, it improves the mechanical strength of hardened products with limited expansion, particularly after wet curing (Berger et al., 2011). Thus, CSA could be a potential material for in situ S/S of Zn-contaminated soil/sediment.

Meanwhile, magnesium phosphate cements (MPCs) exhibit remarkable performance for Zn S/S, including high initial strength, good stability, and short setting time. The raw materials of MPC generally include dead-burned MgO, phosphate, and retarder. MPCs are also named chemically bonded phosphate ceramics. Lai et al. (2016) used MPC for Zn S/S and investigated the influence of Zn ions on the hydration behavior of MPC. The results showed that Zn ions had no effect on the hydration products of the MPC. However, the microstructure of the hydration products MPC is changed by Zn ions. Zn^{2+} is present primarily as hydration products, whereby Zn^{2+} could attain a stable state. Moreover, it is effective for S/S of Zn^{2+}. The leaching toxicity of Zn ions in hardened MPC paste was lower than the standard limits (Lai et al., 2016).

Limestone calcined clay cement (LC^3) is a new ternary blend produced by replacing 50% cement clinker with limestone and calcined clay. This LC^3 exhibits a tremendous potential for use as green and low-cost supplementary cementitious material. In general, calcined clay is an amorphous pozzolanic material obtained from the calcined kaolin between 700°C and 800°C. Research showed that the coupling replacement of limestone and calcined clay can effectively improve the physical and chemical properties (Latifi et al., 2017). In addition, it has been reported that the pozzolanic reaction of calcined clays generated additional calcium–aluminate–silicate–hydrate (C–(A)–S–H) gels in the presence of limestone. Thus, the generation of ettringite achieves better space-filling and densified pore structure (Reddy et al., 2019). Researchers evaluated the potential use of LC^3 for S/S of Zn-contaminated soil. Studies revealed that the leaching Zn concentrations in the stabilized products after 14 d curing are substantially below the standard of corresponding hazardous waste standard. The compressive strength and pH of the S/S treated soil increase along with the curing time. The addition could significantly reduce Zn leaching. Thus, Zn is converted into a stable residual phase, making the stabilized soil a sustainable engineering material. The results also revealed that the precipitates formed included $Ca_6Al_2(SO_4)_3(OH)_{12} \cdot 26 H_2O$ (ettringite) and $3CaO \cdot SiO_2$ (alite). These were observed to be the major mechanisms for Zn immobilization. Moreover, generation of precipitates in LC^3-treated soil, including $Ca(OH)_2$ (portlandite) and $Zn(OH)_2$ (wulfingite) also contributed to Zn immobilization (Reddy et al., 2019).

15.2.3 S/S of Arsenic (As)

Arsenic (As) is widely present in tailings and soil that are contaminated during metallurgical processes. Moreover, biowastes and certain sediments also contain high concentrations of As. This could pose significant hazards to the environment and human body (Li et al., 2017; Beiyuan et al., 2017). As is ubiquitous in the environment and highly toxic to all life forms. Furthermore, most plants and animals could be poisoned by low levels of As. As can accumulate in living organisms and then be excreted gradually. As poisoning can cause keratosis, skin cancer, melanosis, hyperkeratosis, gangrene, and other diseases (Singh et al., 2015). Although the absorption of anionic and soluble species of As by the human body is high, the absorption of As is very low for insoluble species (Roy et al., 2015).

The treatment methods for As include stabilization, fixation, leaching/washing, biological methods, phytoremediation, and calcination. However, S/S is the most used and mature technology for As treatment. OPC-based S/S method for remediating As-containing solid waste is simple and effective (Li et al., 2018). As can be S/S effectively in the OPC-based system by the mechanisms of chemical fixation and physical encapsulation (Liu et al., 2018). The binders react with As during hydration to form stable Ca-arsenate precipitates, sorption into C-S-H gel, or substitution mechanisms within the crystalline lattice of ettringite (Phenrat et al., 2005).

Recent research developed green binders of As-contaminated soil remediation by using kaolinitic clay mineral systems (Wang et al., 2019b). Kaolinitic clays (>40% kaolinite) are inexpensive and widely distributed. Owing to the presence of Al- and Si-rich phases with partially disordered structures, calcined clays have higher pozzolanic reactivity (Antoni et al., 2012). These alumina- and silica-rich phases in clay minerals also exhibit high compatibility with metallic/metalloid elements and soil/clay (Mukhopadhyay et al., 2017). In the S/S process, a part of As(III) was oxidized into the less toxic As(V). As reacted with calcium hydroxide (CH) and formed precipitates of $Ca_3(AsO_4)_2 \cdot 4H_2O$. This accounted for the low leachability of As. Results indicated that these clay minerals immobilized As effectively and ensured satisfactory physical encapsulation.

Certain additives can also be employed for the S/S of As. The S/S efficiency can be enhanced by (i) generating additional cement hydrates or (ii) incorporating As stabilizers. In particular, blast furnace slag (BS, a by-product from the iron and steel-making industry) is an SCM that can react with excessive hydrated CH to generate secondary hydrates via pozzolanic reaction (Prentice et al., 2019). Metakaolin (MK, produced from calcination of kaolinite) is regarded as an inexpensive and low-carbon SCM because it releases less CO_2 during manufacture (0.18 ton CO_2 per ton MK) compared to PC production (0.82 ton CO_2 per ton PC) (Kavitha et al., 2016). Similarly, MK could be activated by CH to generate additional hydrates such as C–S–H and calcium aluminum hydrate (C–A–H) (Avet et al., 2018; Ke et al., 2018). These additional hydrates efficiently encapsulate As and densify the structure of S/S products. In addition, As(III) and As(V) are established to bind strongly with iron oxides/hydroxides by inner-sphere complexation for their effective removal (Sun et al., 2018). Red mud (RM, an alumina refinery residue) demonstrated remarkable compatibility with As by precipitating crystalline Ca–As and amorphous Fe–As compounds. Therefore, waste valorization of red mud as a green and low-carbon additive is effective for sustainable remediation of As-contaminated sediments. Furthermore, the upgraded S/S method is generally applied with a reducing agent such as ferrous sulfate solid sulphur, and nano-zero-valent iron to reduce high-migration pentavalent As to low-migration trivalent As, to achieve a high efficiency of fixation.

15.2.4 S/S of Cadmium (Cd)

Cd is considered as an exceptionally toxic element for aquatic and human life. Exposure to high concentrations of Cd could be harm to the reproductive system, DNA, lungs, and kidneys. Moreover, it could cause deficiencies in cognition, learning, behavior, and neuromotor skills in children (Adelopo et al., 2018). Cd is an important raw material in industries such as the manufacture of coatings, alloys, pigments, batteries, and plastic stabilizers (United States Geological Survey, 2018).

The S/S method is effective for addressing Cd wastes. OPC is a common matrix for the S/S of heavy metal wastes. Researchers investigated distribution of cadmium and lattice replacement in cement kiln co-processing of municipal solid waste and hazardous wastes by using cement clinkers doped

with CdO (0%–2.0% by weight). The results revealed that the concentration of Cd was high in the silicate phases, which resulted in a lower Ca/Si ratio. The presence of the Cd–O bond was observed. This could be attributed to the lattice replacement of Cd^{2+} with Ca^{2+} (Wang et al., 2016).

A few SCMs can be incorporated into a cement system to enhance the physical encapsulation and chemical fixation in the S/S processes. However, because of the weak bonding to this matrix, the stabilization effects of metal are generally ineffective and inefficient. The stabilized product containing heavy metals such as Cd is easily leached in an acidic environment (Quina et al., 2008). Thermal treatment using existing industrial ceramic sintering procedure has been considered to be an effective technology for stabilizing Cd into crystal structures due to rich-Al or Fe in precursors (Tang et al., 2011). Similar crystal products have good acid resistance and can be transformed into eco-friendly construction materials. Cd can also be incorporated into $CdFe_2O_4$ spinel by a ceramic manufacturing process at an achievable temperature with hematite and magnetite.

Research has indicated that the slow setting of OPC prolongs the treatment time of hazardous wastes. However, the long-time exposure of hazardous waste to unstable environments can result in uncertainty and a likely increase in the diffusion of hazardous substances (Anastasiadou et al., 2012). Previous research has also demonstrated that the addition of heavy metals may severely reduce the mechanical strength and the S/S performance of OPC. Other research has shown that MPC has the characteristics of rapid setting time, high early strength, low porosity, and remarkable stability (Lu et al., 2016). It exhibits remarkable performance of the S/S of Cd contaminants and of the rapid reduction of the leaching toxicity of Cd ions. During the hydration process of MPC, no apparent crystal phase was generated in the reaction of phosphate ions and Cd ions. The Cd ions could exist in the form of hydration phases and amorphous exist in the hardened MPC. In the hydration process of MPC, Cd^{2+} does not affect the trend of the pH variation. However, it can postpone the rate of pH variation and affect the morphology of the hydration products. The leaching toxicity of Cd ions decreased with the decrease in the M/P ratio. This is because PO_4^{2-} is the major ion for Cd fixation (He et al., 2019). In summary, Cd can be solidified rapidly in MPC system. Therefore, it is feasible to use MPC for solidification Cd contaminants.

15.2.5 S/S of Mercury (Hg)

Hg is a heavy metallic element with high toxic, long persistent, and high bio-accumulative properties. Hg emissions through water or gas from the treatment of sewage sludge such as landfill and incineration have significantly contributed to human toxicity and eco-toxicity. Hg is used widely in various household products (such as fluorescent lamps, button cells, thermometers, alkaline batteries, and sphygmomanometers) as well as in certain dental, electronic, and fluorescent-lighting manufacturing industries (United States Geological

Survey, 2018; Hu et al., 2018;). The most common forms of Hg include the soluble and reactive Hg(II) forms as well as less reactive volatile Hg(0) species. The former can generate aqueous complexes with numerous surface complexes and ligands on solids such as HgSe and HgS (Bartov et al., 2013).

Toxic elements such as Cr, Pb, and Cd have been immobilized successfully by using S/S technologies. However, the common S/S approaches cannot reduce the leachability of Hg effectively. Studies showed that the S/S process using thiol-functionalized zeolite (TFZ) and OPC is effective for treating and disposing Hg-containing wastes (Ricardo et al., 2006). The grafting of a thiol group to zeolite enhances the Hg adsorption ability of the modified zeolite effectively. This is likely to be a result of the action between the SH group and the Hg on the thiol-functionalized zeolite surface. Moreover, the adsorption of Hg by TFZ was in line with the Freundlich adsorption isotherm. The Hg adsorption ability is substantially improved by thiol grafting. The results revealed that the S/S process for disposing Hg-containing wastes by using TFZ and OPC is feasible (Zhang et al., 2009).

The S/S technology using cement and reactivated carbon has been established to be effective and low-cost for treating and disposing Hg-contaminated wastes. The pretreatment of powder reactivated carbon (PAC) by soaking it in CS_2 enhanced the capacity of Hg adsorption. This is likely to be a result of the generation of mercuric sulphide on surface of the PAC (Zhang and Bishop, 2002). The pH value between 5.0 and 5.5 was regarded as the optimum pH for the Hg stabilization using PAC. The adsorption of Hg by PAC and PACs are in accordance with the Freundlich theory, and the adsorption equilibrium could be attained within 24 h. Among various anions and cations, only chloride ion affected Hg adsorption by PAC or PACs significantly. The combined use of PAC and cement could form a cement hydrate barrier surrounding the PAC particles and reduce the reduced interference by Cl ion (Zhang and Bishop, 2002).

15.2.6 S/S of Lead (Pb)

Pb and its compounds are hazardous, which can cause aquatic, atmospheric, and incineration, and soil pollution (Xu et al., 2014). Pb-bearing materials originate from either natural or unnatural sources. Natural Pb-bearing materials are dispersed into the environment through natural phenomena such as volcanic eruptions and forest fires. Meanwhile, unnatural sources are mainly human activities, e.g., industrial wastes (e.g., slag, water treatment sludge, scrap Pb acid batteries, and tailings), institutional wastes, commercial wastes, and residential wastes. Whereas tailings, smelting waste, coal ash, and clinker are the main sources of Pb emission, automobile exhaust emission is an important source of Pb present in the dust collected on roadsides.

A high concentration of Pb hinders the hydration reactions of conventional OPC systems. The sulfate/phosphate-containing by-products are established as being effective as additives for enhancing Pb S/S treatment (Wang et al., 2018c). The S/S performance was detected in terms of mechanical strength,

static/semi-dynamic leaching, and thermogravimetric/spectroscopic characteristics. In comparison, binary binders (OPC and GGBS) showed remarkable tolerance with the interference of Pb (Yu et al., 2018; Chen et al., 2019). However, there still was significant retarded hydration in the S/S treatment of Pb contaminated waste. This was particularly so for addition of a high amount of Pb-soil where physical encapsulation could not be feasible. Selected additives (incinerated sewage sludge ash, phosphogypsum, and potassium dihydrogen phosphate) reduced the Pb leaching and increased initial strength to a varying extent relative to the molar ratio of Pb to PO_4 or SO_4 for precipitate generation and the amount of hydration products in the monolithic S/S product. Furthermore, the high efficacy of clay minerals with lime and limestone for simultaneous S/S of Pb in field-contaminated soil also has been investigated. In clay mineral systems, the calcined clay was activated by lime effectively to generate additional C–S–H and C–A–H. The presence of Pb interfered significantly with the hydration of clay minerals. Pb reacted with $Ca(OH)_2$ and formed precipitates of $Pb_3(NO_3)(OH)_5$ during the S/S process. Owing to the consumption of $Ca(OH)_2$ by Pb such as inhibitory effects and the inherent soil buffering capacity, a relatively high content of lime was necessary for sufficient hydration of clay minerals. The results indicated that clay minerals immobilized Pb effectively and ensured effective physical encapsulation (Wang et al., 2019b).

The use of MPC for Pb S/S process has been investigated. TCLP results indicated that MPC is more effective for Pb immobilization owing to the chemical precipitation of lead phosphate and pyromorphite (Wang et al., 2018c). The presence of Pb species adversely impacted the mechanical characteristic of cement. The impact was less obvious than that on the compressive strength of MPC. Since Pb interference through reaction with hydroxide ions of $Ca(OH)_2$ in the initial stage to form lead hydroxide, the final setting time of OPC is substantially prolonged. The induction procedures were delayed, and the subsequent cement hydration process was delayed. These studies demonstrated that physical encapsulation was the major mechanism for cement-based S/S process of Pb, whereas physical fixation of Pb in struvite-K component and chemical stabilization by phosphate precipitation are the main mechanisms for MPC-based S/S (Wang et al., 2018c).

15.3 S/S OF RADIOACTIVE METALS

The cement-based S/S method is an effective and mature technology for immobilizing radioactive metals through physical encapsulation and chemical fixation (Sanderson et al., 2015; Collier et al., 2019). The mechanicals and applications of S/S for specific radioactive elements are introduced in this section.

15.3.1 S/S of Uranium (U)

With the rapid development of the nuclear industry, the mining and smelting of U mines have produced a large volume of U-contaminated waste tailings and rock, which has become the most hazardous source of U pollution. Till the present, over 20 billion tons of U tailings have accumulated worldwide. In certain areas, U has caused severe and complex radiation pollution. As the grade of U ores continues to decrease, the number of tailings and the amount of waste rock produced will increase (Bowker et al., 2007). Thus, the treatment of U-containing waste has been a significant challenge.

Hexavalent U(VI) is commonly present as complexed, adsorbed, and precipitated uranyl (UO_2^{2+}) species in oxidizing environments, which exhibit high mobility and toxicity. Meanwhile, tetravalent U(IV) mainly exists in the mineral phase U(IV), which is insoluble and immobile in a reductive environment. Thus, reduction from U(VI) to U(IV) has been considered as a feasible remediation technique for U-containing waste (Sheng et al., 2014).

S/S is regarded as an effective method for treating radioactive waste. Studies have shown that the incorporation of aqueous Fe(II) to OPC slurry has successfully reduced the degradation of organic pollutants during S/S (Hwang and Batchelor, 2002). This suggests that the S/S technology based on Fe(II) has the potential for the *in situ* remediation of contaminated by toxic elements (e.g., As and Cr) and radioactive chemicals (e.g., Th, U, and Pu). The U(VI) removal mechanism of nanoscale zero-valent iron (NZVI) has been widely studied. It presents a stronger ability for U(VI) removal than microscale ZVI due to its higher effective surface area (Riba et al., 2008). The S/S method of reducing and fixing hexavalent U in cement-soil matrix with NZVI-OPC system has been investigated (Sihn et al., 2019). The results established that enhanced U(VI) immobilization in NZVI–OPC binder is due to the reduction of U(VI) to insoluble, and reduction of U species (UO_3 and U_3O_8) by receiving electrons from Fe(0)/Fe(II). The increase of NZVI dosage, pH value, and humic acid promoted the U(VI) reduction in the NZVI-OPC binder systems.

15.3.2 S/S of Thorium (Th)

Th is regarded as a potential nuclear fuel because of its high reserves (three times more than those of U) and good stability. Th has been developed further and utilized by various countries in recent years (ATSDR, 2015). However, the associated radioactive pollution cannot be omitted. Th is a highly radioactive metal element observed mainly in rare earth minerals. A large amount of radioactive Th is associated with rare earth ores. In the smelting process of rare earth ores, a large amount of Th-containing slag is produced annually. The centralized storage of Th-containing slag causes cockroaches to diffuse and migrate to the surrounding environment, enter the soil and groundwater sources, and cause pollution to the ecological environment of the rare earth industrial zone. The Th-rich tailing powder is easily dispersed by wind in a dry state. This pollutes the environment, particularly the soil (Yan and Luo, 2015). In addition, in the process of stacking, the rich tailing powder is transferred to the soil owing to sedimentation and rainfall, which eventually causes soil pollution. As mentioned above, the Th in associated tailings may enter the biosphere under natural conditions and potentially impact

the ecological environment. The chemical toxicity of the Th compound is almost equivalent to that of heavy metals such as Pb, Sb, and Cu. Moreover, it is infeasible to completely discharge into the body and cause adverse effects (Yan and Luo, 2015).

The overwhelming majority (99%) of Th excessed prevalently in the form of Th-232 (^{232}Th) as thorianite and thorite. Meanwhile, after mining processes, ^{232}Th generally existed as ThO_2. OPC-based S/S method was demonstrated to be effective to solidify ^{232}Th-containing soils. However, high ^{232}Th concentration can inhibit its application substantially (Falciglia et al., 2014). The composition of binders employed for S/S approach also presented significant effect on the properties of the S/S materials. Especially, there is a strong relationship between the setting time and Th concentration in the contaminated soil. Falciglia et al. (2014) investigated the effect of various OPC-based and OPC-barite grouts on performance of S/S products and demonstrated that relatively high ^{232}Th-concentrations (>2%) in soil result in a significant delay in setting times. The hydration of cement is significantly affected by insoluble forms of heavy metals. This leads to the formation of a colloidal hydroxide coating on the surface of the cement particles, which determines the particles of the unhydrated cement, resulting in the delay of cement setting and deterioration of the physical properties of the S/S treated matrix (Chen et al., 2009). Therefore, *in situ* S/S approach is not effective for the treatment of ^{232}Th highly contaminated soil owing to the severe delay in the S/S mass curing and the increase of water permeation in the soil that could facilitate the leachability of the pollutants. In cases of low ^{232}Th concentrations (<2%), the OPC-based S/S method shows good applicability. Recent research (Falciglia et al., 2017) demonstrated that an increase in the barite amount in OPC grout leads to an increasing density and a decreasing porosity. This is owing to the dense structure and fine particle size of barite, which functions as a filler to induce porosity reduction and pore refinement. The barite powder addition in OPC could fill the pores and increase the cohesion and compactness of the paste (Shaaban and Assi, 2011). Overall, it could be concluded that barite-modified OPC S/S approach ensures remarkable ^{232}Th immobilization.

15.4 SUMMARY AND FUTURE TRENDS

This chapter provides an overview of green and sustainable S/S approach for specific heavy metals and radioactive waste by using different cement-based materials including SCM-incorporated cement, alkaline activated cement, MgO-based cement, and certain special cements. The S/S mechanisms of various cements for toxic elements (Cr, Zn, As, Cd, Hg, and Pb) and radioactive metals (U and Th) have been systematically discussed. Green remediation using low-carbon cement-based S/S is a novel and significant approach. It reduces environmental impacts and generates valuable products. The use of industrial by-products or green alternative cements for S/S addresses the problems of contaminated soil effectively and mitigates greenhouse gas emission.

For special cases, the single cement-based material is not highly effective for toxic element immobilization. Binary cement or even ternary cement systems are effective for treating certain special wastes. Moreover, certain specified additives (e.g., chelating agent, activated carbon/biochar, zeolite, clay, and oxidizing and reducing agents) could be added to improve the S/S efficiency. Moreover, the S/S method can be used collaboratively with other remediation methods such as adsorption, washing, and extraction.

The requirements of S/S technology vary across applications. For recycling waste into construction materials, leachability and mechanical strength are major concerns. In certain cases, the workability and setting time determined the feasibility of the S/S method. For example, the long setting time of cement may result in exposure to hazardous waste for a long period, resulting in uncertainty risks. It may also influence the quality and progress of engineering projects. Therefore, special cements should be designed to treat special toxic elements in certain special cases. Although the technical and economic feasibility of novel and sustainable S/S approaches should be validated further, green and sustainable S/S technology will become mainstream remediation methods for toxic and radioactive elements.

REFERENCES

Adelopo, A.O., Haris, P.I., Alo, B.I., Huddersman, K., Jenkins, R.O., 2018. Multivariate analysis of the effects of age, particle size and landfill depth on heavy metals pollution content of closed and active landfill precursors. *Waste Manag.* 78, 227–237.

Agency for Toxic Substances and Disease Registry (ATSDR), 2015. U.S. Department of Health and Human Services. Toxic Substances List. http://www.atsdr.cdc.gov.

Anastasiadou, K., Christopoulos, K., Mousios, E., Gidarakos, E., 2012. Solidification/stabilization of fly and bottom ash from medical waste incineration facility. *J. Hazard. Mater.* 207–208, 165–170.

ANS 16.1, 1986. *American National Standard for the Measurement of the Leachability of Solidified Low-level Radioactive Wastes by a Short-term Tests Procedure*, American National Standards Institute, New York.

Antoni, M., Rossen, J., Martirena, F., Scrivener, K., 2012. Cement substitution by a combination of metakaolin and limestone. *Cem. Concr. Res.* 42, 1579–1589.

Avet, F., Li, X., Scrivener, K., 2018. Determination of the amount of reacted metakaolin in calcined clay blends. *Cem. Concr. Res.* 106, 40–48.

Bartov, G., Deonarine, A., Johnson, T.M., Ruhl, L., Vengosh, A., Hsu-Kim, H., 2013. Environmental impacts of the Tennessee Valley Authority Kingston coal ash Spill. 1. Source apportionment using mercury stable isotopes. *Environ. Sci. Technol.* 47, 2092–2099.

Beiyuan, J., Li, J.S., Tsang, D.C.W., Wang, L., Poon, C.S., Li, X.D., Fendorf, S., 2017. Fate of arsenic before and after chemical-enhanced washing of an arsenic-containing soil in Hong Kong. *Sci. Total Environ.* 599–600, 679–688.

Berger, S., Cau Dit Coumes, C., Champenois, J.B., Douillard, T., Le Bescop, P., Aouad, G., Damidot, D., 2011. Stabilization of ZnCl$_2$-containing wastes using calcium sulfoaluminate cement: Leaching behaviour of the solidified waste form, mechanisms of zinc retention. *J. Hazard. Mater.* 194, 268–276.

Bowker, K.A., 2007. Barnett shale gas production, Fort Worth Basin: Issues and discussion. *Am. Assoc. Pet. Geol. Bull.* 91, 523–533.

BS EN 6073, 1981. *Precast Concrete Masonry Units. Specification for Precast Concrete Masonry Units.* British Standards Institution, London, UK.

Chen, L., Wang, L., Cho, D.W., Tsang, D.C.W., Tong, L., Zhou, Y., Yang, J., Hu, Q., Poon, C.S., 2019. Sustainable stabilization/solidification of municipal solid waste incinerator fly ash by incorporation of green materials. *J. Clean. Prod.* 222, 335–343.

Chen, Q.Y., Tyrer, M., Hills, C.D., Yang, X.M., Carey, P., 2009. Immobilisation of heavy metal in cement-based solidification/stabilisation: A review. *Waste Manag.* 29, 390–403.

Claramunt, J., Ventura, H., Toledo Filho, R.D., Ardanuy, M., 2019. Effect of nanocelluloses on the microstructure and mechanical performance of CAC cementitious matrices. *Cem. Concr. Res.* 119, 64–76.

Collier, N.C., Heyes, D.W., Butcher, E.J., Borwick, J., Milodowski, A.E., Field, L.P., Kemp, S.J. et al., 2019. Gaseous carbonation of cementitious backfill for geological disposal of radioactive waste: Nirex Reference Vault Backfill. *Appl. Geochemistry* 106, 120–133.

Dung, N.T., Unluer, C., 2017. Carbonated MgO concrete with improved performance: The influence of temperature and hydration agent on hydration, carbonation and strength gain. *Cem. Concr. Compos.* 82, 152–164.

Duxson, P., Fernández-Jiménez, A., Provis, J.L., Lukey, G.C., Palomo, A., Van Deventer, J.S.J., 2007. Geopolymer technology: The current state of the art. *J. Mater. Sci.* 42, 2917–2933.

Falciglia, P.P., Cannata, S., Romano, S., Vagliasindi, F.G.A., 2014. Stabilisation/solidification of radionuclide polluted soils–Part I: Assessment of setting time, mechanical resistance, γ-radiation shielding and leachate γ-radiation. *J. Geochem. Explor.* 142, 104–111.

Falciglia, P.P., Romano, S., Vagliasindi, F.G.A., 2017. Stabilisation/solidification of soils contaminated by mining activities: Influence of barite powder and grout content on γ-radiation shielding, unconfined compressive strength and ^{232}Th immobilisation. *J. Geochem. Explor.* 174, 140–147.

Goodarzi, A.R., Movahedrad, M., 2017. Stabilization/solidification of zinc-contaminated kaolin clay using ground granulated blast-furnace slag and different types of activators. *Appl. Geochem.* 81, 155–165.

He, Y., Lai, Z., Yan, T., He, X., Lu, Z., Lv, S., Li, F., Fan, X., Zhang, H., 2019. Effect of Cd^{2+} on early hydration process of magnesium phosphate cement and its leaching toxicity properties. *Constr. Build. Mater.* 209, 32–40.

HK EPD, 2011. Practice Guide for Investigation and Remediation of Contaminated Land, Environmental Protection Department, Hong Kong SAR Government.

HK ETWB, 2004. Specification Facilitating the Use of Concrete Paving Units Made of Recycled Aggregates. Technical Circular (Works) No. 24/2004. Environment, Transport and Works Bureau, Hong Kong SAR Government, Hong Kong.

Hu, Y., Cheng, H., Tao, S., 2018. The growing importance of waste-to-energy (WTE) incineration in China's anthropogenic mercury emissions: Emission inventories and reduction strategies. *Renew. Sustain. Energy Rev.* 97, 119–137.

Hwang, I., Batchelor, B., 2002. Reductive dechlorination of chlorinated methanes in cement slurries containing Fe(II). *Chemosphere* 48, 1019–1027.

Ivanov, R.C., Angulski da Luz, C., Zorel, H.E., Pereira Filho, J.I., 2016. Behavior of calcium aluminate cement (CAC) in the presence of hexavalent chromium. *Cem. Concr. Compos.* 73, 114–122.

Karamalidis, A.K., Voudrias, E.A., 2007. Release of Zn, Ni, Cu, SO_4^{2-} and CrO_4^{2-} as a function of pH from cement-based stabilized/solidified refinery oily sludge and ash from incineration of oily sludge. *J. Hazard. Mater.* 141, 591–606.

Kavitha, O.R., Shanthi, V.M., Arulraj, G.P., Sivakumar, V.R., 2016. Microstructural studies on eco-friendly and durable Self-compacting concrete blended with metakaolin. *Appl. Clay Sci.* 124–125, 143–149.

Lai, Z., Lai, X., Shi, J., Lu, Z., 2016. Effect of Zn^{2+} on the early hydration behavior of potassium phosphate based magnesium phosphate cement. *Constr. Build. Mater.* 129, 70–78.

Latifi, N., Vahedifard, F., Ghazanfari, E., Horpibulsuk, S., Marto, A., Williams, J., 2017. Sustainable improvement of clays using low-carbon nontraditional additive. *Int. J. Geomech.* 18, 4017162.

Li, J.S., Beiyuan, J., Tsang, D.C.W., Wang, L., Poon, C.S., Li, X.D., Fendorf, S., 2017. Arsenic-containing soil from geogenic source in Hong Kong: Leaching characteristics and stabilization/solidification. *Chemosphere* 182, 31–39.

Li, J.S., Wang, L., Cui, J.L., Poon, C.S., Beiyuan, J., Tsang, D.C.W., Li, X.D., 2018. Effects of low-alkalinity binders on stabilization/solidification of geogenic As-containing soils: Spectroscopic investigation and leaching tests. *Sci. Total Environ.* 631–632, 1486–1494.

Liu, L.W., Li, W., Song, W.P., Guo, M.X., 2018. Remediation techniques for heavy metal-contaminated soils: Principles and applicability. *Sci. Total Environ.* 633, 206–219.

Lu, Z., Hou, D., Ma, H., Fan, T., Li, Z., 2016. Effects of graphene oxide on the properties and microstructures of the magnesium potassium phosphate cement paste. *Constr. Build. Mater.* 119, 107–112.

Melamed, R., da Luz, A.B., 2006. Efficiency of industrial minerals on the removal of mercury species from liquid effluents. *Sci. Total Environ.* 368, 403–406.

Mo, L., Deng, M., Tang, M., Al-Tabbaa, A., 2014. MgO expansive cement and concrete in China: Past, present and future. *Cem. Concr. Res.* 57, 1–12.

Mukhopadhyay, R., Manjaiah, K.M., Datta, S.C., Yadav, R.K., Sarkar, B., 2017. Inorganically modified clay minerals: Preparation, characterization, and arsenic adsorption in contaminated water and soil. *Appl. Clay Sci.* 147, 1–10.

ÖzverdI, A., Erdem, M., 2010. Environmental risk assessment and stabilization/solidification of zinc extraction residue: I. Environmental risk assessment. *Hydrometallurgy* 100, 103–109.

Paria, S., Yuet, P.K., 2006. Solidification/stabilization of organic and inorganic contaminants using portland cement: A literature review. *Environ. Rev.* 14, 217–255.

Phenrat, T., Marhaba, T.F., Rachakornkij, M., 2005. A SEM and X-ray study for investigation of solidified/stabilized arsenic-iron hydroxide sludge. *J. Hazard. Mater.* 118, 185–195.

Prentice, D.P., Walkley, B., Bernal, S.A., Bankhead, M., Hayes, M., Provis, J.L., 2019. Thermodynamic modelling of BFS-PC cements under temperature conditions relevant to the geological disposal of nuclear wastes. *Cem. Concr. Res.* 119, 21–35.

Quina, M.J., Bordado, J.C., Quinta-Ferreira, R.M., 2008. Treatment and use of air pollution control residues from MSW incineration: An overview. *Waste Manag.* 28, 2097–2121.

Rajapaksha, A.U., Alam, M.S., Chen, N., Alessi, D.S., Igalavithana, A.D., Tsang, D.C.W., Ok, Y.S., 2018. Removal of hexavalent chromium in aqueous solutions using biochar: Chemical and spectroscopic investigations. *Sci. Total Environ.* 625, 1567–1573.

Reddy, V.A., Solanki, C.H., Kumar, S., Reddy, K.R., Du, Y.J., 2019. New ternary blend limestone calcined clay cement for solidification/stabilization of zinc contaminated soil. *Chemosphere* 235, 308–315.

Riba, O., Scott, T.B., Vala Ragnarsdottir, K., Allen, G.C., 2008. Reaction mechanism of uranyl in the presence of zero-valent iron nanoparticles. *Geochim. Cosmochim. Acta* 72, 4047–4057.

Roy, M., Giri, A.K., Dutta, S., Mukherjee, P., 2015. Integrated phytobial remediation for sustainable management of arsenic in soil and water. *Environ. Int.* 75, 190–198.

Shaaban, I., Assi, N., 2011. Measurement of the leaching rate of radionuclide 134Cs from the solidified radioactive sources in Portland cement mixed with microsilica and barite matrixes. *J. Nucl. Mater.* 415, 132–137.

Sheng, G., Shao, X., Li, Y., Li, J., Dong, H., Cheng, W., Gao, X., Huang, Y., 2014. Enhanced removal of uranium(VI) by nanoscale zerovalent iron supported on na-bentonite and an investigation of mechanism. *J. Phys. Chem. A* 118, 2952–2958.

Sihn, Y., Bae, S., Lee, W., 2019. Immobilization of uranium(VI) in a cementitious matrix with nanoscale zerovalent iron (NZVI). *Chemosphere* 626–633.

Singh, R., Singh, S., Parihar, P., Singh, V.P., Prasad, S.M., 2015. Arsenic contamination, consequences and remediation techniques: A review. *Ecotoxicol. Environ. Saf.* 112, 247–270.

Sun, J., Prommer, H., Siade, A.J., Chillrud, S.N., Mailloux, B.J., Bostick, B.C., 2018. Model-based analysis of arsenic immobilization via iron mineral transformation under advective flows. *Environ. Sci. Technol.* 52, 9243–9253.

Tang, Y., Shih, K., Wang, Y., Chong, T.C., 2011. Zinc stabilization efficiency of aluminate spinel structure and its leaching behavior. *Environ. Sci. Technol.* 45, 10544–10550.

Tsang, D.C.W., Yip, A.C.K., Olds, W.E., Weber, P.A., 2014. Arsenic and copper stabilisation in a contaminated soil by coal fly ash and green waste compost. *Environ. Sci. Pollut. Res.* 21, 10194–10204.

United States Geological Survey, 2018. United States Geological Survey Mineral Commodity Summaries. Rare Earth.

Unluer, C., Al-Tabbaa, A., 2014. Enhancing the carbonation of MgO cement porous blocks through improved curing conditions. *Cem. Concr. Res.* 59, 55–65.

Vinter, S., Montanes, M.T., Bednarik, V., Hrivnova, P., 2016. Stabilization/solidification of hot dip galvanizing ash using different binders. *J. Hazard. Mater.* 320, 105–113.

Wang, F.Z., Shang, D.C., Wang, M.G., Hu, S.G., Li, Y.Q., 2016. Incorporation and substitution mechanism of cadmium in cement clinker. *J. Clean. Prod.* 112, 2292–2299.

Wang, L., Chen, L., Tsang, D.C.W., Kua, H.W., Yang, J., Ok, Y.S., Ding, S.M., Hou, D.Y., Poon, C.S., 2019a. The roles of biochar as green admixture for sediment-based construction products. *Cement Concrete Comp.* 104, 103348.

Wang, L., Chen, L., Tsang, D.C.W., Li, J.S., Yeung, T.L.Y., Ding, S., Poon, C.S., 2018a. Green remediation of contaminated sediment by stabilization/solidification with industrial by-products and CO_2 utilization. *Sci. Total Environ.* 631–632, 1321–1327.

Wang, L., Kwok, J.S.H., Tsang, D.C.W., Poon, C.S., 2015. Mixture design and treatment methods for recycling contaminated sediment. *J. Hazard. Mater.* 283, 623–632.

Wang, L., Yeung, T.L.K., Lau, A.Y.T., Tsang, D.C.W., Poon, C.S., 2017. Recycling contaminated sediment into eco-friendly paving blocks by a combination of binary cement and carbon dioxide curing. *J. Clean. Prod.* 164, 1279–1288.

Wang, L., Yu, K., Li, J.S., Tsang, D.C.W., Poon, C.S., Yoo, J.C., Baek, K., Ding, S., Hou, D., Dai, J.G., 2018b. Low-carbon and low-alkalinity stabilization/solidification of high-Pb contaminated soil. *Chem. Eng. J.* 351, 418–427.

Wang, Y.S., Dai, J.G., Wang, L., Tsang, D.C.W., Poon, C.S., 2018c. Influence of lead on stabilization/solidification by ordinary Portland cement and magnesium phosphate cement. *Chemosphere* 190, 90–96.

Xia, S., Song, Z., Jeyakumar, P., Shaheen, S.M., Rinklebe, J., Ok, Y.S., Bolan, N., Wang, H., 2019. A critical review on bioremediation technologies for Cr(VI)-contaminated soils and wastewater. *Crit. Rev. Environ. Sci. Technol.* 49, 1027–1078.

Xiong, X., Liu, X., Yu, I.K.M., Wang, L., Zhou, J., Sun, X., Rinklebe, J. et al., 2019. Potentially toxic elements in solid waste streams: Fate and management approaches. *Environ. Pollut.* 253, 680–707.

Xu, C., Chen, W., Hong, J., 2014. Life-cycle environmental and economic assessment of sewage sludge treatment in China. *J. Clean. Prod.* 67, 79–87.

Yan, X., Luo, X., 2015. Radionuclides distribution, properties, and microbial diversity of soils in uranium mill tailings from southeastern China. *J. Environ. Radioact.* 139, 85–90.

Yu, K., Li, L., Yu, J., Xiao, J., Ye, J., Wang, Y., 2018. Feasibility of using ultra-high ductility cementitious composites for concrete structures without steel rebar. *Eng. Struct.* 170, 11–20.

Zhang, J., Bishop, P.L., 2002. Stabilization/solidification (S/S) of mercury-containing wastes using reactivated carbon and Portland cement. *J. Hazard. Mater.* 92, 199–212.

Zhang, J., Provis, J.L., Feng, D., van Deventer, J.S.J., 2008. The role of sulfide in the immobilization of Cr(VI) in fly ash geopolymers. *Cem. Concr. Res.* 38, 681–688.

Zhang, X.Y., Wang, Q.C., Zhang, S.Q., Sun, X.J., Zhang, Z.S., 2009. Stabilization/solidification (S/S) of mercury-contaminated hazardous wastes using thiol-functionalized zeolite and Portland cement. *J. Hazard. Mater.* 168, 1575–1580.

Zhao, G., Zhang, L., Cang, D., 2018. Pilot trial of detoxification of chromium slag in cyclone furnace and production of slag wool fibres. *J. Hazard. Mater.* 358, 122–128.

16 Treatment Strategies for Wastewater from Hydraulic Fracturing

Yuqing Sun, Di Wang, Linling Wang, and Daniel C.W. Tsang

CONTENTS

16.1 INTRODUCTION

The increasing consumption of traditional fossil fuels, such as coal, oil, and natural gas, has prompted urgent consideration for alternative energy sources. Nevertheless, as far as the main global energy supply is concerned, the contemporary gap between non-renewable (81.6%) and renewable (13.4%) energy sources remains too large to balance in recent decades (Melikoglu, 2014). Thus, shale gas is recognized in the transitional stage as an alternative, non-renewable supply for traditional energy (Howarth et al., 2011a, 2011b). Shale gas is natural gas in formations that is low in permeability and rich in organics (Gregory et al., 2011). The fast development of horizontal drilling and hydraulic fracturing (HF) technologies has made shale gas mining technically and economically viable (Shaffer et al., 2013). In the 1970s, the United States and Canada started full commercial production of shale gas, while it was initiated from 2015 to 2016 in Poland and the United Kingdom with estimated technically recoverable resources of 5.3 and 0.57 trillion m³, respectively

(European Commission, 2014). Apart from European countries, China, Argentina, and Australia with discovered evaluated technically recoverable shale gas of 31, 22, and 11 trillion m³, respectively, are progressively perceived as prospective manufacturers (Mauter et al., 2014; Wang et al., 2014). Accordingly, shale gas can gradually replace the traditional fossil fuels such as oil, coal, and natural gas, and it can be used as a substituted energy source for transition to more environmentally friendly and renewable energy sources (Vidic et al., 2013).

However, the comprehensive impacts of applying shale gas globally remain controversial and subjects of intense debate. For shale gas, the positive effect of diminishing green house gas (GHG) emissions might not be accomplished, which indicates an improved appreciation of the environmental and social implications (Bazilian et al., 2014; McJeon et al., 2014; Newell and Raimi, 2014). Proponents promote shale gas extraction as an economic benefit and a prospective connection to a low-carbon society (Helman, 2012; Moniz et al., 2011). Opponents argue that the exploitation of shale gas has more disadvantages than advantages, and its potential environmental impact is great, such as drinking water pollution, wastewater management, and greenhouse gas leakage, among which water-related issues are of particular concern (Howarth et al., 2011a, 2011b; Jackson et al., 2013; Olmstead et al., 2013; Osborn et al., 2011). HF processes produced time-fluctuating volumes of liquid waste, which involves high flowback rates of fracturing fluid after initial fracturing stage and low flow rates of water during the production stage both leading to the volume fluctuation of liquid waste produced by HF process with time. HF wastewater contains a variety of pollutants, such as solid suspended solids (SS), inorganic salts, heavy metal ions, organic matter, and natural radioactive substances (NORMs) (Barbot et al., 2013). Until recent years, HF wastewater is usually treated by injection into deep underground wells. However, this approach has been difficult to work with because deep wells that are permitted to be discharged near HF sites are limited (Gregory et al., 2011; Lutz et al., 2013). The HF industry is driven by public anxiety, government supervision, local water shortages, and cost savings requirements to find more sustainable solutions, which will promote effective water reuse and surface water discharge (Mauter et al., 2014).

From the perspective of the above background, the effective way to promote public acceptance of shale gas development is to find an effective scheme for HF wastewater management and reduce the environmental impact of this process. This chapter focuses on six typical areas in the global shale gas production: Duvernay (Alberta, CA), Montney (Alberta and British Columbia, CA), Eagle Ford (Texas, US), Barnett (Texas, US), Marcellus (Pennsylvania, US), and Bakken (North Dakota, US). The present chapter depicts the procedures and environmental impacts of HF process, which orientate in wastewater production, characterization, impact, and potential management strategies. This chapter summarizes the recent literature to understand the production of wastewater in HF process, as well as the different strategies for the treatment and reuse of HF wastewater. This chapter is anticipated as a reference for selecting appropriate technology for wastewater treatment and identifying key weaknesses for further research directions.

16.2 WASTEWATER FROM HYDRAULIC FRACTURING

16.2.1 Hydraulic Fracturing Process

The first step of the HF process (Figure 16.1) is to determine the source of water, water consumption, and wastewater management means (API, 2010). During the drilling process, the casing pipe is sealed with cement to prevent the leakage of water, gas, oil, and coal (API, 2010; NDCC, 2012). After completion of construction, water is piped or trucked from the source to the well, then mixed with sand and drilling additives to start the HF process (Lutz et al., 2013). The formation ruptures under high-pressure injection of HF fluid, which releases oil and gas. The initial oil and gas are collected by flowback process before formal oil and gas exploitation. The large amount of water used for flowback mainly comes from injected water and formation water in shale (EPA, 2012). Then comes the production stage of formation water and hydrocarbons (WEF, 2013). The production process lasts until re-fracturing or abandonment of wells, during which oil, gas, and produced water reach the wellbore and are collected to the surface (Nicot et al., 2014; WEF, 2013). Recovered flowback water and produced water need to be disposed of by some means as follows: (a) storage, (b) disposal, (c) treatment and reuse, and (d) treatment and disposal (API, 2010). After production, wells will be abandoned and blocked or sealed with cement barriers (API, 2010). Notably, the process in Figure 16.1 only represents a universal procedure, and some minor alternatives are followed at different locations.

16.2.2 Flowback Process and Produced Water

Inorganic salts, natural compounds, chemical additives, oils and grease, and NORMs in flowback water and produced water are contaminants that require primary attention (NPC, 2011), as shown in Tables 16.1 through 16.3. Spill/leakage may result in surface and groundwater contamination equivalent to chemical mixing and well injection stages (EPA, 2012). However, specific spill/leakage information associated with flowback and produced water is not entirely accessible to the public, which restrains risk assessment corresponding to this stage. In 2012, 25.5 million barrels of brine were produced. Meanwhile, there were 141 pipeline leaks in North Dakota, resulting in a total of 8000 barrels of salt water (Al Jazeera America, 2014). Texas found that the leaking material was not HF fluid, but mainly oil, natural gas, and liquid condensate (EPA, 2012). A recent study examined the effects of flowback water pollution on soil, environment, and human health in typical shale gas areas in China. The results showed that the soil toxicity increased slightly after one month aging with

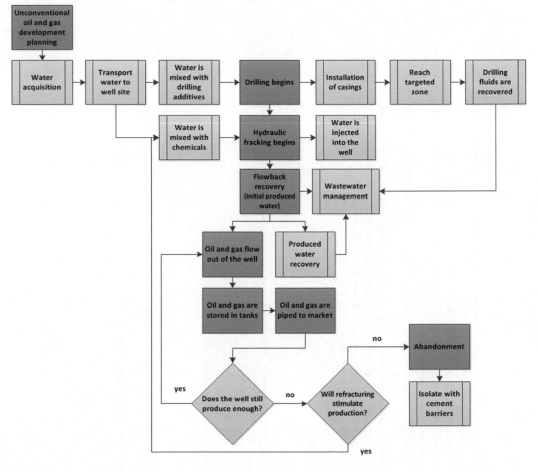

FIGURE 16.1 General process of the HF process. (Adapted from Torres, L. et al., *Sci. Total Environ.*, 39, 478–493, 2016.)

TABLE 16.1
Production and Quality of Wastewater in HF Process

Shale Region	Wastewater Production per Well	TDS (mg L⁻¹)	Organics (mg L⁻¹)	Radioactivity (pCi L⁻¹)
Bakken	1305–3,480 m³ (15%–40% of injected water)	<632,689	COD: 1220	^{226}Ra: 527.1–1211 ^{228}Ra: 259.5–510.9
Barnett	1890–4,725 m³ (10%–25% of injected water)	21,581–300,155	COD: 1052	—
Eagle Ford	~2730 m³ (~15% of injected water)	1033–317,876	TOC: 1089	—
Marcellus	1140–4,725 m³ (10%–25% of injected water)	<476,500	BOD: 1–12,400 DOC: 20–3750 TOC: 1.2–1530 COD: 13–67,383	Alpha, Gross: 0–123,00 Beta, Gross: 0–597.699 ^{137}Cs: 0–9 ^{210}Pb: 0–310 ^{212}Pb: 0–54 ^{214}Pb: 6–7350 ^{226}Ra: 0.0672–17,000 ^{228}Ra: 0–2589 ^{90}Sr: 0.005–0.44 ^{228}Th: 0.881–45.9 ^{230}Th: 0.078–6.9 ^{232}Th: 0.003–0.355 ^{234}U: 0.089–25 ^{235}U: 0–40 ^{238}U: 0–497
Montney	5000–12,500 m³	—	—	—
Duvernay	1000–6000 m³	—	—	—

Source: Sun, Y. et al., *Environ. Int.*, 125, 452–469, 2019a.

TABLE 16.2

Organic Contents of Fracturing Wastewater from Typical Shale Gas Wells, and Their Health Effects and Regulatory Levels

Pollutant	Example Compound	Range (µg L⁻¹)	MCLª (µg L⁻¹)	Health Effect of Different Compounds
Dissolved organic carbon	—	Hydrocarbons found in HF wastewater at levels as high as 5,500 mg L⁻¹	—	• Cyclic octaatomic sulfur: microbiological activity indicator. • Straight chain alkanes/alkenes: mucosal irritation in nasal turbinates and larynx in rat, cystic uterine endometrial hyperplasia in mice, and carcinogenic potential. • Aromatics and aliphatics: hepatic and renal effects, hemolytic anemia, and respiratory irritant effects in animals. • Carboxylic acids: low genotoxic potential.
Added organic chemicals	Chlordane	0.47–2.5	2	• Aliphatic hydrocarbons (solvents): respiratory irritant effects in animals, asphyxia and chemical pneumonitis. • Brominated nitrilopropionamides and hexahydro-1,3,5-trimethyl-1,3,5-triazine-2-thione (biocide): developmental, reproductive, mutagenic, carcinogenic, or neurological effects. • Ethylene glycol and derivatives (cross linker and scale inhibitors): central nervous system depression, cardiopulmonary effects, and renal damage. • Guar gum and diesel fuel (gelling agent): guar gum does not pose a threat but diesel fuel contains known carcinogens. • Ethanol (foaming agent): malnutrition, effects on hepatic metabolism and immunological functions. • Methanol (corrosion inhibitor): visual disturbances, neurological damage, and dermatitis. • Fatty acid phthalate esters (breaker): liver effects.
	Dibromochloromethane	0.25–2000	0.2	
	Dinoseb	1.19–940	7	
	Endrin	0.02–5	2	
	Heptachlor	0.01–2.4	0.4	
	Heptachlor epoxide	0.01–9.5	0.2	
	Hexachlorobenzene	0.5–190	1	
	Hexachlorocyclopentadiene	0.5–940	50	
	Methoxychlor	0.09–47	40	
	Pentachlorophenol	1.25–940	1	
	Toxaphene	0.1–10	3	
	Hexahydro-1,3,5-trimethyl-1,3,5-triazine-2-thione	< 15,000	—	
Benzene, toluene, ethylbenzene, and xylene (BTEX)	**Benzene**	0.25–2000	5	• Cancer risk, neurological effects, primarily central nervous system depression, ototoxicity, hematological, immunological, and lymphoreticular effects.
	Toluene	0.25–6,200	1000	
	Ethylbenzene	0.25–2000	700	
	Xylenes	0.5–6500	10,000	
Polycyclic aromatic hydrocarbons (PAHs)	**Benzo(a)pyrene**	0.5–190	0.2	• Carcinogenic, reproductive problems in mice, and respiratory effects.
	2-Methylnaphthalene	<20,000	—	
	Naphthalene	<1400	—	
	Phenanthrene	<1,400	—	
Volatile fatty acids (VFAs)	1,1,1-Trichloroethane	0.25–2000	200	• Aliphatic acid anion (primarily acetate): induces headache in sensitized rats, corrosive for the skin, eye damage, and mucous membranes irritation. • VFAs are responsible for unpleasant odor in wastewater.
	1,1,2-Trichloroethane	0.25–2000	5	
	1,2-Dichloroethane	0.25–2000	5	
	1,2-Dichloropropane	0.25–2000	5	
	1,2,4-Trichlorobenzene	0.25–2000	70	
	1,2,4-Trimethylbenzene	<4000	—	
	1,3,5-Trimethylbenzene	<1900	—	
	Carbon disulfide	<7.3	—	
	Carbon tetrachloride	0.25–2000	5	
	Styrene	0.25–2000	100	
	Vinyl chloride	0.25–2000	2	

Source: Sun, Y. et al., *Environ. Int.*, 125, 452–469, 2019a.

ª MCL from EPA drinking water standards. The compounds in bold indicate mean/MCL ratio > 1. The compounds in italic indicate a risk quotient (RQ) > 1, which implies potential risk of a compound to the surface water ecosystems and the necessity of its removal prior to wastewater discharge (Butkovskyi et al., 2017).

TABLE 16.3

Inorganic Contents of Fracturing Wastewater from Typical Shale Gas Wells, and Their Health Effects and Regulatory Levels

Pollutant	Range (mg L⁻¹)	MCL[a]	Health Effect of Different Compounds
Aluminum (Al)	0.01–862.8	0.2	• Severe trembling, listlessness, loss of memory, damage to the central nervous system, and dementia.
Ammonia (NH_4^+)	44.8–2,520	—	• Skin, eyes, respiratory tract, and lung irritation.
Arsenic (As)	0.001–1.1	0.01	• Carcinogen and mutagen. • Long term: sometimes can cause fatigue and loss of energy, dermatitis, endocrine disruptor, and reproductive toxicity.
Antimony (Sb)	0.001–0.2	0.006	• Increase in blood cholesterol, and decrease in blood sugar.
Barium (Ba)	0.005–12,000	2	• Flammable at room temperature in powder form. • Long term: increased blood pressure and nerve block.
Beryllium (Be)	0.00021–0.008	0.004	• Kidney damage.
Bromide (Br)	0.24–3340	0.01	• Malfunctioning of the nervous system and disturbances in genetic materials.
Boron (B)	0.02–57.5	—	• Infection in stomach, liver, kidneys, and brain.
Cadmium (Cd)	0.00019–0.1	0.005	• Flammable in powder form, toxic by inhalation of dust or fume, a carcinogen. • Soluble compounds of Cd are highly toxic. • Long term: concentrates in the liver, kidney, pancreas, and thyroid. • Hypertension suspected effect.
Calcium (Ca)	16–39,800	—	• Kidney stones and reproductive toxicity.
Chloride (Cl)	18–200,000	250	• Irritates skin and eyes, chest pain, and water retention in the lungs.
Chromium (Cr)	0.0008–2.2	0.1	• Hexavalent Cr compounds are carcinogens and corrosive on tissue. • Long term: skin sensitization, kidney damage, and reproductive toxicity.
Cobalt (Co)	0.0029–0.169	—	• An essential element for life in minute amounts. • Long term: respiratory problems when inhaled and a major cause of contact dermatitis.
Copper (Cu)	0.0025–116	1.3	• Gastrointestinal distress, liver or kidney damage, and neurological disorder.
Cyanide (CN)	0.0019–0.954	0.2	• Nerve damage or thyroid problem.
Fluoride (F)	0.0009–58.3	4	• Dermal, musculoskeletal, ocular (eyes), and respiratory (from nose to lungs).
Iron (Fe)	0.025–1,390	0.3	• Conjunctivitis, choroiditis, retinitis, and neurological disorder.
Lead (Pb)	0.0005–0.97	0.015	• Toxic by ingestion or inhalation of dust or fumes, long term-brain and kidney damage, birth defects, reproductive toxicity, neurological disorder, and immunological disorder.
Lithium (Li)	0.02–634	—	• Corrosive to the eyes, skin, and respiratory tract.
Manganese (Mn)	0.0104–72.8	0.05	• Hallucinations, forgetfulness, Parkinson's disease, lung embolism, bronchitis, nerve damage, and endocrine disruptor.
Magnesium (Mg)	0.25–3670	—	• Fever, chills, nausea, vomiting, muscle pain, and irritation of upper respiratory tract upon inhalation.
Mercury (Hg)	0–0.065	0.0002	• Kidney damage, reproductive toxicity, immunological disorder.
Molybdenum (Mo)	0.0068–1.98	—	• Hyperbilirubinemia, gout, and joint pains.
Nickel (Ni)	0.05–3.2	—	• Lung cancer, nose cancer, larynx cancer, prostate cancer, asthma and chronic bronchitis, heart disorders, and allergic reactions such as skin rashes.
Nitrate as N	0.02–15.9	10	• Harmful to infants, shortness of breath, and blue-baby syndrome.
Nitrite as N	0.034–146	1	
Phosphorus (P)	0.015–4,222	—	• Kidney damage, osteoporosis, nausea, stomach cramps, and drowsiness.
Potassium (K)	0.07–5,030	—	• Fluid in the lungs, eye irritation, and nose and throat irritation.
Rubidium (Rb)	0.342–1.29	—	• Skin and eye burns, failure to gain weight, ataxia, hyper irritation, skin ulcers, and extreme nervousness.
Selenium (Se)	0.0025–0.35	0.05	• Long term: red staining of fingers, teeth and hair, general weakness, depression, and irritation of nose and mouth.
Silicon (Si)	0.03–117.6	—	• Fibrosis in lung tissue, skin and eye irritation, and renal system diseases.
Silver (Ag)	0.0005–0.1	0.1	• Toxic metal. • Long term: permanent gray discoloration of skin, eyes, and mucous membranes.
Sodium (Na)	8–81,590	—	• Kidney damage; increases in blood pressure; and skin, eye, nose, and throat irritation.
Strontium (Sr)	0.06–7,890	—	• Problems with bone growth, anemia, and carcinogenic.
Sulfate (SO_4^{2-})	0.5–2,920	250	• Salty taste but no health hazard.
Thallium (Tl)	0.001–1	0.002	• Hair loss; changes in blood; and kidney, intestine, or liver problems.
Vanadium (V)	0.148–1.02	—	• Cardiac and vascular disease; damage to the nervous system; dizziness; and eye, nose, and throat irritation.
Zinc (Zn)	0.0025–247	5	• Loss of appetite, decreased sense of taste and smell, slow wound healing and skin sores, endocrine disruptor, and neurological disorders.

Source: Sun, Y. et al., *Environ. Int.*, 125, 452–469, 2019a.

[a] MCL from EPA drinking water standards. The compounds in bold indicate mean/MCL ratio > 1.

flowback solution, and the activities of dehydrogenase and phosphate monoesterase were significantly reduced by ions in flowback solution (Chen et al., 2016, 2017).

16.2.2.1 Flowback Water

The fracture fluid will flow back to the surface after the fracturing process, namely, "flowback water," which must first be recovered in order to obtain the gas (Gregory et al., 2011; Balaba and Smart, 2012). Approximately 10% to 40% of the fracturing fluid returns to the surface during the first two-week flowback period. The total amount of flowback rests with the characteristics of formation and operation factors during the well development. The flowback rate reaches its top on the first day, and declines over time; the original rate can be up to 1000 m^3 d^{-1} (Arthur et al., 2008; GWPC and ALL Consulting 2009). It may happen that the fracturing liquids become trapped and impede the flow of the gas (Kargbo et al., 2010). Flowback water contents are time related, which infers possible mathematical functions and models (Gregory et al., 2011).

16.2.2.2 Gels

Gels may be added to increase the viscosity of the fracturing fluid so as to minimize the loss when it passes through the fracture. After the complete process of the fracturing activity, fracturing fluids are supposed to reduce the viscosity soon, and then the fluids can be recovered from the ground easily. However, sometimes there is residue of gels in the flowback water following partial gel decomposition (Kargbo et al., 2010). Guar gum (a gelling agent) is identified as one of the principal ingredients of flowback water. It could have adverse impacts on advanced flowback water such as membrane separation due to its gel-like nature, though the findings also showed it biodegradable (Lester et al., 2013).

16.2.2.3 Brines

It has long been known that deeper ancient salt formations in shale gas basins can salinize groundwater. The flowback water typically contains brines composed of high concentrations of metals, salts, organics, and radioactive materials from shale that it has contacted with (Vidic et al., 2013; Lutz et al., 2013). Those ion concentrations vary from different formations and regions, and the ionic strength may be raised by the use of different chemical additives in fracturing fluid (Balaba and Smart, 2012). The generation of brine also changes over time, which shows the dominating effect when the amount of drilled wells become stabilized. The corresponding values reported in the first year often represent the partial amount, which means there could be an underestimation (Lutz, et al., 2013).

16.2.2.4 Produced Water

When the gas production is activated, much lower volumes (2–8 m^3 day^{-1}) of liquids are generated at the surface throughout the life cycle of the well. This wastewater, also called "produced water," is constituted of slight levels of fracturing chemicals but substantial TDS concentrations and petroleum hydrocarbons that can be followed by subsequent separation and recovery. Its higher TDS than flowback water is generally associated with the

equilibrium between rock and water, and the mixture of fracturing liquid with subsurface brines (GWPC and ALL Consulting 2009; Bibby et al., 2013). It is hardly economical to conduct an exhaustive examination of all components of flowback and produced water. Standard water quality measurements such as pH, TDS, total suspended solids (TSS), specific conductivity, chemical oxygen demand (COD), total organic carbon (TOC), and alkalinity provide preferable cross-sample comparisons. The presence of radium (Ra) and uranium (U) in some shale formation areas has raised a particular concern of naturally occurring radioactive materials (NORMs). Additionally, precipitations of Ba and Sr can incorporate Ra, which makes a worse case (Bibby et al., 2013; Vidic et al., 2013).

16.2.3 SPATIAL AND TEMPORAL CORRELATION OF WATER QUALITY

16.2.3.1 Injected Water

In the studies of water quality, spatial and temporal information of chemical analyses were used to manifest fundamental relationships. Most surveys reported low concentrations of chloride, TDS, and other water parameters in the injected water, which were conducted between the local surface and shallow groundwater (Hayes, 2008). Some flowback water is reused as injected water, which can be noticed from the increased Ca, Ba, Cl, and other ions. Low concentrations of chloride and sulfate and also the approximately neutral pH of fracturing fluid imply that H_2SO_4 and HCl were not the ingredients of the fracturing fluids. The similar pH of the injected and flowback waters in the range of 6–8 suggests that rock material dissolution due to acidic solutions is not a valid explanation for elevated cation levels. In other words, the injected water attacked by acids cannot elucidate the high contents of cations in the flowback water (Haluszczak et al., 2013).

16.2.3.2 Chloride

Chloride, calcium, and sodium are the dominating ions in flowback water, and then come to calcium, barium, magnesium, and strontium. Their trends with time correspond to that of TDS. The chemistry of flowback water has great variation during the first weeks of collection. As the concentration of Cl varies with time, it is commonly adopted as a reference to compare with the trends of other ions. Cl is also the principal anion in flowback water, which shows high correlation with TDS regardless of the spatial and temporal variations ($R^2 = 0.90169$). In Barbot et al. (2013) research, key ions (i.e., Na, Ba, Mg, Sr, Br) are matched to chloride in the Marcellus Shale case study to compare the change of concentrations. Primary cations (i.e., Ca, Mg, Sr) showed high correlations with Cl concentrations while Ba was significantly affected by geographic location.

16.2.3.3 Bromide

Bromide, known as a conservative element in water solution during evaporation, is normally used to portray the relationships with the salinity in the produced water. Water samples with high salinity fit well with the model of seawater evaporation, which indicates that the origins of the produced water are the condensed seawater. Produced water may get condensed in Ca, Ba, Mg, and Sr gradually, while the contents of sulfate and carbonate become lower with time.

16.2.3.4 Sulfate

The concentration of sulfate reported in the Marcellus brines was at an average level less than 100 mg L^{-1}, and the highest level also did not exceed 500 mg L^{-1}. Likely due to a low level of SO$_4$ in the *in situ* brine, its concentration was also reduced in the flowback water. As a result, the concentrations of alkaline earth constituents (Ca, Ba, Sr, Ra), alkali elements (K, Na, Li), and halides (Br, Cl) are highly accumulated.

16.2.3.5 Brine Variations

Comparing brine variations with nearby formations gives clues to the origin of salinity in produced waters. Produced water from Marcellus Shale demonstrated different tendencies in the proportions of Ca:Cl and Mg:Cl from other shale formations where lower chloride concentrations were found at the early stages of the flowback. On the contrary, the concentrations of barium can reach thousands of mg L^{-1} in the later period of flowback, which shows the potential to be a toxic element in the Marcellus brines. Simple mixing with existing formation brine and the fracturing fluid cannot explain the variations of major ions over the entire life. The constituents of the fracturing liquid and solid-liquid interactions affect the produced water quality, particularly in the early flowback period (Barbot et al., 2013).

16.2.3.6 Naturally Occurring Radioactive Materials (NORMs)

Reported total concentrations of radium (Ra) in the flowback water were 10 to 1000 times higher than the maximum contaminant level (MCL) for potable water. It drew particular attention to this potential radiation hazard. The research pointed out that high levels of Ra also presented in the conventional brines, which suggested that high levels of Ra were caused by the *in situ* brines instead of the dissolution in the fracturing activities (Dresel and Rose, 2010; Haluszczak et al., 2013). However, though it showed relatively high concentrations of uranium (U) in the Marcellus black shale, its concentrations did not demonstrate a corresponding high level in the brines, which may attribute to the reducing character of the pore water, leading to a low solubility of U. Thus, the fates and transport of these NORMs need deep investigation to protect human drinking water sources (Haluszczak et al., 2013).

16.2.3.7 Elevated Salt Concentrations

The spatial and temporal correlations with brine components in Marcellus Shale study suggest that (1) salt dissolution and/or the presence of rock minerals are not the main causes for the elevated concentration of salts in the flowback water; (2) the compositions of flowback water mixed with greatly condensed *in situ* brines have little difference with those in other formations; and (3) that the levels of Ba and Ra in these waters are hundredfold over the US drinking water standards (Barbot et al., 2013; Haluszczak et al., 2013). The existing hypothesis suggests that constituent dissolution increases the salt concentrations in the flowback water, when it is injected into the shale during fracturing activities (Blauch et al., 2009). A more recent study highlights an alternative that the high salinity in flowback/produced water is due to the release of *in situ* brines (Haluszczak et al., 2013).

16.2.4 CURRENT TREATMENT STRATEGIES FOR WASTEWATER FROM HYDRAULIC FRACTURING

Wastewater management is one of the most challenging aspects for the scale up of the fracturing industry, which will unavoidably produce huge amounts of wastewater. Common wastewater management options are: (1) treated within the public sewage treatment facilities, and then discharged to local waterways; (2) treated at private industrial wastewater facilities either for future reuse or discharge into local waterways; (3) transported for underground injection; (4) partial treated and recycled *in situ* (Lutz et al., 2013).

In North Dakota, fracturing wastewater is not disposed through evaporation ponds, instead via on-site storage before underground injection (NDCC, 2012). In the western region, because of the high evaporation rate, the amount of wastewater can be reduced by evaporation before the reserve pit is closed (RRCT, 2015). In Texas, produced water is first stored in a reserve pit and then evaporated in an evaporation pool, which varies across the state (RRCT, 2015). In contrast, due to the low evaporation rate in the east, there are fewer field treatments near gas wells, usually dewatering in pits, as is the case in the Barnett area (Nicot et al., 2014). In Alberta, municipal sewage treatment facilities are not suitable for HF wastewater. When HF wastewater is not suitable for reuse/recovery/treatment, it can only be discharged into approved treatment wells (Rokosh et al., 2012). In British Columbia, HF wastewater can be either injected into approved treatment wells or treated and reused in water treatment facilities (OGC, 2014). The ultimate way out for large quantities of HF wastewater in western Canada is deep-ground injection (Rokosh et al., 2012; Rivard et al., 2014). However, most HF wastewater in Marcelus has been recycled (Brantley et al., 2014), which unfortunately only accounts for a comparatively low proportion. Recently, reusing/recycling through integrated industrial or on-site/mobile wastewater treatment facilities and occasional deep-well injection have been developed into the principal operations for HF wastewater management in Marcellus (Barbot et al., 2013). Therefore, the leakage from reserve pits and pipelines, along with accidents in the transportation and subsequent ground discharge will inevitably lead to HF wastewater leakage. All these conditions may lead to the pollution of drinking water (EPA, 2012).

16.2.4.1 Limited Report Data

It has been reported that compared with conventional natural gas extraction approaches, less wastewater was produced in terms of per unit gas produced (about 35%). Well operators also claimed that approximately one-third of wastewater from Marcellus wells came from flowback water, and the remainder was classified brine, which kept generating for several years. The estimations of wastewater volumes were unclear in Marcellus wells as the categories of wastewater (drilling, flowback, or produced/brine) were never classified (Arthur et al., 2009; Gregory et al., 2011; MSAC, 2011). The amount of fracturing fluid injected into a well was usually in a range from 11.5 to 19 million liters (MSAC, 2011), but the majority of the estimations considered flowback only rather than drilling and brine (Gregory et al., 2011), which may lead to a significant underestimation of total wastewater generation. Despite less wastewater production per unit gas, the rapid expansion of shale gas extraction has greatly raised the wastewater volume in Marcellus by almost 600 times since 2004. Current wastewater treatment facilities are overwhelming (Lutz et al., 2013).

16.2.4.2 Barriers for Water Reuse

There was about 56% (1,763.2 mL) of wastewater in Marcellus Shale recycled in 2011, compared with only 13% recycled before (Lutz et al., 2013). Reuse of wastewater for shale gas well development draws lots of interest due to potential economic benefits. Internal reuse does not require expensive desalination and can reduce the adverse effects of high dissolved solids concentration by simply blending with fresh water (Shaffer et al., 2013). However, there are still some barriers to be overcome for wastewater reuse. High concentrations of Ca, Ba, Sr, and other ions can impede gas flow due to high scaling potential. Formation minerals may release divalent cations into the flowback water, which will lead to the formations of carbonate and sulfate precipitations when reusing the flowback water. Particularly, low solubility of Ba and Sr with sulfate and high concentration of Ca can contribute to calcite formation. The growth of anaerobic bacteria may result in corrosive by-products (e.g., H_2S) and biological fouling. Brine variations can affect well integrity in a disadvantageous way. Moreover, fracturing additives need to maintain their chemical properties in brine solutions in the case of reusing flowback water for reinjection (Montgomery and Smith, 2010; Gregory et al., 2011).

16.3 ADVANCED WASTEWATER TREATMENT IN HYDRAULIC FRACTURING PROCESS

Traditionally, there are several post-treatment methods for HF wastewater in North America: (1) surface treatment (e.g., spreading) (Notte et al., 2016); (2) injection into approved deep wells; (3) reuse and recycling, especially in the water-deficient areas of the United States (Freyman, 2014); and (4) using on-site, industrial, municipal, and other facilities for treatment. Local geology, fracture characteristics, water sources, fracturing fluid compositions, and completion conditions all affect the quantity and characteristics of HF wastewater (Nicot and Scanlon, 2012; Chen and Carter, 2016). In addition to the above factors, when choosing alternatives

for HF wastewater treatment or disposal, HF operators will also consider the treatment capacity of treatment/disposal facilities, the level of infrastructure improvement, legal requirements, the surrounding environment, costs, and other factors (Kargbo et al., 2010; Lutz et al., 2013; Vengosh et al., 2014; Zhang et al., 2016).

16.3.1 DEEP-WELL INJECTION

In the exploitation of Marcelus shale gas, wastewater treatment methods have been changing over time. Wastewater was treated directly by municipal wastewater treatment facilities in the earliest time, then by industrial wastewater treatment plants, then by deep well injection, and finally wastewaters were reused with high reusability (Lutz et al., 2013). In addition, the concentrated salt wastewater obtained from desalination is expected to be injected into deep wells (Lutz et al., 2013; Shaffer et al., 2013). In this case, trucks transport wastewater to storage tanks at the mining site for treatment or disposal by gas companies or other companies (Veil, 2010). However, fewer and fewer injection wells are approved, and wastewater is often transported from one shale area to another in order to be disposed, which poses a risk for HF waste transport (Lutz et al., 2013). Meanwhile, deep well injection complemented by the onshore HF operations is considered as less dangerous activity, which is correlated with no major environment effects. However, there are many mechanisms and long-term factors that need to be further studied and discussed (Ferguson, 2015).

16.3.2 REUSE FOR FUTURE FRACTURING OPERATION

On-site recycling of flowback water exhibits the benefits of reducing water transportation via truck. Reuse is the most economical method. Wastewater can be reused with high quality after simple treatment (Lester et al., 2015). The main problems in water quality include blockage, scaling, contaminants, and soluble solids that can lead to increased resistance (Acharya et al., 2011; Hayes, 2008). However, reuse is only achievable if new fracturing fluid is necessitated, which indicates the shale gas industry has been continuously expanding with new wells to be drilled and fractured (Estrada and Bhamidimarri, 2016). Despite some controversy, it is generally believed that the primary factor affecting wastewater reuse is to determine the discharge water quality suitable for reuse (Estrada and Bhamidimarri, 2016). However, no mutually recognized criterions have been established so far (Blauch et al., 2009). Generally, total dissolved solid (TDS) concentrations in fracturing fluids comprising of recycled wastewater should not exceed 50,000–65,000 mg L^{-1}. Particularly, the chloride concentration should be maintained below 20,000–30,000 mg L^{-1}. The water quality indexes to be investigated are as follows: total SS content is less than 50 mg L^{-1}, total hardness is less than 2500 mg L^{-1}, pH value is within 6 ~ 8, sulfate content is less than 100 mg L^{-1}, iron concentration is less than 20 mg L^{-1}, oil and soluble organic matter are less than 25 mg L^{-1},

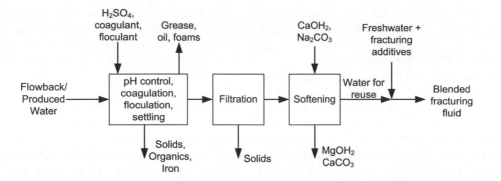

FIGURE 16.2 Diagram of the process proposed for water pretreatment prior to re-use as fracturing fluid including possible inputs and outputs. Note that not all the stages might be needed depending on the quality of the flowback/produced water. (Adapted from Estrada, J.M. and Bhamidimarri, R., *Fuel*, 182, 292–303, 2016.)

and total bacterial count is less than 100/100 mL (Keister et al., 2012; Henderson et al., 2011; Ord, 2014).

Hence, a practical treatment sequence (Figure 16.2) for flowback water recycling in the HF process could be comprised essentially of aeration for iron, with subsequent sedimentation/filtration (with optional flocculation) for SS and colloids, and disinfection for microorganisms (Lester et al., 2015). The pH of the wastewater is first adjusted. SS, colloidal particles, and NORM can be aggregated by coagulation/flocculation process. Degreasing and de-oiling would also be involved at this stage. Subsequently, an easy filtration of the residual SS would take place (Lester et al., 2015). Last, the effluent would be softened contributing to the aggregation of magnesium and calcium salts (Ord, 2014). The content of TDS and chloride in wastewater determines whether freshwater dilution can be directly used for reuse (Lester et al., 2015). Low-salt wastewater can be directly reused after dilution and mixing with fresh water, otherwise deep desalination treatment should be carried out before that (Mantell, 2011). Moreover, solids will be produced in the above processes. These solids should be managed according to their different properties, especially when they contain high concentrations of NORM (Lester et al., 2015).

16.3.3 Treatment for Surface Disposal or Reuse outside the Gas Industry

Fracturing wastewater applied outside the gas industry can be used for irrigation crops, livestock drinking, and indirect potable reuse (IOGCC and ALL Consulting, 2006). However, most of the above reuse methods need to be desalinated first. The allowable concentration of TDS in recovered fracturing fluid is 16,000 ~ 25,000 mg L^{-1} (Vengosh et al., 2014), which can reach 80,000 mg L^{-1} in extreme cases (US EPA, 2011). The US government suggests that TDS content of wastewater for long-term irrigation should be 500 ~ 2000 mg L^{-1}. The TDS standard for drinking water for livestock should be less than 10,000 mg L^{-1} (ALL Consulting, 2003), while the label for surface water discharged into Pennsylvania should be less than 500 mg L^{-1} for TDS (FracFocus, 2014).

Dissolved organic matter (DOM) also affects the reuse of fracturing wastewater (Lester et al., 2015). For example, surface drainage in Colorado requires a five-day biological oxygen demand (BOD$_5$) concentration of less than 45 mg L^{-1} (Colorado Water Quality Control Commission, 2012). In addition, DOM will bring biological pollution, reduce permeation flux, and then decrease the desalination efficiency (Lester et al., 2013; Ozgun et al., 2013). The organic matter, toxicity, and salinity brought about by the use of HF process in shale gas production will affect the treatment of fracturing wastewater (Abualfaraj et al., 2014; Barbot et al., 2013; Chapman et al., 2012; Haluszczak et al., 2013; Thacker et al., 2015). The existence of salinity, trace toxins, and reluctant organics surrender unique challenges necessitating a comprehensive strategy for HF wastewater treatment (Gregory and Mohan, 2015; Shaffer et al., 2013).

16.3.3.1 Membrane Separation (MS)

MS, including microfiltration (MF), ultrafiltration (UF), nanofiltration (NF), and reverse osmosis (RO), is a membrane technology based on differential pressure (Miller et al., 2013). The difference of these techniques is their filtration precision, so they have different target pollutions for treatment of wastewater. In general, MF and NF membranes can be used to remove suspended solids and organic matter in water, while NF and RO membranes can be used to remove soluble solids. It is a traditional treatment process for wastewater and has been used in some studies of fracturing wastewater. More than 99% of TSS in Marcelus shale fracturing wastewater was removed by Jiang et al. (2013) using ceramic membrane. 32.5% ~ 83.3% DOM in Fuling shale fracturing wastewater was removed by UF membrane by Kong et al. (2017). The 99.85% ~ 99.95% of hardness of Barnett shale in the United States was removed by UF-RO process by Miller et al. (2013). The advantages of this technique's small modules and wide usage make MS mobile and efficient for on-site treatment. However, membrane fouling is the main obstacle to the application of MS. The fouling of microfiltration membranes in the treatment of wastewater from Marcelus shale gas production in the United States is mainly caused by ferrous oxide submicron particles and organic coating, which was investigated by He et al. (2014).

16.3.3.2 Membrane Distillation (MD)

MD is a novel and promising desalination technology. The process is driven by thermal difference, and the water is heated to vapor after entering. The vapor enters the cooler through hydrophobic porous membranes due to osmosis (Estrada and Bhamidimarri, 2016), as shown in Figure 16.3. Compared with the traditional method, the process can withstand high temperature, and the steam pressure difference on both sides of the membrane makes the process faster (Estrada and Bhamidimarri, 2016). Normally, the temperature difference on the surface of the membrane is between 10°C and 20°C, which can achieve good separation effect without boiling salt-containing wastewater (Alkhudhiri et al., 2012; Minier-Matar et al., 2014). Pollutants other than TDS in wastewater will affect the operation of the process, so the key to the successful operation of the process is the treatment of wastewater, especially the removal of scaling substances (Alkhudhiri et al., 2012; Minier-Matar et al., 2014; Shaffer et al., 2013). Particularly, small volatile organic compounds (VOCs) can pass through the membrane to pollute the permeates, while alcohols and surfactants can make the membrane pores more hydrophilic, which enables the influent to flow across the membrane (Estrada and Bhamidimarri, 2016). Only in this way can MD play a great potential in HF wastewater treatment. MD process has low energy consumption, because it can use low levels of heat to achieve smooth operation, which provides a good example for low energy treatment of HF wastewater (Shaffer et al., 2013). However, only laboratory scale studies under controlled conditions have been conducted. Therefore, a larger scale test of MD application under similar conditions in the field is the key to comprehensively evaluate the advantages and disadvantages of this technology (Alkhudhiri et al., 2012).

16.3.3.3 Forward Osmosis (FO)

FO is a completely different separation technology from reverse osmosis. Its essence is a membrane separation technology, which uses the concentration difference between the two sides of the membrane to remove the total dissolved solids. The key components of FO are semipermeable membranes. The two sides of the semipermeable membranes are low concentration wastewater to be treated and high concentration solution (drawn solution), respectively. The concentration difference between the two forms the driving force for the influent water flux to pass through the semipermeable membranes (Estrada and Bhamidimarri, 2016). In order to ensure the normal operation of the process, the concentration of specific components in the draw solution should be higher than that in the wastewater to be treated (Shaffer et al., 2015). As shown in Figure 16.4, the FO process consists of two steps: (1) influent passes through a semipermeable membrane and draw solution is diluted; and (2) water and drawn solution are separated by RO or thermal distillation to achieve the purpose of re-concentrating drawn solution and producing high-quality effluent (Coday et al., 2014). The application of FO process in the treatment of fracturing wastewater is less, which is mainly due to the poor performance of semipermeable membranes. In a few studies, the drilling mud and fracturing wastewater in the Haynesville Shale of the United States were treated by this process by Hickenbott et al. (2013), and more than 80%

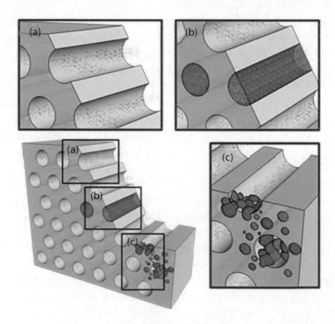

FIGURE 16.3 Conceptual illustration of membrane distillation (MD). (a) The temperature difference between the warmer feed and cooler permeate streams gives rise to a vapor pressure gradient, which drives the transport of water vapor across the membrane pores. (b) Wetting of the pores, where the aqueous solution penetrates the hydrophobic membrane, results in the feed solution flowing directly across the membrane, causing permeate quality to deteriorate. (c) Fouling of the MD membrane leads to performance decline, such as flux decay. (Adapted from Shaffer, D.L. et al., *Environ. Sci. Technol.*, 47, 9569–9583, 2013.)

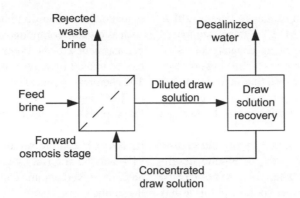

FIGURE 16.4 Simplified diagram for an FO process for desalination including the second stage for draw solution re-concentration. (Adapted from Estrada, J.M. and Bhamidimarri, R., *Fuel*, 182, 292–303, 2016.)

water recovery was obtained. In addition, in the treatment of Fayetteville Shale wastewater, Sardari et al. (2018) used the combination of electrocoagulation and FO process to achieve 70% water recovery. Although FO system is more complex than other methods, it still has some rare advantages. Shaffer et al. (2013) pointed out that FO process operates at low pressure, so the membrane is not susceptible to contamination, which can simplify some pretreatment, reduce maintenance operations, and effectively prolong the overall life of the membrane (Shaffer et al., 2013).

16.3.3.4 Evaporation with Mechanical Vapor Compression (MVC)

Evaporation technology is a common technology to obtain fresh water and concentrated brine with high TDS content by heating some water to vapor. In this technology, heat is continuously transferred to liquid flux. MVC system is a novel technology combination, where the heat is transferred to brine through superheated compressed steam conveyed by tubular evaporator, as shown in Figure 16.5. The system uses thermal condensate and brine flow to preheat wastewater, maximizes energy utilization, and realizes energy integration.

The principal energy input to this system is the electricity exhausted for compressing the vapor. Generally, the temperature of fracturing wastewater is 65°C ~ 100°C, while MVC system can operate under the environment of less than 70°C, which can make HF wastewater be effectively treated under low energy consumption (Henderson et al., 2011; Igunnu and Chen, 2014; Shaffer et al., 2013). Because of the low operating temperature, the system also has the characteristics of high efficiency and high reliability. Vertical tube heat exchangers have also been tried out in MVC to maximize heat transfer and further improve evaporation performance. As shown in Figure 16.5, the tube heat exchangers are arranged vertically. The feed flow is film-like on the inner surface of the heat exchanger tube, and the condensate obtained by superheated evaporation is enriched on the outer surface of the tube (Heins, 2010). For MVC's application research on fracturing wastewater treatment, Rangharajan et al. (2017) studied desalting high salinity Utica Shale flowback water via MVC-FO in the United States and the results showed salt was removed more than 99%. In HF process, wastewater desalination by MVC has lowered treatment complexity and restrained liquid waste emissions in contrast with traditional procedure incorporating

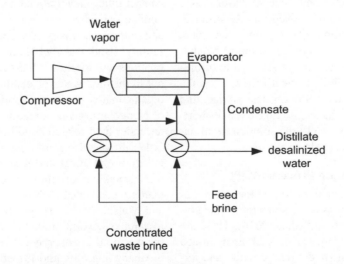

FIGURE 16.5 Simplified flow diagram for an MVC process for desalination. (Adapted from Estrada, J.M. and Bhamidimarri, R., *Fuel*, 182, 292–303, 2016.)

boiling, softening, filtration, and ion exchange (Heins, 2010). Modular design is a highlight of MVC systems compared with traditional methods. The processing capacity of MVC can be significantly improved by stacking evaporator-condenser units (Veza, 1995). This shows that modular design can reduce equipment costs and operating costs. The desalinated water does not need to be treated separately, but can be recycled in the preheater for oil-gas separation (Koren and Naday, 1994). MVC is not easily polluted by oil and grease, which is its remarkable advantage over membrane desalination (Heins, 2010; Veza, 1995). Therefore, MVC can be used to pretreat wastewater. However, sodium chloride has a high solubility in water, so the MVC process still has an upper limit on TDS in the treatment of fracturing wastewater. Under better operating conditions, it is usually limited to 200,000 mg L^{-1}. In large-scale desalination of HF wastewater, the MVC system requires about 20 kWh of energy per cubic meter of wastewater (Thiel et al., 2015). Therefore, compared with other desalination technologies, the main shortage of MVC is the higher energy consumption caused by the vapor compression during operation (Koren and Naday, 1994; Veza, 1995). In addition, Bean et al. (2018) found that evaporative wastewater could cause PM$_{2.5}$ to rise and produce air pollution such as ammonia water.

16.3.3.5 Electrocoagulation (EC)

Compared with the traditional chemical coagulation method, the electrocoagulation method has significant differences. It does not depend on chemical agents and only needs alternating current or direct current. EC is a developing process, but it has shown a good universality, including TSS, heavy metals, oil, vitamins, and inorganic salts with high hardness. Hence, there are some cases of fracturing wastewater treatment via EC. The wastewater produced from Denver-Julesburg basin, USA, had the characteristics of the high turbidity and high TSS, which was effectively treated by Lobo et al. (2016) using biochar-EC method. In the study of Kausley et al. (2017), the fracturing wastewater with high hardness and TOC was simulated, and the organic matter and divalent cations were removed by EC system. The results showed that the removal rates of chloride, inorganic carbon, CaCO$_3$, and TOC were 25%, 63%, 99%, and 50%, respectively. They found biochar indeed improved the removal efficiency of EC so that turbidity and TSS could be removed 99%, respectively. EC has the advantages of less environmental pollution, less sludge, and high water recovery, but it also has some unavoidable shortcomings, such as low efficiency due to the solution desalination and high cost due to the electrode passivation.

16.3.3.6 Advanced Oxidization Process (AOP)

AOP can be applied in HF wastewater treatment for removal of organic and some inorganic composites, disinfection, and elimination of odor and color (Igunnu and Chen, 2014). There are many kinds of oxidants in water treatment, such as hydrogen peroxide, ozone, Fenton reagent (hydrogen peroxide and ferrous), and perchlorate. In practical applications, hydrodynamic cavitation, acoustic cavitation, ozonation, and electrochemical oxidation are often combined to effectively remove additives such as organic compounds, bacteria, and scale inhibitors. After pretreatment with AOP-combined technologies, the wastewater can be treated by reverse osmosis or reused (Ely et al., 2011). The chemical agents used in AOP treatment vary with the types of fracturing wastewater. In the study of Abass et al. (2017), the complex organic compounds and metals in wastewater from south-eastern Ordos Basin were degraded and adsorbed by NZVI-H$_2$O$_2$, the removal rate of polyethylene glycol was 91%, and the removal rate of total petroleum hydrocarbons was 88%. Dissolved organic carbon (DOC) in the flowback water of the southeastern Anatolia Basin, Turkey, was removed by ozone oxidation by Turan et al. (2017). Chemical oxygen demand (COD), color, and total phenol were removed around 87.4%, 89.2%, and 91.8%, respectively. In a synthetic water treatment, Hong et al. (2018) used photolysis technology to remove glutaraldehyde (GA) from synthetic water under ultraviolet light. The removal rate reached 85% in one hour. AOP is commonly considered as an environment friendly process for efficient COD degradation. AOP cannot be used to reduce the dissolved salt content in wastewater, so it needs to be combined with other methods. When ozone oxidation process is used to treat fracturing wastewater containing high concentration chloride and bromide, the formation of chlorine and bromine free radicals will result in more harmful organic chlorides and bromides than the original components (Langlais et al., 1991).

16.3.3.7 Adsorption

According to previous studies, different components will behave differently in HF wastewater. At present, only about one third of these components can be separated by adsorption or attachment to other phases (Camarillo et al., 2016). Previous data show that only 22% of the compounds can be adsorbed on the organic surface, which indicates that the components in wastewater are difficult to be adsorbed by organic matter (Camarillo et al., 2016). Adsorption by activated carbon is efficient for various organic compounds, but strongly depends on many factors, e.g., experiment conditions (i.e., dosages and contact time), inherent characteristic of activated carbon (i.e., particle size, chemical structure, hydrophilicity, and polarity), and solution chemistry (i.e., competing organic matter, pH, and salinity) (Butkovskyi et al., 2017). Notably, most of the investigations on sorption of organic contaminants were carried out in distilled water or distilled water with artificially added organic matter or NaCl. Therefore, information about sorption of particular organic contaminants in real fracturing wastewater under field-related conditions is absent (Butkovskyi et al., 2017). It was also found that the increase of salinity resulted in salting-out effect and weakened the electrostatic exclusion of activated carbon, thus enhancing the adsorption of organic matter on activated carbon (Lamichhane et al., 2016; Younker and Walsh, 2015). There are many different materials for research of adsorption. In the study of Chang et al. (2017), zeolite was used as adsorbent to remove cations from wastewater in Changning city, and the effect of adsorbent properties on the removal of calcium and magnesium ions was also studied. According to the study of Butkovskyi et al. (2018), the removal rates of HOC, DOC, and COD from the Baltic shale fracturing wastewater in Poland are only about 20% ~ 45% after adsorption by granular activated carbon (GAC). The effects of ZVI, NZVI, and aging NZVI on removal of heavy metal ions and chlorine-containing organic solvents from fracturing wastewater were tested (Sun et al., 2017a, 2017b, 2018, 2019b; Lei et al., 2018). After a month of immersion, NZVI removed 58.7%–42.9% zinc(II), 25.4%–80.0% copper(II), 66.7%–75.1% arsenic(V), 65.7%–44.1% chromium(VI), 38.7%–53.5% carbon tetrachloride (CT), 21.7%–71.1% 1,1,2-TCA (1,1,2-trichloroethane), which was the best adsorption material. Interestingly, aging NZVI performs better than NZVI in removing metal ions, because the increase of specific surface area after aging is beneficial to electron transfer. However, aging will lead to a decrease in the actual content of NZVI, making it less capable of removing TCA. In the cases of application of adsorption, almost all removal efficiencies of pollutions are a little low. In a word, the high ion exchange capacity and high specific surface area of adsorbents are conducive to the removal of inorganic and organic matter, which makes the adsorption treatment widely used at present.

16.3.3.8 Biological Treatment

Based on the existing literature, 37% of organic compounds identified in HF wastewater were biodegradable in laboratory tests, suggesting applicability of biological process for HF wastewater treatment (Struijs and Vandenberg, 1995). Preliminary tests showed that the biodegradability of HF wastewater was not affected by high salinity and trace toxicants (Kahrilas et al., 2015; Lester et al., 2013; Strong et al., 2013). HF wastewater has high oxygen content, a large number of microorganisms, and organic matter can be consumed in high salinity environment, which proves the feasibility of its biological treatment (Cluff et al., 2014; Mohan et al., 2013a, 2013b). Under both aerobic and anaerobic conditions, biological treatment of saline wastewaters is achievable, which specifies that a proper halophilic bacterial group exists (Xiao and Roberts, 2010). Biological treatment systems for HF wastewater with total dissolved solids content up to 100,000 mg L^{-1} have been established (Pendashteh et al., 2012). It is noteworthy that the removal efficiency of pollutants does decrease with the increase of salinity. Aerobic treatment can be carried out in submerged membrane bioreactor, which is the preferred method for degradation of organic halogens in HF wastewater. Compared with activated sludge, MBR is more efficient in PAHs removal. Meanwhile, contrasting the side stream filtration, the submerged-filter configuration is more appropriate for eliminating organic compounds with lower volatilization (Kwon et al., 2008; Mozo et al., 2011). Aerobic sludge from sewage treatment plants is often used in biodegradation technology. Some researchers studied the performances of the biological treatment, and found that the removal rate of DOC in Marcellus shale wastewater reached more than 90% (Kekacs et al., 2015), the removal rate of TPH reached 98%–99% when treating Shengli Oilfield fracturing wastewater (Yang et al., 2014), and the removal rate of COD can exceed 90% when treating guar gum synthetic water (Lester et al., 2013). When conducted in the produced water from the Bakken shale, microbial fuel cells (MFCs) and microbial capacitive deionization cells (MCDCs) demonstrated different abilities, according to the study of Shrestha et al. (2018). The COD removal in MFCs and MCDCs were 88% and 76%, respectively. Carbon and energy sources in fracturing wastewater are suitable for microbial growth, making this technology possible. But when the TDS of wastewater is very high, there are some shortcomings such as high risk, long working time, and low efficiency of intermediate metabolites.

16.3.3.9 Summary of Treatment Options

As above presented, different techniques behaved more efficiently for different target pollutions (Table 16.4). AOP, biodegradation, and adsorption technology can be used to remove organic components such as COD, TOC, and BOD, while MF/UF, EC, and coagulation technology can be used to remove inorganic substances such as total suspended solids (TSS) and hardness. Because desalination process requires membrane separation of water and soluble salts, it is the most complex of all wastewater treatment processes. Thus, RO, NF, FO, and MD showed more advantages in desalinate. Based on the discussion from referred researches, we put forward some recommended options of technique combinations for fracturing wastewater treatment. Figure 16.6 is a typical drilling wastewater treatment process. Wastewater is mixed first to achieve the purpose of pH regulation and pre-oxidation. Then wastewater should remove inorganics by combination of coagulation and filtration, because too much TSS and hardness could inhibit radicals' oxidation and microorganisms' degradation that were from organics removal process. After filtration such as MF or UF, the sludge sent to make a disposal, and the solution went to remove organics that could produce some membrane fouling in desalinate process. AOP, biodegradation, and adsorption technologies usually make a

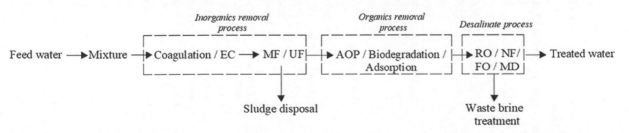

FIGURE 16.6 Options of technique combinations for fracturing wastewater treatment. (Adapted from Sun, Y. et al., *Environ. Int.*, 125, 452–469, 2019a.)

TABLE 16.4

Comparison of Different Options for Fracturing Wastewater Treatment

Treatment	Case							
	Water Source	Water Quality	Method	Efficiency	Cost	Advantages	Disadvantages	References
MS	Fuling Shale China	TOC: 78 mg L^{-1} COD: 472.6 mg L^{-1} Turbidity: 37.1 NTU Conductivity: 2.36 mS cm^{-1}	Coagulation-UF	32.5% ~ 83.3% DOM removal	—	• MF/UF: useful to remove residual TSS and organics;	• Membrane fouling	Kong et al. (2017)
	Marcellus Shale USA	TSS: 98 ~ 776 mg L^{-1} TDS: 38,000 ~ 166,484 mg L^{-1} Turbidity: 32 ~ 111 NTU TOC: 5.2 ~ 19.4 mg L^{-1}	MF	—	—	• NF/RO: useful to desalinate; • Small modules: mobile and efficient for on-site treatment.		He et al. (2014)
	Barnett Shale USA	TSS: 10.8 ~ 3220 mg L^{-1} TDS: 23,600 ~ 238,000 mg L^{-1} TOC: 3.7 ~ 323 mg L^{-1}	RO-UF	99.85% ~ 99.95% Hardness removal	—			Miller et al. (2013)
	Marcellus Shale USA	TSS: 881 mg L^{-1} TDS: 48,000 mg L^{-1} TOC: 720 mg L^{-1} Turbidity: 770 NTU	Ceramic MF	>99%TSS removal	18.4 $/m^3			Jiang et al. (2013)
FO	Haynesville Shale USA	Total cations: 91 mg L^{-1} Total anions: 89 mg L^{-1}	FO	>80% Water recovery	—	• High rejection of inorganic and organic;	• Membrane robustness, permeability, chemical stability, and range need to be improved.	Hickenbottom et al. (2013)
	Fayetteville Shale USA	TSS: 639.1 mg L^{-1} TDS: 23,255 mg L^{-1} Turbidity: 117.1 NTU TOC: 154.7 mg L^{-1}	EC-FO	>70% TSS, TOC, and turbidity removal; ~70% water recovery		• Less irreversible membrane fouling.		Sardari et al. (2018)
MVC	Utica Shale USA	TDS: 184,000 mg L^{-1}	MVC-FO	>99% Salt removal	—	• Low cost; • Less membrane fouling.	• Evaporated wastewater: air pollution (PM2.5 and ammonia); • High energy requirement.	Rangharajan et al. (2017)

(Continued)

TABLE 16.4 (*Continued*)
Comparison of Different Options for Fracturing Wastewater Treatment

Treatment	Water Source	Water Quality	Method	Case Efficiency	Cost	Advantages	Disadvantages	References
EC	Synthetic water	Hardness: 200 mg L^{-1} as CaCO$_3$ Conductivity: 17 mS cm^{-1} TOC: 407 mg L^{-1}	EC	50% TOC, 63% IC, 99% hardness, and 25% chloride removal	4.15 USD/m^3	• High water recovery; • Less sludge production; • Lower environmental impact.	• Not effective for dissolved salts removal; • Electrode passivation; • High cost.	Kausley et al. (2017)
	Denver–Julesburg Basin USA	Turbidity: 400 ± 44 NTU TSS: 514 ± 47 mg L^{-1} COD: 3631 ± 69 mg L^{-1} Conductivity: 46.1 mS cm^{-1}	EC-biochar	99% Turbidity and TSS removal	—			Lobo et al. (2016)
AOP	Southeastern Ordos Basin China	TOC: 700 ~ 1514 mg L^{-1} COD: 1577 ~ 2318 mg L^{-1} Turbidity: 78.7 ~ 90.4 NTU TDS: 8640 ~ 9630 mg L^{-1} TPH: 92.4 ~ 130.6 mg L^{-1} PEGs: 17.8 ~ 19.9 mg L^{-1}	NZVI-H$_2$O$_2$	88% TPH and 91% PEGs removal	200 \$/m^3	• Environment friendly processes; • Useful for COD degradation.	• Not effective for dissolved salts removal.	Abass et al. (2017)
	Southeast Anatolia Basin Turkey	COD: 5375 mg L^{-1} TSS: 525 mg L^{-1} Total phenol: 21.4 mg L^{-1} Conductivity: 104.9 mS cm^{-1}	Ozonation	87.4% COD, 89.2% color, and 91.8% total phenol removal	6.35 €/m^3			Turan et al. (2017)
	Synthetic water	GA: 0.1 mM	UV Photolyzation	52% ~ 85% GA removal	—			Hong et al. (2018)

(*Continued*)

TABLE 16.4 (Continued)
Comparison of Different Options for Fracturing Wastewater Treatment

Treatment	Water Source	Water Quality	Method	Efficiency (Case)	Cost	Advantages	Disadvantages	References
Adsorption	Changning shale China	TDS: 20,520 ~ 20,755 mg L⁻¹	Activated zeolite	40% Ca^{2+} and 70% Mg^{2+} removal	—	• High cation-exchange ability: good for inorganics removal; • High surface area: good for organics removal; • Low cost.	• Low pollutions removal efficiency.	Chang et al. (2017)
	Baltic shale Poland	COD: 1800 mg L⁻¹ DOC: 649 mg L⁻¹	GAC	23% COD, 20% DOC, and 45% HOC removal				Butkovskyi et al. (2018)
	Denver–Julesburg Basin USA	Turbidity: 400 ± 44 NTU TSS: 514 ± 47 mg L⁻¹ COD: 3631 ± 69 mg L⁻¹ Conductivity: 46.1 mS cm⁻¹	Biochar	5% ~ 10% TSS removal by biochar 99% TSS removal by EC-biochar				Lobo et al. (2016)
	Synthetic water	As(V): 150 µg L⁻¹ Se(VI): 350 µg L⁻¹	ZVI	90% As(V) and Se(VI) removal				Sun et al. (2017a)
	Synthetic water	TDS: 233,300 mg L⁻¹ TOC: 670 mg L⁻¹ Cu(II): 116 mg L⁻¹ Zn(II): 247 mg L⁻¹ Cr(VI): 2.2 mg L⁻¹ As(V): 1.1 mg L⁻¹ CT: 2 mg L⁻¹ 1,1,2-TCA: 2 mg L⁻¹	NZVI	25.4% ~ 80.0% Cu(II), 58.7% ~ 42.9% Zn(II), 65.7% ~ 44.1% Cr(VI), 66.7% ~ 75.1% As(V), 38.7% ~ 53.5% CT, and 21.7% ~ 71.1% 1,1,2-TCA removal				Sun et al. (2017b) Lei et al. (2018)
			Aged NZVI	84.2% Cu(II), 70.8% Cr(VI), 31.2% Zn(II), and 39.8% As(V) removal				Sun et al. (2018) Lei et al. (2018)
			Fe-biochar composite	99.1% Cu(II), 88.2% Cr(VI), 45.8% Zn(II), 77.7% As(V), 91.0% 1,1,2-TCA, and 3.2~7.2 g cations/g removal				Sun et al. (2019b)
Biological treatment	Marcellus shale USA	DOC: 250 mg L⁻¹	Aerobic sludge	57% ~ 90% DOC removal	—	• Wastewater with high carbon and energy sources to support microbial growth is suitable for biological treatment.	• Incompletely degraded intermediate of metabolic risks; • Long reaction time; • Low efficiency with high TDS.	Kekacs et al. (2015)
	Shengli Oilfield China	BOD₅: 1525 mg L⁻¹ TSS: 2330 mg L⁻¹ COD: 9360 mg L⁻¹ TDS: 6850 mg L⁻¹ TPH: 14 mg L⁻¹	Aerobic activated sludge	98% ~ 99% TPH removal				Yang et al. (2014)
	Synthetic water	COD: 2500 mg L⁻¹ TDS: 22,000 ~ 45,000 mg L⁻¹	Activated sludge	>90% COD removal				Lester et al. (2013)
	Bakken shale	COD: 20,000 ~ 78,000 mg L⁻¹ TDS: 150,000 ~ 219,000 mg L⁻¹	MFCs MCDCs	88% COD removal 76% COD removal				Shrestha et al. (2018)

good effect on organics removal. The inorganic and organic components are then removed by adsorption or oxidation. Finally, the desalination process was used for deep purification until the salt concentration was reduced to a certain extent. The wastewater treated by the above process can be discharged into the natural environment. Flowing through desalination process, treated water can be discharged into the environment, but the waste brine with high concentration of salt still needs to be treated.

16.4 CONCLUSIONS

The wastewater produced in the process of shale gas exploitation by HF method has seriously hindered the development of shale gas exploitation. With the remarkable growth of the shale gas industry, the aim of wastewater management has become to find the best wastewater treatment scheme. In this process, the following aspects should be considered: (1) the ultimate goal is to achieve zero discharge of HF wastewater; (2) attention should be paid to the treatment of high concentration saline wastewater produced by desalination process; (3) the actual situations determine the choice of centralized treatment plants or on-site treatment via modular plants; and (4) the feasibility of reusing recovered HF wastewater as fracturing fluid should be considered. In the absence of sufficient operational experience under field-related conditions, it is still too early to systematically address these questions for the shale gas industry by means of fracturing. Currently, there remains a desperate necessity to comprehend the time-fluctuating and location-dependent composition of HF wastewaters in various shale reservoirs. Once the compositions of HF wastewater have been characterized, the most suitable treatment technologies will be potentially selected and optimized for different scenarios. No single technology can achieve the requirements for discharging or reusing HF wastewater in/outside the shale gas industry. Presumably, an incorporation of various pretreatment technologies and further exploration of available desalination technologies must be accomplished in order to minimize the environmental impact of wastewaters from shale gas production by HF.

REFERENCES

Abass, O.K., Zhuo, M., Zhang, K., 2017. Concomitant degradation of complex organics and metals recovery from fracking wastewater: Roles of nano zerovalent iron initiated oxidation and adsorption. *Chem. Eng. J.* 328, 159–171.

Abualfaraj, N., Gurian, P.L., Olson, M.S., 2014. Characterization of marcellus shale flowback water. *Environ. Eng. Sci.* 31, 514–524.

Acharya, H.R., Henderson, C., Matis, H., Kommepalli, H., Moore, B., Wang, H., 2011. Cost effective recovery of low-TDS frac flowback water for re-use. U.S. Department of Energy, Washington, DC (DE-FE0000784).

Al Jazeera America, 2014. Cleanup of North Dakota pipeline spill may take weeks. http://america.aljazeera.com/articles/2014/7/10/north-dakota-saltwater.html

Alkhudhiri, A., Darwish, N., Hilal, N., 2012. Membrane distillation: A comprehensive review. *Desalination* 287, 2–18.

ALL Consulting, 2003. Handbook on coal bed methane produced water: Management and beneficial use alternatives, Chapter 5–beneficial use alternatives. Prepared For, Ground Water Protection Research Foundation U.S. Department of Energy National Petroleum Technology Office Bureau of Land Management. http://www.all-llc.com/CBM/BU/index.htm

API, 2010. Water management associated with hydraulic fracturing. http://www.api.org/~/media/files/policy/exploration/hf2_e1.pdf.

Arthur, J., Bohm, B., Coughlin, B.J., Layne, M., Cornue, D., 2009. Evaluating the environmental implications of hydraulic fracturing in shale gas reservoirs, Paper presented at *SPE Americas E&P Environmental and Safety Conference*, San Antonio, TX, March 23–25, 2009.

Arthur, D., Bohm, B., Coughlin, B., Layne, M., 2008. Evaluating the environmental implications of hydraulic fracturing in shale gas reservoirs. http://www.allllc.com/publicdownloads/ArthurHydrFracPaperFINAL.pdf.

Balaba, R.S. and Smart, R.B., 2012. Total arsenic and selenium analysis in Marcellus shale, high-salinity water, and hydrofracture flowback wastewater. *Chemosphere* 89 (11), 1437–1442.

Barbot, E., Vidic, N.S., Gregory, K.B., Vidic, R.D., 2013. Spatial and temporal correlation of water quality parameters of produced waters from Devonian-age shale following hydraulic fracturing. *Environ. Sci. Technol.* 47 (6), 2562–2569.

Bazilian, M., Brandt, A.R., Billman, L., Heath, G., Logan, J., Mann, M., 2014. Ensuring benefits from North American shale gas development: Towards a research agenda. *J. Unconv. Oil Gas Resour.* 7, 71–74.

Bean, J.K., Bhandari, S., Bilotto, A., Hildebrandt Ruiz, L., 2018. Formation of particulate matter from the oxidation of evaporated hydraulic fracturing wastewater. *Environ. Sci. Technol.* 52 (8), 4960–4968.

Bibby, K.J., Brantley, S.L., Reible, D.D., Linden, K.G., Mouser, P.J., Gregory, K.B., Ellis, B.R., Vidic, R.D., 2013. Suggested reporting parameters for investigations of wastewater from unconventional shale gas extraction. *Environ. Sci. Technol.* 47, 13220–13221.

Blauch, M.E., Myers, R.R., Moore, T., Lipinski, B.A., Houston, N.A., 2009. Marcellus Shale post-frac flowback waters-where is all the salt coming from and what are the implications? In: *SPE Eastern Regional Meeting*, Charleston, WV.

Brantley, S.L., Yoxtheimer, D., Arjmand, S., Grieve, P., Vidic, R., Pollak, J., Llewellyn, G.T., Abad, J., Simon, C., 2014. Water resource impacts during unconventional shale gas development: The Pennsylvania experience. *Int. J. Coal Geo.* 126, 140–156.

British Columbia Oil and Gas Commission (OGC), 2014. Application guideline for: Deep well disposal of produced water, deep well disposal of nonhazardous waste. https://www.bcogc.ca/node/8206/download.

Butkovskyi, A., Bruning, H., Kools, S.A., Rijnaarts, H.H., Van, Wezel, A.P., 2017. Organic pollutants in shale gas flowback and produced waters: Identification, potential ecological impact, and implications for treatment strategies. *Environ. Sci. Technol.* 51 (9), 4740–4754.

Butkovskyi, A., Faber, A.H., Wang, Y., Grolle, K., Hofmancaris, R., Bruning, H., Wezel, A.P.V., Rijnaarts H.H.M., 2018. Removal of organic compounds from shale gas flowback water. *Water Res.* 138, 47–55.

Camarillo, M.K., Domen, J.K., Stringfellow, W.T., 2016. Physical-chemical evaluation of hydraulic fracturing chemicals in the context of produced water treatment. *J. Environ. Manage.* 18, 164–174.

Chapman, E.C., Capo, R.C., Stewart, B.W., Kirby, C.S., Hammack, R.W., Schroeder, K.T., Edenborn, H.M., 2012. Geochemical and strontium isotope characterization of produced waters from Marcellus Shale natural gas extraction. *Environ. Sci. Technol.* 46, 3545–3553.

Chang, H., Liu, T., He, Q., Li, D., Crittenden, J., Liu, B., 2017. Removal of calcium and magnesium ions from shale gas flowback water by chemically activated zeolite. *Water Sci. Technol.* 76, 575–583.

Chen, H. and Carter, K.E., 2016. Water usage for natural gas production through hydraulic fracturing in the United States from 2008 to 2014. *J. Environ. Manage.* 170, 152–159.

Chen, S.S., Sun, Y.Q., Tsang, D.C.W., Graham, N.J.D., Ok, Y.S., Feng, Y.J., Li, X.D., 2016. Potential impact of flowback water from hydraulic fracturing on agricultural soil quality: Metal/metalloid bioaccessibility, Microtox bioassay, and enzyme activities. *Sci. Total Environ.* 579, 1419–1426.

Chen, S.S., Sun, Y.Q., Tsang, D.C.W., Graham, N.J.D., Ok, Y.S., Feng, Y.J., Li, X.D., 2017. Insights into the subsurface transport of As(V) and Se(VI) in produced water from hydraulic fracturing using soil samples from Qingshankou Formation, Songliao Basin, China. *Environ. Pollut.* 223, 449–456.

Cluff, M.A., Hartsock, A., MacRae, J.D., Carter, K., Mouser, P.J., 2014. Temporal changes in microbial ecology and geochemistry in produced water from hydraulically fractured Marcellus shale gas wells. *Environ. Sci. Technol.* 48, 6508–6517.

Coday, B.D., Xu, P., Beaudry, E.G., Herron, J., Lampi, K., Hancock, N.T., Cath, T.Y., 2014. The sweet spot of forward osmosis: Treatment of produced water, drilling wastewater, and other complex and difficult liquid streams. *Desalination* 333, 23–35.

Colorado Water Quality Control Commission, 2012. Regulation #62 - Regulations for Effluent Limitations (5CCR 1002-62).

Dresel, P.E. and Rose, A.W., 2010. Chemistry and Origin of Oil and Gas Well Brines in Western Pennsylvania, 4th ser., Pennsylvania Geological Survey, Open-File Report OFOG 10-01.0, 48 pp.

EC Commission Staff Working Document Impact Assessment: Exploration and production of hydrocarbons (such as shale gas) using high volume hydraulic fracturing in the EU. European Commission, SWD/2014/021, 2014.

Ely, J.W., Horn, A., Cathey, R., Fraim, M., Jakhete, S., 2011. Game changing technology for treating and recycling frac water. In SPE Annual Technical Conference and Exhibition, Denver, Colorado, USA, October 30–November 2, 2011.

EPA, 2012. Study of the potential impacts of hydraulic fracturing on drinking water resources progress report. http://www2.epa.gov/sites/production/files/documents/hf-report20121214.pdf

Estrada, J.M. and Bhamidimarri, R., 2016. A review of the issues and treatment options for wastewater from shale gas extraction by hydraulic fracturing. *Fuel* 182, 292–303.

Ferguson, G., 2015. Deep injection of waste water in the western Canada sedimentary basin. *Groundwater* 53 (2), 187–194.

FracFocus, 2014. Why chemicals are used. http://www.fracfocus.org/chemical-use/why-chemicals-are-used

Freyman, M., 2014. Hydraulic fracturing & water stress: Water demand by the numbers. http://www.ceres.org/issues/water/shale-energy/shale-and-water-maps/hydraulic-fracturing-water-stress-water-demand-by-the-numbers

Gregory, K. and Mohan, A.M., 2015. Current perspective on produced water management challenges during hydraulic fracturing for oil and gas recovery. *Environ. Chem.* 12, 261–266.

Gregory, K.B., Vidic, R.D., Dzombak, D.A., 2011. Water management challenges associated with the production of shale gas by hydraulic fracturing. *Elements* 7, 181–186.

GWPC and ALL Consulting, 2009. Modern shale gas development in the United States: A primer, United States Department of Energy, National Energy Technology Laboratory, DE-FG26-04NT15455: 67www.netl.doe.gov/technologies/oil-gas/publications/epreports/shale_gas_primer_2009.pdf

Haluszczak, L.O., Rose, A.W., Kump, L.R., 2013. Geochemical evaluation of flowback brine from Marcellus gas wells in Pennsylvania, USA. *Appl. Geochem.* 28, 55–61.

Hayes, T., 2008. *Proceedings and Minutes of the Hydraulic Fracturing Expert Panel XTO Facilities*, Fort Worth, TX Sept 26, 2007.

He, C., Wang, X., Liu, W., Barbot, E., Vidic, R.D., 2014. Microfiltration in recycling of marcellus shale flowback water: Solids removal and potential fouling of polymeric microfiltration membranes. *J. Mater. Sci.* 462, 88–95.

Heins, W.F., 2010. Is a paradigm shift in produced water treatment technology occurring at SAGD facilities? *J. Can. Petrol. Technol.* 49 (1), 10–15.

Helman, C., 2012. The arithmetic of shale gas. *Forbes* June 22, 2012.

Henderson, C., Acharya, H., Matis, H., Kommepalli, H., Moore, B., Wang, H., 2011. Cost effective recovery of low-TDS frac flowback water for re-use. Department of Energy DE-FE0000784 final report.

Hickenbottom, K.L., Hancock, N.T., Hutchings, N.R., Appleton, E.W., Beaudry, E.G., Xu, P., Cath, T. Y., 2013. Forward osmosis treatment of drilling mud and fracturing wastewater from oil and gas operations. *Desalination* 312, 60–66.

Hong, S., Ratpukdi, T., Sivaguru, J., Khan, E., 2018. Photolysis of glutaraldehyde in brine: A showcase study for removal of a common biocide in oil and gas produced water. *J. Hazard. Mater.* 353, 254–260.

Howarth, R., Santoro, R., Ingraffea, A., 2011a. Methane and the greenhouse-gas footprint of natural gas from shale formations. *Clim. Change.* 106 (4), 679–690.

Howarth, R.W., Ingraffea, A., Engelder, T., 2011b. Natural gas: Should fracking stop? *Nature* 477 (7364), 271–273.

Igunnu, E.T. and Chen, G.Z., 2014. Produced water treatment technologies. *Int. J. Low-Carbon Technol.* 9 (3), 157–177.

Jackson, R.B., Vengosh, A., Darrah, T.H., Warner, N.R., Down, A., Poreda, R.J., Osborn, S.G., Zhao, K., Karr, J.D., 2013. Increased stray gas abundance in a subset of drinking water wells near Marcellus shale gas extraction. *Proc. Natl. Acad. Sci. U.S.A.* 110 (28), 11250–11255.

Jiang, Q., Rentschler, J., Perrone, R., Liu, K., 2013. Application of ceramic membrane and ion-exchange for the treatment of the flowback water from Marcellus shale gas production. *J. Mater. Sci.* 431, 55–61.

Kahrilas, G.A., Blotevogel, J., Stewart, P.S., Borch, T., 2015. Biocides in hydraulic fracturing fluids: A critical review of their usage, mobility, degradation, and toxicity. *Environ. Sci. Technol.* 49, 16–32.

Kargbo, D.M., Wilhelm, R.G., Campbell, D.J., 2010. Natural gas plays in the Marcellus shale: Challenges and potential opportunities. *Environ. Sci. Technol.* 44 (15), 5679–5684.

Kausley, S.B., Malhotra, C.P., Pandit, A.B., 2017. Treatment and reuse of shale gas wastewater: Electrocoagulation system for enhanced removal of organic contamination and scale causing divalent cations. *J. Water Process Eng.* 16, 149–162.

Keister, T., Sleigh, J., Briody, M., Brockway, P.A., 2012. Sequential precipitation-fractional crystallization treatment of Marcellus Shale flowback and production wastewaters. *IWC*, 12, 72.

Kekacs, D., Drollette, B.D., Brooker, M., Plata, D.L., Mouser, P.J., 2015. Aerobic biodegradation of organic compounds in hydraulic fracturing fluids. *Biodegradation* 26, 271–287.

Koren, A. and Nadav, N., 1994. Mechanical vapour compression to treat oil field produced water. *Desalination* 98 (1–3), 41–48.

Kong, F.X., Chen, J.F., Wang, H.M., Liu, X.N., Wang, X.M., Wen, X., Chen, C.M., Xie, Y.F., 2017. Application of coagulation-UF hybrid process for shale gas fracturing flowback water recycling: Performance and fouling analysis. *J. Mater. Sci.* 524, 460–469.

Kwon, S., Sullivan, E.J., Katz, L., Kinney, K., Chen, C.C., Bowman, R., Simpson, J., 2008. Pilot scale test of a produced water-treatment system for initial removal of organic compounds. In *SPE Annual Technical Conference and Exhibition*, Denver, CO, September 21–24, 2008.

Lamichhane, S., Bal Krishna, K.C., Sarukkalige, R., 2016. Polycyclic aromatic hydrocarbons (PAHs) removal by sorption: A review. *Chemosphere* 148, 336–353.

Langlais, B., Reckhow, D.A., Brink, D.R., 1991. *Ozone in Water Treatment: Application and Engineering.* CRC Press, Boca Raton, FL.

Lei, C., Sun, Y., Khan, E., Chen, S.S., Tsang, D.C.W., Graham, N.J.D., Ok, Y.S. et al., 2018. Removal of chlorinated organic solvents from hydraulic fracturing wastewater by bare and entrapped nanoscale zero-valent iron. *Chemosphere* 196, 9–17.

Lester, Y., Yacob, T., Morrissey, I., Linden, K.G., 2013. Can we treat hydraulic fracturing flowback with a conventional biological process? the case of guar gum. *Environ. Sci. Technol. Lett.* 1, 133–136.

Lester, Y., Ferrer, I., Thurman, E.M., Sitterley, K.A., Korak, J.A., Aiken, G., Linden, K.G., 2015. Characterization of hydraulic fracturing flowback water in Colorado: Implications for water treatment. *Sci. Total Environ.* 512, 637–644.

Lobo, F.L., Wang, H., Huggins, T., Rosenblum, J., Linden, K.G., Ren, Z.J., 2016. Low-energy hydraulic fracturing wastewater treatment via ac powered electrocoagulation with biochar. *J. Hazard. Mater.* 309, 180–184.

Lutz, B.D., Lewis, A.N., Doyle, M.W., 2013. Generation, transport, and disposal of wastewater associated with Marcellus Shale gas development. *Water Resour. Res.* 49, 647–656.

Mantell, M.E., 2011. Produced water reuse and recycling challenges and opportunities across major shale plays. In: *US EPA Technical Workshops for the Hydraulic Fracturing Study.* Water Resources Management.

Marcellus Shale Advisory Commission (MSAC), 2011. Governor's Report on the Marcellus Shale, Marcellus Shale Advisory Commission, Harrisburg, PA, January, 2011, pp. 137.

Mauter, M.S., Alvarez, P.J.J., Burton, A., Cafaro, D.C., Chen, W., Gregory, K.B., 2014. Regional variation in water-related impacts of shale gas development and implications for emerging international plays. *Environ. Sci. Technol.* 48, 8298–8306.

McJeon, H., Edmonds, J., Bauer, N., Clarke, L., Fisher, B., Flannery, B.P., 2014. Limited impact on decadal-scale climate change from increased use of natural gas. *Nature* 514, 482–485.

Melikoglu, M., 2014. Shale gas: Analysis of its role in the global energy market. *Renew. Sust. Energ. Rev.* 37, 460–468.

Miller, D.J., Huang, X., Li, H., Kasemset, S., Lee, A., Agnihotri, D., Thomas, H., Paul, D.R., Freeman, B.D., 2013. Fouling-resistant membranes for the treatment of flowback water from hydraulic shale fracturing: A pilot study. *J. Mater. Sci.* 437, 265–275.

Minier-Matar, J., Hussain, A., Janson, A., Adham, S., 2014. Treatment of produced water from unconventional resources by membrane distillation. In: *International Petroleum Technology Conference*, Doha, Qatar. January 2014.

Mohan, A.M., Hartsock, A., Bibby, K.J., Hammack, R.W., Vidic, R.D., Gregory, K.B., 2013a. Microbial community changes in hydraulic fracturing fluids and produced water from shale gas extraction. *Environ. Sci. Technol.* 47, 13141–13150.

Mohan, A.M., Hartsock, A., Hammack, R.W., Vidic, R.D., Gregory, K.B., 2013b. Microbial communities in flowback water impoundments from hydraulic fracturing for recovery of shale gas. *FEMS Microbiol. Ecol.* 86, 567–580.

Moniz, E.J., Jacoby, H.D., Meggs, A.J., Armtrong, R.C., Cohn, D.R., Connors, S.R., Kaufman, G.M., 2011. The future of natural gas. Cambridge, MA: Massachusetts Institute of Technology.

Montgomery, C.T. and Smith, M.B., 2010. Hydraulic fracturing: History of an enduring technology. *J. Pet. Technol.* 12, 26.

Mozo, I., Stricot, M., Lesage, N., Sperandio, M., 2011. Fate of hazardous aromatic substances in membrane bioreactors. *Water Res.* 45 (15), 4551–4561.

NDCC, 2012. Chapter 43-02-03 oil and gas conservation. http://www.legis.nd.gov/information/acdata/pdf/43-02-03.pdf?2015 0306121451.

Newell, R.G. and Raimi, D., 2014. Implications of shale gas development for climate change. *Environ. Sci. Technol.* 48, 8360–8368.

Nicot, J.P. and Scanlon, B.R., 2012. Water use for shale-gas production in Texas, U.S. *Environ. Sci. Technol.* 46 (6), 3580–3586.

Nicot, J.P., Scanlon, B.R., Reedy, R.C., Costley, R.A., 2014. Source and fate of hydraulic fracturing water in the Barnett Shale: A historical perspective. *Environ. Sci. Technol.* 48, 2464–2471.

Notte, C., Allen, D., Gehman, J., Alessi, D., Goss, G., 2016. Comparative analysis of hydraulic fracturing wastewater practices in unconventional shale developments: Regulatory regimes. *Can. Water Resour. J.* 42 (2), 122–137.

NPC, 2011. Management of produced water from oil and gas wells. http://www.npc.org/Prudent_Development-Topic_Papers/2-17_Management_of_Produced_Water_Paper. pdf

Olmstead, S.M., Muehlenbachs, L.A., Shih, J.S., Chu, Z., Krupnick, A.J., 2013. Shale gas development impacts on surface water quality in Pennsylvania. *Proc. Natl. Acad. Sci. U.S.A.* 110 (13), 4962–4967.

Ord, J., 2014. *Strategies for Disposing Waste Water.* London, UK: Shale Gas Environment Summit.

Osborn, S.G., Vengosh, A., Warner, N.R., Jackson, R.B., 2011. Methane contamination of drinking water accompanying gas-well drilling and hydraulic fracturing. *Proc. Natl. Acad. Sci. U.S.A.* 108 (20), 8172–8176.

Ozgun, H., Ersahin, M.E., Erdem, S., Atay, B., Kose, B., Kaya, R., Altinbas, M. et al., 2013. Effects of the pre-treatment alternatives on the treatment of oil-gas field produced water by nano-filtration and reverse osmosis membranes. *J. Chem. Technol. Biotechnol.* 88, 1576–1583.

Pendashteh, A.R., Abdullah, L.C., Fakhru'l-Razi, A., Madaeni, S.S., Abidin, Z.Z., Biak, D.R.A., 2012. Evaluation of membrane bioreactor for hypersaline oily wastewater treatment. *Process Saf. Environ. Prot.* 90, 45–55.

Rangharajan, K., Lochab, V., Prakash, S., 2017. Desalting high salinity shale flowback water via high-flux nanofluidic evaporation-condensation. In *Micro Electro Mechanical Systems (MEMS), 2017 IEEE 30th International Conference on.* IEEE, pp. 60–63.

Rivard, C., Lavoie, D., Lefebvre, R., Séjourné, S., Lamontagne, C., Duchesne, M., 2014. An overview of Canadian shale gas production and environmental concerns. *Int. J. Coal Geo.* 126, 64–76.

Rokosh, C.D., Lyster, S., Anderson, S.D.A., Beaton, A.P.H., Berhane, T., Brazzoni, D., 2012. Summary of Alberta's shale- and siltstone-hosted hydrocarbons. ERCB/AGS Open File Report No. 2012-06. Edmonton: Energy Resources Conservation Board and Alberta Geological Survey, 327.

RRCT, 2015. Chapter III–pollution potential and statewide regulation. http://www.rrc.state.tx.us/oil-gas/applications-and-permits/environmental-permit-types-information/chapter-3-pollution-potential/

Sardari, K., Fyfe, P., Lincicome, D., Wickramasinghe, S.R., 2018. Aluminum electrocoagulation followed by forward osmosis for treating hydraulic fracturing produced waters. *Desalination* 428, 172–181.

Shaffer, D.L., Arias Chavez, L.H., Ben-Sasson, M., Romero-Vargas Castrillón, S., Yip, N.Y., Elimelech, M., 2013. Desalination and reuse of high-salinity shale gas produced water: Drivers, technologies, and future directions. *Environ. Sci. Technol.* 47, 9569–9583.

Shaffer, D.L., Werber, J.R., Jaramillo, H., Lin, S., Elimelech, M., 2015. Forward osmosis: Where are we now? *Desalination* 365, 271–284.

Shrestha, N., Chilkoor, G., Wilder, J., Ren, Z.J., Gadhamshetty, V., 2018. Comparative performances of microbial capacitive deionization cell and microbial fuel cell fed with produced water from the bakken shale. *Bioelectrochemistry* 121, 56–64.

Strong, L.C., Gould, T., Kasinkas, L., Sadowsky, M.J., Aksan, A., Wackett, L.P., 2013. Biodegradation in waters from hydraulic fracturing: Chemistry, microbiology, and engineering. *J. Environ. Eng.* 140 (5), B4013001.

Struijs, J. and Vandenberg, R., 1995. Standardized biodegradability tests: Extrapolation to aerobic environments. *Water Res.* 29, 255–262.

Sun, Y., Chen, S.S., Tsang, D.C.W., Graham, N.J.D., Ok, Y.S., Feng, Y., Li, X.D., 2017a. Zero-valent iron for the abatement of arsenate and selenate from flowback water of hydraulic fracturing. *Chemosphere* 167, 163–170.

Sun, Y., Lei, C., Khan, E., Chen, S.S., Tsang, D.C.W., Ok, Y.S., Lin, D., Feng, Y., 2017b. Nanoscale zero-valent iron for metal/metalloid removal from model hydraulic fracturing wastewater. *Chemosphere* 176, 315–323.

Sun, Y., Lei, C., Khan, E., Chen, S.S., Tsang, D.C.W., Ok, Y.S., Feng, Y., Li, X.D., 2018. Aging effects on chemical transformation and metal(loid) removal by entrapped nanoscale zero-valent iron for hydraulic fracturing wastewater treatment. *Sci. Total Environ.* 615, 498–507.

Sun, Y., Wang, D., Tsang, D. C., Wang, L., Ok, Y. S., Feng, Y., 2019a. A critical review of risks, characteristics, and treatment strategies for potentially toxic elements in wastewater from shale gas extraction. *Environ. Int.* 125, 452–469.

Sun, Y., Yu, I.K.M., Tsang, D.C.W., Cao, X., Lin, D., Wang, L., Graham, N.J.D. et al., 2019b. Multifunctional iron-biochar composites for the removal of potentially toxic elements, inherent cations, and hetero-chloride from hydraulic fracturing wastewater. *Environ. Int.* 124, 521–532.

Thacker, J.B., Carlton, D.D., Hildenbrand, Z.L., Kadjo, A.F., Schug, K.A., 2015. Chemical analysis of wastewater from unconventional drilling operations. *Water* 7, 1568–1579.

The Interstate Oil and Gas Compact Commission (IOGCC) and ALL Consulting, 2006. A guide to practical management of produced water from onshore oil and gas operations in the United States (DE-PS26-04NT15460-02).

Thiel, G.P., Tow, E.W., Banchik, L.D., Chung, H.W., Lienhard, V.J.H., 2015. Energy consumption in desalinating produced water from shale oil and gas extraction. *Desalination* 366, 94–112.

Torres, L., Yadav, O.P., Khan, E., 2016. A review on risk assessment techniques for hydraulic fracturing water and produced water management implemented in onshore unconventional oil and gas production. *Sci. Total Environ.* 39, 478–493.

Turan, N.B., Erkan, H.S., Engin, G.O., 2017. The investigation of shale gas wastewater treatment by electro-Fenton process: Statistical optimization of operational parameters. *Process Saf. Environ.* 109, 203–213.

U.S. EPA, 2011. Proceedings of the technical workshops for the hydraulic fracturing study: Water resources management (EPA 600/R-11/048).

Veil, J.A., 2010. Water management technologies used by Marcellus shale gas producers. Office of Fossil Energy–United States Department of Energy National Energy Technology Laboratory. FWP 49462.

Vengosh, A., Jackson, R.B., Warner, N., Darrah, T.H., Kondash, A., 2014. A critical review of the risks to water resources from unconventional shale gas development and hydraulic fracturing in the United States. *Environ. Sci. Technol.* 48 (15), 8334–8348.

Veza, J.M., 1995. Mechanical vapor compression desalination plants: A case study. *Desalination* 101 (1), 1–10.

Vidic, R.D., Brantley, S.L., Vandenbossche, J.M., Yoxtheimer, D., Abad, J.D., 2013. Impact of shale gas development on regional water quality. *Science* 340 (6134), 1235009.

Wang, Q., Chen, X., Jha, A.N., Rogers, H., 2014. Natural gas from shale formation: The evolution, evidences and challenges of shale gas revolution in United States. *Renew. Sustain. Energy Rev.* 30, 1–28.

WEF, 2013. Considerations for accepting fracking wastewater at water resource recovery facilities. http://www.wef.org/uploadedFiles/Access_Water_Knowledge/Wastewater_Treatment/FrackingFactsheetFinal(1).pdf

Xiao, Y.Y. and Roberts, D.J., 2010. A review of anaerobic treatment of saline wastewater. *Environ. Technol.* 31, 1025–1043.

Yang, J., Hong, L., Liu, Y.H., Guo, J.W., Lin, L.F., 2014. Treatment of oilfield fracturing wastewater by a sequential combination of flocculation, Fenton oxidation and SBR process. *Environ. Technol.* 35, 2878–2884.

Younker, J.M. and Walsh, M.E., 2015. Impact of salinity and dispersed oil on adsorption of dissolved aromatic hydrocarbons by activated carbon and organoclay. *J. Hazard. Mater.* 299, 562–569.

Zhang, X., Sun, A.Y., Duncan. I.J., 2016. Shale gas water management under uncertainty. *J. Environ. Manage.* 165, 188–198.

17 Societal Support for Remediation Technologies

Jason Prior and Deyi Hou

CONTENTS

17.1 INTRODUCTION

In recent decades, the selection of technologies used to remediate contaminated environments is done increasingly through engagement with a multitude of stakeholders, including residents living on or near a contaminated site. Despite this shift, little is known about the diverse factors that affect how these residents support the remediation technologies that are selected. This lack of understanding is complicated by the fact that the technologies that can be used to remediate contaminated environments continue to evolve far beyond the traditional techniques of capping pollution in-place and off-site removal (Kennen and Kirkwood, 2015), to include a growing diversity of biotechnologies, chemical technologies, thermal technologies, and physical technologies (Henry et al., 2013; Kennen and Kirkwood, 2015; Prasad et al., 2010).

The increased importance being placed on involving broader stakeholder buy-in in the selection of remediation technologies has resulted in the adoption of measures of public support of remediation technologies within some remediation guidelines and decision support tools (Harclerode et al., 2016; Schädler et al., 2011; Sorvari and Seppälä, 2010).

This emergence of broader stakeholder support as an evaluation criterion informing remediation technology selection acknowledges, firstly, a growing understanding that technology selection can be improved through the bringing together of the diverse knowledge of all those affected by the application of a technology (Delgado et al., 2011; Irwin, 2006; Hou and Al-Tabbaa, 2014), secondly, that a lack of understanding of broader stakeholder support hinders the ability of those within the remediation industry to effectively engage with stakeholders about the selection of technologies, and thirdly, that ignorance of the public's level of support for technologies can lead to significant sociopolitical risks to technology applications (de Groot et al., 2013; Horst, 2005; Siegrist and Visschers, 2013; Steg et al., 2006; Hou et al., 2014a).

While research has focused on regulatory and practitioner support for the selection of remediation technologies (Fan et al., 2017; Focht and Albright, 2009; Gerhardt et al., 2017; Gillespie and Philp, 2013; Grieger et al., 2010; Hou et al., 2014b; Kocher et al., 2002; Marinovich et al., 2016; Morillo and Villaverde, 2017; Page and Atkinson-Grosjean, 2013; Pollard et al., 1994; Ramirez-Andreotta et al., 2016; Zhang et al., 2016; Zhu et al., 2016), there has been little consideration of the factors that affect the level of support of residents living on or near contaminated sites for technologies that might be used to remediate those sites. This chapter addresses this research gap by presenting a unique study that systematically develops a framework for understanding the diverse factors that affect a resident's level of support for remediation technologies that might be used to address environmental contamination in their local area.

This chapter begins by outlining a conceptual framework for residents' support for remediation technologies that was developed through reference to broader technology research (Allansdottir et al., 2000; Bonfadelli et al., 2002; Clothier et al., 2015; Connor and Siegrist, 2010; Cowell et al., 2011; de Groot et al., 2013; Fischhoff et al., 1978; Frewer et al., 2004; Gilbert, 2007; Gupta et al., 2012; Jenkins-Smith et al., 2011; Krause et al., 2013; Luo et al., 2010; Marques et al., 2015; Siegrist, 2000; Siegrist et al., 2007; Ganesh Pillai, 2017;

Siegrist and Visschers, 2013; Slovic, 1987). The chapter then explains how the framework was tailored to residents' support for remediation technologies using data from a telephone survey of 2009 residents living near 13 contaminated sites across Australia. Regression analysis of closed-ended survey questions and coding of open-ended survey questions were combined to identify key predictors of a resident's level of support for remediation technologies.

The study acknowledges that remediation embodies a suite of potential technology types that can be used to address specific contaminants (Prior et al., 2017). One significant challenge in exploring how residents support the application of technology types is their diversity. To facilitate the study, we worked with industry experts to develop a high-level typology that could be used within the study: chemical technologies, physical technologies, thermal technologies, and biotechnologies (see Table 17.1).

TABLE 17.1
Remediation Technology Types Used within the Study

Bioremediation generally refers to the use of biological technologies in the form of microbes, fungi, and enzymes to clean up contaminated land and groundwater. For example:

- *Microbial bioremediation* (*in situ*) utilizes microbial activity to remove contaminants in groundwater, waste, or soil, and involves delivering something that can stimulate native microorganisms that can degrade contaminants, or a microbial culture to the contaminated medium that is capable of degrading contaminants.
- *Phytoremediation* (*in situ*) uses plants to clean up contaminated soils and groundwater. This process takes advantage of the ability of plants to take up, accumulate, stabilize, and/or degrade contaminants in soil and groundwater.

Thermal remediation generally refers to the use of heat to de-contaminate an area that can be done on-site (*in situ*) (e.g., steam injection, resistance heating, and conductive heating) or carrying out a treatment of excavated soil off-site (*ex situ*). In particular, thermal treatment is used to treat recalcitrant compounds such as persistent organic pollutants. For example:

- *Thermal desorption* (*ex situ* on-site) involves excavating and heating soils so that contaminants are vaporized and the vaporized contaminants are then collected and treated by other means.
- *Incineration* (*ex situ* off-site) involves excavating and heating soils so that the contaminants are destroyed. Thermal desorption differs from incineration in that it does not aim to destroy the organic but rather to change the form to a more treatable one.
- *Thermal vapor extraction* (*in situ*) involves injecting heat into the soil or waste so that contaminants are vaporized and extracting the vapor that is formed by the heat.

Chemical remediation generally involves the use of chemical reagents to oxidize or reduce contaminants, particularly in groundwater, although the method can extend to soils. There are several chemical oxidants that can be used to treat chlorinated solvents, and certain mobile heavy metals. For example:

- *Chemical treatment general* (*in situ*) involves the injection of chemical oxidants or reductants into groundwater or soil, which subsequently leads to the destruction of contaminants of concern or its transformation into something safer.
- *Nanoremediation* (*in situ*) involves introducing chemical substances containing microscopic particles called nanoparticles to destroy or degrade the contaminant in the soil or groundwater to an acceptable level.
- *Permeable reactive barrier* (*in situ*) involves introducing a chemical treatment wall into the groundwater flow; as contaminated groundwater passes through the treatment wall, the contaminants are either trapped by the treatment wall or transformed into harmless substances that flow out of the wall.

Physical remediation generally involves a range of physical techniques such as vacuum extraction (to remove contaminants in vapor form), soil washing, and separation. Excavation and removal of contaminated soil and disposal in a landfill is a very common method of remediation, although the increasing costs of landfill disposal are making this technique less widely used. For example:

- *Encapsulation* (*in situ*) comprises the physical isolation and containment of the contaminated material. In this technique, the impacted soils are isolated by low permeability caps, slurry walls, grout curtains, or cutoff walls.
- *Immobilizing/stabilization* (*in situ*, *ex situ*) generally refers to the process that reduces the risk posed by a waste or soil by converting the contaminant into a less soluble, immobile, and less toxic form.
- *Mining* (*ex situ* on-site, *ex situ* off-site) involves excavation, screening and separation, and recycling of all old landfill material. Unusable or contaminant producing materials are then stored.
- *Dig and dump* (*ex situ* off-site) involves the excavation and removal of the contaminated soil from the site and its transportation to a landfill site where it is stored and monitored.

17.2 CONCEPTUALIZING RESIDENTS' SUPPORT

The conceptual framework for resident's support of remediation technology presented in this section is built upon broader technology research (Clothier et al., 2015; Connor and Siegrist, 2010; Cowell et al., 2011; de Groot et al., 2013; Flynn, 2007; Focht and Albright, 2009; Ganesh Pillai and Bezbaruah, 2017; Siegrist et al., 2007; Siegrist and Visschers, 2013; Todt, 2011) and is a starting point for the study presented within this chapter. Within the context of the framework, residents' support of remediation technology is not equated with the formal approval, direct use, or application of remediation technologies (Rogers, 1995; Venkatesh and Davis, 2000); these are the responsibility of remediation regulators and remediation service providers. Within the context of this framework, resident's level of support for the application of the remediation technology in their local area is understood as occurring along a spectrum that is an expression of the resident's perception of a technology's sufficiency or lack (Eagly and Chaiken, 2007; Whitfield et al., 2009). For example, this spectrum can range from complete support, through to finding a technology tolerable, through to withholding support for the technology, and/or resisting the technology being applied or protesting against the technology's application.

Within the framework, a resident's level of support is understood as being guided by diverse predictors (see Figure 17.1). The conceptual framework groups predictors into four key dimensions, and each contain several attributes (see Figure 17.1). These dimensions include the physical context, institutional context (e.g., trust for organizations communicating with residents), as well as technology characteristics, and demographic and personal characteristics (e.g., values) that have been shown in broader technology research to influence people's support of technologies.

17.2.1 PREDICTOR DIMENSIONS OF RESIDENTS' SUPPORT

In our framework, a resident's level of support for the application of a remediation technology in their local area is understood firstly as being affected by the resident's demographic characteristics and personal values. Demographic characteristics have been found to predict the public's level of support of technologies (Connor and Siegrist, 2010; Flynn et al., 1994; Gupta et al., 2011; Kraus et al., 1992; Lee et al., 2005; Zingg and Siegrist, 2011). Men, in contrast to women, have been found to assess technology as more beneficial, and indicated a greater level of support regardless of a specific technology's applications (Magnusson and Koivisto Hursti, 2002). While findings are mixed, some research suggests that increasing age, education, household tenure, and income has been found to increase levels of support for technology (Gaskell et al., 1999, 2005; Hoban, 1998; Jenkins-Smith et al., 2011; Krause et al., 2013). The framework identifies personal motivational values as an important predictor of residents' support for remediation technology (Peters and Slovic, 1996; Peters et al., 2004; Plant et al., 2016; Prior, 2016; Sjöberg, 2002; Wong, 2015) where values are understood as an enduring foundation for the formation of beliefs that guide a resident's support for the application of those technologies (de Groot and Steg, 2008; Ibtissem, 2010; Nordlund and Garvill, 2002; Schwartz, 1973, 1992, 2012; Schwartz and Bilsky, 1987; Steg et al., 2014; Stern et al., 1999; Thøgersen and Ölander, 2002).

Environmental contaminants from nearby source sites often impact residents' local neighborhoods (Prior et al., 2014). Our conceptual framework therefore recognizes a second set of factors that relate to physical context as influencing a resident's support of technologies that are to be applied to remediate contaminants in their local area. These include the nature, extent, and impact of the environmental contamination on a resident's daily life, and their sense of place. While studies have examined residents' perceptions of contaminants and their impact on their daily life (Prior et al., 2014; Shusterman et al., 1991), no studies examine the way in which the residents' proximity to a nearby contaminated site, and their experience of the contaminant from that site in their everyday life, affects their support of the application of remediation technologies in their local area.

Thirdly, within our framework, institutional context refers to the organizations (e.g., companies, state and federal governments, experts, advocacy groups) that influence or manage the remediation of contaminated sites (Prior, 2016). Within

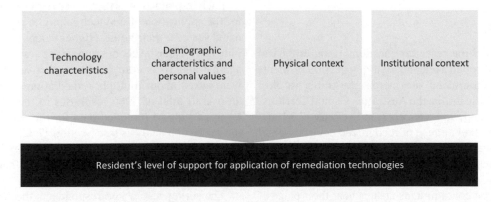

FIGURE 17.1 Conceptual framework of residents' support for the application of remediation technologies in their local area.

the framework, the ways in which these institutions engage with residents during the development of a remedial solution is understood to play a role in mediating the resident's level of support for the application of remediation technologies in their local area. Different dimensions of an organization's engagement – trust or language used – have been shown to be important predictors of levels of support for technology (Besley, 2010; Cvetkovich and Löfstedt, 1999; Flynn et al., 1992, 1994; Flynn, 2007; Freudenburg, 1993; Lindell and Perry, 2004; Siegrist, 2000). In particular, while research has shown that public level of trust of organizations that regulate and apply technologies, such as government, industry, and experts, influence their level of support for those technologies (Besley, 2010; Gupta et al., 2012; Siegrist, 1999, 2000; Siegrist et al., 2012; Terwel and Daamen, 2012; Terwel et al., 2010), no research has specifically explored how trust affects a resident's support for remediation technologies.

Fourthly, within our framework residents' level of support for the application of remediation technologies in their local area is understood as being affected by a technology's perceived characteristics. For example, broader technology research suggests that the public's support levels vary between different technology applications (Connor and Siegrist, 2010; Frewer et al., 1997; Magnusson and Koivisto Hursti, 2002; Siegrist, 2000), and is influenced by a technology's perceived naturalness (Connor and Siegrist, 2010;Miles, 2005; Morillo and Villaverde, 2017; Tenbült et al., 2005) and its perceived safety (Terwel and Daamen, 2012; Wallquist et al., 2009). Finally, a resident's weighing up of a technology's risks and benefits is understood as a strong predictor influencing the resident's support of the technology (Alhakami and Slovic, 1994; Kim, 2016; Poortinga and Pidgeon, 2006). While technologies can deliver perceived benefits to society, they may also have perceived risks (Gunter and Harris, 1998), and an individual's weighing up of these risks and benefits has been shown to influence their level of support for technologies (Connor and Siegrist, 2010; de Groot et al., 2013; Frewer et al., 1996, 1997; Kasperson et al., 1988; Pin and Gutteling, 2009; Renn et al., 1992; Renn and Roco, 2006).

In relation to the conceptual framework described above, this study posed the following research question: (RQ1) What are the main predictors of a resident's perceived level of support for the application of remediation technologies at a nearby contaminated site?

17.3 METHOD

The study addressed the research questions using empirical insights collected through a telephone survey of 2009 residents living near 13 contaminated sites across Australia, in New South Wales, South Australia, the Australian Capital Territory, Tasmania, Queensland, and Victoria (see Sections 17.3.1 and 17.3.2). The 13 sites had a range of recognized environmental contaminants present – solvents, hydrocarbons, heavy metals, asbestos, and putrescible waste – and we explored the magnitude of residents' support for the application of technologies to remediate the contaminants at sites near their place of residence. While this analysis focuses on the residents living

near these different locations across Australia, it aims to provide broader insights into the support levels of residents for the application of remediation technologies (Byrne, 2009). The Human Research Ethics Committee of the University of Technology Sydney approved this research. To protect the confidentiality of survey participants and sites, only generic information is provided.

The telephone survey collected data through closed-format and open-format questions. We began by identifying predictors of residents' level of support for the application of remediation technologies (RQ1) through an ordered logistic regression model analysis in combination with coding of verbatim responses to open-format questions (Sturgis and Allum, 2004; Voils et al., 2008).

The survey was designed through engagement with remediation experts and three focus groups each containing 9–12 residents from the target survey population areas. The survey was piloted using a sample of 50 respondents from the target resident populations. This was used to refine the instruments prior to surveying. The final questionnaire was conducted using a computer-assisted telephone interviewing (CATI) software application. This enabled the direct recording of data, controlled for logically incorrect answers, enabled interim reporting to ensure data was being recorded appropriately, provided built-in logic to enhance data accuracy, and provided built-in branching logic to direct interviewers through the questionnaire. Respondents were selected at random from a database of residential telephone numbers sourced from a commercial electronic telephone directory. Surveying commenced on March 24, 2014, and continued until the September 30, 2014, with a team of 12 researchers calling residents on weekdays (excluding Friday) from 3.30 to 8.00 pm. Residents were screened only to ensure they were aged 18 or over. Follow-up interviews occurred between February 2015 and January 27, 2016, with two researchers calling residents on weekdays (excluding Friday) at an agreed time.

17.3.1 SURVEY MEASURES

Within the survey, a remediation technology was described to each respondent: descriptions were selected from a list of technologies applicable to the contamination at the site in the respondent's local area (see Table 17.1).

Each respondent was then asked to rate their level of support for the application of that technology to remediate a contaminated site near their home. Higher values on the 11-point scale indicate greater levels of support (i.e., where 0 is completely against and 10 is completely supportive). Most respondents were asked about multiple remediation technologies, giving an overall total of 3966 responses to these questions. These levels of support provided data for the dependent variable in the regression analysis (see Section 17.3.3). Respondents were also asked to provide the first thought or image that came to mind for the technology applications that they had rated; these responses were recorded verbatim.

The survey also asked respondents the following question about each technology application: "Overall for this application would you say the benefits outweigh the risks, or the risks outweigh the benefits, or are they equal?" Furthermore, the survey also explored respondents' demographic information, factors associated with their physical and institutional context (e.g., trust), and preferences concerning the regulation and naturalness of technology applications.[1] These constituted the independent variables in the regression analysis.

Sixty randomly selected respondents were engaged in follow-up one-on-one telephone interviews after the survey to obtain further insights into the survey data and collect data on their motivational values (see Section 17.3.3). These interviews used a slightly modified short version of Schwartz's value scale survey previously developed, which has been shown to have good reliability and validity (Lindeman and Verkasalo, 2005). Respondents were asked to rate "the importance of each value as a life-guiding principle in your decision to support the application of remediation technologies at [the site] … Using the 8-point scale in which 0 indicates that the value is opposed to your principle, 1 indicates that the value is not important to you, 4 indicates that the value is important, and 8 indicates that the value is of supreme importance for you." They were urged to vary the scores and to rate only a few values as extremely important.

17.3.2 Sample Characteristics

Of the 2009 completed respondent surveys, four were excluded due to data-entry errors at the analysis stage, leaving 2,005 respondents. Fifty-eight percent ($n = 1175$) of the respondents were female, and 42% ($n = 834$) were male. One respondent did not report their gender. The age distribution was from 18 to 89 years. Of the 2009 respondents, 7% ($n = 144$) were aged between 18 and 34 years, 23% ($n = 579$) were aged between 35 and 54 years, 50% ($n = 1006$) were aged between 55 and 74 years, and 14% ($n = 280$) were more than 75 years old. One respondent did not report age.

17.3.3 Predictor Analysis

Ordered logistic regression analysis was used to assess the effects of a range of likely predictors on the respondents' levels of support for the application of technologies to remove contaminants from a source site in their local area (RQ1). The predictors were identified through our conceptual framework. Since the dependent variable (support) was measured on a Likert scale, they were treated as ordinal variables (see Section 17.3.1). Therefore, ordinal logistic regression was considered more appropriate than linear regression. The independent variables (likely predictors) used in the regression models included demographic information about the respondents, factors associated with the physical context and institutional factors (e.g., trust), and information about the remediation technology.[1] All statistical analyses were conducted using SPSS and R; findings of the regression analyses are presented in Table 17.2. Since some of the independent variables collected through the survey were conceivably measuring the same underlying variable or latent variable, we eliminate highly correlated variables by either eliminating or

TABLE 17.2

Ordered Logistic Regression Coefficients with the Dependent Variable Being the Degree to Which Respondents Are Supportive of the Application of Remediation Technologies

Dimension	Attribute	Dependent Variable Acceptance Value
Demographic and personal characteristics	Male	−0.000
	Education university	−0.053
	Age under 35	0.057
	Age 35–54	0
	Age 55–74	−0.112
	Age 75+	−0.203
	Income unspecified	−0.173
	Income zero to $40k	−0.036
	Income $40k to $80k	−0.142
	Income $80k to $120k	**−0.191***
	Income $120k+	0
	Children in household yes	0.111
	Tenure own or purchasing	**−0.184***
Physical context	Heard of contaminant yes	0.020
	Contamination at site features in my life	−0.005
	Transport contaminant through local area	**−0.124*****
	Sense of place	−0.002
	Lived less than 10 years	0.017
	Move away	−0.035
Institutional context	Trust 1–general	**0.051*****
	Trust 1–central	0.020
	Trust 1–commercial	0.004
	Government should regulate technology	0.007
	Experts know best	0.004
	Language other than English yes	−0.035
Technology characteristics	Technology is out of control	−0.015
	Technology fascinated	−0.002
	Technology solves our problems	**0.033***
	Should use natural methods	0.006
	Technology type–chemical	**−1.149*****
	Technology type–physical	**−1.154*****
	Technology type–thermal	**−0.988*****
	Technology type–biological	0
	On-site in-ground treatment	**0.200*****
	On-site out-of-ground treatment	**0.312*****

(Continued)

TABLE 17.2 (*Continued*)
Ordered Logistic Regression Coefficients with the Dependent Variable Being the Degree to Which Respondents Are Supportive of the Application of Remediation Technologies

Dimension	Attribute	Dependent Variable Acceptance Value
	Off-site treatment	**0.329***
	Risks outweigh benefits	**−3.408***
	Benefits and risks are equal	**−1.809***
	Benefits outweigh risks	0

Note: Positive coefficients indicate variables are associated with higher levels of support.
*** $p < 0.001$ is highly significant.
** $p < 0.01$ is very significant.
* $p < 0.05$ is significant.

combining duplicate or redundant variables. There was evident correlation between the six different "trust" variables, so we used principal components analysis (PCA) to separate out the underlying latent factors; we used PCA to extract three latent factors from the data.[2]

While regression analysis of closed-format questions provided robust quantitative findings of likely predictors, coding analysis of respondents' open-ended responses provided a more heterogeneous set of perspectives than the regression analysis allowed (Voils et al., 2008). This coding was used to identify remediation technology characteristics that affected residents' level of support. This coding used a similar two stage process to that explained in Prior et al. (2017): involving first cycle coding methods (attribute coding, in-vivo coding) and second cycle coding methods (focused coding, elaboration coding) (Auerbach and Silverstein, 2003; Saldaña, 2013), using frequency counts to quantify these codes in due course.

As the modified short version of Schwartz's value scale only collected a small random sample, we only sought to identify the values that respondents identified as important through to extremely important (4 through 8 on the Likert scale) as guiding principles for each respondent. Furthermore, while the analysis collected data on the relative priorities respondents placed on 10 broad universal value types (see Figure 17.2) (Schwartz, 1999, 2011), these were grouped into four higher order domains of values for this study; the domains are organized on two polar dimensions: self-enhancement versus self-transcendence as a basis for supporting technology applications, and openness to change versus conservation as a basis for supporting technology applications (see Figure 17.1) (Schwartz, 2012). These value domains have been found to hold relevance to

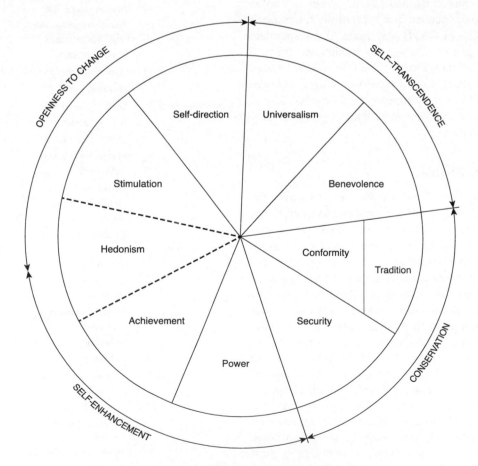

FIGURE 17.2 Schwartz's 10 universal motivational values and four motivational value domains. (From Schwartz, S.H., *J. Cross Cult. Psychol.*, 42, 307–319, 2011.)

environmental management (Ibtissem, 2010; Thøgersen, 1999; Thøgersen and Ölander, 2002). The identified value domains guiding respondents were then cross-referenced with residents' verbatim responses (see Section 17.3.1) to explore how they manifested in respondents' discourse.

Where parts of verbatim responses are presented in Section 17.4, the respondents' basic details are represented in brackets after the verbatim response as follows (gender, age, and interview ID) (e.g., Female, 45, 2015).

17.4 RESULTS

Section 17.4 addresses the study's first research question (RQ1) by presenting findings on the most salient predictors of respondents' level of support for the application of remediation technologies in their local area.

Predictors were found across all four dimensions of the conceptual framework discussed in Section 17.2. These predictors include those that were identified as significant within the regression analysis (see Table 17.2), those most frequently identified through the coding analysis, and those motivational values identified through the Schwartz value system. Collectively, these variables predict the level of residents' support for the application of remediation technologies in their local area (Figure 17.3).

17.4.1 TECHNOLOGY CHARACTERISTICS

Technology type: The type of technology proposed for remediation had a highly significant ($p < 0.001$) effect on the residents' level of support for a technology's application at a nearby site (see Table 17.2).

Respondents are more supportive of the application of biotechnologies than they are of chemical technologies, thermal technologies, and physical technologies.

Weigh-up of a technology's risks and benefits: How a respondent weighed up the perceived risks and benefits of a technology's application was found to have a highly significant effect ($p < 0.001$) on their level of support for a technology's application (see Table 17.2). Those who believed the risks outweighed the benefits of a technology were less supportive of that technology's application at the site than those who believed the risks and benefits of a technology application were equal, who in turn were less supportive than those who believed the benefits outweighed the risks of the technology's application at the site.

Effectiveness: Those agreeing that technologies are effective for solving most problems faced by human beings were significantly ($p < 0.05$) more likely to support remediation technology applications at a nearby site (see Table 17.2). Respondents' verbatim comments provided further insight into concerns about the effectiveness of remediation technologies. For most respondents, support for technologies was connected to their perceived ability to eliminate environmental contaminants, and restrain or block their exposure pathway to receptors. As one respondent noted:

I disagree with the application [of incineration] and do not believe it could possibly be effective. … [it] would be more likely to release more of the fibers. (Male, 66, 3023)

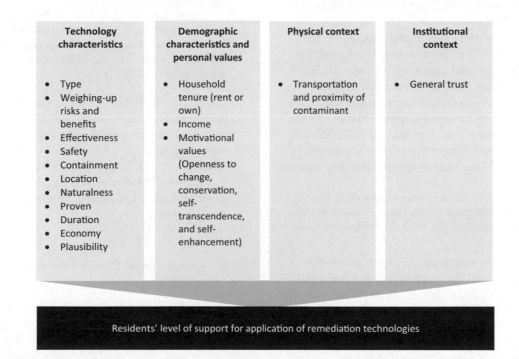

FIGURE 17.3 Developed framework for considering and understanding residents' level of support for the application of remediation technologies in their local area.

Safety: Respondents' verbatim comments frequently referred to a technology's perceived safety in order to assess support for it. Respondents used safety as a measure of support in two principal ways: first, respondents often declared their support for one technology over another based on their perception of the safety of the technology, and secondly, respondents would not support a technology unless it could be shown that that technology was safe: "nothing, unless they could demonstrate [thermal vapor extraction] was *safe*" (Female, 70, 366).

Containment: Respondents frequently highlighted how their support for the application of a technology was contingent on its perceived ability to contain the contaminant during application, as well as both the side-effects of the application and the long-term fate of the technology on the environment. As one respondent noted:

> I don't like … [the idea of using microbial bioremediation] … because it is a very populated area, so I don't think they should be introducing bacteria which could spread into the air. (Female, 43, 2063)

Containment was more frequently reported as a motivation for withholding support for the application of thermal and physical technology types. Respondents associated a relatively consistent set of containment risks with these technologies. For physical technology, the risk was focused on the ability of encapsulation or immobilization techniques to "hold" the contaminated material overtime. For thermal technology, the risk focused on the escape of vapors, as one respondent noted: "How [are] they are going to capture the vapor [from thermal vapor extraction]?" (Male, 67, 359).

Location: Respondents' stated preferences for remediation location (See On-site in-ground treatment, On-site out-of-ground treatment, Off-site treatment in Table 17.2) are consistent with their responses to specific remediation technologies, in the sense that respondents who were more supportive of off-site treatment generally were significantly (<0.001) more likely to be more supportive of the application of off-site remediation technology. As one respondent noted:

> [I would] only [support encapsulation] if they … removed [the contaminated material] to a site especially for toxic chemicals and put it far from inhabited areas. (Female, 56, 92)

Naturalness: While respondents' preference for the use of natural methods for remediation did not significantly relate to level of support for the application of technologies (see Table 17.2), verbatim responses frequently mentioned a technology's naturalness as a motivation for withholding or granting support for the application of a remediation technology (For further discussion of naturalness see Motivational Values.)

Proven: Respondents noted how they often withheld support for technologies that they perceived as unproven, in particular emergent technologies like nanoremediation and phytoremediation, and often only supported technologies that were successfully "trialed" or had "evidence of success." As a respondent explained:

> Nano[remediation] is *unproven*. Nanoparticles do still currently pose a risk as to how they enter the body, I would be more comfortable with a more *proven* application. (Male, 65, 5205)

Duration: Two dimensions of a Technology's duration influenced respondents' level of support for it: the perception that it took a lengthy time to complete the remediation, and the perception that some technologies were only short-term solutions that "transfer[ed] the problem" to future generations. Perceived duration was a more frequently stated reason for respondents to withhold support for physical and biotechnology applications (see Figure 17.3). As a respondent explained:

> This seems to be a temporary solution, we cannot guarantee eternal monitoring, it should be dealt with permanently [encapsulation]. (Female, 58, 3174)

Economy: The economy of the remediation technology was a reason that many respondents chose to withhold or grant their support for the application of a technology. Respondents' perceptions of a technology's economy was not concentrated on the monetary "risks" of the technology alone, but also extended to its economic use of resources and energy. As one respondent asserted:

> I'm rating this as a risk, as I feel [incineration] would be quite an *expensive* endeavor. If I could see the financials described I may be more inclined to be more supportive – that is to say, if it is not as *expensive* as it seems. (Male, 77, 3002)

Plausibility: A further stated reason that many respondents gave either to withhold or grant support was the perceived plausibility of the remediation technology application. As one respondent asserted: "No, I think [chemical treatment general] is ludicrous" (Female, 48, 332). Plausibility was a more frequently stated concern for emergent technologies.

17.4.2 Demographic Characteristics and Personal Values

Income was found to have a significant effect ($p < 0.05$) on respondents' level of support for the application of remediation technology in their local area. Respondents on a moderate income ($80K to $120K) were less likely to support the application of technologies than those on higher incomes (over $120K) (see Table 17.2).

Household tenure was found to have a significant effect ($p < 0.05$) on respondents' level of support for the application of remediation technology in their local area. Respondents who owned or were purchasing their own home in the neighborhoods surrounding the site were less likely to support the application of remediation technologies (see Table 17.2).

The range of *values* found to motivate respondents' decisions to withhold or grant support for the application of technologies spanned the four Schwartz value domains (see Figure 17.2).

Openness to change (self-direction and stimulation): Some respondents' willingness to withhold or grant support was motivated by a belief in the "beneficial," "positive," rather than threatening change that could be brought about by remediation technologies. This motivational belief was reflected in the finding that those who agreed that technologies are effective at solving most problems faced by human beings were significantly ($p < 0.05$) more likely to support remediation technology applications at a nearby site (see Table 17.2).

Conservation (conformity, tradition): Some respondents' willingness to withhold or grant support was motivated by their "conformity" and "commitment" to specific customs and ideas that their culture, belief, or religion provided. *Traditional* values respondents mentioned as reasons for withholding support for technologies, focused on beliefs that "nature knows best" and that human "interference does not improve the situation" (Female, 85, 5529) at the contaminated site. In these beliefs, natural is generally associated with non-human, non-artificial, and unspoiled environment. As one respondent asserted: "No, I think [chemical treatment general] is ludicrous. … leave it alone … Let nature fix itself" (Female, 48, 332). *Conformity* in the form of subordination to "independent experts" who were believed to have "neutral" and "unbiased" approaches to the application of technologies was frequently relied upon by respondents as a means of assessing their support for a technology's application. As one respondent explained what would make them more supportive of a technology: "Well-informed and externally validated advice on the methodology and arrangements …, and independent monitoring [of mining]" (Male, 72, 3053).

Self-transcendence (universalism, benevolence): Respondents' level of support for a technology was dependent on the belief that it enhanced the welfare of their local community, neighbors, or family. This *benevolence* was often made apparent when respondents openly assessed their support for a remediation technology, using pronouns like "we," "residents," "community," "us," as one respondent noted: "Digging up contaminants and exposing the contaminants to *residents* is too big of a risk. [thermal desorption]" (Female, declined, 5303). More embracingly, some respondents assessed their support for a technology

based on its perceived ability to protect the welfare of all people (the public) and the environment. This contrasts with the in-group focus of benevolence, as one respondent explained: "[I need] more information about the bacteria and side effects to *environment and to people* [from microbial remediation before I will support it]" (Female, 49, 2053). In some instances, respondents' universalism values extended to the welfare of distant communities and future generations, as one respondent explained: "I don't like [dig and dump] at all as it will be causing issues for another area" (Male, 51, 372).

Self-enhancement (power, achievement): Some respondents' level of support for a technology was dependent on the belief that the technology was in their own interests, that is, the technology enhanced their own welfare (e.g., health, amenity). This was apparent in frequent use of first person language and the pronouns "my" and "I."

17.4.3 Physical Context

Transportation: Those who agreed that they were concerned about contaminants being transported through local streets to be treated at an off-site location were significantly (< 0.001) less likely to support the application of remediation technologies (see Table 17.2). For example, respondents frequently explained that they withheld support for the much-applied dig and dump technology for removing contamination from the site, because of concern for the risk associated with transporting the contaminated medium through streets in proximity to local homes. As one respondent noted: "I would not like it to be transported at all it should be done onsite [dig and dump]" (Female, 67, 2).

17.4.4 Institutional Context

General trust: A resident's general level of trust (in companies, government, scientific organizations, NGOs, as well as other residents) was found to have a highly significant effect ($p < 0.001$) on their level of support for remediation technologies (see Table 17.2). Those who were more trusting generally were more supportive of the application of technologies, and conversely those who were less trusting generally were less supportive of the application of technologies. As one respondent explained: "If the people doing it were more trustworthy [I would be more supportive of the application of nanoremediation]" (Female, 76, 5425).

17.5 CONCLUDING DISCUSSION

Over recent decades, the selection of technologies for remediating contaminated land and groundwater has evolved from a simple process, dominated usually by cost and few stakeholders, to a process involving a broader range of stakeholders in a multi-criteria decision process that assesses a growing diversity of technology options (Pollard et al., 2004). Despite these evolving processes, attention to how and why residents accept the application of remediation technologies remains fragmented and limited. Within this chapter, we examined residents' level of support for the application of remediation technologies. The significance of the framework developed in this study is that it highlights the multidimensional nature of residents' support, highlighting some of the diverse predictors that affect a resident's level of support for the application of remediation technology options in their local area. While the developed framework is imperfect and will need modifications as research on stakeholder support of remediation technologies develops further, it is a first step toward developing a framework that researchers and remediation practitioners can use to better navigate the complex landscape of residents' support for remediation technologies.

17.5.1 PREDICTORS

The identification of a broad range of predictors for residents' support for remediation technology including demographic, personal, physical, institutional, and technology-specific factors reflects those found in broader technology research (Clothier et al., 2015; Cowell et al., 2011; Jenkins-Smith et al., 2011; Krause et al., 2013; Siegrist and Visschers, 2013). Moreover, the analysis highlights the relative predictive power of many of these factors (see Table 17.2).

Aligning with broader technology research, the analysis highlights how residents' support for a technology depends on a range of technology's characteristics, including technology type (Allansdottir et al., 2000; Bonfadelli et al., 2002; Connor and Siegrist, 2010; Ganesh Pillai and Bezbaruah, 2017; Marinovich et al., 2016; Morillo and Villaverde, 2017; Siegrist, 2000; Siegrist et al., 2007). For example, the application of emergent technologies, such as nanotechnologies, often lacks support because they are perceived to be *unproven* technologies (Gerhardt et al., 2017; Kim, 2016). Furthermore, carbon capture and storage technologies (Terwel and Daamen, 2012; Wallquist et al., 2009), like technologies that encapsulate contaminants, receive less public support because it is perceived that they will inevitably fail to *contain* in the longer-term (Morillo and Villaverde, 2017; Zhang et al., 2016). Finally, the study findings that a resident's weighing up of a technology's risk and benefit influences their support for that remediation technology adds to earlier research which has begun to show how residents' risk and benefit perceptions for remediation technologies are linked (Huynh et al., 2018).

Echoing wider technology research two demographic factors – income and tenure – were found to be significant predictors of residents' level of support for remediation technologies (Krause et al., 2013). On the other hand, the absence of association between a resident's age and education reflects mixed findings in broader research (Cummings et al., 2013). The study's findings on personal values motivating people's level of support for technologies reflect studies into sustainable remediation that have found residents' motivational values about technology decision making to be divergent (Prior, 2016; Hou, 2016). This study suggests that a technology's perceived threat to these motivational values has a key impact on a resident's level of support for that technology, and furthermore that residents holding contrasting motivational values may generate conflict about a technology's selection. For example, a potential conflict may exist between those residents who support a technology's application based on an openness to changes brought about by that technology and those who seek to withhold support for that technology based on the conservative belief that all technologies are unnatural (risky) interventions.

Within the resident's physical context, concerns for transportation through local streets was the only factor that had a significant impact on a resident's level of support for the application of technologies. This reflects broader findings which argue that proximity to a technology (e.g., landfills, incinerators, and nuclear facilities) affects the public's support for that technology (Gawande and Jenkins-Smith, 2001; Jenkins-Smith et al., 2011; Terwel and Daamen, 2012), and in some instances results in greater concerns and opposition to that technology by residents (Schively, 2007; Slovic et al., 1991). Furthermore, within the institutional remediation context, a resident's general level of trust was found to have a significant effect on a resident's level of support for technologies. These results support the hypothesis that trust in institutions or persons is a key factor influencing a person's support of technologies.

17.5.2 IMPLICATIONS

The rare insights provided by this research into residents' support of remediation technology applications arguably afford a means for increasing awareness among policy makers and remediation practitioners of how residents support technologies. The reasoning for engaging with the diverse predictors guiding residents' support of technologies is not that this awareness is necessarily more significant than, for example, guidelines that experts use to select remediation technologies, but because they provide unique knowledge that may be of value in developing remediation approaches that respond to the needs of all those who may be affected by the application of such a remediation technology rather than just the views of experts (Cooperative Research Centre for Contamination Assessment and Remediation of the Environment, 2013; NICOLE, 2008; Delgado et al., 2011; Irwin, 2006; O'Riordan and Cameron, 1994).

Research indicates that there are still evident data gaps in how to engage with residents in the remediation context (Cooperative Research Centre for Contamination Assessment and Remediation of the Environment, 2013). The framework developed in this study provides detailed insights into the complex arrangement of predictors that guide resident support for remediation technologies, which can be used to enhance and develop the simple metrics for broader stakeholder support that are found in current remediation decision support tools (Harclerode et al., 2016; Schädler et al., 2011; Sorvari and Seppälä, 2010). The aim is to enhance engagement with residents during both the planning and implementation of remediation approaches, as well as the development of community engagement strategies associated with the remediation of contaminated sites. This knowledge is valuable in engagement strategies in that it provides insights into how residents in local communities are likely to support specific technologies based on their demographic and physical context, as well as the institutional and technological factors that may affect their level of support.

Beyond providing value to engagement strategies, the study's insights have wider use for remediation planning. In the case of risk management, attunement to how residents support technologies based on subjective understanding of their risks is needed to inform a more holistic understanding of how risks in the context of remediation can be managed (Moussaïd et al., 2015; Wong, 2015). Moreover, in the context of sustainable remediation, being attuned to residents' support can be used to develop a broader appreciation of the benefits of remediation technologies to affected stakeholders (Beck and Mann, 2010; Bubna-Litic and Lloyd-Smith, 2004, p. 269; Hardisty et al., 2008; Hou and Al-Tabbaa, 2014).

17.5.3 LIMITATIONS AND FURTHER RESEARCH

Despite the significance of the study's new insights on residents' support, the study is not without limitations. While coding using the Schwartz value system provides a means of systematically identifying values, they may be subject to misinterpretation and people's reluctance to reveal what motivated their decisions. Also, the ability to use the coding of verbatim responses to access predictors guiding residents' support of technologies was arguably hampered by the fact that many verbatim responses were short, containing just a few words. Limitations also extend to the study's findings. Because the surveying was of a random Australia cohort, caution needs to be taken when applying those findings to other contexts, given that technology support may vary across countries (Connor and Siegrist, 2010; Frewer et al., 1997; Gaskell et al., 1999; Gupta et al., 2012; Ho et al., 2011; Hoban, 1998). Furthermore, this study has focused upon the application of individual remediation technology types. Given that technologies are often applied in combination, there is a need for future research to examine how residents support the application of combined technologies.

ACKNOWLEDGMENTS

This research has been assisted by the New South Wales Government through its Environmental Trust. Furthermore, this research has been funded by the Cooperative Research Centre for Contamination Assessment and Remediation of the Environment (CRC CARE).

NOTES

1 The independent variables included in the regression are described below. Unless otherwise specified, all variables are 11-point Likert scale variables, with higher values indicating stronger agreement. Independent variables, grouped into dimensions, are described below:

Demographic variables and personal motivational values:
- Education university: 0/1 dummy with value 1 if the respondent had a university qualification.
- Male: 0/1 dummy with value 1 if the respondent is male.
- Tenure own or purchasing: 0/1 dummy with value 1 if the respondent owns or is purchasing their home. Other tenures are renting (private), renting (public/social), and other.
- Children in household: 0/1 dummy with value 1 if children younger than 14 are in the household.
- Age under 35: 0/1 dummy with value 1 if the respondent is under 35. Age 35–54: is the control category. Age 55–74: 0/1 dummy for respondents aged 55–74. Age 75+: 0/1 dummy for respondents aged 75+.
- Income unspecified: 0/1 dummy for respondents who did not specify income. Income Zero to $40k: 0/1 dummy for annual household income between $0 and $40k. Income $40k to $80k: 0/1 dummy for annual household income between $40 and $80k. Income $80k to $120k: 0/1 dummy for annual household income between $80 and $120k. Income $120k+: is the control category. Annual household income over $120K.
- Motivational values: Respondents were asked to rate "the importance of each value as a life-guiding principle in their decision to support the application of remediation technologies at [the site] … Using the 8-point scale in which 0 indicates that the value is opposed to your life guiding principle, 1 indicates that the value is not important to you, 4 indicates that the value is important, and 8 indicates that the value is of supreme importance for you." The motivational values were: power, achievement, hedonism, stimulation, self-direction, universalism, benevolence, tradition, conformity, security.

Physical context variables:
- Heard of contaminant yes: 0/1 dummy with value 1 if the respondent had heard of the contaminant in their local area.
- Contamination at site features in my life: Likert-scale response to the statement *"The contamination at [LOCAL CONTAMINATION SITE] has featured strongly in my life or community."*
- Transport material through local area: A 0/1 dummy with value 1 if the respondent was supportive of remediation if the contaminated medium is transported through local streets.
- Sense of place: combined response to the statements *"I feel like I belong to the community where I live"* and *"For me, this is the ideal place to live,"* which were found to be highly (0.87) correlated.

- Lived < 10 years local area: 0/1 dummy with value 1 if the person has lived in the local area less than 10 years.
- Intend move away from local area: 0/1 dummy with value 1 if the respondent indicates they intend to move away from the local area (regardless of intended time-frame for departure).

Institutional context variables:
- Trust: As already described, survey respondents were asked a number of questions about the degree to which they trusted different groups/organizations. Responses to many of these were highly correlated, and principal components analysis was used to derive the variables (first three principal components) including: Trust-general, a numeric variable measuring how trusting the respondent is, generally. Positive/negative values indicate the respondent is more/less trusting; Trust-central, a numeric variable measuring how trusting the respondent is of "centralized" organizations such as companies and government, relative to "local" groups/organizations such as neighbors, community groups and non-government organizations, and media. Positive/negative values indicate the respondent is more/less trusting of central/local organizations; Trust-commercial, a numeric variable measuring how trusting the respondent is of "commercial" organizations such as companies, media, and local residents, relative to "non-commercial" groups such as government, scientific organizations, and non-government organizations. Positive/negative values indicate the respondent is more/less trusting of commercial/non-commercial organizations.
- Government should regulate technology: response to the statement: *"It is important for governments to regulate technology."*
- Experts know best: response to the statement *"Experts know best what is good for the public"*
- Language other than English yes: 0/1 dummy variable, which is 1 if the household speaks a language other than English in the home.

Technology characteristic variables
- Should use natural methods: response to the statement: *"We should use more natural ways to clean up the environment."*
- Technology is out of control: combined response to the statements: *"Technologies are out of control, and beyond the control of governments"* and *"Technological change happens too fast for me to keep up with."*

- Technology fascinated: combined response to the statements: *"I'm fascinated by technology stories in the media"* and *"After I encounter news about a technology, I am likely to seek out further information on it."*
- Technology solves our problems: combined response to the statements: *"Technologies can solve most problems faced by human beings"* and *"Technologies are continuously improving our quality of life."*
- Technology type: As already described, respondents were given details of a particular remediation technology applicable to the contamination in their suburb. The following variables relate to the particular technology discussed with the respondent: *Technology type-chemical*, a 0/1 dummy with value 1 if the remediation technology discussed with the respondent was a chemical technology; *Technology type-physical*, a 0/1 dummy with value 1 if the remediation technology discussed with the respondent was classified as a physical technology; *Technology type-thermal*, a 0/1 dummy with value 1 if the remediation technology discussed with the respondent was classified as a thermal technology; *Technology type-biological*, is the control category. Technology classified as a biological technology.
- Location variables: As already discussed, respondents had a particular remediation technology described to them, and after this description, they were asked to rate their supportiveness of the technology. Specifically, they were asked: *"Using a scale of 0 to 10 can you please rate your support for the use of each technology at [LOCAL CONTAMINATION SITE], where 0 is completely against it and where 10 is completely supportive."* The following variables relate to the response to that question:
  - On-site in-ground treatment: Support (on 11 point Likert scale) for on-site in-ground treatment (which can include storage) of contamination, if the technology discussed with the respondent was an on-site in-ground technology (also known as in situ technology). 0 otherwise.
  - On-site out-of-ground treatment: Support (on 11 point Likert scale) for on-site out of ground treatment of contamination, if the technology discussed with the respondent was an on-site out-of-ground technology (also known as on-site ex situ technology). 0 otherwise.
  - Off-site treatment: Support (on 11 point Likert scale) for off-site treatment (which can include storage) of contamination, if the technology discussed with the respondent

was an off-site technology (Off-site treatment locations can be either *ex situ* or *in situ*). 0 otherwise.

- The above three variables are standard Likert scale responses to support for different treatment locations, interacted with 0/1 dummy variables that specify the actual treatment location for the remediation technology discussed with the respondent. They thus indicate the respondent's level of support for the location at which the remediation technology is applied.

- Risks/benefits variable: Is a categorical variable with three levels: benefits outweigh risks, risks outweigh benefits, and risks and benefits are equal/unsure.

[2] Survey respondents were asked six questions about their trust in various groups and institutions. Specifically, they were asked: *"Can you tell us how much you trust the following sources to give you balanced and reliable information about technologies used to clean up [LOCAL CONTAMINATION SITE] where 0 means you don't trust them at all, and 10 means you trust them completely: companies and industry associations, government agencies or regulators (EPA, Dept of Health etc.), other local residents (neighbors etc.), mass media (TV, radio, newspapers), non-governmental organizations and community advocacy groups, science institutes, and organizations such as the CSIRO and universities."*

There was evident correlation between the six different trust questions, and so we used principal components analysis to separate out the underlying latent factors. The six orthogonal axes are shown in the table below:

	PC1	PC2	PC3	PC4	PC5	PC6
Trust-Companies	0.43590	0.46878	0.54759	0.223211	−0.47858	0.107366
TrustGovt	0.53485	0.43920	−0.37252	0.156407	0.48576	−0.349060
TrustLocal-Residents	0.32471	−0.62717	0.27636	0.581939	0.28493	0.070759
TrustMedia	0.49858	−0.22974	0.25427	−0.759387	0.19651	0.136743
TrustNon-Govt	0.33411	−0.37537	−0.39417	−0.041288	−0.63175	−0.437360
Trust-Science	0.24140	0.01808	−0.51477	0.093202	−0.12684	0.807243
Variance explained	0.369	0.227	0.135	0.122	0.0857	0.0604

We took the first three principal components (PC1, PC2, PC3), and gave them more descriptive names: Trust-General (PC1), Trust-Central (PC2), and Trust-Commercial (PC3). We estimated logistic regression models with the original (untransformed) trust variables, and with the three (PCA-derived) trust variables and compared their AIC scores, and the PCA-derived variables produced a better model.

REFERENCES

Alhakami AS, Slovic P. A psychological study of the inverse relationship between perceived risk and perceived benefit. *Risk Analysis* 1994; 14: 1085–1096.

Allansdottir A, Allum N, Bauer M, Bonfadelli H, Boy D, de Cheveigne S, et al. Biotechnology and the European public. *Nature Biotechnology* 2000; 18: 935+.

Auerbach CF, Silverstein LB. *Qualitative Data : An Introduction to Coding and Analysis*. New York: New York University Press, 2003.

Beck P, Mann B. Technical Report No.15: A technical guide for demonstrating monitored natural attenuation of petroleum hydrocarbons in ground water, Adelaide, Australia, 2010, p. 120.

Besley J. Current research on public perceptions of nanotechnology. *Emerging Health Threats* 2010; 3: 1–25.

Bonfadelli H, Dahinden U, Leonarz M. Biotechnology in Switzerland: High on the public agenda, but only moderate support. *Public Understanding of Science* 2002; 11: 113–130.

Bubna-Litic K, Lloyd-Smith M. The role of public participation in the disposal of HCBs: An Australian case study. *Environmental and Planning Law Journal* 2004; 21: 264–288.

Byrne D. Case-based methods: Why we need them; What they are; How to do them. In: Bryne D, Ragin CC, editors. *The SAGE Handbook of Case-Based Methods*. London, UK: SAGE Publications, 2009.

Clothier RA, Greer DA, Greer DG, Mehta AM. Risk perception and the public acceptance of drones. *Risk Analysis* 2015; 35: 1167–1183.

Connor M, Siegrist M. Factors influencing people's acceptance of gene technology: The role of knowledge, health expectations, naturalness, and social trust. *Science Communication* 2010; 32: 514–538.

Cooperative Research Centre for Contamination Assessment and Remediation of the Environment. National Framework for Remediation and Management of Contaminated Sites in Australia. CRC for Contamination Assessment and Remediation of the Environment, Adelaide, Australia, 2013.

Cowell R, Bristow G, Munday M. Acceptance, acceptability and environmental justice: The role of community benefits in wind energy development. *Journal of Environmental Planning and Management* 2011; 54: 539–557.

Cummings CL, Berube DM, Lavelle ME. Influences of individual-level characteristics on risk perceptions to various categories of environmental health and safety risks. *Journal of Risk Research* 2013; 16: 1277–1295.

Cvetkovich G, Löfstedt R. *Social Trust and the Management of Risk*. London, UK: Earthscan, 1999.

de Groot JIM, Steg L, Poortinga W. Values, Perceived risks and benefits, and acceptability of nuclear energy. *Risk Analysis* 2013; 33: 307–317.

de Groot JIM, Steg L. Value orientations to explain beliefs related to environmental significant behavior: How to measure egoistic, altruistic, and biospheric value orientations. *Environment and Behavior* 2008; 40: 330–354.

Delgado A, Lein Kjølberg K, Wickson F. Public engagement coming of age: From theory to practice in STS encounters with nanotechnology. *Public Understanding of Science* 2011; 20: 826–845.

Eagly A, Chaiken S. The advantages of an inclusive definition of attitude. *Social Cognition* 2007; 25: 582–602.

Fan D, Gilbert EJ, Fox T. Current state of in situ subsurface remediation by activated carbon-based amendments. *Journal of Environmental Management* 2017; 204: 793–803.

Fischhoff B, Slovic P, Lichtenstein S, Read S, Combs B. How safe is safe enough? A psychometric study of attitudes towards technological risks and benefits. *Policy Sciences* 1978; 9: 127–152.

Flynn J, Burns W, Mertz CK, Slovic P. Trust as a determinant of opposition to a high-level radioactive waste repository: Analysis of a structural model. *Risk Analysis* 1992; 12: 417–429.

Flynn J, Slovic P, Mertz CK. Gender, race, and perception of environmental health risks. *Risk Analysis* 1994; 14: 1101–1108.

Flynn R. Risk and the public acceptance of new technologies. In: Flynn R, Bellaby P, editors. *Risk and the Public Acceptance of New Technologies.* New York: Palgrave Macmillan, 2007.

Focht W, Albright M. Enhancing stakeholder acceptance of bioremeidation technologies. U.S. Department of Energy, 2009.

Freudenburg WR. Risk and recreancy: Weber, the division of labor, and the rationality of risk perceptions. *Social Forces* 1993; 71: 909–932.

Frewer L, Howard C, Shepherd R. Public concerns about general and specific applications of genetic engineering: Risk, benefit and ethics. *Science, Technology and Human Values* 1997; 22: 98–124.

Frewer L, Lassen J, Kettlitz B, Scholderer J, Beekman V, Berdal KG. Societal aspects of genetically modified foods. *Food and Chemical Toxicology* 2004; 42: 1181–1193.

Frewer LJ, Howard C, Hedderley D, Shepherd R. What determines trust in information about food-related risks? Underlying psychological constructs. *Risk Analysis* 1996; 16: 473–486.

Ganesh Pillai R, Bezbaruah AN. Perceptions and attitude effects on nanotechnology acceptance: An exploratory framework. *Journal of Nanoparticle Research* 2017; 19: 41.

Gaskell G, Bauer M, Durant J, Allum N. Worlds apart? The reception of genetically modified foods in Europe and the U.S. *Science* 1999; 285: 384–387.

Gaskell G, Einsiedel E, Hallman W, Priest S, Hornig J, Olsthoorn J. Social values and the governance of science. *Science* 2005; 310: 1908–1909.

Gawande K, Jenkins-Smith H. Nuclear waste transport and residential property values: Estimating the effects of perceived risks. *Journal of Environmental Economics and Management* 2001; 42: 207–233.

Gerhardt KE, Gerwing PD, Greenberg BM. Opinion: Taking phytoremediation from proven technology to accepted practice. *Plant Science* 2017; 256: 170–185.

Gilbert C. Crisis analysis: Between normalization and avoidance. *Journal of Risk Research* 2007; 10: 925–940.

Gillespie IMM, Philp JC. Bioremediation, an environmental remediation technology for the bioeconomy. *Trends in Biotechnology* 2013; 31: 329–332.

Grieger KD, Fjordbøge A, Hartmann NB, Eriksson E, Bjerg PL, Baun A. Environmental benefits and risks of zero-valent iron nanoparticles (nZVI) for in situ remediation: Risk mitigation or trade-off? *Journal of Contaminant Hydrology* 2010; 118: 165–183.

Gunter VJ, Harris CK. Noisy winter: The DDT controversy in the years before silent spring. *Rural Sociology* 1998; 63: 179–198.

Gupta N, Fischer AR, van der Lans IA, Frewer LJ. Factors influencing societal response of nanotechnology: An expert stakeholder analysis. *Journal of Nanoparticle Research* 2012; 14: 1–15.

Gupta N, Fischer ARH, Frewer LJ. Socio-psychological determinants of public acceptance of technologies: A review. *Public Understanding of Science* 2011; 21: 782–795.

Harclerode MA, Lal P, Miller ME. Quantifying global impacts to society from the consumption of natural resources during environmental remediation activities. *Journal of Industrial Ecology* 2016; 20: 410–422.

Hardisty P, Ozdemiroglu E, Arch S. Sustainable remediation: Including the external costs of remediation. *Land Contamination & Reclamation* 2008; 16: 307–318.

Henry HF, Burken JG, Maier RM, Newman LA, Rock S, Schnoor JL, et al. Phytotechnologies: Preventing exposures, improving public health. *International Journal of Phytoremediation* 2013; 15: 889–899.

Ho SS, Scheufele DA, Corley EA. Value predispositions, mass media, and attitudes toward nanotechnology: The interplay of public and experts. *Science Communication* 2011; 33: 167–200.

Hoban TJ. Trends in consumer attitudes about agricultural biotechnology. *AgBioForum* 1998; 1: 3–7.

Horst M. Cloning sensations: Mass mediated articulation of social responses to controversial biotechnology. *Public Understanding of Science* 2005; 14: 185–200.

Hou D, Al-Tabbaa A. Sustainability: A new imperative in contaminated land remediation. *Environmental Science & Policy* 2014; 39: 25–34.

Hou D, O'Connor D, Al-Tabbaa A. Comparing the adoption of contaminated land remediation technologies in the United States, United Kingdom, and China. *Remediation Journal* 2014b; 25: 33–51.

Hou D, O'Connor D, Al-Tabbaa A. Modeling the diffusion of contaminated site remediation technologies. *Water, Air, & Soil Pollution* 2014a; 225: 2111.

Hou D. Divergence in stakeholder perception of sustainable remediation. *Sustainability Science* 2016; 11: 215–230.

Ibtissem MH. Application of value beliefs norms theory to the energy conservation behaviour. *Journal of Sustainable Development* 2010; 3: 129–139.

Irwin A. The politics of talk: Coming to terms with the "New" scientific governance. *Social Studies of Science* 2006; 36: 299–320.

Jenkins-Smith HC, Silva CL, Nowlin MC, deLozier G. Reversing nuclear opposition: Evolving public acceptance of a permanent nuclear waste disposal facility. *Risk Analysis* 2011; 31: 629–644.

Kasperson RE, Renn O, Slovic P, Brown HS, Emel J, Goble R, et al. The social amplification of risk: A conceptual framework. *Risk Analysis* 1988; 8: 177–187.

Kennen K, Kirkwood N. *Phyto: Principles and Resources for Site Remediation and Landscape Design.* London, UK: Routledge, 2015.

Kim EJ. Phytoremediation for lightly toxic sites: Hazard perception and acceptance of remediation alternatives. *Human and Ecological Risk Assessment: An International Journal* 2016; 22: 1078–1090.

Kocher S, Levi D, Aboud R. Public attitudes toward the use of bioremediation to clean up toxic contamination. *Journal of Applied Social Psychology* 2002; 32: 1756–1770.

Kraus N, Malmfors T, Slovic P. Intuitive toxicology: Expert and lay judgments of chemical risks. *Risk Analysis* 1992; 12: 215–232.

Krause RM, Carley SR, Warren DC, Rupp JA, Graham JD. "Not in (or Under) My Backyard": Geographic proximity and public acceptance of carbon capture and storage facilities. *Risk Analysis* 2014; 34: 529–540.

Lee C, Scheufele D, Lewenstein B. Public attitudes toward emerging technologies. *Science Communication* 2005; 27: 240–267.

Lindell MK, Perry RW. *Communicating Environmental Risk in Multiethnic Communities.* Thousand Oaks, CA: Sage, 2004.

Lindeman M, Verkasalo M. Measuring values with the Short Schwartz's Value Survey. *Journal of Personality Assessment* 2005; 85: 170–178.

Luo X, Li H, Zhang J, Shim JP. Examining multi-dimensional trust and multi-faceted risk in initial acceptance of emerging technologies: An empirical study of mobile banking services. *Decision Support Systems* 2010; 49: 222–234.

Magnusson MK, Koivisto Hursti U-K. Consumer attitudes towards genetically modified foods. *Appetite* 2002; 39: 9–24.

Marinovich MJ, Funk WA, Kelly S, Elliott C, Hansen VG. Sustainable remediation and decision analysis practices at an onshore gas well site. *Remediation Journal* 2016; 26: 95–115.

Marques MD, Critchley CR, Walshe J. Attitudes to genetically modified food over time: How trust in organizations and the media cycle predict support. *Public Understanding of Science* 2015; 24: 601–618.

Morillo E, Villaverde J. Advanced technologies for the remediation of pesticide-contaminated soils. *Science of the Total Environment* 2017; 586: 576–597.

Moussaïd M, Brighton H, Gaissmaier W. The amplification of risk in experimental diffusion chains. *Proceedings of the National Academy of Sciences* 2015; 112: 5631–5636.

Nordlund AM, Garvill J. Value structures behind proenvironmental behavior. *Environment and Behavior* 2002; 34: 740–756.

O'Riordan T, Cameron J. *Interpreting the Precautionary Principle.* London, UK: Cameron May, 1994.

Page J, Atkinson-Grosjean J. Mines and microbes: Public responses to biological treatment of toxic discharge. *Society & Natural Resources* 2013; 26: 270–284.

Peters E, Slovic P. The role of affect and worldviews as orienting dispositions in the perception and acceptance of nuclear power. *Journal of Applied Social Psychology* 1996; 26: 1427–1453.

Peters EM, Burraston B, Mertz CK. An emotion-based model of risk perception and stigma susceptibility: Cognitive appraisals of emotion, affective reactivity, worldviews, and risk perceptions in the generation of technological stigma. *Risk Analysis* 2004; 24: 1349–1367.

Pin RR, Gutteling JM. The development of public perception research in the genomics field: An empirical analysis of the literature in the field. *Science Communication* 2009; 31: 57–83.

Plant R, Boydell S, Prior J, Chong J, Lederwasch A. From liability to opportunity: An institutional approach towards value-based land remediation. *Environment and Planning C: Politics and Space* 2016; 35: 197–220.

Pollard SJT, Brookes A, Earl N, Lowe J, Kearney T, Nathanail CP. Integrating decision tools for the sustainable management of land contamination. *Science of the Total Environment* 2004; 325: 15–28.

Pollard SJT, Hrudey SE, Fedorak PM. Bioremediation of petroleum- and creosote-contaminated soils: A review of constraints. *Waste Management & Research* 1994; 12: 173–194.

Poortinga W, Pidgeon NF. Exploring the structure of attitudes toward genetically modified food. *Risk Analysis* 2006; 26: 1707–1719.

Prasad MNV, Freitas H, Fraenzle S, Wuenschmann S, Markert B. Knowledge explosion in phytotechnologies for environmental solutions. *Environmental Pollution* 2010; 158: 18–23.

Prior J, Hubbard P, Rai T. Using residents' worries about technology as a way of resolving environmental remediation dilemmas. *Science of the Total Environment* 2017; 580: 882–899.

Prior J, Partridge E, Plant R. "We get the most information from the sources we trust least": Residents' perceptions of risk communication on industrial contamination. *Australasian Journal of Environmental Management* 2014; 21: 346–358.

Prior J. The norms, rules and motivational values driving sustainable remediation of contaminated environments: A study of implementation. *Science of the Total Environment* 2016; 544: 824–836.

Ramirez-Andreotta MD, Lothrop N, Wilkinson ST, Root RA, Artiola JF, Klimecki W, et al. Analyzing patterns of community interest at a legacy mining waste site to assess and inform environmental health literacy efforts. *Journal of Environmental Studies and Sciences* 2016; 6: 543–555.

Renn O, Burns WJ, Kasperson JX, Kasperson RE, Slovic P. The social amplification of risk: Theoretical foundations and empirical applications. *Journal of Social Issues* 1992; 48: 137–160.

Renn O, Roco MC. Nanotechnology and the need for risk governance. *Journal of Nanoparticle Research* 2006; 8: 153–191.

Rogers E. *Diffusion of Innovations.* New York: Free Press, 1995.

Saldaña J. *The Coding Manual for Qualitative Researchers.* Los Angeles, CA: Sage, 2013.

Schädler S, Morio M, Bartke S, Rohr-Zänker R, Finkel M. Designing sustainable and economically attractive Brownfield revitalization options using an integrated assessment model. *Journal of Environmental Management* 2011; 92: 827–837.

Schively C. Understanding the NIMBY and LULU phenomena: Reassessing our knowledge base and informing future research. *Journal of Planning Literature* 2007; 21: 255–266.

Schwartz SH, Bilsky W. Towards a universal psychological structure of human values. *Journal of Personality and Social Psychology* 1987; 53: 550–562.

Schwartz SH. A theory of cultural values and some implications for work. *Applied Psychology* 1999; 48: 23–47.

Schwartz SH. An overview of the Schwartz theory of basic values. *Online Readings in Psychology and Culture* 2012; 2: 1–20.

Schwartz SH. Normative explanations of helping behavior: A critique, proposal, and empirical test. *Journal of Experimental Social Psychology* 1973; 9: 349–364.

Schwartz SH. Studying values: Personal adventure, future directions. *Journal of Cross-Cultural Psychology* 2011; 42: 307–319.

Schwartz SH. Universals in the content and structure of values: Theoretical advances and empirical tests in 20 countries. *Advances in Experimental Social Psychology* 1992; 25: 1–65.

Shusterman D, Lipscomb J, Neutra R, Satin K. Symptom prevalence and odor-worry interaction near hazardous waste sites. *Environmental Health Perspectives* 1991; 94: 25–30.

Siegrist M, Connor M, Keller C. Trust, Confidence, procedural fairness, outcome fairness, moral conviction, and the acceptance of GM field experiments. *Risk Analysis* 2012; 32: 1394–1403.

Siegrist M, Cousin M-E, Kastenholz H, Wiek A. Public acceptance of nanotechnology foods and food packaging: The influence of affect and trust. *Appetite* 2007; 49: 459–466.

Siegrist M, Visschers VHM. Acceptance of nuclear power: The Fukushima effect. *Energy Policy* 2013; 59: 112–119.

Siegrist M. A causal model explaining the perception and acceptance of gene technology. *Journal of Applied Social Psychology* 1999; 29: 2093–2106.

Siegrist M. The influence of trust and perceptions of risks and benefits on the acceptance of gene technology. *Risk Analysis* 2000; 20: 195–203.

Sjöberg L. Attitudes toward technology and risk: Going beyond what is immediately given. *Policy Sciences* 2002; 35: 379–400.

Slovic P, Flynn JH, Layman M. Perceived risk, trust, and the politics of nuclear waste. *Science* 1991; 254: 1603–1607.

Slovic P. Perception of risk. *Science* 1987; 236: 280–285.

Sorvari J, Seppälä J. A decision support tool to prioritize risk management options for contaminated sites. *Science of the Total Environment* 2010; 408: 1786–1799.

Steg L, Dreijerink L, Abrahamse W. Why are energy policies acceptable and effective? *Environment and Behavior* 2006; 38: 92–111.

Steg L, Perlaviciute G, van der Werff E, Lurvink J. The significance of hedonic values for environmentally relevant attitudes, preferences, and actions. *Environment and Behavior* 2014; 46: 163–192.

Stern PC, Dietz T, Abel T, Guagnano GA, Kalof L. A value-belief-norm theory of support for social movements: The case of environmentalism. *Human Ecology Review* 1999; 6: 81–97.

Sturgis P, Allum N. Science in Society: Re-evaluating the deficit model of public attitudes. *Public Understanding of Science* 2004; 13: 55–74.

Tenbült P, de Vries NK, Dreezens E, Martijn C. Perceived naturalness and acceptance of genetically modified food. *Appetite* 2005; 45: 47–50.

Terwel BW, Daamen DDL. Initial public reactions to carbon capture and storage (CCS): Differentiating general and local views. *Climate Policy* 2012; 12: 288–300.

Terwel BW, Harinck F, Ellemers N, Daamen DDL. Voice in political decision-making: The effect of group voice on perceived trustworthiness of decision makers and subsequent acceptance of decisions. *Journal of Experimental Psychology: Applied* 2010; 16: 173–186.

Thøgersen J, Ölander F. Human values and the emergence of a sustainable consumption pattern: A panel study. *Journal of Economic Psychology* 2002; 23: 605–630.

Thøgersen J. Spillover processes in the development of a sustainable consumption pattern. *Journal of Economic Psychology* 1999; 20: 53–81.

Todt O. The limits of policy: Public acceptance and the reform of science and technology governance. *Technological Forecasting and Social Change* 2011; 78: 902–909.

Venkatesh V, Davis FD. A theoretical extension of the technology acceptance model: Four longitudinal field studies. *Management Science* 2000; 46: 186.

Voils CI, Sandelowski M, Barroso J, Hasselblad V. Making sense of qualitative and quantitative findings in mixed research synthesis studies. *Field Methods* 2008; 20: 3–25.

Wallquist L, Visschers VHM, Siegrist M. Lay concepts on CCS deployment in Switzerland based on qualitative interviews. *International Journal of Greenhouse Gas Control* 2009; 3: 652–657.

Whitfield S, Rosa E, Dan A, Dietz T. The future of nuclear power: Value orientations and risk perception. *Risk Analysis* 2009; 29: 425–437.

Wong CML. The mutable nature of risk and acceptability: A hybrid risk governance framework. *Risk Analysis* 2015; 35: 1969–1982.

Zhang C, Zhu M-Y, Zeng G-M, Yu Z-G, Cui F, Yang Z-Z, et al. Active capping technology: A new environmental remediation of contaminated sediment. *Environmental Science and Pollution Research* 2016; 23: 4370–4386.

Zhu X, Li W, Zhan L, Huang M, Zhang Q, Achal V. The large-scale process of microbial carbonate precipitation for nickel remediation from an industrial soil. *Environmental Pollution* 2016; 219: 149–155.

Zingg A, Siegrist M. Lay people's and experts' risk perception and acceptance of vaccination and culling strategies to fight animal epidemics. *Journal of Risk Research* 2011; 15: 53–66.

18 Post-Remediation Site Management

Deyi Hou, David O'Connor, and Yinan Song

CONTENTS

18.1 INTRODUCTION

Where contaminants in soil or groundwater pose a risk to human health or the natural environment, removing all contamination is normally cost prohibitive and uncalled for. Risk-based remediation standards are set at acceptable risk levels for intended land uses (e.g., industrial or residential use) beyond which further remedial action is deemed unnecessary (O'Connor and Hou, 2018). For this purpose, partial source depletion, pathway management, or receptor modifications are each considered appropriate strategies, so long as the risk can be demonstrably mitigated. The disadvantage of this approach is that residual contamination remains a hazard, and can pose a risk if conditions change. Therefore, sound risk-based site management entails long-term care of remediated sites (Vegter, 2001; Vegter et al., 2002). This chapter explores post-remediation site management (PRSM) as a means to uphold this principle, focusing on the following four aspects:

- Engineering controls (ECs)
- Long-term monitoring (LTM)
- Institutional controls (ICs)
- Monitored natural attenuation (MNA)

As a post-remediation action phase, PRSM fits after remediation system curtailment but before site closure (Figure 18.1). It should be noted that some PRSM steps may overlap with other remediation phases. For example, LTM or ICs may commence during remedial action or operation and maintenance phases (O&M). Similarly, MNA and ECs can be viewed as remedial actions in themselves, although they are equally considered PRSM aspects. In addition, different areas/zones at large sites may progress along the remediation phases at different times.

During system curtailment, care should be taken to demonstrate that remediation objectives have been met, contaminant concentrations are stable, and contaminant migration potential is minimal. Afterward, a short-term robust management process may be sufficient to bring rapid site closure after curtailment at some sites, with no need for long-term PRSM. At such sites, it is especially important that a robust verification phase is completed. In all cases, but especially

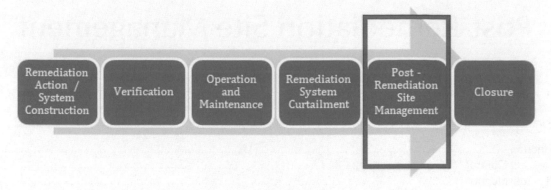

FIGURE 18.1 Post-remediation site management (PRSM) in the progression of remediation phases.

so for rapid site closures, dialogue with stakeholders is crucial, particularly in relation to environmental regulators, who have specific legal duties and powers for environmental protection.

At many sites though, site management will extend over years, or decades, before site closure is achieved. In particular, long-term PRSM becomes necessary whenever remediation measures leave residual contaminants *in situ*, whether intentionally or otherwise, at concentrations that do not allow for unlimited use and unrestricted exposure. Furthermore, PRSM may still be necessary even where contaminant concentrations have been sufficiently reduced, to ensure that contaminant rebound does not occur. For example, this phenomenon occurs within one year of remediation action at approximately one-third of dense non-aqueous phase liquid (DNAPL) cleanup sites (McGuire et al., 2006). In general, if a site necessitates restrictions on humans or ecosystems after remediation, then its fitness for multifunctional use remains hindered and PRSM may need to be considered (US EPA, 2018).

After several decades of remediation activity, reliance on PRSM in the United States is growing. As an estimate, 80% of superfund sites require some form of post-remediation stewardship (US EPA, 2006). Long-term management is also increasingly of importance for brownfield site redevelopment. Redeveloped sites are often located in areas with contamination remaining *in situ*, and so PRSM becomes necessary to ensure receptor protection. In such cases, site redevelopment and the implementation of appropriate PRSM activities can be harmonized, so that development proceeds in good time. PRSM can also provide a number of other advantages, which include (based on EA, 2010):

- It provides scientific and justifiable evidence for regulators, proponents, and other stakeholders that remediation goals are achieved in the long and short term
- Demonstration of remediation contractual requirements
- Documentary evidence for corporate or governmental reporting
- Providing confidence to future landowners/proponents and users that the remediated land is fit for its intended use

- Decreasing unknowns and increasing confidence in the efficacy of innovative remediation technologies
- Identification of unsuccessful remediation activities, where land users would otherwise be exposed to unacceptable risks, or polluters to liability
- Better holistic understanding of the sustainability of different remediation techniques (i.e., economic, social, and environmental performance) (Song et al., 2018).

18.2 ENGINEERING CONTROLS (ECs)

18.2.1 INTRODUCTION

ECs are physical controls or barriers that chemo-physically block contaminants from reaching identified receptors. Examples of ECs include capping systems, vertical barriers (e.g., slurry walls), and vapor intrusion mitigation systems. Collated monitoring data for sites under ECs indicate that, overall, when well designed, constructed, operated, and maintained, they perform at, or above, design objectives.

ECs may be used as part of a site's remediation objectives if they are included in the final remediation action, and not a temporary measure. In such cases, the PRSM phase is equated to the operation and maintenance (O&M) phase of remediation (Figure 18.1), and will require periodic inspections and maintenance of the EC physical state and long-term monitoring (LTM) for contaminant release (NRC, 2007). Because the need for effective LTM and maintenance are crucial, these should be enforced by an IC. It should be noted that the O&M of ECs will typically last decades. For example, the Massachusetts Department of Environmental Protection in the United States requires EC management for at least 30 years unless unlimited use and unrestricted exposure is attained at a site. This period may be extended if action remains necessary to protect public health or the environment.

18.2.2 OPERATION AND MAINTENANCE (O&M)

For vertical hydraulic barriers (e.g., slurry walls), periodic inspections and maintenance should ensure continued operation effectiveness. After construction and verification are completed, vertical barriers tend to be relatively low maintenance, with the main focus being on hydraulic pumping systems, if installed. Specific actions that may need attention include the following (based on US EPA, 1998):

- Operating the hydraulic pumping system to maintain a specified hydraulic gradient across the barrier wall
- Maintenance of the pumping systems
- Physically repairing the barrier as necessary based on period monitoring results

Vertical barriers require monitoring to ensure their continued effectiveness. Monitoring may include quarterly measurements of groundwater levels and water quality in monitoring wells, and any other sampling required by regulatory permits. Other monitoring activities should be based on-site-specific conditions (US EPA, 1998).

For capping systems, O&M will generally address erosion and subsidence effects, as well as maintenance of leachate and gas management systems, where necessary. Specific elements to consider include (based on US EPA, 1998):

- Repair any erosion, surface slumping, or cracking
- Pavement maintenance
- Landscaping measures
- Repair any damaged drainage control structures
- Leachate collection pipe cleanout, as required
- Maintenance of gas management systems
- Clearing obstructions from gas vents

For vapor intrusion mitigation systems, O&M will include equipment maintenance to ensure continued operation and integrity. This should include visible components of venting systems, multi-level gas probes, blowers, etc. If an inspection finds that building foundations or components of the mitigation system have been modified in any way, appropriate testing should be conducted to ensure the system performance. Maintenance frequency should consider site-specific requirements. Some sites will require frequent inspections, but for most sites annual inspections will be appropriate (DTSC, 2011).

18.2.3 Assessment, Modification, and Reporting

Monitoring data assessments should reflect the performance of ECs. Inadequate performance may require further maintenance, repair, or modification. Data assessment and statistical analysis should be carried out, and modification to ECs carried out to maintain long-term performance. A contingency plan should be in place in case of EC failure, including specific repair actions and identification of the parties responsible for implementing these actions. Regular inspections should include a review of the site use, to ensure continued applicability of the EC (US EPA, 2012). Annual reporting is recommended to confirm ECs remain in place and effective. Shorter reporting periods are appropriate when frequent land activities or potential changes in land and/or resource use are anticipated. If site changes are highly unlikely, and monitoring parameters have remained stable for some years, longer monitoring periods may be deemed appropriate.

18.3 LONG-TERM MONITORING (LTM)

18.3.1 Introduction

LTM involves measuring a site's "fitness" over the long-term. However, there is no unanimously accepted definition for the term "long-term monitoring" among different agencies and organizations. It has been defined by the US Environmental Protection Agency (US EPA) as "monitoring conducted after some active, passive, or containment remedy has been selected and put in place, and is used to evaluate the degree to which the remedial measure achieves its objectives" and "it is usually assumed that after a site enters the long-term monitoring phase of remediation, site characterization is essentially complete, and the existing monitoring network can be adapted as necessary, to achieve the objectives of the long-term monitoring program" (US EPA, 2005).

LTM will typically commence during curtailment, after a remediation action and its verification and O&M are completed (shown in dark color on the arrow below in Figure 18.2). It may be necessary for LTM to overlap/form a part of the O&M phase (shown in pink in Figure 18.2) to evaluate whether a site is fit for use during O&M. Monitoring is likely to be needed during verification, but this monitoring will be concerned primarily with whether or not a remediation action has successfully met the remediation objectives (CRC CARE, 2016). LTM, on the other hand, is used to confirm whether the risk from residual contamination remains low; it can also help demonstrate whether remedial goals continue to be met (EA, 2010). LTM may also extend into the closure phase of a project, because closure cannot be agreed until monitoring data allows it.

Contaminants below the water table are more susceptible to contaminant rebound than in the vadose zone. Therefore, LTM is mostly applied to groundwater contamination. LTM should be considered after a remediation action is completed when:

- Contaminants are contained on-site (e.g., using ECs)
- Immobilized contaminants remain on-site (e.g., solidification/stabilization)
- A long-term remedial approach or IC is implemented;
- A source depletion approach is used which will leave some residual contaminant mass (e.g., soil vapor extraction)
- The site involves groundwater contamination

It is good practice for LTM requirements to be recorded in the remediation or verification report, including the requirements for formalizing a long-term monitoring plan.

18.3.2 Long-Term Monitoring Plans

The first stage in developing an LTM plan involves setting clear and quantifiable objectives that set the scope and intent. Objectives should be identified after reviewing the site remediation objectives, the remediation actions, the contaminant exposure pathways and conceptual site model,

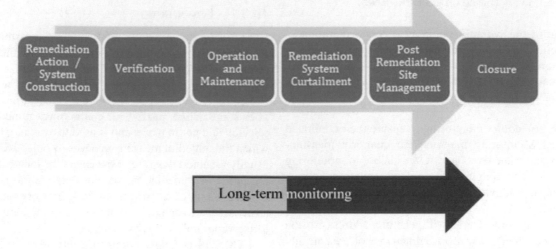

FIGURE 18.2   Long-term monitoring (LTM) in the progression of remediation phases.

and the types of contaminants of concern. Key assumptions and expectations that require LTM to ensure continued protectiveness should be identified at this stage. Stakeholder involvement at this stage will help gain greater acceptance for LTM activity.

Next, the establishment of data quality objectives (DQOs) helps guide effective and economic LTM. The DQO setting process can be flexible and iterative, and applies to both decision-making (e.g., compliance/non-compliance to a standard) and estimation (e.g., estimating contamination levels). The DQO process aims to ensure the data collected are of sufficient quality and quantity to judge the progress of LTM objectives. The approach to DQO planning and the depth of the DQO setting process will, therefore, depend upon the remediation objectives. For example, at sites with multiple phases of remediation, the DQO process will allow the separation of data requirements for each phase (US EPA, 2006).

LTM decision rules are "pass/fail" statements that are used to evaluate monitoring data against criteria as it is generated, which then trigger subsequent courses of action as needed. The LTM plan should include options for contingency and response procedures. Scientific evidence-based decision criteria are necessary for rational and effective decision making (US EPA, 2004; Environment Canada, 2013). Decision rules should be designed to be consistent with the site remediation objectives. Where possible, quantitative "exit criteria" representing the successful completion of objectives should be established in the LTM plan to help progress toward site closure. When an exit criterion is achieved, LTM for a particular objective may be considered complete, although ongoing LTM for other objectives may still be necessary. To avoid unnecessary cost, a specific or stated reason for each sample collected should always be known. The LTM plan should state purposes for sampling and be linked to uncertainties that are identified in the post remediation conceptual site model and data quality objectives (DQO) (EA, 2010).

The LTM plan should document the following:

*   Determinands to be analyzed
*   Sampling methods
*   Analytical methods
*   Limits of quantification

Where possible, these should be consistent with those used previously during the site investigation, remediation, and verification phases of remediation. Determinands will typically include the contaminants of concern and daughter products where applicable. Hydrogeological or geochemical parameters that affect the fate and transport of contaminants may also be measured (e.g., oxidation/reduction potential [ORP], dissolved oxygen, pH, temperature, and electrical conductivity) (US EPA, 2005). Monitoring environmental changes that influence the sampling and/or the interpretation of data should also be included (e.g., fluctuations in groundwater tables).

18.3.3   Sampling and Analysis

Soil sampling and analysis in LTM is used to evaluate residual contamination levels post-remediation. This is particularly applicable to sites where NAPL is present in soil, with respect to post-remediation contaminant rebound in groundwater and soil gas. Soil sampling during long-term monitoring will typically involve location targeted samples. Soil sampling can be used to estimate changes in the remaining contaminant mass with time.

Where necessary, soil gas sampling should be undertaken periodically to monitor the concentrations of contaminants and related daughter compounds. Concentrations of fixed gases such as oxygen, carbon dioxide, and methane in soil gas may be monitored to aid in the interpretation of contaminant attenuation and biodegradation. The utilization of field testing devices such as soil gas detector tubes, or the monitoring of surrogate parameters such as fixed gas concentrations, may be considered suitable. These typically lower the cost of monitoring when used appropriately in the context of the conceptual site model.

Where groundwater sampling is necessary, necessary locations and frequency for monitoring should be identified in the LTM plan; this should take into account the following factors (based on [CRC CARE, 2016]):

- Location of monitoring points relative to the location of residual contamination, contaminant plume(s), and potential receptors
- Contaminant (and daughter products) concentration trends
- Contaminant concentrations compared to other locations within and surrounding the plume
- Vertical hydraulic gradients
- Groundwater table elevation
- Presence of multiple discrete media
- Relative plume size
- Plume shape
- Local ambient groundwater quality and uses

Each sampling location should have a unique identifier with known geographical coordinates to ensure location consistency, and to help find the location in the field. Groundwater sampling date (taking account of, for example, tidal fluctuations, facility operating schedules, or climatic conditions) should be described in the plan to ensure relevant scenarios of contaminant behavior are taken into account. For example, concentration fluctuations may be associated with groundwater table fluctuations. Additionally, long-term environment/climatic changes should be accounted for in the selection and density of monitoring locations, particularly with respect to the potential for the groundwater flow direction to change.

Groundwater sampling frequency is an important aspect of a monitoring program. If groundwater samples are not collected frequently enough, some temporal variability may be overlooked. On the other hand, resources are wasted if more groundwater samples are collected than strictly necessary. Therefore, the rationale for sampling at each location should be documented (US EPA, 2005).

There are various methods that can be used to collect groundwater samples from monitoring locations, which should be performed in accordance with established guidance. It is recommended that consistency be kept with any previous sampling at the site. However, if a demonstrably better sampling technology is available, and it will not cause a significant bias regarding previously collected samples, then it may be considered for an LTM program.

LTM data often relate to contaminant concentrations that are obtained by laboratory analysis of soil, groundwater, or soil gas samples. The laboratory chosen to perform the analysis should be competent to carry out the methods determined in the LTM monitoring plan. Competence can be demonstrated by third party accreditation, for example, CMA certification, or equivalent. It should be noted that laboratory accreditation is often method-specific and not laboratory specific. Accreditation should cover QA/QC procedures, as well as analysis technical factors. Suitable QA/QC measures provide confidence that the data reported is reliable and repeatable. The use of laboratories that participate in external QA schemes is advised, indicating an intention of quality standards. Copies of QA certification and performance measures for the contaminants of concern should be kept on file.

The selection of an appropriate analysis depends on the determinands and sample matrix; analytical laboratories will often provide advice on the selection of appropriate analysis methods. It is also important to ensure that the selected methods can achieve a limit of quantification that is equal to or higher than the relative monitoring criteria (EA, 2010).

18.3.4 Data Assessment and Statistical Analysis

LTM data assessments should first determine whether or not the DQOs are being met. If the data do not meet the DQOs, the underlying reasons should be investigated, and then the remediation strategy or the LTM plan be revised, if necessary. Any changes should be recorded as revisions in the LTM plan. If the data do meet the DQOs, they are then evaluated using the decision rules to determine the next course of action for monitoring objectives. Where appropriate, data trend analysis is used to assess new monitoring data in the context of the results of previous monitoring rounds.

Identifying long-term trends in contaminant concentrations is an important part of LTM. If removal of contaminant mass is occurring due to attenuation processes or the operation of a remediation system, mass removal may be evidenced by this, for example, by (i) a concentration decrease at sampling locations with time or (ii) an observed geochemistry change/the occurrence of contaminant daughter products with time.

Plotting concentrations temporally reveals groundwater plume stability (Wiedemeier and Haas, 2000). Temporal chemical concentration data can be evaluated by plotting contaminant concentrations against time for individual sample points, or by plotting contaminant concentrations against down-gradient distance from the source zone for several monitoring points for different monitoring events. Visual identification trends can be subjective, particularly if the data exhibit noise rather than a uniform trend. In such cases, statistical analysis is advised. Various statistical procedures including regression analyses or the Mann-Kendall test for trends can be applied. Non-parametric tests are often useful as they can be applied to small sample size data sets with no assumptions regarding the underlying statistical distribution of the data. They can also account for seasonal variation. For example, the Mann-Kendall test statistic can be used to evaluate temporal trends for individual wells. Whichever statistical procedure is used to identify temporal trends, it is important to ensure that spurious trends/bias are not caused by externalities such as a change in analytical method or method quantification limits.

18.4 INSTITUTIONAL CONTROLS (ICs)

18.4.1 INTRODUCTION

ICs are non-engineering instruments that are used to: (i) mini-mize exposure to any residual contamination remaining on a site after remediation action, (ii) prevent receptor exposure contaminants until the remediation objectives are met, and/or (iii) offer protection for engineered remediation components (US EPA, 2001, 2012). ICs act by limiting land uses, obli-gating certain actions, or making available information that modifies human behavior. For example, a remediation action at a site may have involved the placement of a capping EC, an IC placed on the site would then prohibit intrusive construc-tion/building action to a certain depth at the site and obligate periodic inspections and repairs to protect the cap. In many countries, ICs are widely used as a relatively straightforward and low-cost device to protect site users. For example, in the United States, it has been estimated that approximately half of the superfund National Priority List (NPL) sites are associ-ated with one or more ICs (US EPA, 2001). It must be recog-nized that the long-term effectiveness of ICs heavily relies on legal and management elements, and will not be effective if these elements are absent or disregarded (ITRC, 2016). Key activities in the institutional control life cycle are summarized in Figure 18.3, which are discussed within the sections below.

18.4.2 TYPES OF INSTITUTIONAL CONTROL

There are a number of IC approaches, which can be divided into three categories: (i) property controls, (ii) government controls, and (iii) information devices.

Property controls are legal instruments based in property law, which are expressly written on the site's property title, planning certificate, or equivalent legal documentation, in order to surrender certain usage rights for that site. When a site is under a property control, it should not be developed for a land use that has been renounced. For example, if site owner agrees that their site is to be restricted to industrial use only, in order that a less stringent remediation target can be applied in the remediation action, a property control would formalize this agreement. For example, the property control would legally ensure that the site is not developed for residen-tial use. Another example of a property control is an arrange-ment that provides access rights to a property so that a party, such as a regulator, may conduct inspections and monitoring (US EPA, 2000).

Government controls impose restrictions on the land use, usually in the form of permits/prohibitions. They are nor-mally employed by local governments or regulatory bod-ies as and when required to impose land use restrictions or enforcement orders to maintain safety. For example, a prohibition on groundwater abstraction may be put in place because of groundwater contamination. At some sites, site-specific health and safety plans that address contaminated soils may be necessitated before issuing a permit to dig below certain depths.

Information devices alert the presence of contamination risk or disclose restrictions on access, use, or development of a site by providing information or a method of notification to interested parties. They differ from property and government controls in that they do not enforce or prohibit actions. As such, an information device by itself will not usually be suf-ficient to address the risk of contaminant exposure and should be used in conjunction with other IC types (US EPA, 2012). Notices can be imparted using various information devices, which may include: (i) local authority records, which may rely on-site owner/practitioner reporting. For example, future site owners or other interested parties can access a databases of contaminated sites made available to the public; (ii) recorded notices on property titles or other documents (see property controls above) can also be considered a form of information

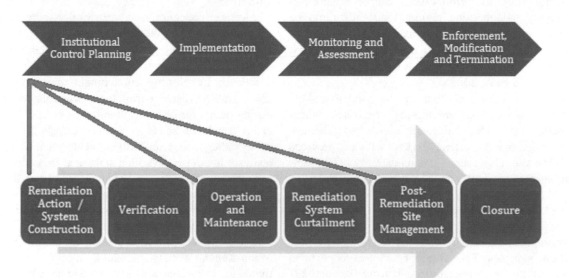

FIGURE 18.3 The institutional control (IC) life cycle and its initiation during different remediation phases.

device in itself. For example, notices on property documents usually provide warning to the reader of the document regarding the contamination status; or (iii) other ways, e.g., warnings to the public of potential risks associated with using the land, surface water, or groundwater, in the form of a public health warning.

The process of identifying appropriate ICs involves an appraisal of all potentially applicable ICs, i.e., a process similar to an appraisal of remediation actions. This will involve evaluating site characteristics to find the most durable and effective IC that can meet the objectives. ICs can be inadequate without full consideration of factors such as feasibility, stakeholder concerns, the intended future land uses, and their long-term monitoring and assessment requirements. When evaluating different types of ICs the following specific factors should be considered (based on US EPA, 2012):

- Will the IC achieve the necessary substantive use restrictions and/or provide adequate notice of site conditions, i.e., what are the source-pathway-receptor pollutant linkages, and how will the IC break the pathway?
- What are the various legal and practical limits for long-term implementation, e.g., are IC life cycle costs prohibitive?
- Who will ultimately be responsible for the different activities through each phase of the IC life cycle?
- Are the parties responsible for activities fully aware of their roles and capable to fulfill their responsibilities?

In addition, a professional legal practitioner should carefully consider ICs being considered. Potentially relevant legal questions for evaluating different possible ICs include the following (based on US EPA, 2012):

- Based on an initial evaluation of land title records, are property controls considered practical?
- Who has authority for implementing and enforcing property controls?
- Do current property laws allow property controls to bind future landowners?
- What is the government's zoning and permitting authority in issuing government controls?
- Which agencies have the legal authority to control potential receptors, e.g., prohibition on groundwater extraction for drinking water supplies, or recreational fishing?
- Do regulatory agencies actively enforce existing environmental regulations?

18.4.3 PLANNING AND IMPLEMENTATION

The intention when planning ICs is to ensure that (i) the objectives of the IC are clearly stated, (ii) the best approach for introducing an IC is applied, and (iii) to coordinate between the site owner, regulatory agency and local governments, and other key stakeholders, for their effective implementation. ICs can be implemented at any stage of the remediation process (Figure 18.1), to achieve various objectives and can be "layered" by using multiple ICs simultaneously, or implemented in series to provide overlapping assurances of protection from the remediation action through PRSM (US EPA, 2012).

Site owners or users will normally set the content of IC with help from professional practitioners/consultants, and then submit the IC contents to the local environmental regulator for approval. ICs should be operated by the site owner, site users, or contractors, and audited by the local government. Local stakeholders/communities should be aware of the relevant information regarding an IC. The implementation stage may include various activities to put ICs in place. This includes drafting, negotiating, signing, and recording specific documents as necessary to establish the IC legally. The planning stage should be completed before the IC implementation is needed (US EPA, 2001).

In the preparation of a property type IC, it is recommended that a legal professional be engaged to ensure compliance with respect to legislation and other legal requirements. It is important that the legal practitioner understand the relevant legislation and whether the IC will be effective in achieving the required objectives. Additionally, environmental practitioners with specialist skills, for example in the design of ECs, can be consulted as subject matter experts. Regulators may have subject matter experts in-house who can also be consulted in the evaluation of ICs. The method of documenting an IC should be considered with regard to communicating with other parties and ensuring its effectiveness and longevity.

An IC implementation and assurance plan should be written in order to: (i) systematically establish and document activities associated with implementing and ensuring long-term implementation of ICs, and (ii) specify the entity/stakeholders responsible for conducting activities. The plan should focus on identifying how ICs will be implemented, monitored, enforced, modified, and terminated. The plan will normally be a stand-alone document, and its enactment may be enforced by the IC.

The monitoring and performance assessment stage of ICs involve monitoring and reporting activities over the full duration of their use, which may last many years, or decades, before they are terminated, except in the case of temporary ICs. This is necessary to evaluate their effectiveness with respect to objectives. The evaluation of ICs should involve a rigorous assessment, as with any other remediation action, to ensure IC effectiveness. The specific monitoring and reporting requirements for each site should be outlined in the implementation and assurance plan.

Reports on ICs should be kept on file for future inspection requirements, including documentation to show that ICs remain in place and are effective. The reporting frequency will depend on the context of the site and risk management needs. Typically, an annual reporting period is deemed appropriate. If it is highly unlikely that site conditions will change, a monitoring period longer than a year may be appropriate (US EPA, 2012).

18.4.4 Enforcement, Modification, and Termination

Enforcement is usually required after they are violated, improperly implemented, or improperly maintained. It is important to identify the responsible party for enforcement, which should be stated within the IC implementation and assurance plan. IC enforcement may first involve informal communication to encourage voluntary compliance, followed, when appropriate, by formal legal steps to enforce compliance.

Modifications to ICs may include administrative or legal activities to adapt ICs in place. Modifications may be necessary when monitoring and assessment indicates that an IC is not achieving its intended objective. It may also be necessary due to changes in site characteristics, or because certain remediation objectives or other IC objectives have been met, for example, by changing the boundary of an IC, or by changing the monitoring requirements. The modification process should involve a thorough evaluation, similar to the planning phase, which allows for maximum beneficial safe use of the property.

An IC should apply to a site in perpetuity, or until the regulatory agency determines that there is no longer a need. At no time should a site be used in a manner inconsistent with an IC in place. Termination of an IC can commence if the remediation objectives and/or other conditions have been fully achieved. A termination report should be prepared and filed for the property subject to an IC before site closure is achieved.

18.5 MONITORED NATURAL ATTENUATION (MNA)

18.5.1 Introduction

Natural contaminant attenuation levels at certain sites may be sufficient to lower the risk of harm to receptors to safe levels without any need for further remediation action. Monitored natural attenuation (MNA) is as an approach to demonstrate whether physical, chemical, and biological processes can reduce the contaminant mass, concentration, flux, or toxicity of polluting substances to acceptable levels within an acceptable time frame (EA, 2000). In this context, physical processes include dispersion and diffusion processes; chemical and abiotic processes include migration retardation or biogeochemical transformations; and biological processes include metabolic processes resulting in the degradation of pollutants to non-toxic daughter products (CRC CARE, 2017). MNA is not a "do nothing" option, but is based on robust site characterization, implementation, and LTM, which are designed to provide multiple line of evidence that sufficient attenuation is occurring to protect human health and the environment.

MNA can also provide a baseline for understanding whether or not ECs are needed, or when they can be terminated because natural attenuation is sufficient. Employing MNA may enhance sustainability of site remediation, as it often has lower associated environmental, social, and economic impacts than other actions. For example, reduced waste generation, reduced energy use, and, in some cases, reduced lifecycle costs. The advantages and disadvantages of selecting MNA as part of a remediation strategy should be evaluated holistically, as with all remediation actions, by a robust remediation evaluation process.

18.5.2 Applicability

When appraising remediation options for a site, MNA should not be considered a sole or principal remediation option in the first place. MNA is most appropriate when used to complement other remedial actions, or as a secondary follow-up process to an active remediation (US EPA, 1999). However, in certain circumstances, MNA may be deemed appropriate as a sole remedy, if an assessment robustly shows that only MNA can meet the remediation objectives.

18.5.2.1 Applicable Contaminants

MNA is applicable for the remediation of contaminated groundwater plumes. MNA has been demonstrated to be potentially applicable for a wide range of plume contaminants, including the following (based on CRC CARE, 2017; Wiedemeier et al., 1995; US EPA, 2007a,b):

- Petroleum hydrocarbons, such as gasoline and diesel range organics and polynucleic aromatic hydrocarbons (PAHs)
- Chlorinated aliphatic hydrocarbons, including chlorinated solvents (e.g., trichloroethene)
- Inorganics, including ammonia, cyanide, and heavy metals

However, it should be regarded that not all possible contaminants that fall within these categories may be suitable. The contaminants present at a site should be evaluated individually for MNA suitability. There are certain chemical properties that make certain contaminants favorable or not. In general, these include:

- Solubility, which influences the potential for high mass loading and concentrations
- Sorption properties, which influence the transport rate and, therefore, plume length
- Potential biodegradation rates (i.e., half-life) relative to the groundwater flow velocity, which influences contaminant migration distance and the prospect for plume steady state conditions
- Toxicity to microbes, which influences contaminant biodegradation potential

18.5.2.2 Favorable Conditions

MNA is most often applied for PRSM, where the risk is first significantly reduced by another remediation action, but not sufficiently for rapid site closure. In such circumstances, MNA can play an important role. Sites where contaminant source zones have first been addressed by source control measures are favorable for MNA. This is because it reduces the potential for source zone contaminants to leach into groundwater at rates exceeding natural attenuate. Source control measures may include source removal (i.e., dig and haul), *in situ* treatment/destruction, or containment by an EC.

18.5.2.3 Practical Requirements

The practicalities of MNA vary site-by-site; some examples are listed below (based on CRC CARE, 2017):

- Access to monitoring points is certain
- Protection of monitoring points between sampling events is ensured
- A budget has been reserved for the proposed implementation time frame
- Future land uses will not adversely impact monitoring points, or they will be replaced with suitable alternatives
- Other remedial action will not adversely affect MNA
- Third-party agreement has been obtained, where necessary, to monitor off-site monitoring points
- Relevant legislation and health and safety requirements are met
- MNA will demonstrably protect receptors for the entirety of the implementation period
- Remedial objectives can be achieved in a reasonable time frame
- A contingency plan is available

18.5.2.4 MNA Prohibitions

There are some circumstances where MNA will normally be considered ineffective or inappropriate. MNA may still be used at such sites if other remediation systems are in place and MNA plays a complementary component of the site remediation, or where MNA is the only viable option to achieve the remediation objectives. Conditions that will normally prohibit the use of MNA include the following (based on NJ DEP, 2012):

- Other viable remediation actions have not been suitably evaluated
- The groundwater plume is expanding, indicating that contaminant release is exceeding natural attenuation capacity
- The site has a relatively complex media/hydrogeological regime. For example, fractured bedrock or karst formations are difficult to monitor for contaminant migration and natural attenuation processes. Therefore, it may not be possible to adequately ensure the protection of potential receptors
- There is an impending threat to receptors. A robust justification is necessary to demonstrate the viability of MNA when the groundwater flow velocity indicates an imminent threat to a receptor
- Free and/or residual non-aqueous phase liquid (NAPL) is present. In general, unaddressed source zones are not recommended to be within the scope of

MNA (see above), unless the risk has robustly been quantified as low due to natural attenuation
- The remediation objectives cannot be reached in a reasonable time frame

18.5.3 Implementation

The term "implementation" here refers to the collection of sufficient evidence to first demonstrate that natural attenuation is taking place. MNA implementation involves a three-tier approach, referred to as the "three lines of evidence." By this approach, detailed information is collected to provide sufficient confidence in MNA. These three tiers of site-specific information are as follows (based on EA, 2000; CRC CARE, 2017):

- Primary lines of evidence – Documented loss of contaminants
- Secondary lines of evidence – Geochemical and analytical data
- Tertiary lines of evidence – Direct microbiological evidence

Primary lines of evidence involve analysis of historical groundwater and/or soil data to evidence decreasing trends in contaminant mass or concentrations at suitable monitoring points, providing evidence for a decreasing plume area, decreasing plume mass/concentration, and a stable/receding plume center of mass. At some sites, the plume status may be apparent from a visual review of plume plots. However, at many sites a statistical evaluation of the historical monitoring data is required to provide an objective measure of the plume status. Contaminant degradation rate constants, calculated using monitoring data from within groundwater plumes, are a valuable way to assess contaminant trends and to help understand long-term plume behaviors.

Secondary lines of evidence involve geochemical and analytical data, which are used to indirectly indicate the natural attenuation at a site. There are three general types of geochemical indicators considered for secondary lines of evidence (based on DEEP, 2014): (i) consumption of electron acceptors used for direct oxidative reactions, that is the apparent loss of dissolved oxygen, nitrate, and sulfate in a plume area; (ii) production of metabolic by-products; (iii) presence of appropriate redox/microbial environments, for example, dissolved hydrogen levels may indicate whether the site is favorable for reductive dechlorination at some sites.

Tertiary lines of evidence include data from field or microcosm studies conducted in or with actual contaminated site media, which evidence the occurrence of specific natural attenuation processes. Such studies are most commonly used to demonstrate and quantify biological degradation processes. This line of evidence typically relates to biodegradable organic contaminants. Tertiary lines may also involve biochemical and isotope data (CRC CARE, 2017). For sites where robust data sets exist, and the uncertainty is low, tertiary lines may be omitted. It is necessary to comprehensively demonstrate

MNA effectiveness at sites with recalcitrant contaminants, including data from microcosm studies or other tertiary lines of evidence (EA, 2000; CRC CARE, 2017).

Long-term monitoring (LTM) is a necessary component of MNA, serving two purposes: (1) to evaluate the long-term performance, and (2) to ensure potential receptors are not impacted. A monitoring well network is necessary for this purpose, including up-gradient wells, a transect of wells along the longitudinal axis of the plume, and transects of wells across the transverse axis of the plume (Missouri, 2007). In addition, LTM at MNA sites will usually require down-gradient wells between the leading plume edge and receptors, where contaminant concentrations are below detection level, or at concentrations considered safe to the receptor. These wells are termed "sentinel wells," as they provide warning in the case of an unexpected expanding plume. The monitoring schedule should incorporate performance and sentinel wells to ensure the continued viability of MNA as a protective remediation action. LTM should be frequent to begin with, to confirm predicted contaminant degradation rates. The frequency may be reduced during later stages if MNA proceeds as predicted.

18.6 SITE CLOSURE

Remediation proceeds along various stages, of which site closure is the final phase (Figure 18.4). It is to be commenced only after all other relevant phases of remediation have been completed. PRSM should be continued until the concentrations of identified contaminants allow for unlimited use and unrestricted exposure. Unlimited use and unrestricted exposure generally refers to a situation when there are no exposure or use limitations (US EPA, 2012). It should be noted that at large sites, zones may be closed at different times.

In general, when a site has any restrictions, PRSM remains necessary. For example, if a site is remediated to a standard suitable for industrial use, and not residential use, then the site cannot be considered for closure. This would require additional remediation to lower the contaminant levels to a level suitable for multifunctional land use. In considering a site for potential closure, the regulator should consider whether the following criteria are met:

- All remediation actions have been carried out as required
- Further remediation action is not required
- An investigation has shown that the site poses no threat to public health or the environment

There may still be residual contamination remaining at the point of site closure, but there are no plausible scenarios where the contamination presents an unacceptable risk for any given land use. PRSM can cease when contaminants are reduced to levels that allow for unlimited use and unrestricted exposure. Lines of evidence supporting this decision should be submitted for approval by the appropriate regulators, which may involve inspections, monitoring, and sampling activities. Robust evidence is necessary to demonstrate that remediation has met pre-defined remediation objectives.

It should be noted that at more complex sites, removal of sufficient mass to achieve unlimited use and unrestricted exposure is unlikely for many decades. Furthermore, the US National Research Council (NRC) has stated that "despite nearly 40 years of intensive efforts in the United States as well as in other industrialized countries worldwide, restoration of groundwater contaminated by releases of anthropogenic chemicals to a condition allowing for unrestricted use and unlimited use and unrestricted exposure remains a significant technical and institutional challenge" (LQM, 2013). Therefore, it should be cautioned that unlimited use and unrestricted exposure is a high standard and difficult to achieve, particularly at sites containing recalcitrant groundwater pollutants, such as chlorinated solvents.

Because of the long time frames that may be required in achieving unlimited use and unrestricted exposure, the regulator may decide that a partial closure is allowable/appropriate, where some aspect of PRSM, such as an IC, remain in place. Such partial closures may be considered closure in itself in certain countries, requiring no further action. Under partial closure, ICs should continue to be enforced. Termination of ICs and full site closure can be considered when unlimited use and unrestricted exposure is achieved.

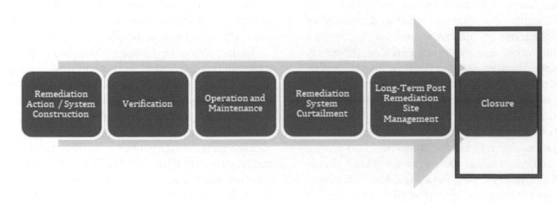

FIGURE 18.4   Site closure in the progression of remediation phases.

REFERENCES

CRC CARE, *Application Guide for Monitored Natural Attenuation.* 2017, Australia: CRC for Contamination Assessment and Remediation of the Environment.

CRC CARE, *National Remediation Framework: Guidelines on Long-Term Monitoring (Draft for consultation).* 1.2.11-14/15. 2016, Australia: CRC for Contamination Assessment and Remediation of the Environment.

DEEP, C., *Public Discussion Draft RSR Wave 2 – Potential Changes to RSRs Monitored Natural Attenuation Class C Cleanup.* Connecticut Department of Energy & Environmental Protection (DEEP 2014).

DTSC, *Vapor Intrusion Mitigation Advisory – Final Revision 1.* 2011: Department of Toxic Substances Control California Environmental Protection Agency

EA, *Guidance on the Assessment and Monitoring of Natural Attenuation of Contaminants in Groundwater.* R&D Publication 95. 2000, Bristol, UK: Environment Agency (EA).

EA, *Verification of Remediation of Land Contamination.* 2010, Bristol, UK: Environment Agency (EA).

Environment Canada, *Executive Summary of the FCSAP Long-Term Monitoring Planning Guidance.* 2013, Gatineau, Canada.

ITRC, *Long-Term Contaminant Management Using Institutional Controls.* 2016, Washington, DC: Interstate Technology & Regulatory Council (ITRC), Long-Term Contaminant Management Using Institutional Controls Team.

LQM, *SP1004 International Processes for Identification and Remediation of Contaminated Land.* Report No.: 1023-0. 2013: Land Quality Management Ltd.

McGuire, T.M., J.M. McDade, and C.J. Newell, Performance of DNAPL source depletion technologies at 59 chlorinated solvent-impacted sites. *Ground Water Monitoring and Remediation,* 2006. **26**(1): pp. 73–84.

Missouri DNR, *Monitored Natural Attenuation of Groundwater Contamination at Brownfields/Voluntary Cleanup Program Sites.* PUB002110. 2007, Jefferson City, MO: Missouri Department of Natural Resources (DNR).

NJ DEP, *Site Remediation Program – Monitored Natural Attenuation Technical Guidance.* Version: 1.0. 2012, Trenton, NJ: New Jersey Department of Environmental Protection.

NRC, *Assessment of the Performance of Engineered Waste Containment Barriers.* 2007, Washington, DC: The National Academies Press, p. 134.

O'Connor, D. and D. Hou, *Targeting Cleanups towards a More Sustainable Future.* Environmental Science: Processes & Impacts, 2018.

Song, Y., D. Hou, J. Zhang, D. O'Connor, G. Li, Q. Gu, S. Li, and P. Liu, Environmental and socio-economic sustainability appraisal of contaminated land remediation strategies: A case study at a mega-site in China. *Science of the Total Environment,* 2018. **610**: pp. 391–401.

US EPA, *Evaluation of Subsurface Engineered Barriers at Waste Sites.* EPA 542-R-98-005. 1998: United States Environmental Protection Agency Office of Solid Waste and Emergency Response.

US EPA, *Guidance for Monitoring at Hazardous Waste Sites: Framework for Monitoring Plan Development and Implementation.* OSWER Directive 9355.4-28. 2004.

US EPA, *Guidance on Systematic Planning Using the Data Quality Objectives Process* EPA QA/G-4. 2006, Washington, DC: United States Environmental Protection Agency Office of Environmental Information.

US EPA, *Institutional Controls: A Guide to Planning, Implementing, Maintaining, and Enforcing Institutional Controls at Contaminated Sites.* EPA-540-R-09-001. 2012, Washington, DC: United States Environmental Protection Agency.

US EPA, *Institutional Controls: A Guide to Planning, Implementing, Maintaining, and Enforcing Institutional Controls at Contaminated Sites.* 2012, Washington, DC: United States Environmental Protection Agency.

US EPA, *Institutional Controls: Site Manager's Guide to Identifying, Evaluating and Selecting Institutional Controls at Superfund and RCRA Corrective Action Cleanups.* OSWER 9355.0-74FS-P/EPA 540-F-00-005. 2000, United States Environmental Protection Agency Office of Solid Waste and Emergency Response.

US EPA, *Monitored Natural Attenuation of Inorganic Contaminants in Ground Water, Volume 1 – Technical Basis for Assessment (EPA/600/R-07/139).* 2007a, Washington, DC: United States Environmental Protection Agency.

US EPA, *Monitored Natural Attenuation of Inorganic Contaminants in Ground Water, Volume 2 – Assessment for Non-Radionuclides Including Arsenic, Cadmium, Chromium, Copper, Lead, Nickel, Nitrate, Perchlorate, and Selenium (EPA/600/R-07/140).* 2007b, Washington, DC: United States Environmental Protection Agency.

US EPA, *Roadmap to Long-Term Monitoring Optimization.* EPA 542-R-05-003, ed. EPA/National Service Center for Environmental Publications. 2005, Cincinnati, OH.

US EPA, *Superfund Post Construction Completion: An Overview.* OSWER 9355.0-79FS – EPA/540/F/01/009. 2001, United States Environmental Protection Agency Office of Solid Waste and Emergency Response.

US EPA, *Use of Monitored Natural Attenuation at Superfund, RCRA Corrective Action, and Underground Storage Tank Sites.* Directive Number 9200.4-17P. 1999: US EPA.

US EPA. *Five-Year Reviews, Frequently Asked Questions (FAQs) and Answers.* OSWER 9355.7-21. 2018. Available from: https://semspub.epa.gov/work/11/174052.pdf.

US EPA. *Long-Term Stewardship Fact Sheet.* EPA 500-F-05. 2006. Available from: https://www.epa.gov/sites/production/files/documents/lts_fact_sheet_1006.pdf.

Vegter, J., J. Lowe, and H. Kasamas, (Eds.), *Sustainable Management of Contaminated Land: An Overview.* 2002, Austrian Federal Environment Agency on behalf of CLARINET.

Vegter, J.J., Sustainable contaminated land management: A risk-based land management approach. *Land Contamination and Reclamation,* 2001. **9**(1): pp. 95–100.

Wiedemeier, T.H. and P.E. Haas, Designing monitoring programs to evaluate the performance of natural attenuation. In *Risk, Regulatory, and Monitoring Considerations: Remediation of Chlorinated and Recalcitrant Compounds,* ed. G.B. Wickramanayake, et al. 2000, Columbus: Battelle Press, pp. 357–367.

Wiedemeier, T.H., J.T. Wilson, D.H. Kampbell, R.N. Miller, and J.E. Hansen, *Technical Protocol for Implementing Intrinsic Remediation with Long Term Monitoring for Natural Attenuation of Fuel Contamination Dissolved in Groundwater.* 1995, Air Force Centre for Environmental Excellence.

19 Electrical Resistivity Tomography Monitoring and Modeling of Preferential Flow in Unsaturated Soils

Debao Lu, Chaosheng Zhang, Ajit K. Sarmah, Yinfeng Xia, Nan Geng, James Tsz Fung Wong, Jochen Bundschuh, and Yong Sik Ok

CONTENTS

19.1 INTRODUCTION

Preferential flow (PF) is a common phenomenon of water movement in unsaturated soils which is generally an unsteady seepage inside the large pores or fissures caused by biological, physical, and chemical factors. PF often occurs with potential impacts on ecological and public health because it can promote pollutant transportation into groundwater without any chemical or biological mitigation in the soil layers. This is prevalent both for water and solute movement within soils (Hendrickx and Flury, 2001). With increasing recognition of the significance of PF, many studies have been conducted during the last few decades associated with its characterization in unsaturated soils. The research on PF consists of predictive equations that forecast water flow and solute migration in porous media (Šimůnek et al., 2003), as well as techniques for monitoring soil structure, breakthrough curves (BTC) that describe soil properties quantitatively, and determination of water movement (Allaire et al., 2009). The most widely used model theories related to PF in structured soils can be

summarized into three categories, namely continuity model theory, discrete model theory, and random fractal model theory (Tompson and Gelhar, 1990; Witten Jr and Sander, 1981; Sheng et al., 2009; Kettering et al., 2013). The techniques commonly applied for PF monitoring include the dye tracing method, non-invasive methods, and geophysical methods. The purpose of this chapter is to review the existing model theories and monitoring technique ERT, which is widely used in laboratory and field, thus to provide pros and cons of the existing model and the information which affects development of PF. Although significant development has been achieved in PF modeling and monitoring techniques through a large number of indoor and outdoor experiments, due to the great variability of the PF in temporal and spatial, it is still impossible to obtain its three-dimensional characteristic parameters accurately. It makes the quantitative description of the PF process difficult. Therefore, in the current and future, research on the PF should focus on a large number of field comprehensive experiments, and obtain enough data to conduct quantitative research on PF. And geophysical tomography technology should be combined to further understand the characteristics of soil PF system, so as to carry out reliable simulation and monitoring of the soil hydrology system.

19.2   MAIN TYPES OF PREFERENTIAL FLOW

The main types of PF include macropore flow and finger flow, which are shown conceptually in Figure 19.1.

19.2.1   Macropore Flow

Macropore flow refers to the phenomenon of inhomogeneous water movement via infiltration whereby water and solutes are rapidly transported to deeper soil profiles and eventually into groundwater directly through macropores, without passing through the soil matrix (Allaire et al., 2009). More than a century ago, a study was carried out to explore the relationship between rainfall and drainage, to determine what is now called "macropore flow" (Lawes, 1882). It was found that a large proportion of water applied to the soil surface would migrate through macropores directly and rapidly. That quick flow could be considered as the first discovery of macropore flow, as well as preferential flow. Since then, macropore flow has been observed in many indoor and outdoor experiments (Germann and Beven, 1981; Weiler and Naef, 2003), but did not receive much attention at the time until it was recognized that macropore flow could cause a serious threat to the living environment of human beings.

Usually, the macropore structure in field soil includes soil cracks and fissures, soil pipelines, wormholes, and larger animal burrows. The diameter of the macropores resulting from ant activities range from 2 to 50 mm and the depth of earthworm related pores could reach 1.2–1.4 m (Moran et al., 1988, 1989). The fissures, which mainly exist in the clay pan, are often caused by freezing-thawing and drying-wetting changes. The existence of fissures destroys the integrity of the soil body resulting in heterogeneity. Thus, obvious differences can be found in the horizontal and vertical permeability coefficients, the difference of initial infiltration rate and the

stable infiltration rate could reach up to 1–2 orders of magnitude (Barenblatt et al., 1960). At present, the definitions of soil macropores and pore size ranges have not been unified in the literature (Koestel and Larsbo, 2014). According to previous studies (Beven and Germann, 1982; Cey and Rudolph, 2009; Chen and Wagenet, 1992; Witten Jr and Sander, 1981), several methods for the definition of soil macropores have been proposed as follows.

1. Size of macropores (1–3 mm)
2. Pressure required to empty all the water from the macropores (<5 kPa)
3. Water conductivity of the soil (1–10 mm h⁻¹)

19.2.2   Finger Flow

In homogeneous soils, due to the instability of the infiltration wetting front during the developing process, parts of the originally homogeneous wetting front could be diverted to form several columnar flow paths, thus forming PF movement (Scanlon et al., 2002). The PF caused by wetting front instability is called finger flow. Early studies of finger flow were conducted (Miller and Gardner, 1962; Hendrickx and Flury, 2001) to investigate the instability of infiltration wetting fronts via experiments. However, the study of finger flow had not attracted much attention until Hill and Parlange (1972) performed experiments to study systematically the development and amount of finger flow, as well as the migration speed of its wetting front.

Finger flow often occurs in two-layer soils in which a fine-grained soil layer covers the coarser-grained soil. The water content of the fine-grained soil layer is higher than that of the coarser-grained soil layer assuming they are both subject to the same matric suction. Therefore, the infiltration water gathers above the interface of the two soil layers, making the infiltration wetting front unstable and resulting in the formation of finger flow in the coarser-grained soil layer. Further experimental studies (Liu et al., 1994; Clark et al., 2015) showed that finger flow could also form in homogeneous soils with infiltration flux lower than the saturated hydraulic conductivity of soils, instead of being confined only to the two-layer soil. Liu et al. (1994) conducted an experiment in a glass chamber to determine the influences of initial water moisture on finger width. However, finger flow was found to occur in homogeneous silica sand. Subsequent experiments and analyses revealed that in both of the soil medium the heterogeneity of soil media and the instability of the wetting front (the soil could be homogeneous) could lead to the formation of finger flow phenomena. The finger flow formed in a two-layer soil is usually observed in clay soil with small or medium pores, while the finger flow formed in a homogeneous soil is often observed in water-repellent sandy soil. It was reported by Raats (1973) that the water repellency of soil is responsible for the formation of preferential flow paths. Raats (1973) also found that unstable infiltration wetting fronts resulting from inhomogeneous tension at the water-repellent soil surface

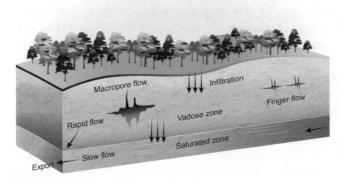

FIGURE 19.1   Conceptual model of macropore flow and finger flow in the unsaturated zone. (Arrows denote the direction of water or solute movement.)

and uneven distribution of capillary force would make moisture and solutes flow in "finger-like" or "tongue-like" forms, resulting in finger flow (Kung, 1990; Ritsema et al., 1993).

Results of some field experiments indicated that the water repellency of soil was also responsible for the instability of the front edge of the water flow. Such instability was found to occur consistently in sandy soil with relatively rough texture where a water-repellent layer could easily form (Xiong, 2014). The water repellency would decrease or even disappear with increasing soil moisture and elevated temperature, and relatively strong water repellency and insignificant hydraulic conductivity could be observed with decreased humidity. These conditions increased the probability of finger flow (De Rooij, 2000; DeBano, 2000; Ritsema and Dekker, 2000). Therefore, there are two main internal drivers for the formation of finger flow: gravity action and the water repellent characteristics of soil. However, the basic principles leading to the occurrence and development of finger flow under different conditions are still not systematically studied. Later studies should focus on analysis how soil characteristics and properties, the geometry of the surface, and the water boundary conditions affect PF processing, accurately measuring the flux of water in the PF channel and determination of the scope and spatial structure of PF.

19.3 FACTORS AFFECTING PREFERENTIAL FLOW

Many factors such as soil structure, texture, initial conditions, infiltration boundary conditions, and unsteady water flow movement could promote the formation of preferential flow and affect its development.

19.3.1 SOIL TEXTURE AND STRUCTURE

Soil macropores formed by the activities of animals (especially soil biota like earthworms), plant root systems, and soil cracks are important paths of preferential flow. Therefore, opening degree, bending degree, and connecting degree of the macropores have significant effects on the formation and development of soil preferential flow (Bishop et al., 2015; Pires et al., 2017; Zhang et al., 2015). Macropore structure plays a major role in the formation and development of the soil preferential flow, which could be summarized as follows (Katuwal et al., 2015; Sheng et al., 2014):

1. Macropores have a capillary shielding effect on the soil-water flow. Therefore, water and solutes could only move along the macropore rather than entering the regular soil matrix
2. Macropores separate the matrix flow from the preferential flow. They restrict stable matrix flow, while supporting formation of high-speed unstable flows at the same time
3. Soil texture usually has a big influence on the development of PF. In sandy soil and others of coarse

texture, the preferential flow could be affected by repellency of soil. Some studies (Seven and Germann, 1981; Clark et al., 2015) have demonstrated that macropore flow often occurs in silt soil and clay soil; unsaturated gravity-driven flow and finger flow often occurs in sandy soil or fine water-repellent soil; and funnel flow usually occurs in fine soil profile mixed with one or more coarse oblique layers

19.3.2 INITIAL AND BOUNDARY CONDITIONS OF INFILTRATION

The initial water content of soil mainly influences the infiltration depth of soil-water and the non-uniformity degree of preferential flow. Merdun et al. (2008) revealed that higher non-uniformity of preferential flow, and more obvious preferential migration, could be achieved under relatively low initial water content. De Rooij (2000) demonstrated that the effect of the initial water content on the preferential flow was obvious only when the initial water content was very low.

Usually, when the irrigation or rainfall intensity exceeds the infiltration rate of soil-water, PF would take place which could be affected greatly by different boundary conditions (e.g., water head, discharge). However, there are different views on how boundary conditions affect preferential flow. Ghodrati and Jury (1990) performed two experiments, respectively, on single ponding infiltration and sprinkler irrigation in loamy-texture sandy soil, showing that the infiltration depth of the stain in the condition of ponding was lower than that under the condition of sprinkler irrigation. Ren et al. (1996) showed that the solute migration velocity under the condition of sprinkler irrigation was higher than that under the condition of ponding. Sheng et al. (2009) pointed out that the non-uniform characteristic of the preferential flow presented regular changes with the increase in the amount of infiltrating water through tracer tests. All their experiments illustrated that when the infiltration capacity is very small, the movement of the non-uniform flow will terminate before it is getting fully developed, making a relatively uniform flow state. When the infiltration capacity volume is getting big, the non-uniform flow will fully develop in both the horizontal and vertical directions, which will increase the degree of non-uniformity of the flow. As the infiltration capacity is sufficiently increased, the flow as a whole becomes relatively uniform due to the expansion and coupling of the preferential flow channels in the lateral direction.

19.3.3 INSTABILITY OF WATER MOVEMENT IN SOIL

In non-structural soils, even if there is no macropore, finger-like preferential flow paths can still be formed in the soil due to the unstable wetting front. Raats (1973) pointed out that when the pressure gradient of the front was opposite to that of the water flow direction, the wetting front would be instable. Philip (1975) further demonstrated that if the migration velocity of the wetting front increased with the infiltration depth, the microdisturbance in the infiltration process would make the front unstable and the wetting front would grow up and eventually become preferential flow.

Nguyen et al. (1999) demonstrated that the finger flow would occur under one or more of the following conditions: (1) the soil was difficult to be wetted and was very dry; (2) the hydraulic conductivity increased with soil depth; and (3) the air pressure increased ahead of the wetting front. Nevertheless, these factors were necessary but not sufficient for the generation of the finger flow. Another condition was the hysteresis of soil-water retention curve. Ritsema et al. (1998) showed that the finger flow was formed by the hysteresis of the soil-water retention curve, and the characteristics of the finger flow were subjected to its soil-water retention curve.

19.4   MODEL THEORY OF PREFERENTIAL FLOW

19.4.1   Continuity Model Theory

The continuity model theory was established based on experimental scale, and the homogeneous medium hypothesis is the main theoretical approach used to describe soil-water movement and solute transfer. According to the continuity model theory, soil hydrodynamic parameters are the keys to influencing and controlling water and solute migration and transfer. Many studies (Bellin et al., 1992; LaBolle et al., 1996; Simmons et al., 2001; Shope et al., 2013, 2014) have been carried out to analyze the spatial variation of hydrodynamic parameters to simulate the movement and transfer of water and solutes under the condition of medium spatial variation. Stochastic simulation methods have also been used to take the anisotropy of nature directly into account to simulate the movement of water flowing through non-uniform media. A key step in the application of the random simulation methods, such as the Monte Carlo method (Sprenger et al., 2016; Vanderborght and Vereecken, 2007), was to determine the statistics of transfer parameters and soil properties of the studied units, namely, the mean, variance, and variation in the spatial structure. Once the structural parameters were determined, the random distribution field of hydrodynamic force would be generated by conditional or non-conditional means. These results are used as the input parameters for the numerical model to simulate the movement of soil-water and solutes in non-uniform media, and to analyze the statistical characteristics of the flow. A large number of experiments and theoretical analyses (García et al., 2014; Peck et al., 1977) have been carried out to reveal the spatial variation and scale characteristics of soil hydrodynamic parameters. However, measurement of soil hydrodynamic parameters under field conditions is rather challenging, expensive, and time consuming, and it is even more difficult to obtain the data needed to inform hydrodynamic parameters at a larger scale.

In the continuity model, the geometric dimension remains the same over the whole of the porous medium and the grid scale is fine. When the 3D model is applied, the geometric dimension is artificially set to "3." However, for a separate preferential flow path, the geometric dimension is close to "1," resulting in the model simulation flow velocity being smaller than the actual flow velocity. This in turn results in increased deviation of the simulation, regarding the characteristics of high-speed water movement within the preferential flow paths. To overcome this defect, a few studies were conducted that revealed that structural pores in the soil were the main factor causing rapid fluid movement (Jarvis, 2007; Köhne et al., 2009). It could be simulated by applying multiporosity models each with a specific flow law (Lewandowska et al., 2004). Otherwise, the study areas could be divided into a preferential flow areas consisting of pores, macropores, and fissures between soil aggregates, and a matrix area of consistently fine pores, and the two areas interacted. The actual flow movement was thought to reflect the coupling of the preferential flow movement and the matrix flow movement, coexisting in the same position to some extent.

To date, Darcy's law and Richard's equation are generally used to calculate the water flow in the matrix area, but no consensus has been reached on describing the water flow in the preferential flow area within the soil matrix. Several researchers have proposed different simplified models and empirical formulae for the water flow movement in the preferential area (Gerke, 2006; Köhne et al., 2009). However, due to the large number of parameters and the lack of standard measurement methods, these models and formulae are still confined to theoretical studies based on controlled indoor conditions (Clark et al., 2015). For example, the van Genuchten-Mualem model (Ippisch et al., 2006; Wallor et al., 2018) utilizes parameters for which it is hard to get data to account for non-capillary flow, and composite hydraulic functions with limited ability for making non-equilibrium predictions. In addition, current continuity models do not explain or predict well the formation and development of finger flow.

Fluid movement in soils involves anisotropy, scale effects, and nonlinearity (instability). For these reasons, the theory and methods adopted by the traditional continuity model for hypothesizing the transfer of soil-water and solutes in homogeneous media are inapplicable to the problem of water flow in inhomogeneous media. Compared with the simulation results of the continuity model, the actual water flow movement and solute transfer in inhomogeneous media exhibit more complex characteristics. Therefore, the continuity model theory cannot be used to describe the behavior of water flow and solute transfer in media at field scale or at a larger observation scale. For example, Wood et al. (2004) suggested that at observation scales of 100 m^2 or greater, non-uniformity and nonlinearity (or instability) of dynamic transmission would be evident in the flow.

19.4.2   Discrete Models

Due to the problems in application of the continuity model, the discrete model has gradually become more often used to describe the movement of water flowing through inhomogeneous media. Unlike the continuity model, the discrete model assumes a water body in soils with a particle-like structure containing certain information and a certain shape. Through simple movement rules, complex spatial distributions can be generated by these "particle" structures (Köhne et al., 2009; Fenicia et al., 2016). Statistically, the distribution patterns produced and the patterns of non-uniform movement of soil-water are similar to each other (to at least some extent). Both the diffusion limited aggregation (DLA) model (Sander, 2000) and invasion percolation (IP) model (Castellano et al., 2009) incorporate these assumptions.

A typical feature of the DLA model is the ability to produce clusters with fractal characteristics. The initial DLA model took the point source as the source particle; then Meakin (1983) introduced the line source DLA model which, in current studies, has mostly been used to simulate the boundary conditions of solute migration tests. By changing the walking probability of particles in different directions, the DLA model could simulate both uniformly distributed chemical substances (e.g., piston flow) and complex distributions triggered by macropores and finger flow (Flury et al., 1995). This model has been successfully applied to describe the test results of preferential flow formation. In addition, the DLA model has been used for viscous fingering analysis of miscible displacement regarding hydrological dynamics in unsaturated soils (Wiekenkamp et al., 2016).

The invasion percolation (IP) model, first proposed by Wilkinson and Willemsen (1983), has been applied to model the permeability of a wetted zone with stable pressure potential. Nooruddin and Blunt (2018) proposed a large-scale model to capture a capillary-controlled Darcy-scale flow in terms of the IP model. A modified invasion percolation (MIP) model was adopted by Glass and Yarrington (1996) to simulate the formation of gravity-driven finger flow and the structure of a wetting front under very small flow velocity, as well as the consideration of gravity only (viscous force neglected) at pore scale. A quick-run algorithm was presented (Masson, 2016) for modeling invasion percolation (IP) with trapping. With the updated algorithm, the computation time was greatly reduced. However, there is no obvious length scale in the IP model when simulating finger flow, while the length scale can be defined sequentially by the width of the finger flow in the MIP model when simulating gravity-driven finger flow. Models based on the IP theory, such as the cellular automata-dynamic model and lattice structure gas model, have also been applied to the study of water flow and solute transfer (Jin et al., 2016; Liu et al., 2016).

At present, the DLA model has been successfully applied to describe experimental results at small scale, but the computational workload of the DLA model is too heavy to solve problems at a large scale. At the same time, a complete theoretical basis is lacking when using the DLA model to describe the flow of water and solute transfer through unsaturated soils. The IP model has mainly been used to describe the fluid movement process at pore scale, for which it has achieved a certain degree of success. However, the physical meaning of this model remains implicit (Köhne et al., 2009), so it is still incapable of solving practical problems at a large scale.

19.4.3 Fractal Models

Fractals are a universal feature of natural phenomena. They are often used to describe and simulate naturally occurring objects using fractal dimensions and mathematical methods. Studies showed that preferential flow had obvious fractal characteristics. For instance, Flury and Flühler (1995) used the DLA model to simulate a solute movement pattern. The results suggested that not only did preferential flow show fractal characteristics, but also that the parameters describing the fractal characteristics were presented with certain regularity. According to one observation (Öhrström et al., 2002), fractal characteristics were observed regarding the path of movement of a stain in soil. Sheng et al. (2009) carried out a series of dye-tracer tests of preferential flow through sandy soil, in which the flow pattern was simulated using fractal theory. The results also revealed that the parameters used to describe the preferential flow characteristics showed certain regular changes associated with the infiltration conditions. Van Genuchten et al. (2004) pointed out that the discrete model could capture detailed characteristics of the preferential flow because the discrete model could produce fractal (multifractal) structure similar to the preferential flow. On this basis, an active fractal model (AFM) was established for describing the process of water and solute transfer in fractured media (Liu et al., 2003).

In order to describe the process of water flow and solute movement in unsaturated zones of soils, an active region model (ARM) (Liu et al., 2005) was established based on AFM, according to the characteristics of the soil medium and fluid movement. The parameters of ARM were functions of the fractal dimension of preferential flow movement; thus, the detailed (fractal) features of the preferential flow became discernible. Sheng et al. (2009, 2014) demonstrated that the overall non-uniform information of the preferential flow movement under a variety of infiltration conditions had been captured effectively with ARM. They also noted that the ARM parameters (which could be used to describe and compare non-uniform features of the preferential flow) also exhibited certain scale invariance. Luo et al. (2018) applied ARM to illustrate that roots are of great importance in processing of solute leaching in soils.

However, until now, the fractal properties of soil preferential flow have not been well developed. Future studies should focus on establishing a theoretical basis for the fractal model, the determination of model parameters, the relationship between the model parameters and the properties of soil media, and the effects of the model on the description of non-uniform flow. Table 19.1 gives a summary of these three theories for modeling PF, as well as their advantages and disadvantages.

TABLE 19.1

Comparison of Three Theories for Preferential Flow Modeling

Model Theory	Advantages	Disadvantages
Continuity model theory	Take spatial variation and scale characteristics into account	Difficulties in obtaining massive parameters Hard to be used in the field Only suitable for macropore flow
Discrete model theory	Can be used to model complex distributed macropore flow and finger flow	Too much calculation Hard to apply in the field Short of systematic theory
Fractal model theory	Able to capture the details of preferential flow	Related theories need to be improved

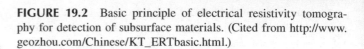

FIGURE 19.2 Basic principle of electrical resistivity tomography for detection of subsurface materials. (Cited from http://www.geozhou.com/Chinese/KT_ERTbasic.html.)

19.5 ELECTRICAL RESISTIVITY TOMOGRAPHY FOR MONITORING PF

Direct measurement and monitoring techniques usually play a big role in calibration and verification of the numerical models. The common techniques used in a laboratory and the field could be classified as dye tracing, micro-tension measurement, and geophysical methods. Advanced technologies (i.e., pertinent to geophysical methods) have enabled us to characterize transport of water and solutes inside the soil with minimal disturbance. The most widely used techniques (e.g., breakthrough curves, dye tracing, and scanning techniques) were already critically reviewed by Allaire et al. (2009). Therefore, this review focuses on the laboratory and field applications of electrical resistivity tomography (ERT). This technology has shown great potential for assessing and quantifying PF as a non-invasive, effective, and efficient tomography tool. ERT method has a flexible scale advantage in PF detection. By adjusting the spacing of the measuring electrodes, the resolution is measured in centimeters or meters. In addition, ERT equipment is relatively inexpensive compared to other specialized detection devices.

19.5.1 THE BASIC CONCEPT FOR ERT MONITORING OF PF

ERT is a tool for characterizing the sub-surface medium according to its electrical properties. The basic principle of electrical resistivity works well for identification of PF paths. Underground fractures and holes cause relative changes in the electrical resistivity that are sensitive to water content, soil porosity, and some other hydraulic parameters (Coscia et al., 2011; Lu et al., 2015). Figure 19.2 shows the basic principle of the high-density electrical resistivity tomography for detection of subsurface materials.

As revealed in the figure, an electric field is established through the direction of current into the ground, and the resulting potential is obtained using counter-electrodes. By placing more electrodes at the measuring points, the data can be quickly and automatically collected using a programmable electrode transfer switch and electrical measuring instrument. Then, the collected data are transferred to a computer for data interpretation. The interpretation of high-density electrical data mainly involves statistical processing, filtration, and inversion. The statistical processing is needed to analyze and evaluate the measured data, and the filtering process is needed to eliminate unreasonable singular values to make the vertical and horizontal transformation trends of the apparent resistivity data fall within a reasonable scale. The inversion process is needed to obtain the dielectric resistivity distribution according to the apparent resistivity, which is the core step of the data processing and interpretation. The most commonly used inversion method is the least squares optimization method. It uses the measured apparent resistivity as the initial model of the dielectric resistivity to solve the non-convergence problem during calculation. The resistivity model is then modified iteratively in terms of the model apparent resistivity and the measured data, until the root mean square (RMS) error is minimized. The electrical resistivity distribution data obtained through the inversion processing can be conveniently used for interpretation of subsurface materials and their characteristics (e.g., water content). Meanwhile, some advantages have been shown in previous research indicating that high-density ERT is capable of PF monitoring. These advantages include:

1. The layout of electrodes is completed all at one time. This not only reduces the malfunctions and interference caused by the electrodes setting, but also makes possible the rapid, long-term, automatic field measurements.
2. Scanning measurements can be performed efficiently using a variety of electrode arrangements, so that a wealth of subsurface information regarding soil structure characteristics can be obtained.
3. The data can be preprocessed and a profile can be displayed visually, allowing various maps of results to be automatically drawn and printed.
4. Compared with traditional methods, ERT is not only more efficient and cheaper, but also provides significant improvement in the capability for detection.

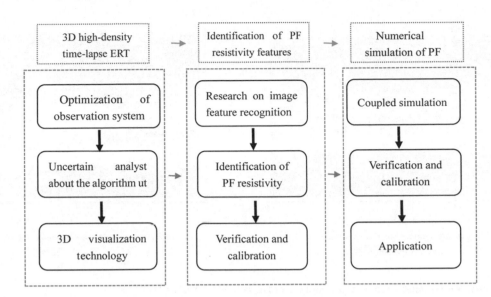

FIGURE 19.3   Flowcharts for how ERT is used to qualify the preferential flow in unsaturated soil.

Figure 19.3 shows simply how ERT is used to qualify the preferential flow in unsaturated soils. Because the measurement of PF systems requires a significant time span and high accuracy, it is necessary to establish a three-dimensional ERT dynamic monitoring system.

This is done in terms of the characteristics of PF by analyzing and calculating the sensitivity of the ERT measurement configuration. Based on this, the characteristics of PF paths (i.e., shape and size) can be identified and extracted using quantitative image identification technology. These results should be verified using the outcomes obtained from other methods that indicate the known structure. After verification, the PF characteristics can be used as the input for preferential flow prediction.

19.5.2 Main Factor Influencing ERT Monitoring of PF

The main factor influencing the use of ERT for PF monitoring might be the sensitivity of the measurement configuration. As it was reported in some studies (Günther et al., 2006), the regions where the absolute value of the sensitivity is higher will show greater responses when there is a change in the soil properties.

Obviously, abnormal responses are more likely in this case. However, in regions where the absolute value of the sensitivity is low, the resistivity response is weak, and detection of changes is also weakened. This means that any changes in the soil properties could not be detected easily. For devices with different specific electrode arrangements, the sensitivity distribution characteristics are also different. A difference in the nature, shape, or spatial position of PF could be regarded as a relative change of the resistivity in the measurement region. Therefore, it is of great significance to improve the understanding of the measurement configuration and to study the response to PF path by analyzing the sensitivity distribution characteristics of the different arrays. Commonly used configurations, such as the pole-pole, Wenner, dipole-dipole, and borehole arrays, were applied in the research calculations on

sensitivity. The sensitivity distribution features of each array can be generated using the codes proposed by Lu et al. (2015).

Figure 19.4 shows the sensitivity distribution for a pole-pole configuration where electrodes are located at the surface. Under ideal conditions, the pole-pole array has only two electrodes, C_1 and P_1. In this case, C_2 and P_2 are generally placed at a distance more than 20 times from C_1 to P_1, which could be considered infinity, so that the measurement is not affected by C_2 and P_2. As shown in Figure 19.4, the area between the C_1 and P_1 is surrounded by a hemispherical negative sensitivity zone, while two positive sensitivity zones appear outside of the two electrodes. The positive sensitivity zone outside encloses the negative sensitivity zone and exhibits a vertical and approximately vertical contour distribution to the ground near the C_1 and P_1 electrodes. Thus, it can be inferred that the pole-pole configuration has better resolution in the horizontal than the vertical direction according to the area of the high-sensitivity zone. Consequently, the configuration could easily detect macropore flow that showed good lateral development.

The sensitivity distribution for the Wenner configuration (involves four electrodes) is also calculated here. Figure 19.5 shows the x–z and x–y sections of the sensitivity distribution located at the surface, which was arranged as: C_1, P_1, P_2, C_2. As is revealed, the near-surface C_1–P_1 and C_2–P_2 pairs are negative sensitivity regions, and the area near the potential electrode pairs P_1–P_2 are positive-sensitivity regions. Generally, the absolute value of the sensitivity increases as it gets closer to the electrode, and the potential electrodes have higher sensitivity to changes in the resistivity. The depth of one-time electrode space zones is a positive value region where the contour shows a parallel, horizontal distribution. The horizontal distribution of sensitivity layers is shown around the electrodes, which implies that the configuration has better resolution for identification of the path with this layer structure than with others. Thus, the Wenner configuration can be used to detect the path of lateral flow in the soil.

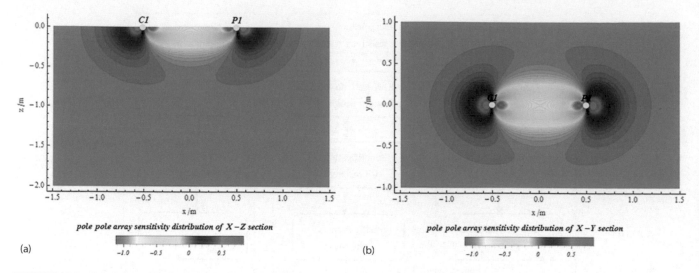

FIGURE 19.4 Sensitivity distribution of a pole-pole array in cross section: (a) *X–Z* and (b) *X–Y*.

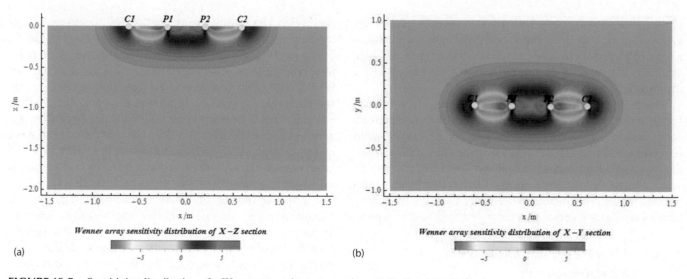

FIGURE 19.5 Sensitivity distribution of a Wenner array in cross section: (a) *X–Z* and (b) *X–Y*.

The dipole-dipole array involves coupled potential and source electrodes at the same intervals. Figure 19.6 shows the distribution of a dipole-dipole array with separate cross sections of *X–Z* and *X–Y*. However, it is different from that of the Wenner and pole-pole arrays described above. The positive and negative sensitivity appears alternatively from C_1 to C_2, inferring a more complex response to the spatial variability of resistivity.

However, the negative sensitivity zone spreads wider in both horizontal and vertical orientations, which causes this configuration to have high sensitivity to the change of resistivity in both directions. Therefore, the dipole-dipole array may be the best choice for the detection of abnormalities that cause big changes in resistivity in both directions.

In contrast to the surface configurations, the cross-hole array is an underground measurement. Because its power supply and potential electrodes are all placed subsurface, it

possesses of a higher-resolution for detecting complex anomalous targets. Hence, as shown in Figure 19.7, the cross-hole measurements have high sensitivity, for which reason, cross hole surveys are widely used. Positive and negative sensitivity scatter zones create a symmetrical spheroid close to the electrodes. A layer-distributed sensitivity distribution appears between the two boreholes and extends only in the vertical direction. Thus, based on its distribution of sensitivity, this method provides considerable potential for the complicated detection of PF path.

19.5.3 MAIN ACHIEVEMENTS OF ERT IN PF MONITORING

Electrical resistivity tomography (ERT) has proven to be a promising non-invasive technique for prospecting of subsurface geological bodies (Shima, 1992; Suzuki and Ohnishi, 1995), assessing the performance of ground structures (Daily and Ramirez, 2000), and imaging water and solute transport

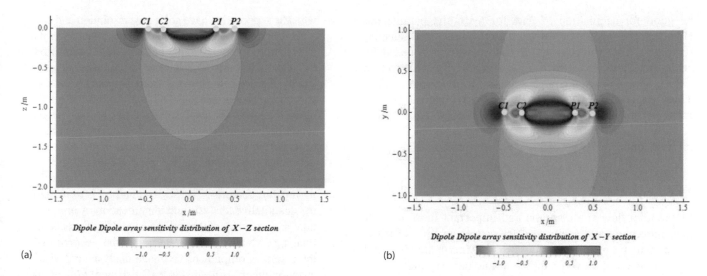

FIGURE 19.6 Sensitivity distribution of a dipole-dipole array in cross section: (a) *X–Z* and (b) *X–Y*.

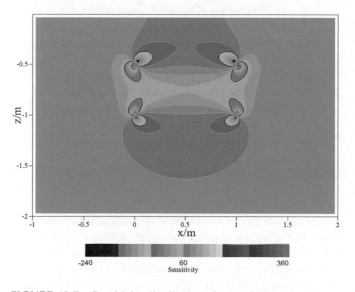

FIGURE 19.7 Sensitivity distribution of a borehole array in cross section: *X–Z*.

through porous media (Binley et al., 1996; Ramirez et al., 1993). Moreover, the technique is also widely applied for gathering data on hydrological parameters in the unsaturated zone (Emch and Yeh, 1998) and in qualifying water content dynamics and tracer breakthrough (Wehrer and Slater, 2015). However, ERT also shows limitations for the identification of the direction of water flow because it has a poor response to liquid flow (Aubert and Atangana, 1996; Revil et al., 2017). Thus, ERT and time domain reflectometry (TDR) have been used to monitor and validate changes in water content in a mountain area (Cassiani et al., 2009; Brunet et al., 2010) and in a cornfield, respectively (Beff et al., 2013).

It has been suggested that one of the ways that solute injection could be applied is to change the electrical resistivity in areas of interest, which would greatly increase the capability for detection of PF movement (Slater et al., 1997; Cassiani et al., 2006; Daily and Ramirez, 1995; Nimmer

et al., 2007). Robert et al. (2012) reviewed several published studies designed to detect water and solute transport in both saturated and unsaturated media under the condition of laboratory and field; however, this aspect is not covered in detail here. The statistics about the type of media, assistance method, and ERT configuration used in the PF research are shown in Figure 19.8 following the work of Robert et al. (2012).

Previous studies clearly show that the ERT is capable to illustrate salt tracer propagation. Most of them adopted the cross-hole ERT method to qualify the preferential flow transport path due to its high vertical resolution. Cross-hole configuration was chosen by Slater et al. (1997), who reported a method to monitor hydrological processes in a limestone environment. The method works best in the detection of hydraulically conductive processes with resistivity changes due to the intrusion of saline water. By using 3-D time lapse electrical resistivity imaging technology, Wehrer and Slater (2015) monitored the infiltration of undisturbed unsaturated soil in a laboratory lysimeter; the results show that the technique can provide quantitative data for the qualifying the transport of water and solutes in unsaturated media. Usually, injected tracers are easier to capture using ERT, which can

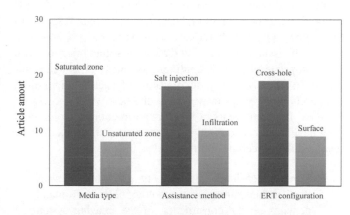

FIGURE 19.8 The statistics about media type, assistance method, and ERT configuration used in PF research.

be good for acquisition of data for hydraulic parameters. However, the method is restricted by the calculated sensitivity matrix. Tracer appearing in a high sensitivity zone can be easily detected, while tracer in low sensitivity areas can be hard to detect or even completely lost (Kemna et al., 2002). Thus, experiments should be prepared with great care to make certain that the tracer propagation occurs in the high sensitivity zone so that it can be detected easily by the ERT instruments.

19.6 SUMMARY

Preferential flow is a common and important form of water flow movement and solute transfer pattern in soils which is difficult to be captured or described. Macropore flow and finger flow are the two preferential flow patterns that are the subjects of most of the studies in the existing literature. Soil structure, texture, initial conditions, infiltration boundary conditions, and unsteady water flow movement are the main factors that promote the formation of preferential flow and affect its development.

To simulate accurately and to predict the spatio-temporal variation characteristics of the soil preferential flow, researchers have put forward many model theories and test methods. Among them, the continuity model theory, discrete model theory, and fractal model theory are the most commonly adopted preferential flow model theories and methods at present.

This review identified the advantages and disadvantages of each model.

Continuity models show potential in PF describing by taking spatial variation and scale character into account while they are limited in describing finger flow. The reliance on parameters which are too hard to get makes it difficult to apply widely in the field. In contrast, the discrete models are useful for both macropore flow and finger flow but are restricted by the need for major calculation resources and lack of a systematic theory. It seems that fractal models might provide the main direction for development in the future, but they are still under development, and relevant theories need to be completed. ERT is a powerful tool for characterizing PF according to its electrical properties. However, it also shows a limitation for the identification of water flow direction because it has a poor response to liquid flow. Based on that, the conclusion on the proper way to resolve the issue appears to be the use of solute injection with an appropriate choice of configuration array. Through comparison of the sensitivity distributions of each configuration, the borehole array has proven to be the best way to access PF features. In recent years, a lot of experiments on preferential flow have been carried out, but some issues still remain to be resolved. These can be summarized as follows:

1. Three-dimensional high-density cross-hole resistivity time-lapse imaging technology: The issue mainly depends on the construction of the imaging system, which is supposed to have good time resolution and can quickly and accurately obtain the required data in terms of variation of the soil in the field. During the inverted calculation, it is not only necessary to compare the sensitivity distribution obtained from homogeneous soil with that of heterogeneous soil, and to find the maximum tolerance for application of the homogeneous sensitivity under heterogeneous conditions. It is also necessary to establish a suitable initial model to solve uncertain problems during the inverted calculation.

2. Quantitative identification technology for PF resistivity imaging features: The key technology is to realize the quantitative analysis and identification of imaging outcomes in terms of three-dimensional high-precision imaging, which directly determines the accuracy of the results. Comprehensive image analysis and visualization theory is needed to establish an optimization algorithm to achieve sufficiently accurate identification and extraction of the PF characteristics.

3. The numerical simulation methods play a significant role in the description of hydrologic processes under unsaturated soils. At present, a natural progression of this work would be to update the models by combining existing multiple-porosity/permeability models with the PF characteristics extracted using ERT. Then studies need to be carried out to validate the accuracy of the updated models.

Overall, the model theory and observation technology of soil preferential flow research are far from being perfect due to numerous formation and influencing factors, various patterns of manifestation, obvious rapid non-equilibrium characteristics, and the highly heterogeneous characteristics of the soil. Therefore, the main issues requiring urgent resolution for the study of preferential flow are to establish uniform criteria for soil preferential flow, to improve the effectiveness of preferential flow model theory, and to develop special technical equipment for the observation and measurement of preferential flow.

19.7 CONCLUSIONS

PF is common phenomenon of water movement that is ubiquitous in soil, which plays a vital role in the transportation of water and pollutants from the surface to the ground. Many scholars have gained a lot of achievement of the PF through a large number of indoor and outdoor experiments; however, due to the great variability of the PF in time and space, it is still impossible to accurately obtain its three-dimensional characteristic parameters. Therefore, in the current and future, long-term, research on the PF issue should focus on a large number of field comprehensive experiments, and obtain enough data to perform quantitative research on the PF. And numerical simulation could be conduct coupling with the parameters which are obtained by using 3-D ERT survey.

ACKNOWLEDGMENT

This work was carried out with the support of "Cooperative Research Program for Agriculture Science and Technology Development (Effects of Plastic Mulch Wastes on crop productivity and Agro-Environment, Project No. PJ01475801)," Rural Development Administration, Republic of Korea, and the National Research Foundation (NRF) of Korea, and Germany–Korea Partnership (GEnKO) Program (2018–2020).

REFERENCES

Allaire SE, Roulier S, Cessna AJ. 2009. Quantifying preferential flow in soils: A review of different techniques. *Journal of Hydrology* **378**: 179–204.

Aubert M, Atangana QY. 1996. Self-potential method in hydrogeological exploration of volcanic areas. *Groundwater* **34**: 1010–1016.

Barenblatt GI, Zheltov IP, Kochina IN. 1960. Basic concepts in the theory of seepage of homogeneous liquids in fissured rocks [strata]. *Journal of Applied Mathematics and Mechanics* **24**: 1286–1303.

Beff L, Günther T, Vandoorne B, Couvreur V, Javaux M. 2013. Three-dimensional monitoring of soil water content in a maize field using electrical resistivity tomography. *Hydrology and Earth System Sciences* **17**: 595.

Bellin A, Salandin P, Rinaldo A. 1992. Simulation of dispersion in heterogeneous porous formations: Statistics, first-order theories, convergence of computations. *Water Resources Research* **28**: 2211–2227.

Beven K, Germann P. 1982. Macropores and water flow in soils. *Water Resources Research* **18**: 1311–1325.

Binley A, Henry-Poulter S, Shaw B. 1996. Examination of solute transport in an undisturbed soil column using electrical resistance tomography. *Water Resources Research* **32**: 763–769.

Bishop JM, Callaghan MV, Cey EE, Bentley LR. 2015. Measurement and simulation of subsurface tracer migration to tile drains in low permeability, macroporous soil. *Water Resources Research* **51**: 3956–3981.

Brunet P, Clément R, Bouvier C. 2010. Monitoring soil water content and deficit using electrical resistivity tomography (ERT)–A case study in the Cevennes area, France. *Journal of Hydrology* **380**: 146–153.

Cassiani G, Bruno V, Villa A, Fusi N, Binley AM. 2006. A saline trace test monitored via time-lapse surface electrical resistivity tomography. *Journal of Applied Geophysics* **59**: 244–259.

Cassiani G, Kemna A, Villa A, Zimmermann E. 2009. Spectral induced polarization for the characterization of free-phase hydrocarbon contamination of sediments with low clay content. *Near Surface Geophysics* **7**: 547–562.

Castellano C, Fortunato S, Loreto V. 2009. Statistical physics of social dynamics. *Reviews of Modern Physics* **81**: 591.

Cey EE, Rudolph DL. 2009. Field study of macropore flow processes using tension infiltration of a dye tracer in partially saturated soils. *Hydrological Processes* **23**: 1768–1779.

Chen C, Wagenet RJ. 1992. Simulation of water and chemicals in macropore soils Part 1. Representation of the equivalent macropore influence and its effect on soilwater flow. *Journal of Hydrology* **130**: 105–126.

Clark MP, Fan Y, Lawrence DM, Adam JC, Bolster D, Gochis DJ, Hooper RP, Kumar M, Leung LR, Mackay DS. 2015. Improving the representation of hydrologic processes in Earth System Models. *Water Resources Research* **51**: 5929–5956.

Coscia I, Greenhalgh SA, Linde N, Doetsch J, Marescot L, Günther T, Vogt T, Green AG. 2011. 3D crosshole ERT for aquifer characterization and monitoring of infiltrating river water. *Geophysics* **76**: G49–G59.

Daily W, Ramirez A. 1995. Electrical resistance tomography during in-situ trichloroethylene remediation at the Savannah River Site. *Journal of Applied Geophysics* **33**: 239–249.

Daily W, Ramirez AL. 2000. Electrical imaging of engineered hydraulic barriers. *Geophysics* **65**: 83–94.

De Rooij GH. 2000. Modeling fingered flow of water in soils owing to wetting front instability: A review. *Journal of Hydrology* **231**: 277–294.

DeBano LF. 2000. Water repellency in soils: A historical overview. *Journal of Hydrology* **231**: 4–32.

Emch PG, Yeh WW. 1998. Management model for conjunctive use of coastal surface water and ground water. *Journal of Water Resources Planning and Management* **124**: 129–139.

Fenicia F, Kavetski D, Savenije HH, Pfister L. 2016. From spatially variable streamflow to distributed hydrological models: Analysis of key modeling decisions. *Water Resources Research* **52**: 954–989.

Flury M, Flühler H. 1995. Modeling solute leaching in soils by diffusion-limited aggregation: Basic concepts and application to conservative solutes. *Water Resources Research* **31**: 2443–2452.

Flury M, Leuenberger J, Studer B, Flühler H. 1995. Transport of anions and herbicides in a loamy and a sandy field soil. *Water Resources Research* **31**: 823–835.

García GM, Pachepsky YA, Vereecken H. 2014. Effect of soil hydraulic properties on the relationship between the spatial mean and variability of soil moisture. *Journal of Hydrology* **516**: 154–160.

Gerke HH. 2006. Preferential flow descriptions for structured soils. *Journal of Plant Nutrition and Soil Science* **169**: 382–400.

Germann P, Beven K. 1981. Water flow in soil macropores III. A statistical approach. *Journal of Soil Science* **32**: 31–39.

Ghodrati M, Jury WA. 1990. A field study using dyes to characterize preferential flow of water. *Soil Science Society of America Journal* **54**: 1558–1563.

Glass RJ, Yarrington L. 1996. Simulation of gravity fingering in porous media using a modified invasion percolation model. *Geoderma* **70**: 231–252.

Günther T, Rücker C, Spitzer K. 2006. Three-dimensional modelling and inversion of DC resistivity data incorporating topography—II. Inversion. *Geophysical Journal International* **166**: 506–517.

Hendrickx JM, Flury M. 2001. Uniform and preferential flow mechanisms in the vadose zone. *Conceptual Models of Flow and Transport in The Fractured Vadose Zone* **4**: 149–187.

Hill DE, Parlange J-Y. 1972. Wetting front instability in layered soils 1. *Soil Science Society of America Journal* **36**: 697–702.

Ippisch O, Vogel H-J, Bastian P. 2006. Validity limits for the van Genuchten–Mualem model and implications for parameter estimation and numerical simulation. *Advances in Water Resources* **29**: 1780–1789.

Jarvis NJ. 2007. A review of non-equilibrium water flow and solute transport in soil macropores: Principles, controlling factors and consequences for water quality. *European Journal of Soil Science* **58**: 523–546.

Jin C, Langston PA, Pavlovskaya GE, Hall MR, Rigby SP. 2016. Statistics of highly heterogeneous flow fields confined to three-dimensional random porous media. *Physical Review E* **93**: 013122.

Katuwal S, Norgaard T, Moldrup P, Lamandé M, Wildenschild D, de Jonge LW. 2015. Linking air and water transport in intact soils to macropore characteristics inferred from X-ray computed tomography. *Geoderma* **237**: 9–20.

Kemna A, Vanderborght J, Kulessa B, Vereecken H. 2002. Imaging and characterisation of subsurface solute transport using electrical resistivity tomography (ERT) and equivalent transport models. *Journal of Hydrology* **267**: 125–146.

Kettering J, Ruidisch M, Gaviria C, Ok YS, Kuzyakov Y. 2013. Fate of fertilizer 15 N in intensive ridge cultivation with plastic mulching under a monsoon climate. *Nutrient cycling in Agroecosystems* **95**: 57–72.

Koestel J, Larsbo M. 2014. Imaging and quantification of preferential solute transport in soil macropores. *Water Resources Research* **50**: 4357–4378.

Köhne JM, Köhne S, Šimůnek J. 2009. A review of model applications for structured soils: a) Water flow and tracer transport. *Journal of Contaminant Hydrology* **104**: 4–35.

Kung KS. 1990. Preferential flow in a sandy vadose zone: 1. Field observation. *Geoderma* **46**: 51–58.

LaBolle EM, Fogg GE, Tompson AF. 1996. Random-walk simulation of transport in heterogeneous porous media: Local mass-conservation problem and implementation methods. *Water Resources Research* **32**: 583–593.

Lawes JB. 1882. *On the Amount and Composition of the Rain and Drainage-Waters Collected at Rothamsted.* W. Clowes and Sons London, UK.

Lewandowska J, Szymkiewicz A, Burzyński K, Vauclin M. 2004. Modeling of unsaturated water flow in double-porosity soils by the homogenization approach. *Advances in Water Resources* **27**: 283–296.

Liu H, Kang Q, Leonardi CR, Schmieschek S, Narváez A, Jones BD, Williams JR, Valocchi AJ, Harting J. 2016. Multiphase lattice Boltzmann simulations for porous media applications. *Computational Geosciences* **20**: 777–805.

Liu H-H, Zhang G, Bodvarsson GS. 2003. The active fracture model: Its relation to fractal flow patterns and an evaluation using field observations. *Vadose Zone Journal* **2**: 259–269.

Liu H-H, Zhang R, Bodvarsson GS. 2005. An active region model for capturing fractal flow patterns in unsaturated soils: Model development. *Journal of Contaminant Hydrology* **80**: 18–30.

Liu Y, Steenhuis TS, Parlange J. 1994. Closed-form solution for finger width in sandy soils at different water contents. *Water Resources Research* **30**: 949–952.

Lu D-B, Zhou Q-Y, Junejo SA, Xiao A-L. 2015. A systematic study of topography effect of ERT based on 3-D modeling and inversion. *Pure and Applied Geophysics* **172**: 1531–1546.

Luo Z, Niu J, Zhang L, Chen X, Zhang W, Xie B, Du J, Zhu Z, Wu S, Li X. 2018. Roots-enhanced preferential flows in deciduous and coniferous forest soils revealed by dual-tracer experiments. *Journal of Environmental Quality.* doi:10.2134/jeq2018.03.0091.

Masson Y. 2016. A fast two-step algorithm for invasion percolation with trapping. *Computers & Geosciences* **90**: 41–48.

Meakin P. 1983. Diffusion-controlled deposition on fibers and surfaces. *Physical Review A* **27**: 2616.

Merdun H, Meral R, Demirkiran AR. 2008. Effect of the initial soil moisture content on the spatial distribution of the water retention. *Eurasian Soil Science* **41**: 1098–1106.

Miller DE, Gardner WH. 1962. Water Infiltration into Stratified Soil 1. *Soil Science Society of America Journal* **26**: 115–119.

Moran CJ, Koppi AJ, Murphy BW, McBratney AB. 1988. Comparison of the macropore structure of a sandy loam surface soil horizon subjected to two tillage treatments. *Soil Use and Management* **4**: 96–102.

Moran CJ, McBratney AB, Koppi AJ. 1989. A rapid method for analysis of soil macropore structure. I. Specimen preparation and digital binary image production. *Soil Science Society of America Journal* **53**: 921–928.

Nguyen HV, Nieber JL, Ritsema CJ, Dekker LW, Steenhuis TS. 1999. Modeling gravity driven unstable flow in a water repellent soil. *Journal of Hydrology* **215**: 202–214.

Nimmer RE, Osiensky JL, Binley AM, Sprenke KF, Williams BC. 2007. Electrical resistivity imaging of conductive plume dilution in fractured rock. *Hydrogeology Journal* **15**: 877–890.

Nooruddin HA, Blunt MJ. 2018. Large-scale invasion percolation with trapping for upscaling capillary-controlled Darcy-scale flow. *Transport in Porous Media* **121**: 479–506.

Öhrström P, Persson M, Albergel J, Zante P, Nasri S, Berndtsson R, Olsson J. 2002. Field-scale variation of preferential flow as indicated from dye coverage. *Journal of Hydrology* **257**: 164–173.

Peck AJ, Luxmoore RJ, Stolzy JL. 1977. Effects of spatial variability of soil hydraulic properties in water budget modeling. *Water Resources Research* **13**: 348–354.

Philip JR. 1975. Stability analysis of infiltration 1. *Soil Science Society of America Journal* **39**: 1042–1049.

Pires LF, Borges JA, Rosa JA, Cooper M, Heck RJ, Passoni S, Roque WL. 2017. Soil structure changes induced by tillage systems. *Soil and Tillage Research* **165**: 66–79.

Raats PAC. 1973. Unstable wetting fronts in uniform and nonuniform soils 1. *Soil Science Society of America Journal* **37**: 681–685.

Ramirez A, Daily W, LaBrecque D, Owen E, Chesnut D. 1993. Monitoring an underground steam injection process using electrical resistance tomography. *Water Resources Research* **29**: 73–87.

Ren G-L, Izadi B, King B, Dowding E. 1996. Preferential transport of bromide in undisturbed cores under different irrigation methods. *Soil Science* **161**: 214–225.

Revil A, Ahmed AS, Jardani A. 2017. Self-potential: A non-intrusive ground water flow sensor. *Journal of Environmental and Engineering Geophysics* **22**: 235–247.

Ritsema CJ, Dekker LW. 2000. Preferential flow in water repellent sandy soils: Principles and modeling implications. *Journal of Hydrology* **231**: 308–319.

Ritsema CJ, Dekker LW, Hendrickx JMH, Hamminga W. 1993. Preferential flow mechanism in a water repellent sandy soil. *Water Resources Research* **29**: 2183–2193.

Ritsema CJ, Dekker LW, Nieber JL, Steenhuis TS. 1998. Modeling and field evidence of finger formation and finger recurrence in a water repellent sandy soil. *Water Resources Research* **34**: 555–567.

Robert T, Caterina D, Deceuster J, Kaufmann O, Nguyen F. 2012. A salt tracer test monitored with surface ERT to detect preferential flow and transport paths in fractured/karstified limestones. *Geophysics* **77**: B55–B67.

Sander LM. 2000. Diffusion-limited aggregation: A kinetic critical phenomenon? *Contemporary Physics* **41**: 203–218.

Scanlon BR, Healy RW, Cook PG. 2002. Choosing appropriate techniques for quantifying groundwater recharge. *Hydrogeology Journal* **10**: 18–39.

Seven K, Germann P. 1981. Water flow in soil macropores II. A combined flow model. *Journal of Soil Science* **32**: 15–29.

Sheng F, Liu H, Wang K, Zhang R, Tang Z. 2014. Investigation into preferential flow in natural unsaturated soils with field multiple-tracer infiltration experiments and the active region model. *Journal of Hydrology* **508**: 137–146.

Sheng F, Wang K, Zhang R, Liu H. 2009. Characterizing soil preferential flow using iodine–starch staining experiments and the active region model. *Journal of Hydrology* **367**: 115–124.

Shima H. 1992. 2-D and 3-D resistivity image reconstruction using crosshole data. *Geophysics* **57**: 1270–1281.

Shope CL, Bartsch S, Kim K, Kim B, Tenhunen J, Peiffer S, Park J-H, Ok YS, Fleckenstein J, Koellner T. 2013. A weighted, multi-method approach for accurate basin-wide streamflow estimation in an ungauged watershed. *Journal of Hydrology* **494**: 72–82.

Shope CL, Maharjan GR, Tenhunen J, Seo B, Kim K, Riley J, Arnhold S, Koellner T, Ok YS, Peiffer S. 2014. Using the SWAT model to improve process descriptions and define hydrologic partitioning in South Korea. *Hydrology and Earth System Sciences* **18**: 539–557.

Šimůnek J, Jarvis NJ, van Genuchten MT, Gärdenäs A. 2003. Review and comparison of models for describing non-equilibrium and preferential flow and transport in the vadose zone. *Journal of Hydrology* **272**: 14–35.

Simmons CT, Fenstemaker TR, Sharp Jr JM. 2001. Variable-density groundwater flow and solute transport in heterogeneous porous media: Approaches, resolutions and future challenges. *Journal of Contaminant Hydrology* **52**: 245–275.

Slater L, Zaidman MD, Binley AM, West LJ. 1997. Electrical imaging of saline tracer migration for the investigation of unsaturated zone transport mechanisms. *Hydrology and Earth System Sciences Discussions* **1**: 291–302.

Sprenger M, Leistert H, Gimbel K, Weiler M. 2016. Illuminating hydrological processes at the soil-vegetation-atmosphere interface with water stable isotopes. *Reviews of Geophysics* **54**: 674–704.

Suzuki K, Ohnishi H. 1995. Application of the electrical method to field survey: 65th Annual International Meeting. *SEG, Expanded Abstracts*, 338–343.

Tompson AF, Gelhar LW. 1990. Numerical simulation of solute transport in three-dimensional, randomly heterogeneous porous media. *Water Resources Research* **26**: 2541–2562.

Van Genuchten MT, Šimůnek J, Feddes RA. 2004. Integrated modeling of vadose zone flow and transport processes. In: *Unsaturated Zone Modelling: Progress, Challenges and Applications, Wageningen UR Frontis Series*, vol. **6**, pp. 37–69 R.A. Feddes, G.H.de Rooij, J.C. van Dam, Editors, Springer Science & Business Media, Wageningen, Netherlands.

Vanderborght J, Vereecken H. 2007. Review of dispersivities for transport modeling in soils. *Vadose Zone Journal* **6**: 29–52.

Wallor E, Herrmann A, Zeitz J. 2018. Hydraulic properties of drained and cultivated fen soils part II—Model-based evaluation of generated van Genuchten parameters using experimental field data. *Geoderma* **319**: 208–218.

Wehrer M, Slater LD. 2015. Characterization of water content dynamics and tracer breakthrough by 3-D electrical resistivity tomography (ERT) under transient unsaturated conditions. *Water Resources Research* **51**: 97–124.

Weiler M, Naef F. 2003. Simulating surface and subsurface initiation of macropore flow. *Journal of Hydrology* **273**: 139–154.

Wiekenkamp I, Huisman JA, Bogena HR, Lin HS, Vereecken H. 2016. Spatial and temporal occurrence of preferential flow in a forested headwater catchment. *Journal of Hydrology* **534**: 139–149.

Wilkinson D, Willemsen JF. 1983. Invasion percolation: A new form of percolation theory. *Journal of Physics A: Mathematical and General* **16**: 3365.

Witten Jr TA, Sander LM. 1981. Diffusion-limited aggregation, a kinetic critical phenomenon. *Physical Review Letters* **47**: 1400.

Wood TR, Glass RJ, McJunkin TR, Podgorney RK, Laviolette RA, Noah KS, Stoner DL, Starr RC, Baker K. 2004. Unsaturated flow through a small fracture–matrix network. *Vadose Zone Journal* **3**: 90–100.

Xiong Y. 2014. Flow of water in porous media with saturation overshoot: A review. *Journal of Hydrology* **510**: 353–362.

Zhang ZB, Peng X, Zhou H, Lin H, Sun H. 2015. Characterizing preferential flow in cracked paddy soils using computed tomography and breakthrough curve. *Soil and Tillage Research* **146**: 53–65.

20 Determination of the Unsaturated Hydraulic Conductivity of Soil
Theoretical and In Situ Approaches

Debao Lu, Yinfeng Xia, Nan Geng, Sang Soo Lee,
Jochen Bundschuh, and Yong Sik Ok

CONTENTS

20.1 INTRODUCTION

The unsaturated zone is comprised of porous media, water, and air that constantly move among particle pores, and is the most active area in the groundwater system (Lehoux et al., 2017).

Its uppermost part, soil, is an import natural resource which is inextricably linked to human life and agricultural production (Mitchell and Soga, 2005; Jury and Stolzy, 2018). The key zone influences surface water, groundwater, and its interaction, thereby contributing to the formation, transformation, and consumption of water resources. Some research of unsaturated zones mainly focused on optimal water management and crop development (Koupai et al., 2008; Kattan, 2018; Martínez-Santos et al., 2018; Riel et al., 2018) and others recently aimed to determine the impact of physical, chemical, and biological parameters and properties on transportation of water through these zones in agricultural and industrial areas (Sprenger et al., 2015; Haverkamp et al., 2016). To ensure accuracy of assessing soil-water movement, modeling and other methods have been developed using the soil hydraulic parameters like unsaturated hydraulic conductivity of soils (Brassington, 2017) which is closely related to the soil particle size, pore distribution, pore shape, and pore continuity (Fredlund et al., 1993; Rahardjo et al., 2018). However, due to high spatial variability, it is relatively difficult to measure

hydraulic conductivity of soils directly in large areas, and it produces residual errors by soil texture (Yustres et al., 2018; Butters et al., 1989; Klute and Dirksen, 1986; Lee et al., 2007).

Currently, the numerical simulation and field test methods are mainly applied to obtain the unsaturated hydraulic conductivity of a soil (Van Genuchten et al., 1992; Lim et al., 2016; Ghanbarian and Hunt, 2017; Haslauer et al., 2017; Ali et al., 2018). However, some articles have only focused on saturated soil and not much on unsaturated conductivity.

In this review, the comprehensive knowledge of unsaturated hydraulic conductivity is discussed in detail, based on soil-water models and test methods related to soil characteristics, soil hydraulic properties, fractal, numerical inversion, and soil morphology for the numerical simulations and double ring infiltration, disc infiltration, and artificial rainfall infiltration for the field test methods.

20.2 NUMERICAL SIMULATIONS

20.2.1 SOIL-WATER CHARACTERISTICS

The soil-water characteristic curve defined the functional relation between pressure head and soil-water moisture. It can determine the movement of water or chemicals in a soil governed by hydraulic conductivity (Ahmadi et al., 2015; Sadeghi et al., 2016). However, it is very difficult to determine the soil-water interaction accurately using the soil-water characteristic curve because this curve is nonlinear and affected by various soil factors (Iden et al., 2015). The lab- or field-scale experiment of soil-water characteristics has high uncertainty due to horizontal and vertical spatial variability of a soil, and is also costly, complex, and time consuming (Bordoni et al., 2017; Klute, 1986; Guimarães et al., 2017; Haverkamp and Parlange, 1986; Dexter and Bird, 2001).

For the numerical simulations, soil-water movement can be calculated quantitatively. Many scholars have established soil-water retention models based on empirical data (Aubertin et al., 2003; Assouline et al., 1998; Tian et al., 2018; Kern, 1995; Rawls et al., 1991). Among them, the power function is the most universal and acceptable as below:

$$\left(1 - S_e^{-a}\right)^b S_e^{-c} = ah \tag{20.1}$$

$$S_e = \frac{\theta - \theta_r}{\theta_s - \theta_r} \tag{20.2}$$

in which S_e is the effective saturation; θ is the water content, θ_r and θ_s are saturated and residual water content, respectively, a, b, and c are empirical parameters; α is scale parameter, inversely proportional to the average pore diameter; $\alpha = 1/hb$, in which hb is value of air-entry suction.

1. The Brooks–Corey model can be applied when b is zero (Brooks and Corey, 1966). This model describes most soil-water retention characteristics under low suction and presents well the effects of homogeneous and isotropic coarse-texture soils with narrow pore spaces. However, under high suction, this model will have some defects such as certain errors, discontinuous curve, and a lack of rapid numerical convergence. This model will also show poor accuracy for fine-texture and undisturbed soils. As the rate of curve is discontinuous at $h = 1/\alpha$, it would not be possible to describe the water characteristics under near saturation conditions. This model has been widely employed the soils in arid and semi-arid regions.
2. The Campbell model is proper if $\theta_r = 0$ and $b = 0$ (Campbell, 1974). However, even for $\theta_r = 0$, this model is certainly irrational in physics.
3. The van Genuchten model (if $a = 1/m$, $b = 1/n$, $c = ab$) (van Genuchten, 1980) is quite universal and can be solved for most soils having the typical range of large water storage potential or water content.
4. The Gardner model (Gardner et al., 1970) is a simple model which needs only a few input parameters:

$$h = a\theta^{-b}$$

where h is pressure head, θ is volumetric water content, a and b are constants (>0). The single parameter is of physical significance and reflects the variability of soil, but correlations between the parameter and soil physical properties (e.g., water content, porosity)

are still not clearly studied. In addition, the interrelation among model parameters for different soil layers (vertical) and different soil textures (horizontal) also needs further study.

Many researchers have attempted to determine the soil-water characteristic curve by using the soil particle distribution curve and soil characteristics. Gregson et al. (1987) established the single parameter model of soil-water characteristics based on the Gardner model. Williams and Ahuja (1992) calculated the entire soil-water characteristic curve by measuring the soil suction-soil-water content relationship based on point soil sample data. Ahujia and Williams (1991) found that the single parameter model was also applicable to determine the soil-water retention curve for heterogeneous soil. The unique parameter of the model can reflect the soil-water characteristic curve at a spatial point. Vogel et al. (1998) showed that minor changes in the water retention curve under near saturation level would cause significant changes in hydraulic conductivity, thus affecting the water flow simulation in unsaturated zones with numerical difficulties.

Arya and Paris (1981) established the soil-water characteristic curve by using soil particle size distribution, bulk density, and specific gravity. Assouline et al. (1998) found the relationship between soil-water characteristic curve and soil mechanical composition, according to capillary pore and water suction. They also obtained van Genuchten model parameters by fitting the soil particle distribution curve.

The soil-water retention value or soil-water retention model's parameters can be estimated through regression analysis of soil physical properties. Vereecken (1990) estimated θ_r, α, and n of the van Genuchten model based on soil characteristics such as sand and clay contents, bulk density, organic carbon content, and porosity. Tian et al. (2018) obtained the soil-water retention curve with different bulk densities based on van Genuchten model and showed the potential of the van Genuchten model to estimate the unsaturated hydraulic conductivity of a soil.

Several methods including tensiometer (Klute and Dirksen, 1986), pressure membrane (Reitemeier and Richards, 1944), and centrifuge (Reatto et al., 2008) have been used for the soil-water characteristic curve. The tensiometer method is mostly applied for *in situ* determination of the soil-water characteristic curve in the field. The soil-water content can commonly be determined by soil oven drying, neutron probe, and time domain reflectometry (TDR) methods (Dalton et al., 1984; Noborio, 2001; Baviskar and Heimovaara, 2017; Martini et al., 2017).

20.2.2 Soil Hydraulic Conductivity

The soil hydraulic conductivity based models are very useful to save time and cost to determine soil-water movement, and provide convenience and excellence in quantitative, theoretical, and mathematical approaches (Wallor et al., 2018).

The soil hydraulic conductivity based models are classified into uniform pore size distribution models and statistical pore size distribution models (Childs and Collis-George, 1950; Fatt, 1956; Allen, 1997). The theoretical uniform pore size distribution models are easy to apply; however, there is still uncertainty in the results because the impact of pore size distribution on hydraulic conductivity is neglected. The statistical pore size distribution models combine porous media into a series of interconnected and randomly distributed pores with different sizes (e.g., Mualem model [Mualem, 1976] and Burdine model [Burdine, 1953]).

Other models are soil hydraulic mathematic models which are based on statistical pore distribution. The proper soil-water characteristic curve might be selected to obtain the analytical expression of soil hydraulic conductivity. For example, the analytical expression of soil hydraulic conductivity may be obtained by van Genuchten's water retention model with the Burdine and Mualem models, namely the Burdine-van Genuchten and Mualem-van Genuchten models. The research by van Genuchten (1980) showed that the Mualem model was more suitable for describing hydraulic conductivity properties of all kinds of soils than Burdine model.

20.2.3 Numerical Inversion

The numerical inversion method is determined by solving the Richards equation according to controllable initial and boundary conditions. The estimating steps of the numerical inversion are: (1) assumed that soil hydraulic properties are described by an analytical model with some unknown parameters, (2) tested that one or more flow control variables such as pressure, water content, or outflow rate is/are simultaneously measured under given initial and boundary conditions and other controllable conditions, (3) unknown parameters of soil hydraulic properties are initially estimated and then substituted into the Richards equation for solution numerically, and (4) the above numerical simulation processes are repeated until the error between measured and forecasted flow variables is minimized so as to optimize the unknown parameters.

The main problem regarding the numerical inversion method is that the solution is not unique. Kool et al. (1985) first applied the inversion method to solve the Richards equation during the instantaneous one-step outflow test. They concluded that uniqueness problems would be minimized if the designed test covers a wide range of water content. Kool and Parker (1988) included pressure and outflow simultaneously in the numerical inversion method. Toorman et al. (1992) revealed that the uniqueness problems could be minimized if the soil-water pressure head is included into one target function of the instantaneous one-step outflow test. In order to satisfy additional soil-water pressure measurement requirements during the outflow test, van Dam et al. (1994) carried out many multistep outflow tests and reported an increase of air pressure. In addition, they simultaneously measured the pressure head, water content, outflow, and other parameters at each air pressure step and the results showed that the uniqueness of soil hydraulic properties was sufficiently estimated based on multistep outflow test data (Eching and Hopmans, 1993; Eching et al., 1994). Inoue et al. (1998) discussed the potential application of soil hydraulic parameters in the field based on the vacuum extraction technique and numerical

simulation/optimization methods. Le Bourgeois et al. (2016) studied the capability of the inversion model on estimating soil and bedrock hydraulic properties only by *situ* soil moisture in different depths; however, this model may not always be good in saturated deep soil.

20.3 FRACTAL MODELS

Since Mandelbrot (1982) proposed the concept of fractal geometry, it has successfully been applied to describe soil structure properties such as aggregate, distribution, and porosity. Many scholars studied the unsaturated hydraulic conductivity on the basis of fractal geometry. Toledo et al. (1990) concluded that, if the water content is low, the relationship between unsaturated soil hydraulic conductivity and water content according to the fractal geometry theory and physical characteristics of the thin water film could be expressed as:

$$k(\theta) = \theta^{\frac{3}{m(3-D_S)}}$$

where m is a parameter for interaction between solid soil and surface liquid ($1 < m < 3$) and D_s is a fractal dimension based on the soil-water characteristic curve. However, this model has low accuracy under a high soil-water content.

Crawford (1994) described the unsaturated hydraulic conductivity of a soil with the mass fractal dimension D_m of a soil and coefficient d characterizing the homogeneity and pore sidewall shape of soil texture. The relationship of soil-water characteristic curve was expressed as $\theta \propto D_m$, and the relationship between unsaturated hydraulic conductivity and soil-water content was expressed as $k(\theta) = \theta^{\left[\frac{1}{D_m-3}\right]\left[(\varepsilon-1)\left[3+\left(\frac{D_m}{d}\right)-D_m\right]\right]}$, in which ε referred to an empirical parameter calibrated.

Rieu and Sposito (1991a) proposed the unsaturated hydraulic conductivity model based on soil particle size distribution, expressed as $k_i = C\beta_r \sum(d_f)_j^2 G^j$, in which C referred to a constant and related to soil pore composition and fluid property, G^j referred to an attenuation coefficient and showed that unsaturated hydraulic conductivity decreased with decreasing the porosity, $(d_f)_j$ referred to vertical pore area of fractal structure and corresponded to the probability value without fractal fragmentation. Rieu and Sposito (1991b) obtained the unsaturated conductivity of a sandy soil by using this model successfully.

Fuentes et al. (1996) believed that the soil-water characteristic curve was expressed as $\theta \propto h^{D-3}$, and derived the following relationship between unsaturated hydraulic conductivity and soil-water content:

$$k(\theta) = \theta^{\frac{2}{3-D} + 2D/3}$$

Tyler and Wheatcraft (1990) also derived an expression similar to the Campbell model in describing the water retention curve based on Sierpinski Carpet's fractal theories, but only considered the fractal properties of pore space, except for soil mass. Kravchenko and Zhang (1998) simplified the models proposed by Perrier et al. (1996) according to particle size distribution and fractal theory. Pachepsky et al. (2001) quantified and simulated the spatial variations of the fine textured soil-water characteristic curve within the entire measurable soil pressure head based on the fractal concept. Perfect et al. (2002) introduced the pressure head on the basis of Rieu and Sposito models, and the study showed that the improved model could better fit to the soil-water characteristic curve without deviation under given pressure head. A fractal-based model has been proposed by Soto et al. (2017) to obtain hydraulic conductivity in the vadose zone. They concluded that the presented model could be a good option if laboratory/field measurements are not available. The unsaturated hydraulic conductivity fractal model is classified into two types:

1. The corresponding parameters of the original unsaturated hydraulic conductivity model are substituted with the fractal dimension of the soil-water characteristic curve. This type of model structure is simple and easy to solve, but the hydraulic conductivity cannot be explained in physics.
2. Poiseuille's law is combined with soil particle size and fractal properties of pore structure. It derived the unsaturated hydraulic conductivity for different scales, with clear physical significance. However, this model is difficult to apply in practice because it has complex structure and is related to many undeterminable parameters. Since the unsaturated hydraulic conductivity is related to soil texture, pore distribution and shape, connectivity among pores of different sizes, and other factors, the various physical factors of the unsaturated hydraulic conductivity might comprehensively be considered to establish a model mechanism.

20.4 SOIL MORPHOLOGICAL MODEL

The morphological method for estimating the soil hydraulic properties is to determine the soil pore size distribution and connectivity through analytical processing of a series of high-resolution soil profiles pictures, converting into parameters, and establishing a network model.

In terms of soil hydraulic properties, the heterogeneity at the minimum scale is caused by the complex pore structure. Soil hydraulic properties depend on the soil pore structure such as pore size distribution and connectivity. Fatt (1956) developed the relationship between soil hydraulic properties and soil pore structure based on the network model. Afterward, this method was developed in the petroleum engineering field. Jerauld and Salter (1990) analyzed the spatial correlation based on different pore size classification. Ferrand and Celia (1992) and Friedman and Seaton (1996) studied the spatial heterogeneity and anisotropy of pore size distribution, respectively. They randomly selected the parameters of the network model, but the unique structure of soil was not considered. In addition, Wise (1992) forecasted the unsaturated hydraulic conductivity based on the network model and obtained the

network model parameters by corresponding the water retention curve of the model. Ewing and Gupta (1993) studied the water retention curve based on the network model so as to accurately and uniquely demonstrate the rationality of the assumption regarding real pore size distribution. In particular, Vogel and Roth (2001) completed a lot of beneficial work. They measured the soil pore size distribution and pore connectivity, and generated the parameters of the network model based on real data for soil hydraulic properties. They also demonstrated that it was feasible to research the soil hydraulic properties according to the soil morphological method by comparing simulation results with the results of the multistep outflow test in the laboratory. An improved pore-network model was presented by Xie et al. (2017). This model had an excellent prediction prospect since the empirical coupling method was involved in the calculation. Table 20.1 gives the comparison of these numerical simulation methods according to their advantages and disadvantages.

TABLE 20.1
The Advantages and Disadvantages of the Numerical Simulation Method for Unsaturated Hydraulic Conductivity Estimating

Models	Advantages	Disadvantages
Brooks–Corey model	1. Multivariate nonlinear regression method 2. The suction of the saturated soil is equal to the intake suction, which is consistent with the characteristics of the dehumidification curve. 3. It is ideal for homogeneous and isotropic coarse ground samples with narrow pore size distribution.	1. Poor precision for fine soil and undisturbed soil 2. It is discontinuous at $h = 1/\alpha$, so the water characteristic data near saturation is not well described.
Campbell model	1. Less parameters, easy to calculate 2. Good precision for sandy soil	1. There are defects in discontinuity and convergence prevention. 2. Hypothetical lack of physical meaning
van Genuchten model	1. The most widely used, almost suitable for all soil texture types 2. Have an intuitive physical meaning 3. Convenient use of statistical pore size distribution model to estimate soil hydraulic conductivity 4. Multiple nonlinear regression method 5. The suction of the saturated soil h is zero, which is consistent with the characteristics of the hygroscopic curve.	1. The model is more complicated, too many parameters are needed, difficult to get the solution 2. When the soil is close to the saturated or close to the dry, the error between the fitted value and the measured value is larger. 3. Poor precision on structural soils with distinct bimodal pore size distribution

(Continued)

TABLE 20.1 (*Continued*)

The Advantages and Disadvantages of the Numerical Simulation Method for Unsaturated Hydraulic Conductivity Estimating

Models	Advantages	Disadvantages
Gardner model	1. Multivariate nonlinear regression method 2. There are fewer parameters involved and the fitting is more convenient. 3. Easy to fit and calculate, can be directly simulated by Excel, the results can basically meet the scientific research requirements.	1. Lack of consideration for retained moisture and saturated water content 2. The model does not converge when there is water vapor thermal coupling in the aeration zone. 3. Cannot meet the requirements of high precision instruments
Soil hydraulic conductivity model	1. Based on the statistical pore size, it has a certain physical basis. 2. High accuracy and easy calculation 3. Suitable for many types of soil	1. Hypothetical lack of physical meaning 2. Influenced by pore size distribution model
Fractal model	1. The parameters in the determined soil moisture characteristic curve have clear physical meaning 2. Utilization of soil texture data	1. There is a lack of physical interpretation of the model's hydraulic conductivity with few parameters. 2. Models with many parameters are difficult to solve.
Numerical inversion method	1. Use the objective function to reduce the error 2. Good applicability to heterogeneous soils	The result of the solution is non-unique.
Soil morphology model	The model that is theoretically closest to the real situation	1. The resolution of the soil imaging device is to be improved. 2. Not easy to operate 3. The cost is relatively high.

20.5 FIELD TEST METHODS

20.5.1 DOUBLE-RING INFILTROMETER

The double-ring infiltration method is one of the most influential methods. In 1986, Bouwer (1986) proposed the method for measuring soil infiltration by a double-ring infiltrometer. After test unit improvement by Prieksat et al. (1992) and Milla and Kish (2006), the commonly used double-ring infiltrometer has been developed. Figure 20.1 shows the components and work setting of the infiltrometer. As it is shown, the infiltrometer is composed of double concentric rings and two Mariotte bottles. The test steps are: pouring water into inner and outer rings and maintaining the water columns in the inner and outer rings at the same height, then calculating the saturated hydraulic conductivity by recording the observation time and amount of infiltrating water and estimating the unsaturated hydraulic conductivity in terms of the saturated hydraulic conductivity.

For a double-ring infiltrometer, it is necessary to determine whether the measured results reflect the infiltration properties within the measured range. Wuest (2005) carried out the double-ring test by comparing the diameters of the inner ring (20, 30, and 45 cm) and concluded that the average infiltration rate was directly proportional to the diameter of the inner ring. Besides, due to spatial variability, the size increment of the test area is not directly equal to the average infiltration increment. Lai and Ren (2007) provided four double-ring infiltrometers with different inner diameters in order to achieve the correlation between inner diameter of double ring and spatial variability of soil infiltration, and further obtained different buffering indicators by setting different ratios between outside and inside diameters. Baiamonte et al. (2017) also discussed some factors which cause great influence on the hydraulic conductivity by using a point based double-ring infiltrometer.

When a diameter of the inner ring is small, the soil infiltration properties of the measured inner ring area would be affected by spatial variability and cannot reflect the real infiltration properties of the measured points. The equipment required for the double-ring measurement method are relatively simple and easy to operate. Compared with the single-ring infiltration method, the impacts of lateral infiltration

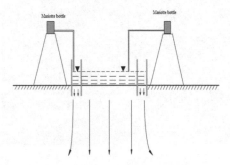

FIGURE 20.1 The components (left) and work setting (right) of the double-ring infiltrometer.

are eliminated so as to improve the measurement accuracy. However, in order to ensure that single test results are not affected by spatial variability, the diameter of the inner ring might not be less than 80 cm so that the double ring infiltration test instruments and equipment are generally large in size and would be inconvenient to carry. Moreover, the test efficiency would be lower in large areas.

20.5.2 Disc Infiltrometer

The disc infiltrometer consists of infiltration disc, water pipe, and constant pressure pipe (Figure 20.2). The infiltration disc is permanently connected to the water tube and the pressure-regulating pipe with a rubber tube so as to control the constant negative water head.

Smettem et al. (1995) proposed the approximate solution for disc infiltration and established the following approximate solution formula with physical significance. When a test is done by a disc infiltrometer, the arrangement scheme of the measuring points would be similar to that of the double-ring infiltrometer. Some problems also existed, for example, whether the measured results reflect the infiltration performance of measured areas and the infiltration performance of the selected measuring points is regionally representative. The impacts of spatial variability on the disc diameter have rarely been researched. According to some studies (Latorre et al., 2015a, 2015b; Kargas et al., 2017), the common diameters of disc infiltrometer were 10 and 20 cm and the results might not be affected by small-scale spatial variability.

20.5.3 Artificial Rainfall Simulation

The indoor artificial rainfall simulation method has been proposed to measure infiltration by Peterson and Bubenzer (1986). Ogden and Saghafian (1997) introduced a device used for measuring infiltration according to the rainfall simulation method, and Singh et al. (1999) and Mao et al. (2016) further improved and used this method. Compared with the double-ring

FIGURE 20.2 The components and work setting of disc infiltrometer.

infiltration measurement method, it is unnecessary to select a measuring point and the measured results directly represented the average infiltration value of the areas. As the measured result represents the average value of the measured area, some problems such as small measurement areas, impacts of spatial variability, and lack of representativeness were avoided. Similarly, when a scale test is carried out in a large area, it is impossible to directly carry out tests in the entire large area. It is also necessary to carry out artificial rainfall simulation tests in small representative areas. Therefore, the selected small areas cannot be regionally representative. Compared with the double-ring method, the artificial rainfall simulation method is characterized by larger single test area so that the impacts of spatial variability on the single test results can be avoided. However, it is impossible to observe the higher soil infiltration capacity at the initial stage because the infiltration rate at the initial stage of soil infiltration is equal to the rainfall intensity. Under high rainfall intensity, a soil is getting rapidly wet and crust formed on the soil surface, namely soil surface sealing, thereby destroying the soil structure and leading to lower soil infiltration. Therefore, the results of this method may have some errors compared to the actual conditions.

Table 20.2 indicates the representativeness, parameters, and single test period for several test methods introduced in this article. The double-ring and disc infiltrometers are obviously characterized by simple test method and short single test period. The disc infiltrometer is more and more widely used to measure the unsaturated hydraulic conductivity because it

TABLE 20.2

The Advantages and Disadvantages of the Field Methods for Unsaturated Hydraulic Conductivity Estimating

Methods	Advantages	Disadvantages
Double-ring infiltrometers method	1. Excludes the influence of lateral infiltration, resulting in higher accuracy of measurement results 2. Infiltration depth and area are large, the outcome is representative.	1. The device is big and not easy to operate. 2. The error is large, especially during the initial infiltration phase.
Disc infiltrometers method	1. High precision, especially the initial infiltration process can be accurately determined. 2. Lightweight, labor-saving, water-saving, easy to operate in the field	1. Infiltration area is small, shallow depth, poor representation. 2. The outcome is influenced by lateral infiltration.
Artificial simulated rainfall method	1. Can avoid single test results from spatial variability 2. The measured result is the mean of the measurement area.	Unable to observe higher soil infiltration capacity during the initial period

is capable of measuring the unsaturated hydraulic conductivity directly. For others, it should be done to convert the saturated hydraulic conductivity into the unsaturated hydraulic conductivity against the model.

20.6 CONCLUSION AND RESEARCH PROSPECT

Many researches have introduced various methodologies and models of soil hydraulic characteristics for quantitatively describing soil-water movement. The empirical, physical, and fractal models have been proposed and employed to be direct measurement or indirect calculation of soil hydraulic characteristics and rapidly developed by computer simulation.

Models based on the soil-water characteristic curve are important to obtain the unsaturated soil conductivity. Models by Brooks–Corey, Campbell, and van Genuchten are relatively well developed and mature. The Brooks–Corey model can be solved by the multivariate nonlinear regression method with assumption that saturated soil suction is equal to the value of air-entry suction accorded with the characteristics of dehumidification curve and is ideal for homogeneous and isotropic coarse-textured soils with narrow pore size. The Campbell model has the defects of discontinuity and non-convergence due to the lack of physical explanation on the parameter assumption. The van Genuchten model can be solved by the multivariate nonlinear regression method and is most widely used, suitable for most types of soil, with intuitive physical significance. The statistical pore size distribution model can also be conveniently used to estimate the soil hydraulic conductivity.

The soil hydraulic conductivity model is generally classified into theoretical and mathematical models. These two kinds of models are based on the pore distribution model. Uniform pore and statistical pore distribution models are commonly used with good theoretical basis, physical significance, and fewer parameters; however, the calculation accuracy of hydraulic conductivity may not be guaranteed due to spatial pore distribution and assumption irrationality.

The fractal method is a powerful tool for indirect estimation of soil hydraulic properties. soil-water characteristics, soil texture, bulk density, clay and mineral structure, and other properties should be considered to clarify physical significance. However, it is not easy to establish the relationship between soil hydraulic properties and real soil heterogeneous systems. For comprehensive research, the combination of theoretical methods and empirical data would be an excellent strategy.

The methods of double-ring infiltration, disc infiltration, and rainfall simulation have commonly been employed to obtain unsaturated conductivity. The double-ring infiltration method is most common and the disc infiltrometer is most capable to simulate hydraulic conductivity in the vadose zone directly. Computer models and simulation methods have broad future development prospects. However, although many methods for directly determining the unsaturated hydraulic conductivity of soils are present, their standardization is urgent for better accuracy, and new experimental techniques and instruments would be necessary. Moreover, various theoretical methods and wide empirical data must be required.

ACKNOWLEDGMENT

This work was carried out with the support of "Cooperative Research Program for Agriculture Science and Technology Development (Effects of Plastic Mulch Wastes on Crop Productivity and Agro-Environment, Project No. PJ01475801)," Rural Development Administration, Republic of Korea, and the National Research Foundation (NRF) of Korea, and Germany–Korea Partnership (GEnKO) Program (2018–2020).

REFERENCES

Ahmadi SH, Sepaskhah AR, Fooladmand HR (2015) A simple approach to predicting unsaturated hydraulic conductivity based on empirically scaled microscopic characteristic length. *Hydrol Sci J* 60:326–335.

Ahujia LR, Williams RD (1991) Scaling Water characteristic and hydraulic conductivity based on Gregson-Hector-McGown approach. *Soil Sci Soc Am J* 55:308–319.

Ali MH, Bhattacharya B, Katimon A (2018) Modelling surface runoff in a large-scale paddy field in Malaysia. *Int J Hydrol Sci Technol* 8:69–90.

Allen T (1997) *Particle Size Measurement.* Chapman and Hall, London, UK.

Arya LM, Paris JF (1981) A physicoempirical model to predict the soil moisture characteristic from particle-size distribution and bulk density data 1. *Soil Sci Soc Am J* 45:1023–1030.

Assouline S, Tessier D, Bruand A (1998) A conceptual model of the soil water retention curve. *Water Resour Res* 34:223–231.

Aubertin M, Mbonimpa M, Bussière B, Chapuis RP (2003) A model to predict the water retention curve from basic geotechnical properties. *Can Geotech J* 40:1104–1122.

Baiamonte G, Bagarello V, D'Asaro F, Palmeri V (2017) Factors influencing point measurement of near-surface saturated soil hydraulic conductivity in a small sicilian basin. *Land Degrad Dev* 28:970–982.

Baviskar SM, Heimovaara TJ (2017) Quantification of soil water retention parameters using multi-section TDR-waveform analysis. *J Hydrol* 549:404–415.

Bordoni M, Bittelli M, Valentino R et al (2017) Improving the estimation of complete field soil water characteristic curves through field monitoring data. *J Hydrol* 552:283–305.

Bouwer H (1986) Intake rate: Cylinder infiltrometer. In: *Methods Soil Analysis Part 1—Physical Mineral Methods*, pp. 825–844, Arnold Klute, Editor, American Society of Agronomy and Soil Science. Society of American, Madison, USA.

Brassington R (2017) *Field Hydrogeology.* Wiley, New York.

Brooks RH, Corey AT (1966) Properties of porous media affecting fluid flow. *J Irrig Drain Div* 92:61–90.

Burdine N (1953) Relative permeability calculations from pore size distribution data. *J Pet Technol* 5:71–78.

Butters GL, Jury WA, Ernst FF (1989) Field scale transport of bromide in an unsaturated soil: 1. Experimental methodology and results. *Water Resour Res* 25:1575–1581.

Campbell GS (1974) A simple method for determining unsaturated conductivity from moisture retention data. *Soil Sci* 117:311–314.

Childs EC, Collis-George N (1950) The permeability of porous materials. *Proc R Soc Lond A* 201:392–405.

Crawford JW (1994) The relationship between structure and the hydraulic conductivity of soil. *Eur J Soil Sci* 45:493–502.

Dalton FN, Herkelrath WN, Rawlins DS, Rhoades JD (1984) Time-domain reflectometry: Simultaneous measurement of soil water content and electrical conductivity with a single probe. *Science* 224:989–990.

Dexter AR, Bird NRA (2001) Methods for predicting the optimum and the range of soil water contents for tillage based on the water retention curve. *Soil Tillage Res* 57:203–212.

Eching SO, Hopmans JW (1993) Optimization of hydraulic functions from transient outflow and soil water pressure data. *Soil Sci Soc Am J* 57:1167–1175.

Eching SO, Hopmans JW, Wendroth O (1994) Unsaturated hydraulic conductivity from transient multistep outflow and soil water pressure data. *Soil Sci Soc Am J* 58:687–695.

Ewing RP, Gupta SC (1993) Modeling percolation properties of random media using a domain network. *Water Resour Res* 29:3169–3178.

Fatt I (1956) The network model of porous media. *Trans AIME* 207:144–181.

Ferrand LA, Celia MA (1992) The effect of heterogeneity on the drainage capillary pressure-saturation relation. *Water Resour Res* 28:859–870.

Fredlund DG, Rahardjo H, Rahardjo H (1993) *Soil Mechanics for Unsaturated Soils*. Wiley, New York.

Friedman SP, Seaton NA (1996) On the transport properties of anisotropic networks of capillaries. *Water Resour Res* 32:339–347.

Fuentes C, Vauclin M, Parlange J-Y, Haverkamp R (1996) A note on the soil-water conductivity of a fractal soil. *Transport Porous Media* 23:31–36.

Gardner WR, Hillel D, Benyamini Y (1970) Post-irrigation movement of soil water: 1. Redistribution. *Water Resour Res* 6:851–861.

Ghanbarian B, Hunt AG (2017) Improving unsaturated hydraulic conductivity estimation in soils via percolation theory. *Geoderma* 303:9–18.

Gregson K, Hector DJ, McGowan M (1987) A one-parameter model for the soil water characteristic. *J Soil Sci* 38:483–486.

Guimarães RM, Lamandé M, Munkholm LJ et al (2017) Opportunities and future directions for visual soil evaluation methods in soil structure research. *Soil Tillage Res* 173:104–113.

Haslauer CP, Bárdossy A, Sudicky EA (2017) Detecting and modelling structures on the micro and the macro scales: Assessing their effects on solute transport behaviour. *Adv Water Resour* 107:439–450.

Haverkamp R, Debionne S, Angulo-Jaramillo R, de Condappa D (2016) Soil properties and moisture movement in the unsaturated zone. In: *The Handbook of Groundwater Engineering*, 3rd edition. CRC Press, Boca Raton, FL, pp. 167–208.

Haverkamp R T, Parlange J-Y (1986) Predicting the water-retention curve from particle-size distribution: 1. Sandy soils without organic matter1. *Soil Sci* 142:325–339.

Iden SC, Peters A, Durner W (2015) Improving prediction of hydraulic conductivity by constraining capillary bundle models to a maximum pore size. *Adv Water Resour* 85:86–92.

Inoue M, Šimunek J, Hopmans JW, Clausnitzer V (1998) In situ estimation of soil hydraulic functions using a multistep soil-water extraction technique. *Water Resour Res* 34:1035–1050.

Jerauld GR, Salter SJ (1990) The effect of pore-structure on hysteresis in relative permeability and capillary pressure: Pore-level modeling. *Transport Porous Media* 5:103–151.

Jury WA, Stolzy LH (2018) Soil physics. In: *Handbook of Soils and Climate in Agriculture*. CRC Press, Boca Raton, FL, pp. 131–158.

Kargas G, Londra PA, Valiantzas JD (2017) Estimation of near-saturated hydraulic conductivity values using a mini disc infiltrometer. *Water Utility J* 16:97–104.

Kattan Z (2018) Using hydrochemistry and environmental isotopes in the assessment of groundwater quality in the Euphrates alluvial aquifer, Syria. *Environ Earth Sci* 77:45.

Kern JS (1995) Evaluation of soil water retention models based on basic soil physical properties. *Soil Sci Soc Am J* 59:1134–1141.

Klute A (1986) Water retention: Laboratory methods. In: *Methods Soil Analysis Part 1—Physical Mineral Methods*, pp. 635–662, Arnold Klute, Editor, American Society of Agronomy and Soil Science. Society of American, Madison, USA.

Klute A, Dirksen C (1986) Hydraulic conductivity and diffusivity: Laboratory methods. In: *Methods Soil Anal Part 1—Physical Mineral Methods*, pp. 687–734, Arnold Klute, Editor, American Society of Agronomy and Soil Science. Society of American, Madison, USA.

Kool JB, Parker JC (1988) Analysis of the inverse problem for transient unsaturated flow. *Water Resour Res* 24:817–830.

Kool JB, Parker JC, Van Genuchten MT (1985) Determining soil hydraulic properties from one-step outflow experiments by parameter estimation: I. Theory and numerical studies 1. *Soil Sci Soc Am J* 49:1348–1354.

Koupai JA, Eslamian SS, Kazemi JA (2008) Enhancing the available water content in unsaturated soil zone using hydrogel, to improve plant growth indices. *Ecohydrol Hydrobiol* 8:67–75.

Kravchenko A, Zhang R (1998) Estimating the soil water retention from particle-size distributions: A fractal approach. *Soil Sci* 163:171–179.

Lai J, Ren L (2007) Assessing the size dependency of measured hydraulic conductivity using double-ring infiltrometers and numerical simulation. *Soil Sci Soc Am J* 71:1667–1675.

Latorre B, Peña C, Lassabatere L et al (2015a) Estimate of soil hydraulic properties from disc infiltrometer three-dimensional infiltration curve. Numerical analysis and field application. *J Hydrol* 527:1–12.

Latorre B, Peña-Sancho C, Angulo-Jaramillo R, Moret-Fernández D (2015b) Soil hydraulic properties estimate based on numerical analysis of disc infiltrometer three-dimensional infiltration curve. In: *EGU General Assembly Conference Abstracts*, EGU General Assembly, Vienna, Austria.

Le Bourgeois O, Bouvier C, Brunet P, Ayral P-A (2016) Inverse modeling of soil water content to estimate the hydraulic properties of a shallow soil and the associated weathered bedrock. *J Hydrol* 541:116–126.

Lee K-S, Kim J-M, Lee D-R et al (2007) Analysis of water movement through an unsaturated soil zone in Jeju Island, Korea using stable oxygen and hydrogen isotopes. *J Hydrol* 345:199–211.

Lehoux AP, Faure P, Lafolie F et al (2017) Combined time-lapse magnetic resonance imaging and modeling to investigate colloid deposition and transport in porous media. *Water Res* 123:12–20.

Lim TJ, Spokas KA, Feyereisen G, Novak JM (2016) Predicting the impact of biochar additions on soil hydraulic properties. *Chemosphere* 142:136–144.

Mandelbrot BB (1982) *The Fractal Geometry of Nature*. WH Freeman and Company, New York.

Mao L, Li Y, Hao W et al (2016) An approximate point source method for soil infiltration process measurement. *Geoderma* 264:10–16.

Martínez-Santos P, Castaño-Castaño S, Hernández-Espriú A (2018) Revisiting groundwater overdraft based on the experience of the Mancha Occidental Aquifer, Spain. *Hydrogeol J* 26:1083–1097.

Martini E, Werban U, Zacharias S et al (2017) Repeated electromagnetic induction measurements for mapping soil moisture at the field scale: Validation with data from a wireless soil moisture monitoring network. *Hydrol Earth Syst Sci* 21:495–513.

Milla K, Kish S (2006) A low-cost microprocessor and infrared sensor system for automating water infiltration measurements. *Comput Electron Agric* 53:122–129.

Mitchell JK, Soga K (2005) *Fundamentals of Soil Behavior*. Wiley, New York.

Mualem Y (1976) A new model for predicting the hydraulic conductivity of unsaturated porous media. *Water Resour Res* 12:513–522.

Noborio K (2001) Measurement of soil water content and electrical conductivity by time domain reflectometry: A review. *Comput Electron Agric* 31:213–237.

Ogden FL, Saghafian B (1997) Green and Ampt infiltration with redistribution. *J Irrig Drain Eng* 123:386–393.

Pachepsky YA, Timlin DJ, Rawls WJ (2001) Soil water retention as related to topographic variables. *Soil Sci Soc Am J* 65:1787–1795.

Perfect E, Dıaz-Zorita M, Grove JH (2002) A prefractal model for predicting soil fragment mass-size distributions. *Soil Tillage Res* 64:79–90.

Perrier E, Rieu M, Sposito G, Marsily GD (1996) Models of the water retention curve for soils with a fractal pore size distribution. *Water Resour Res* 32:3025–3031.

Peterson AE, Bubenzer GD (1986) Intake rate: Sprinkler infiltrometer 1. In: *Methods Soil Analysis Part 1—Physical Mineral Methods*, pp. 845–870, Arnold Klute, Editor, American Society of Agronomy and Soil Science. Society of American, Madison, USA.

Prieksat MA, Ankeny MD, Kaspar TC (1992) Design for an automated, self-regulating, single-ring infiltrometer. *Soil Sci Soc Am J* 56:1409–1411.

Rahardjo H, Thang NC, Kim Y, Leong E-C (2018) Unsaturated elasto-plastic constitutive equations for compacted kaolin under consolidated drained and shearing-infiltration conditions. *Soils Found* 58:534–546.

Rawls WJ, Gish TJ, Brakensiek DL (1991) Estimating soil water retention from soil physical properties and characteristics. In: *Advances in Soil Science*. Springer, New York, pp. 213–234.

Reatto A, da Silva EM, Bruand A et al (2008) Validity of the centrifuge method for determining the water retention properties of tropical soils. *Soil Sci Soc Am J* 72:1547–1553.

Reitemeier RF, Richards LA (1944) Reliability of the pressure-membrane method for extraction of soil solution. *Soil Sci* 57:119–136.

Riel B, Simons M, Ponti D et al (2018) Quantifying ground deformation in the Los Angeles and Santa Ana Coastal Basins due to groundwater withdrawal. *Water Resour Res* 54:3557–3582.

Rieu M, Sposito G (1991a) Fractal fragmentation, soil porosity, and soil water properties: I. Theory. *Soil Sci Soc Am J* 55:1231–1238.

Rieu M, Sposito G (1991b) Fractal fragmentation, soil porosity, and soil water properties: II. Applications. *Soil Sci Soc Am J* 55:1239–1244.

Sadeghi M, Ghahraman B, Warrick AW et al (2016) A critical evaluation of the Miller and Miller similar media theory for application to natural soils. *Water Resour Res* 52:3829–3846.

Singh R, Panigrahy N, Philip G (1999) Modified rainfall simulator infiltrometer for infiltration, runoff and erosion studies. *Agric Water Manag* 41:167–175.

Smettem KRJ, Ross PJ, Haverkamp R, Parlange JY (1995) Three-dimensional analysis of infiltration from the disk infiltrometer: 3. Parameter estimation using a double-disk tension infiltrometer. *Water Resour Res* 31:2491–2495.

Soto MA, Chang HK, Van Genuchten MT (2017) Fractal-based models for the unsaturated soil hydraulic functions. *Geoderma* 306:144–151.

Sprenger M, Volkmann TH, Blume T, Weiler M (2015) Estimating flow and transport parameters in the unsaturated zone with pore water stable isotopes. *Hydrol Earth Syst Sci* 19: 2617–2635.

Tian Z, Gao W, Kool D et al (2018) Approaches for estimating soil water retention curves at various bulk densities with the extended van Genuchten model. *Water Resour Res* 54: 5584–5601.

Toledo PG, Novy RA, Davis HT, Scriven LE (1990) Hydraulic conductivity of porous media at low water content. *Soil Sci Soc Am J* 54:673–679.

Toormann AF, Wierenga PJ, Hills RG (1992) Parameter estimation of soil hydraulic properties from one-step outflow data. *Water Resour Res* 28:3021–3028.

Tyler SW, Wheatcraft SW (1990) Fractal processes in soil water retention. *Water Resour Res* 26:1047–1054.

Van Dam JC (1994) Inverse method for determining soil hydraulic functions from multi-step outflow experiments. *Soil Sci Soc Am J* 56:1042–1050.

Van Genuchten MT (1980) A closed-form equation for predicting the hydraulic conductivity of unsaturated soils 1. *Soil Sci Soc Am J* 44:892–898.

Van Genuchten MT, Leij FJ, Lund LJ (1992) On estimating the hydraulic properties of unsaturated soils. In *Indirect Methods for Estimating the Hydraulic Properties of Unsaturated Soils*, edited by M. T. van Genuchten, F. J. Leij, and L. J. Lund. University of California, Riverside, CA, 718 pp.

Vereecken H, Maes J, Feyen J (1990) Estimating unsaturated hydraulic conductivity from easily measured soil properties. *Soil Sci* 149:1–12.

Vogel HJ, Roth K (1998) A new approach for determining effective soil hydraulic functions. *Eur J Soil Sci* 49:547–556.

Vogel H-J, Roth K (2001) Quantitative morphology and network representation of soil pore structure. *Adv Water Resour* 24:233–242.

Wallor E, Herrmann A, Zeitz J (2018) Hydraulic properties of drained and cultivated fen soils part II—Model-based evaluation of generated van Genuchten parameters using experimental field data. *Geoderma* 319:208–218.

Williams RD, Ahuja LR (1992) Evaluation of similar-media scaling and a one-parameter model for estimating the soil water characteristic. *J Soil Sci* 43:237–248.

Wise WR (1992) A new insight on pore structure and permeability. *Water Resour Res* 28:189–198.

Wuest SB (2005) Bias in ponded infiltration estimates due to sample volume and shape. *Vadose Zone J* 4:1183–1190.

Xie C, Raeini AQ, Wang Y et al (2017) An improved pore-network model including viscous coupling effects using direct simulation by the lattice Boltzmann method. *Adv Water Resour* 100:26–34.

Yustres Á, López-Vizcaíno R, Sáez C et al (2018) Water transport in electrokinetic remediation of unsaturated kaolinite. Experimental and numerical study. *Sep Purif Technol* 192:196–204.

21 New Technologies for Monitoring Contaminated Soil and Groundwater

Kumuduni Niroshika Palansooriya, Xiaomin Dou, Jörg Rinklebe,
Nanthi S. Bolan, and Yong Sik Ok

CONTENTS

21.1 INTRODUCTION

Soil and groundwater pollution occur due to natural and anthropogenic activities. Currently, anthropogenic activities are a major cause of concern. Sources of anthropogenic contaminants include waste disposal, illegal dumping, mineral extraction, abandonment of mines, accidental spills, leaking underground storage tanks, pesticide use, and application of fertilizers (Shaheen et al., 2018; US EPA, 2019). Hurricanes, floods, and weathering of soil parent materials (including igneous and sedimentary rocks, and coal) are natural events that contribute to both soil and groundwater contamination (Bolan et al., 2014). Inorganic contaminants, such as trace elements, phosphates, nitrates, fluorides, and chlorides, and organic contaminants, namely benzene, disinfection by-products (DBPs), pharmaceuticals and personal care products (PPCPs), endocrine disrupting chemicals (EDCs), pesticides, herbicides, and surfactants are some of the typical pollutants that occur in soil and water (Antoniadis et al., 2017; Palansooriya et al., 2019b; Tianlik et al., 2016; Yang et al., 2017). Moreover, microbial contaminants, including bacteria and viruses are also commonly found in water bodies. The inorganic and organic pollutants found in water and soil ultimately enter the food chain and cause major environmental and human health concerns. The US EPA estimates that there are 294,000 contaminated sites in the United States (US EPA, 2004). Moreover, >300,000 ha of land in the United Kingdom are considered potentially contaminated sites (DEFRA, 2006). Contamination of soil may eventually lead to the deterioration of surface and groundwater quality via runoff and leaching. The United Nations World Water Assessment Programme (WWAP) reported that, on a global scale, ~15–18 billion m^3 (15,000–18,000 GL) of freshwater resources are polluted by fossil fuel production and that 768 million people do not have access to improved sources of water (UNESCO, 2017).

Thus, mitigation of soil and groundwater pollution is critical to human and environmental health, and has huge socioeconomic implications (Palansooriya et al., 2019a). To achieve the UN's Sustainable Development Goals by 2030 (e.g., good health and well-being, clean water and sanitation, life on land), it is important to remediate these contaminated sources. In order to address pollution remediation, the type, level, and sources of contaminants, as well as environmental and health risks, and interactions between the contaminants and soil and groundwater sources must be understood. Thus, systematic soil and groundwater monitoring and sensing are required to meet the cleanup goals. However, accurate, real-time, and *in situ* measurements of pollutants in the field remains a technical challenge. Various monitoring and sensing techniques with low-medium and high-resolutions have been used. With an increase in soil and water pollution, there is a need for environmentally friendly and reliable assessment of soil and water quality through efficient techniques. At present, several novel techniques that use advanced methods are being employed to get a greater understanding of the contamination status of soil and water. Information obtained from these techniques could prove to be beneficial to decision-makers to understand, interpret, and use this knowledge in their management activities relating to pollution prevention and safeguarding of soil and water.

Conventional methods used to determine the spatial distribution of pollutants in soil and groundwater typically involve field samplings, laboratory analyses, and geo-statistical interpolation. However, these methods are disadvantageous in terms of high cost, time-consumption, low efficiency when covering large areas, and high environmental interventions due to intensive soil samplings in the field as well as laboratory analyses (Lee et al., 2016; Lorenzo et al., 2018; Pérez-Fernández et al., 2017). Furthermore, they provide limited information with respect to a specific time and location. The data obtained are not adequate to assess the spatial and temporal changes of pollutants over a large area. Thus, these methods are becoming less popular among the scientific community.

In recent years, the development of new technologies has attracted the attention of the scientific community due to their practical advantages, such as the ability to obtain both temporal and spatial data, applicability in both laboratory and field measurements, cost effectiveness, and fast real-time and accurate measurements (Pérez-Fernández et al., 2017; Shi et al., 2014).

In this chapter, the latest advancements and trends in soil and groundwater monitoring and sensing techniques are discussed, including new technologies, their applicability at a field and laboratory level, advantages over conventional methods, and their effectiveness in integrated technologies.

21.2 OCCURRENCE OF CONTAMINANTS IN SOIL AND GROUNDWATER

21.2.1 SOIL CONTAMINATION

Inorganic contaminants, in particular, trace elements such as arsenic (As), cadmium (Cd), chromium (Cr), cobalt (Co), copper (Cu), lead (Pb), mercury (Hg), nickel (Ni), and zinc (Zn) are the major elements in soil that cause adverse impacts on environmental and human health (Bolan et al., 2014; El-Naggar et al., 2019; Hou et al., 2017; Yang et al., 2018). Spatial distribution of trace elements in soil is highly heterogeneous, with higher concentrations in particular localities (El-Naggar et al., 2018). For example, arsenic (As) concentration in igneous rocks and sedimentary rocks can range from 1.5 to 3.0 mg kg^{-1} and 1.7 to 400 mg kg^{-1}, respectively (Mahimairaja et al., 2005). Likewise, weathering of these rocks may release As into the soil at varying concentrations. Apatite (Ca$_5$(F, Cl, OH)(PO$_4$)$_3$), which is a major phosphate mineral in soil, is a source of trace elements. Some trace elements are enriched in apatites via different mechanisms, such as surface adsorption, formation of insoluble compounds, and complexation. Trace elements that are poorly bound are released into the soil (Bolan et al., 2014; Park et al., 2011). The occurrence of inorganic and organic contaminants in soils in various geographical regions is summarized in Table 21.1.

Anthropogenically added trace elements, unlike natural ones, typically have high bioavailability in soils, which lead to food chain contamination. Globally, 800,000 tons of Pb and more than 30,000 tons of Cr were released into the environment during the past five decades (Yang et al., 2018). Organic soil amendments, such as biosolids/sewage sludge and animal manure, and inorganic soil amendments, such as P fertilizers (agrochemicals in particular), pesticides, and herbicides are the major sources of trace elements in soil ecosystems (Bolan et al., 2014; Huang et al., 2019).

Polycyclic aromatic hydrocarbons (PAHs), polychlorinated biphenyls (PCBs), polychlorinated dibenzofurans, polybrominated biphenyls, pesticides, organophosphorus, herbicides, organic fuels, especially gasoline and diesel are the commonly found organic contaminants in soil (Fang et al., 2017; Haddaoui et al., 2016; Lu et al., 2012). Industrial activities, improper waste disposal, and intensive agricultural practices are the main sources of organic contaminants in soils. As reported by Lu et al. (2012), 10.8 µg g^{-1} of PAH was observed in a soil profile near an oil extraction factory in eastern China. However, this concentration was lower in the deeper soils, that is 0.143 µg g^{-1}. Combustion processes, including crude oil refining and traffic emissions, are the main sources of PAH in the soil. Contamination of soil with petroleum is found to be the main source of PAH in deeper soils. Moreover, soil irrigation with treated wastewater has been identified as a source of organic pollutants in soils. Agricultural soil irrigated with treated wastewater for > 30 y were detected with PAHs, PCBs, and organochlorinated pesticides, with 4 rings PAHs exhibiting > 74% of the total concentration (Haddaoui et al., 2016). These observations warn of the potential transfer of these contaminants to plants and their leaching to groundwater; thus, care should be taken while irrigating soils with treated wastewater. Moreover, excessive use of agrochemicals, especially pesticide and herbicides, leads to the contamination and accumulation of these compounds in soils. Although some studies have proposed that the influence of herbicide and pesticide application on soil function is minor and/or temporary, other studies have found that there can be a significant negative impact on soil functions. Rose et al. (2016) reported that

TABLE 21.1

Occurrence of Inorganic and Organic Contaminants in Soils in Various Geographical Regions

Contaminant	Concentration (mg kg⁻¹)	Soil Type	Location	References
		Inorganic Contaminants		
Cd	0.10–2.34	Rice soil	China	Gu et al. (2018)
Cu	11.40–90.20			
Pb	16.00–977.60			
Zn	38.70–201.60			
Pb	80.5	Playground soil	China	Peng et al. (2019)
Pb	204	Residential vegetable garden soil	Australia	Laidlaw et al. (2018)
	102	Community vegetable garden soil		
As	257	Railway corridor soil	Australia	Bolan et al. (2013)
Pb, Cd, Zn	9876, 654, 224	Shooting range soil	Australia	Seshadri et al. (2017)
Cr	81035	Tannery waste disposal site	Australia	Choppala et al. (2016)
As, Ba, Cd, Co, Cr, Cu, Ni, Pb, Zn	39.98, 976.42, 8.34, 19.36, 287.19, 14543.4, 130.24, 1615.8, 4737.74	E-waste recycling sites	India	Singh et al. (2018)
As and Pb	52.58 and 1259.58	Rice soil (near mining site)	South Korea	Igalavithana et al. (2017)
As and Pb	1940.92 and 1445.0	Upland fallowed agricultural soil (near mining site)		
Cr	15.6–525.8	Residential soil	Romania	Mihaileanu et al. (2018)
Pb	25.4–559.5			
Mn	363.1–1389.6			
Cu	87	Avocado orchard soil	New Zealand	Vogeler et al. (2008)
Cd		Pasture soils	New Zealand	Loganathan et al. (2003)
Cd, Cu, Fe, Pb, and Zn	1 1.0, 19.1, 3791.4, 177.0, and 129.0	Roadside soil	Jordan	Alsbou and Al-Khashman (2017)
		Organic Contaminants		
76 types of pesticide	2.87 (total pesticide content)	317 EU agricultural topsoils	European Union	Silva et al. (2019)
Glyphosate	2	317 EU agricultural topsoils	European Union	Silva et al. (2018)
Aminomethylphosphonic acid	2			
Neonicotinoid insecticides	0.0042	Maize grown soil	Southwestern Ontario	Schaafsma et al. (2015)
PAHs	0.235	Field experimental site (0–10 cm layer)	Poland	Kuśmierz et al. (2016)
	0.160	Field experimental site (10–20 cm layer)		
16 PAHs	0.892–3.514	Surface soil: around municipal solid waste landfill	Poland	Melnyk et al. (2015)
7 PCB	0.201–3.033			
PAHs (16 US EPA priority PAHs)	0.96	Industrial, urban, and rural sites soils	South Korea	Kwon and Choi (2014)
23 PAHs	0.164	Industrial area soils	South Korea	Islam et al. (2017)
16 PAHs	0.204	Surface soil around central Himalaya region	Tibetan Plateau, China	Bi et al. (2016)
	0.33	Surface soil around central Himalaya region	Nepal	
Petroleum hydrocarbon	388	Urban open space, agricultural/ rural, road verge, recreational/ municipal, commercial, and light industrial	United Kingdom	Kim et al. (2019)
16 PAHs	32.4			
31 PAHs	45.4			
PCB (tri-hepta)	32400			
16 US EPA PAHs	0.02–0.106	Industrial and agricultural soils	Cuba	Sosa et al. (2017)
7 PCBs	0.0011–0.0076			

the application of herbicides could adversely affect the soil biology and its functions. For instance, exposure to glyphosate and atrazine can damage the earthworm ecology in soils. In addition, sulfonylurea herbicides could inhibit soil N-cycling in alkaline or low organic matter soils (Rose et al., 2016).

21.2.2 GROUNDWATER CONTAMINATION

Inorganic contaminants, such as trace elements, nitrates, nitrites, chloride, bicarbonate, sulfate, and phosphates are commonly found in groundwater (Geng et al., 2019; Palansooriya et al., 2019c). In Bangladesh, groundwater contamination with As has been identified as the biggest threat to its citizens. In some areas of Bangladesh, 2000 $\mu g\ L^{-1}$ of As was detected, which was 200 times higher than the World Health Organization standard (10 $\mu g\ L^{-1}$) for drinking water (Hossain, 2006). Moreover, As found in groundwater was mainly of geological origin. Oxidation of arsenopyrites and reduction of As-associated iron and manganese oxyhydroxide released As into the groundwater (Das et al., 2004). Groundwater samples analyzed in Tamil Nadu, India, revealed the presence of 16 trace elements in the order of Cr < Zn < Cu < Cd < Co < Fe < Al < Ni < Ti < Zr < B < Ag < Mn < Pb < Li < Si (Vetrimurugan et al., 2017). Although Cr and Zn were within the safety limits of the Bureau of Indian Standards for drinking water, Ag, Pb, and Ni concentrations exceeded them. It was found that the sources of trace elements were natural as well as anthropogenic. Moreover, the use of fertilizers and industrial wastewater in agricultural fields, and seawater intrusion as a result of intensive pumping to farm lands have been recognized as the sources of trace elements in the area (Vetrimurugan et al., 2017).

Nitrate is another common inorganic contaminant found in groundwater due to intensive agricultural practices, irrigation of sewage effluents, and poor sanitation in densely populated areas. Consumption of nitrate contaminated water may lead to methemoglobinemia and cancers. Zhai et al. (2017) studied nitrate pollution in groundwater and the potential human health risk in northeast China. Out of the 389 groundwater samples that were tested, >32% of the samples exceeded 20 mg L^{-1} of N, which is the Grade III threshold of the quality standard for groundwater of China. Principal component analysis revealed that a high level of nitrate in groundwater was due to anthropogenic activities, such as excessive use of chemical fertilizers for farming (Zhai et al., 2017). In addition, they also observed other inorganic contaminants in groundwater. Their results showed that the concentrations of the main contaminants were in order of total dissolved solids > HCO_3 > Ca > NO_3 > Cl > Na > SO_4 > Mg > K > NH_4 > NO_2 (Zhai et al., 2017). It is suspected that natural phenomena are the main sources of these contaminants in water. In particular, water-rock interactions play a significant role in controlling these inorganic contaminants in groundwater.

Pesticides, PPCPs, veterinary products (especially antibiotics), industrial by-products, and food additives are the emerging organic contaminants in groundwater (Sui et al., 2015; Yang et al., 2017). In areas where the groundwater table is shallow, there is a higher possibility of groundwater contamination. Urban and storm-water runoff, agricultural runoff, infiltration/leakage from urban sewerage systems, and diffuse aerial deposition are identified as the major pathways for groundwater pollution (Ebele et al., 2017; Gottschall et al., 2012). There are multiple sources of organic contaminants in groundwater; however, wastewater is the most common source. In particular, hospital wastewater, landfill sites, domestic and industrial waste, septic tanks, and animal husbandries are the major point sources of organic contaminants. Figure 21.1 demonstrates the sources and pathways of groundwater contamination by organic contaminants.

PPCPs, such as erythromycin, fluconazole, methyl paraben, sulfamethoxazole, salicylic acid, triclosan, and bisphenol, were frequently detected in groundwater near two multiple landfill sites in Guangzhou, China (Peng et al., 2014). Moreover, 42% of the groundwater samples collected

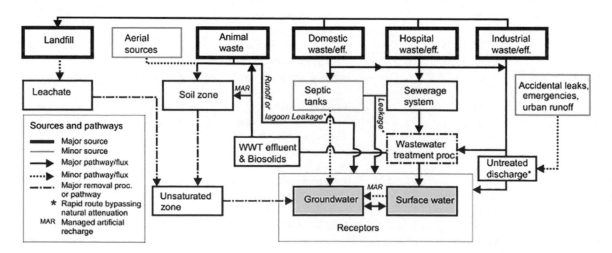

FIGURE 21.1 Sources and pathways of groundwater contamination by organic contaminants. WWT: Wastewater treatment. (Reproduced from Lapworth, D. et al., *Environ. Pollut.*, 163, 287–303, 2012.)

from 164 locations in 23 European countries contained a commonly used analgesic, that is carbamazepine with a maximum concentration of 390 ng L^{-1} (Loos et al., 2010). Veterinary antibiotics used in livestock can end up in waste lagoons and eventually lead to groundwater contamination. Groundwater samples collected from manure applied agricultural areas in Spain contained 11 types of antibiotics (Boy-Roura et al., 2018). Sulfamethoxazole and ciprofloxacin were the major types of antibiotics, with maximum concentrations of 300 ng L^{-1} (Boy-Roura et al., 2018).

According to Sorensen et al. (2015), organic contaminants in groundwater samples collected from Kabwe, Zambia, included N,N-diethyl-meta-toluamide (DEET), bactericide triclosan, trihalomethanes, and surfactant 2,4,7,9-tetramethyl-5-decyne-4,7-diol at concentrations up to 1.8 mg L^{-1}, 0.03 mg L^{-1}, 50 mg L^{-1}, and 0.6 mg L^{-1}, respectively. The diverse and elevated organic compounds in groundwater could be attributed to poor sanitation, inappropriate well protection, and household waste disposal in the area. Table 21.2 shows the occurrence of inorganic and organic contaminants in groundwater in various geographical regions.

21.3 IMPORTANCE OF NOVEL MONITORING AND SENSING TECHNOLOGIES

Various laboratory and field techniques are being employed to monitor soil and groundwater quality. As it is a complex process, many factors need to be considered to obtain better results. Moreover, reliable assessments are required to understand the status of soil and water quality. This would aid decision-makers in recognizing the severity of the pollution, interpreting the data, and designing the remediation techniques to protect resources.

Field samplings, laboratory analyses, and geo-statistical interpolation are the commonly involved steps in conventional soil/water monitoring. In the case of heavy metal contaminated environmental samples, sample storing, preparation, pretreatment, acidic extraction or acidic oxidation, and digestion of the samples are some of the important steps involved. These prepared samples can then be measured by several analytical equipments, such as inductively coupled plasma-atomic emission spectrometry (ICP-AES), inductively coupled plasma-mass spectrometry (ICP-MS), graphite furnace atomic absorption spectrometry (GFAAS), atomic

TABLE 21.2

Occurrence of Inorganic and Organic Contaminants in Groundwater in Various Geographical Regions

Contaminant	Concentration (mg L^{-1})	Source Type	Location	References
Inorganic Contaminants				
Cd	3×10^{-5}	Groundwater	China	Wen et al. (2019)
Fe	0.68692			
As	0.0469	Well	Pakistan	Shah et al. (2019)
As	0.124–0.138	Groundwater	India	Upadhyay et al. (2019)
As and F⁻	0.05 and 5.5	Deep wells	Mexico	Sandoval et al. (2019)
Zn, Pb, and Cd	0.75, 0.632, and 0.00193	Well	Morocco	Yassir et al. (2019)
Fe and As	9.78×10^{-3} and 1.30×10^{-3}	Wells and springs	Iran	Saleh et al. (2018)
NO_3^-	36.8	Groundwater monitoring wells	USA	Atekwana and Geyer (2018)
NO_3^-	12–26	Well	Japan	Kawagoshi et al. (2019)
F⁻	3.8	Well	China	Jia et al. (2019)
Organic Contaminants				
Petroleum hydrocarbon	>0.05	Spring and borehole groundwater samples	China	Liu et al. (2018b)
16 PAHs	0.02–9.5	Well	Nigeria	Ugochukwu and Ochonogor (2018)
PAHs	$5–9.210^{-5}$	Aquifers	North China	Wang et al. (2019)
PAHs	0.0204–1.93	Well	Tunisia	Samia et al. (2018)
PCBs	0.0052–0.196			
Chlorpyrifos organophosphate pesticides	0.0021	Shallow aquifers	Indonesia (Sidoarjo)	Rochaddi et al. (2019)
Endosulfan	4.75×10^{-6}	Aquifer	Argentina	Grondona et al. (2019)
Heptachlors	2.17×10^{-6}			
Sulfamethoxazole	$1\times10^{-7}–3.4\times10^{-5}$	Borehole groundwater samples	India	Lapworth et al. (2018)
Perfluoroalkyl	$1\times10^{-7}–3.3\times10^{-5}$			
Phenoxyacetic acid (pesticides)	$2\times10^{-5}–2.1\times10^{-4}$			
Benzo(a)anthracene	0.01326	Well	Mexico	López-Macias et al. (2019)
Benzo(k)fluoranthene	0.00788			

fluorescence spectroscopy (AFS), and hydride generation atomic absorption spectroscopy (HGAAS). However, these instruments are used for off-site analyses, and the sampling and analyzing are intermittent. It is hard to obtain timely and continuously monitoring results. In addition, they require fully equipped and staffed laboratories to maintain and operate. In general, intensive field sampling and laboratory analysis are time consuming and costly. Moreover, this information may not be applicable for evaluating spatial and temporal changes of contaminants in soil and groundwater.

In recent years, several new measurement technologies have emerged with various advantages. Novel monitoring and sensing technologies (proximal and remote sensing) with sophisticated techniques are being used to overcome limitations. For example, visible and near-infrared reflectance spectroscopy (VNIRS) has received greater attention due to its efficiency in determining concentration of heavy metals in soils. Moreover, some studies have found that physical, chemical, and biological properties of soils can also be estimated using VNIRS. Thus,

these techniques are more advantageous over conventional methods (Rossel et al., 2006; Stenberg et al., 2010).

Moreover, with rapid development of sensor and wireless communication technologies, remote sensing and geographic information system (GIS) techniques have become common (Wang et al., 2018), especially due to their user-friendly nature and accuracy of data. For example, some smart sensors are cheaper and relatively smaller. Moreover, these sensors are ideal for real-time applications and are capable of monitoring a wider range of environmental parameters with continuous-timed monitoring (Jha et al., 2007). Therefore, identification and application of such novel technologies for environmental monitoring are crucial for effective remediation approaches. Strategic plans need to be developed for productive monitoring of soil and groundwater with new techniques. Figure 21.2 is a schematic diagram, which shows the different steps involved in soil and groundwater monitoring program. Through this strategic plan, environmental monitoring can be optimized to generate accurate data while utilizing resources efficiently.

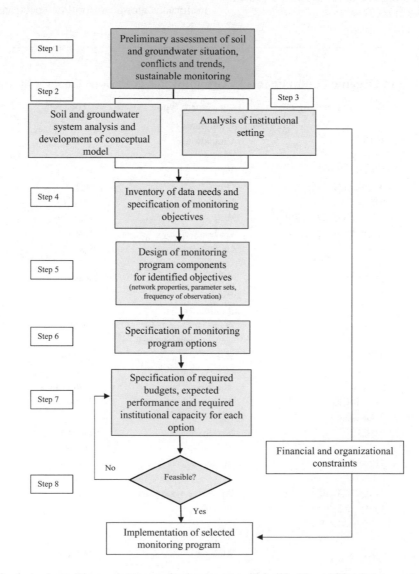

FIGURE 21.2   Scheme for designing soil/groundwater monitoring program. (Modified from IGRAC, International Groundwater Resources Assessment Centre, Guideline on: Groundwater monitoring for general reference purposes, 2008.)

21.4 MONITORING AND SENSING TECHNOLOGIES FOR SOIL AND GROUNDWATER

21.4.1 SOIL MONITORING

Soil monitoring is defined as the *"systematic determination of soil variables so as to record their temporal and spatial changes"* (Morvan et al., 2008). Soil monitoring is important to identify soil contamination levels and to detect changes in soil quality in the early stages. These data are useful for implementing and designing remediation techniques to improve soil quality (Nerger et al., 2016). Moreover, it can help establish policies to safeguard and maintain the sustainable use of soil.

21.4.2 NOVEL SOIL MONITORING AND SENSING TECHNOLOGIES

21.4.2.1 X-Ray Fluorescence (XRF) Spectrometry

X-ray fluorescence spectrometry (XRF) is a commonly used proximal sensing technique to screen a range of contaminants with respect to their elemental composition of complex media (Kalnicky and Singhvi, 2001). XRF has several advantages over the other methods, such as accurate and speedy measurements, non-destructive analysis, multi-element and simultaneous quantitative analysis, low detection limit, cost effectiveness, and ease of use (Hepburn et al., 2018; Lord et al., 2012; Parsons et al., 2013). The energy and intensity of the distinctive fluoresced radiation of a component in the sample are used to identify the component and quantify its concentration, respectively (Kaniu et al., 2012), whereas sample composition and the instrumental parameters may decide the analytical quality of the obtained spectra (Kaniu et al., 2012).

At a laboratory level, XRF spectrometry has been in use for several years. Recently, portable XRF technology (PXRF) has attracted greater attention for field applications due to its feasibility for on-site measurements (Lee et al., 2016; Padilla et al., 2019). As PXRF is equipped with efficient radioisotope source excitation combined with highly sensitive detectors and associated electronics, it becomes more viable for field applications (Kalnicky and Singhvi, 2001; Weindorf and Chakraborty, 2016).

Hepburn et al. (2018) studied heavy metal source separation in groundwater in a reclaimed land using PXRF (Olympus DELTA) together with inductively coupled plasma atomic emission spectroscopy (ICP-AES) analysis. In their study, XRF-derived heavy metal concentrations were cross-checked with the ICP-AES data. The XRF-derived Cr concentrations were found to be three times higher than the ICP-AES data. The authors argued that PXRF calibration (which typically contains high Cr levels) or sample digestion with ICP-AES analysis might have caused variations in the results. More importantly, limit of detection (LOD) played an important role in the analysis as XRF typically has a higher LOD than laboratory tests. As observed by Hepburn et al. (2018), the LOD for Cr using PXRF and ICP-AES was 10 mg kg^{-1} and 2 mg kg^{-1}, respectively. Nevertheless, there was no good correlation between XRF- derived As concentrations and ICP-AES-derived As concentrations. This could be due to the relatively low As concentration in the area, sample heterogeneity, interference by Fe in the sample matrix, and the moisture content in the samples (Hepburn et al., 2018). However, NITON® XLt™ and 700 Series™ field PXRF analyzers were used to measure on-site As concentration in floodplain soils of the Saône in eastern France (Parsons et al., 2013). Even though the As concentration was very low (<20 ppm), they successfully quantified As with minimal sample preparation under field conditions. Moreover, the LOD was recorded between 5.8 and 10.2 ppm, suggesting that PXRF is a useful technique to detect trace amounts of As in soils (Parsons et al., 2013). In addition, the authors proposed that soil moisture is an important parameter in quantitative PXRF analysis to minimize the sources of error and to maximize the precision and accuracy of data.

Rouillon and Taylor (2016) evaluated the analytical capabilities of a field PXRF spectrometer with respect to contaminated soils using a matrix-matched calibration. Their study revealed that PXRF is a better alternative for measuring Cd, Pd, Cu, Zn, Cr, Mn, Fe, Sr, and Ti in metal-contaminated soils, and not for Ni and As. Furthermore, the accuracy of the PXRF can be increased by instrument calibration, as well as proper sample preparation and handling (Rouillon and Taylor, 2016). The development of Niton PXRF analyzers over time is shown in Figure 21.3a. In the early years, Niton PXRF was designed only to quantify Pb. With technological advancements in the field, Niton PXRF has become advanced in terms of sensitivity and capability of detecting various other elements. Figure 21.3b shows Innov-X/Olympus PXRF analyzers with different models.

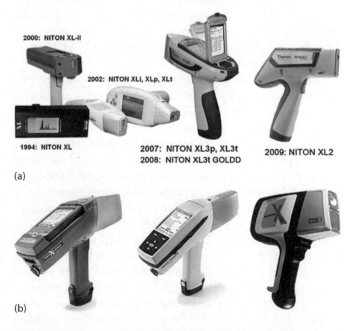

FIGURE 21.3 Various PXRF types used for soil heavy metal monitoring. (a) Evolution of Niton™ PXRF analyzers. (b) Evolution of Innov-X/Olympus PXRF analyzers. From left to right, Alpha, Omega, and Delta models. (Reproduced from Weindorf, D.C. et al., *Adv. Agron.*, 128, 1–45, 2014.)

21.4.2.2 Visible and Near-Infrared Reflectance Spectroscopy

Visible and near-infrared reflectance spectroscopy (350–2500 nm) is a cost-effective, environmentally friendly, and non-invasive proximal sensing technique to determine heavy metal concentrations in soil (Shi et al., 2014). In addition, other benefits of this technique make it more convenient for users. For instance, sample preparation requires only sample drying and crushing, measurements can be taken within a few seconds, a single scan can estimate several soil properties, and both laboratory and on-site measurements can be done (Stenberg et al., 2010). Generally, pure metals lack spectral features in the visible and NIR regions However, these metals are mostly associated with soil constituents, such as clay minerals, iron oxides, sulfides, carbonates, hydroxides, and organic matters (Sun et al., 2018). As these soil properties are spectrally active, heavy metal concentrations in soils can be estimated based on the relationship of heavy metals with spectrally active soil properties (Shi et al., 2016).

Sun and Zhang (2017) evaluated the Zn concentrations in soils using VNIRS technique. In their study, the entire VNIR spectral bands, separate and combined bands of organic matter, and clay minerals were evaluated and compared using genetic algorithm-based partial least square regression (GA-PLSR) model. The observed coefficient of determination (R^2), residual prediction deviation, and root mean square error of prediction for the combined spectral bands were 0.73, 1.96, and 329.65 mg kg^{-1}, respectively; for the entire VNIR spectral bands were 0.71, 1.89, 341.88 mg kg^{-1}, respectively; for organic matter were 0.40, 1.31, 492.65 mg kg^{-1}, respectively; and for clay minerals were 0.54, 1.50, 430.26 mg kg^{-1}, respectively (Sun and Zhang, 2017). Based on these results, it was revealed that the use of a combined band is a reliable and feasible way to estimate soil Zn concentrations.

Even though VNIRS technique exhibited a better performance for predicting soil heavy metal concentration, it has limited potential for determining heavy metals in sediments. For instance, Jiang et al. (2018) studied Cd, Cu, Zn, Pb, Ni, Hg, Cr, and As concentrations in sediments collected from urban lake in Wuhan, China, and quantified the heavy metal(loid) concentrations based on the PLSR calibration model. They observed acceptable model prediction for Cd, Pb, Ni, and Hg (R^2 between 0.32 to 0.40) and unsatisfactory model results for Cu, Zn, Cr, and As (R^2 between 0.01 and 0.06). Authors suggested that the variability in accuracies were possibly due to the diverse relationships between heavy metal(loid) and spectrally active constitutes, such as total organic carbon (Jiang et al., 2018). Thus, the authors proposed that the accuracy of the measurements could be increased by adopting GA-PLSR and CARS-PLSR (competitive adaptive reweighted sampling PLSR) models, as they can remove uninformative spectral variables (Jiang et al., 2018). Figure 21.4 shows the technique for determining heavy metal concentrations in soils using VNIRS.

21.4.2.3 Immunochemical Techniques

Immunochemical techniques/immunoassays are quick, reliable, and cost-effective biochemical techniques to measure the presence or concentration of small molecules and macromolecules (analytes) using specific interactions between antibodies and antigens (Ju et al., 2017). Various constituents ranging from complex substances to simple molecules and industrial contaminants can be measured quantitatively, qualitatively, and semi-quantitatively using these biorecognition agents (Aga and Thurman, 1997). Antibodies are immunoglobulins (e.g., soluble proteins) produced by immune cells in response to stimulation by antigens. Formation of an antibody-antigen complex is measured by labeling (e.g., enzyme label) or using a labeling-free format (Ju et al., 2017).

Evaluation of organic contaminants, such as insecticides in soil was successfully carried out by Watanabe et al. (2016) using a commercially available kit-based enzyme-linked immunosorbent assay (ELISA). The study revealed that ELISA had an excellent analytical sensitivity for the determination of neonicotinoid insecticides in soils, namely clothianidin, dinotefuran, and imidacloprid. A recovery experiment using soil extract spiked with insecticide at a concentration of 2–10 ng mL^{-1} exhibited a high accuracy of 72%–126% and a precision of <16%, respectively, indicating the analytical reliability of ELISA (Watanabe et al., 2016). Chromatographic techniques such as gas chromatography (GC), liquid chromatography-mass spectrometry (LC-MS), and high-performance chromatography (HPLC) are extensively used for the determination of pesticides in soils due to their accuracy and analytical sensitivity (Miyawaki et al., 2018; Tadeo, 2019; Zhao et al., 2018). However, chromatographic techniques require intensive sample preparation prior to the analysis, and they are time consuming and expensive. On the other hand, the ELISA technique is superior compared to chromatographic techniques as it has a high selectivity to particular pesticides, does not require any complicated sample preparation, can measure several samples simultaneously, and requires less labor (Li et al., 2014; Mosallam et al., 2016).

Xu et al. (2012) developed a semi-quantitative one-step strip immunoassay using specific nanocolloidal gold-labeled monoclonal antibodies for rapid detection of neonicotinoid insecticides, such as thiamethoxam and imidacloprid residues in agricultural products. The strip assay that was developed was stable for >5 months under 4°C, which indicates its stability for relatively long-term usage. Moreover, authors suggested that the strip-based immunoassay is a simple and quick technique for the simultaneous detection of pesticides in agricultural products.

ELISA and gold immunochromatographic assay (GICA) were used by Li et al. (2014) for quantifying clothianidin residues in environmental samples (water, soil, and agricultural products) and were compared with HPLC data. ELISA showed an LOD of 3.8 ng mL^{-1} and the GICA showed a visual detection limit of 8 ng mL^{-1}. In the GICA technique, visual evidences were observed within 10 min of using gold nanoparticles as a tracer. The GICA did not require intensive work and equipment, but

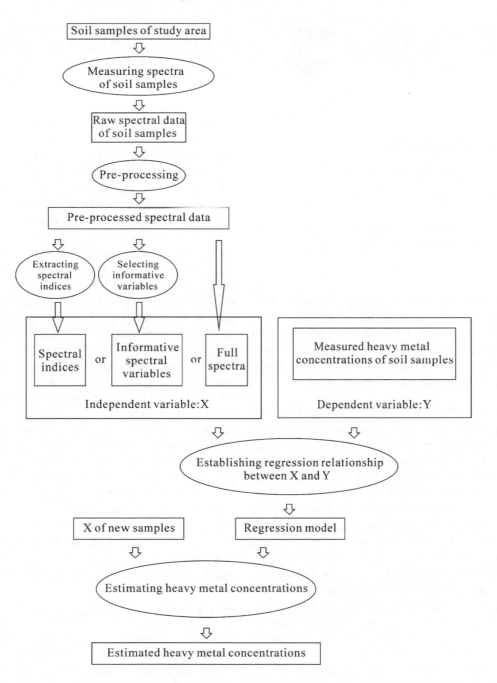

FIGURE 21.4 Technical approaches to estimating heavy metal concentrations in soils using visible and near-infrared reflectance spectroscopy. (Reproduced from Shi, T. et al., *J. Hazard. Mater.*, 265, 166–176, 2014.)

only a simple sample pretreatment. Thus, the authors suggest that GICA is a promising technique for *in situ* semi-quantitative detection of clothianidin (Li et al., 2014). Subsequently, quantitative assessments could be done in laboratories using ELISA. In addition, the used spiked test proved high accuracy of both immunoassays, whereas authentic samples were in good agreement with those of HPLC results (Li et al., 2014).

The US EPA documented the detection of PCB in contaminated soils and solution by competitive binding enzyme immunoassay using EnviroGard PCB Test Kit (US EPA Environmental Technology Verification Report). This field analytical technique can detect PCB rapidly at

specified levels of 1, 5, 10, and 50 ppm. In this technique, PCB in the collected soil samples are extracted using methanol. The extractant and PCB-enzyme conjugate are added to the antibody-coated test tubes, incubated, and rinsed, followed by color development. The color changes in the samples indicate the concentration of PCBs in the samples. Better performance characteristics, such as higher throughput, ease of use, completeness and higher data quality levels make EnviroGard PCB Test Kit ideal for PCB detection in contaminated fields (US EPA Environmental Technology Verification Report). Electrochemical immunosensors (EI) (Figure 21.5) can be used to determine contaminants in both soil and groundwater.

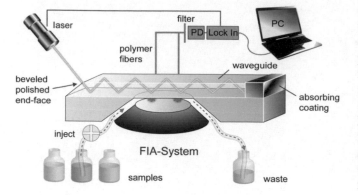

FIGURE 21.5 General scheme of an electrochemical immunosensor. (Reproduced from Zhang, Z. et al., *Trends Anal. Chem.*, 87, 49–57, 2017.)

EI can measure signal responses via changes in electron density or ion concentration resulting from antibody-antigen interaction (Zhang et al., 2017). EI has high sensitivity and high tolerance toward sample turbidity. Moreover, its portability, less power requirement, and low cost of equipment makes it ideal for field measurement (Wang, 2002).

21.4.2.4 Remote Sensing and GIS Technologies for Soil

When compiling distribution maps, spatial distribution of soil contaminants is generally carried out by systematic sampling, laboratory analysis, and interpolation of point data. However, these methods are time consuming, costly, and require more effort. In contrast, remote sensing is a rapid and inexpensive technique, wherein a high-quality imaging spectrometer called a hyperspectral is used for mapping feature events on the Earth (de Araujo Barbosa et al., 2015). Thus, the fundamental basis of this technique involves remotely capturing electromagnetic radiations that are reflected from the target (Aggarwal, 2004). Nevertheless, as soil contaminants (e.g., heavy metals) are spectrally featureless in the VNIR parts of the electromagnetic spectrum, the spectral signatures of features that are associated with soil contaminants (e.g., minerals, heavy metal stressed plants) could be used as indirect detection methods to map the contaminant dispersion. Several researches have shown the applicability of remote sensing spectroscopy in identifying heavy metal contaminated lands by using minerals containing heavy metals as indicators (Ferrier, 1999; Pascucci et al., 2012). Plants are also important indicators of soil heavy metal contamination, as they provide crucial evidence on the degree of soil pollution. Thus, soil heavy metal contaminations can be identified by remote sensing based on spectral responses to changes in chlorophyll content and photosynthesis of a certain plant (Zhuang, 2010). For instance, remote sensing data (e.g., satellite images) together with other integrated techniques (geostatistics, modeling) have been used to identify heavy metal contaminations in soils using rice stress levels (Liu et al., 2016, 2018a, 2019).

GIS is a computer-based technique for data collection, storage, management, analysis, modeling, and transformation of raw data into geospatial data for real world applications

(Sahu et al., 2015; Suh et al., 2017). Table 21.3 summarizes the development of GIS technologies over time. Technological development in techniques, such as remote sensing, global positioning systems (GPS), and GIS have significantly increased the effectiveness of soil survey. Remote sensing and GIS techniques have been used for monitoring various soil properties/phenomenon, namely soil erosion (Ganasri and Ramesh, 2016), changes in soil organic matter (Bhunia et al., 2019), and soil salinity issues (Asfaw et al., 2018; Singh, 2018), and for identifying soil pollutants such as heavy metal(loid)s (Shi et al., 2018). Remote sensing and ancillary data were used for mapping heavy metal(loid)s in contaminated soils in Qatar (Peng et al., 2016). The topsoil, multi-spectral images, spectral indices, and environmental variables were utilized to model and map the spatial distribution of As, Cr, Cu, Ni, Pb, and Zn in the soils. The used prediction model exhibited good predictive performances for all the heavy metal(loid)s, while Cu showed the highest R^2 of 0.74. Moreover, the generated topsoil maps of As, Cr, Cu, Ni, Pb, and Zn could be useful for implementing appropriate remediation actions (Peng et al., 2016).

In addition, GIS and GIS-integrated geostatistical analysis have been identified as an effective method for studying contaminated soil (Chen et al., 2016; Jin et al., 2019; Liang et al., 2017). A review by Hou et al. (2017) showed the potential of integrated GIS and multivariate statistical analysis for evaluating the regional distribution of heavy metals in contaminated soils. They suggested further research to enable the mapping of multivariate results of GIS to identify certain anthropogenic sources and to examine temporal trends besides spatial patterns. With the development of sensing technologies, multiple spectroscopic technologies (i.e., assimilated proximal and remote sensing) can be applied for horizontal and vertical monitoring of contaminants in the soil (Shi et al., 2018). Moreover, these integrated techniques may help to monitor soil contaminants rapidly and accurately at a large scale.

21.4.3 GROUNDWATER MONITORING

Groundwater monitoring is an essential aspect of groundwater management, and it provides information on groundwater quality and quantity (Sundaram et al., 2009). The groundwater volume and groundwater quality are two basic aspects that are concerned during groundwater monitoring. For groundwater volume, the underground water level, the flux, the velocity, and the flow path and direction need to be monitored and identified. For water quality, the basic water quality index (e.g., temperature, salinity, pH, Eh, oxidation-reduction potential (ORP), dissolved oxygen (DO), dissolved organic carbon (DOC), mineral cations and anions) and contamination index (e.g., EDCs, heavy metals, PAHs) need to be monitored. Groundwater monitoring on a regular basis could be beneficial for assessing different water quality parameters, thereby aiding proper groundwater management. In most cases, surface water monitoring is done separately despite the connection between groundwater and surface water monitoring. It should be noted that all these water resources are linked and they influence each other with regard to water quantity and quality

TABLE 21.3

Development of GIS Technology over Time

Development Aspects	Primary Stages of Development and Time Frames		
	Formative Years (1960–1980)	Maturing Technology (1980–Mid-1990s)	GI Infrastructure (Mid-1990s–Present)
Technical environment	Mainframe computer	Mainframe and mini-computers	Workstations and PCs
	Proprietary software	Geo-relational data structure	Network/Internet
	Proprietary data structure	Graphical user interface	Open systems design
	Mainly raster form	New data acquisition technologies (GPS, RS, redefinition of datum)	Multimedia
			Data integration
			Enterprise computing
			Object-relational data model
Major users	Government	Government	Government
	Utilities	Universities	Universities and schools
	Military	Utilities	Business
		Business	Utilities
		Military	Military
			General public
Major application areas	Land and resource management	Land and resource management	Land and resource management
	Census	Census	Census
	Surveying and mapping	Surveying and mapping	Surveying and mapping
		Facilities management	Facilities management
		Market analysis	Market analysis
			Utilities
			Geographic data browsing

Source: Jha, M.K. et al., *Water Resour. Manage.*, 21, 427–467, 2007.

(Hernández et al., 2015), and must be taken into account when planning groundwater monitoring. Such approaches help identify water quality and quantity concerns and could prove beneficial in safeguarding and remediating water sources (Yu et al., 2018). To accomplish such a harmonized monitoring, identifying efficient and novel monitoring techniques are important to optimize the process. Some techniques used for groundwater monitoring and protection are categorized into different groups, as shown in Figure 21.6.

21.4.3.1 Groundwater Sampling

Groundwater sampling is one of the key aspects of groundwater monitoring. Existing bores with natural (e.g., springs) and artificial (e.g., mine shafts or pits) features can be used for groundwater monitoring (Sundaram et al., 2009). Nevertheless, collection of samples from water sources, such as surface water and rainfall, are also needed to integrate those data with groundwater data to comprehend their effects and sources of contaminants. The conventional method of collecting groundwater involves drilling a bore hole to the required depth and then withdrawing water for analysis (Misstear, 2017). Auger drilling, rotary air/mud drilling, vibro-coring, and cable tool

drilling are some of the methods used for constructing bore holes for groundwater monitoring (Sundaram et al., 2009). However, these processes have some drawbacks, such as increase in contaminant concentrations, high disturbances to soil, time consumption, and high cost. In recent years, various new techniques have been employed with higher efficiencies for monitoring groundwater. Table 21.4 summarizes specific sampling procedures and precautions that need to be undertaken for effective groundwater quality monitoring.

21.4.3.1.1 Direct Push Technologies (DPTs)

Direct push technology (DPT) has received greater attention as an alternative to traditional monitoring methods due to its rapid sampling and data collection, and cost-effectiveness (Bourke et al., 2017). DPT can even be installed in unconsolidated materials such as clay, silt, sand, and gravel (Dietrich and Leven, 2009). It involves pushing and/or hammering small-diameter string of steel hollow rods to the desired depth. Usually, these rods can penetrate up to depths of 50 m (Dietrich and Leven, 2009). Once the desired depth is reached, samples are collected, and the obtained water samples are stored in airtight, chemically inert containers until analysis. Moreover, by

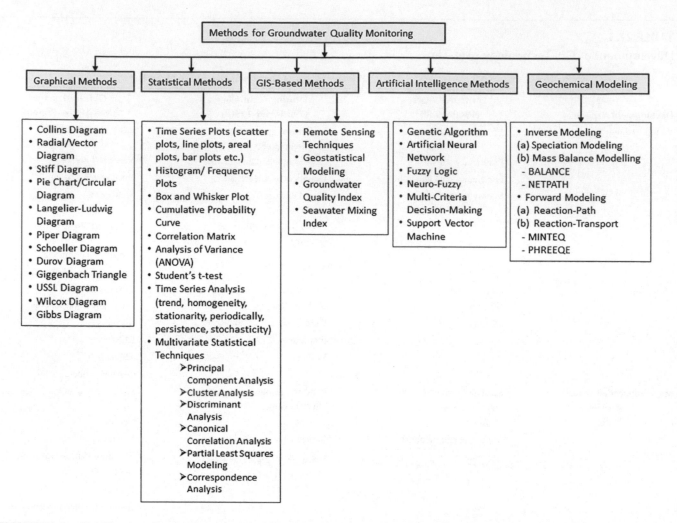

FIGURE 21.6 Classification of groundwater quality monitoring and protection techniques. (Reproduced from Machiwal, D. et al., *Environ. Earth Sci.*, 77, 681, 2018.)

attaching various devices (e.g., sensors) to the end of the steel rods, several groundwater parameters, such as pH, electrical conductivity, redox potential, dissolved oxygen, other gases, and temperature can be measured (Kurup et al., 2017).

21.4.3.1.2   Diffusion Membrane Sampling

Diffusion membrane sampling of contaminants in groundwater is a relatively new and convenient technique, which does not require withdrawing water from contaminated sites (Divine and McCray, 2004). In this technique, a container is filled with deionized water, a chemical reagent, solvent, or a porous adsorbent, and it is sealed with a diffusive membrane and placed in a well-installation apparatus (Verreydt et al., 2010). Then, the apparatus containing samplers is submerged in a monitoring well up to a desired depth and left undisturbed. With time, the contaminants in well water diffuse across the membrane and the contaminant concentrations in the sampler reach an equilibrium with well water (Divine and McCray, 2004; Verreydt et al., 2010). Generally, after 14 days (or less), the sampler is removed from the well and the water in the sampler is analyzed (Tunks et al., 2005). Various

groundwater contaminants, such as volatile organic compounds (VOC) (McHugh et al., 2018; Verreydt et al., 2010), heavy metal(loid)s (Vera et al., 2018), organic contaminants (Orera et al., 2018), PAHs, volatile chlorinated hydrocarbons, and volatile aromatic compounds, namely, benzene, toluene, and xylenes (BTEX) (Martin et al., 2003), can be monitored with the diffusion membrane sampling technique. Moreover, in recent years, several researchers have developed novel membranes materials for sampling contaminants in groundwater successfully (Elias et al., 2019; Franquet-Griell et al., 2017; Orera et al., 2018; Wei et al., 2019).

21.4.3.1.3   The BAT® System

The BAT groundwater monitoring system is supplied and supported by a commercial company. The BAT system is a portable and convenient method for long-term monitoring of groundwater, soil, and gas (INGEO, 2018). It consists of three basic components: a sealed filter tip attached to the extension pipe, an evacuated and sterilized sample vial to collect samples, and a double-ended injection needle (BAT Geosystems AB, 2018). The BAT system is installed at a preferred depth. The sampler is lowered down an extension pipe

TABLE 21.4

Specific Sampling Procedures and Precautions That Are Required for Effective Groundwater Quality Monitoring

Groundwater Contaminant	Sampling Procedure	Preferred Storage Containers	Storage Time/Temperature	Operational Difficulty/Cost
Major Ions Cl, SO$_4$, F, Na, K	• 0.45 µm filter only • No acidification	Any	7 days/4°C	Minimal
Trace Metals Fe, Mn, As, Cu, Zn, Pb, Cr, Cd, etc.	• Sealed 0.45 µm filter • Acidify (pH < 2) • Avoid aeration through splashing/head space	Plastic	150 days	Moderate
N Species NO$_3$, NH$_4$ (NO$_2$)	• Sealed 0.45 µm filter	Any	1 day/4°C	Moderate/low
Microbiological TC, FC, FS	• Sterile conditions • Unfiltered sample • On-site analysis preferred	Dark glass	6 hours/4°C	Moderate/low
Carbonate Equilibria pH, HCO$_3$, Ca, Mg	• Unfiltered well-sealed sample • On-site analysis (pH, HCO$_3$) • (Ca/Mg at base laboratory on acidified sample)	Any	1 hour (150 days)	Moderate
Oxygen Status pE(Eh), DO, T	• On-site in measuring cell • Avoid aeration • Unfiltered	Any	0.1 hour	High/moderate
Organics TOC, VOC, HC, ClHC, etc.	• Unfiltered sample • Avoid volatilization • Direct absorption in cartridges preferred	Dark glass or Teflon	1–7 days (indefinite for cartridges)	High

Source: Tuinhof, A. et al., World Bank. Sustainable groundwater management: Concepts and tools, Groundwater Monitoring Requirements for managing aquifer response and quality threats, Available at http://documents.worldbank.org/curated/en/334871468143053052/pdf/301070BriefingNote9.pdf, 2004.

onto the filter tip. Furthermore, a hypodermic needle makes a hydraulic connection with the groundwater by penetrating the self-sealing flexible septum in the filter tip. Water/gas samples are drawn into the sample tube due to suction in the sample tube and groundwater pressure. This process prevents sample contamination with air and keeps it hermetically sealed until analysis. Moreover, this sampling is not sensitive to human error and maintains a high quality every time.

21.4.3.2 Ion Trap Mass Spectrometry (ITMS)

ITMS is an *in situ* groundwater/soil-contaminant detection technique, which requires direct push wells to detect contaminants in real time using mass spectrometry. The ITMS method is a rapid technique and is ideal for field monitoring as it can detect relatively low concentrations of contaminants (ppb levels) and requires little or no sample preparation (Wise et al., 1995). For instance, an analysis of VOC in groundwater can be done within three minutes (Wise et al., 1995). Water sample containing VOC is introduced by purging the sample with a stream of helium and directing the purge stream into the ITMS (U.S. Department of Energy, 1998). Moreover, ITMS can be performed using tandem mass spectrometry (MS/MS) (Jabali et al., 2018). Thus, specific elements in complex matrices can be analyzed (Niessen and Falck, 2015). To reach high sensitivity and selectivity, the instrumental parameters of ITMS need to be carefully optimized based on the complexity of the groundwater sample. Jabali et al. (2018) successfully determined 16 PAHs in groundwater by a method developed with gas chromatography-ion trap tandem mass spectrometry (GC-ITMS/MS) associated with solid phase micro-extraction. They suggested a non-resonant excitation mode for PAHs analysis by MS/MS. The study found that the developed method was linear, precise, and selective. Moreover, the limits of quantification and LOD were of low concentrations (µg L^{-1}), indicating its high sensitivity (Jabali et al., 2018).

21.4.3.3 Analytical Techniques for Groundwater

Various advanced on-site and off-site techniques have been developed for monitoring groundwater. In many cases, following groundwater sampling, the samples are sent to off-site laboratories for testing. Though the laboratory analysis is time consuming and costly, more accurate results can be obtained with lower LOD. On the other hand, field analysis is rapid, but its sensitivity is not as superior compared to the laboratory analysis. Thus, sending some of the field samples for laboratory analysis and validation is required, as reported by US EPA (2002). Many laboratory techniques are available for analyzing a variety of contaminants in water. To determine heavy metals/trace elements in water samples, ICP-MS, atomic absorption spectroscopy (AAS), atomic emission spectroscopy (AES), and XRF techniques are used (Prasad et al., 2018; Sikdar and Kundu, 2018). Besides the laboratory XRF analysis, PXRF has also been used for analyzing polluted and/or salt affected waters. For instance, Pearson et al. (2017) analyzed 256 of water samples using PXRF, ICP-AES, and digital salinity bridge to assess the salt-affected waters. They found that the PXRF can predict water electrical conductivity by quantifying chlorine in water samples (Pearson et al., 2017). Moreover, the PXRF can detect some elements more effectively than other methods, indicating its potential for analyzing various metal contaminants in waters.

Techniques such as LC-MS and gas chromatography-mass spectrometry (GC-MS) are the commonly used analytical methods for determining organic pollutants in water samples (Lorenzo et al., 2018). However, in recent years, some integrated approaches exhibited relatively better performances for detecting organic contaminants in water. Interestingly, Hernández et al. (2015) developed a method to monitor large numbers of organic compounds in waters. In this work, hybrid quadrupole time-of-flight (QTOF) MS was combined with both LC and GC and used as a single instrument for the analysis. Qualitative validation was done for about 300 compounds, while establishing detection limits. The authors suggested that these types of methods are ideal for the "universal" screening of a large number of organic contaminants in water (Hernández et al., 2015). Similarly, an advanced analytical technique, that is LC and GC coupled to MS with quadrupole-time of flight analyzer (QTOF), was used by Pitarch et al. (2016) for screening ~1500 organic contaminants in surface water and groundwater.

21.4.3.4 Remote Sensing and GIS Technologies for Groundwater

Recent advancements in RS and GIS techniques have facilitated large-scale groundwater monitoring and its associated remediation efforts. Water contamination indicators, such as algal blooms, chlorophyll-a, suspended sediments, turbidity, and mineral content can be monitored by using improved spectral and spatial resolution sensors together with geospatial modeling. Integrated RS, GIS, geospatial modeling, and filed monitoring may further assist in identifying pollution zones and sources of pollutants to develop strategies for remediation. Moreover, production of high-resolution maps of pollution zones may further aid in assigning remediation techniques to the affected areas. Nath et al. (2018) carried out a comprehensive qualitative analysis using integrated GIS and geochemical-based investigations to assess the severity of metal(loid) contamination in the groundwater of 20 districts in India. The spatial distribution of metal(loid)s in groundwater was mapped using the ArcGIS software and a comprehensive groundwater quality index was generated. These outputs could be beneficial to the stakeholders to design appropriate remediation efforts in the area (Nath et al., 2018). Similarly, Tiwari et al. (2016) studied the spatial distribution of heavy metal(loid)s, such as Al, As, Ba, Cr, Cu, Fe, Mn, Ni, Se, and Zn, in 66 groundwater samples of the West Bokaro coalfield in India using the heavy metal pollution index (HPI) and GIS techniques. The sources of heavy metal(loid)s in the groundwater samples were of geological origin and from mining activities. The information generated

through this study would be useful to the locals who rely on groundwater resources, and could aid in future water resource management for the mining area.

Many studies utilized GIS and RS approaches to delineate the spatial distribution of groundwater quality parameters. Among others, Lavanya et al. (2019) interpreted several groundwater quality parameters using Q-GIS software to produce the spatial distribution maps of parameters, including pH, total dissolved solids, total hardness, Ca, Mg, Na, K, Cl, HCO_3, SO_4, and NO_3. They stated that the produced maps will be beneficial for users to assess the groundwater quality in the locality while making decisions for water quality improvement (Lavanya et al., 2019). Recent advances in GIS-integrated statistical techniques for groundwater monitoring were reviewed and documented by Machiwal et al. (2018). They suggested that the integration of various statistical analyses, different time series modeling, and water quality indices in GIS software could prove to be a vital research area for future studies. Moreover, RS and GIS-based groundwater monitoring in combination with field monitoring could be an important strategy for effective monitoring of contaminants in groundwater (Jha et al., 2007).

Ground penetrating radar (GPR) is an emerging geophysical technology for exploring groundwater (Maheswari et al., 2013). In this technique, pulses of ultra-high-frequency radio waves are transmitted into the groundwater using a transmitting antenna. Then, the receiving antenna receives the waves back based on the electrical properties of contact materials. As shown by Benson (1995), the signatures received from GPR corresponded with hydrocarbon contamination in some wells located in Arizona and Utah. In addition to the above-mentioned techniques, modeling techniques are also important for evaluating the level of contaminants and their transportation in an area (Figure 21.6). Modeling packages such as FEFlow (Trefry and Muffels, 2007), Visual Modflow (Rahman et al., 2019), and PHREEQC (Karmegam et al., 2011) are some of the useful modeling tools for describing and predicting the hydrology and chemical behavior of contaminants in groundwater. Furthermore, CXTFIT 2.0 STANMOD v. 2.8 is a novel modeling technique for evaluating contaminant plume transport (Jaramillo et al., 2019).

21.5 CONCLUSIONS

Anthropogenic activities as well as natural phenomena lead to soil and groundwater contamination. Minimizing soil and groundwater contamination is important to secure human and environmental health. Thus, monitoring contaminated sites is crucial to plan remediation techniques and to protect the affected areas. Various conventional methods are available for monitoring contaminants on-site and off-site. However, due to the limitations of these techniques, advanced and novel monitoring approaches are needed for effective and efficient soil and groundwater monitoring. For instance, cost effectiveness and rapid and accurate measurements are some of the benefits of the new techniques over the conventional methods. VNIRS, XRF/PXRF, and biochemical approaches, such as immunoassays, are some of the efficient, rapid, and accurate techniques for monitoring soil contaminants. XRF (or PERF) has been identified as an efficient method for monitoring environmental samples due to its capability of measuring a variety of metals in both soil and groundwater samples. Groundwater monitoring can be carried on-site or off-site. However, unlike soil sampling, groundwater sampling plays an important role in the groundwater monitoring process. DPTs, diffusion membrane sampling, and the BAT system are some of the efficient groundwater sampling methods. The obtained samples could be sent to laboratories to determine contaminants via techniques, such as inductively coupled plasma-mass spectrometry (ICP-MS), atomic absorption spectroscopy (AAS), atomic emission spectroscopy (AES), X-ray fluorescence spectrometry (XRF), liquid chromatography-mass spectrometry (LC-MS), and gas chromatography-mass spectrometry (GC-MS), for precise results. Ion trap mass spectrometry is one of the important methods of groundwater monitoring in real time. Remote sensing of soil and groundwater is beneficial in terms of spatial and temporal monitoring of contaminated sites. Integrated remote sensing and GIS techniques are low-cost and efficient methods and have added advantages over field measurements. Moreover, GIS and geostatistical techniques are closely connected and are useful tools for soil and groundwater monitoring process. Overall, the selection of an "appropriate" monitoring technique for a polluted soil or groundwater site is important for productive outcomes. Finally, the generated data are useful to policy-makers to identify issues and to make decisions on soil and groundwater remediation and pollution prevention, and to safeguard the soil and water.

ACKNOWLEDGMENT

This work was carried out with the support of "Cooperative Research Program for Agriculture Science and Technology Development (Effects of Plastic Mulch Wastes on Crop Productivity and Agro-Environment, Project No. PJ01475801)," Rural Development Administration, Republic of Korea.

REFERENCES

Aga, Diana S., and E. M. Thurman. 1997. Environmental immunoassays: alternative techniques for soil and water analysis. ACS Publications.

Aggarwal, S. 2004. Principles of remote sensing. *Proceedings of the Training Workshop* 7–11 July, 2003, Dehra Dun, India

Alsbou, E.M.E., Al-Khashman, O.A. 2017. Heavy metal concentrations in roadside soil and street dust from Petra region, Jordan. *Environmental Monitoring and Assessment*, **190**(1), 48.

Antoniadis, V., Levizou, E., Shaheen, S.M., Ok, Y.S., Sebastian, A., Baum, C., Prasad, M.N., Wenzel, W.W., Rinklebe, J. 2017. Trace elements in the soil-plant interface: Phytoavailability, translocation, and phytoremediation – A review. *Earth-Science Reviews*, **171**, 621–645.

Asfaw, E., Suryabhagavan, K., Argaw, M. 2018. Soil salinity modeling and mapping using remote sensing and GIS: The case of Wonji sugar cane irrigation farm, Ethiopia. *Journal of the Saudi Society of Agricultural Sciences*, **17**(3), 250–258.

Atekwana, E.A., Geyer, C.J. 2018. Spatial and temporal variations in the geochemistry of shallow groundwater contaminated with nitrate at a residential site. *Environmental Science and Pollution Research*, **25**(27), 27155–27172.

BAT Geosystems AB. 2018. Groundwater Monitoring System. Available at http://www.bat-gms.com/pdf/BAT-GMS-Brochure.pdf, Accessed on May 21, 2019.

Benson, A.K. 1995. Applications of ground penetrating radar in assessing some geological hazards: Examples of groundwater contamination, faults, cavities. *Journal of Applied Geophysics*, **33**(1), 177–193.

Bhunia, G.S., Shit, P.K., Pourghasemi, H.R., Edalat, M. 2019. Prediction of soil organic carbon and its mapping using regression analyses and remote sensing data in GIS and R. In: *Spatial Modeling in GIS and R for Earth and Environmental Sciences*, Elsevier, pp. 429–450 Elsevier, Netherland For more info: https://doi.org/10.1016/B978-0-12-815226-3.00019-3.

Bi, X., Luo, W., Gao, J., Xu, L., Guo, J., Zhang, Q., Romesh, K.Y., Giesy, J.P., Kang, S., de Boer, J. 2016. Polycyclic aromatic hydrocarbons in soils from the Central-Himalaya region: Distribution, sources, and risks to humans and wildlife. *Science of The Total Environment*, **556**, 12–22.

Bolan, N., Kunhikrishnan, A., Gibbs, J. 2013. Rhizoreduction of arsenate and chromate in Australian native grass, shrub and tree vegetation. *Plant and Soil*, **367**(1–2), 615–625.

Bolan, N., Kunhikrishnan, A., Thangarajan, R., Kumpiene, J., Park, J., Makino, T., Kirkham, M.B., Scheckel, K. 2014. Remediation of heavy metal(loid)s contaminated soils–to mobilize or to immobilize? *Journal of Hazardous Materials*, **266**, 141–166.

Bourke, S.A., Hermann, K.J., Hendry, M.J. 2017. High-resolution vertical profiles of groundwater electrical conductivity (EC) and chloride from direct-push EC logs. *Hydrogeology Journal*, **25**(7), 2151–2162.

Boy-Roura, M., Mas-Pla, J., Petrovic, M., Gros, M., Soler, D., Brusi, D., Menció, A. 2018. Towards the understanding of antibiotic occurrence and transport in groundwater: Findings from the Baix Fluvià alluvial aquifer (NE Catalonia, Spain). *Science of the Total Environment*, **612**, 1387–1406.

Chen, T., Chang, Q., Liu, J., Clevers, J., Kooistra, L. 2016. Identification of soil heavy metal sources and improvement in spatial mapping based on soil spectral information: A case study in northwest China. *Science of the Total Environment*, **565**, 155–164.

Choppala, G., Bolan, N., Kunhikrishnan, A., Bush, R. 2016. Differential effect of biochar upon reduction-induced mobility and bioavailability of arsenate and chromate. *Chemosphere*, **144**, 374–381.

Das, H., Mitra, A.K., Sengupta, P., Hossain, A., Islam, F., Rabbani, G. 2004. Arsenic concentrations in rice, vegetables, and fish in Bangladesh: A preliminary study. *Environment International*, **30**(3), 383–387.

de Araujo Barbosa, C.C., Atkinson, P.M., Dearing, J.A. 2015. Remote sensing of ecosystem services: A systematic review. *Ecological Indicators*, **52**, 430–443.

DEFRA. 2006. Defra Circular 01/2006, Environmental Protection Act 1990: Part 2A, Contaminated Land. Defra – Department for Environment, Food and Rural Affairs, London, UK.

Dietrich, P., Leven, C. 2009. Direct push-technologies. In: *Groundwater Geophysics*, Springer, Berlin, Germany, pp. 347–366.

Divine, C.E., McCray, J.E. 2004. Estimation of membrane diffusion coefficients and equilibration times for low-density polyethylene passive diffusion samplers. *Environmental Science & Technology*, **38**(6), 1849–1857.

Ebele, A.J., Abdallah, M.A.-E., Harrad, S. 2017. Pharmaceuticals and personal care products (PPCPs) in the freshwater aquatic environment. *Emerging Contaminants*, **3**(1), 1–16.

El-Naggar, A., Shaheen, S.M., Hseu, Z.-Y., Wang, S.-L., Ok, Y.S., Rinklebe, J. 2019. Release dynamics of As, Co, and Mo in a biochar treated soil under pre-definite redox conditions. *Science of the Total Environment*, **657**, 686–695.

El-Naggar, A., Shaheen, S.M., Ok, Y.S., Rinklebe, J. 2018. Biochar affects the dissolved and colloidal concentrations of Cd, Cu, Ni, and Zn and their phytoavailability and potential mobility in a mining soil under dynamic redox-conditions. *Science of the Total Environment*, **624**, 1059–1071.

Elias, G., Díez, S., Fontàs, C. 2019. System for mercury preconcentration in natural waters based on a polymer inclusion membrane incorporating an ionic liquid. *Journal of Hazardous Materials* 371, 316–322.

Fang, Y., Nie, Z., Die, Q., Tian, Y., Liu, F., He, J., Huang, Q. 2017. Organochlorine pesticides in soil, air, and vegetation at and around a contaminated site in southwestern China: Concentration, transmission, and risk evaluation. *Chemosphere*, **178**, 340–349.

Ferrier, G. 1999. Application of imaging spectrometer data in identifying environmental pollution caused by mining at Rodaquilar, Spain. *Remote Sensing of Environment*, **68**(2), 125–137.

Franquet-Griell, H., Pueyo, V., Silva, J., Orera, V.M., Lacorte, S. 2017. Development of a macroporous ceramic passive sampler for the monitoring of cytostatic drugs in water. *Chemosphere*, **182**, 681–690.

Ganasri, B., Ramesh, H. 2016. Assessment of soil erosion by RUSLE model using remote sensing and GIS-A case study of Nethravathi Basin. *Geoscience Frontiers*, **7**(6), 953–961.

Geng, N., Wu, Y., Zhang, M., Tsang, D.C.W., Rinklebe, J., Xia, Y., Lu, D., Zhu, L., Palansooriya, K.N., Kim, K.-H., Ok, Y.S. 2019. Bioaccumulation of potentially toxic elements by submerged plants and biofilms: A critical review. *Environment International*, **131**, 105015.

Gottschall, N., Topp, E., Metcalfe, C., Edwards, M., Payne, M., Kleywegt, S., Russell, P., Lapen, D. 2012. Pharmaceutical and personal care products in groundwater, subsurface drainage, soil, and wheat grain, following a high single application of municipal biosolids to a field. *Chemosphere*, **87**(2), 194–203.

Grondona, S.I., Gonzalez, M., Martínez, D.E., Massone, H.E., Miglioranza, K.S.B. 2019. Assessment of organochlorine pesticides in phreatic aquifer of Pampean Region, Argentina. *Bulletin of Environmental Contamination and Toxicology*, **102**(4), 544–549.

Gu, Q., Yang, Z., Yu, T., Yang, Q., Hou, Q., Zhang, Q. 2018. From soil to rice – a typical study of transfer and bioaccumulation of heavy metals in China. *Acta Agriculturae Scandinavica, Section B—Soil & Plant Science*, **68**(7), 631–642.

Haddaoui, I., Mahjoub, O., Mahjoub, B., Boujelben, A., Di Bella, G. 2016. Occurrence and distribution of PAHs, PCBs, and chlorinated pesticides in Tunisian soil irrigated with treated wastewater. *Chemosphere*, **146**, 195–205.

Hepburn, E., Northway, A., Bekele, D., Liu, G.-J., Currell, M. 2018. A method for separation of heavy metal sources in urban groundwater using multiple lines of evidence. *Environmental Pollution*, **241**, 787–799.

Hernández, F., Ibáñez, M., Portolés, T., Cervera, M.I., Sancho, J.V., López, F.J. 2015. Advancing towards universal screening for organic pollutants in waters. *Journal of Hazardous Materials*, **282**, 86–95.

Hossain, M.F. 2006. Arsenic contamination in Bangladesh—an overview. *Agriculture, Ecosystems & Environment*, **113**(1–4), 1–16.

Hou, D., O'Connor, D., Nathanail, P., Tian, L., Ma, Y. 2017. Integrated GIS and multivariate statistical analysis for regional scale assessment of heavy metal soil contamination: A critical review. *Environmental Pollution*, **231**, 1188–1200.

Huang, Y., Wang, L., Wang, W., Li, T., He, Z., Yang, X. 2019. Current status of agricultural soil pollution by heavy metals in China: A meta-analysis. *Science of the Total Environment*, **651**, 3034–3042.

Igalavithana, A.D., Lee, S.-E., Lee, Y.H., Tsang, D.C.W., Rinklebe, J., Kwon, E.E., Ok, Y.S. 2017. Heavy metal immobilization and microbial community abundance by vegetable waste and pine cone biochar of agricultural soils. *Chemosphere*, **174**, 593–603.

IGRAC. 2008. International Groundwater Resources Assessment Centre. Guideline on: Groundwater monitoring for general reference purposes.

INGEO. 2018. Geological and Geotechnical Research. Available at http://www.ingeo.com.pl/the-bat-system,34,en.html#kot, Accessed on May 22, 2019.

Islam, M.N., Park, M., Jo, Y.-T., Nguyen, X.P., Park, S.-S., Chung, S.-Y., Park, J.-H. 2017. Distribution, sources, and toxicity assessment of polycyclic aromatic hydrocarbons in surface soils of the Gwangju City, Korea. *Journal of Geochemical Exploration*, **180**, 52–60.

Jabali, Y., Millet, M., El-Hoz, M. 2018. Determination of 16 polycyclic aromatic hydrocarbons (PAHs) in surface and groundwater in North Lebanon by using SPME followed by GC–ITMS/MS. *Euro-Mediterranean Journal for Environmental Integration*, **3**(1), 24.

Jaramillo, M., Grischek, T., Boernick, H., Velez, J.I. 2019. Evaluation of riverbank filtration in the removal of pesticides: An approximation using column experiments and contaminant transport modeling. *Clean Technologies and Environmental Policy*, **21**(1), 179–199.

Jha, M.K., Chowdhury, A., Chowdary, V.M., Peiffer, S. 2007. Groundwater management and development by integrated remote sensing and geographic information systems: Prospects and constraints. *Water Resources Management*, **21**(2), 427–467.

Jia, H., Qian, H., Qu, W., Zheng, L., Feng, W., Ren, W. 2019. Fluoride occurrence and human health risk in drinking water wells from Southern edge of Chinese Loess Plateau. *International Journal of Environmental Research and Public Health*, **16**(10), 1683.

Jiang, Q., Liu, M., Wang, J., Liu, F. 2018. Feasibility of using visible and near-infrared reflectance spectroscopy to monitor heavy metal contaminants in urban lake sediment. *CATENA*, **162**, 72–79.

Jin, Y., O'Connor, D., Ok, Y.S., Tsang, D.C., Liu, A., Hou, D. 2019. Assessment of sources of heavy metals in soil and dust at children's playgrounds in Beijing using GIS and multivariate statistical analysis. *Environment International*, **124**, 320–328.

Ju, H., Lai, G., Yan, F. 2017. *Immunosensing for Detection of Protein Biomarkers*. Elsevier, Amsterdam, the Netherlands.

Kalnicky, D.J., Singhvi, R. 2001. Field portable XRF analysis of environmental samples. *Journal of Hazardous Materials*, **83**(1–2), 93–122.

Kaniu, M., Angeyo, K., Mwala, A., Mangala, M. 2012. Direct rapid analysis of trace bioavailable soil macronutrients by chemometrics-assisted energy dispersive X-ray fluorescence and scattering spectrometry. *Analytica Chimica Acta*, **729**, 21–25.

Karmegam, U., Chidambaram, S., Prasanna, M.V., Sasidhar, P., Manikandan, S., Johnsonbabu, G., Dheivanayaki, V. et al. 2011. A study on the mixing proportion in groundwater samples by using Piper diagram and Phreeqc model. *Chinese Journal of Geochemistry*, **30**(4), 490.

Kawagoshi, Y., Suenaga, Y., Chi, N.L., Hama, T., Ito, H., Van Duc, L. 2019. Understanding nitrate contamination based on the relationship between changes in groundwater levels and changes in water quality with precipitation fluctuations. *Science of the Total Environment*, **657**, 146–153.

Kim, A.W., Vane, C.H., Moss-Hayes, V.L., Beriro, D.J., Nathanail, C.P., Fordyce, F.M., Everett, P.A. 2019. Polycyclic aromatic hydrocarbons (PAHs) and polychlorinated biphenyls (PCBs) in urban soils of Glasgow, UK. *Earth and Environmental Science Transactions of the Royal Society of Edinburgh*, **108**(2–3), 231–247.

Kurup, P., Sullivan, C., Hannagan, R., Yu, S., Azimi, H., Robertson, S., Ryan, D., Nagarajan, R., Ponrathnam, T., Howe, G. 2017. A review of technologies for characterization of heavy metal contaminants. *Indian Geotechnical Journal*, **47**(4), 421–436.

Kuśmierz, M., Oleszczuk, P., Kraska, P., Pałys, E., Andruszczak, S. 2016. Persistence of polycyclic aromatic hydrocarbons (PAHs) in biochar-amended soil. *Chemosphere*, **146**, 272–279.

Kwon, H.-O., Choi, S.-D. 2014. Polycyclic aromatic hydrocarbons (PAHs) in soils from a multi-industrial city, South Korea. *Science of the Total Environment*, **470–471**, 1494–1501.

Laidlaw, M.A., Alankarage, D.H., Reichman, S.M., Taylor, M.P., Ball, A.S. 2018. Assessment of soil metal concentrations in residential and community vegetable gardens in Melbourne, Australia. *Chemosphere*, **199**, 303–311.

Lapworth, D., Baran, N., Stuart, M., Ward, R. 2012. Emerging organic contaminants in groundwater: A review of sources, fate and occurrence. *Environmental Pollution*, **163**, 287–303.

Lapworth, D., Das, P., Shaw, A., Mukherjee, A., Civil, W., Petersen, J., Gooddy, D., Wakefield, O., Finlayson, A., Krishan, G. 2018. Deep urban groundwater vulnerability in India revealed through the use of emerging organic contaminants and residence time tracers. *Environmental Pollution*, **240**, 938–949.

Lavanya, G., Srinivas, P., Sourabh, K.G., Naidu, G.B. 2019. Spatial distribution of groundwater quality parameters in Amaravathi Region—A GIS and remote sensing approach. *Proceedings of International Conference on Remote Sensing for Disaster Management*. Springer, Cham, Switzerland, pp. 779–793.

Lee, H., Choi, Y., Suh, J., Lee, S.-H. 2016. Mapping copper and lead concentrations at abandoned mine areas using element analysis data from ICP–AES and portable XRF instruments: A comparative study. *International Journal of Environmental Research and Public Health*, **13**(4), 384.

Li, M., Hua, X., Ma, M., Liu, J., Zhou, L., Wang, M. 2014. Detecting clothianidin residues in environmental and agricultural samples using rapid, sensitive enzyme-linked immunosorbent assay and gold immunochromatographic assay. *Science of the Total Environment*, **499**, 1–6.

Liang, J., Feng, C., Zeng, G., Gao, X., Zhong, M., Li, X., Li, X., He, X., Fang, Y. 2017. Spatial distribution and source identification of heavy metals in surface soils in a typical coal mine city, Lianyuan, China. *Environmental Pollution*, **225**, 681–690.

Liu, M., Liu, X., Zhang, B., Ding, C. 2016. Regional heavy metal pollution in crops by integrating physiological function variability with spatio-temporal stability using multi-temporal thermal remote sensing. *International Journal of Applied Earth Observation and Geoinformation*, **51**, 91–102.

Liu, M., Skidmore, A.K., Wang, T., Liu, X., Wu, L., Tian, L. 2019. An approach for heavy metal pollution detected from spatio-temporal stability of stress in rice using satellite images. *International Journal of Applied Earth Observation and Geoinformation*, **80**, 230–239.

Liu, M., Wang, T., Skidmore, A.K., Liu, X. 2018a. Heavy metal-induced stress in rice crops detected using multi-temporal Sentinel-2 satellite images. *Science of the Total Environment*, **637–638**, 18–29.

Liu, S., Qi, S., Luo, Z., Liu, F., Ding, Y., Huang, H., Chen, Z., Cheng, S. 2018b. The origin of high hydrocarbon groundwater in shallow Triassic aquifer in Northwest Guizhou, China. *Environmental Geochemistry and Health*, **40**(1), 415–433.

Loganathan, P., Hedley, M., Grace, N., Lee, J., Cronin, S., Bolan, N., Zanders, J. 2003. Fertiliser contaminants in New Zealand grazed pasture with special reference to cadmium and fluorine—a review. *Soil Research*, **41**(3), 501–532.

Loos, R., Locoro, G., Comero, S., Contini, S., Schwesig, D., Werres, F., Balsaa, P., Gans, O., Weiss, S., Blaha, L. 2010. Pan-European survey on the occurrence of selected polar organic persistent pollutants in ground water. *Water Research*, **44**(14), 4115–4126.

López-Macias, R., Cobos-Gasca, V., Cabañas-Vargas, D., Rendón von Osten, J. 2019. Presence and spatial distribution of polynuclear aromatic hydrocarbons (PAHs) in groundwater of Merida City, Yucatan, Mexico. *Bulletin of Environmental Contamination and Toxicology*, **102**(4), 538–543.

Lord, G.G., Karp, K.E., Elmer, J., Hamrick, J. 2012. Portable XRF to guide a groundwater source removal action at the Cañon City, CO, USA Uranium Mill. In: *The New Uranium Mining Boom: Challenge and Lessons Learned* (Eds.) B. Merkel, M. Schipek, Springer, Berlin, Germany, pp. 361–370.

Lorenzo, M., Campo, J., Picó, Y. 2018. Analytical challenges to determine emerging persistent organic pollutants in aquatic ecosystems. *TrAC Trends in Analytical Chemistry*, **103**, 137–155.

Lu, Z., Zeng, F., Xue, N., Li, F. 2012. Occurrence and distribution of polycyclic aromatic hydrocarbons in organo-mineral particles of alluvial sandy soil profiles at a petroleum-contaminated site. *Science of the Total Environment*, **433**, 50–57.

Machiwal, D., Cloutier, V., Güler, C., Kazakis, N. 2018. A review of GIS-integrated statistical techniques for groundwater quality evaluation and protection. *Environmental Earth Sciences*, **77**(19), 681.

Maheswari, K., Senthil Kumar, P., Mysaiah, D., Ratnamala, K., Sri Hari Rao, M., Seshunarayana, T. 2013. Ground penetrating radar for groundwater exploration in granitic terrains: A case study from Hyderabad. *Journal of the Geological Society of India*, **81**(6), 781–790.

Mahimairaja, S., Bolan, N., Adriano, D., Robinson, B. 2005. Arsenic contamination and its risk management in complex environmental settings. *Advances in Agronomy*, **86**, 1–82.

Martin, H., Patterson, B.M., Davis, G.B., Grathwohl, P. 2003. Field trial of contaminant groundwater monitoring: Comparing time-integrating ceramic dosimeters and conventional water sampling. *Environmental Science & Technology*, **37**(7), 1360–1364.

McHugh, T.E., Molofsky, L.J., Adamson, D.T., Newell, C.J. 2018. Long-exposure, time-integrated sampler for groundwater or the like, Google Patents.

Melnyk, A., Dettlaff, A., Kuklińska, K., Namieśnik, J., Wolska, L. 2015. Concentration and sources of polycyclic aromatic hydrocarbons (PAHs) and polychlorinated biphenyls (PCBs) in surface soil near a municipal solid waste (MSW) landfill. *Science of the Total Environment*, **530–531**, 18–27.

Mihaileanu, R.G., Neamtiu, I.A., Fleming, M., Pop, C., Bloom, M.S., Roba, C., Surcel, M., Stamatian, F., Gurzau, E. 2018. Assessment of heavy metals (total chromium, lead, and manganese) contamination of residential soil and homegrown vegetables near a former chemical manufacturing facility in Tarnaveni, Romania. *Environmental Monitoring and Assessment*, **191**(1), 8.

Misstear, B., Banks, D. and Clark, L. 2017. Groundwater sampling and analysis. In: *Water Wells and Boreholes*. John Wiley & Sons, Hoboken, NJ.

Miyawaki, T., Tobiishi, K., Takenaka, S., Kadokami, K. 2018. A rapid method, combining microwave-assisted extraction and gas chromatography-mass spectrometry with a database, for determining organochlorine pesticides and polycyclic aromatic hydrocarbons in soils and sediments. *Soil and Sediment Contamination: An International Journal*, **27**(1), 31–45.

Morvan, X., Saby, N.P.A., Arrouays, D., Le Bas, C., Jones, R.J.A., Verheijen, F.G.A., Bellamy, P.H., Stephens, M., Kibblewhite, M.G. 2008. Soil monitoring in Europe: A review of existing systems and requirements for harmonisation. *Science of the Total Environment*, **391**(1), 1–12.

Mosallam, E., Aly, N., Ahmed, N., El-Gendy, K. 2016. Development of an enzyme immunoassay for detection of fipronil in environmental samples. *International Journal Advanced Science Research and Management*, **1**(12), 13–23.

Nath, B., Chaliha, C., Bhuyan, B., Kalita, E., Baruah, D., Bhagabati, A. 2018. GIS mapping-based impact assessment of groundwater contamination by arsenic and other heavy metal contaminants in the Brahmaputra River valley: A water quality assessment study. *Journal of Cleaner Production*, **201**, 1001–1011.

Nerger, R., Beylich, A., Fohrer, N. 2016. Long-term monitoring of soil quality changes in Northern Germany. *Geoderma Regional*, **7**(2), 239–249.

Niessen, W.M.A., Falck, D. 2015. Introduction to mass spectrometry, a Tutorial. In: *Analyzing Biomolecular Interactions by Mass Spectrometry*, Wiley Online Library, New Jersey.

Orera, V.M., Silva, J., Franquet-Griell, H., Lacorte, S. 2018. Design and characterization of macroporous alumina membranes for passive samplers of water contaminants. *Journal of the European Ceramic Society*, **38**(4), 1853–1859.

Padilla, J.T., Hormes, J., Selim, H.M. 2019. Use of portable XRF: Effect of thickness and antecedent moisture of soils on measured concentration of trace elements. *Geoderma*, **337**, 143–149.

Palansooriya, K.N., Ok, Y.S., Awad, Y.M., Lee, S.S., Sung, J.-K., Koutsospyros, A., Moon, D.H. 2019a. Impacts of biochar application on upland agriculture: A review. *Journal of Environmental Management*, **234**, 52–64.

Palansooriya, K.N., Wong, J.T.F., Hashimoto, Y., Huang, L., Rinklebe, J., Chang, S.X., Bolan, N., Wang, H., Ok, Y.S. 2019b. Response of microbial communities to biochar-amended soils: A critical review. *Biochar*, **1**(1), 3–22.

Palansooriya, K.N., Yang, Y., Tsang, Y.F., Sarkar, B., Hou, D., Cao, X., Meers, E., Rinklebe, J., Kim, K.-H., Ok, Y.S. 2019c. Occurrence of contaminants in drinking water sources and the potential of biochar for water quality improvement: A review. *Critical Reviews in Environmental Science and Technology*. doi.org/10.1080/10643389.2019.1629803.

Park, J.H., Lamb, D., Paneerselvam, P., Choppala, G., Bolan, N., Chung, J.-W. 2011. Role of organic amendments on enhanced bioremediation of heavy metal(loid) contaminated soils. *Journal of Hazardous Materials*, **185**(2–3), 549–574.

Parsons, C., Grabulosa, E.M., Pili, E., Floor, G.H., Roman-Ross, G., Charlet, L. 2013. Quantification of trace arsenic in soils by field-portable X-ray fluorescence spectrometry: Considerations for sample preparation and measurement conditions. *Journal of Hazardous Materials*, **262**, 1213–1222.

Pascucci, S., Belviso, C., Cavalli, R.M., Palombo, A., Pignatti, S., Santini, F. 2012. Using imaging spectroscopy to map red mud dust waste: The Podgorica Aluminum Complex case study. *Remote Sensing of Environment*, **123**, 139–154.

Pearson, D., Chakraborty, S., Duda, B., Li, B., Weindorf, D.C., Deb, S., Brevik, E., Ray, D. 2017. Water analysis via portable X-ray fluorescence spectrometry. *Journal of Hydrology*, **544**, 172–179.

Peng, T., O'Connor, D., Zhao, B., Jin, Y., Zhang, Y., Tian, L., Zheng, N., Li, X., Hou, D. 2019. Spatial distribution of lead contamination in soil and equipment dust at children's playgrounds in Beijing, China. *Environmental Pollution*, **245**, 363–370.

Peng, X., Ou, W., Wang, C., Wang, Z., Huang, Q., Jin, J., Tan, J. 2014. Occurrence and ecological potential of pharmaceuticals and personal care products in groundwater and reservoirs in the vicinity of municipal landfills in China. *Science of the Total Environment*, **490**, 889–898.

Peng, Y., Kheir, R., Adhikari, K., Malinowski, R., Greve, M., Knadel, M., Greve, M. 2016. Digital mapping of toxic metals in Qatari soils using remote sensing and ancillary data. *Remote Sensing*, **8**(12), 1003.

Pérez-Fernández, V., Rocca, L.M., Tomai, P., Fanali, S., Gentili, A. 2017. Recent advancements and future trends in environmental analysis: Sample preparation, liquid chromatography and mass spectrometry. *Analytica Chimica Acta*, **983**, 9–41.

Pitarch, E., Cervera, M.I., Portolés, T., Ibáñez, M., Barreda, M., Renau-Pruñonosa, A., Morell, I., López, F., Albarrán, F., Hernández, F. 2016. Comprehensive monitoring of organic micro-pollutants in surface and groundwater in the surrounding of a solid-waste treatment plant of Castellón, Spain. *Science of the Total Environment*, **548**, 211–220.

Prasad, M., Anil, K., Ramola, R. 2018. Radiological and chemical risk assessment from the exposure to uranium and heavy metals in drinking groundwater. *Proceedings of the Eight Biennial Symposium on Emerging Trends in Separation Science and Technology: Abstract Book*.

Rahman, T.U., Ahsan, N.U., Habib, A., Ara, A. 2019. Assessment of Saline Water Intrusion in Southwest Coastal Aquifer, Bangladesh Using Visual MODFLOW. In: *Advances in Sustainable and Environmental Hydrology, Hydrogeology, Hydrochemistry and Water Resources*, Springer, Cham, Switzerland, pp. 175–177.

Rochaddi, B., Sabdono, A., Zainuri, M. 2019. Preliminary study on the contamination of organophosphate pesticide (chlorpyrifos) in shallow coastal groundwater aquifer of Surabaya and Sidoarjo, East Java Indonesia. *IOP Conference Series: Earth and Environmental Science*, **246**, 012079.

Rose, M.T., Cavagnaro, T.R., Scanlan, C.A., Rose, T.J., Vancov, T., Kimber, S., Kennedy, I.R., Kookana, R.S., Van Zwieten, L. 2016. Impact of herbicides on soil biology and function. In: *Advances in Agronomy*, Vol. 136, Elsevier, Amsterdam, the Netherlands, pp. 133–220.

Rossel, R.V., Walvoort, D., McBratney, A., Janik, L.J., Skjemstad, J. 2006. Visible, near infrared, mid infrared or combined diffuse reflectance spectroscopy for simultaneous assessment of various soil properties. *Geoderma*, **131**(1–2), 59–75.

Rouillon, M., Taylor, M.P. 2016. Can field portable X-ray fluorescence (pXRF) produce high quality data for application in environmental contamination research? *Environmental Pollution*, **214**, 255–264.

Sahu, N., Reddy, G., Kumar, N., Nagaraju, M. 2015. High resolution remote sensing, GPS and GIS in soil resource mapping and characterization–A review. *Agricultural Reviews*, **36**(1) pp. 14–25.

Saleh, H.N., Panahande, M., Yousefi, M., Asghari, F.B., Conti, G.O., Talaee, E. and Mohammadi, A.A., 2019. Carcinogenic and non-carcinogenic risk assessment of heavy metals in groundwater wells in Neyshabur Plain, Iran. *Biological trace element research*, **190**(1), 251–261.

Samia, K., Dhouha, A., Anis, C., Ammar, M., Rim, A., Abdelkrim, C. 2018. Assessment of organic pollutants (PAH and PCB) in surface water: Sediments and shallow groundwater of Grombalia watershed in northeast of Tunisia. *Arabian Journal of Geosciences*, **11**(2), 34.

Sandoval, M.A., Fuentes, R., Nava, J.L., Coreño, O., Li, Y., Hernández, J.H. 2019. Simultaneous removal of fluoride and arsenic from groundwater by electrocoagulation using a filter-press flow reactor with a three-cell stack. *Separation and Purification Technology*, **208**, 208–216.

Schaafsma, A., Limay-Rios, V., Baute, T., Smith, J., Xue, Y. 2015. Neonicotinoid insecticide residues in surface water and soil associated with commercial maize (Corn) fields in Southwestern Ontario. *PLoS One*, **10**(2), e0118139.

Seshadri, B., Bolan, N., Choppala, G., Kunhikrishnan, A., Sanderson, P., Wang, H., Currie, L., Tsang, D.C., Ok, Y.S., Kim, G. 2017. Potential value of phosphate compounds in enhancing immobilization and reducing bioavailability of mixed heavy metal contaminants in shooting range soil. *Chemosphere*, **184**, 197–206.

Shah, A.H., Shahid, M., Khalid, S., Natasha, Shabbir, Z., Bakhat, H.F., Murtaza, B. et al. 2019. Assessment of arsenic exposure by drinking well water and associated carcinogenic risk in peri-urban areas of Vehari, Pakistan. *Environmental Geochemistry and Health*, pp. 1–13 DOI: https://doi.org/10.1007/s10653-019-00306-6.

Shaheen, S.M., Niazi, N.K., Hassan, N.E., Bibi, I., Wang, H., Tsang, D.C., Ok, Y.S., Bolan, N., Rinklebe, J. 2018. Wood-based biochar for the removal of potentially toxic elements in water and wastewater: A critical review. *International Materials Reviews*, 1–32 64(4).

Shi, T., Chen, Y., Liu, Y., Wu, G. 2014. Visible and near-infrared reflectance spectroscopy—An alternative for monitoring soil contamination by heavy metals. *Journal of Hazardous Materials*, **265**, 166–176.

Shi, T., Guo, L., Chen, Y., Wang, W., Shi, Z., Li, Q., Wu, G. 2018. Proximal and remote sensing techniques for mapping of soil contamination with heavy metals. *Applied Spectroscopy Reviews*, **53**(10), 783–805.

Shi, T., Wang, J., Chen, Y., Wu, G. 2016. Improving the prediction of arsenic contents in agricultural soils by combining the reflectance spectroscopy of soils and rice plants. *International Journal of Applied Earth Observation and Geoinformation*, **52**, 95–103.

Sikdar, S., Kundu, M. 2018. A review on detection and abatement of heavy metals. *ChemBioEng Reviews*, **5**(1), 18–29.

Silva, V., Mol, H.G.J., Zomer, P., Tienstra, M., Ritsema, C.J., Geissen, V. 2019. Pesticide residues in European agricultural soils – A hidden reality unfolded. *Science of the Total Environment*, **653**, 1532–1545.

Silva, V., Montanarella, L., Jones, A., Fernández-Ugalde, O., Mol, H.G.J., Ritsema, C.J., Geissen, V. 2018. Distribution of glyphosate and aminomethylphosphonic acid (AMPA) in agricultural topsoils of the European Union. *Science of the Total Environment*, **621**, 1352–1359.

Singh, A. 2018. Managing the salinization and drainage problems of irrigated areas through remote sensing and GIS techniques. *Ecological Indicators*, **89**, 584–589.

Singh, M., Thind, P.S., John, S. 2018. Health risk assessment of the workers exposed to the heavy metals in e-waste recycling sites of Chandigarh and Ludhiana, Punjab, India. *Chemosphere*, **203**, 426–433.

Sorensen, J., Lapworth, D., Nkhuwa, D., Stuart, M., Gooddy, D., Bell, R., Chirwa, M., Kabika, J., Liemisa, M., Chibesa, M. 2015. Emerging contaminants in urban groundwater sources in Africa. *Water Research*, **72**, 51–63.

Sosa, D., Hilber, I., Faure, R., Bartolomé, N., Fonseca, O., Keller, A., Schwab, P., Escobar, A., Bucheli, T.D. 2017. Polycyclic aromatic hydrocarbons and polychlorinated biphenyls in soils of Mayabeque, Cuba. *Environmental Science and Pollution Research*, **24**(14), 12860–12870.

Stenberg, B., Rossel, R.A.V., Mouazen, A.M., Wetterlind, J. 2010. Visible and near infrared spectroscopy in soil science. In: *Advances in Agronomy*, Vol. 107, Elsevier, Amsterdam, the Netherlands, pp. 163–215.

Suh, J., Kim, S.-M., Yi, H., Choi, Y. 2017. An overview of GIS-based modeling and assessment of mining-induced hazards: Soil, water, and forest. *International Journal of Environmental Research and Public Health*, **14**(12), 1463.

Sui, Q., Cao, X., Lu, S., Zhao, W., Qiu, Z., Yu, G. 2015. Occurrence, sources and fate of pharmaceuticals and personal care products in the groundwater: A review. *Emerging Contaminants*, **1**(1), 14–24.

Sun, W., Zhang, X. 2017. Estimating soil zinc concentrations using reflectance spectroscopy. *International Journal of Applied Earth Observation and Geoinformation*, **58**, 126–133.

Sun, W., Zhang, X., Sun, X., Sun, Y., Cen, Y. 2018. Predicting nickel concentration in soil using reflectance spectroscopy associated with organic matter and clay minerals. *Geoderma*, **327**, 25–35.

Sundaram, B., Feitz, A., de Caritat, P., Plazinska, A., Brodie, R., Coram, J., Ransley, T. 2009. Groundwater sampling and analysis – a field guide. *Geoscience Australia, Record*, **27**(95), 104.

Tadeo, J.L. 2019. *Analysis of Pesticides in Food and Environmental Samples*, CRC Press, Hoboken, NJ.

Tianlik, T., Norulaini, N.A.R.N., Shahadat, M., Yoonsing, W., Omar, A.K.M. 2016. Risk assessment of metal contamination in soil and groundwater in Asia: A review of recent trends as well as existing environmental laws and regulations. *Pedosphere*, **26**(4), 431–450.

Tiwari, A.K., Singh, P.K., Singh, A.K., De Maio, M. 2016. Estimation of heavy metal contamination in groundwater and development of a heavy metal pollution index by using GIS technique. *Bulletin of Environmental Contamination and Toxicology*, **96**(4), 508–515.

Trefry, M.G., Muffels, C. 2007. FEFLOW: A finite-element ground water flow and transport modeling tool. *Groundwater*, **45**(5), 525–528.

Tuinhof, A., Foster, S., Kemper, K., Garduno, H., Nanni, M. 2004. World Bank. Sustainable groundwater management: Concepts and tools. Groundwater monitoring requirements for managing aquifer response and quality threats. Available at http://documents.worldbank.org/curated/en/334871468143053052/pdf/301070BriefingNote9.pdf.

Tunks, J., Parsons, J.H., Vazquez, R., Vroblesky, D. 2005. Passive diffusion sampling for metals. In: *Contaminated Soils, Sediments and Water: Science in the Real World Volume 9*, (Eds.) E.J. Calabrese, P.T. Kostecki, J. Dragun, Springer, Boston, MA, pp. 265–285.

U.S. Department of Energy. 1998. Direct sampling ion trap mass spectrometry (DSITMS), Innovative technology summary report. Available at https://frtr.gov/pdf/itsr69.pdf. Accessed May 2, 2019.

Ugochukwu, U.C., Ochonogor, A. 2018. Groundwater contamination by polycyclic aromatic hydrocarbon due to diesel spill from a telecom base station in a Nigerian City: Assessment of human health risk exposure. *Environmental Monitoring and Assessment*, **190**(4), 249.

UNESCO. 2017. United Nations World Water Assessment Programme. United Nations world water development report. Facts and Figures. Available at http://www.unesco.org/new/en/natural-sciences/environment/water/wwap/, Accessed on February 25, 2019.

Upadhyay, M.K., Majumdar, A., Barla, A., Bose, S., Srivastava, S. 2019. An assessment of arsenic hazard in groundwater–soil–rice system in two villages of Nadia district, West Bengal, India. *Environmental Geochemistry and Health*, 41(6), pp. 2381–2395.

US EPA. 2004. *Cleaning Up the Nation's Waste Sites: Markets and Technology Trends, 2004 Edition*. United States Environmental Protection Agency, Washington, DC.

US EPA. 2019. United States Environmental Protection Agency. Contaminated Land: What are the trends in contaminated land and their effects on human health and the environment? Report on the environment. Available at https://www.epa.gov/report-environment/contaminated-land, Accessed on February 25, 2019.

US EPA. 2002. United States Environmental Protection Agency. Guidance on environmental data verification and data validation. Available at https://www.epa.gov/sites/production/files/2015-06/documents/g8-final.pdf. Accessed on May 23, 2019.

US EPA Environmental Technology Verification Report. United States Environmental Protection Agency. Immunoassay Kit, Strategic Diagnostics Inc. EnviroGard PCB Test Kit. Available at https://archive.epa.gov/nrmrl/archive-etv/web/pdf/01_vr_stratdiag_egard.pdf.

Vera, R., Anticó, E., Fontàs, C. 2018. The use of a polymer inclusion membrane for arsenate determination in groundwater. *Water*, **10**(8), 1093.

Verreydt, G., Bronders, J., Van Keer, I., Diels, L., Vanderauwera, P. 2010. Passive samplers for monitoring VOCs in groundwater and the prospects related to mass flux measurements. *Groundwater Monitoring & Remediation*, **30**(2), 114–126.

Vetrimurugan, E., Brindha, K., Elango, L., Ndwandwe, O.M. 2017. Human exposure risk to heavy metals through groundwater used for drinking in an intensively irrigated river delta. *Applied Water Science*, **7**(6), 3267–3280.

Vogeler, I., Vachey, A., Deurer, M., Bolan, N. 2008. Impact of plants on the microbial activity in soils with high and low levels of copper. *European Journal of Soil Biology*, **44**(1), 92–100.

Wang, F., Gao, J., Zha, Y. 2018. Hyperspectral sensing of heavy metals in soil and vegetation: Feasibility and challenges. *ISPRS Journal of Photogrammetry and Remote Sensing*, **136**, 73–84.

Wang, J. 2002. Electrochemical nucleic acid biosensors. *Analytica Chimica Acta*, **469**(1), 63–71.

Wang, J., Zhao, Y., Sun, J., Zhang, Y., Liu, C. 2019. The distribution and sources of polycyclic aromatic hydrocarbons in shallow groundwater from an alluvial-diluvial fan of the Hutuo River in North China. *Frontiers of Earth Science*, **13**(1), 33–42.

Watanabe, E., Seike, N., Motoki, Y., Inao, K., Otani, T. 2016. Potential application of immunoassays for simple, rapid and quantitative detections of phytoavailable neonicotinoid insecticides in cropland soils. *Ecotoxicology and Environmental Safety*, **132**, 288–294.

Wei, M., Yang, X., Watson, P., Yang, F., Liu, H. 2019. A cyclodextrin polymer membrane-based passive sampler for measuring triclocarban, triclosan and methyl triclosan in rivers. *Science of the Total Environment*, **648**, 109–115.

Weindorf, D.C., Bakr, N., Zhu, Y. 2014. Advances in portable X-ray fluorescence (PXRF) for environmental, pedological, and agronomic applications. In: *Advances in Agronomy*, Vol. 128, Elsevier, the Netherlands, pp. 1–45.

Weindorf, D. C., and S. Chakraborty. 2016. Portable X-ray Fluorescence Spectrometry Analysis of Soils. In: *Methods of Soil Analysis, SSSA Book Ser. 5*. doi:10.2136/methods-soil.2015.0033.

Wen, X., Lu, J., Wu, J., Lin, Y., Luo, Y. 2019. Influence of coastal groundwater salinization on the distribution and risks of heavy metals. *Science of the Total Environment*, **652**, 267–277.

Wise, M.B., Merriweather, R., Guerin, M., Thompson, C.V. 1995. Comparison of direct sampling ion trap mass spectrometry to GC/MS for monitoring VOCs in groundwater. Air and Waste Management Association, Pittsburgh, PA.

Xu, T., Xu, Q.G., Li, H., Wang, J., Li, Q.X., Shelver, W.L., Li, J. 2012. Strip-based immunoassay for the simultaneous detection of the neonicotinoid insecticides imidacloprid and thiamethoxam in agricultural products. *Talanta*, **101**, 85–90.

Yang, Q., Li, Z., Lu, X., Duan, Q., Huang, L., Bi, J. 2018. A review of soil heavy metal pollution from industrial and agricultural regions in China: Pollution and risk assessment. *Science of the Total Environment*, **642**, 690–700.

Yang, Y., Ok, Y.S., Kim, K.-H., Kwon, E.E., Tsang, Y.F. 2017. Occurrences and removal of pharmaceuticals and personal care products (PPCPs) in drinking water and water/sewage treatment plants: A review. *Science of the Total Environment*, **596**, 303–320.

Yassir, B., Sana El, F., Mohyeddine El, K., Abdelaziz Ait, M., Alain, P. 2019. Study of the effect of climate changes on the well water contamination by some heavy metals at a mining extract region in Marrakech City, Morocco. In: *Handbook of Research on Global Environmental Changes and Human Health*, (Eds.) K. Kholoud, H. Moulay Abdelmonaim El, H. Omar El, S. Denis, B. Lahouari, IGI Global, Hershey, PA, pp. 121–128.

Yu, L., Rozemeijer, J., Van Breukelen, B.M., Ouboter, M., Van Der Vlugt, C., Broers, H.P. 2018. Groundwater impacts on surface water quality and nutrient loads in lowland polder catchments: Monitoring the greater Amsterdam area. *Hydrology & Earth System Sciences*, **22**(1), 487–508.

Zhai, Y., Zhao, X., Teng, Y., Li, X., Zhang, J., Wu, J., Zuo, R. 2017. Groundwater nitrate pollution and human health risk assessment by using HHRA model in an agricultural area, NE China. *Ecotoxicology and Environmental Safety*, **137**, 130–142.

Zhang, Z., Zeng, K., Liu, J. 2017. Immunochemical detection of emerging organic contaminants in environmental waters. *TrAC Trends in Analytical Chemistry*, **87**, 49–57.

Zhao, P., Zhao, J., Lei, S., Guo, X., Zhao, L. 2018. Simultaneous enantiomeric analysis of eight pesticides in soils and river sediments by chiral liquid chromatography-tandem mass spectrometry. *Chemosphere*, **204**, 210–219.

Zhuang, D.-F. 2010. Study on canopy spectral characteristics of paddy polluted by heavy metals. *Spectroscopy and Spectral Analysis*, **30**(2), 430–434.

22 Selected Examples of Remediation and Reactivation of Old Sites in the Ruhr Region

Volker Selter

CONTENTS

22.1 INTRODUCTION

In the course of industrialization in the nineteenth and twentieth centuries, the Ruhr region was one of the economically most important regions of Germany and Central Europe (Figure 22.1). Its central location in the middle of Germany, the numerous coal reserves, and the good traffic links, especially by shipping, offered optimal conditions.

In particular, the mining, steel, and chemical industries were the cornerstones of an emerging industrial region. However, these processes inevitably led to a widespread environmental impact due to the raw materials or intermediate and end products used. In addition to the intensive temporary air pollution, there was also a high level of contamination of soil and ground or surface water. A prime example is the task of the river Emscher, which flows through the Ruhr area from east to west, and which was converted during the course of industrialization into a sewer for industrial waste.

For the renaturation of the Emscher alone, a high level of funding was made available by the state and the EU in recent years in order to correct the environmental sins of industrialization.

Ultimately, structural change since the 1960s has led to a rethink in environmental policy and the orientation of urban development in the region. The Ruhr region is therefore in a long-term process of transformation from an industrial region to a central location for services and logistics.

22.2 STRUCTURAL CHANGE

The first industrial crisis in the sixties and seventies led to the decommissioning and closure of major industrial sites in the mining industry, such as coal mines or steel sites and chemical plants.

The historic urban development meant that the production sites were in the center or in the immediate vicinity of inner cities or neighborhoods; the working population settled in close proximity to these areas, and even executives had their residences close to the works. Furthermore, a corresponding infrastructure such as shopping, cultural meeting places, and

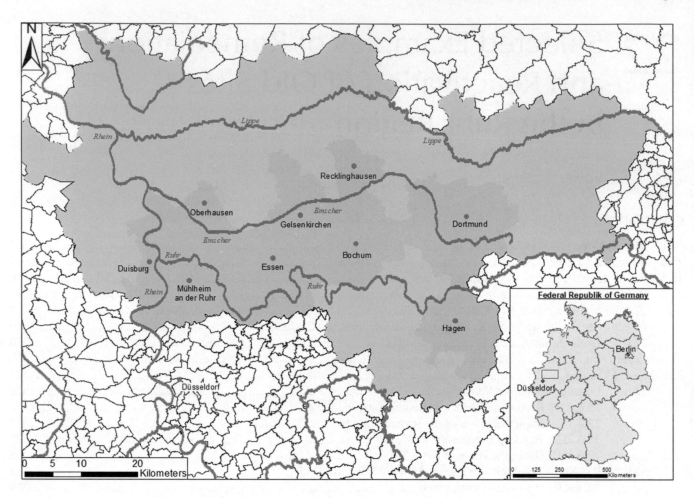

FIGURE 22.1 Map of Germany and NRW showing the location of the Ruhr region.

recreational facilities and the famous Ruhr region pub was created. As a result of the plant closures, large areas of land in an attractive location lay fallow and significantly influenced the cityscape and the further urban development in a negative sense. The previous owners were generally not interested in redeveloping the space, since in particular high expenses for the dismantling of the old facilities and buildings as well as the remediation of the contaminated sites were to be expected. Also a new development of the fallow industrial land was not in the interest of the owners.

In order to promote urban development, the Land Fund was founded by the state of North Rhine-Westphalia. The purpose of this project was to acquire space in cooperation with the affected municipalities and then to develop urban planning in the sense of structural change. One of the political goals of structural change was the creation of qualified and highly qualified jobs in order to counteract the region's sometimes negative "grimy" image.

22.3 FALLOW LAND RECYCLING IN THE RUHR REGION

The concentration of industrial sites such as coal mines, coking plants, and steel mills in the Ruhr region led to a gradual retreat of the mining industry in structural change and thus

to the uncovering of numerous areas, especially near the city centers. The intention was, after rehabilitation and treatment, for fallow land to achieve a higher-value subsequent use, such as for craft workshops, educational institutions, technology centers, and residential areas, but also leisure facilities.

A groundbreaking step was the establishment of the International Building Exhibition Emscher Park (IBA), during which beacon projects on former fallow sites were developed. These included:

- Landscape Park Duisburg-North (Meiderich Steelworks, Figure 22.2a)
- Science Park Rheinelbe in Gelsenkirchen (Cast Steelworks Gelsenkirchen)
- Zeche Zollverein (Colliery Association) in Essen (Figure 22.2b)
- Mont Cenis Academy in Herne
- Gasometer Oberhausen
- Century Hall in Bochum (Figure 22.2c)

Here, a modern residential, cultural, and leisure landscape with ecological requirements was developed on the fallow land throughout the Ruhr area. The new buildings are characterized by a high architectural quality due to the urban planning and political influences of the IBA Emscher Park.

(a)

(b)

(c)

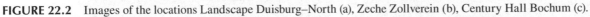

FIGURE 22.2 Images of the locations Landscape Duisburg–North (a), Zeche Zollverein (b), Century Hall Bochum (c).

The end of the IBA in the presentation year 1999 also meant a standstill in the area development in the Ruhr region; the structural change 1.0 phase was over. A continuation of the IBA took place to a much smaller extent in the form of REGIONALS.

A renewed push for the restoration and refurbishment of old sites has brought about an increasing demand for large areas of more than 4 hectares in size through the emerging logistics industry. Distinctive locations, where the logistics industry mainly settled, are:

- Logport site (former Krupp Steelworks) in Duisburg–Rheinhausen
- Opel plants 1 and 2 in Bochum
- Westfalenhütte (formerly Hoesch) in Dortmund
- Schalke Club in Gelsenkirchen

For this up-and-coming branch of industry, large, flat parcels of land (>4 ha) and good traffic links through immediate proximity to a motorway junction or in individual cases also the presence of rail network links are all important. Due to the simple founding of the buildings, the requirements for the surface preparation of the areas appear rather low. The generally large lightweight halls of logistics companies are mainly founded on individual foundations and large supports; deep foundations with high demands on the degree of compaction and special methods, such as vibro displacement compaction, are the exception.

The reactivation of the large industrial wasteland to create logistic locations heralded the start of the phase of structural change 2.0.

In contrast to the first phase during the period of the IBA Emscher Park, however, there is no architectural innovation compared to neighboring countries such as the Netherlands; for the most part, relatively modest functional hall complexes with large open spaces and parking spaces for trucks have been and are being built.

22.4 REMEDIATION PROCESS

With the beginning of the treatment of fallow land at the beginning of the 1990s, the applied remediation processes developed continuously in an emerging new industry: the remediation of contaminated sites.

At the beginning, a complete remediation of the areas by soil replacement was generally favored. The contaminated soil materials were completely excavated and sent for external disposal or treatment. Preferred methods were storage in an approved landfill site or removal of pollutants by soil washing or thermal treatment at external sites.

For the treatment and remediation of old sites, however, these methods proved to be too costly in the long term, and the large amounts of contaminated soils also led to a blockage of valuable landfill space and were therefore classified as environmentally harmful.

In coordination with the technical and regulatory authorities, redevelopment concepts were increasingly developed which made it possible to secure the retention of contaminated soil materials or demolition masses on the former site. The implementation was usually planned by specialist offices in a so-called remediation plan and approved by the authorities in a public law procedure.

22.4.1 REMEDIATION OF HOT SPOTS

Even though, in contrast to the beginnings of land remediation, there has been a reduction in the dumping of contaminated soil materials into approved landfills, this method often remains the only option for certain pollutants or geological conditions. This applies, among others, to waste containing asbestos or organic pollutants such as mineral oil hydrocarbons or tar oils, i.e. pollutants in pasty to liquid form or in phase.

For all this remediation of so-called hot spots, i.e. limited localized areas of soil contamination, the principle of mass control or minimization applies by a separation of contamination and uncontaminated materials, in particular by qualified expert support and construction monitoring.

After the separation of the pollutants in Germany, a so-called waste legislation declaration with a corresponding approval procedure for waste requiring monitoring takes place. These include highly contaminated soil or rubble materials. The procedure is now being implemented as part of electronic verification. Following regulatory approval, the monitored transport may be moved to an approved recovery facility or landfill.

In the case of hot spot remediation, the contaminated soil materials are generally completely excavated and disposed of. In cooperation with the regulatory authorities, usually ground protection, they are assisted by authorization procedures, such as a remediation plan, which sets so-called remediation targets. These describe, for selected parameters, the residual pollutant load that can remain in the soil. The concentration is determined individually by the geological or hydro-geological conditions (groundwater table, proximity to drinking water protection zones) as well as background pollution and is checked or documented by the accompanying expert through wall and floor sampling in the remediation excavation (Figure 22.3).

FIGURE 22.3   Images of remediation hot spots.

22.4.2 Secure Storage

One instrument of the remediation plan according to the BBodSchG (Federal Soil Protection Law) is the secure storage of contaminated soil and demolition materials in a technical structure at a defined location on the former site. This has the advantage, alongside economic aspects such as the elimination of material transport, that external landfill space can be better used both ecologically and economically.

In the case of secure storage, a site is usually selected on the respective fallow land, where soil contamination already exists; uncontaminated partial areas are not provided for this purpose (prohibition of deterioration). A so-called safety structure for storing the contaminated materials from the site can then be erected on these areas. The storage of externally delivered contaminated materials is not permitted and must be excluded.

As a rule, the technical framework conditions for this structure were fixed in the remediation plan. A key element is the surface sealing system, which completely prevents the penetration of precipitation or surface waters through the stored contaminated materials. In the first phase of the construction of the structures, plastic sealing membranes from landfill construction were generally used. Their production and installation on-site is comparatively complicated and therefore also expensive, because the individual membranes must be sealed together, for example.

Therefore, bentonite mats have been recently used which are significantly cheaper and almost show an identical sealing effect. These consist mainly of the mineral sodium bentonite, which is of volcanic origin and is used in the mats as granules (Figure 22.4). Another advantage is the simpler

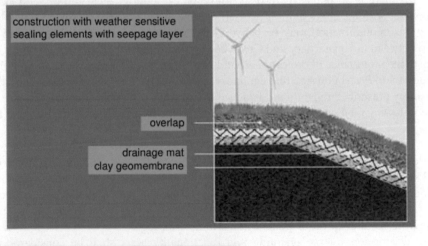

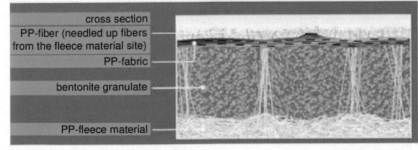

FIGURE 22.4 Structure of bentonite tracks.

way of laying the mats over geosynthetic plastics. The bentonite mats can simply be laid overlapping and used on steeper slopes.

In individual cases, a base seal is built for highly contaminated stored material, which has the same general structure as the surface seal.

In the Ruhr area, there are currently securing structures at the following old sites where soil and building rubble materials contaminated by the respective location have been stored:

- Colliery Consolidation I/VI in Gelsenkirchen
- Schalker Verein West in Gelsenkirchen
- Schalker Verein East ("Spitzbergen") in Gelsenkirchen
- Phoenix East – site (current Lake Phoenix) in Dortmund
- Mining Colliery West in Kamp–Linfort
- Coking plant Hansa in Dortmund
- Graf Bismarck Colliery in Gelsenkirchen

22.4.3 GROUNDWATER

The contaminants of the unsaturated soil zone have been transferred to groundwater on a variety of old sites and industrial wastelands. In the Ruhr area, the first groundwater horizon represents the quaternary, which is predominantly designed as a pore groundwater aquifer. The chalk is a fractured aquifer of the second horizon in large areas.

Predominantly organic pollutants have led to contamination of the first quaternary aquifer at the old sites or their effluent. If there is an intensive vertical migration of pollutants, contact with the chalk aquifer is often present. Another cause of its contamination may be a hydro-geological connection between the quaternary and Cretaceous groundwater.

Due to the importance of protected groundwater in terms of the Federal Soil Protection Act, the regulatory authorities have increasingly prescribed rehabilitation and safety measures.

The following methods are used at the old sites in the Ruhr region:

- Pump and treat
- Deep drainage
- Sealing walls

For the process of deep drainage and the sealing wall, more detailed descriptions of the planning and functionality are shown in the project examples below.

In the pump-and-treat process, the groundwater is pumped through wells located in the center or downstream of the contamination and purified by appropriate systems or system modules adapted to the pollutant spectrum in the groundwater. In the case of the coke-specific pollutants which are frequently found in the Ruhr area, such as polycyclic aromatic hydrocarbons (PAHs), mineral oil hydrocarbons, or BTEX compounds, these systems are predominantly water-activated carbon systems. As a rule, the systems consist of a mechanical stage in which cloudy and suspended solids are filtered off. The groundwater

to be cleared is then passed through one or more activated carbon filters in the case of organic pollutants and then partly discharged into surface water, but usually into the public sewage system. In exceptional cases, a re-infiltration or discharge also takes place into source of the groundwater pollution.

22.5 LEGAL FOUNDATIONS

As part of the development of contaminated site management, legal foundations for evaluation and implementation were gradually created.

At the beginning they were completely missing; as an aid for the evaluation of soil pollution and possible remediation requirements, the so-called Holland list from the neighboring country was used. However, this was based on other geological conditions, i.e. it took into account the low groundwater table in the Netherlands, which was much higher in large parts of the Ruhr area, for example. Thus, the application of this list led to a review of overdrawn remediation measures in this region.

In North Rhine-Westphalia, the first legal and formal foundations for soil protection were developed, which also took into account the regional geological and hydrogeological conditions or soil and groundwater pollution in the environment. On this basis, the Federal Soil Protection Act (BBodSchG) was passed in 1999 and subsequently the Federal Soil Protection Ordinance (BBodSchV), on the basis of which a site-specific investigation of contaminated sites and remediation planning was possible.

One instrument in practical implementation is the so-called remediation plan. It will be approved or made binding by the local soil protection authority after it has been submitted by the person responsible for remediation or by a specialist office appointed by him. This implies complete legal certainty for all parties involved in the process, such as former owners, buyers, or municipalities.

In the course of the planning process, site-specific refurbishment targets are being developed for each existing site. The respective local boundary conditions, such as

- Groundwater conditions
- Distance to groundwater protection zones or areas
- Pollution situation in the area surrounding the site
- Planned subsequent uses
- Geogenic existing contamination

shall be considered. The remediation target values for the respective parameter or substance group determine how far contaminated materials are to be removed from the remediation zones or disposed of. The remediation plan, which in the actual sense represents a redevelopment concept and planning, is examined and approved in a public procedure by the participating authorities, such as soil protection, nature and landscape protection, civil engineering offices, and water protection, and public participation (affected citizens and residents) is also planned.

Another instrument for the assessment of soil contamination and the determination of remediation target values is the recommendation of the state working group (Länderarbeitsgemeinschaft [LAGA]) guideline M 20. In contrast to the BBodSchG, however, this has no legislative character and is used by many regulatory authorities as an orientation tool in the assessment. The LAGA points out recommendable replacement values under certain technical boundary conditions or environmental aspects:

- Z 0: Unrestricted installation due to the pollutant content of the material
- Z 1.1: Unrestricted open installation
- Z 1.2: Restricted open installation under certain hydro-geological conditions
- Z 2: Restricted installation with defined technical safety measures
- >Z 2: No installation, if necessary in safety structures with BBodSchG approval

The unrestricted installation in category Z 0 means that the resulting ground can be reused or installed without technical restrictions. On sensitive land, such as children's playgrounds, gardens, agricultural land, or sports facilities, ground installation, which comes from refurbishment projects or soil treatment facilities, should be omitted because there is still the residual risk of possible contamination.

When installing materials in category Z 1, specific usage restrictions arise due to the pollutant contents. Thus, in hydrologically favorable areas, material up to Z 1.2 with an appropriate distance to the groundwater floor can be installed if there is already a pre-load. In any case, for the assured emplacement of Z 1.2 material erosion protection, e.g., in the form of a vegetation cover. As a rule, 1 m distance to the highest expected groundwater level should be maintained for installation. Installation is not possible in drinking water protection areas (Zone I to IIIa) and mineral springs (Zone I to III), areas with frequent flooding (e.g., flood retention basins, diked areas), nature reserves, and biosphere reserves of sensitive areas.

Almost unlimited installation of Z 1.1 or Z 1.2 is possible in the reactivation of old sites to industrial and commercial areas, if a seal is provided by halls and building complexes or storage and parking areas. On green areas, a cover is made with cohesive or culturable soil. These measures are either regulated in the remediation plan (see above) or approved in the course of the building application.

Storage of material Z 2 is associated with well-defined technical safety measures. Similar to the installation of Z 1 material, installation with a minimum distance of 1 m is connected to the highest expected groundwater level.

The following safety measures or production of a qualified surface seal are associated with the installation of Z 2 material:

- Noise barriers with mineral surface sealing $k_f < 10^{-8}$ m s^{-1} with a minimum thickness of 0.5 m
- Road surfaces with a corresponding mineral surface seal (bituminous layer)
- In the construction of paved areas in industrial and commercial areas as well as other traffic areas (e.g., airfields,

port areas, freight transport centers) as a base course under a waterproof cover layer and as a bound base course under a slightly permeable cover layer
- Building-related use of soil materials in the landfill body, e.g., as a leveling layer between waste body and surface seal

The limit values of the respective installation class are specified separately for each parameter.

The recommendations of the LAGA and the resulting guideline values are used in particular in the re-utilization of recycled building materials. These are prepared in appropriate plants approved under waste legislation and are then used in the refurbishment of commercial sites on former old sites. There is also an expert review, analysis, and documentation of the material flows.

In assessing groundwater pollution, the so-called LAWA list ("Derivation of Insignificance Thresholds for Groundwater") from 2004 is generally applied in North Rhine-Westphalia. The objective was to preserve groundwater for drinking water use and to protect the ecological environment, such as surface waters or wetlands. With this, nationwide standardized criteria for the classification of groundwater pollution were created. It was important to assess the respective substance concentrations with regard to local or regional framework conditions such as anthropogenic or geogenic influences (background pollution). A so-called insignificance threshold (Geringfügigkeits-Schwellenwert – GFS) was defined; this is defined as the concentration at which, despite increased contaminant levels, no relevant toxicological effects on the groundwater are to be expected and the requirements of the Drinking Water Ordinance are complied with.

22.6 CASE EXAMPLES

22.6.1 LOCATION "SCHALKER VEREIN"

The site of the former steelworks "Thyssen Schalke Club" is located just east of the city center of Gelsenkirchen and covers an area of approximately 100 ha. The city of Gelsenkirchen lies in the center of the Ruhr region and is characterized by numerous colliery and coking plants as well as steelworks in the center and the surrounding districts.

The development of the industrial site began at the end of the nineteenth century, and mainly cast and steel products were produced. In addition there were plants for the production of energy and of coke-specific substances. Production at the site was gradually discontinued at the end of the 1990s.

The area was one of the most important industrial sites in the Ruhr area where the first coking plant in the region originated, and production facilities such as blast furnaces, ore bunkers, a gasoline factory, and several tar plants made their mark. The more than 150-year-old production history led to an almost comprehensive contamination with organic and inorganic pollutants in soil and groundwater.

After decommissioning, the western part was sold to a state-owned development company; the eastern part was acquired by the French conglomerate Saint Gobain PAM. After an intensive exploration phase, which was always carried out in coordination with the competent soil protection authority, the development and approval of a remediation plan according to the Federal Soil Protection Act (BBodSchG) for the western part was first carried out. In this, both target values for remediation and limit values for the re-installation of soil materials with regard to the planned commercial reuse were bindingly stipulated. The aim was to dispose of the least contaminated soil materials from the remediation externally; instead, a so-called safety structure was planned according to landfill standard and executed.

The remediation and building preparation was implemented with EU funding (Structural Aid Fund). These were limited in time, so that a nationwide treatment and renovation were implemented to create commercial space ready for construction.

In contrast to this, the eastern part was prepared for investment-related purposes without subsidies, renovated, and prepared for construction. Again, after the exploration phase, a remediation concept or a remediation plan was developed. Also, a safety structure for the storage of contaminated soil and building rubble materials was built from the site.

In investor-related processing, the planning of the resettlement is taken into account in the implementation. For example, for traffic or green areas at the new location, significantly lower expenses for the preparation of building are required than for the building preparation of the building structure. This leads to a more economical implementation of earthworks and processing.

22.6.1.1   Risk Assessment and Remediation Planning

The risk assessment studies were planned and executed on the basis of intensive historical research. In the research, several archives, e.g., the archives of the city of Gelsenkirchen, the archive of the previous owner Thyssen AG, and aerial photographs from various sources, were analyzed.

As a result of the research, partial areas and former production facilities with an increased risk potential were designated (Figure 22.5) and subsequently developed and implemented in consultation with the regulatory authorities of the city of Gelsenkirchen in an investigation program featuring core sounding, dredging, and groundwater investigations.

The results of the investigations at several locations of former production sites, especially tar plants, revealed increased pollution, mainly due to organic pollutants such as PAHs, BTEX compounds, or hydrocarbons (Figure 22.6). In particular, tar oils were in phase, which occurred both in the soil samples from the soundings as well as the dredging. For these so-called hot spots, remediation concepts or a remediation plan according to the Federal Soil Protection Act (BBodSchG) were developed and planned.

22.6.1.2   Implementation of the Remediation Plan

In 2009, the renovation of the first hot spot at the former petroleum factory began. In the process, organically contaminated soil materials from anthropogenic fillings and quaternary high-tide clays were dumped and disposed of down to the groundwater fluctuation range.

In the further course of the remediation and preparation of the site, further "hot spots" were remediated on the site, whereby primarily tar oil-contaminated soil materials were found and mainly disposed of:

* Tar hall AFG
* Sand casting plant
* Pipe foundry 1
* Chill casting workshop

Overall, there was an external disposal of highly contaminated with organic pollutants soil masses in the order of 50,000 tons.

Less heavily contaminated soil masses, in particular slag and foundry sands, were placed in a safety structure on the eastern part of the "Spitzbergen." The partial area has been consciously selected because

* The area outside the planned industrial park is separated from it by a public road
* The site in the underground has a preload due to heavy metal and PAH-containing slag and thus the deterioration prohibition is not violated
* The area should be used as a biotope and public green space after completion (Figure 22.7).

In the safety structure, heavy metals and cyanide-contaminated slag were stored that originated from a dam that had to be dismantled during the process of building preparation. On this, a factory siding was operated, which ran from the connection of the public railway network in the east to the former ore bunkers in the western part of the area. The dam was sometimes up to 10 m high and had a capacity of 80,000 m³, which were almost completely built into the safety structure "Spitzbergen" in the course of the surface preparation.

An essential aspect in the implementation of remedial measures is the detailed and extensive documentation by the expert support of one or more independent engineering firms.

These include:

* Assignment of remediation areas to the implementing company
* Separation of polluted and unpolluted excavation masses
* Implementation of wall and floor sampling in the remediation pits
* Declaration analysis of the soil materials to be disposed of
* Analysis and classification of reusable soil masses in fixed batches, generally 500–1000 m³
* Analytics and guarantees of origin of externally supplied materials
* Proof of proper compaction of installed materials by compaction controls (load plate tests)

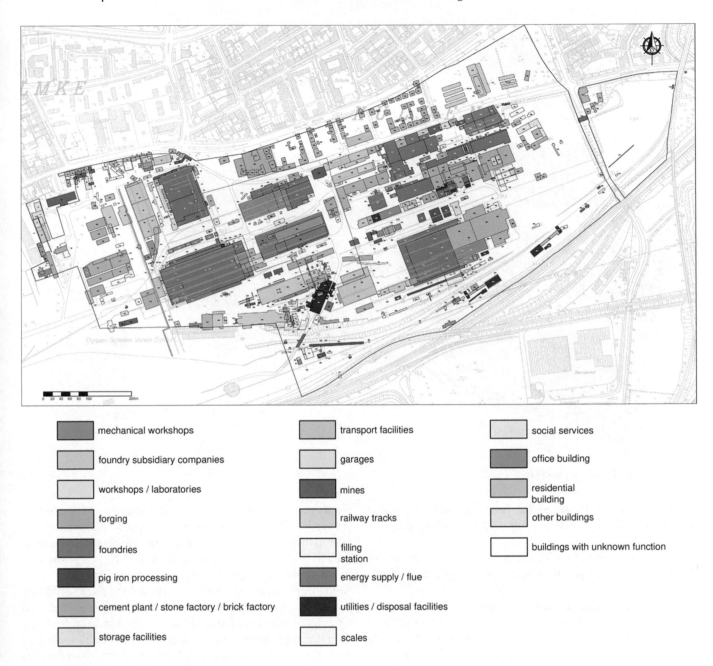

■	mechanical workshops	■	transport facilities	■	social services
■	foundry subsidiary companies	■	garages	■	office building
■	workshops / laboratories	■	mines	■	residential building
■	forging	■	railway tracks	■	other buildings
■	foundries	■	filling station	□	buildings with unknown function
■	pig iron processing	■	energy supply / flue		
■	cement plant / stone factory / brick factory	■	utilities / disposal facilities		
■	storage facilities	□	scales		

FIGURE 22.5   Historical research map.

- Waste declaration of the soil mass stored in a safety structure including documentation of the situation in the structure

The results of these investigations are generally recorded in a final documentation, which is submitted to the authorities as well as to the person responsible for remediation ("state disruptors") or potential land purchasers.

22.6.1.3   Building Preparation

In addition to the remediation of contaminated sites, the construction of the subsoil on the rehabilitated land, the so-called building preparation, plays a decisive role in the reactivation of fallow land. For this purpose, investor-related processing concepts for each settlement are developed on the grounds of the Schalke Club.

Here, the planning of the investors and the associated demands on the subsoil play an essential role. A primary goal is to compact the existing unloaded soil or building waste masses in layers after their preparation in order to achieve a restriction-free ground.

In order to make the entire project economically viable, a soil management concept was developed for the building preparation. This has the objective of

1. If possible, do not dispose of or reutilize any unloaded soil materials externally
2. In return, not delivering any third-party materials
3. Soil storage of uncontaminated soils for installation in the groundwater fluctuation range (LAGA Z 0) and for green areas through economic, external delivery

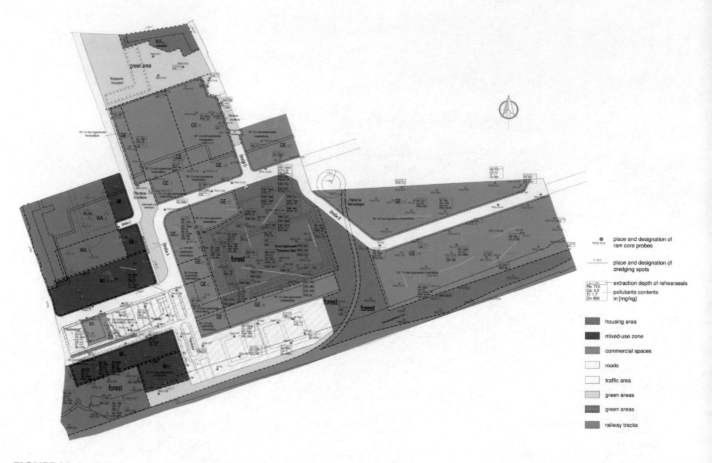

FIGURE 22.6 Pollution plan map.

The soil management concept has been constantly monitored and updated through a GIS-controlled processing program (Figure 22.8).

In close consultation with the planners and architects of the investors or land purchasers, area-specific elevation models were developed.

Except for individual cases, the endeavor was generally to apply filling material of about 2–3 m, in order to produce a restriction-free and load-bearing ground. The aim was to achieve a degree of compaction of 45 MN m^{-2} for the geotechnically simple foundation of the logistics halls, which in individual cases has been improved to 100 MN m^{-2}. The upper 0.5 m was built with recycled material (RC material) as standard, which has been prepared grain-graded for higher degrees of compaction. The RC material was produced on-site from the mineral demolition masses of the building demolition or the foundation removal with the aid of a processing or crushing plant. This measure is also part of the soil management concept.

This approach had the advantage that below the original upper edge of the building remains, such as floor slabs or foundations, could remain underground. It was only the remediation of contaminated sites and a backfilling of cavities and cellars.

A special feature in the project implementation was the establishment of a deep drainage on the western part of the site for groundwater protection. Due to the existing production facilities such as coke furnaces and secondary production plants, significant evidence of soil contamination by carcinogenic, coke, and metallurgical-specific pollutants, such as heavy metals, cyanides, PAH, BTEX, and so on, was found. Since the impurities, in particular the cyanides originating from old sewage ponds and the coking plant secondary production, have already transferred to the groundwater and a discharge into southern or western plots was to be feared, a deep drainage with delivery shafts was built on the southwest edge of the plot to prevent a pollutant discharge from the area as a hydraulic safety measure. The location was chosen deliberately, as the highest pollutant concentrations were found here in the quaternary groundwater.

The underground structure is characterized by up to 8 m thick fillings, which are underlain by the groundwater-bearing quaternary sediments of silts, sands, and gravels. These can reach widths of 3–12 m, with the table between 2 and 6 m. The quaternary sediments are followed by the Upper Cretaceous with the Emscher marl, which is generally clayey-weathered on the surface.

In the center to the eastern part of the terrain, a groundwater elevation was identified, from which the water flows to the north, west, and south, at times also to the east. In the vicinity of the site there are two pumping stations, whose operation affects the groundwater outflow to the north and south of the site.

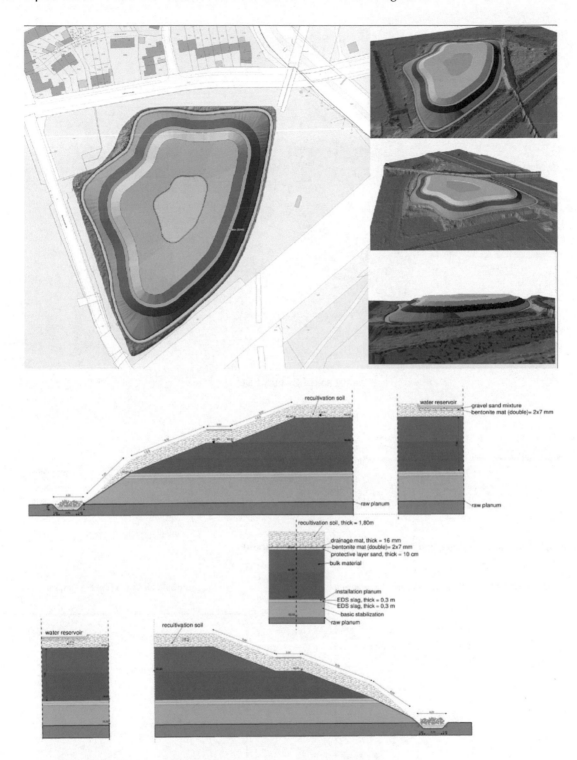

FIGURE 22.7 Location plan of safety structure "Spitzbergen" and profile section.

The deeper groundwater floor is located in the gap and joint system of chalk deposits at approximately 8–12 m below the top ground edge (Emscher marl). The upper part of these sediments is usually clayey weathered and separates the upper groundwater level (quaternary) hydraulically largely from the Cretaceous Quercus, which is why difficult groundwater conditions occur frequently in the Cretaceous layer. The Cretaceous pressure water level is not significantly lower than the free upper groundwater level of the quaternary.

22.6.1.4 Hydraulic Basis of Remediation

A short pumping test was conducted at a quaternary survey site to determine hydraulic parameters to determine the containment conditions at the potential site for hydraulic containment/

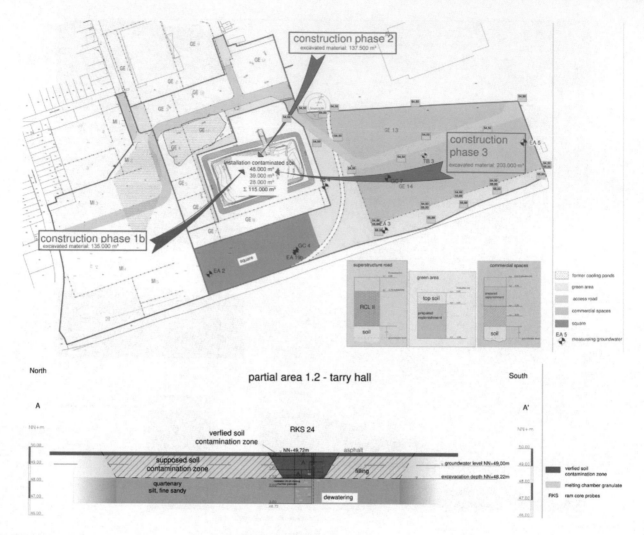

FIGURE 22.8   Soil management plan and profile section of surface preparation.

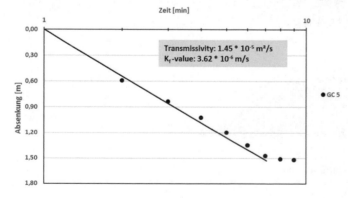

FIGURE 22.9   Evaluation graph of the time reduction at a measuring point (according to Cooper and Jacob).

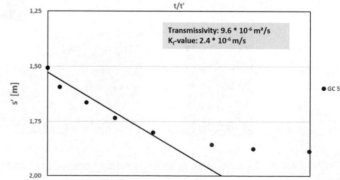

FIGURE 22.10   Evaluation graph of the rise at a measuring point (according to THEIS).

remediation. Important parameters were the transmissivity (**T-value**) and the permeability coefficient (**k_f-value**).

The pumping test was carried out with a delivery rate of 0.5 m³ h⁻¹ over a period of 30 minutes. The rise was measured over 90 minutes. During the production, no reaction was detected at a nearby Cretaceous (TB1) measuring point. From the pumping experiment, the transmissivity (**T-value**) and the permeability coefficient (**k_f-value**) were analyzed graphically for unsteady flow of wells using the time reduction according to Cooper and Jacob (Figure 22.9) and the rise according to THEIS (Figure 22.10).

From the subsidence and rise, an averaged T-value of 1.2×10^{-5} m² s⁻¹ was calculated and, assuming a groundwater-filled thickness of 4 m, a k_f-value of 3×10^{-6} m s⁻¹ was calculated, which characterizes the pore aquifer at this site as weak to moderately permeable.

The main hydraulic parameters for the site of the hydraulic safety measure are compiled in the following:

General parameters

• Formation of the aquifer	U, fs, ms', g' – Fs, u, ms', g', currently mixed grain filling
• Type of aquifer	Pore aquifer
• Groundwater conditions	Unstressed
• Groundwater intensity (H)	i.m. 5.5 m
• Hydraulic gradient (J_0)	Approx. 1.6% (in the southwest)
• Useful porosity (P*)	Approx. 10%; (estimated)
• Clearance speed (v_a)	4.76×10^{-7} m s⁻¹ = 0.04 m d⁻¹ = 15 m a⁻¹

Parameters determined from pumping tests

• Pumping rate (Q)	0.5 m³ h⁻¹
• Subsidence (s)	3.24 m
• Performance quotient (Lq)	0.043 L s⁻¹ m⁻¹
• Transmissivity (T-value)	1.2×10^{-5} m² s⁻¹
• Groundwater thickness (H)	3.9 m
• Permeability coefficient (k_f-value)	3×10^{-6} m s⁻¹
• Range (R) according to Sichardt	16.84 m

Based on these results, the deep drainage was planned and executed in detail. This was positioned as a hydraulic protection on the southwest edge of the old site and led down to the sediments of the chalk. For this a drainage length of approximately 150 m in west-east extension and approximately 25 m in approximately NW-SE extension as well as a depth of approximately 8–10 m was required. The deep drainage was made by overlapping bored piles, which were filled with a filter gravel adapted to the grain size of the surrounding rock (gravel pile wall).

To promote the amount of water in the drainage, three wells were created within the bored pile wall, their arrangement take into account both the hydraulic requirements and the detection of preferred pollutant discharge zones. To monitor the hydraulic conditions in the gravel pile wall, additional measuring points were integrated into the drainage. After a mechanical cleaning of the raw water via settling tanks and gravel filters, the pumped water was discharged into the public sewage system.

Due to the fact that old building substance and strongly solidified slag were encountered in the planned route area within the anthropogenic filling, which interfered with the drilling work, a preliminary excavation was carried out. Subsequently, the drainage was carried out by means of overcut bored piles down to depths of 8–10 m to the surface of the circular surface and executed as a bored pile wall (diameter 900 mm) with gravel filling. The bored piles were produced in two work steps. In the first step, the production of the so-called primary piles took place, followed by the overlapping secondary piles. The exact placement of the holes was ensured by the use of a drilling template. The surface of the filter gravel filling was protected at about 1 m under the planned ground top edge with a filter fleece. On top of that, 0.5 m of cohesive soil was applied to reduce the infiltration of leachate and finally 0.5 m of mixed-grained culturable soil for the purpose of planting.

As delivery wells, removal tubes (filter and extension tubes) were set in the drainage wall. The backfilling of the pipes was done with the material of the drainage wall (filter gravel 2–3 mm). The well was completed with a well space made of shaft rings, which were set up to 1 m under the ground top edge and sealed with a manhole cover, so that a regular inspection and maintenance of the well was guaranteed.

In order to check the hydraulic functionality, a total of 4 groundwater measuring points (PEHD DN 125) were placed in the drainage wall at a distance of 25 m each from the pumping wells.

Before carrying out a continuous production operation, a 2-month trial operation was carried out for optimal control of the plant. There hydraulically required flow rate for securing the groundwater outflow was hereby determined by different operating modes of the individual wells under observation of the groundwater measuring points. Following the trial operation, the optimal operation of the plant was determined for continuous operation.

During commissioning of the rehabilitation facility, the current hydraulic and load situation was determined as part of a zero test by measuring the deadline at all measuring points on and around the site and by collecting groundwater samples from the drainage wall and selected groundwater measuring points. In order to be able to determine an optimal mode of operation of the groundwater pumping and treatment plant, a 2-month pilot phase with appropriate sampling campaigns was initially carried out. During the operation of the plant, regular checks were carried out to check the functioning of the conveyor system and mechanical cleaning. At least twice a week, the flow rates and meter readings were read, the well water levels and water levels in the drainage wall were sounded, and the plant control was adjusted. The documentation of all measurement results, the adjustments made to the system, and other observations and findings are made in an operating journal.

The planned decontamination measures in the area of the cooling/treatment ponds and the sewage treatment plant in connection with the overbuilding or planting of the area caused a considerable reduction of the pollutant availability – in particular of cyanides – for the groundwater via the leachate path. Significantly increased levels of cyanide were predominantly present in the unsaturated zone in the potential sources of damage to the cooling/treatment pond and purification plant/gas purification, with the result that the hydraulic measures at the SW edge of the site in conjunction with the previous decontamination measures improved groundwater quality.

All remediation measures or the control of any pollutant emissions from the safety structures or the deep drainage are monitored by groundwater monitoring, which is usually carried out four times a year. For this purpose, pump samples are taken from groundwater measuring points in the area of the former entry points as well as in groundwater outflow and analyzed for parameters stipulated in the remediation plan.

22.6.2 BSI-Site – Friedrich – Wilhelms Hütte in Mülheim/Ruhr

The site of the Bergische Stahlindustrie (BSI) is located northwest of the city center of Mülheim an der Ruhr and, with more than 100 years of industrial history, is part of the Friedrich-Wilhelms-Hütte of Thyssen-Krupp AG (Figure 22.11). The city of Mülheim is located on the western edge of the Ruhr region between the three cities of Düsseldorf, Duisburg, and Essen.

The former fallow land was acquired by Aldi Grundstücksgesellschaft mbH & Co. KG for the purpose of expanding its logistics center in Mülheim an der Ruhr.

In the period from 1997 to 2011, various contaminated site investigations were carried out on the BSI site. On the basis of the reorganization investigation, the reorganization plan, and a project-related development plan, an implementation contract (urban construction contract) was concluded between the city of Mülheim and the company ALDI in 2018, which regulates the handling of the contaminated sites and provides the necessary legal and cost security for all participants with regard to the measures to be implemented.

22.6.2.1 Historical Development

The company was founded in 1811 and was always characterized by steady growth. There was a specialization in the production and processing of molten iron. One aspect was the commissioning of the first coke-based blast furnace in the Ruhr area in 1849. Through numerous mergers of steel plants and metallurgical enterprises, the company grew steadily until the mid-twentieth century.

After 1901, essentially the following production areas could be outsourced:

- Blast furnace/pig iron production
- Coking plant/secondary production
- Foundry/steel finishing
- Mechanical production, apparatus construction
- Cement factory (stone factory)

FIGURE 22.11 Map of the BSI site.

At the beginning of the twentieth century, a cement and stoneware factory was built on the southern edge of the BSI site, which was intended, among other things, for the supply of blast furnaces and coking plants. Bricks were also made in this factory, for which floodplain clay was mined north of the factory.

For stable filling of the excavations, materials from a so-called agglomeration plant with sintered belt were used, especially in the deepest areas. The fine ore obtained by sieving in the ore deposit was mixed with additives from the foundry molding processes and with blast furnace dusts, coal residues, and dusts from the coke ovens, compacted in the agglomerating plant and sintered using converter belts. The agglomerated fine ore was hot-cast as a melt in the excavations, as indicated by the banded and partially silicified grain structure with rust-brown slag crust.

22.6.2.2 Pollution Situation

In the context of several contaminated site investigations from 1997 to the present, severe soil contamination was determined by inorganic and organic pollutants, which are bound to the old deposits and have led to groundwater pollution. Closer investigation of the main pollution in the soil revealed that the old deposits mainly contaminated with heavy metals, cyanides, PAHs, and MKW in the central and western part of the site came from sintered fine ore (**sinter ore**) material with aggregates from waste of various production sectors (coking plant, blast furnaces, steel refining, foundry). These deposits are present as a compact block, some of which is already less than 1 m below ground level and extends at the lowest points in the western area into the groundwater fluctuation range. The distribution area of the contaminated sintered ore body can be stated as approximately 11,000 m² and the thickness as 2 to 5 m. The banded, slate-clay-like layers are earth-moist and proved to be virtually impermeable to water in a laboratory test. Upon destruction of the grain structure and storage in the air, the sand and clay-marl-like fractions disintegrate.

The anthropogenic filling above and outside the sinter ore distribution consists of a mixture of foundry sands with ashes, slag, building rubble, rearranged soils, building materials, and various waste. Increased levels of heavy metals and PAHs are also found in this filling mixture. Use-related increased mineral oil hydrocarbons occur in places. Cyanides are also detectable in the fillings. The metallurgic-specific deposits have a thickness of between 3 and 6 m as well as pollutant loads that extend to the terrain surface. The highest pollutant content in groundwater results from the direct elution of pollutants from the sintered ore body during temporary contact with the groundwater. Long-term groundwater observations and a groundwater model have shown that an almost stationary state has formed in the spread of pollutants, i.e. there has been a balance between the solution of pollutants and their transport processes. According to the prognosis calculations made by the model, such an equilibrium state would have occurred under unfavorable propagation conditions, e.g., an unrestrained spread, discontinued after 40 years at the latest. The old deposit has been underground for 100 years, so that it could be assumed that the equilibrium, i.e. a stationary state between solution and transport of pollutants, has existed for some time.

Since a change in the influencing factors in the future could not be excluded, e.g., environmental changes in the subsoil, a sharp rise in groundwater levels, a significant increase in groundwater due to groundwater abstraction, or additional pollutant sources, a change in the spread of pollutants was expected. Therefore, appropriate remediation/protection measures have been planned, which take into account the protection of drinking water facilities in the surrounding area.

22.6.2.3 Groundwater-Hydraulic Measures

In combination with the overbuilding of the damage source in the form of a dry storage and extensive sealing of surrounding open spaces, the source of damage was secured by a sealing wall on the western edge of the BSI site to secure the outflow and protect an adjacent waterworks. The sealing wall is a passive hydraulic system, which is intended to divert the groundwater in the form of a baffle to the north away from the waterworks area.

There are different types of sealing wall systems:

- Trench wall (one and two-phase process): a lot of excavated soil, high disposal costs, obstacles in the underground
- Overcut bored pile walls: a lot of excavated soil, high production costs
- MIP walls: mortaring of the existing soil, low-vibration production, reduced ground attack, deflection of the boreholes through high drilling resistance possible, limited control of wall-tightness, cost-effective
- Sheet piling: ramming obstacles in the underground, high workload for back anchoring, advance excavation of obstacles required, high vibration
- Narrow wall: less excavated soil, less space required, lower vibration than sheet piling, cost-effective

After an assessment of the technical feasibility on-site, the execution of the sealing wall was recommended as a narrow wall.

Narrow walls generally have a thickness between 3 to 30 cm and can be rammed or shaken to a depth of up to 20 m. The ramming process hardly produces any excavated soil, and the disposal costs are low. The prerequisite for this procedure is a soil that can be rammed or shaken. For the narrow wall, the present soil is displaced by a vibrated steel profile (vibration plate) and when shaking and drawing the paving screed, a sealing material is injected into the resulting cavity. The individual lines are produced overlapped, whereby the width of these paving screeds varies between 800 and 1200 mm. The sealing material exits at the lower end of the paving screed under pressure via nozzles, so that the resulting cavity and the pores of the remaining soil are pressed with the sealing material. This material is a bentonite-cement suspension to which a certain amount of additives (stone powder, fly ash) is added. The suspension must have a high density, a fast setting capacity, and a low permeability.

FIGURE 22.12 Picture of a drilling machine for the narrow wall.

This method was suitable for the remediation of the BSI site, since the soil can be rammed and shaken and the seal through the narrow wall is a sufficient safeguard for the groundwater on the site. The planned sealing wall has a total length of about 350 m and was rammed to a depth of about 10 m. It was thereby incorporated about 1 m in the sediments of the Upper Cretaceous. The wall thickness of the vibrating narrow wall is 10 cm. On the sealing wall track, initial ordnance probes were performed at a distance of 1.5 m. These holes also served as loosening holes for the insertion of the narrow wall, and were reduced to a depth of 10 m for this purpose. Perforation bores were made later when producing the shaken narrow wall complementary to the ordnance drilling only in case of insufficient ramming progress. After the release of the sealing wall track by the ordnance disposal service, the advance excavation was carried out to create a 0.6 m deep intake ditch.

Prior to the construction of the narrow wall, the sealing wall suspension selected for use was tested for various factors, e.g., the resistance to organic and alkaline waters, as these could produce sulphuric acid, for example, and could possibly affect the tightness of the wall. Likewise, the aggressiveness of concrete and steel was tested in order to be able to exclude an impairment of the sealing wall. Furthermore, a 3 m long sample section of the narrow wall was produced on-site to demonstrate the functionality/suitability of the method under the prevailing soil and groundwater conditions. The suspension used had a permeability coefficient of $\leq 10^{-9}$ m s^{-1} and a compressive strength of ≈ 1000 kN m^{-2} in order to meet the requirements of the planned narrow wall (Figure 22.12).

22.7 COMPARISON OF THE REMEDIATION METHODS BETWEEN THE AREAS SCHALKER VEREIN AND BSI

In principle, both former industrial sites were production facilities of the steel industry with similar production facilities and the resulting comparable contamination situations. However, due to the different surrounding areas and environments, various remediation and treatment strategies have been developed.

In addition to treatment, the protection of a drinking water extraction plant in the immediate vicinity of the BSI site is at the forefront of the planned remediation measures.

On the other hand, at the location "Schalker Verein" there is no acute danger to the groundwater outside the lot or an influence on drinking water protection areas. The circumstances mentioned thus have a considerable influence on the remediation plan or the remedial and safety measures to be carried out. For example, a safety instrument for general groundwater protection was installed on the "Schalker Verein" site by the deep drainage system. Due to the intensive input of pollutants into the saturated soil zone, despite the remediation of the "hot spots," that is, dumps of pollutants, there are still residual pollutants that can only be intercepted and remediated via the groundwater path and the precautionary measure of deep drainage.

In contrast, the sealing wall on the BSI site in Mülheim represents a precautionary safety instrument for the protection of a drinking water extraction plant. A soil remediation in terms of pollutant removal is not planned; however, model calculations and simulations have shown that pollutant leakage via the groundwater path cannot be completely ruled out.

22.7.1 Difference between Deep Drainage and Narrow Wall

The deep drainage intercepts contaminated groundwater in the outflow as a hydraulic barrier, whereby the resulting groundwater is actively raised or pumped and then channeled to the sewage system or the stream.

In contrast, the narrow or sealing wall serves as a passive safety system which deflects the contaminated groundwater, e.g., to secure a drinking water protection zone.

22.8 OUTLOOK

An exciting question for the future is the further handling of the still existing, often also problematic old sites and industrial wasteland. Their remediation and treatment is in addition to possible risk aspects for the protected soil and groundwater dependent on the further economic development and the resulting space requirement.

As of spring 2019, there is a high demand for new settlements and business expansions, especially in the logistics industry. According to the Regionalverband Ruhr (RVR), the following forecasts have been issued for the cities in the Ruhr region:

- Hagen 120 ha
- Bochum 175 ha
- Essen 195 ha

In the entire Ruhr region there will be 2800 hectares of land over the next 20 years – i.e. until 2039.

In the interests of environmental sustainability, settling commercial and industrial areas with a high demand for land on the "green belt" should be avoided. Instead, intelligent and area-specific, i.e. remediation and treatment concepts relevant to the site and respective investor, should be developed and planned to enable the reactivation in terms of the environment but also the respective economic aspects. At present, the pressure of local authorities to design industrial sites is so great that, instead of ecologically sound treatment of industrial wasteland, green areas and agricultural areas should be converted into industrial areas, especially near motorway junctions.

An important long-term aspect in the remediation and treatment of old sites is groundwater protection. Not only the environmental sins of industrialization, but also the effects of climate change on our ecosystem and water conservation play a role here. The impending shortage of drinking water reservoirs and reserves will lead to an appreciation of groundwater resources even in regions such as the Ruhr region, where the groundwater has already been "abandoned" due to increased levels of pollutants, in particular sulfate.

Central European countries, such as the Netherlands or Germany, have taken on a pioneering role in the treatment and remediation of brownfield sites and the development of remediation technologies due to their industrial development and geographic location. In the future, the topic of surface treatment will also affect other regions such as Southern Europe (drinking water protection), North America, and Asia with its emerging industrialization. Here, a suitable transfer of know-how to the countries is desirable.

Index